A Textbook of Physical Chemistry

Volume 2

Thermodynamics and Chemical Equilibrium
(SI Units)

Sixth Edition

A Textbook of Physical Chemistry

Volume 1 : States of Matter and Ions in Solution
Volume 2 : Thermodynamics and Chemical Equilibrium
Volume 3 : Applications of Thermodynamics
Volume 4 : Quantum Chemistry, Molecular Spectroscopy and Molecular Symmetry
Volume 5 : Dynamics of Chemical Reactions, Statistical Thermodynamics, Macromolecules, and Irreversible Processes
Volume 6 : Computational Aspects in Physical Chemistry
Volume 7 : Experimental Aspects in Physical Chemistry

A Textbook of Physical Chemistry

Volume 2

Thermodynamics and Chemical Equilibrium (SI Units)

Sixth Edition

K L KAPOOR
Former Associate Professor
Hindu College
University of Delhi
New Delhi

McGraw Hill Education (India) Private Limited

Published by McGraw Hill Education (India) Private Limited,
Registered Office: Anjana Complex No: 5/90 A, Butt Road, St. Thomas Mount, Chennai – 600016, Tamil Nadu, India.

A Textbook of Physical Chemistry, Volume 2

Seventh reprint 2024

Print Edition:
ISBN-13: 978-93-5316-520-8
ISBN-10: 93-5316-520-2

E-Book Edition:
ISBN-13: 978-93-5316-523-9
ISBN-10: 93-5316-523-7

08 09 10 11 12 RAJ 28 27 26 25 24

Typeset at The Composers, 260, C.A. Apt,, Paschim Vihar, New Delhi 110 063, and Printed and bound in India at Rajkamal Electric Press, Plot No. 2, Phase-IV, Kundli, Haryana

Cover Designer: APS Compugraphics
Cover Image Source: Thinkstock

Visit us at: www.mheducation.co.in
Toll free in India: 1800 103 5875
Write to us at: info.india@mheducation.com
CIN: U80302TN2010PTC111532

*To the memory
of my parents*

Preface

This edition extends the derivations of expressions relating Boyle temperature and inversion temperature for Berthelot, Dieterici, and Redlich-Kwong equations of state for gases. Also, the expressions relating reduced inversion temperature and reduced pressure of gases are derived for van der Waals, Berthelot, Dieterici and Redlich-Kwong equations. These are displayed diagrammatically and compared with the generalized variations as suggested by Robert C. Hendricks, Ildiko C. Peller and Anne K. Baron in their paper in Nasa Technical Note D-6807.

Salient Features

- Strictly in accordance with latest IUPAC recommendations and SI units being adopted throughout the text
- Detailed discussion on expressions relating Reduced Inversion Temperature and Reduced Pressure
- Perfect blend of both theoretical and application-based concepts
- Comprehensive coverage of problems on Joule-Thomson Coefficient
- Extensive chapter-end numericals including Revisionary Problems, Try Yourself Problems and Numerical Problems

K L Kapoor

Publisher's Note

McGraw-Hill Education (India) invites suggestions and comments from you, all of which can be sent to *info.india@mheducation.com* (kindly mention the title and author name in the subject line).
Piracy-related issues may also be reported.

Preface to the Fifth Edition

In recent years, the teaching curriculum of Physical Chemistry in many Indian universities has been restructured with a greater emphasis on a theoretical and conceptual methodology and the applications of the underlying basic concepts and principles. This shift in the emphasis, as I have observed, has unduly frightened undergraduates whose performance in Physical Chemistry has been otherwise generally far from satisfactory. This poor performance is partly because of the non-availability of a comprehensive textbook which also lays adequate stress on the logical deduction and solution of numericals and related problems. Naturally, the students find themselves unduly constrained when they are forced to refer to various books to collect the necessary reading material.

It is primarily to help these students that I have ventured to present a textbook which provides a systematic and comprehensive coverage of the theory as well as of the illustration of the applications thereof.

The present volumes grew out of more than a decade of classroom teaching through lecture notes and assignments prepared for my students of BSc (General) and BSc (Honours). The schematic structure of the book is assigned to cover the major topics of Physical Chemistry in six different volumes. **Volume 1** discusses the states of matter and ions in solutions. It comprises five chapters on the gaseous state, physical properties of liquids, solid state, ionic equilibria and conductance. **Volume 2** describes the basic principles of thermodynamics and chemical equilibrium in seven chapters, viz., introduction and mathematical background, zeroth and first laws of thermodynamics, thermochemistry, second law of thermodynamics, criteria for equilibrium and A and G functions, systems of variable composition, and thermodynamics of chemical reactions. **Volume 3** seeks to present the applications of thermodynamics to the equilibria between phases, colligative properties, phase rule, solutions, phase diagrams of one-, two- and three-component systems, and electrochemical cells. **Volume 4** deals with quantum chemistry, molecular spectroscopy and applications of molecular symmetry. It focuses on atomic structure, chemical bonding, electrical and magnetic properties, molecular spectroscopy and applications of molecular symmetry. **Volume 5** covers dynamics of chemical reactions, statistical and irreversible thermodynamics, and macromolecules in six chapters, viz., adsorption, chemical kinetics, photochemistry, statistical thermodynamics, macromolecules and introduction to irreversible processes. **Volume 6** describes computational aspects in physical chemistry in three chapters, viz., synopsis of commonly used statements in BASIC language, list of programs, and projects.

The study of Physical Chemistry is incomplete if students confine themselves to the ambit of theoretical discussions of the subject. They must grasp the practical significance of the basic theory in all its ramifications and develop a clear perspective to appreciate various problems and how they can be solved.

It is here that these volumes merit mention. Apart from having a lucid style and simplicity of expression, each has a wealth of carefully selected examples and solved illustrations. Further, three types of problems with different objectives in view are listed at the end of each chapter: (1) Revisionary Problems, (2) Try Yourself Problems, and (3) Numerical Problems. Under *Revisionary Problems*, only those problems pertaining to the text are included which should afford an opportunity to the students in self-evaluation. In *Try Yourself Problems*, the problems related to the text but not highlighted therein are provided. Such problems will help students extend their knowledge of the chapter to closely related problems. Finally, unsolved *Numerical Problems* are pieced together for students to practice.

Though the volumes are written on the basis of the syllabi prescribed for undergraduate courses of the University of Delhi, they will also prove useful to students of other universities, since the content of physical chemistry remains the same everywhere. In general, the SI units (*Systeme International d'unite's*), along with some of the common non-SI units such as atm, mmHg, etc., have been used in the books.

Salient Features

- Comprehensive coverage to basic principles of thermodynamics and chemical equilibrium in seven chapters, viz., introduction and mathematical background, zeroth and first laws of thermodynamics, thermochemistry, second law of thermodynamics, equilibrium criteria A and G functions, systems of variable composition, and thermodynamics of chemical reactions
- Emphasis given to applications and principles
- Explanation of equations in the form of solved problems and numericals
- IUPAC recommendations and SI units have been adopted throughout
- Rich and illustrious pedagogy

Acknowledgements

I wish to acknowledge my greatest indebtedness to my teacher, late Prof. R P Mitra, who instilled in me the spirit of scientific inquiry. I also record my sense of appreciation to my students and colleagues at Hindu College, University of Delhi, for their comments, constructive criticism and valuable suggestions towards improvement of the book. I am grateful to late Dr Mohan Katyal (St. Stephen's College), and late Prof. V R Shastri (Ujjain University) for the numerous suggestions in improving the book. I would like to thank Sh. M M Jain, Hans Raj College, for his encouragement during the course of publication of the book.

I wish to extend my appreciation to the students and teachers of Delhi University for the constructive suggestions in bringing out this edition of the book. I also wish to thank my children, Saurabh-Urvashi and Surabhi-Jugnu, for many useful suggestions in improving the presentation of the book.

Finally, my special thanks go to my wife, Pratima, for her encouragement, patience and understanding.

Feedback Request

The author takes the entire responsibility for any error or ambiguity, in fact or opinion, that may have found its way into this book. Comments and criticism from readers will, therefore, be highly appreciated and incorporated in subsequent editions.

K L Kapoor

Publisher's Note

McGraw-Hill Education (India) invites suggestions and comments from you, all of which can be sent to *info.india@mheducation.com* (kindly mention the title and author name in the subject line).
Piracy-related issues may also be reported.

Contents

1 Introduction to Thermodynamics

1.1 SCOPE OF THERMODYNAMICS

The subject of thermodynamics deals basically with the interaction of one body with another in terms of the quantities of heat and work.[†] The entire formulation of thermodynamics is based on two fundamental laws which have been established on the basis of the experimental behaviour of macroscopic aggregates of matter, collected over a long period of time. There is no known example which contradicts the two fundamental laws of thermodynamics. With the help of mathematical tools, it is possible to apply thermodynamic principles in every possible field of science and engineering.

The science which deals with the macroscopic properties of matter is known as classical thermodynamics. Here, the entire formulation can be developed without the knowledge that matter consists of atoms and molecules. Statistical thermodynamics is another branch of science which is based on statistical mechanics and which deals with the calculation of thermodynamic properties of matter from the classical or quantum mechanical behaviour of a large congregation of atoms or molecules.

With the help of thermodynamic principles, the experimental criteria for equilibrium or for the spontaneity of processes are readily established. The

---

[†] The concepts of heat and work are of fundamental importance in thermodynamics. Both these quantities change the internal energy of the system. Heat is best understood in terms of increase or decrease in temperature of a system when it is added to or removed from the system. The convenient unit of heat is calorie (non-SI unit) which is the heat required to raise the temperature of 1 g of water at 15 °C by 1 degree Celsius. The most common work involved in thermodynamics is the work of expansion or compression of a system. This work is best understood in terms of lifting up or lowering down a mass (say, m) through a distance (say, h) in the surroundings; the magnitude of work involved is mgh (see also sections 1.4 and 1.5). Both heat and work have common characteristics of (i) appearing at the boundary of the system, (ii) causing a change in the state of system, and (iii) producing equivalent and opposite effects in the surroundings. The experiments of Joule have established a definite fact (known as *mechanical equivalent of heat*) involving the work and heat. This fact states that the expenditure of a given amount of work, no matter whatever is its origin, always produces the same quantity of heat; 4.184 joules of work is equivalent to 1 calorie of heat. In SI units, both heat and work are expressed in joules. Since heat given to the system and work done on the system increase the internal energy of the system, these two operations are assigned positive values. The converse of these two operations, viz., heat given out and work done by the system are assigned negative values.

equilibrium conditions for any system, in equilibrium state or otherwise, may be calculated. The result of such calculations will indicate the direction the system will take to achieve equilibrium. However, time is not a thermodynamic variable and so thermodynamics cannot give any information about the length of time which would be required for any process to be completed.

The following examples may be helpful.

(1) Liquid water at –10 °C and 0.1 MPa pressure is unstable with respect to ice at the same temperature and pressure. However, water can be supercooled to –10 °C and 0.1 MPa pressure and be maintained at that temperature and pressure for a long time.

(2) Acetylene gas is thermodynamically unstable with respect to graphite and hydrogen gas. However, no one has observed acetylene decompose spontaneously into graphite and hydrogen. Thus, acetylene may take very long time to decompose into graphite and hydrogen gas. The only thing that is predicted by thermodynamics is that had acetylene been in equilibrium with graphite and hydrogen, the concentration of acetylene would have been extremely small and thus essentially only graphite and hydrogen would be present.

(3) Combination of H_2 and O_2 to give water is thermodynamically possible. Nevertheless, both gases can co-exist without combining for a long time.

For chemical reactions, thermodynamics can be used to predict the *extent of reaction at equilibrium*, that is, the equilibrium concentrations of all the active species. In addition, we can predict whether changes in the experimental conditions will increase or decrease the quantity of a product at equilibrium.

1.2 BASIC DEFINITIONS

In thermodynamics, a few terms with their specific definitions are involved. We give below some of the terms along with their definitions.

System

The system is any region of space being investigated.

A system, in general, can be of three types:

(a) *Closed system* Matter can neither be added to nor removed from it.

(b) *Open system* To this system, matter can be added or removed.

(c) *Isolated system* This type of system has no interaction with its surroundings. Neither energy nor matter can be transferred to or from it.

Surroundings

The surroundings are considered to be all other matter that can interact with the system.

Boundary

Anything which separates system and surroundings is called *boundary* (envelope or wall). The envelope may be imaginary or real; it may be rigid or non-rigid; it may be a conductor of heat (*diathermic* wall) or a non-conductor of heat (*adiabatic* wall).

State Variables

The state of a system is defined by ascribing values to a sufficient number of state variables. Such variables are macroscopic properties such as pressure, volume, temperature, mass, composition, surface area, etc. Normally, specifying the values of only a few state variables is necessary for fixing or defining the state of a system.

Take, for example, a system consisting of an ideal gas. In order to define this system completely, we need to state the values of only three variables, namely, p, V and T. The values of other variables (say, for example, amount of the gas, density, etc.) will be definite and thus need not be stated.

Intensive and Extensive Variables

Variables are classified as *intensive* or *extensive*. The classification can be explained by taking a system at a fixed state and dividing it into two or more parts without alternating the state of the entire system. Those variables whose values on division remain the same in any part of the system are called *intensive variables*. Those variables whose values in any part of the divided system are different from the values of the entire system are called *extensive variables*. The magnitudes of extensive variables are proportional to the mass of the system provided the values of all the intensive variables are kept constant.

Examples of Intensive and Extensive Variables

Examples of intensive and extensive variables are given in the following.

Intensive variables Temperature, pressure, concentration, density, dipole moment, refractive index, viscosity, surface tension, molar volume, gas constant, specific heat capacity, vapour pressure, specific gravity, dielectric constant, and emf of a dry cell.

Extensive variables Volume, energy, heat capacity, enthalpy, entropy, free energy, length and mass.

Process

A process is the path along which a change of state takes place. The process can occur under a variety of conditions which must be defined because many things may depend on the nature of the process.

Isothermal process This occurs under constant temperature condition.

Isobaric process This occurs under constant pressure condition.

Isochoric process This occurs under constant volume condition.

Adiabatic process This occurs under the condition that heat can neither be added to nor removed from the system.

Cyclic process It is a process in which a system undergoes a series of changes and ultimately comes back to the initial state.

Quasi-static (or reversible) process If a process is carried out in such a way that at every moment the system departs only infinitesimally from an equilibrium state, the process is called a quasi-static process. At every instant, the system remains virtually in a state of equilibrium.

1.3 MATHEMATICAL BACKGROUND

A great part of thermodynamics is concerned with the change of a thermodynamic property with a change of some independent variable. The mathematical operations used in such derivations are simple differentiations, partial differentiations and integration. In addition, the concepts of exact differentials, inexact differentials and line integrals are commonly used.

Partial Derivatives

Such type of derivatives arise when a function having two or more independent variables is differentiated. A partial derivative is defined as the derivative of a function with respect to one of the independent variables when all other independent variables are kept constant.

First Derivatives

Consider a single-valued function Z of two independent variables x and y; this is usually written as $Z = f(x, y)$ or $Z(x, y)$. If one of the independent variables is held constant, then Z becomes a function of the other variable alone. Partial derivatives can thereby be defined as

$$\left(\frac{\partial Z}{\partial x}\right)_y = \lim_{\Delta x \to 0} \frac{Z(x + \Delta x, y) - Z(x, y)}{\Delta x}$$

and
$$\left(\frac{\partial Z}{\partial y}\right)_x = \lim_{\Delta y \to 0} \frac{Z(x, y + \Delta y) - Z(x, y)}{\Delta y}$$

Partial derivatives are evaluated by the rules for ordinary differentiation, treating the appropriate variables as constants. For example, the volume of one mole of an ideal gas, given by $V_m = RT/p$, is a function of temperature and pressure, i.e. $V_m = f(T, p)$. Thus

$$\left(\frac{\partial V_m}{\partial p}\right)_T = -\frac{RT}{p^2} \quad \text{and} \quad \left(\frac{\partial V_m}{\partial T}\right)_p = \frac{R}{p}$$

Second Derivatives

Since partial derivatives are themselves functions of the independent variables, they can be differentiated again to yield second (and higher) derivatives. If $Z = f(x, y)$, then the first derivatives are $(dZ/dx)_y$ and $(dZ/dy)_x$ and the second derivatives are

$$\frac{\partial^2 Z}{\partial x^2} \equiv \frac{\partial}{\partial x}\left\{\left(\frac{\partial Z}{\partial x}\right)_y\right\}_y; \quad \frac{\partial^2 Z}{\partial y^2} \equiv \frac{\partial}{\partial y}\left\{\left(\frac{\partial Z}{\partial y}\right)_x\right\}_x$$

$$\frac{\partial^2 Z}{\partial y\,\partial x} \equiv \frac{\partial}{\partial y}\left\{\left(\frac{\partial Z}{\partial x}\right)_y\right\}_x; \quad \frac{\partial^2 Z}{\partial x\,\partial y} \equiv \frac{\partial}{\partial x}\left\{\left(\frac{\partial Z}{\partial y}\right)_x\right\}_y$$

Euler's Reciprocity Relation

When a function and its derivative are *single valued* and *continuous*, the order of differentiation in the mixed derivatives is immaterial. Thus

$$\frac{\partial^2 Z}{\partial x\,\partial y} = \frac{\partial^2 Z}{\partial y\,\partial x} \tag{1.3.1}$$

Equation (1.3.1) is known as *Euler's reciprocity relation* (or *cross-derivative rule*). It is applicable to the thermodynamic functions. For an ideal gas. We have

$$\left(\frac{\partial^2 V_m}{\partial p^2}\right)_T = \frac{2\,RT}{p^3}; \quad \left(\frac{\partial^2 V_m}{\partial T^2}\right)_p = 0$$

and
$$\left\{\frac{\partial\,(\partial V_m/\partial p)_T}{\partial T}\right\}_p = \left\{\frac{\partial\,(\partial V_m/\partial T)_p}{\partial p}\right\}_T = -\frac{R}{p^2}$$

Total Differentials

We have considered so far changes in $Z(x, y)$ brought about by changing one of the independent variables at a time. The more general case involves simultaneous variations of x and y. Let ΔZ be the small change in Z brought by simultaneous increments x and y in the independent variables. Thus

$$\Delta Z = Z(x + \Delta x, y + \Delta y) - Z(x, y)$$

Adding and subtracting the quantity $Z(x, y + \Delta y)$, we get

$$\Delta Z = [Z(x + \Delta x, y + \Delta y) - Z(x, y + \Delta y)] + [Z(x, y + \Delta y) - Z(x, y)]$$

Multiplying and dividing the expression within the first bracket by a factor Δx and that within the second bracket by Δy, we get

$$\Delta Z = \left[\frac{Z(x+\Delta x, y+\Delta y) - Z(x, y+\Delta y)}{\Delta x}\right]\Delta x + \left[\frac{Z(x\ y+\Delta y) - Z(x, y)}{\Delta y}\right]\Delta y$$

Approaching the limit $\Delta x \to 0$ and $\Delta y \to 0$ the two bracketed quantities become partial derivatives, while the increments Δx, Δy, ΔZ can be replaced by the differentials dx, dy, dZ, respectively. Thus, the total differential of the function $Z(x, y)$ is

$$\mathrm{d}Z = \left(\frac{\partial Z}{\partial x}\right)_y \mathrm{d}x + \left(\frac{\partial Z}{\partial y}\right)_x \mathrm{d}y$$

For a function Z of n independent variables $Z = f(x_1, x_2, ..., x_n)$, there are n first partial derivatives. The total differential is given by

$$\mathrm{d}Z = \left(\frac{\partial Z}{\partial x_1}\right)\mathrm{d}x_1 + \left(\frac{\partial Z}{\partial x_2}\right)\mathrm{d}x_2 + \cdots + \left(\frac{\partial Z}{\partial x_n}\right)\mathrm{d}x_n = \sum_{i=1}^{n}\left(\frac{\partial Z}{\partial x_i}\right)\mathrm{d}x_i$$

Relations Between Partial Derivatives

To determine the change in the value of the thermodynamic function caused by a change in one or more state variables, it is necessary to express the partial derivatives of the function in terms of experimentally observable quantities. Certain relation between partial derivatives which facilitate obtaining the required expressions are derived below.

(i) Let u be a function of x and y; its differential is

$$\mathrm{d}u = \left(\frac{\partial u}{\partial x}\right)_y \mathrm{d}x + \left(\frac{\partial u}{\partial y}\right)_x \mathrm{d}y \qquad (1.3.2)$$

If $u = f(x, y)$, then $x = f(u, y)$ and its differential is

$$\mathrm{d}x = \left(\frac{\partial x}{\partial u}\right)_y \mathrm{d}u + \left(\frac{\partial x}{\partial y}\right)_u \mathrm{d}y \qquad (1.3.3)$$

Substituting Eq. (1.3.2) in Eq. (1.3.3), we get

$$\mathrm{d}x = \left(\frac{\partial x}{\partial u}\right)_y \left\{\left(\frac{\partial u}{\partial x}\right)_y \mathrm{d}x + \left(\frac{\partial u}{\partial y}\right)_x \mathrm{d}y\right\} + \left(\frac{\partial x}{\partial y}\right)_u \mathrm{d}y$$

or

$$\left[1 - \left(\frac{\partial u}{\partial x}\right)_y \left(\frac{\partial x}{\partial u}\right)_y\right]\mathrm{d}x = \left[\left(\frac{\partial x}{\partial y}\right)_u + \left(\frac{\partial x}{\partial u}\right)_y \left(\frac{\partial u}{\partial y}\right)_x\right]\mathrm{d}y \qquad (1.3.4)$$

The variables x and y are independent. If y is held constant, i.e. dy = 0, then Eq. (1.3.4) becomes

$$\left[1-\left(\frac{\partial u}{\partial x}\right)_y\left(\frac{\partial x}{\partial u}\right)_y\right]\mathrm{d}x = 0$$

But dx may have any value and therefore the term within the bracket must be zero. Thus

$$1-\left(\frac{\partial u}{\partial x}\right)_y\left(\frac{\partial x}{\partial u}\right)_y = 0 \qquad \text{or} \qquad \left(\frac{\partial u}{\partial x}\right)_y = \frac{1}{(\partial x/\partial u)_y} \tag{1.3.5}$$

that is, the partial derivative is equal to the reciprocal of the partial derivative between the same two variables taken in opposite order, provided the same variables are held constant.

Cyclic Rule

If x is held constant, i.e. dx = 0, then Eq. (1.3.4) yields

$$\left(\frac{\partial x}{\partial y}\right)_u + \left(\frac{\partial x}{\partial u}\right)_y\left(\frac{\partial u}{\partial y}\right)_x = 0 \tag{1.3.6}$$

This equation can be written in several different forms such as

$$\left(\frac{\partial x}{\partial y}\right)_u = -\frac{(\partial u/\partial y)_x}{(\partial u/\partial x)_y} \tag{1.3.7a}$$

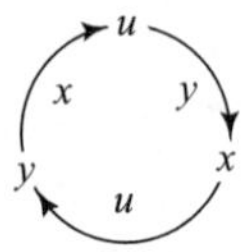

or

$$\left(\frac{\partial u}{\partial x}\right)_y\left(\frac{\partial x}{\partial y}\right)_u\left(\frac{\partial y}{\partial u}\right)_x + 1 = 0 \tag{1.3.7b}$$

Equation (1.3.7b) is known as a *cyclic rule* and is applicable for any three variables of which only two are independent.

(ii) Consider again the function $u = f(x, y)$. Let $y = f(x, s)$. The differential of y in terms of x and s is

$$\mathrm{d}y = \left(\frac{\partial y}{\partial x}\right)_s \mathrm{d}x + \left(\frac{\partial y}{\partial s}\right)_x \mathrm{d}s \tag{1.3.8}$$

But if $u = f(x, y)$ and $y = f(x, s)$, then $u = f(x, s)$. Writing the differential of u in terms of x and s, we have

$$\mathrm{d}u = \left(\frac{\partial u}{\partial x}\right)_s \mathrm{d}x + \left(\frac{\partial u}{\partial s}\right)_x \mathrm{d}s \tag{1.3.9}$$

The differential of u in terms of x and y is

$$\mathrm{d}u = \left(\frac{\partial u}{\partial x}\right)_y \mathrm{d}x + \left(\frac{\partial u}{\partial y}\right)_x \mathrm{d}y \tag{1.3.10}$$

Substituting dy from Eq. (1.3.8) into this, we get

$$\mathrm{d}u = \left[\left(\frac{\partial u}{\partial x}\right)_y + \left(\frac{\partial u}{\partial y}\right)_x\left(\frac{\partial y}{\partial x}\right)_s\right]\mathrm{d}x + \left(\frac{\partial u}{\partial y}\right)_x\left(\frac{\partial y}{\partial s}\right)_x \mathrm{d}s \tag{1.3.11}$$

Equations (1.3.9) and (1.3.11) are identical. Therefore, the coefficients of dx and ds in them must be the same, i.e.

$$\left(\frac{\partial u}{\partial x}\right)_s = \left(\frac{\partial u}{\partial x}\right)_y + \left(\frac{\partial u}{\partial y}\right)_x \left(\frac{\partial y}{\partial x}\right)_s \tag{1.3.12a}$$

and
$$\left(\frac{\partial u}{\partial s}\right)_x = \left(\frac{\partial u}{\partial y}\right)_x \left(\frac{\partial y}{\partial s}\right)_x \tag{1.3.12b}$$

Equations (1.3.12a) and (1.3.12b) can be evaluated directly from Eq. (1.3.10). Dividing Eq. (1.3.10) by dx and introducing the condition of s being constant gives Eq. (1.3.12a). Similarly, dividing Eq. (1.3.10) by ds and introducing the condition of x being constant gives Eq. (1.3.12b).

(iii) If the two independent variables in a function $u = f(x, y)$ are also functions of two other independent variables $x = f(s, t)$, and $y = f(s, t)$, then the function u also becomes a function of s and t. The differentials of these functions are

$$\mathrm{d}u = \left(\frac{\partial u}{\partial x}\right)_y \mathrm{d}x + \left(\frac{\partial u}{\partial y}\right)_x \mathrm{d}y \tag{1.3.13}$$

$$\mathrm{d}x = \left(\frac{\partial x}{\partial s}\right)_t \mathrm{d}s + \left(\frac{\partial x}{\partial t}\right)_s \mathrm{d}t \tag{1.3.14}$$

$$\mathrm{d}y = \left(\frac{\partial y}{\partial s}\right)_t \mathrm{d}s + \left(\frac{\partial y}{\partial t}\right)_s \mathrm{d}t \tag{1.3.15}$$

$$\mathrm{d}u = \left(\frac{\partial u}{\partial s}\right)_t \mathrm{d}s + \left(\frac{\partial u}{\partial t}\right)_s \mathrm{d}t \tag{1.3.16}$$

Substituting dx and dy from Eqs (1.3.14) and (1.3.15) in Eq. (1.3.13), we get

$$\mathrm{d}u = \left[\left(\frac{\partial u}{\partial x}\right)_y \left(\frac{\partial x}{\partial s}\right)_t + \left(\frac{\partial u}{\partial y}\right)_x \left(\frac{\partial y}{\partial s}\right)_t\right] \mathrm{d}s + \left[\left(\frac{\partial u}{\partial x}\right)_y \left(\frac{\partial x}{\partial t}\right)_s + \left(\frac{\partial u}{\partial y}\right)_x \left(\frac{\partial y}{\partial t}\right)_s\right] \mathrm{d}t \tag{1.3.17}$$

Comparing Eqs (1.3.16) and (1.3.17), we get

$$\left(\frac{\partial u}{\partial s}\right)_t = \left(\frac{\partial u}{\partial x}\right)_y \left(\frac{\partial x}{\partial s}\right)_t + \left(\frac{\partial u}{\partial y}\right)_x \left(\frac{\partial y}{\partial s}\right)_t \tag{1.3.18}$$

and
$$\left(\frac{\partial u}{\partial t}\right)_s = \left(\frac{\partial u}{\partial x}\right)_y \left(\frac{\partial x}{\partial t}\right)_s + \left(\frac{\partial u}{\partial y}\right)_x \left(\frac{\partial y}{\partial t}\right)_s \tag{1.3.19}$$

Equations (1.3.18) and (1.3.19) can also be obtained directly from Eq. (1.3.13). Dividing Eq. (1.3.13) by ds and introducing the conditions of constant t, we get Eq. (1.13.18). Similarly, Eq. (1.3.19) can be derived by dividing Eq. (1.3.13) by dt and introducing the condition of constant s.

The following equations can also be derived from Eq. (1.3.13).

$$\left(\frac{\partial u}{\partial x}\right)_y\left(\frac{\partial x}{\partial u}\right)_\upsilon+\left(\frac{\partial u}{\partial y}\right)_x\left(\frac{\partial y}{\partial u}\right)_\upsilon=1 \qquad (1.3.20)$$

$$\left(\frac{\partial u}{\partial x}\right)_y\left(\frac{\partial x}{\partial \upsilon}\right)_u+\left(\frac{\partial u}{\partial y}\right)_x\left(\frac{\partial y}{\partial \upsilon}\right)_u=0 \qquad (1.3.21)$$

where υ is a function of x and y.

Problem 1.3.1 Derive the cyclic rule

$$\left(\frac{\partial p}{\partial T}\right)_V\left(\frac{\partial T}{\partial V}\right)_p\left(\frac{\partial V}{\partial p}\right)_T+1=0$$

Solution Since $p = f(V, T)$, we have

$$\mathrm{d}p=\left(\frac{\partial p}{\partial T}\right)_V\mathrm{d}T+\left(\frac{\partial p}{\partial V}\right)_T\mathrm{d}V$$

For a cyclic process, $\mathrm{d}p = 0$, so that

$$\left(\frac{\partial p}{\partial T}\right)_V(\partial T)_p+\left(\frac{\partial p}{\partial V}\right)_T(\partial V)_p=0$$

Dividing by $(\partial V)_p$, we have

$$\left(\frac{\partial p}{\partial T}\right)_V\left(\frac{\partial T}{\partial V}\right)_p+\left(\frac{\partial p}{\partial V}\right)_T=0 \quad \text{or} \quad \left(\frac{\partial p}{\partial T}\right)_V\left(\frac{\partial T}{\partial V}\right)_p=-\left(\frac{\partial p}{\partial V}\right)_T$$

or

$$\left(\frac{\partial p}{\partial T}\right)_V\left(\frac{\partial T}{\partial V}\right)_p\left(\frac{\partial V}{\partial p}\right)_T+1=0$$

Problem 1.3.2 Test the cyclic rule of Problem 1.3.1 for $pV_\mathrm{m} = RT$.

Solution Differentiating the given equation $pV_\mathrm{m} = RT$, we have

$$p\,\mathrm{d}V_\mathrm{m}+V_\mathrm{m}\,\mathrm{d}p=R\,\mathrm{d}T$$

Dividing this equation by $\mathrm{d}T$ and introducing the condition of constant molar volume, we get

$$V_\mathrm{m}\left(\frac{\partial p}{\partial T}\right)_{V_\mathrm{m}}=R \qquad \text{i.e.} \qquad \left(\frac{\partial p}{\partial T}\right)_{V_\mathrm{m}}=\frac{R}{V_\mathrm{m}}$$

Similarly, we have

$$\left(\frac{\partial T}{\partial V_\mathrm{m}}\right)_p=\frac{p}{R} \qquad \text{and} \qquad \left(\frac{\partial V_\mathrm{m}}{\partial p}\right)_T=-\frac{V_\mathrm{m}}{p}$$

Now substituting these in the cyclic rule of Problem 1.3.1, we get

$$\left(\frac{\partial p}{\partial T}\right)_{V_\mathrm{m}}\left(\frac{\partial T}{\partial V_\mathrm{m}}\right)_p\left(\frac{\partial V_\mathrm{m}}{\partial p}\right)_T+1=\left(\frac{R}{V_\mathrm{m}}\right)\left(\frac{p}{R}\right)\left(-\frac{V_\mathrm{m}}{p}\right)+1=-1+1=0$$

Problem 1.3.3 Test the cyclic rule for

$$\left(p+\frac{a}{V_\mathrm{m}^2}\right)(V_\mathrm{m})=RT$$

Solution Writing the given equation as

$$\left(pV_m + \frac{a}{V_m}\right) = RT$$

and then differentiating, we have

$$p\,dV_m + V_m dp - \frac{a}{V_m^2} dV_m = R\,dT$$

Dividing by dT and introducing the condition of constant volume, i.e. $dV_m = 0$, we get

$$V_m\left(\frac{\partial p}{\partial T}\right)_{V_m} = R \quad \text{i.e.} \quad \left(\frac{\partial p}{\partial T}\right)_{V_m} = \frac{R}{V_m}$$

Similarly, we have

$$\left(\frac{\partial T}{\partial V_m}\right)_p = \frac{p - a/V_m^2}{R} \quad \text{and} \quad \left(\frac{\partial V_m}{\partial p}\right)_T = -\frac{V_m}{p - a/V_m^2}$$

Substituting these in the cyclic rule, we get

$$\left(\frac{\partial p}{\partial T}\right)_{V_m}\left(\frac{\partial T}{\partial V_m}\right)_p\left(\frac{\partial V_m}{\partial p}\right)_T + 1 = \left(\frac{R}{V_m}\right)\left(\frac{p - a/V_m^2}{R}\right)\left(-\frac{V_m}{p - a/V_m^2}\right) + 1 = -1 + 1 = 0$$

Ordinary Integration

The definite integral of a continuous function is defined by the limit

$$\int_a^b f(x)\,dx = \lim_{n\to\infty} \lim_{\Delta x_i \to 0} \sum_{i=1}^{n} f(x_i)\Delta x_i \tag{1.3.22}$$

where $\Delta x_i = x_{i+1} - x_i$ with $x_1 = a$ and $x_{n+1} = b$.

The geometrical interpretation of the above integral as an area is illustrated in Fig. 1.3.1.

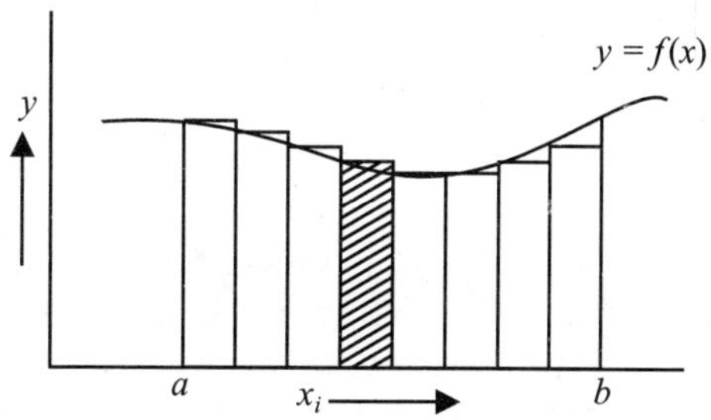

Fig. 1.3.1 Geometrical interpretation of the integral

The operation of integration is the inverse of that of differentiation. Thus

$$\int_a^b f(x)\,dx = F(b) - F(a) \tag{1.3.23}$$

where

$$\frac{dF(x)}{dx} = f(x) \tag{1.3.24}$$

This follows easily from the definition of the derivative

$$\lim_{\Delta x_i \to 0} \frac{F(x_{i+1}) - F(x_i)}{\Delta x_i} = f(x_i) \tag{1.3.25}$$

Substituting this in Eq. (1.3.22), we get

$$\sum_{i=1}^{n} [F(x_{i+1}) - F(x_i)] = F(b) - F(a) \tag{1.3.26}$$

which establishes Eq. (1.3.23). Using Eq. (1.3.24) in Eq. (1.3.23), we have

$$\int_a^b \mathrm{d}F(x) = F(b) - F(a) \tag{1.3.27}$$

showing that a definite integral can be expressed as a difference between two boundary values of a function.

Indefinite Integral

If the integration is done without the limit of integration, it is then called an indefinite *integral.* In this case, we have

$$F(x) = \int f(x)\,\mathrm{d}x \tag{1.3.28}$$

If the function $F(x)$ contains a constant term, the term does not affect the derivative $f(x)$, because the derivative of a constant is zero. Consequently, on integrating the function $f(x)$, the constant term must be added to the integral. Thus, Eq. (1.3.28) must be written as

$$F(x) = \int f(x)\,\mathrm{d}x + I \tag{1.3.29}$$

The value of I (constant of integration) can be determined if the value of $F(x)$ is known at some value of x, say x_i.

$$I = F(x_i) - \left[\int f(x)\,\mathrm{d}x\right]_{x_i} \tag{1.3.30}$$

where the subscript on the last term is used to indicate that the integral is to be evaluated at x_i.

Line Integrals

Differential expressions of the form

$$\mathrm{d}\phi = P(x, y)\,\mathrm{d}x + Q(x, y)\,\mathrm{d}y \tag{1.3.31}$$

for two independent variables are often met in physical sciences and engineering. When $\mathrm{d}x$ and $\mathrm{d}y$ are small, the quantity $\mathrm{d}\phi$ is a small increment of some quantity ϕ, which may or may not be a function of x and y. The integral of such expressions between two points (x_1, y_1) and (x_2, y_2) can be determined along some particular path connecting the two points, since $\mathrm{d}\phi$ can be calculated from Eq. (1.3.31) for each part of a specific path. The integral is the summation of the quantity $\mathrm{d}\phi$, obtained as we move along the curve. Such integrals are called *line* or *contour integrals.*

The value of a line integral between two points depends, in general, upon the path followed in determining the integral. As an example, let us evaluate the line integral

$${}_{\mathrm{L}}\int (y\,\mathrm{d}x - x\,\mathrm{d}y)$$

from A to B in Fig. 1.3.2 along two different paths

(i) A(0, 0) to B(2, 2)

(ii) A(0, 0) to D(2, 0) to B(2, 2)

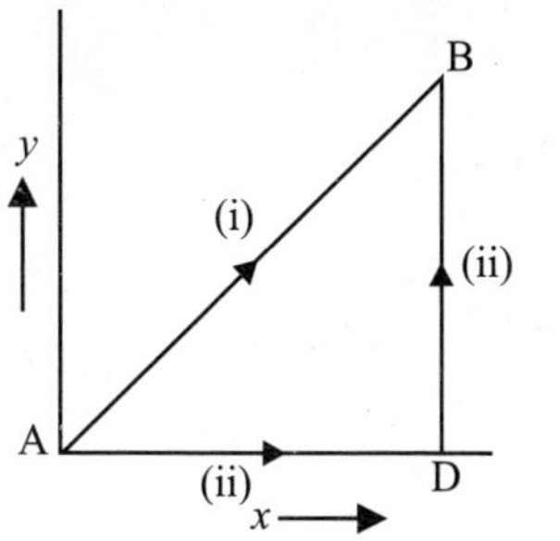

Fig. 1.3.2 Two different paths employed in going from A to B

Path (i) Along the line AB, we have

$$y = x$$

Therefore, $y\,dx - x\,dy = 0$

Hence, $$\int_{AB} (y\,dx - x\,dy) = 0$$

Path (ii) Along AD, we have

$$y = 0 \quad \text{and} \quad dy = 0$$

Thus, $y\,dx - x\,dy = 0$

Along DB, we have

$$x = 2 \quad \text{and} \quad dx = 0$$

Thus, $y\,dx - x\,dy = -2\,dy$

Hence, $$\int_{ADB} (y\,dx - x\,dy) = \int_{AD} (y\,dx - x\,dy) + \int_{DB} (y\,dx - x\,dy)$$

$$= \int_{DB} -2\,dy = \int_0^2 -2\,dy = -4$$

The line integral can be reduced to an ordinary integral with one independent variable, if y is a function of x and

$$dy = (dy/dx)\,dx$$

With this Eq. (1.3.31) becomes

$$_L\int_{x_1}^{x_2} d\phi = \int_{x_1}^{x_2} \left[P(x, y(x)) + Q(x, y(x)) \frac{dy}{dx} \right] dx \qquad (1.3.32)$$

The value of such integral depends upon the particular function chosen for $y(x)$.

Line Integral and Green's Theorem

A line integral of special interest occurs when the path of integration is a closed curve, that is, the initial and final points are identical. Such integrals are called cyclic integrals and are denoted by the symbol $\oint$. Thus, the cyclic integral of the differential expression given by Eq. (1.3.31) is represented as

$$\oint d\phi = \oint [P(x, y)\,dx + Q(x, y)\,dy] \tag{1.3.33}$$

The value of this integral is determined by traversing the closed curve, usually in a counter clockwise direction (Fig. 1.3.3).

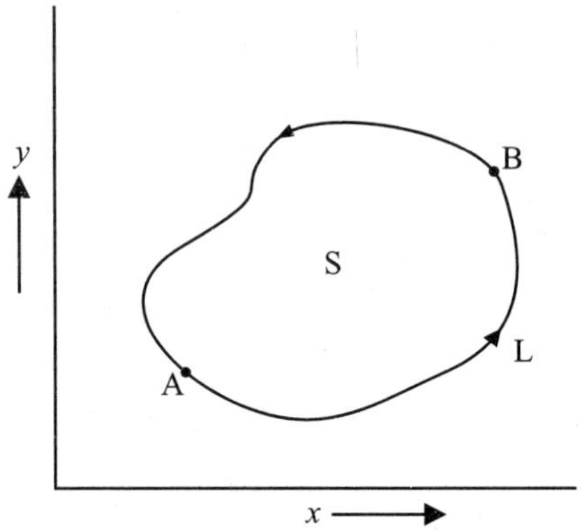

Fig. 1.3.3 Cyclic integration

Green's theorem states that under certain conditions*

$$\oint [P(x, y)dx + Q(x, y)dy] = \iint_S \left[\left(\frac{\partial Q}{\partial x}\right)_y - \left(\frac{\partial P}{\partial y}\right)_x \right] dx\,dy \tag{1.3.34}$$

The right hand side of Eq. (1.3.34) represents the double integral over the surface enclosed by the closed curve.

Exact and Inexact Differentials

A special case occurs when the cyclic integral of a differential expression given by Eq. (1.3.33) equals zero for every closed curve. According to Green's theorem (Eq. 1.3.34), we have

$$\left(\frac{\partial Q}{\partial x}\right)_y = \left(\frac{\partial P}{\partial y}\right)_x \tag{1.3.35}$$

When the condition of Eq. (1.3.35) holds, the differential expression is said to be *exact* and $d\phi$ is said to be an *exact differential;* otherwise, the differential expression is said to be *inexact.*

If $d\phi$ of Eq. (1.3.31) is to be an exact differential, then

$$\oint d\phi = \oint [P(x, y)\,dx + Q(x, y)\,dy] = 0 \tag{1.3.36}$$

From Fig. 1.3.3 the cyclic integration can be replaced by two line integrals (i) from A to B in the counter clockwise direction and (ii) from B to A in the same direction, so that

$$\oint d\phi = {}_L\int_A^B d\phi + {}_L\int_B^A d\phi$$

* If $P(x, y)$, $Q(x, y)$, $(dP/dx)_y$ and $(dQ/dy)_x$ are continuous functions of x and y along the curve L and over the surface S (Fig. 1.3.3).

Since $\oint d\phi = 0$, it follows that

$$\int_{L\,A}^{B} d\phi = -\int_{L\,B}^{A} d\phi \tag{1.3.37}$$

If this condition is true for any cyclic path, the line integral from A to B, $\left(\int_{L\,A}^{B} d\phi\right)$ must be independent of the path and its value depends only on two points. A and B (as in the case of ordinary integration).

It may be readily proved that the condition of Eq. (1.3.35) as derived from Green's theorem is equivalent to Euler's reciprocity relation. If ϕ is a function of x and y, the total differential of ϕ is given by

$$d\phi = \left(\frac{\partial\phi}{\partial x}\right)_y dx + \left(\frac{\partial\phi}{\partial y}\right)_x dy$$

If $d\phi$ is given by the differential expression

$$d\phi = P(x, y)\,dx + Q(x, y)\,dy$$

it follows that

$$P(x, y) = \left(\frac{\partial\phi}{\partial x}\right)_y \quad \text{and} \quad Q(x, y) = \left(\frac{\partial\phi}{\partial y}\right)_x$$

The condition of exactness, as given by Eq. (1.3.35), is

$$\left(\frac{\partial P}{\partial y}\right)_x = \left(\frac{\partial Q}{\partial x}\right)_y$$

Therefore $$\left[\frac{\partial}{\partial y}\left(\frac{\partial\phi}{\partial x}\right)_y\right]_x = \left[\frac{\partial}{\partial x}\left(\frac{\partial\phi}{\partial y}\right)_x\right]_y$$

which is Euler's reciprocity relation (Eq. 1.3.1).

Summary of Exact and Inexact Differentials

The concept of line integral, exact differential and inexact differential may be summarized as follows:

We are concerned with the differential expression

$$d\phi = P(x, y)\,dx + Q(x, y)\,dy$$

The integration of such an expression is carried out along a designated path between two points (x_1, y_1) and (x_2, y_2) or along a closed curve.

- If the line integral $\int_{L\,x_1,y_1}^{x_2,y_2} d\phi$ depends upon the path along which the integration is performed, or, if $\oint d\phi$ is not equal to zero, then $d\phi$ is an inexact differential. There is no function $\phi(x, y)$ which exists whose total differential is given by $P(x, y)\,dx + Q(x, y)\,dy$.

- If the integral ${}_L\int_{x_1,y_1}^{x_2,y_2} \mathrm{d}\phi$ does not depend upon the path along which the integration is performed, or it $\oint \mathrm{d}\phi$ equals zero for every cyclic path, then $\mathrm{d}\phi$ is an exact differential. A function $\phi(x, y)$ does exist and its total differential is equal to $P(x, y)\,\mathrm{d}x + Q(x, y)\,\mathrm{d}y$ with the condition that $(\partial P/\partial y)_x$ is equal to $(\partial Q/\partial x)_y$.
- Symbolically, an inexact differential is distinguished from an exact differential by use of đ (i.e. d cross) instead of d, i.e. đϕ denotes the inexact differential of ϕ while $\mathrm{d}\phi$ is the exact differential of ϕ.

Problem 1.3.4

(i) Given the differential

$$\mathrm{d}\phi = \frac{RT}{p}\,\mathrm{d}p - R\,\mathrm{d}T$$

(i) Carry out the line integration between the limits T_0, p_0 to T_1, p_1 along the following three paths (shown in Fig. 1.3.4).

(a) $T_0, p_0 \rightarrow T_1, p_0 \rightarrow T_1, p_1$

(b) $T_0, p_0 \rightarrow T_0, p_1 \rightarrow T_1, p_1$

(c) $T_0, p_0 \rightarrow T_1, p_1$

(ii) Show that $\mathrm{d}\phi$ is an inexact differential.

(iii) Is it possible to define the function (ϕ) explicitly in terms of T and p?

Solution

(i) Carrying out the line integration along the given paths, we have

Path (a)

$$\Delta\phi_a = \int_{A(T_0,p_0)}^{B(T_1,p_0)} \left[\frac{RT}{p}\mathrm{d}p - R\,\mathrm{d}T\right] + \int_{B(T_1,p_0)}^{C(T_1,p_1)} \left[\frac{RT}{p}\mathrm{d}p - R\,\mathrm{d}T\right]$$

$$= -R\int_{T_0}^{T_1}\mathrm{d}T + RT_1\int_{p_0}^{p_1}\frac{\mathrm{d}p}{p} = -R(T_1 - T_0) + RT_1 \ln\frac{p_1}{p_0}$$

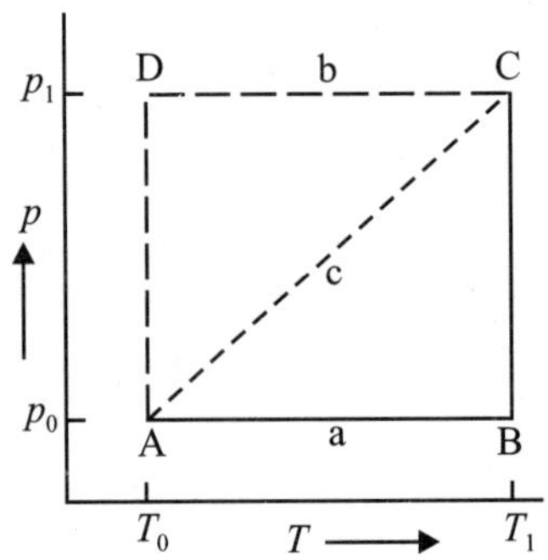

Fig. 1.3.4 Three paths a, b, and c

Path (b)

$$\Delta\phi_b = \int_{A(T_0,p_0)}^{D(T_0,p_1)} \left[\frac{RT}{p}\mathrm{d}p - R\,\mathrm{d}T\right] + \int_{D(T_0,p_1)}^{C(T_1,p_1)} \left[\frac{RT}{p}\mathrm{d}p - R\,\mathrm{d}T\right]$$

$$= RT_0\int_{p_0}^{p_1}\frac{1}{p}\mathrm{d}p - R\int_{T_0}^{T_1}\mathrm{d}T = RT_0 \ln\frac{p_1}{p_0} - R(T_1 - T_0)$$

Path (c) Temperature and pressure along the path c are related by the expression

$$T = T_0 + \left(\frac{T_1 - T_0}{p_1 - p_0}\right)(p - p_0)$$

Hence $$dT = \left(\frac{T_1 - T_0}{p_1 - p_0}\right)dp$$

Substituting T and dT in the given relation and carrying out the integration over p, we have

$$\Delta\phi_c = \int_{p_0}^{p_1}\left\{\frac{R}{p}\left[T_0 + \left(\frac{T_1 - T_0}{p_1 - p_0}\right)(p - p_0)\right] - R\left(\frac{T_1 - T_0}{p_1 - p_0}\right)\right\}dp$$

$$= RT_0 \ln\frac{p_1}{p_0} - Rp_0\left(\frac{T_1 - T_0}{p_1 - p_0}\right)\ln\frac{p_1}{p_0} = \frac{R(T_0p_1 - T_1p_0)}{(p_1 - p_0)}\ln\frac{p_1}{p_0}$$

(ii) We observe that

$$\Delta\phi_a \neq \Delta\phi_b \neq \Delta\phi_c$$

that is, the line integral depends on the path of integration and, hence, $d\phi$ is not an exact differential. This also follows from the fact that Euler's reciprocity relation does not hold good.

(iii) Since $d\phi$ is an inexact differential, the function ϕ cannot be explicitly expressed in terms of T and p.

Problem 1.3.5 Given the differential

$$d\phi = \frac{R}{p}dT - \frac{RT}{p^2}dp$$

(i) Carry out the line integration between the limits T_0, p_0 and T_1, p_1 following the three different paths of the proceeding problem. (ii) Show that $d\phi$ is an exact differential.

Solution **(i)** Carrying out the line integration along the given paths, we have

Path (a) $$\Delta\phi_a = \int_{A(T_0,p_0)}^{B(T_1,p_0)}\left[\frac{R}{p}dT - \frac{RT}{p^2}dp\right] + \int_{B(T_1,p_0)}^{C(T_1,p_1)}\left[\frac{R}{p}dT - \frac{RT}{p^2}dp\right]$$

$$= \frac{R}{p_0}(T_1 - T_0) + RT_1\left(\frac{1}{p_1} - \frac{1}{p_0}\right) = R\left(\frac{T_1}{p_1} - \frac{T_0}{p_0}\right)$$

Path (b) $$\Delta\phi_b = \int_{A(T_0,p_0)}^{D(T_0,p_1)}\left[\frac{R}{p}dT - \frac{RT}{p^2}dp\right] + \int_{D(T_0,p_1)}^{C(T_1,p_1)}\left[\frac{R}{p}dT - \frac{RT}{p^2}dp\right]$$

$$= RT_0\left(\frac{1}{p_1} - \frac{1}{p_0}\right) + \frac{R}{p_1}(T_1 - T_0) = R\left(\frac{T_1}{p_1} - \frac{T_0}{p_0}\right)$$

Path (c) Since for this path,

$$T = T_0 + \left(\frac{T_1 - T_0}{p_1 - p_0}\right)(p - p_0)$$

Therefore $$\mathrm{d}T = \left(\frac{T_1 - T_0}{p_1 - p_0}\right)\mathrm{d}p$$

Substituting T and $\mathrm{d}T$ in the given relation and carrying out the integration over p, we get

$$\Delta\phi_\mathrm{c} = \int_{p_0}^{p_1}\left[\frac{R}{p}\left(\frac{T_1 - T_0}{p_1 - p_0}\right)\mathrm{d}p - \frac{R}{p^2}\left\{T_0 + \left(\frac{T_1 - T_0}{p_1 - p_0}\right)(p - p_0)\right\}\mathrm{d}p\right]$$

$$= \int_{p_0}^{p_1}\left[\frac{R}{p}\left(\frac{T_1 - T_0}{p_1 - p_0}\right)\mathrm{d}p - \frac{RT_0}{p^2}\mathrm{d}p - \frac{R}{p}\left(\frac{T_1 - T_0}{p_1 - p_0}\right)\mathrm{d}p + Rp_0\left(\frac{T_1 - T_0}{p_1 - p_0}\right)\frac{\mathrm{d}p}{p^2}\right]$$

$$= \int_{p_0}^{p_1}\left[-\frac{RT_0}{p^2}\mathrm{d}p + Rp_0\left(\frac{T_1 - T_0}{p_1 - p_0}\right)\frac{\mathrm{d}p}{p^2}\right] = \left\{-RT_0 + Rp_0\left(\frac{T_1 - T_0}{p_1 - p_0}\right)\right\}\int_{p_0}^{p_1}\frac{\mathrm{d}p}{p^2}$$

$$= R\left(\frac{T_1 p_0 - T_0 p_1}{p_1 - p_0}\right)\left(-\frac{1}{p_1} + \frac{1}{p_0}\right) = R\left(\frac{T_1}{p_1} - \frac{T_0}{p_0}\right)$$

(ii) We see that

$$\Delta\phi_\mathrm{a} = \Delta\phi_\mathrm{b} = \Delta\phi_\mathrm{c}$$

and hence the given differential is an exact differential. This also follows from the Euler's reciprocity relation.

Comparing the given differential with the expression

$$\mathrm{d}\phi = P(T, p)\,\mathrm{d}T + Q(T, p)\,\mathrm{d}p$$

we find that

$$P = \frac{R}{p} \quad \text{and} \quad Q = -\frac{RT}{p^2}$$

For $\mathrm{d}\phi$ to be exact differential, we must have

$$\left(\frac{\partial P}{\partial p}\right)_T = \left(\frac{\partial Q}{\partial T}\right)_p$$

which is true since

$$\left(\frac{\partial P}{\partial p}\right)_T = \left\{\frac{\partial(R/p)}{\partial p}\right\}_T = -\frac{R}{p^2} \quad \text{and} \quad \left(\frac{\partial Q}{\partial T}\right)_P = \left\{\frac{\partial(-RT/p^2)}{\partial T}\right\}_p = -\frac{R}{p^2}$$

Exact Differential and State Function

Let us review some properties of the exact differential $\mathrm{d}\phi(x, y)$ and the function $\phi(x, y)$. First, the integral of $\mathrm{d}\phi(x, y)$ between any two points will be a function of the end points only and will be independent of the path. This may be represented as

$$\int_\mathrm{A}^\mathrm{B} \mathrm{d}\phi(x, y) = \phi(x, y)\Big|_\mathrm{A}^\mathrm{B} = \phi_\mathrm{B} - \phi_\mathrm{A} = \Delta\phi \qquad (1.3.38)$$

The term $\Delta\phi$ is then just $(\phi_B - \phi_A)$, that is, its values is dependent on the difference $(\phi_B - \phi_A)$ and not on the path in between. Further, the cyclic integral becomes

$$\oint d\phi(x,y) = \int_A^B d\phi(x,y) + \int_B^A d\phi(x,y) = 0$$

or $$\int_A^B d\phi(x,y) = -\int_B^A d\phi(x,y) \tag{1.3.39}$$

A function which satisfies above requirements is called a *state function.*

Essential Criteria of a State Function

The essential criteria of a state function are as follows:

- The change in the value of a state function depends only the initial and final states and not on the path of the process carried out in going from initial state to final state.
- The cyclic integration involving a state function is zero.
- The state function has an exact differential, i.e. if $p = f(T, V)$ is a state function then

$$dp = \left(\frac{\partial p}{\partial T}\right)_V dT + \left(\frac{\partial p}{\partial V}\right)_T dV$$

with the condition that

$$\frac{\partial}{\partial V}\left\{\left(\frac{\partial p}{\partial T}\right)_V\right\}_T = \frac{\partial}{\partial T}\left\{\left(\frac{\partial p}{\partial V}\right)_T\right\}_V$$

- All thermodynamic properties satisfy the requirements of state function. A few of them are

$\Delta U = q + w$	Change in thermodynamic energy
$S = \dfrac{q_{\text{rev}}}{T}$	Entropy
$H = U + pV$	Enthalpy
$G = H - TS$	Gibb's free energy
$A = U - TS$	Helmholtz free energy
$\mu_i = \left(\dfrac{\partial G}{\partial n_i}\right)_{T,p,n_j,j\neq i}$	Chemical potential

Problem 1.3.6 Show that the volume of a fixed amount of an ideal gas is a state function.

Solution For an ideal gas $V = nRT/p$, therefore

$$dV = \left(\frac{\partial V}{\partial T}\right)_p = \frac{nR}{p} \quad \text{and} \quad \left(\frac{\partial V}{\partial p}\right)_T = -\frac{nRT}{p^2}$$

Substituting these in the relation

$$dV = \left(\frac{\partial V}{\partial p}\right)_T dp + \left(\frac{\partial V}{\partial T}\right)_p dT$$

we have $$dV = -\frac{nRT}{p^2}dp + \frac{nR}{p}dT$$

If V is to be a state function, $\mathrm{d}V$ must be an exact differential, for which the Euler's reciprocity condition states that

$$\frac{\partial}{\partial T}\left\{\left(\frac{\partial V}{\partial p}\right)_T\right\}_p = \frac{\partial}{\partial p}\left\{\left(\frac{\partial V}{\partial T}\right)_p\right\}_T$$

Evaluating these, we get

$$\frac{\partial}{\partial T}\left\{\left(\frac{\partial V}{\partial p}\right)_T\right\}_p = \frac{\partial}{\partial T}\left(-\frac{nRT}{p^2}\right)_p = -\frac{nR}{p^2}$$

$$\frac{\partial}{\partial p}\left\{\left(\frac{\partial V}{\partial T}\right)_p\right\}_T = \frac{\partial}{\partial p}\left(\frac{nR}{p}\right)_T = -\frac{nR}{p^2}$$

Since $(\partial^2 V/\partial T\,\partial p)$ and $(\partial^2 V/\partial p\,\partial T)$ are identical, the volume of an ideal gas is a state function.

Problem 1.3.7

Show that pressure is a state function for a gas obeying

$$\left(p + \frac{a}{V_\mathrm{m}^2}\right)(V_\mathrm{m}) = RT$$

Solution

Rewriting the given equation as

$$p = \frac{RT}{V_\mathrm{m}} - \frac{a}{V_\mathrm{m}^2}$$

Therefore $\left(\frac{\partial p}{\partial V_\mathrm{m}}\right)_T = -\frac{RT}{V_\mathrm{m}^2} + \frac{2a}{V_\mathrm{m}^3}$; $\left(\frac{\partial p}{\partial T}\right)_{V_\mathrm{m}} = \frac{R}{V_\mathrm{m}}$

$$\frac{\partial^2 p}{\partial T\,\partial V_\mathrm{m}} = -\frac{R}{V_\mathrm{m}^2}; \quad \text{and} \quad \frac{\partial^2 p}{\partial V_\mathrm{m}\,\partial T} = -\frac{R}{V_\mathrm{m}^2}$$

Hence $\frac{\partial^2 p}{\partial T\,\partial V_\mathrm{m}} = \frac{\partial^2 p}{\partial V_\mathrm{m}\,\partial T}$

Therefore, $\mathrm{d}p$ is an exact differential and p is a state function.

Differential Expression in Three or More Variables

Many application of thermodynamics involve more than two independent variables. A differential expression involving more than two variables (say, for example, three variables x, y, z) will be of the type

$$\mathrm{d}\phi = M(x, y, z)\,\mathrm{d}x + N(x, y, z)\,\mathrm{d}y + P(x, y, z)\,\mathrm{d}z \tag{1.3.40}$$

As in the case of two variables, $\mathrm{d}\phi$ can be either exact or inexact. If it is exact, then ϕ is a state function and its total differential will also be given by

$$\mathrm{d}\phi = \left(\frac{\partial\phi}{\partial x}\right)_{y,z}\mathrm{d}x + \left(\frac{\partial\phi}{\partial y}\right)_{x,z}\mathrm{d}y + \left(\frac{\partial\phi}{\partial z}\right)_{x,y}\mathrm{d}z \tag{1.3.41}$$

Comparing Eq. (1.3.40) with Eq. (1.3.41), we get

$$M(x, y, z) = \left(\frac{\partial \phi}{\partial x}\right)_{y,z} \;;\; N(x,y,z) = \left(\frac{\partial \phi}{\partial y}\right)_{x,z} \;;\; P(x,y,z) = \left(\frac{\partial \phi}{\partial z}\right)_{x,y}$$

Equation (1.3.40) can be tested for its exactness with the help of reciprocity relations (one between each pair of variables). These are

$$\left(\frac{\partial M}{\partial y}\right)_{x,z} = \left(\frac{\partial N}{\partial x}\right)_{y,z};\; \left(\frac{\partial N}{\partial z}\right)_{x,y} = \left(\frac{\partial P}{\partial y}\right)_{x,z};\; \left(\frac{\partial P}{\partial x}\right)_{y,z} = \left(\frac{\partial M}{\partial z}\right)_{x,y} \qquad (1.3.42)$$

Problem 1.3.8 For an ideal gas $pV = nRT$. Taking $V = f(n, p, T)$, verify that $\mathrm{d}V$ is an exact differential.

Solution For $V = f(n, p, T)$, the total differential of V is

$$\mathrm{d}V = \left(\frac{\partial V}{\partial n}\right)_{p,T} \mathrm{d}n + \left(\frac{\partial V}{\partial p}\right)_{n,T} \mathrm{d}p + \left(\frac{\partial V}{\partial T}\right)_{n,p} \mathrm{d}T$$

For $\mathrm{d}V$ to be an exact differential, we must have

$$\frac{\partial}{\partial p}\left\{\left(\frac{\partial V}{\partial n}\right)_{p,T}\right\}_{n,T} = \frac{\partial}{\partial n}\left\{\left(\frac{\partial V}{\partial p}\right)_{n,T}\right\}_{p,T} \qquad (1.3.43)$$

$$\frac{\partial}{\partial T}\left\{\left(\frac{\partial V}{\partial n}\right)_{p,T}\right\}_{n,p} = \frac{\partial}{\partial n}\left\{\left(\frac{\partial V}{\partial T}\right)_{n,p}\right\}_{p,T} \qquad (1.3.44)$$

and

$$\frac{\partial}{\partial T}\left\{\left(\frac{\partial V}{\partial p}\right)_{n,T}\right\}_{n,p} = \frac{\partial}{\partial p}\left\{\left(\frac{\partial V}{\partial T}\right)_{n,p}\right\}_{n,T} \qquad (1.3.45)$$

Now for the given gas $pV = nRT$, we have

$$\left(\frac{\partial V}{\partial n}\right)_{p,T} = \frac{RT}{p};\; \left(\frac{\partial V}{\partial p}\right)_{n,T} = -\frac{nRT}{p^2};\; \left(\frac{\partial V}{\partial T}\right)_{n,p} = \frac{nR}{p}$$

and

$$\frac{\partial}{\partial p}\left\{\left(\frac{\partial V}{\partial n}\right)_{p,T}\right\}_{n,T} = \left\{\frac{\partial(RT/p)}{\partial p}\right\}_{n,T} = -\frac{RT}{p^2}$$

$$\frac{\partial}{\partial n}\left\{\left(\frac{\partial V}{\partial p}\right)_{n,T}\right\}_{p,T} = \left\{\frac{\partial(-nRT/p^2)}{\partial n}\right\}_{p,T} = -\frac{RT}{p^2}$$

$$\frac{\partial}{\partial T}\left\{\left(\frac{\partial V}{\partial n}\right)_{p,T}\right\}_{n,p} = \left\{\frac{\partial(RT/p)}{\partial T}\right\}_{n,p} = \frac{R}{p}$$

$$\frac{\partial}{\partial n}\left\{\left(\frac{\partial V}{\partial T}\right)_{n,p}\right\}_{p,T} = \left\{\frac{\partial(nR/p)}{\partial n}\right\}_{p,T} = \frac{R}{p}$$

$$\frac{\partial}{\partial T}\left\{\left(\frac{\partial V}{\partial p}\right)_{n,T}\right\}_{n,p} = \left\{\frac{\partial(-nRT/p^2)}{\partial T}\right\}_{n,p} = -\frac{nR}{p^2}$$

$$\frac{\partial}{\partial p}\left\{\left(\frac{\partial V}{\partial T}\right)_{n,p}\right\}_{n,T} = \left\{\frac{\partial(nR/p)}{\partial p}\right\}_{n,T} = -\frac{nR}{p^2}$$

From the expressions of second differentials, we find that Eqs (1.3.43) to (1.3.45) are satisfied. Hence, dV is an exact differential for an ideal gas.

Integrating Factor

An inexact differential expression $P\mathrm{d}x + Q\,\mathrm{d}y$ (with $\partial P/\partial y \neq \partial Q/\partial x$) can be converted into an exact one by means of an integrating factor $G(x, y)$. In that case, $G(P\,\mathrm{d}x + Q\,\mathrm{d}y)$ becomes exact, that is

$$\frac{\partial(GP)}{\partial y} = \frac{\partial(GQ)}{\partial x}$$

Take, for example, the differential expression

$$đ\phi = y\,\mathrm{d}x - x\,\mathrm{d}y$$

It can be converted into an exact differential by choosing $G(x, y) = 1/x^2$. Thus, we have

$$\frac{1}{x^2}\,đ\phi = \mathrm{d}f = \frac{y}{x^2}\mathrm{d}x - \frac{1}{x}\mathrm{d}y$$

It can be seen that

$$\frac{\partial}{\partial y}\left\{\frac{y}{x^2}\right\}_x = \frac{\partial}{\partial x}\left(-\frac{1}{x}\right)_y$$

Thus, $\mathrm{d}f$ is an exact differential. It can be verified that the function $f = -y/x$. Alternatively, the integrating factors $1/y^2$, $1/xy$ and $1/(x^2 + y^2)$ convert the above relation to $\mathrm{d}(x/y)$, $\mathrm{d}[\ln(x/y)]$ and $\mathrm{d}[\arctan(x/y)]$, respectively. Evidently the choice of $G(x, y)$ is not unique.

Another case of integrating factor is cited by the differentials given in Problems (1.3.4) and (1.3.5). Differential of Problem (1.3.5) can be obtained by multiplying the differential of Problem (1.3.4) by the factor $(-1/p)$. Since differential of Problem (1.3.4) is inexact whereas that of Problem (1.3.5) is exact, it is obvious that $(-1/p)$ is an integrating factor.

The mathematical formulation of the first and second laws of thermodynamics is based on the construction of exact differentials from inexact ones. Thus the first law postulates that even though $\mathrm{d}q$, the heat exchanged by a system, and $\mathrm{d}w$, the work involved in the system, are individually inexact differentials, but the sum of these two (i.e. $\mathrm{d}q + \mathrm{d}w = \mathrm{d}U$) is an exact differential. This constitutes a definition of the internal energy U. The second law postulates that $1/T$ is an integrating factor for $\mathrm{d}q_{\mathrm{rev}}$. Thus $\mathrm{d}S = \mathrm{d}q_{\mathrm{rev}}/T$ is exact, which defines the state function entropy S.

Legendre Transformation

The principle of Legendre transformation can be used to modify a differential expression so as to change its independent variables. For example, take the following exact differential expression:

$$dF(x, y) = M(x, y)\, dx + N(x, y)\, dy \tag{1.3.46}$$

Let a function ϕ be defined as

$$\phi = F - Mx \tag{1.3.47}$$

Its differential is given by

$$d\phi = dF - M\, dx - x\, dM$$

Substituting dF from Eq. (1.3.46), we get

$$d\phi = N\, dy - x\, dM \tag{1.3.48}$$

The differential is appropriate for a function $\phi = f(y, M)$. In transforming a function of x and y into a function of M and y, the independent variable x and the corresponding coefficient M have thereby exchange roles with the change of sign. Pairs of variables which can be interchanged by a Legendre transformation such as M and x or N and y are said to be conjugate to one another. The prescription for one of these transformations is given by Eq. (1.3.47) in which a new function is defined by subtracting from the original one the appropriate product of conjugate variables.

The most important application of Legendre transformations in thermodynamics is based upon the differential relation obtained by union of the first and second laws:

$$dU = T\, dS - p\, dV \tag{1.3.49}$$

The three possible transformations of this relation are

$$dH = T\, dS + V\, dp$$

$$dA = -S\, dT - p\, dV$$

$$dG = -S\, dT + V\, dp$$

where $H = U + pV$; $A = U - TS$; and $G = U + pV - TS = A + pV$

Problem 1.3.9

From the following thermodynamic relation

$$G = H - TS\,; \quad H = U + pV\,; \quad dq_{rev} = T\, dS\,; \quad dU = dq_{rev} - p\, dV$$

Show that, $(\partial G/\partial T)_p = -S$; $(\partial G/\partial p)_T = V$ and $\left(\dfrac{\partial V}{\partial T}\right)_p = -\left(\dfrac{\partial S}{\partial p}\right)_T$

Solution

We start with

$$G = H - TS$$

Since $H = U + pV$, therefore

$$G = U + pV - TS$$

Differential of this expression is

$$dG = dU + p\, dV + V\, dp - T\, dS - S\, dT$$

Now $T\,dS = dq_{rev} = dU + p\,dV,$ therefore

$$dG = V\,dp - S\,dT \tag{1.3.50}$$

Thus, we establish that G is a function of T and p. Moreover since G is a state function, therefore, we have

$$dG = \left(\frac{\partial G}{\partial p}\right)_T dp + \left(\frac{\partial G}{\partial T}\right)_p dT \tag{1.3.51}$$

Comparing Eqs (1.3.50) and (1.3.51), we get

$$\left(\frac{\partial G}{\partial p}\right)_T = V \quad \text{and} \quad \left(\frac{\partial G}{\partial p}\right)_p = -S$$

Applying Euler's reciprocity relation to Eq. (1.3.50), we get

$$\left(\frac{\partial V}{\partial T}\right)_p = -\left(\frac{\partial S}{\partial p}\right)_T$$

Problem 1.3.10 From the following thermodynamic relations

$$A = U - TS; \qquad dq_{rev} = T\,dS; \qquad dU = dq_{rev} - p\,dV$$

show that

$$\left(\frac{\partial A}{\partial V}\right)_T = -p; \qquad \left(\frac{\partial A}{\partial T}\right)_V = -S \quad \text{and} \quad \left(\frac{\partial p}{\partial T}\right)_V = \left(\frac{\partial S}{\partial V}\right)_T$$

Solution Since $A = U - TS,$ we get

$$dA = dU - T\,dS - S\,dT$$

Now $T\,dS = dq_{rev} = dU + p\,dV,$ therefore

$$dA = -p\,dV - S\,dT \tag{1.3.52}$$

Thus, we establish that A is a function of T and V. Moreover, since A is a state function. Therefore, we have

$$dA = \left(\frac{\partial A}{\partial V}\right)_T dV + \left(\frac{\partial A}{\partial T}\right)_V dT \tag{1.3.53}$$

Comparing Eqs (1.3.52) and (1.3.53), we obtain

$$\left(\frac{\partial A}{\partial V}\right)_T = -p \quad \text{and} \quad \left(\frac{\partial A}{\partial T}\right)_V = -S$$

Applying Euler's reciprocity relation to Eq. (1.3.52), we get

$$\left(\frac{\partial p}{\partial T}\right)_V = \left(\frac{\partial S}{\partial V}\right)_T$$

Miscellaneous Problems

Problem 1.3.11 Using the definition $H = U + pV$ and, when necessary, obtaining conversion relationship by considering H (or U) as a function of any two of the variables p, V and T, derive the following relationships:

(i) $$\left(\frac{\partial H}{\partial T}\right)_p = \left(\frac{\partial U}{\partial T}\right)_V + \left[p + \left(\frac{\partial U}{\partial V}\right)_T\right]\left(\frac{\partial V}{\partial T}\right)_p$$

(ii) $$\left(\frac{\partial H}{\partial T}\right)_p = \left(\frac{\partial U}{\partial T}\right)_V + \left[V - \left(\frac{\partial H}{\partial p}\right)_T\right]\left(\frac{\partial p}{\partial T}\right)_V$$

(iii) $$\left(\frac{\partial H}{\partial T}\right)_p = \left(\frac{\partial U}{\partial T}\right)_V + \left[V + \left(\frac{\partial H}{\partial p}\right)_p\left(\frac{\partial T}{\partial p}\right)_H\right]\left(\frac{\partial p}{\partial T}\right)_V$$

Solution **(i)** Differentiating the given relation $H = U + pV$, we get

$$\mathrm{d}H = \mathrm{d}U + p\,\mathrm{d}V + V\,\mathrm{d}p$$

Taking $H = f(T, p)$ and $U = f(T, V)$ and replacing $\mathrm{d}H$ and $\mathrm{d}U$ in the above equation by

$$\mathrm{d}H = \left(\frac{\partial H}{\partial T}\right)_p \mathrm{d}T + \left(\frac{\partial H}{\partial p}\right)_T \mathrm{d}p$$

$$\mathrm{d}U = \left(\frac{\partial U}{\partial T}\right)_V \mathrm{d}T + \left(\frac{\partial U}{\partial V}\right)_T \mathrm{d}V$$

we have $$\left(\frac{\partial H}{\partial T}\right)_p \mathrm{d}T + \left(\frac{\partial H}{\partial p}\right)_T \mathrm{d}p = \left(\frac{\partial U}{\partial T}\right)_V \mathrm{d}T + \left(\frac{\partial U}{\partial V}\right)_T \mathrm{d}V + p\,\mathrm{d}V + V\,\mathrm{d}p \tag{1.3.54}$$

Dividing by $\mathrm{d}T$, keeping p constant, we have

$$\left(\frac{\partial H}{\partial T}\right)_p = \left(\frac{\partial U}{\partial T}\right)_V + \left(\frac{\partial U}{\partial V}\right)_T\left(\frac{\partial V}{\partial T}\right)_p + p\left(\frac{\partial V}{\partial T}\right)_p$$

or $$\left(\frac{\partial H}{\partial T}\right)_p = \left(\frac{\partial U}{\partial T}\right)_V + \left[p + \left(\frac{\partial U}{\partial V}\right)_T\right]\left(\frac{\partial V}{\partial T}\right)_p \tag{1.3.55}$$

(ii) Dividing Eq. (1.3.54) by $\mathrm{d}T$, keeping V constant, we have

$$\left(\frac{\partial H}{\partial T}\right)_p + \left(\frac{\partial H}{\partial p}\right)_T\left(\frac{\partial p}{\partial T}\right)_V = \left(\frac{\partial U}{\partial T}\right)_V + V\left(\frac{\partial p}{\partial T}\right)_V$$

or $$\left(\frac{\partial H}{\partial T}\right)_p = \left(\frac{\partial U}{\partial T}\right)_V + \left[V - \left(\frac{\partial H}{\partial p}\right)_T\right]\left(\frac{\partial p}{\partial T}\right)_V \tag{1.3.56}$$

(iii) The cyclic rule for $H = f(T, p)$ is

$$\left(\frac{\partial H}{\partial p}\right)_T\left(\frac{\partial p}{\partial T}\right)_H\left(\frac{\partial T}{\partial H}\right)_p + 1 = 0$$

Rearranging this in the form

$$\left(\frac{\partial H}{\partial p}\right)_T = -\left(\frac{\partial T}{\partial p}\right)_H\left(\frac{\partial H}{\partial T}\right)_p$$

and then substituting in Eq. (1.3.56), we have

$$\left(\frac{\partial H}{\partial T}\right)_p = \left(\frac{\partial U}{\partial T}\right)_V + \left[V + \left(\frac{\partial H}{\partial T}\right)_p\left(\frac{\partial T}{\partial p}\right)_H\right]\left(\frac{\partial p}{\partial T}\right)_V \tag{1.3.57}$$

Problem 1.3.12 Considering U as a function of any two of the variables p, V and T, prove that

(i) $$\left(\frac{\partial U}{\partial T}\right)_p\left(\frac{\partial T}{\partial p}\right)_V = -\left(\frac{\partial U}{\partial V}\right)_p\left(\frac{\partial V}{\partial p}\right)_T$$

(ii) $$\left(\frac{\partial U}{\partial p}\right)_T = \left(\frac{\partial U}{\partial V}\right)_T\left(\frac{\partial V}{\partial p}\right)_T$$

Solution **(i)** Taking $U = f(T, p)$ we get

$$\mathrm{d}U = \left(\frac{\partial U}{\partial T}\right)_p \mathrm{d}T + \left(\frac{\partial U}{\partial p}\right)_T \mathrm{d}p$$

Dividing by dp and introducing the condition of constant V, we have

$$\left(\frac{\partial U}{\partial p}\right)_V = \left(\frac{\partial U}{\partial T}\right)_p\left(\frac{\partial T}{\partial p}\right)_V + \left(\frac{\partial U}{\partial p}\right)_T$$

or $$\left(\frac{\partial U}{\partial p}\right)_V - \left(\frac{\partial U}{\partial p}\right)_T = \left(\frac{\partial U}{\partial T}\right)_p\left(\frac{\partial T}{\partial p}\right)_V \qquad (1.3.58)$$

Taking now $U = f(p, V)$, we get

$$\mathrm{d}U = \left(\frac{\partial U}{\partial p}\right)_V \mathrm{d}p + \left(\frac{\partial U}{\partial V}\right)_p \mathrm{d}V$$

Dividing by dp and introducing the condition of constant T, we have

$$\left(\frac{\partial U}{\partial p}\right)_T = \left(\frac{\partial U}{\partial p}\right)_V + \left(\frac{\partial U}{\partial V}\right)_p\left(\frac{\partial V}{\partial p}\right)_T$$

or $$\left(\frac{\partial U}{\partial p}\right)_T - \left(\frac{\partial U}{\partial p}\right)_V = \left(\frac{\partial U}{\partial V}\right)_p\left(\frac{\partial V}{\partial p}\right)_T \qquad (1.3.59)$$

Comparing Eq. (1.3.58) and Eq. (1.3.59), we have the required relation

$$\left(\frac{\partial U}{\partial T}\right)_p\left(\frac{\partial T}{\partial p}\right)_V = -\left(\frac{\partial U}{\partial V}\right)_p\left(\frac{\partial V}{\partial p}\right)_T$$

(ii) Since $U = f(T, V)$, we get

$$\mathrm{d}U = \left(\frac{\partial U}{\partial T}\right)_V \mathrm{d}T + \left(\frac{\partial U}{\partial V}\right)_T \mathrm{d}V$$

Dividing by dp and introducing the condition of constant temperature (dT = 0), we have

$$\left(\frac{\partial U}{\partial p}\right)_T = \left(\frac{\partial U}{\partial V}\right)_T\left(\frac{\partial V}{\partial p}\right)_T$$

Problem 1.3.13 Cubic expansion coefficient, α (or expansivity in short, also formerly known as coefficient of thermal expansion) and isothermal compressibility, κ_T (formerly known as compressibility factor) are defined as:

$$\alpha = \frac{1}{V}\left(\frac{\partial V}{\partial T}\right)_p; \qquad \kappa_T = -\frac{1}{V}\left(\frac{\partial V}{\partial p}\right)_T$$

Show that

(i) $\dfrac{\alpha}{\kappa_T} = \left(\dfrac{\partial p}{\partial T}\right)_V$ and (ii) $\left(\dfrac{\partial \alpha}{\partial p}\right)_T + \left(\dfrac{\partial \kappa_T}{\partial T}\right)_p = 0$

Solution **(i)** Taking $V = f(p, T)$, we get

$$dV = \left(\frac{\partial V}{\partial p}\right)_T dp + \left(\frac{\partial V}{\partial T}\right)_p dT$$

Dividing this by dT and introducing the constant volume condition, we have

$$0 = \left(\frac{\partial V}{\partial p}\right)_T \left(\frac{\partial p}{\partial T}\right)_V + \left(\frac{\partial V}{\partial T}\right)_p$$

that is $$\left(\frac{\partial p}{\partial T}\right)_V = -\frac{(\partial V/\partial T)_p}{(\partial V/\partial p)_T} = \frac{V\alpha}{V\kappa_T} = \frac{\alpha}{\kappa_T} \qquad (1.3.60a)$$

(ii) Since V is a state function, and dV is an exact differential, using Euler's reciprocity relation, we have

$$\frac{\partial}{\partial T}\left\{\left(\frac{\partial V}{\partial p}\right)_T\right\}_p = \frac{\partial}{\partial p}\left\{\left(\frac{\partial V}{\partial T}\right)_p\right\}_T$$

or $$\frac{\partial}{\partial T}(-V\kappa_T)_p = \frac{\partial}{\partial p}(V\alpha)_T$$

Carrying out the differentiation, we get

$$-\kappa_T\left(\frac{\partial V}{\partial T}\right)_p - V\left(\frac{\partial \kappa_T}{\partial T}\right)_p = \alpha\left(\frac{\partial V}{\partial p}\right)_T + V\left(\frac{\partial \alpha}{\partial p}\right)_T$$

that is $$\left(\frac{\partial \alpha}{\partial p}\right)_T + \left(\frac{\partial \kappa_T}{\partial T}\right)_p = -\frac{\kappa_T}{V}\left(\frac{\partial V}{\partial T}\right)_p - \frac{\alpha}{V}\left(\frac{\partial V}{\partial p}\right)_T = -\frac{\kappa_T}{V}(\alpha V) - \frac{\alpha}{V}(-\kappa_T V) = 0$$

(1.3.60b)

Problem 1.3.14 Derive the expressions for α and κ_T for one mole of (i) an ideal gas and (ii) a van der Waals gas.

Solution **(i)** For one mole of an ideal gas

$$pV_m = RT$$

On differentiating, we get

$$p\, dV_m + V_m\, dp = R\, dT \qquad (1.3.61)$$

Dividing by dT at constant p, we have

$$p\left(\frac{\partial V_m}{\partial T}\right)_p = R$$

that is $$\alpha = \frac{1}{V_m}\left(\frac{\partial V_m}{\partial T}\right)_p = \frac{R}{V_m p} = \frac{1}{T}$$

Dividing Eq. (1.3.61) by dp at constant T, we have

$$p\left(\frac{\partial V_m}{\partial p}\right)_T + V_m = 0 \quad \text{or} \quad \left(\frac{\partial V_m}{\partial p}\right)_T = -\frac{V_m}{p}$$

Hence
$$\kappa_T = -\frac{1}{V_m}\left(\frac{\partial V_m}{\partial p}\right)_T = \left(-\frac{1}{V_m}\right)\left(-\frac{V_m}{p}\right) = \frac{1}{p}$$

(ii) For one mole of a van der Waals gas

$$\left(p + \frac{a}{V_m^2}\right)(V_m - b) = RT$$

or
$$pV_m + \frac{a}{V_m} - pb - \frac{ab}{V_m^2} = RT$$

or
$$pV_m^3 + aV_m - pV_m^2 b - ab = V_m^2 RT$$

Differentiation gives

$$V_m^3\,dp + 3V_m^2 p\,dV_m + a\,dV_m - V_m^2 b\,dp - 2pV_m b\,dV_m = 2V_m RT\,dV_m + V_m^2 R\,dT \quad (1.3.62)$$

Dividing by dT and introducing the condition of constant p (i.e. dp = 0), we have

$$3V_m^2 p\left(\frac{\partial V_m}{\partial T}\right)_p + a\left(\frac{\partial V_m}{\partial T}\right)_p - 2pV_m b\left(\frac{\partial V_m}{\partial T}\right)_p = 2V_m RT\left(\frac{\partial V_m}{\partial T}\right)_p + V_m^2 R$$

Therefore
$$\left(\frac{\partial V_m}{\partial T}\right)_p = \frac{V_m^2 R}{3V_m^2 p + a - 2pV_m b - 2V_m RT}$$

and
$$\alpha = \frac{1}{V_m}\left(\frac{\partial V_m}{\partial T}\right)_p = \frac{R}{3pV_m + \dfrac{a}{V_m} - 2pb - 2RT} = \frac{R}{pV_m - \dfrac{a}{V_m} + \dfrac{2ab}{V_m^2}}$$

Dividing Eq. (1.3.62) by dp and introducing the condition of constant temperature, we have

$$V_m^3 + 3pV_m^2\left(\frac{\partial V_m}{\partial p}\right)_T + a\left(\frac{\partial V_m}{\partial p}\right)_T - V_m^2 b - 2pV_m b\left(\frac{\partial V_m}{\partial p}\right)_T = 2V_m RT\left(\frac{\partial V_m}{\partial p}\right)_T$$

Therefore
$$\left(\frac{\partial V_m}{\partial p}\right)_T = \frac{V_m^2 b - V_m^3}{3pV_m^2 + a - 2pV_m b - 2V_m RT} = \frac{V_m(b - V_m)}{pV_m - \dfrac{a}{V_m} + \dfrac{2ab}{V_m^2}}$$

and
$$\kappa_T = -\frac{1}{V_m}\left(\frac{\partial V_m}{\partial p}\right)_T = \frac{(V_m - b)}{\left(pV_m - \dfrac{a}{V_m} + \dfrac{2ab}{V_m^2}\right)}$$

Problem 1.3.15

Taking V as a state function, derive the equation of state for which

(i) $V = k_1/p$, keeping T constant and $V = k_2T$, keeping p constant

(ii) $\alpha = (V - a)/TV$ and $\kappa_T = 3(V - a)/4pV$, where a is constant

Solution **(i)** Since $V = k_1/p$ for constant temperature, from where we can get

$$\left(\frac{\partial V}{\partial p}\right)_T = -\frac{k_1}{p^2} = -\frac{pV}{p^2} = -\frac{V}{p}$$

Also $V = k_2T$ for constant pressure, from where we can get

$$\left(\frac{\partial V}{\partial T}\right)_p = k_2 = \frac{V}{T}$$

Since $V = f(p, T)$, therefore

$$\mathrm{d}V = \left(\frac{\partial V}{\partial p}\right)_T \mathrm{d}p + \left(\frac{\partial V}{\partial T}\right)_p \mathrm{d}T$$

Substituting the values of $(\partial V/\partial p)_T$ and $(\partial V/\partial T)_p$, we get

$$\mathrm{d}V = -\frac{V}{p}\,\mathrm{d}p + \frac{V}{T}\,\mathrm{d}T$$

Dividing by V and rearranging, we get

$$\frac{\mathrm{d}V}{V} + \frac{\mathrm{d}p}{p} = \frac{\mathrm{d}T}{T}$$

Integrating the above expression, we get

$$\ln\left(\frac{V_2}{V_1}\right) + \ln\left(\frac{p_2}{p_1}\right) = \ln\left(\frac{T_2}{T_1}\right) \quad \text{or} \quad \ln\left(\frac{p_2V_2}{p_1V_1}\right) = \ln\left(\frac{T_2}{T_1}\right)$$

or $$\frac{p_2V_2}{T_2} = \frac{p_1V_1}{T_1} = \text{constant} \quad \text{or} \quad pV = RT \qquad \text{(where } R \text{ is a constant)}$$

(ii) Taking $V = f(p, T)$ we have

$$\mathrm{d}V = \left(\frac{\partial V}{\partial p}\right)_T \mathrm{d}p + \left(\frac{\partial V}{\partial T}\right)_p \mathrm{d}T$$

Replacing $(\partial V/\partial p)_T$ and $(\partial V/\partial T)_p$ in terms of κ_T and α, respectively, we have

$$\mathrm{d}V = -V\kappa_T\,\mathrm{d}p + V\alpha\,\mathrm{d}T$$

Substituting the expressions of α and κ_T, we get

$$\mathrm{d}V = -V\frac{3(V-a)}{4pV}\mathrm{d}p + V\frac{V-a}{TV}\mathrm{d}T$$

which on rearrangement gives

$$\frac{\mathrm{d}V}{V-a} = -\frac{3}{4}\frac{\mathrm{d}p}{p} + \frac{\mathrm{d}T}{T}$$

Integrating both sides, we get

$$\ln\left(\frac{V_2-a}{V_1-a}\right) + \frac{3}{4}\ln\left(\frac{p_2}{p_1}\right) = \ln\left(\frac{T_2}{T_1}\right) \quad \text{or} \quad \ln\frac{p_2^{3/4}(V_2-a)}{p_1^{3/4}(V_1-a)} = \ln\frac{T_2}{T_1}$$

or $$\frac{p_2^{3/4}(V_2-a)}{T_2} = \frac{p_1^{3/4}(V_1-a)}{T_1} = \text{constant} \quad \text{or} \quad p^{3/4}(V-a) = AT;$$

(where A is constant)

Homogeneous Functions

A function $f(x, y, ...)$ is said to be a homogeneous function of degree n if the following condition is satisfied:

$$f(\lambda x, \lambda y, \ldots) = \lambda^n f(x, y, \ldots) \tag{1.3.63}$$

where λ is an arbitrary parameter and n has a constant integer value. Take, for example

(i) $f(x, y) = ax + by$

which as a homogeneous function is

$$f(\lambda x, \lambda y) = a(\lambda x) + b(\lambda y) = \lambda(ax + by) = \lambda f(x, y)$$

(ii) $f(x, y) = ax^2 + bxy + cy^2$

which yields

$$f(\lambda x, \lambda y) = a(\lambda x)^2 + b(\lambda x)(\lambda y) + c(\lambda y)^2$$
$$= \lambda^2 (ax^2 + bxy + cy^2) = \lambda^2 f(x, y)$$

If $n = 0$, we have a homogeneous function of zero degree; for $n = 1$, homogeneous function is of the first degree, etc.

Degree of Intensive and Extensive Variables

All functions of zero degree with respect to the amounts of substances are intensive variables whereas those of the first degree are extensive variables.

Take, for example, an ideal binary liquid solution of benzene and toluene. The total volume of the solution is given by

$$V = n_b V_b + n_t V_t \tag{1.3.64}$$

where n_b and n_t are the amounts of benzene and toluene, respectively, and V_b and V_t are their respective molar volumes. Suppose that the values of n_b and n_t are doubles, then the total volume will be given by

$$V' = (2n_b)V_b + (2n_t)V_t = 2^1(n_b V_b + n_t V_t) = 2^1 V$$

Since the exponent of the parameter 2 is 1, therefore, volume of a solution is a homogeneous function of the first degree with respect to the amounts of its constituents. The relation as given in Eq. (1.3.64) is true not only for ideal solutions but also for nonideal solutions. However, for the latter, partial molar volumes should be used instead of the molar volumes of the pure components (*see* also Eq. 1.3.66).

Euler's Theorem

If a function is homogeneous of degree n, according to Euler's theorem, it should satisfy the following relation

$$x\left(\frac{\partial f}{\partial x}\right)_{y,z,\ldots} + y\left(\frac{\partial f}{\partial y}\right)_{x,z,\ldots} + \cdots = n\, f(x, y, \cdots) \tag{1.3.65}$$

The Euler's theorem can be proved as follows.

Differentiating Eq. (1.3.63) with respect to λ, we get

$$\frac{\mathrm{d}f(\lambda x, \lambda y, \cdots)}{\mathrm{d}\lambda} = \frac{\mathrm{d}\{\lambda^n f(x, y, \cdots)\}}{\mathrm{d}\lambda}$$

i.e. $$\left(\frac{\partial f}{\partial(\lambda x)}\right)_{y,z,\ldots}\left(\frac{\partial(\lambda x)}{\partial\lambda}\right)+\left(\frac{\partial f}{\partial(\lambda y)}\right)_{x,z,\ldots}\left(\frac{\partial(\lambda y)}{\partial\lambda}\right)+\cdots=n\lambda^{n-1}f$$

Simplifying this, we get

$$x\left(\frac{\partial f}{\partial(\lambda x)}\right)_{y,z,\ldots}+y\left(\frac{\partial f}{\partial(\lambda y)}\right)_{x,z,\ldots}+\cdots=n\lambda^{n-1}f$$

For the special case where $\lambda = 1$, we get

$$x\left(\frac{\partial f}{\partial x}\right)_{y,z,\ldots}+y\left(\frac{\partial f}{\partial y}\right)_{x,z,\ldots}+\cdots=nf$$

which is the Euler's theorem.

Applying the Euler's theorem to the volume function, which is of the first degree in respect to the amounts of the constituents, we have

$$V=n_1\left(\frac{\partial V}{\partial n_1}\right)_{n_2,n_3,\ldots}+n_2\left(\frac{\partial V}{\partial n_2}\right)_{n_1,n_3,\ldots}+\cdots \tag{1.3.66}$$

Thus, the total volume of a solution of known composition (i.e. n_1, n_2, ... are known) is the sum of the products of the amount of each component with the slope of the plot of V versus n at the given amount of the component in the solution, when the amounts of all other components of the solution are kept constant.[†]

1.4 IUPAC CONVENTIONS OF WORK AND HEAT

According to IUPAC[*] convention, heat absorbed by the system is regarded as the positive quantity whereas the heat released by the system is regarded as the negative quantity. Conversely speaking, if the numerical value of the heat is positive, it is absorbed by the system and if it is negative, the heat is released by the system. We may spell q as the heat involved instead of specifically mentioning the phrase 'heat absorbed by the system' or 'heat released by the system'. The numerical value of q will automatically imply either of these two phrases. To be more specific, the following alternative statements bringing out the same sense may be noted down.

(i)	Heat involved	=	20 kJ
	Heat absorbed by the system	=	20 kJ
	Heat released by the system	=	–20 kJ
(ii)	Heat involved	=	–20 kJ
	Heat absorbed by the system	=	–20 kJ
	Heat released by the system	=	20 kJ

[†] The slope $(\partial V/\partial n)$ is known as the partial molar volume of the said component. See also Section. 6.1.

[*] International Union of Pure and Applied Chemistry.

For work, the IUPAC convention is to take the work done on the system as the positive quantity whereas the work done by the system is taken as the negative quantity. Conversely speaking, if the numerical value of work is positive, it implies that the work is done on the system and if it carries a negative sign, it implies that the work is done by the system. We may spell w as the work involved instead of specifically mentioning the phrase 'work done on the system' or 'work done by the system'. The numerical value of w will automatically imply either of these two phrases. To be more specific, the following alternative statements bringing out the same sense may be noted down.

(i) Work involved = 20 kJ
   Work done on the system = 20 kJ
   Work done by the system = – 20 kJ

(ii) Work involved = – 20 kJ
   Work done on the system = – 20 kJ
   Work done by the system = 20 kJ

Another convention (non-IUPAC) which was in use earlier assigns a negative sign to the work done on the system and a positive sign to work done by the system. The IUPAC convention puts energy and work on the same footing. The work done on the system, like heat added to the system, increases the internal energy of the system and thus is assigned a positive sign. The treatment followed in the text is based on IUPAC convention and can be converted to the non-IUPAC convention by replacing w by $-w$ and the phrases like maximum, minimum, greater than and lesser than by their opposite phrases, i.e. minimum, maximum, lesser than and greater than, respectively. The above replacements do not affect the defining equations of thermodynamic functions.

1.5 WORK INVOLVED IN EXPANSION AND COMPRESSION PROCESSES

In most thermodynamic calculations we will be dealing with the evaluation of work involved in the expansion or compression of gases. If the volume of the system is increased against some pressure (constant or varying), then the work is done by the system on the surroundings (or the work is produced) and is, by convention, given a negative sign. On the other hand, if the volume of the system is decreased, then the work is done by the surroundings on the system (or the work is destroyed) and is given a positive sign.

Essential Criterion of Expansion/ Compression

The essential criterion of expansion/compression is that there should exist a difference between the internal pressure of the system and the external pressure. Suppose a gas is contained in a cylinder fitted with a piston. The latter is assumed to be weightless and frictionless. The piston can be held anywhere against a set of stops. Let the piston be initially held at stops S_1 (Fig. 1.5.1). If the stops are now removed, then the position of the piston will be decided by the external pressure which can be controlled by putting different masses on the piston. If m is the mass that is put on the piston, then the force F acting downwards is

$$F = mg$$

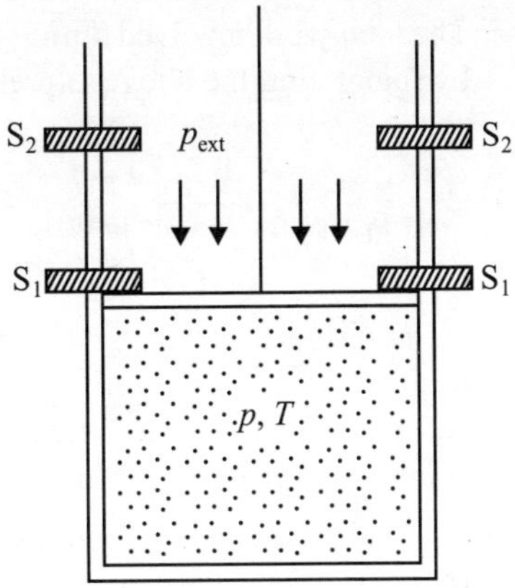

Fig. 1.5.1 Process of expansion of a gas

and the pressure acting downwards

$$p_{ext} = \frac{F}{A} = \frac{mg}{A}$$

where A is the area of cross-section of the piston.

If the external pressure is greater than the internal pressure of the system, the piston moves downward. The volume of the system will decrease and it continues to decrease till the external pressure becomes equal to the internal pressure of the system. If the external pressure is smaller than the internal pressure, the piston moves upward resulting in expansion. Again, the volume will continue to increase till the external pressure becomes equal to the internal pressure. We can stop the expansion or compression in between by providing a set of stops at that stage. During the process of expansion or compression, we can change the external pressure. Thus, the process can be achieved either in one-stage or in multistage. In the latter, the external pressure may be different in each stage.

Mathematical Expression of Work

By definition, the work involved is given by:

$$w = -\text{(External force) (Distance through which piston moves)}$$

or $$w = -\left(\frac{\text{Force}}{\text{Area of cross-section of piston}}\right)\text{(Area of cross-section of piston)} \times \text{(Distance through which piston moves)}$$

or $$w = -p_{ext}\,\mathrm{d}V \qquad (1.5.1)^{\dagger}$$

where ΔV is the change in volume of the system. If the piston moves by an infinitesimal amount, the work involved is given by

$$\mathrm{d}w = -p_{ext}\,\mathrm{d}V$$

† The negative sign in Eq. (1.5.1) is due to the IUPAC convention. In expansion (work is done by the system), ΔV has a positive value and thus work carry a negative sign. In compression (work is done on the system), ΔV has a negative value and thus work carry a positive sign.

The total work involved during the change of volume from V' to V'' can be obtained by integrating the above expression

$$w = -\int_{V'}^{V''} p_{ext}\, dV$$

If p_{ext} remains constant during the volume change, then

$$w = -p_{ext}(V'' - V') \qquad (1.5.2)$$

It can be seen that, if $V'' > V'$ then w is negative and if $V'' < V'$ then w is positive, i.e. if w is negative it automatically implies that expansion has taken place and that work is done by the system on the surroundings, and if w is positive, compression has taken place and work is done by the surroundings on the system.

If the expansion or compression is done in many stages, the total work involved is equal to the sum of the work involved in each stage, i.e.

$$\begin{aligned} w_{total} &= w_1 + w_2 + \cdots \\ &= \{-p_{ext}(V_2 - V_1)\} + \{-p'_{ext}(V_3 - V_2)\} + \cdots \end{aligned} \qquad (1.5.3)$$

Graphical Representation of Isothermal Expansion of a Gas

The work involved during the expansion or compression can be represented by the graphs drawn between p and V.

Expansion in One Stage

Let an isothermal expansion take place from V' to V'' against a constant external pressure p_{ext}, and let p' and p'' be the respective pressures of the system at these two stages. The magnitude of the work involved is given by the shaded area in the $p - V$ diagram (Fig. 1.5.2).

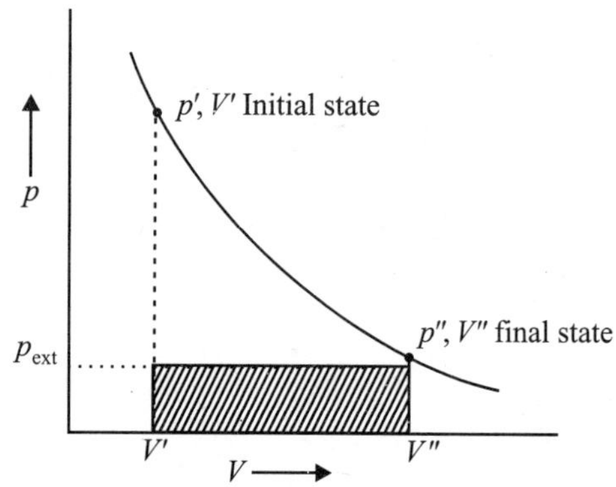

Fig. 1.5.2 Expansion of a gas against a constant pressure

Note that the external pressure has to satisfy the following condition for expansion

$$p_{ext} \leq p''$$

It $p_{ext} < p''$ the piston can be stopped by a set of stops and when $p_{ext} = p''$, it will stop automatically at the final state. Thus, we can see that the magnitude of the work involved during the expansion can vary and the range of magnitude of the work involved will be

$$0 \leq |w_{min}| \leq p''(V'' - V') \qquad (1.5.4)$$

It will be zero when the external pressure is zero (free expansion) and will be maximum when the external pressure is equal to the pressure of the system at the final state.

Expansion in Two Stages

If the above expansion is done in many stages, the magnitude of the work involved will be more than that involved in the one-stage expansion. This is evident from the graph in Fig. 1.5.3 where the magnitude of the work involved in a two-stage expansion is shown by the shaded area.

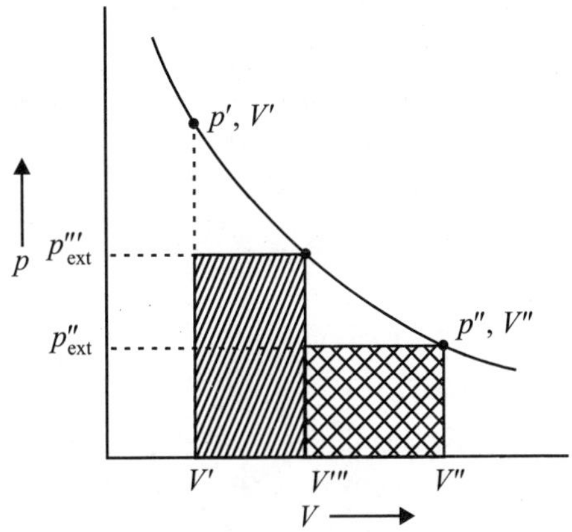

Fig. 1.5.3 Expansion in two stages

Expansion in Multistages

If the expansion is carried out involving larger number of stages and if each stage involves a constant external pressure of

$$p_{ext} = p_{int} - \Delta p$$

then the magnitude of the work involved will be given by

$$|w| = \sum_{i=1}^{n} (p_{i-1} - \Delta p_i)(\Delta V_i)$$

with $\qquad p_0 = p', \quad p_n = p'' \qquad$ and $\qquad p_i = p_{i-1} - \Delta p_i$

This work will be equal to the shaded area of Fig. 1.5.4. It is obvious that the magnitude of the work involved in this case is much larger than in the case of a one-stage or a two-stage expansion.

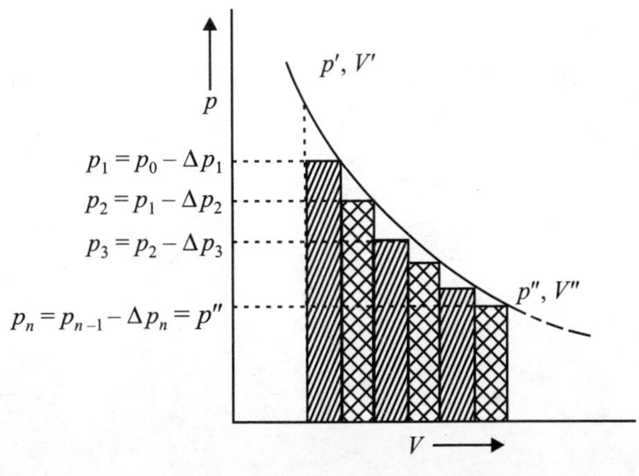

Fig. 1.5.4 Expansion in many stages

Expansion in Infinite Stages

The magnitude of the work involved goes on increasing as Δp_i becomes smaller and smaller. In the limit when $\Delta p_i \rightarrow 0$ the magnitude of the work involved will have a maximum value. In this case

$$p_{\text{ext}} = p_{\text{int}} - \mathrm{d}p$$

and the magnitude of the work involved is given by

$$|w| = \int_{V'}^{V''} p_{\text{ext}} \mathrm{d}V = \int_{V'}^{V''} (p_{\text{int}} - \mathrm{d}p)\, \mathrm{d}V \simeq \int_{V'}^{V''} p_{\text{int}}\, \mathrm{d}V \qquad (1.5.5)$$

The second integral of Eq. (1.5.5) has been ignored since it involves the product of two infinitesimally small values. For an ideal gas, the above integral can be evaluated directly since

$$P_{\text{int}} = \frac{nRT}{V}$$

$$|w| = \int_{V'}^{V''} \frac{nRT}{V} \mathrm{d}V = nRT \ln \frac{V''}{V'} \qquad (1.5.6)$$

The magnitude of the work involved in this case will be the area under the isothermal curve shown in Fig. 1.5.5 and it will be the maximum work that can be obtained by the system during the expansion from volume V' to V''.

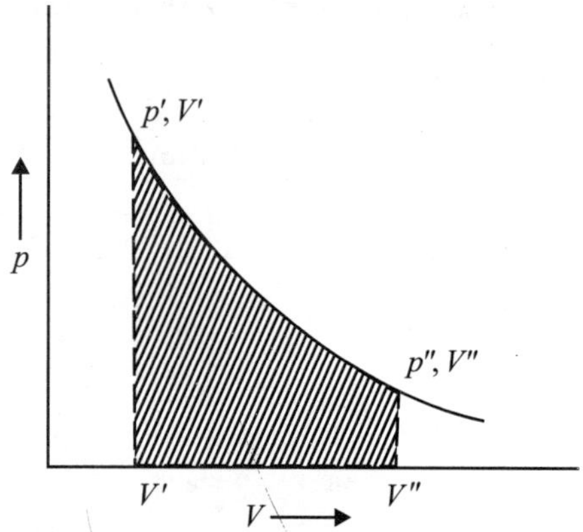

Fig. 1.5.5 Expansion involving infinite number of stages

Graphical Representation of Isothermal Compression of a Gas

Compression in One Stage

In compression processes, the essential condition which should be fulfilled is that the external pressure must be larger than the internal pressure of the gas. If the compression is being done from p'', V'' to p', V' in one stage, the minimum value of external pressure should be p'. If the external pressure is greater than this, the desired compression can be achieved by arresting the position of the piston by a set of stops. Thus, the work involved in compressing the gas from p'', V'' to p', V' against the external pressure p' is

$$w = -\int p_{\text{ext}}\, \mathrm{d}V = -p'(V' - V'')$$

Its value is equal to the area of the shaded rectangle in Fig. 1.5.6. If we use larger external pressure the surroundings do more work in bringing about the desired compression.

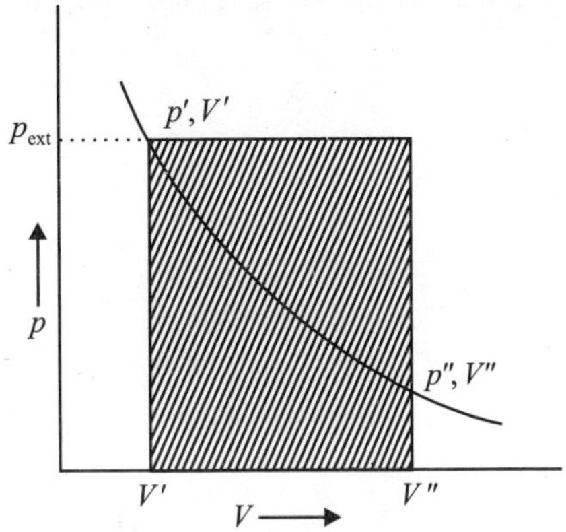

Fig. 1.5.6 Compression in one stage

Compression in Two Stages

If the compression is done in many stages, lesser amount of work is done by the surroundings. For example, in a two-stage compression with external pressures p''' and p'. The work involved is given by

$$w = -p'''(V''' - V'') - p'(V' - V''')$$

Its value is shown by the shaded area of Fig. 1.5.7.

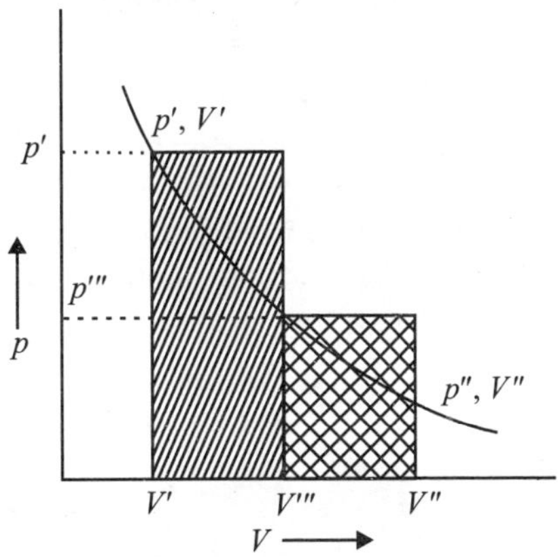

Fig. 1.5.7 Compression in two stages

Compression in Multistages

If the compression is done in multistages, still lesser and lesser work is involved. If at any stage of compression

$$p_{ext} = p_{int} + \Delta p \tag{1.5.7}$$

the expression for the work is

$$w = -\int p_{ext}\, dV = -\sum_{i=1}^{n}(p_{i-1} + \Delta p_i)(\Delta V_i)$$

with $p_0 = p''$, $p_n = p'$ and $p_i = p_{i-1} + \Delta p_i$. The value of this work will be equal to the shaded area of Fig. 1.5.8.

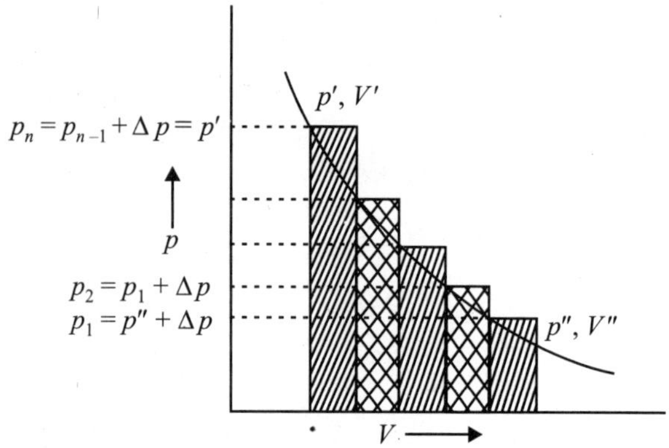

Fig. 1.5.8 Compression in many stages

Compression in Infinite Stages

If the number of stages of compression is infinite, then $\Delta p_i \rightarrow 0$ and the external pressure at [illegible] of compression is given by

$$p_{\text{ext}} = p_{\text{int}} + \mathrm{d}p$$

The work involved will be given by

$$w = -\int_{V''}^{V'} p_{\text{ext}}\, \mathrm{d}V = -\int_{V''}^{V'} (p_{\text{int}} + \mathrm{d}p)\mathrm{d}V \simeq \int_{V''}^{V'} p_{\text{int}}\, \mathrm{d}p \qquad (1.5.8)$$

Here again the second integral of Eq. (1.5.8) is ignored because it involves the product of two infinitesimally small values. Equation (1.5.8) can be evaluated for an ideal gas, for which $p_{\text{int}} = nRT/V$, and therefore, we have

$$w = -\int_{V''}^{V'} \frac{nRT}{V}\mathrm{d}V = -nRT \ln \frac{V'}{V''}$$

whose value is equal to the area under the isothermal curve shown in Fig. 1.5.9. This represents the limiting minimum value of the work done by the surrounding on the system.

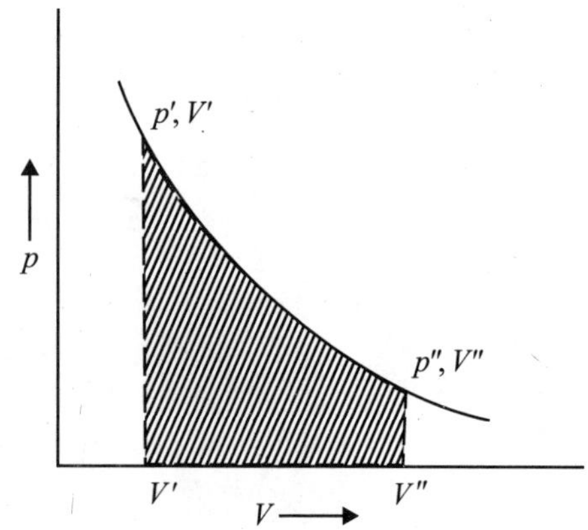

Fig. 1.5.9 Compression involving infinite number of stages

Work is not a State Function

The above analysis very clearly indicates that work is not a state function, since its value depends on the path which has been followed in order to achieve the required expansion or compression.

1.6 REVERSIBLE AND IRREVERSIBLE PROCESSES

The changes of a system from one state to another may occur either in a reversible or in an irreversible way.

Characteristics of Reversible Process

The reversible processes are characterized by the fact that when the system is restored to its original state by traversing the forward sequence of steps in the reverse order, then not only the system but also the surroundings are restored to their original states.

Take, for example, the limiting multistage isothermal expansion of an ideal gas from volume V_1 to V_2 involving an infinite number of steps. As seen earlier, the work involved (actually the work is done by the system on surroundings) is given by

$$w_{\text{exp}} = -nRT \ln \frac{V_2}{V_1} \tag{1.6.1}$$

Let the system be now restored to its initial volume V_1 by following again the limiting multistage isothermal compression involving infinite number of steps (i.e. the forward sequence of steps is being reversed), then the work involved (actually the work is done by the surroundings on the system) is given by

$$w_{\text{comp}} = -nRT \ln \frac{V_1}{V_2} \tag{1.6.2}$$

The net work involved in the above two processes (cyclic process) is

$$w_{\text{cyc}} = w_{\text{exp}} + w_{\text{comp}} = 0 \tag{1.6.3}$$

that is, whatever work has been done by the system on the surroundings during expansion is nullified by the work done by the surrounding on the system during compression and thus no net work effect is produced. Thus in the above example, besides the system, surroundings are also restored to their initial states and hence the above process is reversible in nature.

Hence, the criterion of reversibility is

$$\oint w_{\text{cyc}} = 0 \tag{1.6.4}$$

Characteristics of Irreversible Process

In an irreversible process, the cyclic integral of work is not zero, i.e.

$$\oint w_{\text{cyc}} \neq 0 \tag{1.6.5}$$

One of the examples of irreversible process is a single-stage expansion (or multistage expansion involving a finite number of steps) and then bringing the system beck to its original state by following a single-stage compression (or multistage compression involving a finite number of steps) in the reverse order. In a single-stage process, we have

Work involved during expansion, $w_{exp} = -p_2(V_2 - V_1)$ (1.6.6)

Work involved during compression, $w_{comp} = -p_1(V_1 - V_2)$ (1.6.7)

The net work involved is

$$w_{exp} + w_{comp} = -p_2(V_2 - V_1) + \{-p_1(V_1 - V_2)\}$$
$$= (p_1 - p_2)(V_2 - V_1) \quad (1.6.8)$$

Since $p_1 > p_2$ and $V_2 > V_1$, therefore, the network involved is positive. This means that the surroundings have to do more work in bringing the system black to the original state than the work done by the system during expansion. Thus, we see that here $w_{cyc} \neq 0$ and, therefore, the process is irreversible in nature.

Operation of a Reversible Process

In a reversible process, at any stage, the external condition responsible for the process to occur differs from the internal condition by an infinitesimal amount. For example, during expansion or compression $p_{ext} = p_{int} \pm dp$ and the corresponding change in volume of the system is also infinitesimally small. During this stage, the internal equilibrium is disturbed only infinitesimally and in the limit it is not disturbed at all. Thus, virtually, the system always remains in the equilibrium state during the process. Another example of reversible process is the heat exchanges between the system and the surroundings when the temperature of the latter differs from that of the former by an infinitesimal amount, i.e. $T_{surr} = T_{sys} \pm dT$. Obviously, a reversible process cannot be conducted in actual practice because the external condition has to differ from the internal condition by an infinitesimal amount and moreover, even if it is possible to monitor this, an infinite time would be required to complete the process. Thus, *reversible processes are not real but only ideal which can be carried out only theoretically.* Nevertheless, they are very important as they give the limit of the effect, whether maximum or minimum, that can be produced. Although reversible processes cannot be carried out, the goal of reversibility can be approached by adjusting the conditions with patience and skill to those of reversible processes.

Operation of an Irreversible Process

An irreversible process is one which occurs suddenly or spontaneously without the restrictions of occurring in successive stages of infinitesimal quantities. The system need not remain in equilibrium during the process. Examples are the sudden expansion of gases, the heating of water in a beaker over a Bunsen flame, the dissolution of sugar, the flow of water from a higher to lower level, etc. In the above examples properties may not be uniform throughout the system, e.g., during heating, water near the bottom of the beaker will have a higher temperature and during dissolution of sugar, the concentration of the solution immediately in contact with the sugar will be higher. The fundamental characteristic of any irreversible process is that more work is done by the surroundings in bringing the system back to its original stage than by the system during the forward direction. Thus, if a system is kept at a constant temperature and subjected to a cyclic transformation by an irreversible process, a net amount of work is done by the surroundings. This is, in fact, a statement of *second law of thermodynamics.* The greatest work effect that can be produced is in the reversible process, and, as given by Eq. (1.6.8), it is $w_{cyc} = 0$. Therefore, we cannot expect to get a positive amount of work in the surroundings for cyclic transformation of a system kept at a constant temperature.

Irreversible processes can be completed in a finite time and can be carried out in actual practice. Therefore, *irreversible processes are real processes.* All spontaneous transformations which occur in nature are real processes and hence also irreversible processes.

Problem 1.6.1 Show mathematically that the magnitude of the work involved in a reversible expansion of an ideal gas from volume V_1 and V_2 is larger than the corresponding work involved in an irreversible expansion against a constant pressure of p_2.

Solution We have the relation

$$|w_{\text{rev}}| = nRT \ln \frac{V_2}{V_1} \tag{1.6.9}$$

$$= nRT \ln \left\{1+\left(\frac{V_2}{V_1}-1\right)\right\}$$

Expanding the logarithmic term, we have

$$|w_{\text{rev}}| = nRT\left\{\left(\frac{V_2}{V_1}-1\right)+\text{higher terms}\right\} = \frac{nRT}{V_1}(V_2-V_1)+\text{higher terms}$$

$$= p_1(V_2 - V_1) + \text{higher terms}$$

and $$|w_{\text{irr}}| = p_2(V_2 - V_1) \tag{1.6.10}$$

Therefore $$|w_{\text{rev}}| - |w_{\text{irr}}| = \{p_1(V_2 - V_1) + \text{higher terms}\} - p_2(V_2 - V_1)$$

$$= (V_2 - V_1)(p_1 - p_2) + \text{higher terms}$$

Since, in expansion $V_2 > V_1$ and $p_1 > p_2$, therefore

$$|w_{\text{rev}}| - |w_{\text{irr}}| = \text{positive}$$

that is, the magnitude of the work involved in a reversible expansion is larger than the corresponding work involved in an irreversible expansion.

REVISIONARY PROBLEMS

1.1 Explain, with examples, the following terms:

(i) System (closed, open and isolated); (ii) Surroundings; (iii) Boundary

(iv) State variables (intensive and extensive);

(v) Process (isothermal, adiabatic, isobaric, isochoric, cyclic and quasi-static);

(vi) Reversible and irreversible paths.

1.2 What are intensive and extensive variables? Classify the following into intensive and extensive variables:

(i) energy, (ii) dipole moment, (iii) refractive index, (iv) viscosity, (v) volume, (vi) density, (vii) surface tension, (viii) molar volume, (ix) kinetic energy, (x) heat capacity, (xi) temperature, (xii) gas constant, (xiii) critical density, (xiv) specific heat capacity, (xv) vapour pressure, (xvi) internal energy, (xvii) enthalpy, (xviii) entropy, (xix) free energy, and (xx) chemical potential.

1.3 (a) What do you understand by 'exact and inexact differentials'?

(b) State the Euler's reciprocity relation for an exact differential.

(c) What are state functions? Do the state functions have exact differentials?

(d) Test the following differential expressions for exactness:

(i) $d\phi = (x^2 + y^2)\,dx + 2xy\,dy$ (ii) $d\phi = \frac{p}{T^2}dp - \frac{p^2}{T^3}dT$

(iii) $d\phi = \frac{RT}{p}\,dp - R\,dT$

(e) What is an integrating factor? Determine the integrating factors for the following differentials.

$$d\phi = y\,dx - x\,dy \quad \text{and} \quad d\phi = \frac{RT}{p}dp - R\,dT$$

(f) What is a cyclic rule? Derive this rule for a function $Z = f(x, y)$.

(g) Show that the volume of the following equations is a state function:

(i) Ideal gas equation

(ii) Real gas at low pressure for which van der Waals equation reduces to

$$\left(p + \frac{n^2 a}{V^2}\right)(V) = nRT$$

1.4. An arbitrary variable ϕ is found to have the following relationship

$$d\phi = \frac{RT}{p}dp - R dT$$

Are the following statements correct? Justify your answers.

(a) ϕ is a state function

(b) $d\phi$ is an exact differential

(c) $\oint d\phi \neq 0$

(d) It is possible to write down ϕ explicitly in terms of T and p as the independent variables, i.e. $\phi = f(T, p)$.

1.5 What is the Legendre transformation? Given the relation

$$dU = T\,dS - p\,dV$$

transform this to

(i) $dH = T\,dS + V\,dp$ (ii) $dA = -S\,dT - p\,dV$

(iii) $dG = -S\,dT + V\,dp$

What functional relations H, A and G will have with U?

1.6 (a) From the following thermodynamic relations

$$G = H - TS; \quad H = U + pV; \quad dq = T\,dS; \quad dU = dq - p\,dV$$

Show that

(i) $\left(\frac{\partial G}{\partial T}\right)_p = -S$; (ii) $\left(\frac{\partial G}{\partial p}\right)_T = V$ (iii) $\left(\frac{\partial V}{\partial T}\right)_p = -\left(\frac{\partial S}{\partial p}\right)_T$

(b) From the following thermodynamic relations

$$A = U - TS; \quad dq = T\,dS; \quad dU = dq - p\,dV$$

Show that

$$\text{(i)}\ \left(\frac{\partial A}{\partial V}\right)_T = -p; \quad \text{(ii)}\ \left(\frac{\partial A}{\partial T}\right)_V = -S; \quad \text{(iii)}\ \left(\frac{\partial p}{\partial T}\right)_V = \left(\frac{\partial S}{\partial V}\right)_T$$

1.7 (a) Using the definition $H = U + pV$, and, wherever necessary, obtaining conversion relationships by considering H (or U) as a function of any two variables amongst p, V and T, derive the following relationships:

$$\text{(i)}\ \left(\frac{\partial H}{\partial T}\right)_p = \left(\frac{\partial U}{\partial T}\right)_V + \left[p + \left(\frac{\partial U}{\partial V}\right)_T\right]\left(\frac{\partial V}{\partial T}\right)_p$$

$$\text{(ii)}\ \left(\frac{\partial H}{\partial T}\right)_p = \left(\frac{\partial U}{\partial T}\right)_V + \left[V - \left(\frac{\partial H}{\partial p}\right)_T\right]\left(\frac{\partial p}{\partial T}\right)_V$$

$$\text{(iii)}\ \left(\frac{\partial U}{\partial T}\right)_V = \left(\frac{\partial H}{\partial T}\right)_p + \left[V - \left(\frac{\partial H}{\partial T}\right)_p\left(\frac{\partial T}{\partial p}\right)_H\right]\left(\frac{\partial p}{\partial T}\right)_V$$

1.8 (a) For $p = f(V, T)$ and $V = f(p, T)$, derive the following cyclic rule:

$$\left(\frac{\partial p}{\partial T}\right)_V\left(\frac{\partial T}{\partial V}\right)_p\left(\frac{\partial V}{\partial p}\right)_T + 1 = 0$$

(b) Test the validity of the above equation using

(i) $pV = nRT$ (ii) $\left(p + \frac{n^2 a}{V^2}\right)V = nRT$

1.9 (a) What are cubic expansion coefficient α and isothermal compressibility factor κ_T? Show that

(i) $\frac{\alpha}{\kappa_T} = \left(\frac{\partial p}{\partial T}\right)_V$ (ii) $\left(\frac{\partial \alpha}{\partial p}\right)_T + \left(\frac{\partial \kappa_T}{\partial T}\right)_p = 0$

(b) Derive the expressions for κ_T and α for (i) an ideal gas and (ii) a van der Waals gas.

1.10 Starting from V as a state function derive the eqution of state for which

(i) $V = k_1 p$, keeping T constant, and $V = k_2 T$, keeping p constant.

(ii) $\alpha = \frac{V - a}{TV}$ and $\kappa_T = \frac{3(V - a)}{4pV}$, where a is constant.

1.11 (a) What do you understand by the term 'homogeneous function of the nth degree'? Show that 'a homogeneous function of zero degree with respect to the amounts of substances is an intensive variable while that of the first degree is an extensive variable'.

(b) State Euler's theorem as applicable to homogeneous function. Give its proof.

(c) If we write, for a system containing more than one component

$V = f(n_1, n_2, \ldots)$

then, using Euler's theorem, show that

$$V = \sum_i n_i\left(\frac{\partial V}{\partial n_i}\right)_{T,p,n_j(j \neq i)}$$

1.12 What is the work of expansion or compression? Show that:

(a) The magnitude of the work involved during a multistage expansion is larger than that involved during a single-stage expansion.

(b) The magnitude of the work involved during expansion is maximum if the process is quasi-static.

(c) The work involved during a multistage compression is smaller than that involved in a single-stage compression.

(d) Work involved during compression is minimum if the process is quasi-static.

1.13 (a) Let the given system of an ideal gas be expanded isothermally from p_1, V_1 to p_2, V_2 following a single-stage expansion. What should the external pressure be so that the work done by the system is maximum? What is the amount of work?

(b) Now, suppose that the system is to be restored isothermally to p_1, V_1 following a single-stage compression. What should the external pressure be so that the work done by the surroundings is minimum? What is the amount of work?

(c) Will there by any difference in the work involved in steps (a) and (b)? if so, why?

1.14 Suppose the work of expansion and compression in Question 1.13 is carried out quasi-statically. Show that 'the net work that can be obtained by the system in the above cyclic process is zero, i.e. the maximum amount of the work that can be obtained by a system if its temperature is to be kept constant is zero. If the above cyclic process is not carried out quasi-statically, then the amount of work that can be obtained by system is negative, i.e. there is a net work that is destroyed'.

1.15 Comment on the following:

(a) All natural processes (i.e. spontaneous processes) are irreversible in nature.

(b) Reversible processes are an ideal conception and can be carried out only theoretically.

(c) In a reversible cyclic process, not only the system but also the surroundings are restored to their initial states.

TRY YOURSELF PROBLEMS

1.1 Given that

$$z = ax^2y + by + cxy^3$$

where a, b, c are constants. Show that

(i) dz is an exact differential and z is a state function.

(ii) $\left(\frac{\partial z}{\partial x}\right)_y \left(\frac{\partial x}{\partial y}\right)_z \left(\frac{\partial y}{\partial z}\right)_x = -1$

(iii) $\oint dz = 0$

1.2 Given that

$$dq = dU + p\,dV$$

(a) Assuming $U = f(T, V)$, show that

$$dq = \left(\frac{\partial U}{\partial T}\right)_V dT + \left\{\left(\frac{\partial U}{\partial V}\right)_T + p\right\} dV$$

Following Euler's reciprocity condition, show that dq is an inexact differential.

(b) Assuming that for an ideal gas $(\partial U/\partial V)_T$ *is* zero, show that $1/T$ is an integrating factor of the equation $dq = dU + p\,dV$.

(*Hint:* See Sections 2.3 and 4.12.)

1.3 Given that $u = f(x, y)$ where x and y are themselves functions of s and t, i.e. $x = f(s, t)$ and $y = f(s, t)$. Show that

$$\left(\frac{\partial u}{\partial x}\right)_y \left(\frac{\partial x}{\partial u}\right)_\upsilon + \left(\frac{\partial u}{\partial y}\right)_x \left(\frac{\partial y}{\partial u}\right)_\upsilon = 1$$

$$\left(\frac{\partial u}{\partial x}\right)_y \left(\frac{\partial x}{\partial \upsilon}\right)_u + \left(\frac{\partial u}{\partial y}\right)_x \left(\frac{\partial y}{\partial \upsilon}\right)_u = 0$$

where υ is also a function of x and y.

1.4 (a) Given A as a function of any two of the variable B, C and D, prove that

$$\left(\frac{\partial A}{\partial C}\right)_B \left(\frac{\partial C}{\partial B}\right)_D = -\left(\frac{\partial A}{\partial D}\right)_B \left(\frac{\partial D}{\partial B}\right)_C$$

(b) Given the function $S = f(T, p)$, show that

$$(\partial p/\partial T)_S = -\,(\partial S/\partial T)_p\,/(\partial S/\partial p)_T$$

(c) Given the function of $S = f(T, p)$ and $p = f(T, V)$, show that

$$\left(\frac{\partial S}{\partial T}\right)_V = \left(\frac{\partial S}{\partial T}\right)_p + \left(\frac{\partial S}{\partial p}\right)_T \left(\frac{\partial p}{\partial T}\right)_V$$

(d) If G is a function *of* p, V and T, prove that

$$(\partial G/\partial p)_T = (\partial G/\partial V)_p\,(\partial V/\partial p)_T + (\partial G/\partial p)_V$$

(e) Show that the function $Z = pV$ is a state function.

1.5 A cylinder with a movable piston contains the amount n of an ideal gas. Consider the path indicated in the following diagram.

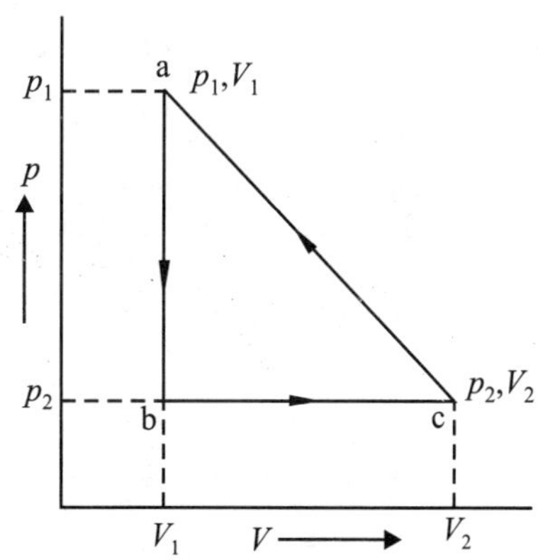

(a) Develop an expression in terms *of* p_1, V_1, p_2 and V_2 for the work that is carried out quasi-statically over the cyclic path, i.e. process going from a to b to c and back to a.

(b) What relation does the area of the triangle has to the magnitude of this work?

1.6 (a) An ideal gas undergoes a single-stage expansion against a constant opposing pressure from p_1, V_1, T to p_2, V_2, T. What is the largest mass m which can be lifted through a height h in this expansion?

(b) The system in (a) is restored to its initial state by a single-stage compression. What is the smallest mass m which must fall through the height h to restore the system?

(c) What is the net mass lowered through height h in the cyclic transformation in (a) and (b)?

(d) If h = 10 cm, p_1 = 1 MPa, p_2 = 0.5 MPa, T = 300 K and one mole of the gas is involved, calculate the numerical values of the mass m required in (a) and (b).

(*Ans.* (a) $m = (nRT/gh)(1 - p_2/p_1)$; (b) $m' = (nRT/gh)(p_1/p_2 - 1)$;
(c) $m' - m = (nRT/gh)(p_1 - p_2)^2/p_1 p_2$; (d) 1.27 Mg and 2.54 Mg)

1.7 Explain which of the following processes are reversible/irreversible:

(a) One teaspoon of a salt is dissolved in water.

(b) A gas is expanded into an evacuated vessel.

(c) Diffusion of a gas into another gas at constant T and p.

(d) Two blocks of iron at different temperatures are brought into closer contact.

(e) One mole of liquid water in equilibrium with its vapour at 100 °C and 0.1 MPa.

(f) Vaporization of benzene into a vacuum at 60 °C.

1.8 (a) Starting from V as a state function, derive the equation of state for a fluid which has an isothermal compressibility $\kappa_T = K[1 + M(T - T_0)]$ and a cubic expansion coefficient $\alpha = A[1 - Np]$, where K, M, A and N are all constants. Assume α and κ_T to be independent of T and p.

(b) Show that for the above fluid $AN = KM$.

(c) If the above fluid fills a container of constant volume at zero pressure at 273 K, calculate how much must the temperature be raised in order to increase the pressure to 10.132 5 MPa. Given that

$K = 2.487 \times 10^{-4}\ \mathrm{MPa^{-1}}$ $\qquad M = 2 \times 10^{-3}\ \mathrm{K^{-1}}$

$T_0 = 273$ K $\qquad A = 4.2 \times 10^{-4}\ \mathrm{K^{-1}}$

$N = 1.184 \times 10^{-3}\ \mathrm{MPa^{-1}}$

(*Ans.* $\ln \dfrac{V}{V_0} = A(1 - Np)(T - T_0) - \{K - AN(T - T_0)\}(p - p_0)$;
$\Delta T = T - T_0 = 6.1$ K)

(*Hint:* Write dV in terms of dT and dp. Substitute $(\partial V/\partial T)_p$ and $(\partial V/\partial p)_T$ and integrate the resultant expression under the assumption that for a fluid p is independent of temperature and vice versa.)

1.9 (a) Show that

$$p = T\left(\frac{\partial p}{\partial T}\right)_V - \left(\frac{\partial U}{\partial V}\right)_T$$

can be written in the form

$$\left(\frac{\partial U}{\partial V}\right)_T = T^2\left[\frac{\partial (p/T)}{\partial T}\right]_V = -\left[\frac{\partial (p/T)}{\partial (1/T)}\right]_V$$

(b) Show that

$$V = T\left(\frac{\partial V}{\partial T}\right)_p + \left(\frac{\partial H}{\partial p}\right)_T$$

can be written as

$$\left(\frac{\partial H}{\partial p}\right)_T = -T^2\left(\frac{\partial (V/T)}{\partial T}\right)_p = \left[\left(\frac{\partial (V/T)}{\partial (1/T)}\right)_p\right]$$

1.10 Test the following function for homogeneity and show by direct test the validity of Euler's theorem on homogeneous function

$$f(x, y) = \frac{ax^2}{y^2} + \frac{b(x^{3/2} + y^{3/2})}{xy^{1/2}}$$

1.11 (a) Show that, if the molar volume of a substance is a homogeneous function of zeroth degree in pressure and temperature, the energy of the substance must be a function of temperature only.

(*Hint:* Using Euler's theorem on homogeneous function, show that $(\partial p/\partial T)_V = p/T$. Write dU in terms of dT and dV and show that $(dU/dV)_T$ is zero. Make use of thermodynamic equation of state $(\partial U/\partial V)_T = T\,(\partial p/\partial T)_V - p$.)

(b) Similarly, show that, if the pressure is a homogeneous function of zeroth degree with respect to volume and temperature, the enthalpy must be a function of temperature only. Make use of the equation

$$\left(\frac{\partial H}{\partial p}\right)_T = V - T\left(\frac{\partial V}{\partial T}\right)_p$$

1.12 Taking U to be a function of T and V and $1/T$ as an integrating factor of $dq = dU + p\,dV$, derive the relation

$$\left(\frac{\partial U}{\partial V}\right)_T = T\left(\frac{\partial p}{\partial T}\right)_V - p$$

(*Hint:* Write the differential of U and substitute in the given relation $dq = dU + p\,dV$. Divide the resultant expression by T and apply the Euler's reciprocity relation. *See* also Section 4.12.)

1.13 Taking H to be a function of T and p and $1/T$ as an integrating factor of $dq = dH - V\,dp$, derive the relation

$$\left(\frac{\partial H}{\partial p}\right)_T = V - T\left(\frac{\partial V}{\partial T}\right)_p$$

1.14 For an ideal gas, show that $1/T$ is an integrating factor for the relation $dw = -p\,dV$. (*Hint:* Write dV in terms of dT and dp. Substitute $(\partial V/\partial T)_p$ and $(\partial V/\partial p)_T$ *from* $pV = RT$ and then substitute dV in the relation $dw = -p\,dV$. Divide by T and apply Euler's reciprocity relation.)

1.15 Using Euler's theorem on homogeneous function, show that the volume of an ideal gas is a homogeneous function of zeroth degree in pressure and temperature.

1.16 For the differential $dz = y\,dx - x\,dy$, show that $x^m y^n$ is the integrating factor provided $m + n = -2$.

2 Zeroth and First Laws of Thermodynamics

2.1 ZEROTH LAW OF THERMODYNAMICS

The zeroth law of thermodynamics, also known as the *law of thermal equilibrium*, was put forward much after the establishment of the first and second laws of thermodynamics. It is placed before the first and second laws as it provides a logical basis for the concept of temperature of a system. The law states that

Two systems in thermal equilibrium with a third system are also in internal equilibrium with each other.

According to this law, if systems A and B separately are in thermal equilibrium with another system C, then systems A and B will also be in thermal equilibrium with each other. In other words, if systems A and B are placed in contact with each other, no exchange of heat will take place. Recording of temperature of a system by a thermometer is also based on this law. When a thermometer is placed in the system, it comes to thermal equilibrium with the latter and thus records a constant value.

2.2 FIRST LAW OF THERMODYNAMICS

Establishment of First Law of Thermodynamics

All experiences have shown that if a system is subjected to any cyclic transformation, the sum of heat and work involved is zero, that is

$$\oint \mathrm{d}q + \oint \mathrm{d}w = 0 \quad \text{or} \quad \oint (\mathrm{d}q + \mathrm{d}w) = 0 \qquad (2.2.1)^{\dagger}$$

Thus, the cyclic integral of the quantity ($\mathrm{d}q + \mathrm{d}w$) is zero. This equation is a mathematical expression of the first law of thermodynamics.

Interpretation of First Law of Therodynamics

Let us represent

$$q = \oint \mathrm{d}q \quad \text{and} \quad w = \oint \mathrm{d}w$$

From Eq. (2.2.1), if follows that for a system undergoing cyclic transformation

$$q + w = 0 \quad \text{or} \quad q = -w \qquad (2.2.2)$$

Two cases may be distinguished.

1. If q is positive then w is negative.
   This implies that the net heat absorbed by the system is equal to the net work done by the system.

---

†The work involved may be of any type. For instance, it may be work of expansion or compresson of gases, work against a magnetic field or against a gravitational field or against an electrical field, etc.

2. If q is negative then w is positive

This implies that the heat released form the system is equal to the net work done on the system.

Identification of Energy Function

The fact that the cyclic integral expressed in Eq. (2.2.1) is zero for all cyclic processes indicates that $dq + dw$ is an exact differential, and therefore, is a differential of a state function. We call this function the *energy function* (internal energy or intrinsic energy) and use the symbol U to represent the function. The differential of the energy is given by

$$dU = dq + dw \qquad (2.2.3)$$

The energy function is a function of the state variables which are used to define the system. For example, the energy function for one mole of a gas may be written as $U = f(T, V)$ or $U = f(T, p)$ or $U = f(p, V)$. The value of the energy function for a given state is determined by the values assigned to the state variables. The function is determined only in terms of its differential, and the absolute value of energy function for any given state of the system may not be precisely known.

Change in Energy Function for a Finite Process

For a finite process, the difference between the values of energy function in two states can be determined by integrating dU as the latter is an exact differential. This difference is independent of the path followed in going form the initial to the final state. Thus, for a process

State A → State B

we have $\int_A^B dU = \int_A^B dq + \int_A^B dw$

$$U_B - U_A = q + w \quad \text{or} \quad \Delta U = q + w \qquad (2.2.4)$$

It may be noted that the values of both q and w depend upon the path followed in going form state A to state B, but the algebraic sum of these two quantities is independent of the path. This follows immediately from the fact that U is a state function.

Change in Energy Function Under a Constant Volume Condition

If a system can undergo only p-V work, then under the condition of constant volume we will have

$$dw = -\,p_{ext}\, dV = 0$$

Hence, Eq. (2.2.3) reduces to

$$dU = dq_V \qquad (2.2.5)$$

that is, when a system undergoes a change at constant volume condition, the heat involved merely changes it internal energy. To be specific,

The heat absorbed (or released) by a system under constant volume condition increases (or decreases) its internal energy.

Law of Conservation of Energy

For an isolated system, the values of both q and w are zero since no interaction of the system with the surroundings can take place. From Eq. (2.2.4) it follows that the value of ΔU must be zero for such a process. We, therefore, conclude that the value of the *energy function of an isolated system is constant.* This is another statement of the first law and is known as the *law of conservation of energy.* No matter what changes of state may occur in an isolated system, the value of the energy function is always a constant.

Denial of Perpetual Motion Machine of First Kind

For a cyclic process, $\Delta U = 0$ since the system returns to its initial state. Thus, according to Eq. (2.2.4), we have

$$q = -w \tag{2.2.6}$$

One of the implications of Eq. (2.2.6) is that the heat absorbed by the system from the surroundings is equal to the work done by the system on the surroundings. A perpetual motion machine of the first kind is one which would do work on the surroundings without absorbing an equivalent amount of heat. The first law denies the possibility of such a machine.

2.3 MATHEMATICAL PROOF OF HEAT AND WORK BEING INEXACT FUNCTIONS

Proof of Heat Being an Inexact Function

According to the first law

$$\mathrm{d}q = \mathrm{d}U - \mathrm{d}w \tag{2.3.1}$$

If the work is restricted to pressure-volume work performed quasi-statically, then $\mathrm{d}w = -p\,\mathrm{d}V$. With this Eq. (2.3.1) becomes

$$\mathrm{d}q = \mathrm{d}U + p\,\mathrm{d}V \tag{2.3.2}$$

Taking U as a function of temperature and volume, we write its differential as

$$\mathrm{d}U = \left(\frac{\partial U}{\partial V}\right)_T \mathrm{d}V + \left(\frac{\partial U}{\partial T}\right)_V \mathrm{d}T$$

On substituting this Eq. (2.3.2), we have

$$\mathrm{d}q = \left(\frac{\partial U}{\partial V}\right)_T \mathrm{d}V + \left(\frac{\partial U}{\partial T}\right)_V \mathrm{d}T + p\,\mathrm{d}V \tag{2.3.3}$$

Dividing this by $\mathrm{d}V$ and introducing the condition of constant T, we get

$$\left(\frac{\partial q}{\partial V}\right)_T = \left(\frac{\partial U}{\partial V}\right)_T + p \tag{2.3.4}$$

Similarly, on dividing by $\mathrm{d}T$ and introducing the condition of constant V, we get

$$\left(\frac{\partial q}{\partial T}\right)_V = \left(\frac{\partial U}{\partial T}\right)_V \tag{2.3.5}$$

If q is to be a state function, then according to Euler's reciprocity relation, we must have

$$\left(\frac{\partial^2 q}{\partial T\,\partial V}\right) = \left(\frac{\partial^2 q}{\partial V\,\partial T}\right) \tag{2.3.6}$$

Evaluating these second differentials form Eqs (2.3.4) and (2.3.5), we get

$$\left(\frac{\partial^2 U}{\partial T\,\partial V}\right) + \left(\frac{\partial p}{\partial T}\right)_V = \left(\frac{\partial^2 U}{\partial V\,\partial T}\right) \tag{2.3.7}$$

Since U is a state function, we have

$$\frac{\partial^2 U}{\partial T\,\partial V} = \frac{\partial^2 U}{\partial V\,\partial T}$$

With this, Eq. (2.3.7) reduces to

$$\left(\frac{\partial p}{\partial T}\right)_V = 0 \tag{2.3.8}$$

Equation (2.3.8) implies that the pressure of a system at constant volume is independent of temperature. When applied to ideal gases, we find that it is contrary to Charles law, which states that *at constant volume, the pressure of a fixed mass of a gas is directly proportional to its kelvin temperature, i.e.*

$$p \propto T$$

or $\quad p = KT$

Thus $(\partial p/\mathrm{d}T)_V = K$ and is not equal to zero as given by Eq. (2.3.8). Hence, the assumption that q is a state function must be wrong, and thus $\mathrm{d}q$ is an inexact differential.

Proof of Work Being an Inexact Function

We have

$$\mathrm{d}w = -p\,\mathrm{d}V \tag{2.3.9}$$

Taking V as a function of T and p, we write its differential as

$$\mathrm{d}V = \left(\frac{\partial V}{\partial T}\right)_p \mathrm{d}T + \left(\frac{\partial V}{\partial p}\right)_T \mathrm{d}p$$

Substituting this in Eq. (2.3.9), we get

$$-\,\mathrm{d}w = p\left\{\left(\frac{\partial V}{\partial T}\right)_p \mathrm{d}T + \left(\frac{\partial V}{\partial p}\right)_T \mathrm{d}p\right\}$$

From this, we have

$$-\left(\frac{\partial w}{\partial p}\right)_T = p\left(\frac{\partial V}{\partial p}\right)_T \tag{2.3.10}$$

and

$$-\left(\frac{\partial w}{\partial T}\right)_p = p\left(\frac{\partial V}{\partial T}\right)_p \tag{2.3.11}$$

If w is to be a state function, $\mathrm{d}w$ must be an exact differential, and then Euler's reciprocity relation states that

$$\frac{\partial^2 w}{\partial T\,\partial p} = \frac{\partial^2 w}{\partial p\,\partial T} \tag{2.3.12}$$

Evaluating these second differentials form Eqs (2.3.10) and (2.3.11), we have

$$p\frac{\partial^2 V}{\partial T\,\partial p} = p\frac{\partial^2 V}{\partial p\,\partial T} + \left(\frac{\partial V}{\partial T}\right)_p \tag{2.3.13}$$

Since V is a state function, we get

$$\frac{\partial^2 V}{\partial T\,\partial p} = \frac{\partial^2 V}{\partial p\,\partial T}$$

With this Eq. (2.3.7) reduces to

$$\left(\frac{\partial V}{\partial T}\right)_p = 0 \tag{2.3.14}$$

Equation (2.3.14) implies that the volume of a system at constant pressure is independent of temperature. When applied to ideal gases, we again find that it is contrary to Charles law which states that *at constant pressure, the volume of a fixed mass of a gas is directly proportional to its kelvin temperature, i.e.*

$$V \propto T$$

or $$V = KT$$

Thus $(\partial V/\mathrm{d}T)_p = K$ and is not equal to zero as given by Eq. (2.3.14). Hence, the assumption that w is a state function must be wrong and thus dw is an inexact differential.

Conditions for d*q* and d*w* to be Exact Differentials

If the work in Eq. (2.2.3) is restricted to pressure-volume work, the differential of the work will be given by $-p\,\mathrm{d}V$, where p is the external pressure against which expansion is carried out and dV is the change in volume in an infinitesimal stage. With this, Eq. (2.2.3) becomes

$$\mathrm{d}U = \mathrm{d}q - p\,\mathrm{d}V \tag{2.3.15}$$

For an adiabatic process $\mathrm{d}q = 0$ and thus $\mathrm{d}U = -p\,\mathrm{d}V$. Hence, for this condition, the differential of work is exact. Similarly, for an isochoric process (i.e. when V is held constant), $\mathrm{d}w = -p\,\mathrm{d}V = 0$ and $\mathrm{d}U = \mathrm{d}q$, and thus; under this condition, the differential of heat becomes exact.

2.4 CHANGE IN ENERGY FUNCTION WITH TEMPERATURE

Experiences have shown that the energy for a closed system (i.e. when n is constant) can be written as a function of temperature and volume, i.e. $U = f(T, V)$ and, therefore

$$\mathrm{d}U = \left(\frac{\partial U}{\partial T}\right)_V \mathrm{d}T + \left(\frac{\partial U}{\partial V}\right)_T \mathrm{d}V \tag{2.4.1}$$

Since for an isochoric process, $\mathrm{d}V = 0$, therefore

$$\mathrm{d}U = \left(\frac{\partial U}{\partial T}\right)_V \mathrm{d}T \qquad \text{(constant volume)} \tag{2.4.2}$$

Also, for an isochoric process

$$\mathrm{d}U = \mathrm{d}q_V \tag{2.4.3}$$

Comparing Eqs (2.4.2) and (2.4.3), we get

$$\mathrm{d}U = \mathrm{d}q_V = \left(\frac{\partial U}{\partial T}\right)_V \mathrm{d}T \tag{2.4.4}$$

or $$\frac{dq_V}{dT} = \left(\frac{\partial U}{\partial T}\right)_V \tag{2.4.5}$$

Definition of Heat Capacities

The heat capacity of a system is defined as the limit of $(q/\Delta T)$ as ΔT goes to zero and is written as (dq/dT), such that

$$C = \lim_{\Delta T \to 0}\left(\frac{q}{\Delta T}\right) = \left(\frac{dq}{dT}\right) \tag{2.4.6}$$

Heat capacity is an extensive quantity as the heat q required to raise the temperature by ΔT depends upon the mass of the substance. The specific heat capatcity is the heat capacity per unit mass of a substance and thus is an intensive quantity. It has the unit of J K^{-1} kg^{-1}. The molar heat capacity is the heat capacity per unit amount of the substance and hence is also an intensive quantity. It is written as C_m and has the unit of J K^{-1} mol^{-1}. Since dq is an inexact differential, therefore, heat capacity C is, in general, also path dependent. The following two heat capacities are commonly used, especially for gases.

(i) Heat capacity at constant volume (*Isochoric Process*)

$$C_V = \frac{dq_V}{dT} \tag{2.4.7}$$

(ii) Heat capacity at constant pressure (*Isobaric Process*)

$$C_p = \frac{dq_p}{dT} \tag{2.4.8}$$

Comparing Eq. (2.4.7) with Eq. (2.4.5), we find that for an isochoric process

$$C_V = \left(\frac{\partial U}{\partial T}\right)_V \tag{2.4.9}$$

With this, Eq. (2.4.4) becomes

$$dU = dq_V = C_V\, dT \tag{2.4.10}$$

For a finite change, we have

$$\Delta U = q_V = \int_{T_1}^{T_2} C_V\, dT \tag{2.4.11}$$

If C_V is considered to be independent of temperature in the range T_1 to T_2, then

$$\Delta U = q_V = C_V\, \Delta T \tag{2.4.12}$$

where ΔT is the change in temperature of the system and is equal to $(T_2 - T_1)$. If C_V is temperature dependent, and its dependence on temperature is known in the form of an analytic expression, then this expression can be substituted in Eq. (2.4.11) which on carrying out the integration gives the value of q_V or ΔU.

Problem 2.4.1 Prove that if $(\partial U/\partial V)_T = 0$, it follows that $(\partial U/\partial p)_T = 0$.

Solution Since $U = f(T, V)$, we have

$$dU = \left(\frac{\partial U}{\partial T}\right)_V dT + \left(\frac{\partial U}{\partial V}\right)_T dV$$

Dividing by $\mathrm{d}p$ and introducing the condition of constant temperature, we have

$$\left(\frac{\partial U}{\partial p}\right)_T = \left(\frac{\partial U}{\partial V}\right)_T \left(\frac{\partial V}{\partial p}\right)_T$$

Since $(\partial U/\partial V)_T = 0$, therefore $(\partial U/\partial p)_T = 0$.

2.5 ENTHALPY FUNCTION

Identification of Enthalpy Function

Most chemical processes which take place in a laboratory are carried out under the condition of constant pressure and the evaluation of energy changes in these processes are often required. The expressions relating the energy change at constant pressure condition can be derived by defining a new state function, known as *enthalpy function*. This can be done as follows.

For a closed system involving only the reversible work of expansion or compression, we have

$$\mathrm{d}U = \mathrm{d}q - p\,\mathrm{d}V \tag{2.5.1}$$

where p is the pressure of the system and is a function of temperature and volume. Adding the differential $\mathrm{d}(pV) = p\,\mathrm{d}V + V\,\mathrm{d}p$ to Eq. (2.5.1), we get

$$\mathrm{d}U + \mathrm{d}(pV) = \mathrm{d}q + V\,\mathrm{d}p \tag{2.5.2}$$

or

$$\mathrm{d}(U + pV) = \mathrm{d}q + V\,\mathrm{d}p \tag{2.5.3}$$

Since p and V are state functions, the product pV is also a state function. This may be verified by using Euler's reciprocity relation.† Thus, the sum $(U + pV)$, known as enthalpy function, is also a state function. This is represented by the symbol H. Thus

$$H = U + pV \tag{2.5.4}$$

Substituting Eq. (2.5.4) in Eq. (2.5.3), we have

$$\mathrm{d}H = \mathrm{d}q + V\,\mathrm{d}p \tag{2.5.5}$$

Physical Significance of Enthalpy Function

Enthalpy is a thermodynamic function of the state variables. For a closed system, these variables are usually the temperature and pressure. The change in enthalpy for a given change of state is independent of the path and the change is zero over any cyclic path. It is an extensive property.

If to a closed system, heat is supplied at constant pressure conditions, then according to Eq. (2.5.5), we have

$$\mathrm{d}H = \mathrm{d}q_p \tag{2.5.6}$$

Thus, *the change in enthalpy function of the system is equal to the heat exchanged by the system at constant pressure.*

Since H is a state function and $\mathrm{d}H$ is an exact differential, the quantity $\mathrm{d}q_p$ under the condition of constant pressure becomes an exact differential. It may be pointed out here that the absolute value of H need not be determined precisely as the absolute value of U may not be known. But as is usual in thermodynamics, we

† See Q. 2.5 in Try Yourself Problems.

will be interested only in knowing the change in the enthalpy function, which can be evaluated by employing Eq. (2.5.6).

Defining Heat Capacity in terms of Enthalpy

If the enthalpy of a closed system is taken as a function of temperature and pressure, the differential of the enthalpy is given by

$$\mathrm{d}H = \left(\frac{\partial H}{\partial T}\right)_p \mathrm{d}T + \left(\frac{\partial H}{\partial p}\right)_T \mathrm{d}p \tag{2.5.7}$$

At constant pressure, the above expression reduces to

$$\mathrm{d}H = \left(\frac{\partial H}{\partial T}\right)_p \mathrm{d}T \qquad \text{(constant pressure)} \tag{2.5.8}$$

Comparing Eq. (2.5.8) and Eq. (2.5.6), we have

$$\mathrm{d}H = \mathrm{d}q_p = \left(\frac{\partial H}{\partial T}\right)_p \mathrm{d}T \tag{2.5.9}$$

Using Eq. (2.4.8), we have

$$C_p = \frac{\mathrm{d}q_p}{\mathrm{d}T} = \left(\frac{\partial H}{\partial T}\right)_p \tag{2.5.10}$$

which relates the temperature dependence of the enthalpy at constant pressure to the heat capacity at constant pressure.

Change in Enthalpy with Temperature

Substituting Eq. (2.5.10) in Eq. (2.5.9), we get

$$\mathrm{d}q_p = \mathrm{d}H = C_p\,\mathrm{d}T \tag{2.5.11}$$

For a finite change, we have

$$q_p = \Delta H = \int_{T_1}^{T_2} C_p\,\mathrm{d}T \tag{2.5.12}$$

C_p independent of T If C_p is considered independent of temperature in the range T_1 to T_2, then

$$q_p = \Delta H = C_p\,(T_2 - T_1) \tag{2.5.13}$$

C_p dependent on T Heat capacity at constant pressure is usually temperature dependent. An analytical expression is available which relates C_p and temperature as

$$C_p = a + bT + cT^2 + \dots \tag{2.5.14}$$

where a, b, c, … are constants. Substituting this in Eq. (2.5.12), we have

$$q_p = \Delta H = \int_{T_1}^{T_2} (a + bT + cT^2 + \dots)\mathrm{d}T$$

which on carrying out the integration gives

$$q_p = \Delta H = \left| aT + \frac{bT^2}{2} + \frac{cT^3}{3} + \cdots \right|_{T_1}^{T_2}$$

$$= a(T_2 - T_1) + \frac{b}{2}(T_2^2 - T_1^2) + \frac{c}{3}(T_2^3 - T_1^3) + \ldots \tag{2.5.15}$$

Example 2.5.1 10.0 dm^3 of O_2 at 101.325 kPa and 298 K is heated to 348 K. Assuming ideal behaviour, calculate the heat absorbed, ΔH and ΔU of this process at (a) constant pressure, and (b) constant volume. Given: $C_{p,\,m} = C_{V,\,m} + R$.
Given: $C_{p,\,m}/\mathrm{J\,K^{-1}\,mol^{-1}} = 25.72 + 0.013\,(T/\mathrm{K}) - 3.86 \times 10^{-6}\,(T/\mathrm{K})^2$. Assume ideal behaviour.

Solution Amount of the gas, $n = \dfrac{pV}{RT} = \dfrac{(101.325\ \mathrm{kPa})\,(10.0\ \mathrm{dm^3})}{(8.314\ \mathrm{dm^3\ kPa\ K^{-1}\ mol^{-1}})\,(298\ \mathrm{K})}$

$$= 0.409\ \mathrm{mol}$$

(a) *Constant pressure*

$$q_p = \Delta H = n\int_{T_1}^{T_2} C_{p,\,\mathrm{m}}\,\mathrm{d}T$$

$$= n\left[(25.72\ \mathrm{J\,K^{-1}\,mol^{-1}})(T_2 - T_1) + (0.013\ \mathrm{J\,K^{-2}\,mol^{-1}})\left(\frac{T_2^2}{2} - \frac{T_1^2}{2}\right) - (3.86 \times 10^{-6}\ \mathrm{J\,K^{-3}\,mol^{-1}})\left(\frac{T_2^3}{3} - \frac{T_1^3}{3}\right)\right]$$

Substituting the given data, we get

$$q_p = \Delta H = (0.409)\left[(25.72\ \mathrm{J})\,(348 - 298) + (0.013\ \mathrm{J})\left(\frac{348^2}{2} - \frac{298^2}{2}\right) - (3.86 \times 10^{-6}\ \mathrm{J})\left(\frac{348^3}{3} - \frac{298^3}{3}\right)\right]$$

$$= (0.409)\,(1\,475.77\ \mathrm{J}) = 603.59\ \mathrm{J}$$

$$\Delta U = \Delta H - \Delta(pV) = \Delta H - nR\,(\Delta T)$$

$$= 603.59\ \mathrm{J} - (0.409\ \mathrm{mol})\,(8.314\ \mathrm{J\,K^{-1}\,mol^{-1}})\,(50\ \mathrm{K})$$

$$= 603.59\ \mathrm{J} - 170.02\ \mathrm{J} = 433.57\ \mathrm{J}$$

(b) *Constant volume*

$$q_V = \Delta U = \int_{T_1}^{T_2} nC_{V,\mathrm{m}}\,\mathrm{d}T = \int_{T_1}^{T_2} nC_{p,\mathrm{m}}\,\mathrm{d}T - \int_{T_1}^{T_2} nR\,\mathrm{d}T$$

$$= 603.59\ \mathrm{J} - 170.02\ \mathrm{J} = 433.57\ \mathrm{J}$$

$$\Delta H = \Delta U + \Delta(pV) = \Delta U + nR(\Delta T) = 603.59\ \mathrm{J}$$

(**Note:** The values of ΔU and ΔH are the same in both the cases as these are state functions. However, the value of q differs as it is a path function.)

Problem 2.5.1 Prove that it does not necessarily follow that if $(\partial U/\partial V)_T = 0$ then $(\partial H/\partial p)_T = 0$.

Solution Since $H = U + pV$, we have

$$\left(\frac{\partial H}{\partial p}\right)_T = \left(\frac{\partial U}{\partial p}\right)_T + p\left(\frac{\partial V}{\partial p}\right)_T + V$$

or $$\left(\frac{\partial H}{\partial p}\right)_T = \left(\frac{\partial U}{\partial V}\right)_T \left(\frac{\partial V}{\partial p}\right)_T + p\left(\frac{\partial V}{\partial p}\right)_T + V \qquad (2.5.16)$$

If $(\partial U/\partial V)_T = 0$, then

$$\left(\frac{\partial H}{\partial p}\right)_T = p\left(\frac{\partial V}{\partial p}\right)_T + V \qquad (2.5.17)$$

Hence if $(\partial U/\partial V)_T = 0$, then $(\partial H/\partial p)_T$ need not be zero.

Problem 2.5.2 Show that for a fixed amount of an ideal gas (for which $(\partial U/\partial V)_T = 0$)

(i) $\left(\frac{\partial H}{\partial p}\right)_T = 0$ (ii) $\left(\frac{\partial U}{\partial V}\right)_p = C_V \frac{p}{nR} = C_p\left(\frac{\partial T}{\partial V}\right)_p - p$

(iii) $\left(\frac{\partial U}{\partial p}\right)_V = C_V\left(\frac{\partial T}{\partial p}\right)_V = \left(\frac{\kappa_T}{\alpha}\right) C_V$

(iv) $\left(\frac{\partial C_V}{\partial V}\right)_T = 0$ and $\left(\frac{\partial C_p}{\partial p}\right)_T = 0$

(v) $\left(\frac{\partial H}{\partial V}\right)_T = 0$ (vi) $\left(\frac{\partial U}{\partial T}\right)_p = C_p - p\left(\frac{\partial V}{\partial T}\right)_p$

Solution **(i)** From Eq. (2.5.17), we have

$$\left(\frac{\partial H}{\partial p}\right)_T = p\left(\frac{\partial V}{\partial p}\right)_T + V \qquad \text{(since for an ideal gas } (\partial U/\partial V)_T = 0)$$

For an ideal gas $V = nRT/p$, therefore $\left(\frac{\partial V}{\partial p}\right)_T = -\frac{nRT}{p^2}$

Substituting this in the previous equation, we get

$$\left(\frac{\partial H}{\partial p}\right)_T = p\left(-\frac{nRT}{p^2}\right) + V = -\frac{nRT}{p} + V = -V + V = 0$$

(ii) Since $U = f(T, V)$, therefore

$$dU = \left(\frac{\partial U}{\partial T}\right)_V dT + \left(\frac{\partial U}{\partial V}\right)_T dV$$

For an ideal gas, $(\partial U/\partial V)_T = 0$, therefore

$$dU = \left(\frac{\partial U}{\partial T}\right)_V dT = C_V\, dT \qquad (2.5.18)$$

Dividing by dV and introducing the condition of constant p, we get

$$\left(\frac{\partial U}{\partial V}\right)_p = \left(\frac{\partial U}{\partial T}\right)_V \left(\frac{\partial T}{\partial V}\right)_p = C_V\left(\frac{\partial T}{\partial V}\right)_p \qquad (2.5.19)$$

Now $pV = nRT$, therefore $T = pV/nR$. Differentiating this with respect to V, keeping p constant, we have

$$\left(\frac{\partial T}{\partial V}\right)_p = \frac{p}{nR}$$

With this, Eq. (2.5.19) becomes

$$\left(\frac{\partial U}{\partial V}\right)_p = C_V \frac{p}{nR} = C_{V,\,\mathrm{m}} \frac{p}{R}$$

We have the relation

$$U = H - pV$$

Differentiating this with respect to V, keeping p constant, we have

$$\left(\frac{\partial U}{\partial V}\right)_p = \left(\frac{\partial H}{\partial V}\right)_p - p = \left(\frac{\partial H}{\partial T}\right)_p \left(\frac{\partial T}{\partial V}\right)_p - p = C_p \left(\frac{\partial T}{\partial V}\right)_p - p \tag{2.5.20}$$

(iii) Dividing Eq. (2.5.18) by dp and introducing the condition of constant V, we get

$$\left(\frac{\partial U}{\partial p}\right)_V = \left(\frac{\partial U}{\partial T}\right)_V \left(\frac{\partial T}{\partial p}\right)_V = C_V \left(\frac{\partial T}{\partial p}\right)_V$$

From Problem (1.3.13), we get

$$\left(\frac{\partial p}{\partial T}\right)_V = \frac{\alpha}{\kappa_T}$$

Thus $$\left(\frac{\partial U}{\partial p}\right)_V = C_V \left(\frac{\partial T}{\partial p}\right)_V = C_V \left(\frac{\kappa_T}{\alpha}\right) \tag{2.5.21}$$

(iv) We have

$$C_V = \left(\frac{\partial U}{\partial T}\right)_V$$

Differentiating with respect to V, keeping T constant, we have

$$\left(\frac{\partial C_V}{\partial V}\right)_T = \frac{\partial}{\partial V}\left\{\left(\frac{\partial U}{\partial T}\right)_V\right\}_T$$

Since U is a state function, it follows that

$$\frac{\partial}{\partial V}\left\{\left(\frac{\partial U}{\partial T}\right)_V\right\}_T = \frac{\partial}{\partial T}\left\{\left(\frac{\partial U}{\partial V}\right)_T\right\}_V$$

Hence $$\left(\frac{\partial C_V}{\partial V}\right)_T = \frac{\partial}{\partial T}\left\{\left(\frac{\partial U}{\partial V}\right)_T\right\}_V$$

But for an ideal gas, $(\partial U/\partial V)_T = 0$. Therefore

$$\left(\frac{\partial C_V}{\partial V}\right)_T = 0$$

In other words, at constant temperature, C_V of an ideal gas is independent of volume and depends only on temperature.

Now $$C_p = \left(\frac{\partial H}{\partial T}\right)_p$$

Differentiating this with respect to p, keeping T constant, we have

$$\left(\frac{\partial C_p}{\partial p}\right)_T = \frac{\partial}{\partial p}\left\{\left(\frac{\partial H}{\partial T}\right)_p\right\}_T$$

Since H is a state function, it follows that

$$\frac{\partial}{\partial p}\left\{\left(\frac{\partial H}{\partial T}\right)_p\right\}_T = \frac{\partial}{\partial T}\left\{\left(\frac{\partial H}{\partial p}\right)_T\right\}_p$$

Hence $$\left(\frac{\partial C_p}{\partial p}\right)_T = \frac{\partial}{\partial T}\left\{\left(\frac{\partial H}{\partial p}\right)_T\right\}_p$$

But, for an ideal gas $(\partial H/\partial p)_T = 0$. Therefore

$$\left(\frac{\partial C_p}{\partial p}\right)_T = 0$$

In other words, at constant temperature C_p of an ideal gas is independent of pressure and depends only on temperature.

(v) Differential of the relation $H = U + pV$ is

$$dH = dU + p\,dV + V\,dp$$

Dividing by dV and introducing the condition of constant temperature, we have

$$\left(\frac{\partial H}{\partial V}\right)_T = \left(\frac{\partial U}{\partial V}\right)_T + p + V\left(\frac{\partial p}{\partial V}\right)_T \qquad (2.5.22)$$

Now, for an ideal gas

$$\left(\frac{\partial U}{\partial V}\right)_T = 0 \qquad \text{and} \qquad p = \frac{nRT}{V}$$

Therefore $$\left(\frac{\partial p}{\partial V}\right)_T = -\frac{nRT}{V^2} = -\frac{pV}{V^2} = -\frac{p}{V}$$

Substituting these in Eq. (2.6.22), we have

$$\left(\frac{\partial H}{\partial V}\right)_T = 0$$

(vi) Differential of the relation $U = H - pV$ is

$$dU = dH - p\,dV - V\,dp$$

Dividing by dT and introducing the condition of constant p, we get

$$\left(\frac{\partial U}{\partial T}\right)_p = \left(\frac{\partial H}{\partial T}\right)_p - p\left(\frac{\partial V}{\partial T}\right)_p = C_p - p\left(\frac{\partial V}{\partial T}\right)_p$$

2.6 RELATION BETWEEN HEAT CAPACITIES

Heat capacity at constant pressure is given by

$$C_p = \left(\frac{\partial H}{\partial T}\right)_p \qquad (2.6.1)$$

and that at constant volume by the relation

$$C_V = \left(\frac{\partial U}{\partial T}\right)_V \tag{2.6.2}$$

Earlier in Problem 1.3.11, we have derived the following three relations between $(\partial H/\partial T)_p$ and $(\partial U/\partial T)_V$.

$$\left(\frac{\partial H}{\partial T}\right)_p = \left(\frac{\partial U}{\partial T}\right)_V + \left[p + \left(\frac{\partial U}{\partial V}\right)_T\right]\left(\frac{\partial V}{\partial T}\right)_p$$

$$\left(\frac{\partial H}{\partial T}\right)_p = \left(\frac{\partial U}{\partial T}\right)_V + \left[V - \left(\frac{\partial H}{\partial p}\right)_T\right]\left(\frac{\partial p}{\partial T}\right)_V$$

$$\left(\frac{\partial H}{\partial T}\right)_p = \left(\frac{\partial U}{\partial T}\right)_V + \left[V + \left(\frac{\partial H}{\partial T}\right)_p\left(\frac{\partial T}{\partial p}\right)_H\right]\left(\frac{\partial p}{\partial T}\right)_V$$

Replacing $(\partial H/\partial T)_p$ and $(\partial U/\partial T)_V$ by C_p and C_V, respectively, we get

$$C_p = C_V + \left[p + \left(\frac{\partial U}{\partial V}\right)_T\right]\left(\frac{\partial V}{\partial T}\right)_p \tag{2.6.3}$$

$$C_p = C_V + \left[V - \left(\frac{\partial H}{\partial p}\right)_T\right]\left(\frac{\partial p}{\partial T}\right)_V \tag{2.6.4}$$

$$C_p = C_V + \left[V + \left(\frac{\partial H}{\partial T}\right)_p\left(\frac{\partial T}{\partial p}\right)_H\right]\left(\frac{\partial p}{\partial T}\right)_V \tag{2.6.5}$$

Difference of Heat Capacities in Terms of Easily Determinable Derivatives

The difference between C_p and C_V may be expressed in terms of easily determinable derivatives by making use of the thermodynamic equation of state†

$$\left(\frac{\partial U}{\partial V}\right)_T = T\left(\frac{\partial p}{\partial T}\right)_V - p$$

Substituting this in Eq. (2.6.3), we get

$$C_p - C_V = T\left(\frac{\partial p}{\partial T}\right)_V\left(\frac{\partial V}{\partial T}\right)_p \tag{2.6.6}$$

Using definitions of α and κ_T and that $(\partial p/\partial T)_V = \alpha/\kappa_T$ (Problem 1.3.13), we get

$$C_p - C_V = TV\,\frac{\alpha^2}{\kappa_T} \tag{2.6.7}$$

Making use of the cyclic relation

$$\left(\frac{\partial V}{\partial T}\right)_p\left(\frac{\partial T}{\partial p}\right)_V\left(\frac{\partial p}{\partial V}\right)_T + 1 = 0$$

† See Section 4.14.

we get $\left(\dfrac{\partial V}{\partial T}\right)_p = -\dfrac{1}{(\partial T/\partial p)_V\,(\partial p/\partial V)_T} = -\left(\dfrac{\partial p}{\partial T}\right)_V\left(\dfrac{\partial V}{\partial p}\right)_T$

Substituting this in Eq. (2.6.6), we get

$$C_p - C_V = -T\left(\frac{\partial p}{\partial T}\right)_V^2\left(\frac{\partial V}{\partial p}\right)_T \tag{2.6.8}$$

Evaluation of $C_{p,\,m} - C_{V,\,m}$ for an Ideal Gas

For one mole of an ideal gas $pV_m = RT$, therefore

$$\left(\frac{\partial p}{\partial T}\right)_V = \frac{R}{V_m} \quad \text{and} \quad \left(\frac{\partial V_m}{\partial T}\right)_p = \frac{R}{p}$$

Substituting these in Eq. (2.6.6), we get

$$C_{p,\,m} - C_{V,\,m} = T\left(\frac{R}{V_m}\right)\left(\frac{R}{p}\right) = \frac{R^2T}{pV_m} = \frac{R^2T}{RT} = R \tag{2.6.9}$$

Alternatively, Eq. (2.6.3) can be used to determine $C_p - C_V$ for an ideal gas. The term $(\partial U/\partial V)_T$ represents the change in internal energy when the volume of the system is changed at constant temperature. Since there exist no forces of attraction between the molecules of an ideal gas (i.e. the molecules move independent of each other), there will not be any expense or gain of energy on expansion (when the molecules are pulled apart) or compression (when the molecules are brought near to each other) of such a gas at constant temperature. This amounts to $(\partial U/\partial V)_T = 0$ for an ideal gas (Joule's law, see Section 2.7).

Thus, Eq. (2.6.3) reduces to

$$C_{p,\,m} - C_{V,\,m} = p\left(\frac{\partial V_m}{\partial T}\right)_p$$

But $\left(\dfrac{\partial V_m}{\partial T}\right)_p = \dfrac{R}{p}$

Hence $C_{p,\,m} - C_{V,\,m} = p\left(\dfrac{R}{p}\right) = R$

Evaluation of $C_{p,\,m} - C_{V,\,m}$ for a van der Waals Gas

For one mole of a van der Waals gas, we have

$$\left(p + \frac{a}{V_m^2}\right)(V_m - b) = RT$$

or $$pV_m - pb + \frac{a}{V_m} - \frac{ab}{V_m^2} = RT \tag{2.6.10}$$

Differentiating this with respect to T at constant V_m, we get

$$\left(\frac{\partial p}{\partial T}\right)_{V_m} V_m - \left(\frac{\partial p}{\partial T}\right)_{V_m} b = R$$

i.e. $$\left(\frac{\partial p}{\partial T}\right)_{V_m} = \frac{R}{V_m - b} \qquad (2.6.11)$$

Differentiating Eq. (2.6.10) with respect to T at constant p, we get

$$p\left(\frac{\partial V_m}{\partial T}\right)_p - \frac{a}{V_m^2}\left(\frac{\partial V_m}{\partial T}\right)_p + \frac{2ab}{V_m^3}\left(\frac{\partial V_m}{\partial T}\right)_p = R$$

i.e. $$\left(\frac{\partial V_m}{\partial T}\right)_p = \frac{R}{p - \dfrac{a}{V_m^2} + \dfrac{2ab}{V_m^3}} \qquad (2.6.12)$$

Substituting Eqs (2.6.11) and (2.6.12) in Eq. (2.6.6), we get

$$C_{p,\,m} - C_{V,\,m} = T\left(\frac{R}{V_m - b}\right)\left(\frac{R}{p - \dfrac{a}{V_m^2} + \dfrac{2ab}{V_m^3}}\right)$$

$$= \left(\frac{RT}{V_m - b}\right)\left(\frac{R}{p + \dfrac{a}{V_m^2} - \dfrac{2a}{V_m^3}(V_m - b)}\right)$$

$$= \frac{R}{1 - \dfrac{2a}{V_m^3\,RT}(V_m - b)^2} \qquad (2.6.13)$$

Under the conditions of high temperature and low pressure, Eq. (2.6.13) may be approximated as

$$C_{p,\,m} - C_{V,\,m} \simeq \frac{R}{1 - 2a/V_m RT} \simeq R\left(1 + \frac{2a}{V_m RT}\right) \simeq R\left(1 + \frac{2ap}{R^2T^2}\right) \qquad (2.6.14)$$

Equation (2.6.14) is not applicable under the conditions of low temperature and high pressure. From Eq. (2.6.14), if follows that the difference $C_{p,\,m} - C_{V,\,m}$ increases linearly with pressure. This difference also depends on the nature of the gas through the constant a. Thus larger the value of a, larger the difference between $C_{p,\,m}$ and $C_{V,\,m}$.

Problem 2.6.1 For a van der Waals gas, show that

$$C_{p,\,m} - C_{V,\,m} = R\left[1 + \frac{2ap}{R^2T^2} - \frac{4abp^2}{R^3T^3}\right]$$

Solution From Eq. (2.6.12), we have

$$\left(\frac{\partial V_m}{\partial T}\right)_p = \frac{R}{p - \dfrac{a}{V_m^2} + \dfrac{2ab}{V_m^3}}$$

$$= \frac{R}{\left(p + \frac{a}{V_m^2}\right) - \frac{2a}{V_m^3}(V_m - b)} = \frac{R}{\frac{RT}{V_m - b} - \frac{2a}{V_m^3}(V_m - b)}$$

$$= \frac{R(V_m - b)}{RT - \frac{2a}{V_m^3}(V_m - b)^2} = \left(\frac{V_m - b}{T}\right)\left[1 - \frac{2a}{V_m^3 RT}(V_m - b)^2\right]^{-1}$$

$$= \left(\frac{V_m - b}{T}\right)\left[1 + \frac{2a}{V_m^3 RT}(V_m - b)^2\right] \simeq \left(\frac{V_m - b}{T}\right)\left[1 + \frac{2a}{V_m^3 RT}(V_m^2 - 2V_m b)\right]$$

$$= \left(\frac{V_m - b}{T}\right)\left[1 + \frac{2ap}{R^2T^2} - \frac{4abp^2}{R^3T^3}\right] \qquad (2.6.15)$$

From Eq. (2.6.11), we have

$$\left(\frac{\partial p}{\partial T}\right)_V = \frac{R}{V_m - b}$$

Hence $C_{p,\,m} - C_{V,\,m} = T\left(\frac{\partial p}{\partial T}\right)_V \left(\frac{\partial V_m}{\partial T}\right)_p = T\left(\frac{R}{V_m - b}\right)\left[\left(\frac{V_m - b}{T}\right)\left\{1 + \frac{2ap}{R^2T^2} - \frac{4abp^2}{R^3T^3}\right\}\right]$

$$= R\left[1 + \frac{2ap}{R^2T^2} - \frac{4abp^2}{R^3T^3}\right] \qquad (2.6.16)$$

Example 2.6.1 The coefficient of cubical expansion α of a metal at 293 K is $21.3 \times 10^{-6}\ \text{K}^{-1}$ and the compressibility coefficient κ_T is $1.56 \times 10^{-11}\ \text{Pa}^{-1}$. The molar mass of metal is $63.55\ \text{g mol}^{-1}$ and its density is $0.97\ \text{g cm}^{-3}$. Calculate $C_{p,\,m} - C_{V,\,m}$ of the metal at 293 K.

Solution We are provided with

$$\alpha = 21.3 \times 10^{-6}\ \text{K}^{-1}$$

$$\kappa_T = 1.56 \times 10^{-11}\ \text{Pa}^{-1} = 1.56 \times 10^{-11}\ (\text{N m}^{-2})^{-1}$$

$$\rho = 0.97\ \text{g cm}^{-3}$$

Now $\quad$ Molar volume $= \frac{\text{Molar mass}}{\text{Density}} = \frac{63.55\ \text{g mol}^{-1}}{0.97\ \text{g cm}^{-3}} = 65.515\ \text{cm}^3\ \text{mol}^{-1}$

Substituting the above data in the expression

$$C_{p,\,m} - C_{V,\,m} = \frac{\alpha^2 T V_m}{\kappa_T} \qquad \text{(Eq. 2.6.7)}$$

we get $\quad C_{p,\,m} - C_{V,\,m} = \frac{(21.3 \times 10^{-6}\ \text{K}^{-1})^2\ (293\ \text{K})\ (65.515 \times 10^{-6}\ \text{m}^3\ \text{mol}^{-1})}{(1.56 \times 10^{-11}\ \text{N}^{-1}\ \text{m}^2)}$

$$= 0.558\ \text{N m K}^{-1}\ \text{mol}^{-1} = 0.558\ \text{J K}^{-1}\ \text{mol}^{-1}$$

Example 2.6.2 Calculate $C_{p,\,m} - C_{V,\,m}$ for nitrogen at 298 K and 100 bar pressure, the van der Waals constant a is $141\ \text{dm}^6\ \text{kPa mol}^{-2}$.

Solution We have

$$p = 100 \text{ bar} = 10^4 \text{ kPa}$$

$$a = 141 \text{ dm}^6 \text{ kPa mol}^{-2}$$

For a gas obeying van der Waals equation, we have

$$C_{p,\,\mathrm{m}} - C_{V,\,\mathrm{m}} \simeq R\left(1 + \frac{2a}{R^2T^2}p\right) \quad \text{(Eq. 2.6.14)}$$

Substituting the given values, we get

$$C_{p,\,\mathrm{m}} - C_{V,\,\mathrm{m}} \simeq (8.314 \text{ J K}^{-1} \text{ mol}^{-1}) \times \left(1 + \frac{2(141 \text{ dm}^6 \text{ kPa mol}^{-2})}{(8.314 \text{ dm}^3 \text{ kPa K}^{-1} \text{ mol}^{-1})^2 (298 \text{ K})^2} \times 10^4 \text{ kPa}\right)$$

$$\simeq (8.314 \text{ J K}^{-1} \text{ mol}^{-1})(1 + 0.459)$$

$$= 12.13 \text{ J K}^{-1} \text{ mol}^{-1}$$

2.7 JOULE'S EXPERIMENT

Experimentation

The evaluation of the differential $(\partial U/\partial V)_T$ for gases, in principle, can be done with the help of Joule's experiment (Fig. 2.7.1). Two vessels A and B are connected via a stopcock. The vessel A is filled with a gas at a certain pressure and the vessel B is completely evacuated. The entire assembly is immersed in a large vat of water and is allowed to come to thermal equilibrium with the water. The stopcock is opened and the gas is allowed to expand till both the vessels are uniformly occupied. After some time, when the vessel has again come to thermal equilibrium, temperature of the water is recorded. The result shows that the temperature of water after the experiment is the same as that before the expansion.

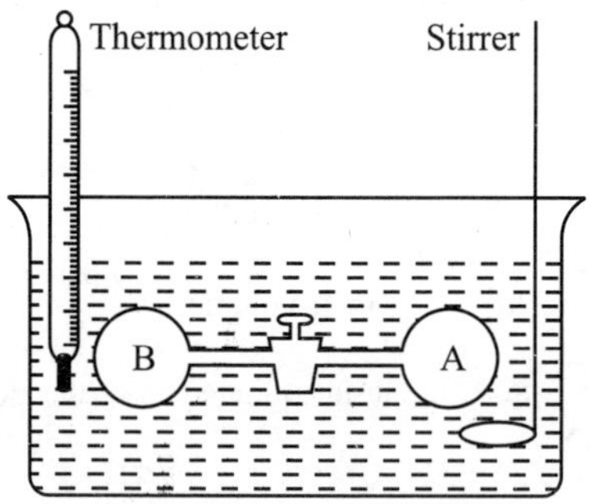

Fig. 2.7.1 Joule expansion experiment

Interpretation of the Results

Before expansion, the entire gas is present in the vessel A. On opening the stopcock, the boundary of the gas which always encloses the entire mass of gas expands against a zero opposing pressure. Such an expansion is called *free expansion*. Since $\mathrm{d}w = -p_{\mathrm{ext}}\,\mathrm{d}V$, it is obvious that the work involved in the above expansion is zero. Now, according to the first law of thermodynamics $\mathrm{d}U = \mathrm{d}q + \mathrm{d}w$, we have

$$\mathrm{d}U = \mathrm{d}q$$

Further, since there occurs no change in temperature of the bath, it follows that the heat $\mathrm{d}q$ involved must also be equal to zero, Thus, we conclude that

$$dU = 0$$

that is, Joule expansion is an isenergetic expansion.

Taking U to be a function to T and V, we write its differential as

$$dU = \left(\frac{\partial U}{\partial T}\right)_V dT + \left(\frac{\partial U}{\partial V}\right)_T dV$$

Since $dT = 0$ and $dU = 0$, we must have

$$\left(\frac{\partial U}{\partial V}\right)_T dV = 0$$

Now since $dV \neq 0$, if follows that

$$\left(\frac{\partial U}{\partial V}\right)_T = 0 \qquad (2.7.2)$$

Equation (2.7.2) is known as Joule's law and it implies the following.

The change in energy of a gas with change in volume at a constant temperature is zero.

In other words, the energy of the gas is a function of temperature only. Mathematically, it is written as $U = f(T)$. Joule's law is strictly applicable only for ideal gases and not for real gases.

Joule Coefficient

The results of Joule's experiment involving any gas (either ideal or real) can be expressed in terms of Joule coefficient η, defined as

$$\eta = \left(\frac{\partial T}{\partial V}\right)_U \qquad (2.7.3)$$

The Joule coefficient can be expressed in terms of the derivative $(\partial U/\partial V)_T$ by employing the first law of thermodynamics. Taking $U = f(T, V)$, we have

$$dU = \left(\frac{\partial U}{\partial T}\right)_V dT + \left(\frac{\partial U}{\partial V}\right)_T dV$$

Since, the Joule experiment is a process of constant energy, we have

$$0 = \left(\frac{\partial U}{\partial T}\right)_V \left(\frac{\partial T}{\partial V}\right)_U + \left(\frac{\partial U}{\partial V}\right)_T$$

Rearranging this, we get

$$\left(\frac{\partial U}{\partial V}\right)_T = -\left(\frac{\partial U}{\partial T}\right)_V \left(\frac{\partial T}{\partial V}\right)_U = -C_V\,\eta$$

or $$\eta = -\frac{1}{C_V}\left(\frac{\partial U}{\partial V}\right)_T \qquad (2.7.4)$$

Joule Coefficient in Terms of Easily Determinable Derivatives

The Joule coefficient η can be expressed in terms of quantities obtainable from the equation of state of the gas by employing the thermodynamic equation of state, according to which (*see* Section 4.14), we have

$$\left(\frac{\partial U}{\partial V}\right)_T = T\left(\frac{\partial p}{\partial T}\right)_V - p \tag{2.7.5}$$

Therefore $$\eta = -\frac{1}{C_V}\left(\frac{\partial U}{\partial V}\right)_T = -\frac{1}{C_V}\left\{T\left(\frac{\partial p}{\partial T}\right)_V - p\right\} \tag{2.7.6}$$

If ΔT and ΔV are small, Joule coefficient can be written as

$$\eta = \left(\frac{\partial T}{\partial V}\right)_U \simeq \frac{(\Delta T)_U}{(\Delta V)_U} \tag{2.7.7}$$

Substituting this in the previous expression, we get

$$(\Delta T)_U \simeq \frac{1}{C_V}\left[p - T\left(\frac{\partial p}{\partial T}\right)_V\right]\Delta V \tag{2.7.8}$$

Joule Coefficient for an Ideal Gas

For an ideal gas $p = nRT/V$, therefore

$$\left(\frac{\partial p}{\partial T}\right)_V = \frac{nR}{V}$$

This implies that

$$\eta = -\frac{1}{C_V}\left\{T\left(\frac{\partial p}{\partial T}\right)_V - p\right\} = -\frac{1}{C_V}\left(\frac{nRT}{V} - p\right) = -\frac{1}{C_V}\,(p - p) = 0 \tag{2.7.9}$$

that is, Joule coefficient for an ideal gas is zero.

Joule Coefficient for a Real Gas

For a real gas, the sign of η depends on the relative magnitudes of p and $(\partial p/\partial T)_V$ as can be seen from Eq. (2.7.6). Thus we have

(*i*) *When* $p > T\,(\partial p/\partial T)_V$ then η is positive. Since $\mathrm{d}V$ is always positive in Joule's experiment, it follows from Eq. (2.7.3) that $(\partial T)_U$ is positive, that is, there occurs an increase in temperature of the gas. In this case, as can be seen from Eq. (2.7.4), $(\partial U/\partial V)_T$ will have a negative sign.

(*ii*) *When* $p < T\,(\partial p/\partial T)_V$ then η is negative. Hence, it follows that $(\partial T)_U$ is negative, that is, there occurs a decrease in temperature of the gas. In this case, $(\partial U/\partial V)_T$ will have a positive sign.

The fact that $(\partial U/\partial V)_T$ is negative at very high pressures and is positive at ordinary pressures may be explained on the basis that the energy U actually consists of two types of energies, namely, kinetic energy and potential energy. When the volume is increased at constant temperatures, the kinetic energy remains constant but potential energy varies. At ordinary pressures where the attractive molecular forces predominate, the potential energy increases on expansion and hence $(\partial U/\partial V)_T$ is positive. At very high pressures, where the excluded volume plays the dominating role, the potential energy decreases on expansion with the result that $(\partial U/\partial V)_T$ is negative.

2.8 JOULE-THOMSON EXPERIMENT

Experimentation

The Joule-Thomson experiment involves the expansion of a gas form one fixed pressure to another fixed pressure under adiabatic conditions. The apparatus for the Joule-Thomson experiment is shown in Fig 2.8.1. It consists of a cylinder divided into two parts by a porous plug, a needle valve or any other throttling device. A fixed quantity of gas is contained in the cylinder by frictionless pistons placed on either sides of the porous plug. The entire apparatus is surrounded by an adiabatic wall. The gas to the left of the porous plug is at a high pressure p_1 and that to the right is at a low pressure p_2 ($p_1 > p_2$). The gas is allowed to pass through the porous plug from left to right, maintaining the pressures p_1 and p_2 at constant values. This can be achieved by removing the left-hand and right-hand pistons to the right quasi-statically.

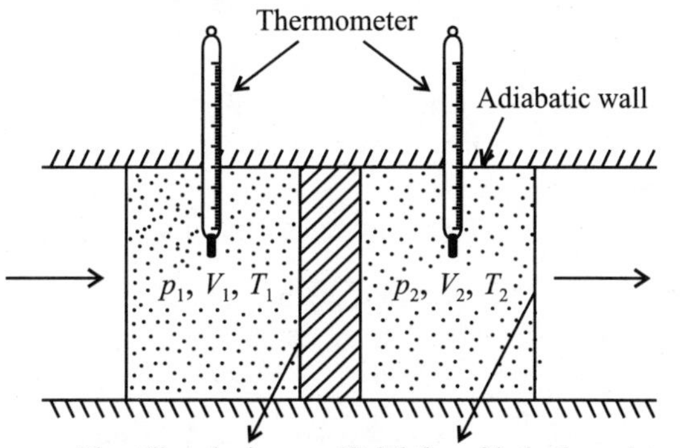

Fig. 2.8.1 Joule-Thomson Experiment

Nature of Joule-Thomson Expansion

Suppose a certain amount of gas is passed through the porous plug. We then have

Change in volume on the left hand side $= -V_1$
Work involved on the left hand side $= p_1V_1$
Change in volume on right hand side $= V_2$
Work involved on the right hand side $= -p_2V_2$
Net work involved in the system $= -p_2V_2 + p_1V_1$

Because the process is adiabatic, dq = 0, and thus from the first law, we have

$$\Delta U = w$$

or $$U_2 - U_1 = -p_2V_2 + p_1V_1$$

or $$U_2 + p_2V_2 = U_1 + p_1V_1$$

or $$H_2 = H_1 \tag{2.8.1}$$

that is, the enthalpy of the gas which has moved across the porous plug remains unchanged and thus the *Joule-Thomson expansion is an isenthalpic process.*

Joule-Thomson Coefficient

The measured change in temperature $-\Delta T$ and the measured change in pressure $-\Delta p$ are combined in the ratio

$$\left(\frac{-\Delta T}{-\Delta p}\right)_H = \left(\frac{\Delta T}{\Delta p}\right)_H$$

The Joule-Thomson coefficient μ_{JT} is defined as the limiting value of this ratio as Δp approaches zero

$$\mu_{JT} = \lim_{\Delta p \to 0} \left(\frac{\Delta T}{\Delta p}\right)_H = \left(\frac{\partial T}{\partial p}\right)_H \tag{2.8.2}$$

Example 2.8.1

At 573 K and pressures of 0 – 6.0 MPa, the Joule-Thomson coefficient of N_2(g) can be represented as

$$\mu_{JT}/\text{K MPa}^{-1} = 0.140 - 2.533 \times 10^{-2}\ (p/\text{MPa})$$

Assuming this equation to be temperature-independent near 573 K, find the temperature drop which may be expected in the Joule-Thomson expansion of the gas from 6.0 MPa to 2.0 MPa pressure.

Solution

Since $\mu_{JT} = (\partial T/\partial p)_H$, therefore

$$\mathrm{d}T = \mu_{JT}\,\mathrm{d}p$$

$$= (0.140\ \text{K MPa}^{-1})\,\mathrm{d}p - (2.533 \times 10^{-2}\ \text{K MPa}^{-1})\,(p/\text{MPa})\,\mathrm{d}p$$

Intergrating within the limits of pressure, we get

$$\Delta T = (0.140\ \text{K MPa}^{-1})(p_2 - p_1) - (2.533 \times 10^{-2}\ \text{K MPa}^{-1})\left(\frac{p_2^2}{2\ \text{MPa}} - \frac{p_1^2}{2\ \text{MPa}}\right)$$

Substituting the given values of p, we get

$$\Delta T = (0.140\ \text{K MPa}^{-1})\,(2.0\ \text{MPa} - 6.0\ \text{MPa}) - (2.533 \times 10^{-2}\ \text{K MPa}^{-1}) \times \left(\frac{(2.0\ \text{MPa})^2}{2\ \text{MPa}} - \frac{(6.0\ \text{MPa})^2}{2\ \text{MPa}}\right)$$

$$= -\,0.560\ \text{K} + 0.405\ \text{K}$$

$$= -\,0.155\ \text{K}$$

Joule-Thomson Coefficient in Terms of Easily Determinable Derivative

Since $H = f(T, p)$, its differential is

$$\mathrm{d}H = \left(\frac{\partial H}{\partial T}\right)_p \mathrm{d}T + \left(\frac{\partial H}{\partial p}\right)_T \mathrm{d}p$$

Dividing by dp and introducing the condition of constant H, we get

$$0 = \left(\frac{\partial H}{\partial T}\right)_p \left(\frac{\partial T}{\partial p}\right)_H + \left(\frac{\partial H}{\partial p}\right)_T$$

or

$$\left(\frac{\partial T}{\partial p}\right)_H = -\frac{(\partial H/\partial p)_T}{(\partial H/\partial T)_p} = -\frac{1}{C_p}\left(\frac{\partial H}{\partial p}\right)_T \tag{2.8.3}$$

According to the thermodynamic equation of state (see Section 4.15), we have

$$\left(\frac{\partial H}{\partial p}\right)_T = -T\left(\frac{\partial V}{\partial T}\right)_p + V$$

Substituting this in Eq. (2.8.3), we get

$$\mu_{JT} = \left(\frac{\partial T}{\partial p}\right)_H = \frac{T\,(\partial V/\partial T)_p - V}{C_p} \tag{2.8.4}$$

or
$$\mu_{JT} = \frac{V}{C_p}\left[\frac{T}{V}\left(\frac{\partial V}{\partial T}\right)_p - 1\right] = \frac{V}{C_p}\,(T\,\alpha - 1) \tag{2.8.5}$$

Example 2.8.2

Calculate the Joule-Thomson coefficient of carbon monoxide at 298 K and 40.53 MPa pressure, given that $(T/V)\ (\partial V/\partial T)_p$ is 0.984, the molar volume is 76 $cm^3\ mol^{-1}$ and $C_{p,\,m} = 37.28\ J\ K^{-1}\ mol^{-1}$.

Solution

Substituting the given values in the expression

$$\mu_{JT} = \frac{V_m}{C_{p,m}}\left[\frac{T}{V}\left(\frac{\partial V}{\partial T}\right)_p - 1\right]$$

we get
$$\mu_{JT} = \frac{(76\times10^{-6}\ m^3\ mol^{-1})}{(37.28\ J\ K^{-1}\ mol^{-1})}\ (0.984 - 1)$$

$$= -\ 0.032\ 6\times10^{-6}\ K\ J^{-1}\ m^3 = -\ 0.032\ 6\times10^{-6}\ K\ (Pa\ m^3)^{-1}\ m^3$$

$$= -\ 0.032\ 6\times10^{-6}\ K\ (1\ atm/101.325\times10^3)^{-1} = -\ 0.003\ 3\ K\ atm^{-1}$$

Joule-Thomson Coefficient for an Ideal Gas

For an ideal gas $(\partial V/\partial T)_p = nR/p$, therefore

$$\mu_{JT} = \frac{T\,(\partial V/\partial T)_p - V}{C_p} = \frac{nRT/p - V}{C_p} = 0 \tag{2.8.6}$$

that is, *Joule-Thomson coefficient for an ideal gas is zero* and thus the temperature of such a gas will not suffer any change in the Joule-Thomson experiment.

Joule-Thomson Coefficient for a Real Gas

For real gases, the numerator in Eq. (2.8.4) (and thus Joule-Thomson coefficient) may be positive, negative or zero depending upon the temperature and pressure of the gas. Since the change in pressure, dp, in Joule-Thomson experiment is always negative, it follows from Eq. (2.8.2) that dT is negative when μ_{JT} is positive, and vice versa, These effects are summarized in Table 2.8.1.

Table 2.8.1 Relative Signs of μ_{JT} and $(\partial H/\partial p)_T$

Experimental result	μ_{JT} *from Eq.* (2.8.2)	$(\partial H/\partial p)_T$ *from Eq.* (2.8.3)	*Comments*
dT negative, i.e. $T_2 < T_1$ Cooling	(+)	(–)	Most gases, H_2 and He at low temperature
dT zero, i.e. $T_2 = T_1$ Neither heating nor cooling	Zero	Zero	Ideal gas, real gas at inversion temperature
dT positive, i.e. $T_2 > T_1$ Heating	(–)	(+)	Most gases at high temperature, H_2 and He at ordinary temperature

Real Gas Undergoing Joule-Thomson Expansion

The effects produced in the Joule-Thomson experiment involving a real gas can be discussed very neatly with the help of Eq. (2.8.3). We have

$$\mu_{JT} = \left(\frac{\partial T}{\partial p}\right)_H = -\frac{1}{C_p}\left(\frac{\partial H}{\partial p}\right)_T$$

Since $H = U + pV$, therefore

$$\text{or} \qquad \left(\frac{\partial H}{\partial p}\right)_T = \left(\frac{\partial U}{\partial p}\right)_T + \left(\frac{\partial (pV)}{\partial p}\right)_T$$

$$\text{or} \qquad \left(\frac{\partial H}{\partial p}\right)_T = \left(\frac{\partial U}{\partial V}\right)_T\left(\frac{\partial V}{\partial p}\right)_T + \left(\frac{\partial (pV)}{\partial p}\right)_T \qquad (2.8.7)$$

Substituting this in the previous expression, we get

$$\mu_{JT} = \left(\frac{\partial T}{\partial p}\right)_H = -\frac{1}{C_p}\left[\left(\frac{\partial U}{\partial V}\right)_T\left(\frac{\partial V}{\partial p}\right)_T + \left(\frac{\partial (pV)}{\partial p}\right)_T\right] \qquad (2.8.8)$$

The first term on the right hand side of the above equation is negative since $(\partial U/\partial V)_T$ is positive (unless the pressure is very high where $(\partial U/\partial V)_T$ is negative) and the fact that $(\partial V/\partial p)_T$ is always negative.

The second term $(\partial(pV)/\partial p)_T$ which, however, represents the deviation from the ideal gas, can have either sign.

Three cases may be distinguished.

(i) *When $\{\partial(pV)/\partial p)\}_T$ is negative*

This condition is usually valid at low temperatures. In this case,

$$\mu_{JT} = -\frac{1}{C_p}\left[\left(\frac{\partial U}{\partial V}\right)_T\left(\frac{\partial V}{\partial p}\right)_T + \left(\frac{\partial (pV)}{\partial p}\right)_T\right]$$

$$= -\frac{1}{(+\text{ve})}\ [(-\text{ve}) + (-\text{ve})] = \text{positive}$$

Since dp in Joule-Thomson expansion is always negative, we will have

$$\mu_{JT} = \left(\frac{\partial T}{\partial p}\right)_H = \frac{?}{(-\text{ve})} = \text{positive}$$

Therefore, $(\Delta T)_H = -\text{ve}$

that is, the gas undergoes cooling in Joule-Thomson expansion.

(ii) *When $\{\partial(pV)/\partial p\}_T$ is positive and is greater than the magnitude of $(\partial U/\partial V)_T\,(\partial V/\partial p)_T$*

This condition is usually valid at high temperatures. In this case

$$\mu_{JT} = -\frac{1}{C_p}\left[\left(\frac{\partial U}{\partial V}\right)_T\left(\frac{\partial V}{\partial p}\right)_T + \left(\frac{\partial (pV)}{\partial p}\right)_T\right]$$

$$= -\frac{1}{(+\text{ve})}\ [(-\text{ve}) + (\text{more } +\text{ve})] = \text{negative}$$

and $\mu_{JT} = \left(\frac{\partial T}{\partial p}\right)_H = \frac{?}{(-ve)}$ = negative

Therefore, $(\Delta T)_H$ = +ve

that is, the gas undergoes heating in Joule-Thomson expansion.

(iii) *When $\{\partial(pV)/\partial p\}_T$ is positive and is equal to the magnitude of $(\partial U/\partial V)_T\,(\partial V/\partial p)_T$*

In this case

$$\mu_{JT} = 0 \quad \text{and thus} \quad (\Delta T)_H = 0$$

that is, the gas does not exhibit any cooling or heating during the Joule-Thomson expansion.

Inversion Temperature

The temperature of the gas at which $\mu_{JT} = 0$ is known as the *inversion temperature.*

From Eq. (2.8.5), it follows that

$$\mu_{JT} = \frac{V}{C_p}\left[\frac{T_i}{V}\left(\frac{\partial V}{\partial T}\right)_p - 1\right] = 0$$

or
$$T_i = \frac{1}{\frac{1}{V}\left(\frac{\partial V}{\partial T}\right)_p} = \frac{1}{\alpha_i} \tag{2.8.9}$$

where α_i is the isobaric expansivity of the gas at the temperature T_i.

Evaluation of $(\partial H/\partial p)_T$ and $(\partial U/\partial V)_T$ from μ_{JT}

The evaluation of the differential $(\partial H/\partial p)_T$ for gases can be done by utilizing Eq. (2.8.3). Thus

$$\left(\frac{\partial H}{\partial p}\right)_T = -C_p\left(\frac{\partial T}{\partial p}\right)_H = -C_p\,\mu_{JT} \tag{2.8.10}$$

For an ideal gas, $\mu_{JT} = 0$ and thus the differential $(\partial H/\partial p)_T$ is zero. This also follows from Eq. (2.8.7) since $(\partial U/\partial V)_T$ and $(\partial(pV)/\partial p)_T$ are both zero for an ideal gas. For real gases, $(\partial H/\partial p)_T$ can be positive, negative or zero depending upon the temperature of the gas. Once $(\partial H/\partial p)_T$ is evaluated, the differential $(\partial U/\partial V)_T$ can be calculated from Eq. (2.8.7). Thus

$$\left(\frac{\partial H}{\partial p}\right)_T = \left(\frac{\partial U}{\partial V}\right)_T\left(\frac{\partial V}{\partial p}\right)_T + \left(\frac{\partial (pV)}{\partial p}\right)_T$$

$$= \left(\frac{\partial U}{\partial V}\right)_T\left(\frac{\partial V}{\partial p}\right)_T + p\left(\frac{\partial V}{\partial p}\right)_T + V$$

Rearranging this, we get

$$\left(\frac{\partial U}{\partial V}\right)_T = \frac{(\partial H/\partial p)_T - V}{(\partial V/\partial p)_T} - p \tag{2.8.11}$$

Hence, substitution of $(\partial H/\partial p)_T$, $(\partial V/\partial p)_T$, V and p gives the value of $(\partial U/\partial V)_T$.

Problem 2.8.1 Show that the value of $(\partial H/\partial p)_T = \simeq V$ for solids and liquids.

Solution Since $H = U + pV$, we get on differentiation

$$\mathrm{d}H = \mathrm{d}U + p\,\mathrm{d}V + V\,\mathrm{d}p$$

Dividing by dp and introducing the condition of constant temperature, we get

$$\left(\frac{\partial H}{\partial p}\right)_T = \left(\frac{\partial U}{\partial p}\right)_T + p\left(\frac{\partial V}{\partial p}\right)_T + V = \left(\frac{\partial U}{\partial V}\right)_T\left(\frac{\partial V}{\partial p}\right)_T + p\left(\frac{\partial V}{\partial p}\right)_T + V$$

For solids and liquids, the variation of volume with pressure is negligibly small, i.e. $(\partial V/\partial p)_T \simeq 0$, and therefore

$$\left(\frac{\partial H}{\partial p}\right)_T \simeq V$$

2.9 JOULE-THOMSON COEFFICIENT AND VAN DER WAALS EQUATION OF STATE

Expression of μ_{JT} The expression for Joule-Thomson coefficient as given by Eq. (2.8.4) is

$$\mu_{\text{JT}} = \frac{1}{C_p}\left[T\left(\frac{\partial V}{\partial T}\right)_p - V\right] \tag{2.9.1}$$

The expression for $(\partial V/\partial T)_p$ for one mole of a van der Waals gas as given by Eq. (2.6.12) is

$$\left(\frac{\partial V_{\text{m}}}{\partial T}\right)_p = \frac{R}{p - \dfrac{a}{V_{\text{m}}^2} + \dfrac{2ab}{V_{\text{m}}^3}}$$

Substituting this in Eq. (2.9.1), we get

$$\mu_{\text{JT}} = \frac{1}{C_{p,\text{m}}}\left[\frac{RT}{p - \dfrac{a}{V_{\text{m}}^2} + \dfrac{2ab}{V_{\text{m}}^3}} - V_{\text{m}}\right] = \frac{1}{C_{p,\text{m}}}\left[\frac{\left(p + \dfrac{a}{V_{\text{m}}^2}\right)(V_{\text{m}} - b)}{p - \dfrac{a}{V_{\text{m}}^2} + \dfrac{2ab}{V_{\text{m}}^3}} - V_{\text{m}}\right]$$

$$= \frac{1}{C_{p,\text{m}}}\left[\frac{\dfrac{2a}{V_{\text{m}}} - \dfrac{3ab}{V_{\text{m}}^2} - pb}{p - \dfrac{a}{V_{\text{m}}^2} + \dfrac{2ab}{V_{\text{m}}^3}}\right] \tag{2.9.2}$$

Equation (2.9.2) may be simplified under the following approximations.

(i) $-\dfrac{a}{V_{\text{m}}^2} + \dfrac{2ab}{V_{\text{m}}^3}$ may be ignored in comparison to p

(ii) $pV_{\text{m}} \simeq RT$

With these, Eq. (2.9.2) becomes

$$\mu_{\text{JT}} = \frac{1}{C_{p,\text{m}}}\left[\frac{2a}{RT} - \frac{3abp}{R^2T^2} - b\right] \tag{2.9.3}$$

If temperature of the gas is not too low and its pressure is not too high, Eq. (2.9.3) can be reduced to a more simple form by neglecting the term $3abp/R^2T^2$, so that

$$\mu_{JT} = \frac{1}{C_{p,\mathrm{m}}}\left(\frac{2a}{RT} - b\right) \tag{2.9.4}$$

Results of Joule-Thomson Expansion

Values of μ_{JT} for H_2, O_2 and CO_2 at 0 °C as calculated from Eq. (2.9.4) are given in Table 2.9.1. The agreement between observed and calculated values seems reasonable keeping in mind the assumptions made in deriving Eq. (2.9.4) and the applicability of the van der Waals equation.

Table 2.9.1 Values of μ_{JT} at 0 °C for some Gases

	a	b	$C_{p,\mathrm{m}}$	μ_{JT}/K atm^{-1}	
Gas	dm^6 kPa mol^{-2}	dm^3 mol^{-1}	J K^{-1} mol^{-1}	*Calculated*	*Observed*
H_2	25.74	2.67×10^{-2}	29.38	– 0.014	– 0.03
O_2	133.75	3.12×10^{-2}	30.29	0.29	0.31
CO_2	364.77	4.28×10^{-2}	38.50	0.73	1.30

Explanation of Results

The Joule-Thomson coefficient has a positive value when $(2a/RT)$ is greater than b and the gas exhibits cooling on expansion. The cooling of the gas can be explained on the basis that the forces of attraction between molecules of the gas predominate over the excluded volume b. Consequently, on expansion the gas has to do some work against these forces of attraction. The energy required for this comes from within the system as the expansion takes place under adiabatic conditions and thus cooling is observed. On increasing the temperature, the term $(2a/RT)$ decreases and eventually at the temperature known as the *inversion temperature*, it becomes equal to b with the result that μ_{JT} becomes equal to zero. At this temperature, neither heating nor cooling is observed. At still higher temperature, the factor $(2a/RT)$ becomes lesser than b and thus now the excluded volume b predominates over forces of attraction. Here μ_{JT} is negative and there occurs a rise in temperature of the gas after it has passed through the porous plug. These effects are also shown in Fig. 2.9.1.

Inversion Temperature

The inversion temperature of a gas as defined above is the temperature at which $\mu_{JT} = 0$. Thus equating Eq. (2.9.3) to zero, we get

$$\frac{2a}{RT_\mathrm{i}} - \frac{3abp}{R^2T_\mathrm{i}^2} - b = 0 \tag{2.9.5}$$ [†]

Equation (2.9.5) implies that a gas has two inversion temperatures at each pressure. Experimental results are in agreement with this fact as illustrated in Fig. 2.9.2.

Under the conditions of low pressure and high temperature inversion temperature may be determined from Eq. (2.9.4). Thus, we have

$$T_\mathrm{i} = \frac{2a}{Rb} \tag{2.9.6}$$

[†] See Annexure I at the end of this Chapter for the inversion temperature related to pressure and reduced inversion temperature related to reduced pressure for a real gas following van der Waals, Berthelot, Dieterici and Redlich-Kwong equations of state.

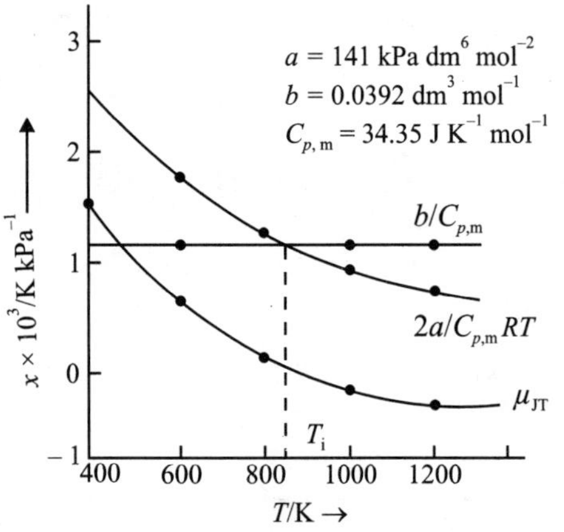

Fig. 2.9.1 The variation of μ_{JT} and its two components of N_2 (a van der Waals gas) with temperature:

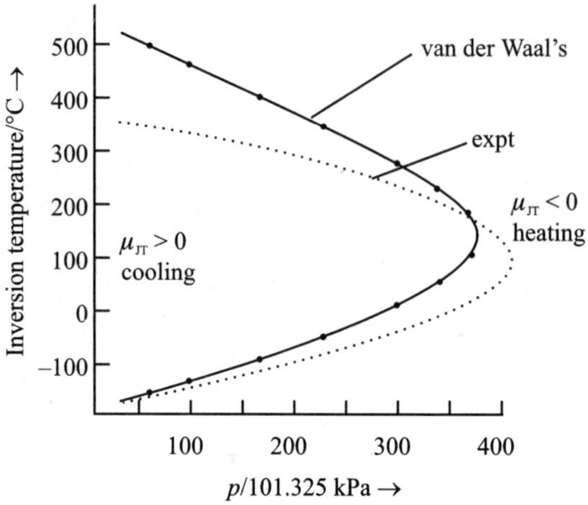

Fig. 2.9.2 A Typical Joule-Thomson inversion curve

Example 2.9.1

Calculate μ_{JT} for nitrogen gas at 293 K and 10.133 MPa pressure, taking $C_{p,\mathrm{m}}$ as 34.35 J K^{-1} mol^{-1}, a = 0.141 dm^6 MPa mol^{-2} and b = 3.92 × 10^{-2} dm^3 mol^{-1}.

Solution

The expression of μ_{JT} is

$$\mu_{JT} = \frac{1}{C_{p,\mathrm{m}}}\left(\frac{2a}{RT} - b - \frac{3abp}{R^2T^2}\right)$$

Now $$\frac{2a}{RT} = \frac{2(0.141\,\mathrm{dm^6\ MPa\ mol^{-2}})}{(8.314\times10^{-3}\,\mathrm{dm^6\ MPa\ mol^{-1}})\,(293\,\mathrm{K})} = 0.115\ 7\ \mathrm{dm^3\ mol^{-1}}$$

$$\frac{3abp}{R^2T^2} = \frac{3(0.141\,\mathrm{dm^6\ MPa\ mol^{-2}})(3.92\times10^{-2}\,\mathrm{dm^3\ mol^{-1}})(10.133\,\mathrm{MPa})}{(8.314\times10^{-3}\,\mathrm{dm^6\ MPa\ K^{-1}mol^{-1}})^2\,(293\,\mathrm{K})^2}$$

$$= 0.028\ 3\ \mathrm{dm^3\ mol^{-1}}$$

Thus $$\mu_{JT} = \frac{1}{(34.35\,\mathrm{J\ K^{-1}\ mol^{-1}})}\ [(0.115\ 7 - 0.039\ 2 - 0.028\ 3)\mathrm{dm^3\ mol^{-1}}]$$

$$= 1.403 \times 10^{-3}\ \mathrm{K\ J^{-1}\ dm^3} = 1.403 \times 10^{-3}\ \mathrm{K\ (kPa\ dm^3)^{-1}\ dm^3}$$

$$= 1.403 \times 10^{-3}\ \mathrm{K\ kPa^{-1}} = 1.403 \times 10^{-3}\ \mathrm{K\ (1\ atm/101.325)^{-1}} = 0.142\ \mathrm{K\ atm^{-1}}$$

Example 2.9.2 Calculate the inversion temperature of nitrogen. Given : a = 1.408 dm^6 bar mol^{-2} and b = 0.039 13 dm^3 mol^{-1}.

Solution Substituting the given data in the expression $T_i = 2a/Rb$, we get

$$T_i = \frac{2(1.408\ \text{dm}^6\ \text{bar mol}^{-2})}{(0.083\ 14\ \text{dm}^3\ \text{bar K}^{-1}\ \text{mol}^{-1})(0.039\ 13\ \text{dm}^3\ \text{mol}^{-1})}$$

$$= 865.6\ \text{K}$$

Evaluation of $(\partial H/\partial p)_T$ from μ_{JT} of a van der Waals Gas

From Eq. (2.8.10), we have

$$\left(\frac{\partial H}{\partial p}\right)_T = -C_p\,\mu_{JT}$$

Also from Eq. (2.9.4), we have

$$\mu_{JT} = \frac{1}{C_{p,m}}\left(\frac{2a}{RT} - b\right)$$

From these two expressions, we get

$$\left(\frac{\partial H_m}{\partial p}\right)_T = b - \frac{2a}{RT} \tag{2.9.7}$$

For a gas undergoing isothermal expansion or compression, we will have

$$\Delta H_m = \left(b - \frac{2a}{RT}\right)\Delta p \tag{2.9.8}$$

Example 2.9.3 Calculate the value of ΔH for the isothermal compression at 300 K of 1 mol of nitrogen from 10^5 Pa to 500 × 10^5 Pa. Given: a = 135.78 (dm^3)^2 kPa mol^{-2} and b = 0.039 dm^3 mol^{-1}.

Solution We have

$$\Delta p = 500 \times 10^5\ \text{Pa} - 1 \times 10^5\ \text{Pa} = 499 \times 10^5\ \text{Pa} = 499 \times 10^2\ \text{kPa}$$

$$a = 135.78\ (\text{dm}^3)^2\ \text{kPa mol}^{-2}$$

$$b = 0.039\ \text{dm}^3\ \text{mol}^{-1}$$

Substituting these values in the expression

$$\Delta H_m = \left(b - \frac{2a}{RT}\right)\Delta p$$

we get

$$\Delta H_m = \left[0.039\ \text{dm}^3\ \text{mol}^{-1} - \frac{2 \times 135.78\ \text{dm}^6\ \text{kPa mol}^{-2}}{(8.314\ \text{dm}^3\ \text{kPa K}^{-1}\ \text{mol}^{-1})\ (300\ \text{K})}\right](499 \times 10^2\ \text{kPa})$$

$$= (0.039\ \text{dm}^3\ \text{mol}^{-1} - 0.108\ 9\ \text{dm}^3\ \text{mol}^{-1})\ (499 \times 10^2\ \text{kPa})$$

$$= -3\ 488\ \text{dm}^3\ \text{kPa mol}^{-1} = -3.488\ \text{kJ mol}^{-1}$$

Example 2.9.4 In a single-stage Joule-Thomson expansion, one mole of nitrogen gas suffers a change in temperature from 25 °C to – 196 °C. If the final pressure of the gas is 101.325 kPa, calculate its initial pressure. Given: a = 135.78 dm^6 kPa mol^{-2}, $C_{p,m}$ = 20.92 J K^{-1} mol^{-1} and b = 0.039 dm^3 mol^{-1}.

Solution We have

$T_1 = 25\ °C = 298\ K;$ $\qquad T_2 = -196\ °C = 77\ K$

$p_1 = ?;$ $\qquad p_2 = 101.325\ kPa$

$a = 135.78\ dm^6\ kPa\ mol^{-2};$ $\qquad b = 0.039\ dm^3\ mol^{-1}$

We know that

$$\mu_{JT} = \left(\frac{\partial T}{\partial p}\right)_H = \frac{1}{C_{p,m}}\left(\frac{2a}{RT} - b\right)$$

Therefore $$dp = \frac{C_{p,m}}{(2a/RT) - b}\,dT = C_{p,m}\left(\frac{RT}{2a - bRT}\right)dT$$

Changing the variable T to θ with the help of equation $2a - bRT = \theta$, we have

$$-bR\ dT = d\theta$$

Thus $$dT = \frac{d\theta}{(-bR)} \quad \text{and} \quad RT = \frac{2a - \theta}{b}$$

With these, the expression for dp becomes

$$dp = C_{p,m}\left(\frac{(2a-\theta)/b}{\theta}\right)\cdot\frac{d\theta}{(-bR)} \quad \text{or} \quad dp = -\frac{C_{p,m}}{b^2R}\left(\frac{2a}{\theta} - 1\right)d\theta$$

Integrating, we have

$$\int_{p_1}^{p_2} dp = -\frac{C_{p,m}}{b^2R}\int_{\theta_1}^{\theta_2}\left(\frac{2a}{\theta} - 1\right)d\theta$$

or $$p_2 - p_1 = -\frac{C_{p,m}}{b^2R}\{2a\ \ln(\theta_2/\theta_1) - (\theta_2 - \theta_1)\}$$

Now the values of θ_1 and θ_2 corresponding to temperatures 25 °C (298 K) and –196 °C (77 K) are

$$\theta_1 = 2a - bRT_1 = 2(135.78\ dm^6\ kPa\ mol^{-2}) - (0.039\ dm^3\ mol^{-1}) \times (8.314\ dm^3\ kPa\ K^{-1}\ mol^{-1})\ (298\ K)$$
$$= 174.93\ dm^6\ kPa\ mol^{-2}$$

$$\theta_2 = 2a - bRT_2 = 2(135.78\ dm^6\ kPa\ mol^{-2}) - (0.039\ dm^3\ mol^{-1}) \times (8.314\ dm^3\ kPa\ K^{-1}\ mol^{-1})\ (77\ K)$$
$$= 246.59\ dm^6\ kPa\ mol^{-2}$$

Substituting these values along with $p_2 = 101.325$ kPa in the previous expression, we get

$$101.325\ kPa - p_1 = -\frac{(20.92\ J\ K^{-1}\ mol^{-1})}{(0.039\ dm^3\ mol^{-1})^2\ (8.314\ J\ K^{-1}\ mol^{-1})}$$
$$\times [2(135.78\ dm^6\ kPa\ mol^{-2})\ \{\ln(246.59/174.93)\} - (246.59 - 174.93)]\ dm^6\ kPa\ mol^{-2}$$
$$= -(1\,654.33\ dm^{-6}\ mol^2)\ (21.58\ dm^6\ kPa\ mol^{-2})$$
$$= -35\ 700\ kPa$$

Thus $p_1 = 35\ 801\ kPa = 353.3\ atm$

Problem 2.9.1 (a) For the equation

$$pV_m = RT\left(1 + \frac{B}{V_m}\right); \quad \text{where} \quad B = b - a/RT$$

Show that the expression of Joule-Thomson coefficient for this gas is given by

$$\mu_{JT} = \frac{1}{C_{p,\mathrm{m}}}\left[\frac{-B + T(dB/dT)_p}{1 + 2B/V_\mathrm{m}}\right]$$

(b) At the inversion temperature, $\mu_{JT} = 0$. Hence, $B - T_\mathrm{i}(dB/dT)_p = 0$. Show that it leads to the expression $T_\mathrm{i} = 2a/Rb$. Calculate the inversion temperature of nitrogen gas for which $a = 141$ kPa dm^6 mol^{-2} and $b = 0.039\ 2$ dm^3 mol^{-1}.

(c) Draw a graph between B and T for the nitrogen and show that the inversion temperature is equal to the temperature of contact of the tangent to B versus T curve drawn from the origin of the graph.

Solution

(a) The expression of Joule-Thomson coefficient is

$$\mu_{JT} = \frac{1}{C_{p,\mathrm{m}}}\left[T\left(\frac{\partial V_\mathrm{m}}{\partial T}\right)_p - V_\mathrm{m}\right]$$

From the given equation of the gas, we get

$$p\left(\frac{\partial V_\mathrm{m}}{\partial T}\right)_p = R\left(1 + \frac{B}{V_\mathrm{m}}\right) + RT\left[\frac{1}{V_\mathrm{m}}\frac{\mathrm{d}B}{\mathrm{d}T} - \frac{B}{V_\mathrm{m}^2}\left(\frac{\partial V_\mathrm{m}}{\partial T}\right)_p\right]$$

Form this expression, we get

$$\left(\frac{\partial V_\mathrm{m}}{\partial T}\right)_p = \frac{R(1 + B/V_\mathrm{m}) + (RT/V_\mathrm{m})(\mathrm{d}B/\mathrm{d}T)}{(p + BRT/V_\mathrm{m}^2)}$$

Replacing p by (RT/V_m) $(1 + B/V_\mathrm{m})$ and multiplying numerator and denominator by V_m/RT, we get

$$\left(\frac{\partial V_\mathrm{m}}{\partial T}\right)_p = \frac{(V_\mathrm{m}/T)(1 + B/V_\mathrm{m}) + (\mathrm{d}B/\mathrm{d}T)}{(1 + 2B/V_\mathrm{m})}$$

Hence, $$\mu_{JT} = \frac{1}{C_{p,\mathrm{m}}}\left[T\left(\frac{\partial V_\mathrm{m}}{\partial T}\right)_p - V_\mathrm{m}\right]$$

$$= \frac{1}{C_{p,\mathrm{m}}}\left[\frac{V_\mathrm{m}(1 + B/V_\mathrm{m}) + T(\mathrm{d}B/\mathrm{d}T)}{(1 + 2B/V_\mathrm{m})} - V_\mathrm{m}\right]$$

$$= \frac{1}{C_{p,\mathrm{m}}}\left[\frac{-B + T(\mathrm{d}B/\mathrm{d}T)}{(1 + 2B/V_\mathrm{m})}\right]$$

(b) Since $B = b - a/RT$, we get

$$\frac{\mathrm{d}B}{\mathrm{d}T} = \frac{a}{RT^2}$$

Hence, at inversion temperature, we have

$$B - T_\mathrm{i}\frac{\mathrm{d}B}{\mathrm{d}T} = b - \frac{a}{RT_\mathrm{i}} - T_\mathrm{i}\left(\frac{a}{RT_\mathrm{i}^2}\right) = b - \frac{2a}{RT_\mathrm{i}}$$

Equating this to zero, we get

$$T_i = \frac{2a}{Rb}$$

For N_2, we have

$$T_i = \frac{2(141\ \text{kPa dm}^6\ \text{mol}^{-2})}{(8.314\ \text{kPa dm}^3\ \text{K}^{-1}\ \text{mol}^{-1})(0.039\,2\ \text{dm}^3\ \text{mol}^{-1})} = 865.3\ \text{K}$$

(c) Given values of a and b for N_2 are $a = 141$ kPa dm^6 mol^{-2} and $b = 0.039\,2$ mol^{-1} dm^3. The values of B (= $b - a / RT$) at different temperatures are as follows.

T/K	B / dm^3 mol^{-1}	T/K	B / dm^3 mol^{-1}
100	– 0.130	600	0.011
200	– 0.046	700	0.015
300	– 0.017	800	0.018
400	– 0.003	900	0.020
500	0.005	1000	0.022

The graph between B and T is shown in Fig. 2.9.3. A slope line passing through the origin (B = 0, T = 0) is also shown in Fig. 2.9.3. It is obvious that this slope line meets B verses T graph at $T = T_i$, because only at this point the values of B calculated via slope line and from the expression $B = b - a/RT$ have the same value. The equation of the slope line is $B = (\mathrm{d}B/\mathrm{d}T)\ T = (B_i/T_i)\ T$.

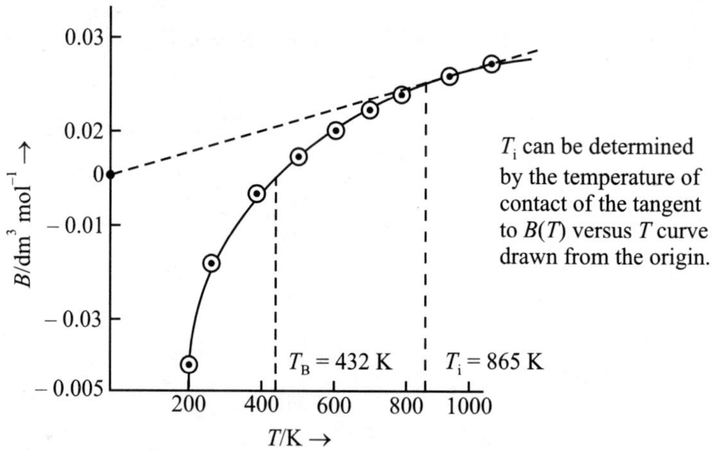

Fig. 2.9.3 Graph between *B* and *T* for nitrogen

Problem 2.9.2

(a) A gas follows the equation $Z = 1 + B/V_m$. The Joule-Thomson coefficient of the gas is given by the expression

$$\mu_{JT} = \frac{V_m}{C_{p,m}}\left[\frac{T(\mathrm{d}B/\mathrm{d}T) - B}{(pV_m^2/RT) + B}\right] \tag{2.9.9}$$

where Z represents compression factor and B represents second virial coefficient of a gas.

(b) Reduce (i) van der Waals, (ii) Berthelot, (iii) Dieterici and (iv) Redlich-Kwong equations to the form $Z = 1 + B/V_m$ and derive the expressions of Boyle and inversion temperatures.

Solution **(i) Van der Waals Equation** We have

$$p = \frac{RT}{V_m - b} - \frac{a}{V_m^2}$$

Multiplying throughout by V_m/RT, we get

$$Z = \frac{pV_m}{RT} = \frac{V_m}{V_m - b} - \frac{a}{V_m RT}$$

$$= \left(1 - \frac{b}{V_m}\right)^{-1} - \frac{a}{V_m RT} = \left[1 + \frac{b}{V_m} + \left(\frac{b}{V_m}\right)^2 + \cdots\right] - \frac{a}{V_m RT}$$

$$= 1 + \left(b - \frac{a}{RT}\right)\frac{1}{V_m} + \cdots$$

Under the condition of low pressure, we write

$$Z = 1 + \left(b - \frac{a}{RT}\right)\frac{1}{V_m}$$

The second virial coefficient is

$$B = b - \frac{a}{RT}$$

At Boyle temperature, $B = 0$. This gives

$$T_B = \frac{a}{Rb} \tag{2.9.10}$$

At inversion temperature, $\mu_{JT} = 0$. Hence, from Eq. (2.9.9), we get

$$T\left(\frac{dB}{dT}\right) - B = 0$$

i.e. $T\left[\frac{d(b - a/RT)}{dT}\right] - \left(b - \frac{a}{RT}\right) = 0$ i.e. $T\left(\frac{a}{RT^2}\right) - \left(b - \frac{a}{RT}\right) = 0$

Hence, $$T_i = \frac{2a}{Rb} \tag{2.9.11}$$

On comparing Eqs (2.9.10) and (2.9.11), we get

$$T_i = 2T_B \tag{2.9.12}$$

(ii) Berthelot Equation We have

$$p = \frac{RT}{V_m - b} - \frac{a}{V_m^2 T}$$

Multiplying throughout by V_m/RT, we get

$$Z = \frac{pV_m}{RT} = \frac{V_m}{V_m - b} - \frac{a}{V_m RT^2} = \left(1 - \frac{b}{V_m}\right)^{-1} - \frac{a}{V_m RT^2}$$

$$= \left[1 + \frac{b}{V_m} + \left(\frac{b}{V_m}\right)^2 + \cdots\right] - \frac{a}{V_m RT^2}$$

$$\approx 1 + \left(b - \frac{a}{RT^2}\right)\frac{1}{V_m} \qquad \text{(applicable at low pressure)}$$

Hence, the second virial coefficient is

$$B = b - \frac{a}{RT^2}$$

At Boyle temperature, $B = 0$. This gives

$$T_B = \left(\frac{a}{Rb}\right)^{1/2} \tag{2.9.13}$$

At inversion temperature

$$T\left(\frac{dB}{dT}\right) - B = 0$$

i.e. $$T\left[\frac{d(b - a/RT^2)}{dT}\right] - \left(b - \frac{a}{RT^2}\right) = 0$$

i.e. $$\frac{2a}{RT^2} - b + \frac{a}{RT^2} = 0$$

This gives $$T_i = \left(\frac{3a}{Rb}\right)^{1/2} \tag{2.9.14}$$

Comparing Eqs (2.9.13) and (2.9.14), we get

$$T_i = \sqrt{3}\, T_B \tag{2.9.15}$$

(iii) Dieterici Equation We have

$$p = \left(\frac{RT}{V_m - b}\right) e^{-a/V_m RT}$$

Multiplying throughout by V_m/RT, we get

$$Z = \frac{pV_m}{RT} = \left(\frac{V_m}{V_m - b}\right) e^{-a/V_m RT}$$

$$= \left(1 - \frac{b}{V_m}\right)^{-1} e^{-a/V_m RT} = \left[1 + \frac{b}{V_m} + \cdots\right]\left[1 - \frac{a}{V_m RT} + \cdots\right]$$

$$= 1 + \left(b - \frac{a}{RT}\right)\frac{1}{V_m}; \qquad \text{(applicable at low pressure)}$$

The second virial coefficient is

$$B = b - \frac{a}{RT}$$

At Boyle temperature, $B = 0$. This gives

$$T_B = \frac{a}{Rb} \tag{2.9.16}$$

At inversion temperature,

$$T\left(\frac{dB}{dT}\right) - B = 0 \qquad \text{i.e.} \qquad T\left[\frac{d(b - a/RT)}{dT}\right] - \left(b - \frac{a}{RT}\right) = 0$$

This gives $$T_i = \frac{2a}{Rb} \qquad (2.9.17)$$

On comparing Eqs (2.9.16) and (2.9.17), we get

$$T_i = 2T_B \qquad (2.9.18)$$

(iv) Redlich-Kwong Equation We have

$$p = \frac{RT}{V_m - b} - \frac{a}{T^{1/2}V_m(V_m + b)}$$

Multiplying throughout by V_m/RT, we get

$$\begin{aligned}
Z = \frac{pV_m}{RT} &= \frac{V_m}{V_m - b} - \frac{a}{RT^{3/2}(V_m + b)} \\
&= \left(1 - \frac{b}{V_m}\right)^{-1} - \frac{a}{RT^{3/2}V_m}\left(1 + \frac{b}{V_m}\right)^{-1} \\
&= \left(1 + \frac{b}{V_m} + \cdots\right) - \frac{a}{RT^{3/2}V_m}\left(1 - \frac{b}{V_m} + \cdots\right) \\
&\approx 1 + \left(b - \frac{a}{RT^{3/2}}\right)\frac{1}{V_m} \qquad \text{; (applicable at low pressure)}
\end{aligned}$$

The second virial coefficient is

$$B = b - \frac{a}{RT^{3/2}}$$

At Boyle temperature, $B = 0$. This gives

$$T_B = \left(\frac{a}{Rb}\right)^{2/3} \qquad (2.9.19)$$

At inversion temperature,

$$T\left(\frac{dB}{dT}\right) - B = 0$$

i.e. $$T\left[\frac{d(b - a/RT^{3/2})}{dT} - \left(b - \frac{a}{RT^{3/2}}\right)\right] = 0$$

i.e. $$\frac{3}{2}\frac{a}{RT^{3/2}} - b + \frac{a}{RT^{3/2}} = 0$$

This gives $$T_i = \left(\frac{5a}{2Rb}\right)^{2/3} \qquad (2.9.20)$$

On comparing Eqs (2.9.19) and (2.9.20), we get

$$T_i = \left(\frac{5}{2}\right)^{2/3} T_B \qquad (2.9.21)$$

Example 2.9.5

Given are the following data for N_2 gas.

	Equation	*a*	*b*
(i)	Van der Waals	136 kPa dm^6 mol^{-2}	0.0386 dm^3 mol^{-1}
(ii)	Berthelot	17226.1 kPa dm^6 K mol^{-2}	0.0386 dm^3 mol^{-1}
(iii)	Dieterici	175.3 kPa dm^6 mol^{-2}	0.0418 dm^3 mol^{-1}
(iv)	Redlich-Kwong	1554.4 kPa dm^6 K$^{1/2}$ mol^{-2}	0.0268 dm^3 mol^{-1}

Calculate the Boyle and inversion temperatures for the given equations of state.

Solution

(i) Van der Waals equation of state

$$T_B = \frac{a}{Rb} = \frac{(136 \text{ kPa dm}^6 \text{ mol}^{-2})}{(8.314 \text{ kPa dm}^3 \text{ K}^{-1} \text{ mol}^{-1})(0.0386 \text{ dm}^3 \text{ mol}^{-1})}$$

$$= 423.8 \text{ K}$$

$$T_i = 2\, T_B = 2 \times 423.8 \text{ K} = 847.6 \text{ K}$$

(ii) Berhelot equation of state

$$T_B = \left(\frac{a}{Rb}\right)^{1/2} = \frac{(1722.61 \text{ kPa dm}^6 \text{ K mol}^{-2})}{(8.314 \text{ kPa dm}^3 \text{ K}^{-1} \text{ mol}^{-1})(0.0386 \text{ dm}^3 \text{ mol}^{-1})}$$

$$= 231.7 \text{ K}$$

$$T_i = \sqrt{3}\ T_B = (1.732)(231.7 \text{ K}) = 401.3 \text{ K}$$

(iii) Dieterici equation of state

$$T_B = \frac{a}{Rb} = \frac{(175.3 \text{ kPa dm}^6 \text{ mol}^{-2})}{(8.314 \text{ kPa dm}^3 \text{ K}^{-1} \text{ mol}^{-1})(0.0418 \text{ dm}^3 \text{ mol}^{-1})}$$

$$= 504.4 \text{ K}$$

$$T_i = 2\, T_B = 2 \times 504.4 \text{ K} = 1008.8 \text{ K}$$

(iv) Redlich-Kwong equation of state

$$T_B = \left(\frac{a}{Rb}\right)^{2/3} = \left[\frac{(1554.4 \text{ kPa dm}^6 \text{ K}^{1/2} \text{ mol}^{-2})}{(8.314 \text{ kPa dm}^3 \text{ K}^{-1} \text{ mol}^{-1})(0.0268 \text{ dm}^3 \text{ mol}^{-1})}\right]^{2/3}$$

$$= 365.1 \text{ K}$$

$$T_i = (5/2)^{2/3}\ T_B = (2.5)^{2/3}(365.1 \text{ K}) = 672.5 \text{ K}$$

Experimental values $T_B = 328$ K; $T_i = 621$ K. Redlich-Kwong equation provides the values near to the experimental values.

2.10 THERMODYNAMIC CHANGES IN ISOTHERMAL VARIATION IN VOLUME OF AN IDEAL GAS

Expressions for q, w, ΔU and ΔH

We shall use the first law of thermodynamics to calculate the changes in thermodynamic properties when an ideal gas undergoes the process of expansion or compression.

In an isothermal expansion or compression process, the temperature of the system remains constant throughout the expansion or compression process. Since for an ideal gas, U depends only on temperature (Joule's law, $(\partial U/\partial V)_T = 0$), it follows that

$$dU = 0 \quad \text{or} \quad \Delta U = 0 \tag{2.10.1}$$

Substituting the above expression in the first law of thermodynamics, we get

$$dq = -\,dw \quad \text{or} \quad q = -\,w \tag{2.10.2}$$

Hence in an isothermal expansion or compression process, heat is converted into work and vice versa. Thus, if heat is supplied to the system (q positive), the equivalent amount of work is done by the system (w negative), and if some work is done on the system (w positive), the equivalent amount of heat is given out (q negative).

The enthalpy change of the system is also zero as

$$\begin{aligned}\Delta H &= \Delta(U + pV) = \Delta U + \Delta(pV)\\ &= \Delta U + \Delta(nRT) = \Delta U + nR(\Delta T)\\ &= 0 + 0 = 0\end{aligned} \tag{2.10.3}$$

The magnitude of w (or q) depends on how the expansion or compression process is carried out. Two different types of processes may be distinguished, namely, reversible and irreversible. The changes in q, w, ΔU and ΔH in these two types of processes are given below.

Reversible Expansion or Compression Process

The expression for the work involved in an isothermal expansion or compression from volume V_1 to V_2 can be worked out as follows:

Since $\quad dw = -\,p_{ext}\,dV$, therefore

$$w = -\int_{V_1}^{V_2} p_{ext}\,dV$$

In the reversible change of volume, the external pressure differs from the pressure of the gas by an infinitesimal amount, i.e.

$$p_{ext} = p_{int} \pm dp$$

where +ve and –ve signs are meant for compression and expansion processes, respectively.

Substituting this is the previous expression, we get

$$w = -\int_{V_1}^{V_2}(p_{int} \pm dp)\,dV = -\int_{V_1}^{V_2} p_{int}\,dV \mp \int_{V_1}^{V_2} dp\,dV$$

The second integration can be neglected, since it is the product of two infinitesimal small differentials. For an ideal gas, p_{int} is given by

$$p_{int} = \frac{nRT}{V}$$

With this the previous expression becomes

$$w = -\int_{V_1}^{V_2}\frac{nRT}{V}\,dV$$

Since temperature remains constant in an isothermal process, we have

$$w = -\,nRT\int_{V_1}^{V_2}\frac{dV}{V} = -\,nRT\ln\frac{V_2}{V_1} \tag{2.10.4}$$

In terms of pressure, we have

$$w = -\,nRT \ln \frac{p_1}{p_2} \tag{2.10.5}$$

Hence, for an isothermal reversible expansion or compression, we have

$$q = -\,w = nRT \ln \frac{V_2}{V_1} = nRT \ln \frac{p_1}{p_2}$$

$$\Delta U = \Delta H = 0$$

Example 2.10.1 Calculate the work which could be obtained from an isothermal reversible expansion of 1 mol of Cl_2 from 1 dm^3 to 50 dm^3 at 273 K using ideal gas behaviour.

Solution For an ideal gas

$$w = -\,RT \ln \frac{V_2}{V_1}$$

$$= -\,(8.314 \text{ J K}^{-1} \text{ mol}^{-1})\,(273 \text{ K}) \times 2.303 \times \log\left(\frac{50 \text{ dm}^3}{1 \text{ dm}^3}\right)$$

$$= -\,8\,880.8 \text{ J mol}^{-1} \simeq -\,8.88 \text{ kJ mol}^{-1}$$

Example 2.10.2 One dm^3 of an ideal gas at a pressure of 1.013 3 MPa expands reversibly and isothermally from its volume to 10 dm^3. How much of heat is absorbed and how much of work is done in expansion?

Solution For an ideal gas undergoing reversible volume change, we have

$$q = -\,w = nRT \ln \frac{V_2}{V_1}$$

The temperature in the above expression can be replaced in terms of p_1 and V_1 by using the ideal gas equation. Thus

$$q = (p_1 V_1) \ln \frac{V_2}{V_1}$$

Substituting the value of p_1, V_1 and V_2, we have

$$q = (1.013\,3 \text{ MPa})\,(1 \text{ dm}^3) \times 2.303 \times \log\left(\frac{10 \text{ dm}^3}{1 \text{ dm}^3}\right)$$

$$= 2.333\,6 \text{ MPa dm}^3 = 2.333\,6 \text{ kJ}$$

Irreversible Expansion or Compression Process

Two types of irreversible change in volume may be distinguished, namely, expansion against a zero pressure (free expansion) and expansion or compression against a constant pressure (intermediate expansion or compression). The change in q, w, ΔU and ΔH for these processes are given below.

Free Expansion

Here

$$p_{\text{ext}} = 0$$

Therefore $$w = \int \mathrm{d}w = -\int p_{\text{ext}}\, \mathrm{d}V = 0 \tag{2.10.6}$$

Thus, $q = -w = 0$

$\Delta U = \Delta H = 0$

Intermediate Expansion or Compression

Here, the work is done against a constant external pressure.

Therefore $w = -\int_{V_1}^{V_2} p_{ext}\, dV = -p_{ext}(V_2 - V_1)$ (2.10.7)

Hence, in the present case, we have

$$q = -w = p_{ext}(V_2 - V_1)$$
$$\Delta U = \Delta H = 0$$

It may be mentioned here that the magnitude of work involved in an intermediate expansion process will be less than that involved in the reversible expansion. Consequently, less heat will be absorbed in the case of intermediate expansion. On the other hand, the work involved in the intermediate compression process is larger than that involved in the reversible compression. Consequently, more heat will be released in the case of intermediate compression.

Example 2.10.3

(a) Five moles of an ideal gas at 293 K are expanded isothermally from an initial pressure of 0.405 3 MPa to a final pressure of 0.101 3 MPa against a constant external pressure of 0.101 3 MPa. Calculate q, w, ΔU and ΔH. (b) Calculate the corresponding values of q, w, ΔU and ΔH if the above process is carried out reversibly.

Solution

(a) For an isothermal expansion against a constant pressure, we have

$$w = -p_{ext}(V_2 - V_1) = -p_{ext}\left(\frac{nRT}{p_2} - \frac{nRT}{p_1}\right) = -nRT\, p_{ext}\left(\frac{1}{p_2} - \frac{1}{p_1}\right)$$

Substituting the values, we get

$$w = -(5.0\text{ mol})\,(8.314\text{ J K}^{-1}\text{ mol}^{-1})\,(293\text{ K})\,(0.101\ 3\text{ MPa}) \times \left(\frac{1}{0.101\,3\text{ MPa}} - \frac{1}{0.405\,3\text{ MPa}}\right)$$

$$= -9\ 135.77\text{ J} \simeq -9.136\text{ kJ}$$

Since temperature is constant

$$\Delta U = 0, \quad \Delta H = 0 \quad \text{and} \quad q = -w$$

(b) For an isothermal reversible expansion, we have

$$w = -nRT \ln\frac{V_2}{V_1} = -nRT \ln\frac{p_1}{p_2}$$

Substituting the values, we get

$$w = -(5.0\text{ mol})\,(8.314\text{ J K}^{-1}\text{ mol}^{-1})\,(293\text{ K}) \times 2303 \times \log\left(\frac{0.405\,3\text{ MPa}}{0.101\,3\text{ MPa}}\right)$$

$$= -16\ 889\text{ J} = -16.889\text{ kJ}$$

Comparison of Isothermal Expansion, Compression and Cyclic Processes

The changes in the thermodynamic properties in the expansion processes and those involved in compressing back to the original state and in the cyclic processes are listed in the Table 2.10.1.

Table 2.10.1 Comparison of q, w ΔU and ΔH for Different Types of Isothermal Processes†

Process	*Expansion*	*Compression*	*Cyclic*
Reversible			
q, $(-w)$	$nRT \ln \frac{V_2}{V_1}$	$nRT \ln \frac{V_1}{V_2}$	0
ΔU, ΔH	0	0	0
Free			
q, $(-w)$	0	$p_{ext}(V_1 - V_2) < nRT \ln \frac{V_1}{V_2}$	$p_{ext}(V_1 - V_2) < 0$
ΔU, ΔH	0	0	0
Intermediate			
q, $(-w)$	$0 < p'_{ext}(V_2 - V_1) < nRT \ln \frac{V_2}{V_1}$	$p_{ext}(V_1 - V_2) < nRT \ln \frac{V_1}{V_2}$	$(p'_{ext} - p_{ext})(V_2 - V_1) < 0$
ΔU, ΔH	0	0	0
	(i) $p_{ext} \geq p_1$	(ii) $0 < p'_{ext} \leq p_2$	(iii) $p'_{ext} < p_{ext}$

†**Notes:** The work involved during expansion is negative as V_1 is less then V_2. This negative work means that the work is done by the system on the surroundings. It may be noted that the magnitude of work involved in the irreversible expansion will be smaller than the corresponding magnitude of work involved in the reversible expansion. The pressure p_{ext} can have any value ranging from zero to p_2. On the other hand, the work involved during the compression is positive since V_1 is less then V_2. The positive work means that the work is done on the system. The work done during irreversible compression will be larger than the corresponding work done in the reversible compression. The pressure p_{ext} can have a value either equal to or greater than p_1.

2.11 THERMODYNAMIC CHANGES IN ADIABATIC VARIATION IN VOLUME OF AN IDEAL GAS

In adiabatic expansion or compression process heat is neither allowed to enter nor leave the system. Therefore

$$q = 0 \tag{2.11.1}$$

So, according to the first law of thermodynamics, $\Delta U = q + w$, we have

$$\Delta U = w \tag{2.11.2}$$

If there is an expansion, w will be negative and, therefore, ΔU will also be negative, i.e. there will occur a decrease in the internal energy of the system and hence a decrease in temperature will be observed. This amounts to the fact that the work is done by the system at the expense of internal energy.

If there is a compression, w will be positive and, therefore, ΔU will also be positive, i.e. there will occur an increase in the internal energy, and hence an increase in temperature will be observed. Here, the work is done by the surroundings on the system, which is stored as the internal energy.

Expressions for q, w, ΔU and ΔH

We have seen that

$$q = 0 \quad \text{and} \quad w = \Delta U$$

The change in the energy can be calculated as follows.

Writing the differential of U, we get

$$\mathrm{d}U = \left(\frac{\partial U}{\partial T}\right)_V \mathrm{d}T + \left(\frac{\partial U}{\partial V}\right)_T \mathrm{d}V$$

But for an ideal gas $(\partial U/\partial V)_T = 0$. Therefore

$$\mathrm{d}U = \left(\frac{\partial U}{\partial T}\right)_V \mathrm{d}T$$

Hence $\mathrm{d}U = nC_{V,\mathrm{m}}\,\mathrm{d}T$

For a finite change, we have

$$\Delta U = U_2 - U_1 = nC_{V,\mathrm{m}}\,(T_2 - T_1) \tag{2.11.3}$$

Thus, we have

$$w = \Delta U = nC_{V,\mathrm{m}}\,(T_2 - T_1) \tag{2.11.4}$$

The change in enthalpy is given by

$$\begin{aligned}\Delta H &= \Delta(U + pV)\\ &= \Delta U + \Delta(pV) = \Delta U + nR\,\Delta T\\ &= nC_{V,\mathrm{m}}\,(T_2 - T_1) + nR(T_2 - T_1) = n\,(C_{V,\mathrm{m}} + R)\,(T_2 - T_1)\end{aligned}$$

or
$$\Delta H = nC_{p,\mathrm{m}}\,(T_2 - T_1) \tag{2.11.5}$$

Expressions as given by Eqs (2.11.1) to (2.11.5) are applicable to any type of adiabatic processes (reversible, irreversible, free). However, the final temperature will be different in each case and hence, w, ΔU and ΔH will be different. We proceed now to calculate these quantities for the two types of expansion or compression processes, namely, reversible and irreversible.

Reversible Expansion or Compression Process

We have

$$w = \int \mathrm{d}w = -\int p_{\mathrm{ext}}\,\mathrm{d}V = -\int p_{\mathrm{int}}\,\mathrm{d}V$$

The integration in the above expression cannot be performed directly, because p_{int} is a function of both T and V and both variables are changing in an adiabatic expansion or compression process. However, if the final temperature is available, the values of q, w, ΔU and ΔH can be determined using Eqs (2.11.1) to (2.11.5). However, if the final temperature is not available, but the final pressure or volume is available, the first step is to calculate the final temperature from the expression which relates the initial and final temperatures to the respective volumes or pressures. Such expressions can be derived as follows.

Relationship between T and V

From Eq. (2.11.2), we have

$$\mathrm{d}U = \mathrm{d}w$$

Since $\mathrm{d}U = nC_{V,\mathrm{m}}\,\mathrm{d}T$ and $\mathrm{d}w = -\,p_{\mathrm{ext}}\,\mathrm{d}V$, therefore

$$nC_{V,\,m}\,\mathrm{d}T = -\,p_{\mathrm{ext}}\,\mathrm{d}V$$

For a reversible process

$$p_{\mathrm{ext.}} = p_{\mathrm{int}} \pm \mathrm{d}p \simeq p_{\mathrm{int}}$$

Thus $\quad nC_{V,\,m}\,\mathrm{d}T = -\,p_{\mathrm{int}}\,\mathrm{d}V$

For an ideal gas $p = (nRT/V)$, therefore

$$nC_{V,\,m}\,\mathrm{d}T = -\,\frac{nRT}{V}\,\mathrm{d}V \quad \text{or} \quad C_{V,\,m}\,\mathrm{d}T = -\,\frac{RT}{V}\,\mathrm{d}V$$

(**Note** that $C_{V,\,m}$ is molar heat capacity. The above equation will also be true for more than 1 mole of an ideal gas.)

Thus for a finite change, we have

$$\int_{T_1}^{T_2} C_{V,\mathrm{m}}\frac{\mathrm{d}T}{T} = -\int_{V_1}^{V_2} R\frac{\mathrm{d}V}{V}$$

If $C_{V,\,m}$ is considered independent of temperature, then

$$C_{V,\,\mathrm{m}}\ln\left(\frac{T_2}{T_1}\right) = -\,R\ln\left(\frac{V_2}{V_1}\right) \tag{2.11.6}$$

$$\text{or} \quad \ln\left(\frac{T_2}{T_1}\right)^{C_{V,\mathrm{m}}/R} = \ln\left(\frac{V_2}{V_1}\right)^{-1} \quad \text{or} \quad \left(\frac{T_2}{T_1}\right)^{C_{V,\mathrm{m}}/R} = \left(\frac{V_1}{V_2}\right)$$

Therefore $(T_2)^{C_{V,\,\mathrm{m}}/R}\,(V_2) = (T_1)^{C_{V,\,\mathrm{m}}/R}\,(V_1)$

The above expression implies that

$$T^{C_{V,\,\mathrm{m}}/R}\,V = \text{constant} \tag{2.11.7}$$

The alternative forms of this expression are

$$\left.\begin{aligned} &T\,V^{R/C_{V,\,\mathrm{m}}} = \text{constant} \\ \text{or}\quad &T\,V^{\gamma-1} = \text{constant} \end{aligned}\right\} \quad (\text{where } \gamma = C_{p,\mathrm{m}}/C_{V,\mathrm{m}}) \tag{2.11.8}$$

Example 2.11.1 20 g of N_2 at 300 K is compressed reversibly and adiabatically from 20 dm^3 to 10 dm^3. Calculate the final temperature, q, w, ΔU and ΔH.

Solution

$$\text{Amount of } N_2 = \frac{(20\ \mathrm{g})}{(28\ \mathrm{g\ mol^{-1}})}\ 0.714\ \mathrm{mol}$$

$$T_1 = 300\ \mathrm{K}; \quad V_1 = 20\ \mathrm{dm^3}; \quad V_2 = 10\ \mathrm{dm^3}$$

For an adiabatic reversible process

$$T_2 V_2^{R/C_{V,\,\mathrm{m}}} = T_1 V_1^{R/C_{V,\,\mathrm{m}}}$$

Thus $T_2 = T_1 \left(\dfrac{V_1}{V_2}\right)^{R/C_{V,\mathrm{m}}} = (300\ \mathrm{K})\left(\dfrac{20\ \mathrm{dm}^3}{10\ \mathrm{dm}^3}\right)^{2/5} = (300\ \mathrm{K})\ (1.32) = 396\ \mathrm{K}$

Hence, $\Delta U = w = nC_{V,\,\mathrm{m}}\ (T_2 - T_1) = (0.714\ \mathrm{mol})\left(\dfrac{5}{2}\times 8.314\ \mathrm{J\ K^{-1}\ mol^{-1}}\right)(96\ \mathrm{K})$

$= 1\ 424.69\ \mathrm{J}$

$\Delta H = nC_{p,\,\mathrm{m}}\ (T_2 - T_1) = (0.714\ \mathrm{mol})\left(\dfrac{7}{2}\times 8.314\ \mathrm{J\ K^{-1}\ mol^{-1}}\right)(96\ \mathrm{K})$

$= 1\ 994.56\ \mathrm{J}$

Relationship Between *T* and *p*

Replacing V in Eq. (2.11.7) by nRT/p, we get

$$T^{C_{V,\,\mathrm{m}}/R}\left(\frac{nRT}{p}\right) = \text{constant}$$

or $T^{-C_{p,\,\mathrm{m}}/R}\ p = \text{constant}$ (since n is constant) (2.11.9)

The alternative forms of this expression are

$$\left.\begin{aligned} T\,p^{-R/C_{p,\mathrm{m}}} &= \text{constant}\\ T\,p^{(1-\gamma)/\gamma} &= \text{constant}\\ T^{\gamma/(1-\gamma)}\,p &= \text{constant}\end{aligned}\right\}\qquad (2.11.10)$$

Relationship Between *p* and *V*

Replacing T in Eq. (2.11.7) by pV/nR, we get

$$\left(\frac{pV}{nR}\right)^{C_{V,\mathrm{m}}/R} V = \text{constant}$$

or $p^{C_{V,\,\mathrm{m}}/R}\ V^{C_{p,\,\mathrm{m}}/R} = \text{constant}$ (since n is constant) (2.11.11)

The alternative form of this expression is

$$pV^{\gamma} = \text{constant} \qquad \text{where} \qquad \gamma = (C_{p,\,\mathrm{m}}/\ C_{V,\,\mathrm{m}}) \qquad (2.11.12)$$

After calculating the final temperature from either Eq. (2.11.7) or Eq. (2.11.9) we can use Eqs (2.11.1) to (2.11.5) to calculate q, w, ΔU and ΔH.

Problem 2.11.1

Show that the work involved in a reversible adiabatic expansion of an ideal gas from p_i and V_i to p_f and V_f is given by

(i) $w = \dfrac{p_\mathrm{f}V_\mathrm{f} - p_\mathrm{i}V_\mathrm{i}}{\gamma - 1}$ (ii) $w = -\,nC_{V,\,\mathrm{m}}\ T_\mathrm{i}\left[1 - \left(\dfrac{p_\mathrm{f}}{p_\mathrm{i}}\right)^{R/C_{p,\mathrm{m}}}\right]$

Solution

(i) For an adiabatic reversible process

$$w = nC_{V,\,\mathrm{m}}(T_\mathrm{f} - T_\mathrm{i}) = C_{V,\,\mathrm{m}}\left(\frac{p_\mathrm{f}V_\mathrm{f}}{R} - \frac{p_\mathrm{i}V_\mathrm{i}}{R}\right)$$

$$= \frac{C_{V,\mathrm{m}}}{R}(p_\mathrm{f}V_\mathrm{f} - p_\mathrm{i}V_\mathrm{i}) = \frac{C_{V,\mathrm{m}}}{C_{p,\mathrm{m}} - C_{V,\mathrm{m}}}(p_\mathrm{f}V_\mathrm{f} - p_\mathrm{i}V_\mathrm{i}) = \frac{p_\mathrm{f}V_\mathrm{f} - p_\mathrm{i}V_\mathrm{i}}{\gamma - 1}$$

(ii) For an adiabatic reversible process

$$w = nC_{V,\mathrm{m}}(T_\mathrm{f} - T_\mathrm{i}) = -\,nC_{V,\mathrm{m}}(T_\mathrm{i} - T_\mathrm{f})$$

Using the expression $T_\mathrm{i}(p_\mathrm{i})^{-R/C_{p,\mathrm{m}}} = T_\mathrm{f}(p_\mathrm{f})^{-R/C_{p,\mathrm{m}}}$, the above equation becomes

$$w = -nC_{V,\mathrm{m}}\left[T_\mathrm{i} - T_\mathrm{i}\left(\frac{p_\mathrm{i}}{p_\mathrm{f}}\right)^{-R/C_{p,\mathrm{m}}}\right] = -\,nC_{V,\mathrm{m}}T_\mathrm{i}\left[1 - \left(\frac{p_\mathrm{f}}{p_\mathrm{i}}\right)^{R/C_{p,\mathrm{m}}}\right]$$

Example 2.11.2 0.410 mol of a monatomic gas fills a 1 $\mathrm{dm^3}$ container to a pressure of 1.013 3 MPa. It is expanded reversibly and adiabatically until a pressure of 0.101 33 MPa is reached. What are the final volume and temperature? What is the work done in the expansion?

Solution The final volume V_f of gas after adiabatic and reversible expansion can be obtained by using the expression

$$p_\mathrm{i}V_\mathrm{i}^{\gamma} = p_\mathrm{f}\,V_\mathrm{f}^{\gamma}$$

Substituting the values of p_i, V_i, p_f and γ we get

$$(1.013\ 3\ \mathrm{MPa})\ (1\ \mathrm{dm^3})^{5/3} = (0.101\ 33\ \mathrm{MPa})\ V_\mathrm{f}^{5/3}$$

or $$V_\mathrm{f} = 10^{3/5}\ \mathrm{dm^3} = 3.98\ \mathrm{dm^3}$$

The final temperature T_f after the expansion is

$$T_\mathrm{f} = \frac{p_\mathrm{f}V_\mathrm{f}}{nR} = \frac{(0.101\,33 \times 10^3\ \mathrm{kPa})\,(3.98\ \mathrm{dm^3})}{(0.410\ \mathrm{mol})\,(8.314\ \mathrm{dm^3\ kPa\ mol^{-1}\ K^{-1}})} = 118.3\ \mathrm{K}$$

The work done during the expansion is

$$w = -\frac{p_\mathrm{i}V_\mathrm{i} - p_\mathrm{f}V_\mathrm{f}}{\gamma - 1} = -\frac{(1.013\,3\ \mathrm{MPa})\,(1\ \mathrm{dm^3}) - (0.101\,33\ \mathrm{MPa})\,(3.98\ \mathrm{dm^3})}{(5/3) - 1}$$

$$= -\,0.915\ \mathrm{dm^3\ MPa} = -\,915\ \mathrm{J}$$

Irreversible Expansion or Compression Process

Here again, we consider two cases, namely, free expansion and intermediate expansion or compression processes.

Free Expansion

In a free expansion, we have

$$p_\mathrm{ext} = 0$$

Thus, $$\mathrm{d}w = -\,p_\mathrm{ext}\,\mathrm{d}V = 0$$

Now according to Eq. (2.11.2), we have

$$\mathrm{d}U = 0$$

Since, for an ideal gas, $U = f(T)$, it follows that the temperature of the gas after expansion remains unchanged.

Now $$\mathrm{d}H = \mathrm{d}U + \mathrm{d}(pV) = \mathrm{d}U + \mathrm{d}(nRT) = \mathrm{d}U + nR\,\mathrm{d}T$$

$$= 0 + 0 = 0$$

For a finite change, we will have

$$w = 0, \qquad \Delta U = 0, \qquad \Delta T = 0, \qquad \Delta H = 0$$

Comparing these changes of properties with those of the isothermal free expansion, we find that the adiabatic irreversible free expansion of an ideal gas is identical with the isothermal free expansion.

Intermediate Expansion or Compression

Here the work is involved against a constant external pressure and is given by

$$w = -p_{ext}(V_2 - V_1) \tag{2.11.13}$$

Substituting this in Eq. (2.11.2), we get

$$w = \Delta U = -p_{ext}(V_2 - V_1) \tag{2.11.14}$$

For an ideal gas, ΔU is given by

$$\Delta U = nC_{V,\mathrm{m}}\,\Delta T = nC_{V,\mathrm{m}}(T_2 - T_1) \tag{2.11.15}$$

The expression for ΔH is

$$\Delta H = nC_{p,\mathrm{m}}\,\Delta T = nC_{p,\mathrm{m}}(T_2 - T_1) \tag{2.11.16}$$

Eliminating ΔT from Eqs (2.10.15) and (2.10.16), we get

$$\Delta H = nC_{p,\mathrm{m}}\frac{\Delta U}{nC_{V,\mathrm{m}}} = \frac{C_{p,\mathrm{m}}}{C_{V,\mathrm{m}}}\Delta U = \gamma\Delta U \tag{2.11.17}$$

Equation (2.11.14) can be employed to calculate w and ΔU from the given values of p_{ext}, V_2 and V_1. Knowing ΔU, we can calculate ΔH by using Eq. (2.11.17).

Calculation of Temperature

In some expansion or compression processes, the values of p_{ext}, p_1, p_2 and T_1 are provided. In such a case, first of all we calculate the temperature of the gas after the expansion or compression. This can be done as follows.

Equating Eqs (2.11.14) and (2.11.15), we get

$$nC_{V,\mathrm{m}}(T_2 - T_1) = -p_{ext}(V_2 - V_1) = p_{ext}(V_1 - V_2)$$

Replacing V_2 and V_1 in terms of T and p, we get

$$nC_{V,\mathrm{m}}(T_2 - T_1) = p_{ext}\left(\frac{nRT_1}{p_1} - \frac{nRT_2}{p_2}\right)$$

or

$$C_{V,\mathrm{m}}(T_2 - T_1) = p_{ext}\left(\frac{RT_1}{p_1} - \frac{RT_2}{p_2}\right) \tag{2.11.18}$$

For a special case, where $p_{ext} = p_2$ (the pressure of the gas after expansion), we get

$$C_{V,\mathrm{m}}(T_2 - T_1) = p_2\left(\frac{RT_1}{p_1} - \frac{RT_2}{p_2}\right) \tag{2.11.19}$$

Rearranging this, we get

$$T_2 = T_1\left(\frac{C_{V,\mathrm{m}} + R\,p_2/p_1}{C_{p,\mathrm{m}}}\right) \tag{2.11.20}$$

After knowing T_2 from Eq. (2.11.20) (or from Eq. 2.11.18 if p_{ext} is not equal to p_2), we can calculate ΔU and ΔH from Eqs (2.11.15) and (2.11.16), respectively.

Example 2.11.3 2.0 mol of an ideal diatomic gas at 300 K and 0.507 MPa are expanded adiabatically to a final pressure of 0.203 MPa against a constant pressure of 0.101 MPa. Calculate the final temperature, q, w, ΔU and ΔH.

Solution For an adiabatic process, $q = 0$

$$w = -p_{\text{ext}}(V_2 - V_1) = -p_{\text{ext}}\left(\frac{nRT_2}{p_2} - \frac{nRT_1}{p_1}\right)$$

$$\Delta U = w$$

i.e. $$nC_{V,\text{m}}(T_2 - T_1) = p_{\text{ext}}\, nR\left(\frac{T_1}{p_1} - \frac{T_2}{p_2}\right)$$

$C_{V,\text{m}}$ for diamotic molecule $= 3\left(\frac{R}{2}\right) + 2\left(\frac{R}{2}\right) = \frac{5}{2}R$; (assuming no contribution from vibration)

Substituting the expression of $C_{V,\text{m}}$ in the previous expression, we get

$$n\left(\frac{5}{2}R\right)(T_2 - T_1) = p_{\text{ext}}\, nR\left(\frac{T_1}{p_1} - \frac{T_2}{p_2}\right)$$

or $$\frac{5}{2}(T_2 - T_1) = p_{\text{ext}}\left(\frac{T_1}{p_1} - \frac{T_2}{p_2}\right)$$

Substituting the values, we get

$$\frac{5}{2}(T_2 - 300\ \text{K}) = (0.101\ \text{MPa})\left(\frac{300\ \text{K}}{0.507\ \text{MPa}} - \frac{T_2}{0.203\ \text{MPa}}\right)$$

Solving for T_2, we get

$$T_2 = 270\ \text{K}$$

Thus $$\Delta U = nC_{V,\text{m}}(T_2 - T_1) = (2.0\ \text{mol})\left(\frac{5}{2}\times 8.314\ \text{J K}^{-1}\ \text{mol}^{-1}\right)(270\ \text{K} - 300\ \text{K})$$

$$= -1\,247.1\ \text{J}$$

$$\Delta H = \Delta U + nR(\Delta T) = -1\,247.1\ \text{J} + (2.0\ \text{mol})(8.314\ \text{J K}^{-1}\ \text{mol}^{-1})(-30\ \text{K})$$

$$= -1\,745.9\ \text{J}$$

Example 2.11.4 Two moles of an ideal monatomic gas ($C_{V,\text{m}} = 12.55\ \text{J K}^{-1}\ \text{mol}^{-1}$) expands irreversibly and adiabatically from an initial pressure of 1.013 MPa against a constant external pressure of 0.101 3 MPa, until the temperature drops from the initial value of 325 K to a final value of 275 K. How much work is done and what is the final volume?

Solution We have

$$\Delta U = nC_{V,\text{m}}(T_2 - T_1) = (2.0\ \text{mol})(12.55\ \text{J K}^{-1}\ \text{mol}^{-1})(275\ \text{K} - 325\ \text{K})$$

$$= -1\,255\ \text{J}$$

$$w = \Delta U = -p_{\text{ext}}(V_2 - V_1) = -(0.101\ 3\ \text{MPa})(V_2 - V_1) = -1\,255\ \text{J}$$

Now $$V_1 = \frac{nRT_1}{p} = \frac{(2.0\ \text{mol})(8.314\ \text{J K}^{-1}\ \text{mol}^{-1})(325\ \text{K})}{(1.013\ \text{MPa})} = 5{,}334.75\ \text{cm}^3$$

Therefore $$V_2 = \frac{1\,255\ \text{J}}{0.101\,3\ \text{MPa}} + V_1 = 12\ 388.94\ \text{cm}^3 + 5\ 334.75\ \text{cm}^3$$

$$= 17\ 723.69\ \text{cm}^3 = 17.72\ \text{dm}^3$$

Example 2.11.5

An ideal monatomic gas ($C_{V,\text{m}} = 1.5\ R$) initially at 298 K and 1.013 MPa pressure expands adiabatically and irreversibly until it is equilibrium with a constant external pressure of 0.101 3 MPa. What is the final temperature of the gas?

Solution

For an adiabatic irreversible process

$$\text{d}U = -\,p_{\text{ext}}\ \text{d}V$$

$$nC_{V,\text{m}}\,(T_2 - T_1) = -\,p_{\text{ext}}(V_2 - V_1) = -\,p_{\text{ext}}\left(\frac{nRT_2}{p_2} - \frac{nRT_1}{p_1}\right)$$

Substituting $C_{V,\text{m}} = 1.5\ R$ and simplifying, we get

$$1.5(T_2 - T_1) = -\,p_{\text{ext}}\left(\frac{T_2}{p_2} - \frac{T_1}{p_1}\right)$$

Substituting the values of p_2, p_1 and T_1, we get

$$1.5\ (T_2 - 298\ \text{K}) = -\ (0.101\,3\ \text{MPa})\left(\frac{T_2}{0.101\,3\ \text{MPa}} - \frac{298\ \text{K}}{1.013\ \text{MPa}}\right)$$

$$2.5\ T_2 = 1.5\ (298\ \text{K}) + \frac{298\ \text{K}}{10}$$

$$T_2 = \frac{447.0\ \text{K} + 29.8\ \text{K}}{2.5} = 190.7\ \text{K}$$

Example 2.11.6

With what minimum pressure must a given volume of nitrogen, originally at 373 K and 0.101 3 MPa pressure, be adiabatically compressed in order to raise its temperature to 673 K. Given: $C_{V,\text{m}} = (5/2)R$.

Solution

For an adiabatic irreversible compression, we have

$$\Delta U = -\int p_{\text{ext}}\,\text{d}V = -p_{\text{ext}}\,(V_2 - V_1)$$

or $$C_{V,\text{m}}(T_2 - T_1) = -p_{\text{ext}}\left(\frac{RT_2}{p_2} - \frac{RT_1}{p_1}\right)$$

Here $$p_{\text{ext}} = p_2$$

Thus $$C_{V,\text{m}}\,(T_2 - T_1) = -\,p_2\left(\frac{RT_2}{p_2} - \frac{RT_1}{p_1}\right)$$

Substituting the values of $C_{V,\text{m}}$, T_2, T_1 and p_1, we get

$$\frac{5}{2}\ R\ (673\ \text{K} - 373\ \text{K}) = -\,p_2\left(\frac{R(673\ \text{K})}{p_2} - \frac{R(373\ \text{K})}{0.101\,3\ \text{MPa}}\right)$$

or $$\frac{5}{2} \times 300 = -673 + \frac{373\, p_2}{0.1013\ \text{MPa}}$$

Hence, $$p_2 = \frac{750 + 673}{373} \times 0.1013\ \text{MPa} = 0.386\ 5\ \text{MPa}$$

Comparison of Adiabatic Expansion Processes

The Changes in thermodynamic properties in the adiabatic expansion processes are listed in Table 2.11.1.

Table 2.11.1 Comparison of w, ΔU and ΔH for Different Types of Adiabatic Expansion Processes†

Expansion	w	T_2	ΔU	ΔH
Reversible	$w = nC_{V,\text{m}}(T_2 - T_1) < 0$	$T_2 < T_1$	$\Delta U = nC_{V,\text{m}}(T_2 - T_1) < 0$	$\Delta H = nC_{p,\text{m}}(T_2 - T_1) < 0$
Intermediate	$w' = nC_{V,\text{m}}(T_2' - T_1) < 0$	$T_2' < T_1$	$\Delta U' = nC_{V,\text{m}}(T_2' - T_1) < 0$	$\Delta H' = nC_{p,\text{m}}(T_2' - T_1) < 0$
Free	$w'' = nC_{V,\text{m}}(T_2'' - T_1) = 0$	$T_2'' = T_1$	$\Delta U'' = nC_{V,\text{m}}(T_2'' - T_1) = 0$	$\Delta H'' = nC_{p,\text{m}}(T_2'' - T_1) = 0$

(i) $T_2 < T_2' < T_2'' = T_1$

(ii) $w < w' < w'' = 0$ and $|w| > |w'| > |w''| = 0$

(iii) $\Delta U < \Delta U' < \Delta U'' = 0$ and $|\Delta U| > |\Delta U'| > |\Delta U''| = 0$

(iv) $\Delta H < \Delta H' < \Delta H'' = 0$ and $|\Delta H| > |\Delta H'| > |\Delta H''| = 0$

†*Note:* Since the magnitude of work involved in an irreversible expansion is less than the corresponding magnitude of the work involved in a reversible expansion, it, therefore, follows from Eq. (2.11.15) that the temperature decrease in an irreversible expansion will be less than the corresponding temperature decrease in the reversible expansion. On the other side, since the work involved in an irreversible compression is larger than the corresponding work involved in the reversible compression, the increase in temperature in the former will be larger than in the latter.

2.12 COMPARISON BETWEEN REVERSIBLE ISOTHERMAL AND ADIABATIC EXPANSIONS OF AN IDEAL GAS

Final Volumes are Same

Let us consider that a gas from initial stage p_1 and V_1 undergoes isothermal reversible and adiabatic reversible expansions such that the final volumes are the same. Let it be represented by V_f. Let p_iso and p_adi be the final pressures in the isothermal and adiabatic expansions, respectively. For an isothermal expansion, initial p_1 and V_1 will be related to the final p_iso and V_f by the relation

$$p_1 V_1 = p_\text{iso} V_\text{f}$$

or $$\frac{V_\text{f}}{V_1} = \frac{p_1}{p_\text{iso}} \tag{2.12.1}$$

And for an adiabatic expansion, these variables will be related by

$$p_1 V_1^\gamma = p_\text{adi}\, V_\text{f}^\gamma$$

or $$\left(\frac{V_\text{f}}{V_1}\right)^\gamma = \frac{p_1}{p_\text{adi}} \tag{2.12.2}$$

Since for an expansion $V_f > V_1$ and the fact that $\gamma > 1$, we have

$$\left(\frac{V_f}{V_1}\right)^{\gamma} > \left(\frac{V_f}{V_1}\right) \quad \text{so that} \quad \frac{p_1}{p_{adi}} > \frac{p_1}{p_{iso}}$$

Thus, $p_{adi} < p_{iso}$ (2.12.3)

The two expansions are shown in Fig. 2.12.1. Since the magnitude of work involved is equal to the area under the curve, it can be seen from this figure that the magnitude of the work involved in the isothermal expansion is larger than that involved in the adiabatic expansion. This also follows from the fact that in the adiabatic expansion, temperature decreases whereas in the isothermal expansion, temperature remains the same. Since the final volumes are the same, according to Charles law ($p \propto T$, V constant) p_{adi} will be smaller than p_{iso}, and thus the magnitude of the reversible work involved in going from volume V_1 to V_f ($V_f > V_1$) for and ideal gas is greater in the isothermal process than that involved in the adiabatic process.

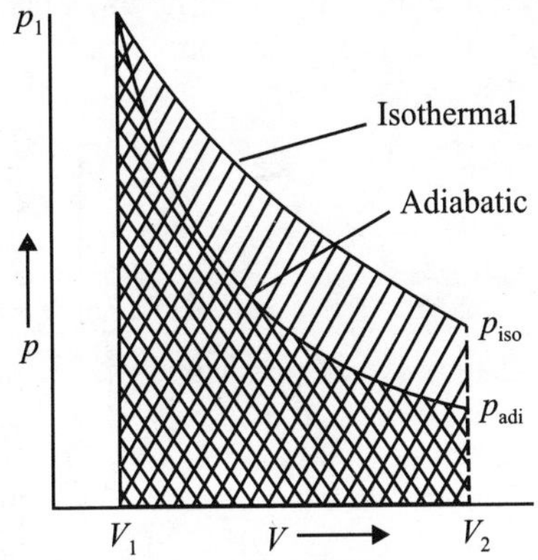

Fig. 2.12.1 Comparison of the magnitude of the work involved in isothermal and adiabatic expansions

Final Pressures are Same

Let us consider now the expansions in which the final pressure p_f is the same. Let V_{iso} and V_{adi} be the final volumes in isothermal and adiabatic expansions, respectively.

For an isothermal expansion

$$p_1V_1 = p_fV_{iso}$$

or $$\left(\frac{p_1}{p_f}\right) = \frac{V_{iso}}{V_1} \tag{2.12.4}$$

For an adiabatic expansion

$$p_1V_1^{\gamma} = p_f\,V_{adi}^{\gamma}$$

or $$\left(\frac{p_1}{p_f}\right) = \left(\frac{V_{adi}}{V_1}\right)^{\gamma} \tag{2.12.5}$$

Comparing Eqs (2.12.4) and (2.12.5), we have

$$\left(\frac{V_{\text{iso}}}{V_1}\right) = \left(\frac{V_{\text{adi}}}{V_1}\right)^{\gamma} \quad \text{or} \quad \ln \frac{V_{\text{iso}}}{V_1} = \gamma \ln\left(\frac{V_{\text{adi}}}{V_1}\right)$$

Since γ is greater than one, it follows that

$$\frac{V_{\text{iso}}}{V_1} > \frac{V_{\text{adi}}}{V_1}$$

or $$V_{\text{adi}} < V_{\text{iso}}$$

that is, the final volume in the case of an adiabatic expansion is lesser than the corresponding volume in an isothermal expansion. This also follows from the fact that in adiabatic expansion, temperature decreases whereas in isothermal expansion, it remains the same and since the final pressures are the same, therefore, according to Charles law ($V \propto T$, p constant), V_{adi} will be smaller than V_{iso}.

The two expansions are shown in Fig. 2.12.2. Since the magnitude of work involved is equal to the area under the curve, it follows that the magnitude of the work involved in an isothermal expansion is greater than that involved in an adiabatic expansion.

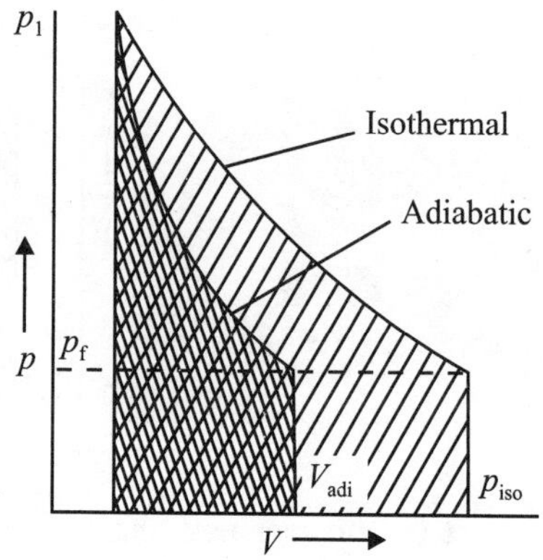

Fig. 2.12.2 Comparison of the magnitude of the work involved in isothermal and adiabatic expansions

2.13 THERMODYNAMIC CHANGES IN ISOTHERMAL VARIATION IN VOLUME OF A VAN DER WAALS GAS

Reversible Expansion or Compression Process

In this section, we derive expressions for q, w, ΔU and ΔH for an isothermal expansion or compression process involving a van der Waals gas.

The various expressions can be derived as fallows:

Expression for *w*

We have

$$\mathrm{d}w = -p_{\text{ext}}\,\mathrm{d}V \tag{2.13.1}$$

For the reversible change in volume, we have

$$p_{\text{ext}} = p_{\text{int}} \pm \mathrm{d}p \simeq p_{\text{int}}$$

where p_{int} is the pressure of the gas and is given by

$$p_{\text{int}} = \frac{nRT}{V - nb} - \frac{n^2 a}{V^2}$$

Substituting this in Eq. (2.13.1), we get

$$dw = -\left(\frac{nRT}{V-nb} - \frac{n^2a}{V^2}\right) dV$$

Hence

$$w = -\int_{V_1}^{V_2}\left(\frac{nRT}{V-nb} - \frac{n^2a}{V^2}\right) dV = -nRT \ln\frac{V_2-nb}{V_1-nb} - n^2a\left(\frac{1}{V_2}-\frac{1}{V_1}\right) \tag{2.13.2}$$

Expression for ΔU The expression for ΔU can be derived by using the thermodynamic equation of state

$$\left(\frac{\partial U}{\partial V}\right)_T = T\left(\frac{\partial p}{\partial T}\right)_V - p$$

For a van der Waals gas

$$\left(\frac{\partial p}{\partial T}\right)_V = \frac{nR}{V-nb}$$

Thus $$\left(\frac{\partial U}{\partial V}\right)_T = \frac{nRT}{V-nb} - p = \frac{nRT}{V-nb} - \left(\frac{nRT}{V-nb} - \frac{n^2a}{V^2}\right) = \frac{n^2a}{V^2} \tag{2.13.3}$$

Now $$dU = \left(\frac{\partial U}{\partial T}\right)_V dT + \left(\frac{\partial U}{\partial V}\right)_T dV$$

For an isothermal process, $dT = 0$, therefore

$$dU = \left(\frac{\partial U}{\partial V}\right)_T dV = \frac{n^2a}{V^2} dV$$

For a finite process, we have

$$\Delta U = \int_{V_1}^{V_2}\frac{n^2a}{V^2} dV = -n^2a\left(\frac{1}{V_2}-\frac{1}{V_1}\right) \tag{2.13.4}$$

Expression for q From the first law of thermodynamics, we have

$$q = \Delta U - w$$

Substituting the expressions for ΔU and w, we have

$$q = -n^2a\left(\frac{1}{V_2}-\frac{1}{V_1}\right) + nRT\ln\frac{V_2-nb}{V_1-nb} + n^2a\left(\frac{1}{V_2}-\frac{1}{V_1}\right)$$

$$= nRT\ln\frac{V_2-nb}{V_1-nb} \tag{2.13.5}$$

Expression for ΔH Since $\Delta H = \Delta U + \Delta(pV)$, we get

$$\Delta H = -n^2a\left(\frac{1}{V_2}-\frac{1}{V_1}\right) + \Delta\left\{\left(\frac{nRT}{V-nb} - \frac{n^2a}{V^2}\right)V\right\}$$

$$= -n^2a\left(\frac{1}{V_2}-\frac{1}{V_1}\right)+nRT\left(\frac{V_2}{V_2-nb}-\frac{V_1}{V_1-nb}\right)-n^2a\left(\frac{1}{V_2}-\frac{1}{V_1}\right)$$

$$= -2n^2a\left(\frac{1}{V_2}-\frac{1}{V_1}\right)+nRT\left\{\frac{V_2-nb+nb}{V_2-nb}-\frac{V_1-nb+nb}{V_1-nb}\right\}$$

$$= -2n^2a\left(\frac{1}{V_2}-\frac{1}{V_1}\right)+n^2RTb\left\{\frac{1}{V_2-nb}-\frac{1}{V_1-nb}\right\} \qquad (2.13.6)$$

Example 2.13.1 Calculate the work which could be obtained from an isothermal reversible expansion of 1 mol of Cl_2 from 1 dm^3 to 50 dm^3 at 273 K using van der Waals gas behaviour (a = 0.655 dm^6 MPa mol^{-2}, b = 0.055 dm^3 mol^{-1}).

Solution For one mole of a van der Waals gas

$$p = \frac{RT}{V_m - b} - \frac{a}{V_m^2}$$

Thus $$w = -\int_{V_1}^{V_2} p\,dV_m = -\int_{V_1}^{V_2}\left(\frac{RT}{V_m-b}-\frac{a}{V_m^2}\right)dV_m = -RT\ln\frac{V_2-b}{V_1-b} - a\left(\frac{1}{V_2}-\frac{1}{V_1}\right)$$

Substituting the values, we get

$$w = -(8.314 \text{ J K}^{-1}\text{ mol}^{-1})(273 \text{ K}) \times 2.303 \times \log\left(\frac{50-0.055}{1-0.055}\right)$$

$$-(0.655 \text{ dm}^6 \text{ MPa mol}^{-2})\left(\frac{1}{50 \text{ dm}^3 \text{ mol}^{-1}}-\frac{1}{1 \text{ dm}^3 \text{ mol}^{-1}}\right)$$

$$w = -9\,006.73 \text{ J mol}^{-1} + 0.642 \text{ dm}^3 \text{ MPa mol}^{-1}$$

$$= -9.006\,7 \text{ kJ mol}^{-1} + 0.642 \text{ kJ mol}^{-1}$$

$$= -8.\,365\,7 \text{ kJ mol}^{-1}$$

Example 2.13.2 One mole of a van der Waals gas at 300 K expands isothermally and reversibly from a volume of 10 dm^3 to 30 dm^3. Calculate q, w, ΔU and ΔH. Given; a = 556 dm^6 kPa mol^{-2} and b = 0.064 dm^3 mol^{-1}.

Solution For one mole of a van der Waals gas, we have

$$w = -RT\ln\left(\frac{V_2-b}{V_1-b}\right) - a\left(\frac{1}{V_2}-\frac{1}{V_1}\right)$$

Substituting the given data, we have

$$w = -(8.314 \text{ J K}^{-1}\text{ mol}^{-1})(300 \text{ K}) \times 2.303 \times \log\left(\frac{30-0.064}{10-0.064}\right)$$

$$-(556 \text{ dm}^6 \text{ kPa mol}^{-2})\left(\frac{1}{30 \text{ dm}^3 \text{ mol}^{-1}}-\frac{1}{10 \text{ dm}^3 \text{ mol}^{-1}}\right)$$

$$= - (8.314 \text{ J mol}^{-1})\ (300)\ (2.303)\ (0.4790) + (556 \text{ dm}^3 \text{ kPa mol}^{-1}) \left(\frac{20}{30 \times 10}\right)$$

$$= - 2\,751.4 \text{ J mol}^{-1} + 37.07 \text{ J mol}^{-1}$$

$$= - 2\,714.33 \text{ J mol}^{-1}$$

$$\Delta U = - a \left(\frac{1}{V_2} - \frac{1}{V_1}\right) = - (556 \text{ dm}^6 \text{ kPa mol}^{-2}) \left(\frac{1}{30 \text{ dm}^3 \text{ mol}^{-1}} - \frac{1}{10 \text{ dm}^3 \text{ mol}^{-1}}\right)$$

$$= 37.07 \text{ J mol}^{-1}$$

$$q = RT \ln \left(\frac{V_2 - b}{V_1 - b}\right) = 2\,751.4 \text{ J mol}^{-1} \qquad (= \Delta U - w)$$

$$\Delta H = RTb \left(\frac{1}{V_2 - b} - \frac{1}{V_1 - b}\right) - 2a\left(\frac{1}{V_2} - \frac{1}{V_1}\right) = RTb\left(\frac{1}{V_2 - b} - \frac{1}{V_1 - b}\right) + 2\,\Delta U$$

$$= (8.314 \text{ J K}^{-1} \text{ mol}^{-1})\ (300 \text{ K})\ (0.064) \left(\frac{1}{30 - 0.060} - \frac{1}{(10 - 0.064)}\right) + 2\ (37.07 \text{ J mol}^{-1})$$

$$= - 10.733 \text{ J mol}^{-1} + 74.14 \text{ J mol}^{-1}$$

$$= 63.407 \text{ J mol}^{-1}$$

Comparison of a Reversible Isothermal Change in Volume of a van der Waals Gas with that of an Ideal Gas

Table 2.13.1 compares the expressions of q, w, and ΔU for a reversible isothermal change in volume of a van der Waals gas with those of an ideal gas.

Table 2.31.1 Expressions of q, $(-w)$ and ΔU for Isothermal Reversible Change in Volume of van der Waals and Ideal Gases

Variables	*van der Waals gas*	*Ideal gas*
q	$nRT \ln \left(\frac{V_2 - b}{V_1 - b}\right)$	$nRT \ln \frac{V_2}{V_1}$
$(- w)$	$nRT \ln \left(\frac{V_2 - b}{V_1 - b}\right) + n^2 a \left(\frac{1}{V_2} - \frac{1}{V_1}\right)$	$nRT \ln \frac{V_2}{V_1}$
ΔU	$- n^2 a \left(\frac{1}{V_2} - \frac{1}{V_1}\right)$	0

Comments to Table 2.13.1

- From the expressions of q, $(- w)$ and ΔU for a real gas, we find that

$$(- w)_{\text{real}} = q_{\text{real}} - \Delta U$$

Thus, in expansion, whole of heat absorbed is not available for the work. A part of the absorbed heat (usually a small fraction) is utilized increasing the internal energy of the system. This energy is stored in the system as the potential energy as it is utilized in overcoming the forces of attraction between the molecules. (*See*, Example 2.13.2). In compression, the effects are just reversed of the effects described above.

- Since in expansion,

$$V_1 < V_2 \Rightarrow V_1nb < V_2nb \Rightarrow -V_1nb > -V_2nb$$
$$\Rightarrow V_2V_1 - V_1nb > V_2V_1 - V_2nb$$
$$\Rightarrow V_1(V_2 - nb) > V_2(V_1 - nb)$$

we will have

$$\frac{V_2 - nb}{V_1 - nb} > \frac{V_2}{V_1}$$

Hence $q_{real} > q_{ideal}$

Thus, to carry out the same increase in volume, heat absorbed in case of a real gas is greater than the corresponding heat absorbed in case of an ideal gas.

- In most real gases at ordinary temperature and pressure, increase in potential energy during the expansion process is such that

$$q_{real} - \Delta U < q_{ideal}$$

indicating that

$$(-w)_{real} < (-w)_{ideal}$$

that is, the net work of expansion in case of a real gas is smaller than the corresponding work involved in the expansion of an ideal gas.

Irreversible Expansion or Compression

Intermediate Expansion

The various expressions are as follows:

Expression for w We have

$$dw = -p_{ext}\,dV$$

$$w = -p_{ext}(V_2 - V_1) \qquad (2.13.7)$$

Expression for ΔU The expression for ΔU will be the same as that given by Eq. (2.13.4), i.e.

$$\Delta U = -n^2a\left(\frac{1}{V_2} - \frac{1}{V_1}\right) \qquad (2.13.8)$$

Expression for q From the first law of thermodynamics, $\Delta U = q + w$, we have

$$q = \Delta U - w$$

$$= -n^2a\left(\frac{1}{V_2} - \frac{1}{V_1}\right) + p_{ext}(V_2 - V_1) \qquad (2.13.9)$$

Free Expansion

For an isothermal free expansion, we will have

$$dw = 0 \qquad (2.13.10)$$

$$\Delta U = -n^2a\left(\frac{1}{V_2} - \frac{1}{V_1}\right) \qquad (2.13.11)$$

$$q = -n^2a\left(\frac{1}{V_2} - \frac{1}{V_1}\right) \qquad (2.13.12)$$

Note that during the isothermal free expansion, ΔU and q are not equal to zero. There will be an absorption of heat which is utilized in overcoming the forces of attraction of the molecules and is stored as the potential energy.

2.14 THERMODYNAMIC CHANGES IN ADIABATIC VARIATION IN VOLUME OF A VAN DER WAALS GAS

Reversible Expansion or Compression Process

Expression for Change in Energy

In the adiabatic process, we have

$$q = 0 \quad \text{and} \quad \Delta U = w \tag{2.14.1}$$

The change in energy can be calculated as follows.
Writing the differential of U, we get

$$\mathrm{d}U = \left(\frac{\partial U}{\partial T}\right)_V \mathrm{d}T + \left(\frac{\partial U}{\partial V}\right)_T \mathrm{d}V$$

Making use of Eq. (2.13.3), we get

$$\mathrm{d}U = nC_{V,\mathrm{m}}\,\mathrm{d}T + \frac{n^2a}{V^2}\mathrm{d}V \tag{2.14.2}$$

For a finite process, we have

$$\Delta U = \int_{T_1}^{T_2} nC_{V,\mathrm{m}}\,\mathrm{d}T + \int_{V_1}^{V_2} \frac{n^2a}{V^2}\mathrm{d}V$$

Assuming $C_{V,\mathrm{m}}$ to be independent of temperature, we get

$$\Delta U = nC_{V,\mathrm{m}}(T_2 - T_1) - n^2a\left(\frac{1}{V_2} - \frac{1}{V_1}\right) \tag{2.14.3}$$

Thus knowing T_1, T_2, V_1 and V_2 for a given process, we can calculate ΔU from Eq. (2.14.3).

Calculation of Temperature

In some expansion processes, the value of T_2 is not provided. In such a case, first of all, we calculate the temperature of the gas after the expansion (or compression) process. This can be done as follows.
For a van der Waals gas

$$p = \frac{nRT}{V - nb} - \frac{n^2a}{V^2} \tag{2.14.4}$$

Substituting Eqs (2.14.2) and (2.14.4) in the expression

$$\mathrm{d}U = \mathrm{d}w = -p\,\mathrm{d}V$$

we get $$nC_{V,\mathrm{m}}\,\mathrm{d}T + \frac{n^2a}{V^2}\mathrm{d}V = -\frac{nRT}{V - nb}\mathrm{d}V + \frac{n^2a}{V^2}\mathrm{d}V$$

or $$C_{V,\mathrm{m}}\,\mathrm{d}T = -\frac{RT}{V - nb}\mathrm{d}V$$

Separating the variables, we get

$$C_{V,\mathrm{m}}\frac{\mathrm{d}T}{T} = -R\frac{\mathrm{d}V}{V - nb}$$

Integrating the above expression, we have

$$C_{V,\mathrm{m}} \ln\left(\frac{T_2}{T_1}\right) = -R\ln\left(\frac{V_2-nb}{V_1-nb}\right) \quad \text{or} \quad \left(\frac{T_2}{T_1}\right)^{C_{V,\mathrm{m}}/R} = \left(\frac{V_1-nb}{V_2-nb}\right)$$

or $\quad T_2^{C_{V,\mathrm{m}}/R}(V_2-nb) = T_1^{C_{V,\mathrm{m}}/R}(V_1-nb)$ or $T^{C_{V,\mathrm{m}}/R}(V-nb) = \text{constant}$

or $\quad T(V-nb)^{R/C_{V,\mathrm{m}}} = \text{constant} \qquad (2.14.5)$

Thus for a given adiabatic reversible expansion, we will have

$$T_2(V_2-nb)^{R/C_{V,\mathrm{m}}} = T_1(V_1-nb)^{R/C_{V,\mathrm{m}}}$$

or $$T_2 = T_1\left(\frac{V_1-nb}{V_2-nb}\right)^{R/C_{V,\mathrm{m}}} \qquad (2.14.6)$$

Hence knowing T_1, V_1 and V_2, the value of T_2 can be calculated from Eq. (2.14.6). Finally, the substitution of T_2 in Eq. (2.14.3) gives the value of ΔU.

Expression of Change in Enthalpy

The enthalpy change in the present case can be calculated as given below.

$$\Delta H = \Delta U + \Delta(pV)$$

$$= \Delta U + \Delta\left\{\left(\frac{nRT}{V-nb} - \frac{n^2a}{V^2}\right)V\right\}$$

$$= \Delta U + nR\left(\frac{V_2T_2}{V_2-nb} - \frac{V_1T_1}{V_1-nb}\right) - n^2a\left(\frac{1}{V_2} - \frac{1}{V_1}\right) \qquad (2.14.7)$$

Substituting ΔU from Eq. (2.14.3), we get

$$\Delta H = nC_{V,\mathrm{m}}(T_2-T_1) - 2n^2a\left(\frac{1}{V_2}-\frac{1}{V_1}\right) - nR\left(\frac{V_2T_2}{V_2-nb} - \frac{V_1T_1}{V_1-nb}\right) \qquad (2.14.8)$$

Example 2.14.1

One mole of chlorine undergoes adiabatic reversible expansion from 1 dm^3 to 10 dm^3 with initial temperature of 273 K. Calculate q, w, ΔU and ΔH if the gas is considered a van der Waals gas. Given: $a = 655\ \mathrm{dm^6\ kPa\ mol^{-2}}$, $b = 0.055\ \mathrm{dm^3\ mol^{-1}}$. $C_{V,\mathrm{m}} = 33.91\ \mathrm{J\ K^{-1}\ mol^{-1}}$.

Solution

Since the gas undergoes adiabatic reversible expansion,

$$q = 0$$

$$w = \Delta U = C_{V,\mathrm{m}}(T_2-T_1) - a\left(\frac{1}{V_2}-\frac{1}{V_1}\right)$$

$$\Delta H = \Delta U - a\left(\frac{1}{V_2}-\frac{1}{V_1}\right) + R\left(\frac{V_2T_2}{V_2-b} - \frac{V_1T_1}{V_1-b}\right)$$

To calculate ΔU and ΔH, we need to know T_2, which can be determined from the expression

$$T_2 = T_1\left(\frac{V_1 - b}{V_2 - b}\right)^{R/C_{V,\mathrm{m}}} = (273\ \mathrm{K})\left(\frac{1-0.055}{10-0.055}\right)^{8.314/33.91}$$

$$= (273\ \mathrm{K})\,(0.562) = 153.3\ \mathrm{K}$$

Hence, $$w = \Delta U = \left[33.91\,(153.3-273) - (655)\left(\frac{1}{10}-\frac{1}{1}\right)\right]\ \mathrm{J\ mol^{-1}}$$

$$= (-\,4\ 059.0 + 589.5\ )\ \mathrm{J\ mol^{-1}}$$

$$= -\,3\ 469.5\ \mathrm{J\ mol^{-1}}$$

$$\Delta H = \left[-\,3\ 469.5 - (655)\left(\frac{1}{10}-\frac{1}{1}\right) + (8.314)\left(\frac{(10)(153.3)}{10-0.055} - \frac{(1)(273)}{1-0.055}\right)\right]\mathrm{J\ mol^{-1}}$$

$$= (-\,3\ 469.5 + 589.5 - 1\ 120.2)\ \mathrm{J\ mol^{-1}}$$

$$= -\,4\ 000.2\ \mathrm{J\ mol^{-1}}$$

Irreversible Expansion or Compression Process

Intermediate Expansion or Compression

The various expressions are as follows.

$$w = -\,p_{\mathrm{ext}}\,(V_2 - V_1) \tag{2.14.9}$$

$$q = 0 \qquad \text{(adiabatic process)} \tag{2.14.10}$$

$$\Delta U = q + w = -\,p_{\mathrm{ext}}\,(V_2 - V_1) \tag{2.14.11}$$

The temperature of the gas after expansion may be determined as follows. For an adiabatic process

$$\mathrm{d}U = \mathrm{d}w$$

Writing dU in terms of dT and dV, we have

$$\left(\frac{\partial U}{\partial T}\right)_V \mathrm{d}T + \left(\frac{\partial U}{\partial V}\right)_T \mathrm{d}V = \mathrm{d}w$$

or $$nC_{V,\mathrm{m}}\,\mathrm{d}T + \frac{n^2 a}{V^2}\,\mathrm{d}V = -\,p_{\mathrm{ext}}\,\mathrm{d}V$$

or $$nC_{V,\mathrm{m}}\,\mathrm{d}T = -\,\frac{n^2 a}{V^2}\,\mathrm{d}V - p_{\mathrm{ext}}\,\mathrm{d}V$$

or $$nC_{V,\mathrm{m}}\,(T_2' - T_1) = n^2 a\left(\frac{1}{V_2} - \frac{1}{V_1}\right) - p_{\mathrm{ext}}\,(V_2 - V_1)$$

or $$T_2' = T_1 + \frac{1}{nC_{V,\mathrm{m}}}\left[n^2 a\left(\frac{1}{V_2} - \frac{1}{V_1}\right) - p_{\mathrm{ext}}\,(V_2 - V_1)\right] \tag{2.14.12}$$

Since for expansion $V_2 > V_1$, it follows that

$$T_2' < T_1$$

Knowing T_2', ΔH can be determined using Eq. (2.14.7).

Example 2.14.2

One mole of chlorine undergoes adiabatic expansion from 1 dm^3 to 10 dm^3 against an external pressure of 1 bar. The initial temperature of the gas was 273 K. Calculate q, w, ΔU and ΔH if the gas is considered a van der Waals gas. Given: a = 655 dm^6 kPa mol^{-2}, b = 0.055 dm^3 mol^{-1}, $C_{V,\text{m}}$ = 33.91 J K^{-1} mol^{-1}.

Solution

Since the gas undergoes adiabatic irreversible expansion,

$$q = 0$$

$$\Delta U = w = -p_{\text{ext}}(V_2 - V_1)$$

$$\Delta H = \Delta U - a\left(\frac{1}{V_2} - \frac{1}{V_1}\right) + R\left(\frac{V_2T_2}{V_2 - b} - \frac{V_1T_1}{V_1 - b}\right)$$

Thus $\Delta U = w = -(100\ \text{kPa})\,(10\ \text{dm}^3 - 1\ \text{dm}^3)\ \text{mol}^{-1} = -900\ \text{J mol}^{-1}$

To calculate ΔH, we need to know T_2, which can be determined from the expression

$$T_2 = T_1 + \frac{1}{C_{V,\text{m}}}\left[a\left(\frac{1}{V_2} - \frac{1}{V_1}\right) - p_{\text{ext}}(V_2 - V_1)\right]$$

$$= (273\ \text{K}) + \frac{1}{33.91}\left[655\left(\frac{1}{10} - \frac{1}{1}\right) - 100\,(10 - 1)\right]\text{K}$$

$$= (273\ \text{K}) + \frac{1}{33.91}\,[-589.5 - 900]\ \text{K}$$

$$= (273 - 43.93)\ \text{K} = 229.1\ \text{K}$$

Hence $$\Delta H = \left[-900 - 655\left(\frac{1}{10} - \frac{1}{1}\right) + 8.314\left(\frac{10\times 229.1}{10 - 0.055} - \frac{1\times 273}{1 - 0.055}\right)\right]\text{J mol}^{-1}$$

$$= (-900 + 589.5 - 486.6)\ \text{J mol}^{-1}$$

$$= -797.1\ \text{J mol}^{-1}$$

Free Expansion

For a free expansion, we will have

$$w = 0\ ; \qquad q = 0\ ; \qquad \Delta U = 0$$

The temperature of the gas after expansion may be derived from Eq. (2.14.12) by substituting $p_{\text{ext}} = 0$. Thus,

$$T_2'' = T_1 + \frac{1}{nC_{V,\text{m}}}\left[n^2a\left(\frac{1}{V_2} - \frac{1}{V_1}\right)\right] \tag{2.14.13}$$

Since for expansion $V_2 > V_1$, it follows that $T_2'' < T_1$. The decrease in temperature in the present case will be smaller than that for a non-free expansion, i.e., $T_2'' > T_1'$. It is worth noticing that ΔU for the free expansion process is zero in spite of the fact that there occurs a decrease in temperature of the gas. This decrease in temperature is due to the fact that some heat is required to overcome the forces of attraction between the molecules. The work done in this process is stored as the potential energy. Thus, during expansion, kinetic energy of the molecules is converted into potential energy without changing the overall energy of the system. In other words, the decrease in energy due to the temperature fall is exactly equal to the increase in energy due to the increase in volume of the gas and hence sum of these two is equal to zero.

Example 2.14.3 One mole of chlorine undergoes adiabatic free expansion from 1 dm^3 to 10 dm^3. If the initial temperature of the gas was 273 K, calculate T_{final}, q, w, ΔU and ΔH if the gas is considered a van der Waals gas. Given: a = 655 dm^6 kPa mol^{-2}, b = 0.055 dm^3 mol^{-1}, $C_{V,m}$ = 33.91 J K^{-1} mol^{-1}.

Solution Since the gas undergoes adiabatic free expansion

$$q = 0$$

$$\Delta U = w = 0$$

$$T_2 = T_1 + \frac{1}{C_{V,m}}\left[a\left(\frac{1}{V_2} - \frac{1}{V_1}\right)\right]$$

$$\Delta H = -a\left(\frac{1}{V_2} - \frac{1}{V_1}\right) + R\left(\frac{V_2T_2}{V_2 - b} - \frac{V_1T_1}{V_1 - b}\right)$$

Thus, we have

$$T_2 = 273\ \text{K} + \frac{1}{33.91}\left[655\left(\frac{1}{10} - 1\right)\right]\text{K}$$

$$= \left(273 - \frac{589.5}{33.91}\right)\ \text{K} = 255.6\ \text{K}$$

$$\Delta H = \left[-655\left(\frac{1}{10} - \frac{1}{1}\right) + 8.314\left(\frac{10\times 255.6}{10-0.055} - \frac{1\times 273}{1-0.055}\right)\right]\text{J mol}^{-1}$$

$$= (589.5 - 256.0)\ \text{J mol}^{-1} = 324.5\ \text{J mol}^{-1}$$

Problem 2.14.1 One mole of a certain gas obeys the equation of state $p(V - b) = RT$ and has a constant molar heat capacity $C_{V,m}$ which is independent of temperature. The parameter b is a constant. For one mole, find ΔU, w, q, and ΔH for the following processes:

(a) Isothermal reversible process

(b) Isobaric reversible process

(c) Isochoric reversible process

(d) Adiabatic reversible process (in terms of T_1, p_1, p_2, V_1, V_2, $C_{p,m}$ and $C_{V,m}$)

(e) Adiabatic irreversible process (in terms of p_1, p_2, T_1) against a constant pressure p_2.

Solution **(a) Isothermal reversible process, dT = 0**

w: $\quad \mathrm{d}w = -p\,\mathrm{d}V$

$$w = -\int_{V_1}^{V_2} p\,\mathrm{d}V = -\int_{V_1}^{V_2}\frac{RT}{(V-b)}\mathrm{d}V = -RT\ln\frac{V_2 - b}{V_1 - b} = -RT\ln\frac{p_1}{p_2}$$

ΔU: Change in the energy can be evaluated by employing the thermodynamic equation of state:

$$\left(\frac{\partial U}{\partial V}\right)_T = T\left(\frac{\partial p}{\partial T}\right)_V - p = T\left\{\frac{\partial(RT/(V-b))}{\partial T}\right\}_V - p = \frac{RT}{V-b} - p = 0$$

Thus $\quad \Delta U = 0 \quad$ (this is to be expected, since a = 0)

q: According to first law:

$$q = \Delta U - w = 0 + RT \ln \frac{V_2 - b}{V_1 - b} = RT \ln \frac{p_1}{p_2}$$

ΔH : $\Delta H = \Delta U + \Delta(pV) = 0 + \Delta(pV) = \Delta(pV)$

Since $p = \dfrac{RT}{V - b}$, therefore $pV = \dfrac{RTV}{V - b}$

Hence $$\Delta H = \Delta(pV) = RT \left[\frac{V_2}{V_2 - b} - \frac{V_1}{V - b} \right] = RT \left[\frac{V_2 - b + b}{V_2 - b} - \frac{V_1 - b + b}{V_1 - b} \right]$$

$$= RTb \left[\frac{1}{V_2 - b} - \frac{1}{V - b} \right] = b\,(p_2 - p_1)$$

(b) Isobaric reversible process, dp = 0

w: $w = -p \int dV = -p\,(V_2 - V_1)$

ΔU: $$\Delta U = \int_{T_1}^{T_2} C_{V,\mathrm{m}}\, dT = C_{V,\mathrm{m}}\,(T_2 - T_1)$$

$$= C_{V,\mathrm{m}} \left[\frac{p(V_2 - b)}{R} - \frac{p(V_1 - b)}{R} \right] = \frac{C_{V,\mathrm{m}} p}{R}\,(V_2 - V_1)$$

q_p and ΔH: $q_p = \Delta H = \Delta U + \Delta(pV) = \Delta U + p(\Delta V)$

$$= \frac{C_{V,\mathrm{m}} p}{R}\,(V_2 - V_1) + p(V_2 - V_1) = \frac{(C_{V,\mathrm{m}} + R)}{R}\, p(V_2 - V_1)$$

For this gas $$C_{p,\mathrm{m}} - C_{V,\mathrm{m}} = \left[p + \left(\frac{\partial U}{\partial V} \right)_T \right] \left(\frac{\partial V}{\partial T} \right)_p$$

Since $(\partial U/\partial V)_T = 0$, we get

$$C_{p,\mathrm{m}} - C_{V,\mathrm{m}} = p \left(\frac{\partial V}{\partial T} \right)_p = p \frac{R}{p} = R$$

or $C_{V,\mathrm{m}} + R = C_{p,\mathrm{m}}$

Thus $$q_p = \frac{C_{p,\mathrm{m}}}{R}\, p(V_2 - V_1)$$

(c) Isochoric reversible process, dV = 0

w: $w = -\int p\, dV = 0$

q_V and ΔU: $$q_V = \Delta U = \int_{T_1}^{T_2} C_{V,\mathrm{m}}\, dT = C_{V,\mathrm{m}}\,(T_2 - T_1)$$

$$= C_{V,\mathrm{m}} \left(\frac{p_2\,(V - b)}{R} - \frac{p_1\,(V - b)}{R} \right) = \frac{C_{V,\mathrm{m}}}{R}\,(V - b)\,(p_2 - p_1)$$

ΔH: $\Delta H = \Delta U + \Delta(pV) = \Delta U + \Delta(RT + pb) = \Delta U + R(\Delta T) + b(\Delta p)$

$$= \frac{C_{V,\mathrm{m}}}{R}(V-b)(p_2-p_1) + R(T_2-T_1) + b(p_2-p_1)$$

$$= \frac{C_{p,\mathrm{m}}}{R}(V-b)(p_2-p_1) + b(p_2-p_1)$$

Alternatively, we may proceed as follows. Employing the thermodynamic equation

$$\left(\frac{\partial H}{\partial p}\right)_T = V - T\left(\frac{\partial V}{\partial T}\right)_p$$

we find that

$$\left(\frac{\partial H}{\partial p}\right)_T = V - T\frac{R}{p} = b$$

Thus $$\mathrm{d}H = \left(\frac{\partial H}{\partial T}\right)_p \mathrm{d}T + \left(\frac{\partial H}{\partial p}\right)_T \mathrm{d}p$$

$$= C_{p,\mathrm{m}}\,\mathrm{d}T + b\,\mathrm{d}p = C_{p,\mathrm{m}}(T_2-T_1) + b(p_2-p_1)$$

$$= \frac{C_{p,\mathrm{m}}}{R}(V-b)(p_2-p_1) + b(p_2-p_1)$$

(Note that $C_{p,\mathrm{m}}$ is also independent of temperature.)

(d) Adiabatic reversible process, dq = 0

q: $q = 0$

w and ΔU: $\Delta U = w = C_{V,\mathrm{m}}(T_2 - T_1)$

ΔH: $\Delta H = C_{p,\mathrm{m}}(T_2 - T_1) + b(p_2 - p_1)$

Since the results are to be expressed in T_1, p_1, p_2, V_1, V_2, $C_{p,\mathrm{m}}$ and $C_{V,\mathrm{m}}$, we will have to eliminate T_2 in the above expressions. This can be done if a relation between V and T is known for this gas. This relation can be derived as follows:
we have

$$\mathrm{d}U = -p\,\mathrm{d}V$$

i.e. $$C_{V,\mathrm{m}}\,\mathrm{d}T = -\frac{RT}{V-b}\mathrm{d}V$$

Integrating this expression, we get

$$C_{V,\mathrm{m}}\ln\frac{T_2}{T_1} = -R\ln\left(\frac{V_2-b}{V_1-b}\right)$$

or $$\ln\frac{T_2}{T_1} = \ln\left(\frac{V_2-b}{V_1-b}\right)^{-R/C_{V,\mathrm{m}}}$$

or $$\frac{T_2}{T_1} = \left(\frac{V_2-b}{V_1-b}\right)^{-R/C_{V,\mathrm{m}}} = \left(\frac{V_2-b}{V_1-b}\right)^{-(C_{p,\mathrm{m}}-C_{V,\mathrm{m}})/C_{V,\mathrm{m}}} = \left(\frac{V_2-b}{V_1-b}\right)^{1-\gamma}$$

or $$T_2 = T_1\left(\frac{V_1-b}{V_2-b}\right)^{\gamma-1}$$

Therefore, $\Delta U = w = C_{V,\mathrm{m}}(T_2 - T_1) = C_{V,\mathrm{m}}\,T_1\left[\left(\dfrac{V_1 - b}{V_2 - b}\right)^{\gamma - 1} - 1\right]$

and $\quad \Delta H = C_{p,\mathrm{m}}\,T_1\left[\left(\dfrac{V_1 - b}{V_2 - b}\right)^{\gamma - 1} - 1\right] + b\,(p_2 - p_1)$

(e) Adiabatic irreversible process, dq = 0

q: $\quad q = 0$

ΔU and w: $\Delta U = w = \int - p_{\mathrm{ext}}\,\mathrm{d}V = - p_{\mathrm{ext}}\,(V_2 - V_1)$

Now $p_{\mathrm{ext}} = p_2$, the final pressure, and so

$$\Delta U = w = - p_2\,[(V_2 - b) - (V_1 - b)]$$

$$= - p_2\left(\frac{RT_2}{p_2} - \frac{RT_1}{p_1}\right)$$

ΔH: $\quad \Delta H = C_{p,\mathrm{m}}\,(T_2 - T_1) + b(p_2 - p_1)$

Since the results are to be expressed in T_1 we have to eliminate T_2 in the above expressions which can be done if a relation between p and T is known. The required relation can be worked out as follows:

Since $\Delta U = w$, we have

$$\Delta U = - p_2\Delta V = - p_2\,(V_2 - V_1)$$

$$C_{V,\mathrm{m}}\,(T_2 - T_1) = - p_2\left(\frac{RT_2}{p_2} - \frac{RT_1}{p_1}\right)$$

or $\quad C_{V,\mathrm{m}}\,(T_2 - T_1) = - RT_2 + p_2\dfrac{RT_1}{p_1}$

or $\quad T_2(C_{V,\mathrm{m}} + R) = C_{V,\mathrm{m}}\,T_1 + p_2\dfrac{RT_1}{p_1}$

or $\quad T_2 C_{p,\mathrm{m}} = T_1\left(C_{V,\mathrm{m}} + \dfrac{Rp_2}{p_1}\right)$ or $T_2 = \dfrac{T_1}{C_{p,\mathrm{m}}}\left(C_{V,\mathrm{m}} + \dfrac{Rp_2}{p_1}\right)$

Eliminating T_2 from the expression of ΔU and ΔH, we get

$$\Delta U = w = \frac{RT_1}{\gamma}\left(\frac{p_2 - p_1}{p_1}\right)$$

$$\Delta H = RT_1\left(\frac{p_2 - p_1}{p_1}\right)\ b(p_2 - p_1)$$

2.15 MISCELLANEOUS NUMERICALS

1. 20.0 dm^3 of an ideal gas (diatomic, $C_{V,\mathrm{m}} = 5R/2$) at 673 K and 0.7 MPa expands until pressure of the gas is 0.2 MPa. Calculate q, w, ΔU and ΔH for the process if the expansion is:

   (i) Isothermal and reversible $\qquad$ (ii) Adiabatic and reversible

(iii) Isothermal and adiabatic (iv) Against 0.2 MPa and adiabatic

(v) Against 0.2 MPa and isothermal

Solution From the given data of $V_i = 20.0\ \text{dm}^3$, $T_i = 673$ K, and $p_i = 0.7$ MPa, we can determine the amount of the ideal gas as given below:

$$n = \frac{pV}{RT} = \frac{(0.7 \times 10^3\ \text{kPa})(20.0\ \text{dm}^3)}{(8.314\ \text{dm}^3\ \text{kPa K}^{-1}\ \text{mol}^{-1})(673\ \text{K})} = 2.502\ \text{mol}$$

(i) Isothermal and reversible For an isothermal reversible expansion, we have

$$w = -\,nRT \ln \frac{V_f}{V_i} = -\,nRT \ln \frac{p_i}{p_f}$$

$$= -\,(2.502\ \text{mol})\ (8.314\ \text{J K}^{-1}\ \text{mol}^{-1})\ (673\ \text{K}) \times 2.303 \times \log \left(\frac{0.7\ \text{MPa}}{0.2\ \text{MPa}}\right)$$

$$= -\,17\ 541.2\ \text{J} = -\,17.54\ \text{kJ}$$

Since T remains constant, $\Delta U = 0$ and $\Delta H = 0$.

According to the first law of thermodynamics,

$$q = -\,w = 17.54\ \text{kJ}$$

(ii) Adiabatic and reversible For an adiabatic process $q = 0$ and $\Delta U = w$, therefore

$$w = \Delta U = nC_{V,\,\text{m}}\,(T_f - T_i)$$

T_f can be calculated using the formula

$$p_i T_i^{-\,C_{p,\,\text{m}}/R} = p_f T_f^{-\,C_{p,\,\text{m}}/R}$$

or $$\log(p_i/\text{MPa}) - \frac{C_{p,\text{m}}}{R} \log\,(T_i/\text{K}) = \log\,(p_f/\text{MPa}) - \frac{C_{p,\text{m}}}{R} \log\,(T_f/\text{K})$$

Therefore

$$\log(T_f/\text{K}) = \frac{R}{C_{p,\text{m}}}\left\{\log\,(p_f/\text{MPa}) - \log\,(p_i/\text{MPa}) + \frac{C_{p,\text{m}}}{R}\log\,(T_i/\text{K})\right\}$$

$$= \frac{2}{7}\left\{\log\,(0.2) - \log\,(0.7) + \frac{7}{2}\log\,(673)\right\}$$

$$= 2.672\ 5$$

Hence $T_f = 470.4$ K

Thus $$w = nC_{V,\,\text{m}}\,(T_f - T_i)$$

$$= (2.502\ \text{mol})\ (2.5 \times 8.314\ \text{J K}^{-1}\ \text{mol}^{-1})\ (470.4\ \text{K} - 673\ \text{K})$$

$$= -\,10\ 536\ \text{J} = -\,10.536\ \text{kJ}$$

$$\Delta U = w = -\,10.536\ \text{kJ}$$

$$\Delta H = \Delta U + \Delta(pV) = \Delta U + nR(\Delta T)$$

$$= -10\ 536\ \text{J} + (2.502\ \text{mol})\ (8.314\ \text{J K}^{-1}\ \text{mol}^{-1})\ (-202.6\ \text{K})$$

$$= -\,10\ 536\ \text{J} - 4\ 214.4\ \text{J}$$

$$= -\,14\ 750.4\ \text{J} = -\,14.75\ \text{kJ}$$

(iii) Isothermal and adiabatic For a process which is isothermal as well as adiabatic, we have

$$q = 0, \quad \Delta T = 0, \quad \Delta U = 0 \quad \text{and} \quad \Delta H = 0$$

All these values can be zero only if the expansion is a free expansion.

(iv) Against 0.2 MPa and adiabatic For an adiabatic irreversible process, the temperature T_f after the expansion of the gas can be calculated as follows.

$$\Delta U = w = -p_{ext}\,\Delta V = -p_{ext}\,(V_f - V_i)$$

$$nC_{V,m}\,(T_f - T_i) = -p_{ext}\left(\frac{nRT_f}{p_f} - \frac{nRT_i}{p_i}\right)$$

or

$$C_{V,m}\,(T_f - T_i) = -p_{ext}\left(\frac{RT_f}{p_f} - \frac{RT_i}{p_i}\right)$$

Substituting the values of $C_{V,m}$, p_{ext}, p_f, p_i and T_i, we have

$$\frac{5}{2}\,(T_f - 673\text{ K}) = -(0.2\text{ MPa})\left(\frac{T_f}{0.2\text{ MPa}} - \frac{673\text{ K}}{0.7\text{ MPa}}\right)$$

or

$$\frac{5}{2}T_f - \frac{5}{2}\times 673\text{ K} = -T_f + \frac{0.2\times 673\text{ K}}{0.7}$$

or

$$T_f = \frac{2}{7}\left(\frac{5}{2}\times 673\text{ K} + \frac{0.2\times 673\text{ K}}{0.7}\right)$$

$$= 535.7\text{ K}$$

Thus

$$\Delta U = nC_{V,m}\,(T_f - T_i)$$

$$= (2.502\text{ mol})\,(2.5\times 8.314\text{ J K}^{-1}\text{ mol}^{-1})\,(535.7\text{ K} - 673\text{ K})$$

$$= -7\,140.16 = -7.14\text{ kJ}$$

$$w = -7.14\text{ kJ}$$

$$\Delta H = \Delta U + nR(\Delta T) = -7.14\text{ kJ} + (2.502\text{ mol})\,(8.314\text{ K}^{-1}\text{ mol}^{-1})\times(535.7\text{ K} - 673\text{ K})$$

$$= -7.14\text{ kJ} - 2\,856.1\text{ J} = -9.996\text{ kJ}$$

(v) Against 0.2 MPa and isothermal For an irreversible isothermal expansion, we have

$$w = -p_{ext}\,\Delta V = -p_{ext}\,(V_f - V_i)$$

The final volume of the gas can be determined using the ideal gas equation,

$$V_f = \frac{p_iV_i}{p_f} = \frac{(0.7\text{ MPa})\,(20\text{ dm}^3)}{(0.2\text{ MPa})} = 70\text{ dm}^3$$

Thus

$$w = -(0.2\text{ MPa})\,(70\text{ dm}^3 - 20\text{ dm}^3) = -(200\text{ kPa})\,(50\text{ dm}^3)$$

$$= -10\,000\text{ J} = -10.0\text{ kJ}$$

$$\left.\begin{aligned}\Delta U &= 0\\ \Delta H &= 0\end{aligned}\right\}\ \text{since } \Delta T = 0$$

$$q = -w = 10.0\text{ kJ}$$

---

2. A 32 g sample of CH_4 gas initially at 101.325 kPa and 300 K is heated to 550 K. $C_{p,m}/\text{J K}^{-1}\text{ mol}^{-1} = 12.552 + 8.368\times 10^{-2}\,(T/\text{K})$. Assuming CH_4 behaves ideally, compute w, q, ΔU and ΔH for (a) an isobaric reversible process, and (b) an isochoric reversible process.

Solution

$$\text{Amount of the gas} = \frac{32\text{ g}}{16\text{ g mol}^{-1}} = 2\text{ mol}$$

(a) Isobaric reversible process (dp = 0):

$$q_p = \Delta H = \int_{T_1}^{T_2} nC_{p,\,\mathrm{m}}\,\mathrm{d}T$$

$$= n\int_{T_1}^{T_2}\{(12.552\ \mathrm{J\,K^{-1}\,mol^{-1}})\,\mathrm{d}T + (8.368\times10^{-2}\ \mathrm{J\,K^{-2}\,mol^{-1}})\,T\,\mathrm{d}T\}$$

$$= n\left\{(12.552\ \mathrm{J\,K^{-1}\,mol^{-1}})(T_2 - T_1) + (8.368\times10^{-2}\ \mathrm{J\,K^{-2}\,mol^{-1}})\left(\frac{T_2^2}{2} - \frac{T_1^2}{2}\right)\right\}$$

Substituting the values, we get

$$q_p = \Delta H = (2\ \mathrm{mol})\left\{(12.552\ \mathrm{J\,K^{-1}\,mol^{-1}})(550\ \mathrm{K} - 300\ \mathrm{K}) + (8.368\times10^{-2}\ \mathrm{J\,K^{-2}\,mol^{-1}})\left(\frac{(550\ \mathrm{K})^2}{2} - \frac{(300\ \mathrm{K})^2}{2}\right)\right\}$$

$$= 2\,(3\ 138\ \mathrm{J} + 8\ 891\ \mathrm{J}) = 24\ 058\ \mathrm{J}$$

$$w = -\int p\,\mathrm{d}V = -nR\int_{T_1}^{T_2}\mathrm{d}T = -\,(2\ \mathrm{mol})\,(8.314\ \mathrm{J\,K^{-1}\,mol^{-1}})\,(550\ \mathrm{K} - 300\ \mathrm{K})$$

$$= -\,4\ 157\ \mathrm{J} = -\,4.157\ \mathrm{kJ}$$

$$\Delta U = \Delta H - \Delta(pV) = \Delta H - nR(\Delta T)$$

$$= (24\ 058\ \mathrm{J}) - (2\ \mathrm{mol})\,(8.314\ \mathrm{J\,K^{-1}\,mol^{-1}})\,(250\ \mathrm{K})$$

$$= 24\ 058\ \mathrm{J} - 4\ 157\ \mathrm{J} = 19\ 901\ \mathrm{J}$$

(b) Isochoric reversible process (dV = 0):

$$C_{V,\,\mathrm{m}} = C_{p,\,\mathrm{m}} - R = (4.238\ \mathrm{J\,K^{-1}\,mol^{-1}}) + (8.368\times10^{-2}\ \mathrm{J\,K^{-2}\,mol^{-1}})\,T$$

$$q_V = \Delta U = \int_{T_1}^{T_2} nC_{V,\,\mathrm{m}}\,\mathrm{d}T = n\left\{(4.238\ \mathrm{J\,K^{-1}\,mol^{-1}})(T_2 - T_1) + (8.368\times10^{-2}\ \mathrm{J\,K^{-2}\,mol^{-1}})\left(\frac{T_2^2}{2} - \frac{T_1^2}{2}\right)\right\}$$

$$= (2\ \mathrm{mol})\left\{(4.238\ \mathrm{J\,K^{-1}\,mol^{-1}})\,(550\ \mathrm{K} - 300\ \mathrm{K}) + (8.368\times10^{-2}\ \mathrm{J\,K^{-2}\,mol^{-1}})\left(\frac{(500\ \mathrm{K})^2}{2} - \frac{(300\ \mathrm{K})^2}{2}\right)\right\}$$

$$= (2\ \mathrm{mol})\,(1\ 059.5\ \mathrm{J\,mol^{-1}} + 8\ 891\ \mathrm{J\,mol^{-1}})$$

$$= 19\ 901\ \mathrm{J}$$

$$\Delta H = \Delta U + \Delta(pV) = \Delta U + nR(\Delta T)$$

$$= 19\ 901\ \mathrm{J} + 4\ 157\ \mathrm{J} = 24\ 058\ \mathrm{J}$$

$w = 0$ since dV = 0

---

3. One mole of an ideal monatomic gas ($C_{V,\,\mathrm{m}} = 1.5\,R$) is subjected to the following sequence of steps: (a) The gas is heated reversibly at constant pressure of 101.325 kPa from 298 K to 373 K. (b) Next, the gas is expanded reversibly and isothermally to double its volume. (c) Finally, the gas is cooled reversibly and adiabatically to 308 K. Calculate q, w, ΔU and ΔH for the overall process.

Solution **Step (a): Isobaric reversible process**

$$q_p = \Delta H = \int_{298\,\text{K}}^{373\,\text{K}} C_{p,\text{m}}\,\text{d}T = \int_{298\,\text{K}}^{373\,\text{K}} (C_{V,\text{m}} + R)\,\text{d}T$$

$$= (2.5 \times 8.314 \text{ J K}^{-1} \text{ mol}^{-1})(373 \text{ K} - 298 \text{ K})$$

$$= 1\ 558.88 \text{ J mol}^{-1}$$

$$\Delta U = C_{V,\text{m}}\,\Delta T = (1.5 \times 8.314 \text{ J K}^{-1} \text{ mol}^{-1})(373 \text{ K} - 298 \text{ K})$$

$$= 935.33 \text{ J mol}^{-1}$$

$$w = -p(\Delta V) = \Delta U - \Delta H = (935.33 - 1\ 558.88) \text{ J mol}^{-1} = -623.55 \text{ J mol}^{-1}$$

Step (b): Isothermal reversible expansion

$$V_1 = V \quad \text{and} \quad V_2 = 2V$$

$$w = -RT \ln \frac{V_2}{V_1} = -(8.314 \text{ J K}^{-1} \text{ mol}^{-1})(373 \text{ K}) \times 2.303 \times \log(2)$$

$$= -2\ 149.71 \text{ J mol}^{-1}$$

Since temperature does not change

$$\Delta U = 0 \qquad \text{and} \qquad \Delta H = 0$$

and according to the first law

$$q = -w = 2\ 149.71 \text{ J mol}^{-1}$$

Step (c): Adiabatic reversible cooling

$$q = 0$$

$$w = \Delta U = C_{V,\text{m}}(T_2 - T_1) = (1.5 \times 8.314 \text{ J K}^{-1} \text{ mol}^{-1})(308 \text{ K} - 373 \text{ K})$$

$$= -810.62 \text{ J mol}^{-1}$$

$$\Delta H = C_{p,\text{m}}(T_2 - T_1) = (2.5 \times 8.314 \text{ J K}^{-1} \text{ mol}^{-1})(308 \text{ K} - 373 \text{ K})$$

$$= -1\ 351.03 \text{ J mol}^{-1}$$

For the overall process, the values of q, w, ΔU and ΔH are

$$q/\text{J mol}^{-1} = 1\ 558.88 + 2\ 149.71 = 3\ 708.59$$

$$w/\text{J mol}^{-1} = -623.55 - 2\ 149.71 - 810.62 = -3\ 583.88$$

$$\Delta U/\text{J mol}^{-1} = 935.33 - 810.62 = 124.71$$

$$\Delta H/\text{J mol}^{-1} = 1\ 558.88 - 1\ 351.03 = 207.85$$

---

4. One mole of an ideal gas (not necessarily monatomic) is subjected to the following sequence of steps.

(a) It is heated at constant volume from 298 K to 373 K.

(b) It is expanded freely into vacuum to double volume.

(c) It is cooled reversibly at constant pressure to 298 K.

Calculate q, w, ΔU and ΔH for the overall process.

Solution **Step (a): Isochoric process**

$$q_V = \Delta U = \int_{298\text{K}}^{373\text{K}} C_{V,\text{m}}\,\text{d}T = C_{V,\text{m}}(373 \text{ K} - 298 \text{ K}) = C_{V,\text{m}}(75 \text{ K})$$

$$w = 0$$

$$\Delta H = \Delta U + \Delta(pV) = \Delta U + \Delta(RT)$$
$$= C_{V,\,\mathrm{m}}\,(75\ \mathrm{K}) + (75\ \mathrm{K})\,R = (75\ \mathrm{K})\,C_{p,\,\mathrm{m}}$$

Step (b): Free expansion

$$q = 0, \quad w = 0, \quad \Delta U = 0 \quad \text{and} \quad \Delta H = 0$$

Step (c): Isobaric process

$$q_p = \Delta H = \int_{373\mathrm{K}}^{298\mathrm{K}} C_{p,\,\mathrm{m}}\,\mathrm{d}T = (-\,75\ \mathrm{K})\,C_{p,\,\mathrm{m}}$$
$$w = -\int p\,\mathrm{d}V = -R\,\Delta T = (75\ \mathrm{K})\,R$$
$$\Delta U = \int C_{V,\,\mathrm{m}}\,\mathrm{d}T = (-75\ \mathrm{K})\,C_{V,\,\mathrm{m}}$$

For the overall process

$$q = (75\ \mathrm{K})\,C_{V,\,\mathrm{m}} - (75\ \mathrm{K})\,C_{p,\,\mathrm{m}} = -\,(75\ \mathrm{K})\,(C_{p,\,\mathrm{m}} - C_{V,\,\mathrm{m}})$$
$$= -\,(75\ \mathrm{K})\,R = -\,(75\ \mathrm{K})\,(8.314\ \mathrm{J\,K^{-1}\,mol^{-1}})$$
$$= -\,623.55\ \mathrm{J\,mol^{-1}}$$
$$w = (75\ \mathrm{K})\,R = (75\ \mathrm{K})\,(8.314\ \mathrm{J\,K^{-1}\,mol^{-1}})$$
$$= 623.55\ \mathrm{J\,mol^{-1}}$$
$$\Delta U = (75\ \mathrm{K})\,C_{V,\,\mathrm{m}} - (75\ \mathrm{K})\,C_{V,\,\mathrm{m}} = 0$$
$$\Delta H = (75\ \mathrm{K})\,C_{p,\,\mathrm{m}} - (75\ \mathrm{K})\,C_{p,\,\mathrm{m}} = 0$$

---

5. The cubic expansion coefficient of water at 293 K and 101.325 kPa is $2.1 \times 10^{-4}\ \mathrm{K^{-1}}$. Calculate approximately, the work attending the heating at 101.325 kPa pressure of 1 mol of water from 288 K to 298 K. Compare this with work involved in heating 1 mol of an ideal gas from 288 K to 298 K.

Solution

From the cubic expansion coefficient

$$\alpha = \frac{1}{V}\left(\frac{\partial V}{\partial T}\right)_p$$

we get $\mathrm{d}V = \alpha V\,\mathrm{d}T \qquad (p \text{ constant})$

Thus $\mathrm{d}w = -\,p\,\mathrm{d}V = -\,p\,(\alpha V\,\mathrm{d}T) \quad \text{or} \quad w = -\,p\alpha V\,\Delta T$

Now $p = 101.325\ \mathrm{kPa}\ ; \qquad \alpha = 2.1 \times 10^{-4}\ \mathrm{K^{-1}}$

V = volume occupied by 1 mole of water = $18 \times 10^{-3}\ \mathrm{dm^3}$

$\Delta T = 298\ \mathrm{K} - 288\ \mathrm{K} = 10\ \mathrm{K}$

Substituting these values in the previous expression, we get

$$w = -\,(101.325\ \mathrm{kPa})\,(2.1 \times 10^{-4}\ \mathrm{K^{-1}})\,(18 \times 10^{-3}\ \mathrm{dm^3})\,(10\ \mathrm{K})$$
$$= -\,3.83 \times 10^{-3}\ \mathrm{kPa\ dm^3} = -\,3.83 \times 10^{-3}\ \mathrm{J}$$

Work involved in case of an ideal gas

$$w = -\int p\,\mathrm{d}V = -\int R\,\mathrm{d}T = -\,(8.314\ \mathrm{J\,K^{-1}\,mol^{-1}})\,(10\ \mathrm{K}) = -\,83.14\ \mathrm{J\,mol^{-1}}$$

---

6. The isothermal compressibility of water at 293 K is $4.9 \times 10^{-6}\ \mathrm{atm^{-1}}$ over the range 1 to 25 atm. Calculate the work attending the compression of 1 mol of liquid water from a pressure of 1 atm to 25 atm at 293 K. Compare this with work involved when 1 mole of an ideal gas is compressed from 1 atm to 25 atm at 298 K.

Solution From the isothermal compressibility

$$\kappa_T = -\frac{1}{V}\left(\frac{\partial V}{\partial p}\right)_T$$

we get $\quad dV = -\kappa_T V\,dp \qquad (T \text{ constant})$

Thus $\quad dw = -p\,dV = +\kappa_T V p\,dp$

Integrating this expression within the limits of p_1 and p_2, we get

$$w = \kappa_T V\left(\frac{p_2^2}{2} - \frac{p_1^2}{2}\right)$$

Substituting the values of κ_T, V, p_1 and p_2, we get

$$w = \frac{(4.9\times10^{-6}\ \text{atm}^{-1})(18\times10^{-3}\ \text{dm}^3)}{2}\{(25\ \text{atm})^2 - (1\ \text{atm})^2\}$$

$$= 2.752\times10^{-5}\ \text{dm}^3\ \text{atm} = 2.752\times10^{-5}\ \text{dm}^3\ (101.325\ \text{kPa})$$

$$= 0.002\ 79\ \text{J}$$

For an ideal gas, the work involved is

$$w = -\int p\,dV = -RT\ln\frac{V_2}{V_1} = -RT\ln\frac{p_1}{p_2}$$

$$= -(8.314\ \text{J K}^{-1}\ \text{mol}^{-1})\ (298\ \text{K})\times 2.303\times\log\left(\frac{1\ \text{atm}}{25\ \text{atm}}\right)$$

$$= 7.976\times10^3\ \text{J mol}^{-1} = 7.976\ \text{kJ mol}^{-1}$$

REVISIONARY PROBLEMS

2.1 State the zeroth law of thermodynamics and discuss its necessity in the laws of thermodynamics.

2.2 The first law of thermodynamics is essentially a law of conservation of energy and is written as

$$dU = dq + dw$$

(i) Explain the symbols involved in this expression.

(ii) Prove mathematically that q and w are not state functions.

2..3 Show that the heat absorbed at constant volume condition is equal to the increase in the internal energy of the system whereas that at constant pressure is equal to the increase in the enthalpy of the system.

2.4 (a) Heat capacity of a system is defined as

$$C = \lim_{\Delta T\to 0}\left(\frac{q}{\Delta T}\right) = \frac{dq}{dT}$$

Using the first law of thermodynamics, show that

$$C_V = \left(\frac{\partial U}{\partial T}\right)_V \qquad \text{and} \qquad C_p = \left(\frac{\partial H}{\partial T}\right)_p$$

(b) Derive the following relations:

$$C_p - C_V = \left[p + \left(\frac{\partial U}{\partial V}\right)_T\right]\left(\frac{\partial V}{\partial T}\right)_p; \qquad C_p - C_V = \left[V - \left(\frac{\partial H}{\partial p}\right)_T\right]\left(\frac{\partial p}{\partial T}\right)_V$$

$$C_p - C_V = \left[V + \left(\frac{\partial H}{\partial T}\right)_p \left(\frac{\partial T}{\partial p}\right)_H\right]\left(\frac{\partial p}{\partial T}\right)_V$$

$$C_p - C_V = T\left(\frac{\partial p}{\partial T}\right)_V \left(\frac{\partial V}{\partial T}\right)_p = TV\alpha^2/\kappa_T = -T\frac{(\partial p/\partial T)_V^2}{(\partial p/\partial V)_T}$$

(c) Derive the value of $C_{p,\mathrm{m}} - C_{V,\mathrm{m}}$ for an ideal gas.

(d) Show that $C_{p,\mathrm{m}} - C_{V,\mathrm{m}}$ for a van der Waals gas is equal to

$$R\left\{1 - \frac{2a}{V}\frac{(V-b)}{(pV^2+a)}\right\}^{-1}$$

2.5 Prove that if $(\partial U/\partial V)_T = 0$, it follows that $(\partial U/\partial p)_T = 0$.

2.6 Prove that it does not necessarily follow that if $(\partial U/\partial V)_T = 0$ then $(\partial H/\partial p)_T = 0$.

2.7 Show that for an ideal gas

(i) $\left(\frac{\partial U}{\partial V}\right)_p = C_V\frac{p}{nR} = C_p\left(\frac{\partial T}{\partial V}\right)_p - p$

(ii) $\left(\frac{\partial U}{\partial p}\right)_V = C_V\left(\frac{\partial T}{\partial p}\right)_V = \left(\frac{\kappa_T}{\alpha}\right)C_V$

(iii) $\left(\frac{\partial C_V}{\partial V}\right)_T = 0$ and $\left(\frac{\partial C_p}{\partial p}\right)_T = 0$

(iv) $\left(\frac{\partial H}{\partial V}\right)_T = 0$ (v) $\left(\frac{\partial U}{\partial T}\right)_p = C_p - p\left(\frac{\partial V}{\partial T}\right)_p$

2.8 (a) Describe the Joule's expansion experiment. What conclusion do you draw from this experiment? What is the physical significance of the derivative $(\partial U/\partial V)_T$? Will it have a zero value for a real gas?

(b) Starting from the definition of Joule coefficient, derive the relation

$$\eta = -\frac{1}{C_V}\left(\frac{\partial U}{\partial V}\right)_T = \frac{1}{C_V}\left\{p - T\left(\frac{\partial p}{\partial T}\right)_V\right\}$$

and hence show that a real gas exhibits:

(i) heating if $p > T(\partial p/\partial T)_V$

and (ii) cooling if $p < T(\partial p/\partial T)_V$

2.9 (a) Describe the Joule-Thomson experiment. Show that the expansion in this experiment is an isenthalpic process.

(b) Starting from the definition of Joule-Thomson coefficient, derive the relations

$$\mu_{\mathrm{JT}} = -\frac{1}{C_p}\left(\frac{\partial H}{\partial p}\right)_T = \frac{1}{C_p}\left[T\left(\frac{\partial V}{\partial T}\right)_p - V\right]$$

$$= -\frac{1}{C_V}\left[\left(\frac{\partial U}{\partial V}\right)_T\left(\frac{\partial V}{\partial p}\right)_T + \left(\frac{\partial (pV)}{\partial p}\right)_T\right]$$

and hence show that a real gas exhibits:

(i) heating if $V > T\,(\partial V/\partial T)_p$

and (ii) cooling if $V < T\,(\partial V/\partial T)_p$

(c) Show that $\mu_{JT} = 0$ for an ideal gas. What do you comment about the liquefaction of such a gas on the basis of its μ_{JT} value?

(d) What is an inversion temperature? Show that for a van der Waals gas

$$\mu_{JT} = \frac{1}{C_{p,\mathrm{m}}}\left[\frac{2a}{RT} - \frac{3abp}{R^2T^2} - b\right]$$

(e) Justify the statement that at moderately low pressure and high temperature the expression given in part (d) reduces to

$$\mu_{JT} = \frac{1}{C_{p,\mathrm{m}}}\left[\frac{2a}{RT} - b\right]$$

and hence show that $T_\mathrm{i} = 2a/Rb$.

2.10 Show that $(\partial H/\partial p)_T = V$ for solids and liquids.

2.11 For the amount n of an ideal gas, derive the expressions for q, w, ΔU and ΔH for the following processes (involving expansion, compression and the cyclic processes):

(i) Isothermal reversible process.

(ii) Isothermal irreversible process against a constant p.

(iii) Free expansion.

(iv) Adiabatic reversible process.

(v) Adiabatic irreversible process.

(vi) Adiabatic free expansion.

2.12 Derive the following relations for an ideal gas undergoing adiabatic reversible process:

(i) $TV^{\gamma-1}$ = constant

(ii) $Tp^{-R/C_{p,\mathrm{m}}}$ = constant

(iii) pV^{γ} = constant, where $\gamma = C_{p,\mathrm{m}}/C_{V,\mathrm{m}}$

Do these re[illegible]ions hold good for an adiabatic irreversible process?

2.13 Show that th[illegible] work done by an ideal gas in a reversible adiabatic expansion from p_i and V_i to p_f and V_f is given by

(i) $w = (p_\mathrm{f}V_\mathrm{f} - p_\mathrm{i}V_\mathrm{i})/(\gamma - 1)$ (ii) $w = -C_VT_\mathrm{i}\left[1 - \frac{p_\mathrm{f}}{p_\mathrm{i}}\right]^{R/C_{p,\mathrm{m}}}$

2.14 Justify the following statement:

(a) The magnitude of the reversible work involved in going from volume V_1 to V_2 ($V_2 > V_1$) for an ideal gas is greater in isothermal process than in adiabatic process.

(b) The magnitude of the reversible work involved in going from pressure p_1 to p_2 ($p_2 < p_1$) is greater in isothermal process than in adiabatic process.

(c) The reversible work of compression from volume V_2 to V_1 for an ideal gas is less than the corresponding work involved in the irreversible compression.

(d) Final pressure in an adiabatic expansion of an ideal gas is less than that of the isothermal expansion to the same final volume.

2.15 A certain gas obeys the equation of state $p(V - nb) = nRT$ and has a constant volume heat capacity C_V which is independent of temperature. The parameter b is constant. For one mole, determine w, q, ΔU and ΔH for the following processes:

(i) Isothermal reversible expansion.

(ii) Isobaric reversible process.

(iii) Isochoric reversible process.

(iv) Adiabatic reversible expansion (in terms of T_1, V_1, V_2, p_1, p_2 and C_V).

(v) Adiabatic irreversible expansion (in terms of p_1, p_2, T_1) against a constant pressure p_2.

2.16 Show that for a van der Waals gas involved in an isothermal reversible expansion

$$w = -nRT\ln\left(\frac{V_2 - nb}{V_1 - nb}\right) - an^2\left(\frac{1}{V_2} - \frac{1}{V_1}\right)$$

$$q = nRT\ln\left(\frac{V_2 - nb}{V_1 - nb}\right); \qquad \Delta U = -an^2\left(\frac{1}{V_2} - \frac{1}{V_1}\right)$$

$$\Delta H = -2an^2\left(\frac{1}{V_2} - \frac{1}{V_1}\right) + n^2 b\,RT\left(\frac{1}{V_2 - nb} - \frac{1}{V_1 - nb}\right)$$

2.17 For a van der Waals gas

$$U = C_{V,\mathrm{m}}T - \frac{a}{V}$$

Show that $T^{C_{V,\mathrm{m}}/R}(V - b)$ = constant in a reversible adiabatic expansion.

2.18 Derive the following expressions for the isothermal irreversible expansion of a van der Waals gas against a constant pressure of p_{ext}.

$$w = -p_{\text{ext}}(V_2 - V_1)$$

$$\Delta U = -n^2 a\left(\frac{1}{V_2} - \frac{1}{V_1}\right); \qquad q = -n^2 a\left(\frac{1}{V_2} - \frac{1}{V_1}\right) + p_{\text{ext}}(V_2 - V_1)$$

2.19 (a) Derive the following expressions for the adiabatic irreversible expansion of a van der Waals gas against a constant pressure of p_{ext}.

$$w = -p_{\text{ext}}(V_2 - V_1); \qquad q = 0; \qquad \Delta U = -p_{\text{ext}}(V_2 - V_1)$$

(b) Show that the temperature of the gas after carrying out the expansion process of part (a) is given by

$$T_2 = T_1 + \frac{1}{nC_{V,\mathrm{m}}}\left[-p_{\text{ext}}(V_2 - V_1) + n^2 a\left(\frac{1}{V_2} - \frac{1}{V_1}\right)\right]$$

(c) Show that the temperature of a van der Waals gas after carrying out the adiabatic free expansion from V_1 to V_2 is given by

$$T_2' = T_1 + \frac{1}{nC_{V,\mathrm{m}}}\left[n^2 a\left(\frac{1}{V_2} - \frac{1}{V_1}\right)\right]$$

Interpret the fact that $T_2' \neq T_1$ in spite of the fact that $\Delta U = 0$.

TRY YOURSELF PROBLEMS

2.1 Since $C_V = (\partial U/\partial T)_V$ by definition, one often writes without any restriction, $\Delta U = C_V \Delta T$. This is not generally true. Explain why?

(*Hint:* $\mathrm{d}U = (\partial U/\partial T)_V\,\mathrm{d}T + (\partial U/\partial V)_T\,\mathrm{d}V$

$= C_V\,\mathrm{d}T + (\partial U/\partial V)_T\,\mathrm{d}V$; true only for ideal gases.)

2.2 Show that the enthalpy of an ideal gas is a function of temperature only.

(*Hint:* because $(\partial H/\partial p)_T = 0$.)

2.3 For a van der Waals gas the Joule-Thomson coefficient is given by

$$\mu_{\mathrm{JT}} = \frac{1}{C_{p,\mathrm{m}}}\left[\frac{2a}{RT} - b - \frac{3abp}{R^2T^2}\right]$$

Show that the temperature of inversion as a function of pressure is given by the relation

$$T_{\mathrm{i}} = \frac{a \pm \sqrt{a^2 - 3ab^2 p}}{bR}$$

2.4 Show that for a system of constant heat capacity whose energy along a given isotherm is constant, the energy depends only on temperature.

(*Hint:* $\mathrm{d}U = C_V\,\mathrm{d}T + (\partial U/\partial V)_T\,\mathrm{d}V$; isotherm means $\mathrm{d}T = 0$, constant energy means $\mathrm{d}U = 0$. Hence $(\partial U/\partial V)_T = 0$, i.e. $U = f(T)$.)

2.5 Show mathematically that H is a state function

[*Hint:* $\mathrm{d}H = \mathrm{d}(U + pV) = \mathrm{d}U + p\,\mathrm{d}V + V\,\mathrm{d}p$

Take $U = f(T, V)$ and $p = f(T, V)$, then

$$\mathrm{d}(U + pV) = \left\{\left(\frac{\partial U}{\partial V}\right)_T \mathrm{d}V + \left(\frac{\partial U}{\partial T}\right)_V \mathrm{d}T\right\} + p\,\mathrm{d}V + V\left\{\left(\frac{\partial p}{\partial T}\right)_V \mathrm{d}T + \left(\frac{\partial p}{\partial V}\right)_T \mathrm{d}V\right\}$$

$$= \left\{\left(\frac{\partial U}{\partial V}\right)_T + p + V\left(\frac{\partial p}{\partial V}\right)_T\right\}\mathrm{d}V + \left\{\left(\frac{\partial U}{\partial T}\right)_V + V\left(\frac{\partial p}{\partial T}\right)_V\right\}\mathrm{d}T$$

$$= \left(\frac{\partial (U + pV)}{\partial V}\right)_T \mathrm{d}V + \left(\frac{\partial (U + pV)}{\partial T}\right)_V \mathrm{d}T\Bigg]$$

2.6 Show that $(\partial U/\partial p)_V = \kappa_T C_V/\alpha$.

2.7 Show that both $(\partial H/\partial p)_T$ and $(\partial H/\partial V)_T$ are zero for an ideal gas, starting with the condition that $(\partial U/\partial V)_T$ is zero.

2.8 Show that

$$\left(\frac{\partial C_V}{\partial V}\right)_T = T\left(\frac{\partial^2 p}{\partial T^2}\right)_V \quad \text{and} \quad \left(\frac{\partial C_p}{\partial p}\right)_T = -T\left(\frac{\partial^2 V}{\partial T^2}\right)_p$$

Use these results to show that C_p and C_V for an ideal gas depend only on temperature.

(*Hint:* Make use of the thermodynamic equations of state.)

2.9 Derive the relationship

$$\left(\frac{\partial C_p}{\partial p}\right)_T = -\mu_{\mathrm{JT}}\left(\frac{\partial C_p}{\partial T}\right)_p - C_p\left(\frac{\partial \mu_{\mathrm{JT}}}{\partial T}\right)_p$$

2.10 Given the following information:

$p, V, T \xrightarrow{\text{Isothermal reversible expansion}} p_1, V_1, T$

$p, V, T \xrightarrow{\text{Adiabatic reversible expansion}} p_1, V_2, T_1$

$p, V, T \xrightarrow{\text{Adiabatic reversible expansion}} p_2, V_1, T_2$

predict qualitatively, whether:

(i) $T_2 > T$ or $T_2 = T$ or $T_2 < T$

$p_2 > p_1$ or $p_2 = p_1$ or $p_2 < p_1$

(ii) $T_1 > T$ or $T_1 = T$ or $T_1 < T$

$V_2 > V_1$ or $V_2 = V_1$ or $V_2 < V_1$

2.11 (a) What is Joule coefficient? Show that it is given by the relation

$$\eta = -\frac{1}{C_V}\left(\frac{\partial U}{\partial V}\right)_T = \frac{1}{C_V}\left\{p - T\left(\frac{\partial p}{\partial T}\right)_V\right\}$$

(b) Form the relation of η given above derive the following facts:

(i) η is zero for an ideal gas.

(ii) η is positive if $(\partial U/\partial V)_T$ is negative and vice versa for real gases.

2.12 The temperature at which the Joule coefficient is zero is called the Joule inversion temperature. For a gas obeying the equation of state

$$\frac{pV}{RT} = 1 + \frac{B}{V}$$

where B is function of temperature alone, show that the Joule inversion temperature is that temperature for which B is a maximum. If B of a certain gas is given by the equation

$$B = b_1 - \frac{b_2}{T^2} - \frac{b_3}{T^3}$$

where b_1, b_2 and b_3 are positive constants, show that such a gas has no Joule inversion temperature.

2.13 (a) Show that for a van der Waals gas, the Joule coefficient is given by

$$\eta = -\frac{1}{C_{V,\mathrm{m}}}\left(\frac{RT}{V_\mathrm{m} - b} - p\right) = -\frac{1}{C_{V,\mathrm{m}}}\frac{a}{V_\mathrm{m}^2}$$

(b) Show that a van der Waals gas in Joule expansion exhibits:

(i) Heating if a is negative and (ii) cooling if a is positive What is the physical significance of a being a negative quantity? Under what conditions is the gas expected to have negative value of a?

2.14 (a) Show that

$$\mu_\mathrm{JT} = -\frac{1}{C_p}\left(\frac{\partial H}{\partial p}\right)_T = -\frac{1}{C_p}\left[\left(\frac{\partial U}{\partial p}\right)_T + \left(\frac{\partial (pV)}{\partial p}\right)_T\right]$$

$$\mu_\mathrm{JT} = -\frac{V}{C_p}\left(1 - T\frac{\partial \ln V}{\partial T}\right)$$

(b) Discuss under what conditions heating and cooling effects are produced in the Joule-Thomson experiment involving real gases.

2.15 (a) Show that an adiabatic process in which no work is performed is an example of constant energy process.

(b) Show that the change in enthalpy attending an isothermal change in state of an ideal gas must be equal to the change in energy.

2.16 The internal energy of a certain gas depends on volume as well as on temperature, and obeys the relation $(\partial U/\partial V)_T = a/V^2$, where a is constant. Prove that C_V for this gas depends only on temperature.

2.17 Show that for a gas obeying van der Waals equation of state

(a) $C_V = f(T)$ only and (b) $C_p = f(T, V)$

$$\left[\textit{Hint}: \text{Use the formulae } \left(\frac{\partial C_V}{\partial V}\right)_T = T\left(\frac{\partial^2 p}{\partial T^2}\right)_V \text{ and } \left(\frac{\partial C_p}{\partial p}\right)_T = -T\left(\frac{\partial^2 V}{\partial T^2}\right)_p\right]$$

2.18 Show that a process of expansion of an ideal gas which is isothermal as well as adiabatic must be a free-expansion process.

(b) Show that a reversible isothermal process can also be adiabatic only at $T = 0$ and that an irreversible isothermal process can also be adiabatic at $T \neq 0$.

2.19 (a) Show that, if a gas obeys the equation of state

$$pV = RT + \alpha p, \qquad \text{where } \alpha = f(T)$$

then

$$C_p - C_V = R\left(1 + \frac{p\,\mathrm{d}\alpha}{R\,\mathrm{d}T}\right)^2 \simeq R + 2p\left(\frac{\mathrm{d}\alpha}{\mathrm{d}T}\right)$$

(b) For hydrogen, $\alpha = 0.014\,2\ \mathrm{dm^3}$ at 0 °C and $0.014\,8\ \mathrm{dm^3}$ at 20 °C. What is $C_p - C_V$ per mole of hydrogen at 0.101 3 MPa pressure in the neighborhood of 10 °C?

(*Ans.* 8.320 J K^{-1} mol^{-1})

2.20 (a) Show that at moderate and low pressures the van der Waals equation for one mole of a gas may be written in the form

$$pV_\mathrm{m} = RT(1 - Bp) \ \text{ where } B = \frac{1}{RT}\left(\frac{a}{RT} - b\right)$$

(*Hint:* Multiply out the van der Waals equation, neglect the term ab/V_m^2 and replace V_m in a/V_m by RT/p.)

(b) Show that Eq. (2.6.14) can be derived directly from the equation of state given in (a).

(c) The equation of state given in (a) may be written as $pV_\mathrm{m} = RT + Ap$, where A is a function of temperature. Show that for this equation of state

$$\left(\frac{\partial C_p}{\partial p}\right)_T = -T\left(\frac{\partial^2 A}{\partial T^2}\right)_p$$

(d) Making use of the thermodynamic equation of state

$$\left(\frac{\partial U}{\partial V}\right)_T = T\left(\frac{\partial p}{\partial T}\right)_V - p$$

Show that for one mole of a van der Waals gas at moderate and low pressures

$$\left(\frac{\partial U}{\partial p}\right)_T = -\frac{a}{RT} \quad \text{and} \quad \left(\frac{\partial (pV_\mathrm{m})}{\partial p}\right)_T = b - \frac{a}{RT}$$

Further show that

$$\left(\frac{\partial H}{\partial p}\right)_T = b - 2\frac{a}{RT} \quad \text{and} \quad \mu_\mathrm{JT} = \frac{1}{C_{p,\mathrm{m}}}\left(\frac{2a}{RT} - b\right)$$

2.21 Show that the expression of $C_{p,\mathrm{m}} - C_{V,\mathrm{m}}$ derived in Problem 2.6.1 can be reduced to a form as given by Eq. (2.6.14).

2.22 If in Eq. (2.6.13), $(V_\mathrm{m} - b)^2$ is replaced by $V_\mathrm{m}^2 - 2V_\mathrm{m}b$, one gets

$$C_{p,\mathrm{m}} - C_{V,\mathrm{m}} = R + \frac{2ap}{RT^2} - \frac{4\,abp^2}{R^2T^3}$$

In the lower pressure range, the term $2ap/RT^2$ plays a predominant role with the result that $C_{p,\mathrm{m}} - C_{V,\mathrm{m}}$ increases linearly with pressure. At higher pressure, the term $-4abp^2/R^2T^3$ also becomes of increasing importance with the result that at a sufficiently high pressure $C_{p,\mathrm{m}} - C_{V,\mathrm{m}}$ attains a maximum and subsequently starts decreasing. The pressure at which $C_{p,\mathrm{m}} - C_{V,\mathrm{m}}$ has a maximum value can be obtained by setting the derivative $(\partial(C_{p,\,m} - C_{V,\,m})/\partial p)_T$ equal to zero. Determine the pressure at which $C_{p,m} - C_{V,m}$ is maximum for nitrogen at 25 °C. Also determine the value of $C_{p,m} - C_{V,\,m}$ at this pressure. Given: $a = 141$ dm^6 kPa mol^{-2} and $b = 39.1$ cm^3 mol^{-1}.

(*Ans.* 15.84 MPa)

2.23 One mole of an ideal gas undergoes the following transformations one after the other:

(a) Isothermal reversible expansion at T_2 from V_1 to V_2.

(b) Adiabatic reversible expansion from V_2 to V_3. Temperature drops from T_2 to T_1.

(c) Isothermal reversible compression at T_1 from V_3 to V_4.

(d) Adiabatic reversible compression from V_4 to V_1. Temperature rises from T_1 to T_2.

Derive the expressions for q, w, ΔU and ΔH for each process and for the overall process.

2.24 Show that

$$\left(\frac{\partial T}{\partial p}\right)_U = \frac{T\,\alpha V - p\kappa_T V}{C_p - p\alpha V}$$

Given that

$$\left(\frac{\partial H}{\partial p}\right)_T - V = -T\left(\frac{\partial V}{\partial T}\right)_p; \qquad H = U + pV$$

NUMERICAL PROBLEMS

Change in Energy with Temperature

2.1 Calculate the heat absorbed, ΔH and ΔU when one mole of nitrogen is heated from 298 K to 348 K at constant volume condition. Considering (a) $C_{p,\,m} = 29.13$ J K^{-1} mol^{-1} and (b) $C_{p,\,m}$/J K^{-1} mol^{-1} = 28.45 + 2.26 × 10^{-3} (T/K).

(*Ans.* (a) 1 040.8 J, 1 456.5 J, 1 040.8 J (b) 1 043.3 J, 1 459.0 J, 1 043.3 J)

Change in Enthalpy with Temperature

2.2 Calculate the heat absorbed, ΔH and ΔU when one mole of nitrogen is heated from 298 K to 348 K at 1 bar pressure considering (a) $C_{p,\,m} = 29.13$ J K^{-1} mol^{-1}, and (b) $C_{p,\,m}$/J K^{-1} mol^{-1} = 28.45 + 2.26 × 10^{-3} (T/K).

(*Ans.* (a) 1 456.5 J, 1 456.5 J, 1 040.8 J (b) 1 459.0 J, 1 459.0 J, 1 043.3 J)

Difference of Heat Capacities

2.3 The coefficient of cubic expansion α of sodium at 25 °C is 21.3 × 10^{-6} K^{-1} and the isothermal compressibility κ_T is 1.56 × 10^{-11} Pa^{-1}. The density is 0.97 g cm^{-3}. Calculate $C_p - C_V$ per mole of solid sodium at 25 °C. (*Ans.* 0.206 J K^{-1} mol^{-1})

2.4 The coefficient of cubic expansion α and the isothernal compressibility κ_T for metallic copper at 25 °C have values 49.2 × 10^{-6} K^{-1} and 7.747 × 10^{-6} MPa^{-1}, respectively. Density of Cu at 25 °C is 8.93 g cm^{-3}. Calculate $C_p - C_V$ per mole for Cu.

(*Ans.* 0.665 J K^{-1} mol^{-1})

2.5 Calculate the difference between $C_{p,m}$ and $C_{V,m}$ for CO_2 at 298 K and 10 bar pressure; the van der Waals constant, $a = 3.64$ dm^6 bar mol^{-2}. (*Ans.* 9.30 J K^{-1} mol^{-1})

Joule-Thomson Coefficient

2.6 For N_2, the van der Waals constants are $a = 141$ dm^6 kPa mol^{-2} and $b = 39.1$ cm^3 mol^{-1}; $C_{p,\,m} = 28.91$ J K^{-1} mol^{-1} and can be assumed to be independent of the temperature. Calculate

(i) μ_{JT} at 298 K and 0.101 3 MPa

(ii) $(\partial H/\partial p)_T$ at 298 K and 0.101 3 MPa

[*Ans.* (i) 2.58 K MPa^{-1} (ii) –74.7 J MPa^{-1}]

2.7 Compute the Joule-Thomson coefficient for carbon dioxide at 6.08 MPa pressure and 10 °C. Assume the gas to be ideal for the purpose of calculation of molar volume. The values of α and C_p are 1.3 × 10^{-2} K^{-1} and 3.82 J K^{-1} g^{-1}. (*Ans.* 6.32 K MPa^{-1})

2.8 (a) For CO_2(g) at 300 K and 0.101 MPa pressure $(\partial H/\partial p)_T = 421.19$ J MPa^{-1} mol^{-1} and $C_p = 37.42$ J K^{-1} mol^{-1}, calculate μ_{JT} of the gas for the given temperature and pressure conditions. (*Ans.* 11.25 K MPa^{-1})

(b) At 300 K and at pressures 0-6.06 MPa, the μ_{JT} of N_2(g) can be represented by the equation

$$\mu_{JT}/\text{K MPa}^{-1} = 0.140 - 2.556 \times 10^{-2}\ (p/\text{MPa})$$

Assuming this equation to be independent of temperature near 300 K, find the temperature drop which may be expected in Joule-Thomson expansion of the gas from 6.06 MPa to 1.01 MPa. (*Ans.* –0.255 K)

2.9 Calculate the inversion temperature for CO_2 gas. Given: $a = 3.64$ dm^6 bar mol^{-2} and $b = 0.042\ 67$ dm^3 mol^{-1}. (*Ans.* 2 052 K)

2.10 Calculate the value of ΔH for the isothermal expansion at 300 K of 1 mol of carbon dioxide from 1 bar to 50 bar. Given: $a = 3.64$ dm^6 bar mol^{-2} and $b = 0.042\ 67$ dm^3 mol^{-1}. (*Ans.* –1.221 kJ mol^{-1})

Isothermal Volume Change of an Ideal Gas

2.11 (a) 7.0 g of N_2 at 25 °C is expanded isothermally from an initial pressure of 0.505 MPa to a final pressure of 0.202 MPa against a constant external pressure of 0.101 MPa. Calculate q, w, ΔU and ΔH.

(b) The same amount of N_2 is expanded isothermally between the same initial and final volumes, but this time the expansion is carried out reversibly. Calculate q, w, ΔU and ΔH.

(*Ans.* (a) $\Delta U = \Delta H = 0$; $q = -w = 185.9$ J, (b) $\Delta U = \Delta H = 0$; $q = -w = 567.8$ J)

2.12 Calculate the work done when 1 mol of zinc dissolves in hydrochloric acid at 273.15 K in (i) an open beaker and (ii) a closed beaker at 300 K. (*Ans.* –2 271.1 J, zero)

(*Hint:* Evolved gas is driven against a constant atmospheric pressure.)

Adiabatic Reversible Volume Change of an Ideal Gas

2.13 Hydrogen gas is expanded reversibly and adiabatically from a volume of 1.43 dm^3 at a pressure of 0.303 MPa and temperature of 25 °C, until the volume is 2.86 dm^3. The heat capacity C_p of hydrogen can be taken to be 28.87 J K^{-1} mol^{-1}.

(a) Calculate the pressure and temperature of the gas, assumed to be ideal, after the expansion.

(b) Calculate q, w ΔU and ΔH for the gas.

(*Ans.* (a) 0.115 MPa, 226 K; (b) $q = 0$; $w = -259.4$ J; $\Delta U = -259.4$ J; $\Delta H = -364.4$ J)

2.14 0.35 mol of an ideal monatomic gas is expanded adiabatically from a volume of 1 dm^3 at 400 K to a volume of 5 dm^3 against a constant external pressure of 50.35 kPa. What is the final temperature of the gas, and what is its enthalpy change in the process?

(*Ans.* 353.84 K, –335.8 J)

Isothermal Volume Change of a van der Waals Gas

2.15 (a) For a certain gas the van der Waals constants are

$$a = 677.86 \text{ dm}^6 \text{ kPa mol}^{-2} \quad \text{and} \quad b = 0.057 \text{ dm}^3 \text{ mol}^{-1}$$

What will be the maximum work performed in the expansion of 2 mol of the gas from 4 to 40 dm^3 at 300 K? (*Ans.* $w_{max} = -11.05$ kJ)

(b) If C_V of the gas is 29.29 J K^{-1}, mol^{-1}, what will be ΔU and ΔH for a process involving the compression of 5 mol of the above gas from a volume of 100 dm^3 at 300 K to a volume of 10 dm^3 at 400 K? (*Ans.* $\Delta U = 13.138$ kJ)

2.16 One mole of a van der Waals gas expands isothermally and reversibly from a volume of 1 dm^3 to 30 dm^3 at 0 °C. Calculate q, w ΔU and ΔH. Given: a = 658.6 dm^6 kPa mol^{-2} and b = 0.056 dm^3 mol^{-1}.

(*Ans.* w = –7 209.68 J; ΔU = 636.647 J; q = 7 846.33 J; ΔH = 1 142.9 J)

Adiabatic Volume Change of a van der Waals Gas

2.17 One mole of a van der Waals gas undergoes an adiabatic free expansion from 1 dm^3 to 10 dm^3. Calculate the change in temperature of the gas. Given: $C_{V,\mathrm{m}}$ = 20.8 J K^{-1} mol^{-1} and a = 0.138 N m^4 mol^{-2}. (*Ans.* –5.97 K)

2.18 One mole of nitrogen undergoes adiabatic reversible expansion from 1 dm^3 to 10 dm^3 with initial temperature of 273 K. Calculate q, w, ΔU and ΔH if the gas is considered a van der Waals gas. Given: a = 141.8 dm^6 kPa mol^{-2}, b = 0.039 dm^3 mol^{-1}, $C_{V,\mathrm{m}}$ = 20.81 J K^{-1} mol^{-1}. (*Ans.* 0, –3 321.5 J mol^{-1}, –3 321.5 J mol^{-1}, –4 661.0 J mol^{-1})

2.19 One mole of nitrogen undergoes adiabatic expansion from 1 dm^3 to 10 dm^3 against an external pressure of 1 bar. The initial temperature of the gas was 273 K. Calculate q, w, ΔU and ΔH if the gas behaves like a van der Waals gas. Given: a = 140.8 dm^6 kPa mol^{-2}, b = 0.039 dm^3 mol^{-1} and $C_{V,\mathrm{m}}$ = 20.81 J K^{-1} mol^{-1}.

(*Ans.* 0. –900 J mol^{-1}, –900 J mol^{-1}, –1 268.3 J K^{-1} mol^{-1})

Miscellaneous Problems

2.20 One mole of a liquid is confined in a piston and cylinder at 101.33 kPa pressure. The coefficient of cubic expansion of the liquid is 10^{-3} K^{-1}. Calculate the work done when the temperature is raised from 300 K to 500 K. The pressure is kept constant and no liquid evaporates. The molar volume of the liquid at 0 °C is 100 cm^3 mol^{-1}.

(*Ans.* 2.026 J)

2.21 (a) A system with an initial volume of 22.4 dm^3 is compressed adiabatically until its volume is 11.2 dm^3. During this process 1 350 joules of work is performed on the system, and the temperature rises from 0 °C to 160 °C. What is the change in the internal energy during this process?

(b) The same initial system is heated as constant volume to 160 °C whereby 1 320 joules of heat must be added. What is the change in the internal energy of the system during this process?

(c) The system is now compressed at this constant temperature of 160 °C to a volume of 11.2 dm^3, whereby 2 550 J of heat flow out of the system. What is the change in internal energy? How much work is done on the system?

[*Ans.* (a) ΔU = 1 350 J, (b) ΔU = 1 320 J, (c) ΔU = 30 J, w = 2 580 J]

2.22 The equation of state for one mole of gas is $pV = RT + Bp$ where B is a constant and is independent of temperature. Starting with one mole of gas at 300 K and a pressure of 1.013 MPa, consider the following quasi-static processes:

(a) an adiabatic expansion to 20 dm^3

(b) an isobaric expansion to 20 dm^3

(c) an isothermal expansion to 20 dm^3

(d) an isochoric decrease in pressure to 0.101 3 MPa

If the value of B is 1.5 dm^3, C_p = 29.29 J K^{-1} mol^{-1} and C_V = 20.92 J K^{-1} mol^{-1} for this gas, find:

(i) Relations for calculating q, w, ΔU and ΔH for each of the above processes.

(ii) The numerical values of q, w, ΔU and ΔH for each of the above processes.

[*Ans.* (a) q = 0; $\Delta U = w$ = –3.46 kJ; ΔH = –6.274 kJ
(b) w = –16.25 kJ; q = 57.094 kJ; ΔU = 40.846 kJ; ΔH = 57.095 kJ
(c) q = 5.032 kJ; w = –5.032 kJ; ΔU = 0; ΔH = –1.32 kJ
(d) q = –5.638 kJ; w = 0; ΔU = –5.638 kJ; ΔH = –9.248 kJ]

Annexure I Expressions Relating Reduced Inversion Temperature and Reduced Pressure

By definition, we have

1. Reduced pressure, $p_r = p/p_c$
2. Reduced inversion temperature, $T_{i(r)} = T_i/T_c$

where p_c and T_c are critical pressure and critical temperature, respectively.

The inversion temperature is defined as

$$T_i = \frac{1}{\alpha_i} = \frac{1}{(1/V_m)(\partial V_m/\partial T_i)_p} = \frac{V_m}{(\partial V_m/\partial T_i)_p} \qquad \text{(Eq. 2.8.9)}$$

The expression of maximum inversion temperature is derived under the approximation $p \to 0$ (i.e. $V_m \to \infty$).

The expressions of T_i and $T_{i,\,max}$ are derived for each of the following equations:

(i) Van der Waals equation
(ii) Berthelot equation
(iii) Dieterici equation
(iv) Redlich-Kwong equation

The expression between p_r and $T_{i(r)}$ is derived by eliminating V_r in the reduced equation of state.

Van der Waals Equation of State

The van der Waals equation of state is

$$\left(p + \frac{a}{V_m^2}\right)(V_m - b) = RT \qquad \text{(AI.1)}$$

i.e.

$$pV_m - pb + \frac{a}{V_m} - \frac{ab}{V_m^2} = RT \qquad \text{(AI.2)}$$

Expression of Inversion Temperature

Differentiating Eq. (AI.2) with respect to T at constant p, we get

$$p\left(\frac{\partial V_m}{\partial T}\right)_p - \frac{a}{V_m^2}\left(\frac{\partial V_m}{\partial T}\right)_p + \frac{2ab}{V_m^3}\left(\frac{\partial V_m}{\partial T}\right)_p = R$$

This gives

$$\left(\frac{\partial V_m}{\partial T}\right)_p = \frac{R}{p - (a/V_m^2) + (2ab/V_m^3)} \qquad \text{(AI.3)}$$

Substituting Eq. (AI.3) in the expression of inversion temperature $[T_i = V_m/(\partial V_m/\partial T)_p]$, we get

$$T_i = \frac{pV_m - (a/V_m) + (2ab/V_m^2)}{R}$$

Eliminating pV_m by using Eq. (AI.2), we get

$$T_i = \frac{RT_i + pb - (2a/V_m) + (3ab/V_m^2)}{R}$$

This is reduced to

$$pb = \frac{2a}{V_m} - \frac{3ab}{V_m^2} \quad \text{;(applicable at inversion temperature)} \qquad \text{(AI.4a)}$$

or

$$\frac{2a}{V_m p} - \frac{3abp}{V_m^2 p^2} - b = 0 \qquad \text{(AI.4b)}$$

Under the condition of low pressure, we may replace $pV_m \simeq RT_i$. Hence

$$\frac{2a}{RT_i} - \frac{3abp}{R^2 T_i^2} - b = 0 \qquad \text{(AI.5a)}$$

Note: Equation (AI.5a) is identical to Eq. (2.9.5). Also, under the condition of low pressure, the expression $3abp/R^2T_i^2$ may be ignored in comparison to $2a/RT_i$. Hence, Eq. (AI.5a) is reduced to

$$\frac{2a}{RT_i} - b = 0 \quad \text{i.e.} \quad T_i = \frac{2a}{Rb} \qquad \text{(AI.5b)}$$

Expression of Maximum Reduced Inversion Temperature

Replacing van der Waals constants a and b in Eq. (AI.5b) in terms of critical constants, we get

$$T_i = \frac{2(3p_c V_c^2)}{R(V_c/3)} = 18\left(\frac{p_c V_c}{R}\right)$$

But $p_c V_c / RT_c = 3/8$ for van der Waals gas, hence

$$T_i = 18\left(\frac{3T_c}{8}\right) = \frac{27}{4} T_c$$

or

$$T_{i(r)} = \frac{T_i}{T_c} = \frac{27}{4} \qquad \text{(AI.6)}$$

Expression Relating p_r and $T_{i(r)}$

We start with reduced van der Waals equation at reduced inversion temperature:

$$\left(p_r + \frac{3}{V_r^2}\right)(3V_r - 1) = 8T_{i(r)} \qquad \text{(AI.7)}$$

Elimination of V_r in terms of $T_{i(r)}$ in Eq. (AI.7) provides the required relation between p_r and $T_{i(r)}$.

Relation Between V_r and $T_{i(r)}$

We start with Eq. (AI.4a):

$$pb = \frac{2a}{V_m} - \frac{3ab}{V_m^2} \quad \text{;(applicable at inversion temperature)}$$

Eliminating p by using van der Waals equation of state at inversion temperature, we get

$$\left(\frac{RT_\mathrm{i}}{V_\mathrm{m}-b}-\frac{a}{V_\mathrm{m}^2}\right)b=\frac{2a}{V_\mathrm{m}}-\frac{3ab}{V_\mathrm{m}^2}$$

This gives

$$\frac{RT_\mathrm{i}b}{V_\mathrm{m}-b}=\frac{2a}{V_\mathrm{m}}-\frac{2ab}{V_\mathrm{m}^2}\quad \text{i.e.}\quad RT_\mathrm{i}b\simeq 2a\left(1-\frac{b}{V_\mathrm{m}}\right)^2 \tag{AI.8}$$

Replacing a and b in terms of critical constants and expressing V_m and T_i in terms of reduced variables, we get

$$R(T_\mathrm{i(r)}T_\mathrm{c})\left(\frac{V_\mathrm{c}}{3}\right)=2(3p_\mathrm{c}V_\mathrm{c}^2)\left(1-\frac{V_\mathrm{c}/3}{V_\mathrm{r}V_\mathrm{c}}\right)^2$$

Cancelling the common terms and setting $p_\mathrm{c}V_\mathrm{c}/RT_\mathrm{c}=3/8$, we get

$$\frac{T_\mathrm{i(r)}}{3}=\frac{9}{4}\left(1-\frac{1}{3V_\mathrm{r}}\right)^2$$

which on rearranging gives

$$V_\mathrm{r}=\frac{\sqrt{3}}{3\sqrt{3}-2\sqrt{T_\mathrm{i(r)}}} \tag{AI.9}$$

Required Relation Between p_r and $T_\mathrm{i(r)}$

Substituting Eq. (AI.9) in Eq. (AI.7), we get

$$\{p_\mathrm{r}+(3\sqrt{3}-2\sqrt{T_\mathrm{i(r)}})^2\}\left(\frac{3\sqrt{3}}{3\sqrt{3}-2\sqrt{T_\mathrm{i(r)}}}-1\right)=8T_\mathrm{i(r)}$$

or
$$p_\mathrm{r}+(3\sqrt{3}-2\sqrt{T_\mathrm{i(r)}})^2=8T_\mathrm{i(r)}\left(\frac{3\sqrt{3}-2\sqrt{T_\mathrm{i(r)}}}{2\sqrt{T_\mathrm{i(r)}}}\right)$$

or
$$p_\mathrm{r}+\{27+4T_\mathrm{i(r)}-12(\sqrt{3T_\mathrm{i(r)}})\}=12(\sqrt{3T_\mathrm{i(r)}})-8T_\mathrm{i(r)}$$

or
$$p_\mathrm{r}=24(\sqrt{3T_\mathrm{i(r)}})-12T_\mathrm{i(r)}-27 \tag{AI.10}$$

Equation (AI.10) is the required expression relating p_r and $T_\mathrm{i(r)}$.

Maximum and Minimum Reduced Inversion Temperatures

The expression of $T_\mathrm{i(r)}$ in terms of p_r is

$$T_\mathrm{i(r)}-2\sqrt{3}\sqrt{T_\mathrm{i(r)}}+(27+p_\mathrm{r})/12=0 \tag{AI.11}$$

This gives

$$\sqrt{T_\mathrm{i(r)}}=\frac{2\sqrt{3}\pm\sqrt{(2\sqrt{3})^2-4(27+p_\mathrm{r})/12}}{2}$$
$$=\sqrt{3}\pm\sqrt{3-(27+p_\mathrm{r})/12} \tag{AI.12}$$

Maximum value The value of maximum $\sqrt{T_{i(r)}}$ at $p_r = 0$ is

$$\sqrt{T_{i(r)}} = \sqrt{3} + \sqrt{3 - 27/12} = 1.732 + 0.866 = 2.598$$

Hence $T_{i(r)} = (2.598)^2 = 6.75$ (Same as Eq. A1.6) (AI.13)

Minimum value The value of minimum $\sqrt{T_{i(r)}}$ at $p_r = 0$ is

$$\sqrt{T_{i(r)}} = \sqrt{3} - \sqrt{3 - 27/12} = 1.732 - 0.866 = 0.866$$

Hence $T_{i(r)} = (0.866)^2 = 0.75$ (AI.14)

Peak Reduced Pressure and Corresponding Reduced Inversion Temperature

For each reduced pressure, Eq. (AI.11) provides two reduced inversion temperatures. The difference between these two temperatures decreases with increase in the reduced pressure. Eventually, a stage is reached where two temperatures become identical. The reduced pressure corresponding to these two identical temperatures is known as peak reduced pressure. The value of the latter and the corresponding reduced temperature can be obtained by setting $dp_r/dT_{i(r)} = 0$.

From Eq. (AI.10), we get

$$\frac{dp_r}{dT_{i(r)}} = 24\sqrt{3}\left(\frac{1}{2}T_{i(r)}^{-1/2}\right) - 12 = 0$$

This gives $T_{i(r)} = 3$ (AI.15a)

Substituting this $T_{i(r)}$ in Eq. (AI.10), we get

$$p_r = 24\,(3 \times 3)^{1/2} - 12 \times 3 - 27 = 72 - 36 - 27 = 9 \qquad \text{(AI.15b)}$$

Display of Plot of $T_{i(r)}$ Versus p_r

Figure AI.1 displays the plot of T_r versus p_r.

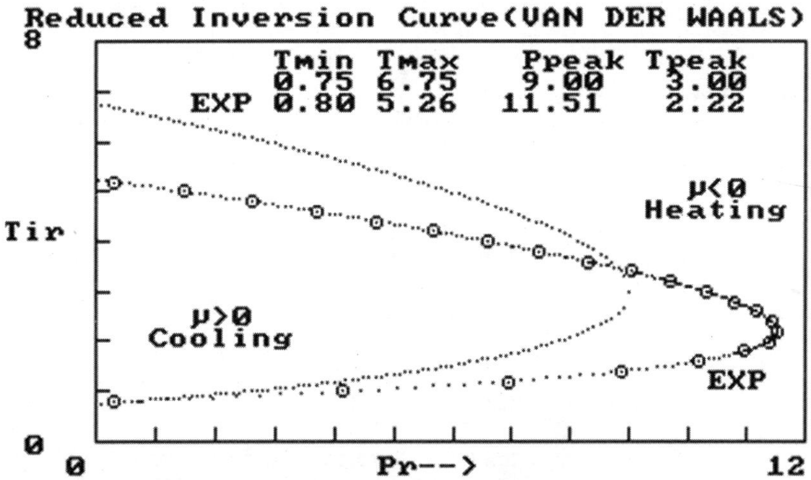

Fig. AI.1

Berthelot Equation of State

The Berthelot equation of state is

$$\left(p+\frac{a}{TV_\mathrm{m}^2}\right)(V_\mathrm{m}-b)=RT \tag{AI.16}$$

i.e.
$$pV_\mathrm{m}-pb+\frac{a}{TV_\mathrm{m}}-\frac{ab}{TV_\mathrm{m}^2}=RT \tag{AI.17}$$

Expression of Inversion Temperature

Differentiating Eq. (AI.17) with respect to T at constant p, we get

$$p\left(\frac{\partial V_\mathrm{m}}{\partial T}\right)_p-\frac{a}{T^2V_\mathrm{m}}-\frac{a}{TV_\mathrm{m}^2}\left(\frac{\partial V_\mathrm{m}}{\partial T}\right)_p+\frac{ab}{T^2V_\mathrm{m}^2}+\frac{2ab}{TV_\mathrm{m}^3}\left(\frac{\partial V_\mathrm{m}}{\partial T}\right)_p=R$$

This gives

$$\left(\frac{\partial V_\mathrm{m}}{\partial T}\right)_p=\frac{R+(a/T^2V_\mathrm{m})-(ab/T^2V_\mathrm{m}^2)}{p-(a/TV_\mathrm{m}^2)+(2ab/TV_\mathrm{m}^3)} \tag{AI.18}$$

Substituting Eq. (AI.18) in the expression of inversion temperature ($T_\mathrm{i}=V_\mathrm{m}/(\partial V_\mathrm{m}/\partial T)_p$), we get

$$T_\mathrm{i}=\frac{pV_\mathrm{m}-(a/T_\mathrm{i}V_\mathrm{m})+(2ab/T_\mathrm{i}V_\mathrm{m}^2)}{R+(a/T_\mathrm{i}^2V_\mathrm{m})-(ab/T_\mathrm{i}^2V_\mathrm{m}^2)} \tag{AI.19}$$

Eliminating pV_m by using Eq. (AI.17), we get

$$T_\mathrm{i}=\frac{RT_\mathrm{i}+pb-2(a/T_\mathrm{i}V_\mathrm{m})+(3ab/T_\mathrm{i}V_\mathrm{m}^2)}{R+(a/T_\mathrm{i}^2V_\mathrm{m})-(ab/T_\mathrm{i}^2V_\mathrm{m}^2)}$$

This is reduced to

$$pb=\frac{3a}{T_\mathrm{i}V_\mathrm{m}}-\frac{4ab}{T_\mathrm{i}V_\mathrm{m}^2}=\frac{a}{T_\mathrm{i}V_\mathrm{m}}\left(3-\frac{4b}{V_\mathrm{m}}\right) \tag{AI.20}$$

Maximum Value of T_i

Under the condition of low pressure (i.e. large volume), the term $4b/V_\mathrm{m}$ may be ignored in comparison to 3. Thus, we have

$$pb=\frac{3a}{T_\mathrm{i}V_\mathrm{m}}$$

Replacing $V_\mathrm{m}\simeq RT_\mathrm{i}/p$, we get

$$pb=\frac{3ap}{RT_\mathrm{i}^2}$$

This gives
$$T_\mathrm{i}=\left(\frac{3a}{Rb}\right)^{1/2} \tag{AI.21}$$

Equation (AI.21) provides the maximum value of T_i.

Expression of Maximum Reduced Inversion Temperature

Replacing Berthelot constants a and b in Eq. (AI.21) in terms of critical constants, we get

$$T_\mathrm{i}^2=\frac{3(9RT_\mathrm{c}^2V_\mathrm{c}/8)}{R(V_\mathrm{c}/3)}=\frac{81}{8}T_\mathrm{c}^2$$

Hence, $T^2_{i(r)} = \left(\frac{T_i}{T_c}\right)^2 = \frac{81}{8}$

or $$T_{i(r)} = \frac{9}{2\sqrt{2}} \quad \text{(AI.22)}$$

Expression Relating p_r and $T_{i(r)}$

We start with reduced Berthelot equation of state at reduced inversion temperature

$$\left(p_r + \frac{3}{V_r^2 T_{i(r)}}\right)(3V_r - 1) = 8T_{i(r)} \quad \text{(AI.23)}$$

Elimination of V_r in terms of $T_{i(r)}$ in Eq. (AI.23) provides the required relation between p_r and $T_{i(r)}$.

Relation Between V_r and $T_{i(r)}$

We start with Eq. (AI.20)

$$pb = \frac{a}{T_i V_m}\left(3 - \frac{4b}{V_m}\right)$$

Eliminating p by using Berthelot equation of state at inversion temperature, we get

$$\left(\frac{RT_i}{V_m - b} - \frac{a}{T_i V_m^2}\right)b = \frac{a}{T_i V_m}\left(3 - \frac{4b}{V_m}\right)$$

or $$\frac{RT_i b}{V_m - b} = \frac{3a}{T_i V_m}\left(1 - \frac{b}{V_m}\right)$$

or $$RT_i b = \frac{3a}{T_i}\left(1 - \frac{b}{V_m}\right)^2 \quad \text{(AI.24)}$$

Replacing a and b in terms of critical constants and expressing V_m and T_i in terms of reduced variables, we get

$$RT_{i(r)}T_c\left(\frac{V_c}{3}\right) = \frac{3(9RT_c^2V_c/8)}{T_{i(r)}T_c}\left(1 - \frac{V_c/3}{V_r V_c}\right)^2$$

Cancelling the common terms, we get

$$\frac{T_{i(r)}}{3} = \frac{27}{8T_{i(r)}}\left(1 - \frac{1}{3V_r}\right)^2$$

which on rearranging, we get

$$V_r = \frac{1}{3 - (\sqrt{8}/3)T_{i(r)}} \quad \text{(AI.25)}$$

Required Relation Between P_r and $T_{i(r)}$

Substituting Eq. (AI.25) in Eq. (AI.23), we get

$$\left[p_r + \frac{3}{T_{i(r)}}\left(3 - \frac{\sqrt{8}}{3}T_{i(r)}\right)^2\right]\left[\frac{3}{3 - (\sqrt{8}/3)T_{i(r)}} - 1\right] = 8T_{i(r)}$$

or $$p_\mathrm{r} + \frac{3}{T_\mathrm{i(r)}}\left(9 + \frac{8}{9}T_\mathrm{i(r)}^2 - 2\sqrt{8}T_\mathrm{i(r)}\right) = 8T_\mathrm{i(r)}\left(\frac{3 - (\sqrt{8}/3)T_\mathrm{i(r)}}{(\sqrt{8}/3)T_\mathrm{i(r)}}\right)$$

or $$p_\mathrm{r} + \frac{27}{T_\mathrm{i(r)}} + \frac{8}{3}T_\mathrm{i(r)} - 6\sqrt{8} = 9\sqrt{8} - 8T_\mathrm{i(r)}$$

or $$p_\mathrm{r} = -\frac{32}{3}T_\mathrm{i(r)} - \frac{27}{T_\mathrm{i(r)}} + 15\sqrt{8} \qquad \text{(AI.26)}$$

Equation (AI.26) is the required expression relating p_r and $T_\mathrm{i(r)}$.

Maximum and Minimum Reduced Inversion Temperature

Equation (AI.26) may be rewritten as

$$T_\mathrm{i(r)}^2 - \frac{3}{32}(15\sqrt{8} - p_\mathrm{r})T_\mathrm{i(r)} + \frac{3}{32} \times 27 = 0$$

i.e. $$T_\mathrm{i(r)}^2 - xyT_\mathrm{i(r)} + \frac{3}{32} \times 27 = 0 \qquad \text{(AI.27)}$$

where $x = 3/32$ and $y = 15\sqrt{8} - p_\mathrm{r}$

Solution of Eq. (AI.27) is

$$T_\mathrm{i(r)} = \frac{xy \pm \sqrt{(xy)^2 - 81/8}}{2} \qquad \text{(AI.28)}$$

The maximum and reduced inversion temperatures are obtained by setting $p_\mathrm{r} = 0$ in Eq. (AI.28). The values are:

$$T_\mathrm{i(r)} = \frac{45/(4\sqrt{8}) \pm \sqrt{(45/4\sqrt{8})^2 - 81/8}}{2}$$

$$= \frac{1}{2} \cdot \frac{1}{\sqrt{8}}\left[\frac{45}{4} \pm \sqrt{\frac{45^2}{16} - 81}\right]$$

Thus

Maximum value: $T_\mathrm{i(r)} = \frac{1}{2} \cdot \frac{1}{\sqrt{8}}[11.25 + 6.75] = \frac{9}{\sqrt{8}} = 3.18$ (Same as Eq. AI.22)

Minimum value: $T_\mathrm{i(r)} = \frac{1}{2} \cdot \frac{1}{\sqrt{8}}(11.25 - 6.75] = \frac{2.25}{\sqrt{8}} = 0.80$

Peak Reduced Pressure and the Corresponding Reduced Inversion Temperature

By setting $\mathrm{d}p_\mathrm{r}/\mathrm{d}T_\mathrm{i(r)} = 0$ from Eq. (AI.26), we have

$$\frac{\mathrm{d}p_\mathrm{r}}{\mathrm{d}T_\mathrm{i(r)}} = -\frac{32}{3} + \frac{27}{T_\mathrm{i(r)}^2} = 0$$

This gives

$$T_\mathrm{i(r)} = \frac{9}{4\sqrt{2}} = 1.59 \qquad \text{(AI.27)}$$

Substituting this $T_{i(r)}$ in Eq. (AI.26), we get

$$p_r = -\frac{32}{3}\left(\frac{9}{4\sqrt{2}}\right) - \frac{27}{(9/4\sqrt{2})} + 15\sqrt{8}$$

$$= -12\sqrt{2} - 12\sqrt{2} + 30\sqrt{2} = 6\sqrt{2} = 8.48 \qquad \text{(AI.28)}$$

Figure (AI.2) displays the plot of $T_{i(r)}$ versus p_r.

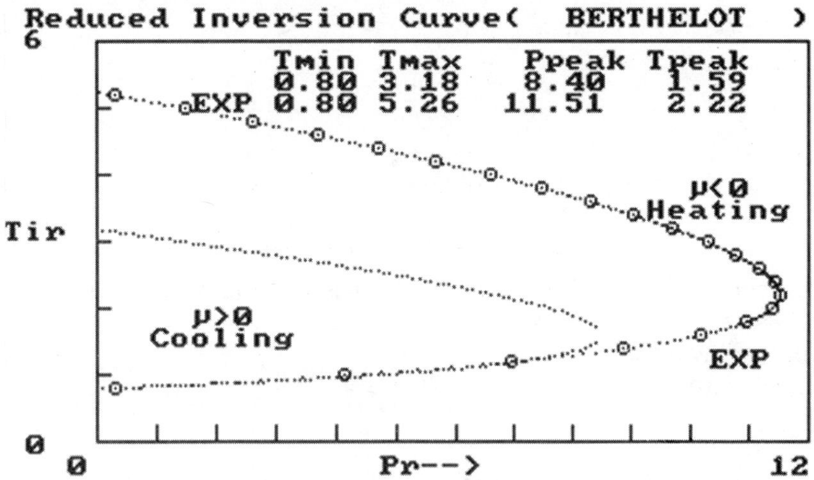

Fig. AI.2

Dieterici Equation of State

The Dieterici equation of state is

$$p(V_m - b) = RT\,e^{-a/V_m RT} \qquad \text{(AI.29)}$$

Expression of Inversion Temperature

Differentiating Eq. (AI.29) with respect to T at constant p, we get

$$p\left(\frac{\partial V_m}{\partial T}\right)_p = \left[R + RT\left\{\frac{a}{V_m^2 RT}\left(\frac{\partial V_m}{\partial T}\right)_p + \frac{a}{V_m RT^2}\right\}\right] e^{-a/V_m RT}$$

This gives

$$\left(\frac{\partial V_m}{\partial T}\right)_p = \frac{(R + a/V_m T)e^{-a/V_m RT}}{p - (a/V_m^2)e^{-a/V_m RT}} \qquad \text{(AI.30)}$$

Substituting Eq. (AI.30) in the expression of inversion temperature ($T_i = V_m/(\partial V_m/\partial p)_T$), we get

$$T_i = \frac{pV_m - (a/V_m)e^{-a/V_m RT_i}}{(R + a/V_m T_i)e^{-a/V_m RT_i}} \qquad \text{(AI.31)}$$

Eliminating pV_m by using Eq. (AI.29), we get

$$T_i = \frac{pb + (RT_i - a/V_m)e^{-a/V_m RT_i}}{(R + a/V_m T_i)e^{-a/V_m RT_i}}$$

which on rearranging, we get

$$pb = \left(\frac{2a}{V_\mathrm{m}}\right)\mathrm{e}^{-a/V_\mathrm{m}RT_\mathrm{i}} \tag{AI.32}$$

Under the condition of low pressure, Eq. (AI.29) at inversion temperature may be written as

$$pV_\mathrm{m} \approx RT_\mathrm{i}\,\mathrm{e}^{-a/V_\mathrm{m}RT_\mathrm{i}}$$

With this, Eq. (AI.32) becomes

$$b = \frac{2a}{RT_\mathrm{i}}$$

or $$T_\mathrm{i} = 2\left(\frac{a}{Rb}\right) \tag{AI.33}$$

Equation (AI.33) provides the maximum value of inversion temperature.

Expression of Maximum Reduced Inversion Temperature

Replacing Dieterici constants a and b in Eq. (AI.33) in terms of critical constants, we get

$$T_\mathrm{i} = 2\left(\frac{2V_\mathrm{c}RT_\mathrm{c}}{RV_\mathrm{c}/2}\right) = 8T_\mathrm{c}$$

or $$T_\mathrm{i(r)} = \frac{T_\mathrm{i}}{T_\mathrm{c}} = 8 \tag{AI.34}$$

Expression Relating p_r and $T_\mathrm{i(r)}$

We start with reduced Dieterici equation at reduced inversion temperature:

$$p_\mathrm{r}\,\mathrm{e}^{2/V_\mathrm{r}T_\mathrm{i(r)}}\left(V_\mathrm{r} - \frac{1}{2}\right) = \mathrm{e}^2\,\frac{T_\mathrm{i(r)}}{2} \tag{AI.35}$$

Elimination of V_r in terms of $T_\mathrm{i(r)}$ in Eq. (AI.35) provides the required relation between p_r and $T_\mathrm{i(r)}$.

Relation Between V_r and $T_\mathrm{i(r)}$

We start with Eq. (AI.32)

$$pb = \left(\frac{2a}{V_\mathrm{m}}\right)\mathrm{e}^{-a/V_\mathrm{m}RT_\mathrm{i}}$$

Eliminating p by using Dieterici equation of state at inversion temperature, we get

$$\left(\frac{RT_\mathrm{i}}{V_\mathrm{m} - b}\mathrm{e}^{-a/V_\mathrm{m}RT_\mathrm{i}}\right)b = \left(\frac{2a}{V_\mathrm{m}}\right)\mathrm{e}^{-a/V_\mathrm{m}RT_\mathrm{i}}$$

This gives

$$RT_\mathrm{i}b = 2a\left(1 - \frac{b}{V_\mathrm{m}}\right)$$

Replacing a and b in terms of critical constants and expressing V_m and T_i in terms of reduced variables, we get

$$R(T_\mathrm{i(r)}T_\mathrm{c})\left(\frac{V_\mathrm{c}}{2}\right) = 2(2RV_\mathrm{c}T_\mathrm{c})\left(1 - \frac{V_\mathrm{c}/2}{V_\mathrm{r}V_\mathrm{c}}\right)$$

This gives

$$\frac{T_{i(r)}}{2} = 4\left(1 - \frac{1}{2V_r}\right)$$

which on rearranging gives

$$V_r = \frac{4}{8 - T_{i(r)}} \tag{AI.36}$$

Required Relation Between p_r and $T_{i(r)}$

Substituting Eq. (AI.36) in Eq. (AI.35), we get

$$p_r e^{(8 - T_{i(r)}/2T_{i(r)})}\left(\frac{4}{8 - T_{i(r)}} - \frac{1}{2}\right) = e^2 \frac{T_{i(r)}}{2}$$

This gives

$$p_r = (8 - T_{i(r)})e^{(5T_{i(r)} - 8)/2T_{i(r)}} \tag{AI.37}$$

Equation (AI.37) is the required expression relating p_r and $T_{i(r)}$.

Maximum and Minimum Reduced Inversion Temperature

The maximum and minimum reduced inversion temperature is obtained by setting $p_r = 0$. Thus, we have

$$(8 - T_{i(r)})e^{(5T_{i(r)} - 8)/2T_{i(r)}} = 0$$

Hence

Maximum $T_{i(r)} = 8$ (Same as Eq. AI.34)

and Minimum $T_{i(r)} = 0$

Peak Reduced Pressure and Corresponding Reduced Inversion Temperature

By setting $dp_r/dT_{i(r)} = 0$ from Eq. (AI.37), we get

$$\frac{dp_r}{dT_{i(r)}} = \left[-1 + (8 - T_{i(r)})\left(\frac{4}{T_{i(r)}^2}\right)\right]e^{(5T_{i(r)} - 8)/2T_{i(r)}}$$

Hence $-1 + (8 - T_{i(r)})\left(\dfrac{4}{T_{i(r)}^2}\right) = 0$

or $T_{i(r)}^2 + 4T_{i(r)} - 32 = 0$

Solving for $T_{i(r)}$, we get

$$T_{i(r)} = \frac{-4 + \sqrt{16 + 4 \times 32}}{2} = \frac{-4 + 12}{2} = 4 \tag{AI.38}$$

Substituting $T_{i(r)} = 4$ in Eq. (AI.37), we get

$$p_r = (8 - 4)e^{(5 \times 4 - 8)/(2 \times 4)}$$

$$= 4e^{3/2} = 17.93 \tag{AI.39}$$

Display of Graph of $T_{i(r)}$ Versus p_r

Figure (AI.3) displays the plot of $T_{i(r)}$ versus p_r.

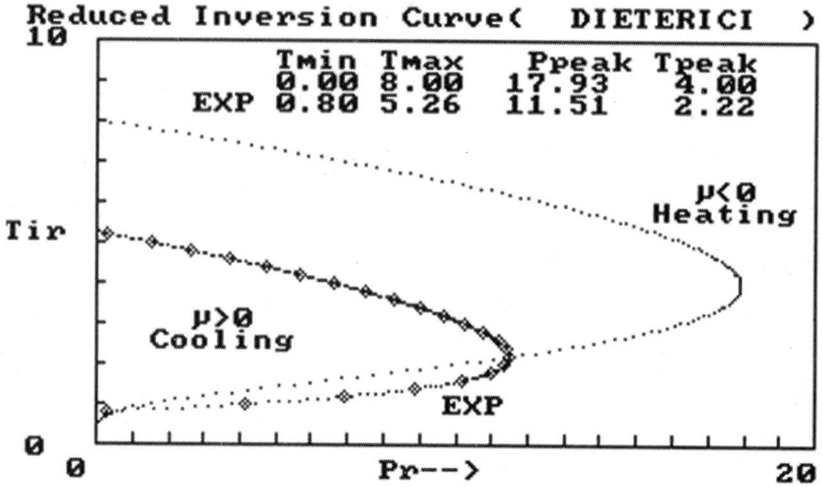

Fig. AI.3

Redlich-Kwong Equation of State

The Redlich-Kwong equation is

$$p = \frac{RT}{V_m - b} - \frac{a}{T^{1/2}V_m(V_m + b)} \tag{AI.40}$$

i.e.
$$p(V_m - b) = RT - \frac{a}{T^{1/2}}\left(-\frac{1}{V_m} + \frac{2}{V_m + b}\right) \tag{AI.41}$$

Expression of Inversion Temperature

Differentiating Eq. (AI.41) with respect to T at constant p, we get

$$p\left(\frac{\partial V_m}{\partial T}\right)_p = R + \frac{a}{2T^{3/2}}\left(-\frac{1}{V_m} + \frac{2}{V_m + b}\right) - \frac{a}{T^{1/2}}\left[\frac{1}{V_m^2} - \frac{2}{(V_m + b)^2}\right]\left(\frac{\partial V_m}{\partial T}\right)_p$$

This gives

$$\left(\frac{\partial V_m}{\partial T}\right)_p = \frac{R + \dfrac{a}{2T^{3/2}}\left(-\dfrac{1}{V_m} + \dfrac{2}{V_m + b}\right)}{p + \dfrac{a}{T^{1/2}}\left[\dfrac{1}{V_m^2} - \dfrac{2}{(V_m + b)^2}\right]} \tag{AI.42}$$

Substituting Eq. (AI.42) in the expression of inversion temperature ($T_i = V_m/(\partial V_m/\partial T)_p$), we get

$$T_i = \frac{pV_m + \dfrac{a}{T^{1/2}}\left[\dfrac{1}{V_m} - \dfrac{2V_m}{(V_m + b)^2}\right]}{R + \dfrac{a}{2T_i^{3/2}}\left(-\dfrac{1}{V_m} + \dfrac{2}{V_m - b}\right)} \tag{AI.43}$$

Eliminating pV_m by using Eq. (AI.41) and rearranging the resultant expression, we get

$$pb = \frac{a}{T_i^{1/2}}\left[-\frac{5}{2}\frac{1}{V_m} + \frac{3}{V_m + b} + \frac{2V_m}{(V_m + b)^2}\right] \quad \text{(AI.44)}$$

Under the condition of low pressure (i.e. large volume), we may approximate $V_m + b \simeq V_m$. With this, Eq. (AI.44) modifies to

$$pb = \frac{a}{T_i^{1/2}}\left(\frac{5}{2V_m}\right)$$

Replacing $V_m \simeq RT/p$, we get

$$pb = \frac{a}{T_i^{1/2}}\left(\frac{5p}{2RT_i}\right) \quad \text{i.e.} \quad b = \frac{5}{2}\frac{a}{RT_i^{3/2}}$$

This gives

$$T_i = \left(\frac{5}{2}\frac{a}{Rb}\right)^{2/3} \quad \text{(AI.45)}$$

Expression of Maximum Reduced Inversion Temperature

Replacing Redlich-Kwong constants a and b in terms of critical constants, we get

$$T_i = \left[\frac{5}{2}\cdot\frac{R^2T_c^{5/2}/9(2^{1/3}-1)p_c}{(2^{1/3}-1)V_cR}\right]^{2/3} = \left[\frac{5}{2}\cdot\frac{1}{9(2^{1/3}-1)^2}\left(\frac{RT_c^{5/2}}{p_cV_c}\right)\right]^{2/3}$$

Replacing $p_cV_c = RT_c/3$, we get

$$T_i = \left[\frac{5}{2}\cdot\frac{1}{9(2^{1/3}-1)^2}\left(\frac{RT_c^{5/2}}{RT_c/3}\right)\right]^{2/3} = \left[\frac{5}{2}\cdot\frac{1}{3(2^{1/3}-1)^2}\right]^{2/3} T_c$$

Hence $$T_{i(r)} = \frac{T_i}{T_c} = \left[\frac{5}{2}\cdot\frac{1}{3(2^{1/3}-1)^2}\right]^{2/3} = 5.34 \quad \text{(AI.46)}$$

Expression Relating $T_{i(r)}$ and V_r

We start with Eq. (AI.44)

$$pb = \frac{a}{T_i^{1/2}}\left[-\frac{5}{2}\cdot\frac{1}{V_m} + \frac{3}{V_m + b} + \frac{2V_m}{(V_m + b)^2}\right]$$

Eliminating p by using Redlich-Kwong equation of state at inversion temperature, we get

$$b\left[\frac{RT_i}{V_m - b} - \frac{a}{T_i^{1/2}V_m(V_m + b)}\right] = \frac{a}{T_i^{1/2}}\left[-\frac{5}{2V_m} + \frac{3}{V_m + b} + \frac{2V_m}{(V_m + b)^2}\right]$$

This gives

$$\frac{RT_ib}{V_m - b} = \frac{a}{T_i^{1/2}}\left[\frac{b}{V_m(V_m + b)} - \frac{5}{2V_m} + \frac{3}{V_m + b} + \frac{2V_m}{(V_m + b)^2}\right]$$

$$= \frac{a}{T_i^{1/2}}\left[\frac{(5V_m + 3b)(V_m - b)}{2V_m(V_m + b)^2}\right] \quad \text{(AI.47)}$$

Replacing Redlich-Kwong constants a and b in terms of critical constants and expressing V_m and T_i in terms of reduced variables, we get

$$\frac{R(T_{i(r)}T_c)(V_c x)}{V_r V_c - V_c x} = \frac{R^2 T_c^{5/2}/9xp_c}{(T_{i(r)}T_c)^{1/2}}\left[\frac{(5V_r V_c + 3V_c x)(V_r V_c - V_c x)}{2V_r V_c (V_r V_c + V_c x)}\right]$$

where $x = (2^{1/3} - 1) = 0.26$.

Cancelling the common terms, we get

$$\frac{RT_{i(r)}T_c x}{V_r - x} = \frac{R^2 T_c^2}{9x(p_c V_c)T_{i(r)}^{1/2}}\left[\frac{(5V_r + 3x)(V_r - x)}{2V_r(V_r + x)^2}\right]$$

Since $p_c V_c / RT_c = 1/3$, we get

$$T_{i(r)}^{3/2} = \frac{1}{3x^2}\left[\frac{(5V_r + 3x)(V_r - x)^2}{2V_r(V_r + x)^2}\right] \tag{AI.48}$$

Equation (AI.48) is the required relation between $T_{i(r)}$ and V_r.

Maximum Reduced Inversion Temperature

The maximum reduced inversion temperature is obtained when $p_r = 0$ (i.e. $V_r = \infty$). Writing Eq. (AI.48) as

$$T_{i(r)}^{3/2} = \frac{1}{3x^2}\left[\frac{(5 + 3x/V_r)(1 - x/V_r)^2}{2(1 + x/V_r)^2}\right]$$

Setting $V_r = \infty$, we get

$$T_{i(r)}^{3/2} = \frac{5}{6x^2} = \frac{5}{6(0.26)^2} = 12.33$$

Hence $T_{i(r)} = (12.33)^{2/3} = 5.34$ (Same as Eq. AI.46) (AI.49)

Expression Relating p_r and $T_{i(r)}$

The relation between p_r and $T_{i(r)}$ is obtained by considering reduced Redlich-Kwong equation:

$$p_r = \frac{3T_{i(r)}}{V_r - x} - \frac{1}{xT_{i(r)}^{1/2}V_r(V_r + x)} \tag{AI.50}$$

where $x = 2^{1/3} - 1 = 0.26$.

Procedure for Plotting $T_{i(r)}$ versus p_r

From Eq. (AI.50), we find that $p_r = \infty$ for $V_r = x = (2^{1/3} - 1) = 0.26$. Starting from $V_r > 0.26$, the value of $T_{i(r)}$ is computed by using Eq. (AI.48). This value of $T_{i(r)}$ along with V_r is substituted in Eq. (AI.50) to compute the value of p_r. The computed p_r is plotted against the computed value of $T_{i(r)}$. The above procedure is repeated with a suitable increase in the value of V_r.

Display of Graph of $T_{i(r)}$ versus p_r

Figure AI.4 displays the graph of $T_{i(r)}$ versus p_r

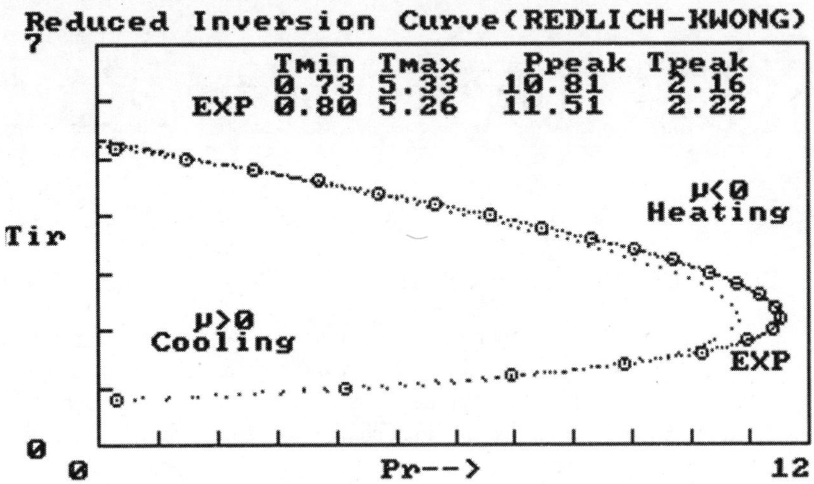

Fig. AI.4 Graph of $T_{i(r)}$ versus p_r

Comment The inversion curve marked as 'EXP' in Figs AI.1 to AI.4 is based on the generalized polynomial:

$$p_r = \sum_{j=0}^{6} \beta_j (T_{i(r)})_j$$

where

$\beta_0 = -32.520\ 9374$ $\beta_1 = 65.692\ 2312$ $\beta_2 = -39.738\ 4300$

$\beta_3 = 12.930\ 0299$ $\beta_4 = -2.461\ 769\ 04$ $\beta_5 = 0.253\ 785\ 53$

$\beta_6 = -0.010\ 986\ 565$

This polynomial is based on the least-square fitting of the inversion data of gases argon, methane, oxygen, nitrogen, fluorine, xenon and krypton. This expression was proposed by Robert C. Hendricks, Ildiko C. Peller and Anne K. Baron published in Nasa Technical note D-6807.

Of the four equations of state described in this section, Redlich-Kwong equation provides the best fitting between the expected and calculated values of reduced inversion temperatures.

3 Thermochemistry

3.1 SCOPE OF THERMOCHEMISTRY

Thermochemistry primarily deals with the transfer of heat between a chemical system and its surroundings when a change of phase or a chemical reaction takes place within the system. Depending upon the conditions under which the reaction is carried out, the quantity of heat transferred is related to energy or enthalpy change due to changes of states which occur in the system. However, in the laboratory, the majority of chemical reactions are carried out under the condition of constant pressure, therefore, the heat transferred is equal to the change in enthalpy of the system.

3.2 ENTHALPY OF A SUBSTANCE

Each substance has a fixed quantity of enthalpy. For one mole of a substance B, the enthalpy is represented as H_m(B). The main characteristics of a substance, if need to be specified, is usually stated along with the symbol B within the parenthesis. For example, the molar enthalpy of water vapours at 398 K and 1 atm pressure may be stated as $H_m(H_2O, g, 398\ K, 1\ atm)$.

Standard State of a Substance The molar enthalpy of a substance is a function of temperature and pressure, i.e. $H_m = H_m(T, p)$. In thermodynamics, the pressure dependence is removed by defining the standard state of a substance listed in the following.

1. For a pure gaseous substance, the standard state at a given temperature is the (hypothetical) ideal gas at one bar pressure.
2. For a pure liquid substance, the standard state at a given temperature is the pure liquid at one bar pressure.
3. For a pure crystalline substance, the standard state at a given temperature is the pure crystalline substance at one bar pressure.
4. For a substance or ion in solution, the standard state at a given temperature is the unit molality of the species in ideal solution at a one bar pressure.

The standard molar enthalpy of a substance is represented by placing the superscript degree to the symbol H. For example, the standard molar enthalpy of liquid water at 273 K is represented as $H_m^\circ(H_2O, 1, 273\ K)$.

3.3 CHANGE IN ENTHALPY DURING THE PROGRESS OF A REACTION

Since the enthalpy of a substance can change with the variation of temperature and pressure, it is essential that the reactants and products of a reaction are present at the same temperature and pressure while computing the enthalpy change during the progress of a chemical reaction.

Extent of Reaction

Let the reaction

$$2\ N_2O_5(g) \rightarrow 4\ NO_2(g) + O_2(g)$$

be started with the amount n_0 of N_2O_5. The progress of the reaction is stated by defining physical quantity known as extent of reaction (Symbol: ξ, Greek word pronounced as xi). By definition,

$$\xi = \frac{\text{Amount of a reactant consumed or product formed}}{\text{Stoichiometric number of the reactant or product}}$$

Enthalpy Change of the Reaction

Enthalpy of the system to start with, where $\xi = 0$, is given by

$$H_{\text{initial}} = n_0\, H_{\text{m}}(N_2O_5)$$

When the reaction has proceeded to the extent ξ, the amounts of reactants and products will be

$$\begin{array}{ccc} 2N_2O_5(g) \rightarrow & 4\ NO_2(g) + & O_2(g) \\ n_0 - 2\xi & 4\,\xi & \xi \end{array}$$

(**Note:** The unit of ξ is that of amount of species, i.e. mol. The stoichiometric numbers are dimensionless quantities and thus carry no units.)

Enthalpy of the system at this stage is

$$H_{\text{final}} = (n_0 - 2\xi)\, H_{\text{m}}(N_2O_5) + 4\xi\, H_{\text{m}}(NO_2) + \xi\, H_{\text{m}}(O_2)$$

Enthalpy change of the system is

$$\begin{aligned} \Delta H &= H_{\text{final}} - H_{\text{initial}} \\ &= [(n_0 - 2\xi)\, H_{\text{m}}(N_2O_5) + 4\xi\, H_{\text{m}}(NO_2) + \xi\, H_{\text{m}}(O_2)] - n_0\, H_{\text{m}}(N_2O_5) \\ &= -2\xi\, H_{\text{m}}(N_2O_5) + 4\,\xi\, H_{\text{m}}(NO_2) + \xi\, H_{\text{m}}(O_2) \qquad (3.3.1) \end{aligned}$$

The value of ΔH of a given chemical equation depends on its extent of reaction. It will vary as the reaction progresses.

From Eq. (3.3.1), it follows that

$$\begin{aligned} \text{unit of } \Delta H &= (\text{unit of } \xi)\ (\text{unit of } H_{\text{m}}) \\ &= (\text{mol})\ (\text{kJ mol}^{-1}) = \text{kJ} \end{aligned}$$

3.4 ENTHALPY OF REACTION

Definition of Enthalpy of Reaction

By definition, enthalpy of reaction is the enthalpy change for the unit extent of reaction. It is represented by the symbol $\Delta_r H$. Thus

$$\Delta_r H = \frac{\Delta H}{\xi} \qquad (3.4.1)$$

For the reaction

$$2N_2O_5(g) \rightarrow 4\ NO_2(g) + O_2(g)$$

$$\Delta H = 4\,\xi\, H_{\text{m}}(NO_2) + \xi\, H_{\text{m}}(O_2) - 2\xi\, H_{\text{m}}(N_2O_5)$$

$$\Delta_r H = \frac{\Delta H}{\xi} = 4H_{\text{m}}(NO_2) + H_{\text{m}}(O_2) - 2H_{\text{m}}(N_2O_5) \qquad (3.4.2)$$

Unit of Enthalpy of Reaction The unit of enthalpy of reaction is

$$\text{unit of } \Delta_r H = \frac{\text{unit of } \Delta H}{\text{Unit of } \xi} = \frac{\text{kJ}}{\text{mol}} = \text{kJ mol}^{-1}$$

Comment on Enthalpy of Reaction Since the enthalpy of reaction is defined as the enthalpy change for unit extent of reaction, the amounts of reactants consumed and products formed will be equal to the corresponding stoichiometric numbers expressed in mol. For example, for the reaction

$$2N_2O_5(g) \rightarrow 4NO_2(g) + O_2(g)$$

the enthalpy of reaction is the enthalpy change when 2 mol of N_2O_5 dissociates to give 4 mol of NO_2 and 1 mol of O_2. It may be noted that

Enthalpy of reaction refers to the entire chemical equation and not to any particular reactant or product.

Expression of Enthalpy of Reaction The enthalpy of a reaction may be computed by using the expression

$$\Delta_r H = \sum_{\text{products}} \nu_B H_m(B) - \sum_{\text{reactants}} |\nu_B| H_m(B) \qquad (3.4.3)$$

where the symbol Σ represent summation over the indicated substances (product or reactant) and ν_B is the stoichiometric number of the substance B in the balanced chemical equation. For example, for the reaction

$$Fe_2O_3(s) + 3H_2(g) \rightarrow 2\,Fe(s) + 3\,H_2O(l)$$

we have

$$\Delta_r H = \sum_{\text{products}} \nu_B H_m(B) - \sum_{\text{reactants}} |\nu_B| H_m(B)$$

$$= [2H_m(Fe, s) + 3H_m(H_2O, l)] - [H_m(Fe_2O_3, s) + 3H_m(H_2, g)]$$

3.5 EXOTHERMIC AND ENDOTHERMIC NATURE OF A REACTION

Since,

$$\Delta_r H = \sum_{\text{products}} \nu_B H_m(B) - \sum_{\text{reactants}} |\nu_B| H_m(B)$$

two cases may be distinguished.

Exothermic Reaction In a case where the enthalpy of products is less than that of reactants, we have

$$\Delta_r H = \sum_{\text{products}} \nu_B H_m(B) - \sum_{\text{reactants}} |\nu_B| H_m(B)$$

$$= \text{negative}$$

that is, there occurs a decrease in enthalpy when reactants are converted into products. This decrease is brought about by the release of heat from the system and the reaction is said to be an exothermic reaction.

For example, for the reaction

$$CH_4(g) + 2\,O_2(g) \rightarrow CO_2(g) + 2H_2O(l) \qquad \Delta_r H^\circ = -\,890.4 \text{ kJ mol}^{-1}$$

The above data may be represented in the form of an enthalpy level diagram, as shown in Fig. 3.5.1

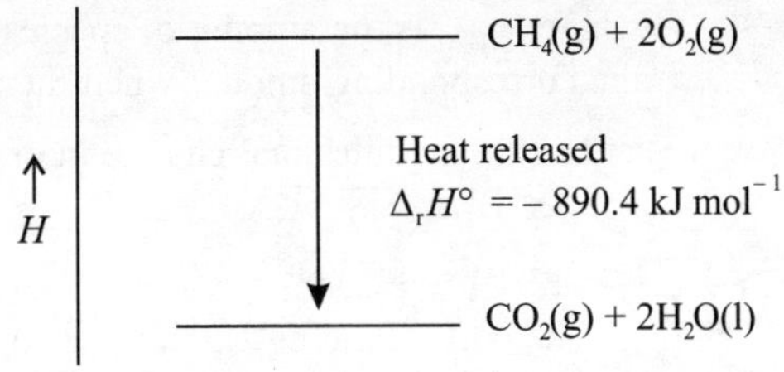

Fig. 3.5.1 Enthalpy-level diagram for an exothermic reaction

Endothermic Reaction

In the case where the enthalpy of products is greater than that of reactants, we have

$$\Delta_r H = \sum_{\text{products}} \nu_B H_m(B) - \sum_{\text{reactants}} |\nu_B| H_m(B)$$

$$= \text{positive}$$

that is, there occurs an increase in enthalpy when reactants are converted into products. This increase is brought about by the absorption of heat by the system from the surroundings and the reaction is said to be an endothermic reaction. For example, for the reaction

$$H_2(g) + I_2(g) \rightarrow 2HI(g) \qquad \Delta_r H° = 52.5 \text{ kJ mol}^{-1}$$

The enthalpy level diagram of this reaction is shown in Fig. 3.5.2

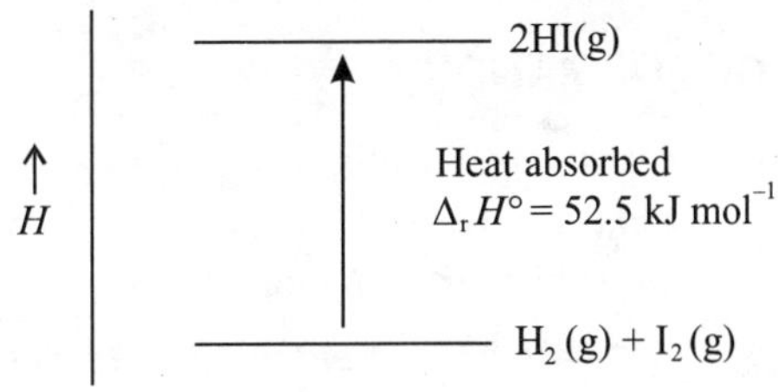

Fig. 3.5.2 Enthalpy-level diagram for an endothermic reaction

3.6 IUPAC RECOMMENDATION OF WRITING CHEMICAL EQUATION AND DEFINITION OF ENTHALPY OF REACTION

Representation of a Chemical Equation

A chemical equation may be conveniently represented as

$$0 = \sum_B \nu_B B \tag{3.6.1}$$

where the summation is carried over all the species (reactants and products) in chemical equation and ν_B is the stoichiometric number of the species B (assumed to be positive for products and negative for reactants).

For example, the chemical equation

$$2N_2O_5(g) \rightarrow 4NO_2(g) + O_2(g)$$

may be represented as

$$0 = 4NO_2(g) + O_2(g) - 2N_2O_5(g)$$

Change in Amount of Substance

The amount of species with the progress of reaction is given by the expression

$$n_B = (n_B)_0 + \nu_B \xi \tag{3.6.2}$$

where $(n_B)_0$ is the amount of species B in the beginning of the reaction and n_B is the corresponding amount when the reaction has proceeded to the extent ξ.

For the infinitesimal change in extent of reaction, the change in the amount of species B is

$$dn_B = \nu_B \, d\xi \tag{3.6.3}$$

The corresponding enthalpy change of the reaction is

$$dH = \sum_B H_m(B)\, dn_B = \sum_B H_m(B)(\nu_B \, d\xi) \tag{3.6.4}$$

Definition of Enthalpy of Reaction

By definition, enthalpy of reaction is given as

$$\Delta_r H = (\partial H / \partial \xi)_{T,p} = \sum_B \nu_B H_m(B) \tag{3.6.5}$$

that is, *the enthalpy of reaction is the rate of change of enthalpy of system with the extent of reaction at constant T and p.*

Standard Enthalpy of Reaction

When all the chemical species in a chemical equation are present in the respective standard states, the enthalpy of reaction is spelled as standard enthalpy of reaction. It is defined as

$$\Delta_r H^\circ = (\partial H^\circ / \partial \xi)_{T,p} = \sum_B \nu_B H^\circ_m(B) \tag{3.6.6}$$

3.7 ENTHALPY OF FORMATION

Accepted Convention

It is not possible to determine the absolute value of the molar enthalpy of a substance. However, based on the following convention, the relative values of standard molar enthalpies of formation of various substances can be built.

The standard enthalpy of formation of every element in its stable state of aggregation at one bar pressure and at specified temperature is assigned a zero value.

The specified temperature is usually taken as 25 °C.

A few example are

$\Delta_f H^\circ(O_2, g) = 0$

$\Delta_f H^\circ(C, \text{graphite}) = 0$ $\qquad$ $\Delta_f H^\circ(C, \text{diamond}) \neq 0$

$\Delta_f H^\circ(Br_2, l) = 0$

$\Delta_f H^\circ(S, \text{rhombic}) = 0$ $\qquad$ $\Delta_f H^\circ(S, \text{monoclinic}) \neq 0$

$\Delta_f H^\circ(P, \text{white}) = 0$ $\qquad$ $\Delta_f H^\circ(P, \text{black}) \neq 0$

Definition of Enthalpy of Formation

The standard enthalpy of formation is defined as follows:

The standard enthalpy of formation of a compound is the change in the standard enthalpy when one mole of the compound is formed starting from the requisite amounts of elements in their stable states of aggregation.

The formation of one mole of the compound implies that the compound appears as product with stoichiometric number equal to one.

The chemical equations corresponding to enthalpy of formation of a few substances are given below.

Enthalpy of formation of HBr(g) The chemical equation to be referred is

$$\tfrac{1}{2}\, H_2(g) + \tfrac{1}{2}\, Br_2(l) \rightarrow HBr(g)$$

$$\begin{aligned} \Delta_f H°(HBr, g) &= \Sigma \nu_B H^\circ_m(B) \\ &= H^\circ_m(HBr, g) - \tfrac{1}{2} H^\circ_m(H_2, g) - \tfrac{1}{2} H^\circ_m(Br_2, l) \end{aligned} \qquad (3.7.1)$$

Enthalpy of formation of SO_2(g) The chemical equation to be referred is

$$S(\text{rhombic}) + O_2(g) \rightarrow SO_2(g)$$

$$\Delta_f H°(SO_2, g) = H^\circ_m(SO_2, g) - H^\circ_m(S, \text{rhombic}) - H^\circ_m(O_2, g) \qquad (3.7.2)$$

Enthalpy of formation of SO_3(g) The chemical equation to be referred is

$$S(\text{rhombic}) + \tfrac{3}{2} O_2(g) \rightarrow SO_3(g)$$

$$\Delta_f H°(SO_3, g) = H^\circ_m(SO_3, g) - H^\circ_m(S, \text{rhombic}) - \tfrac{3}{2} H^\circ_m(O_2, g) \quad (3.7.3)$$

Enthalpy of Reaction from Enthalpies of Formation

Consider the reaction

$$SO_2(g) + \tfrac{1}{2} O_2(g) \rightarrow SO_3(g)$$

Its enthalpy of reaction is

$$\Delta_r H° = H^\circ_m(SO_3, g) - H^\circ_m(SO_2, g) - \tfrac{1}{2} H^\circ_m(O_2, g)$$

Expressing the molar enthalpies of SO_2 and SO_3 in terms of the corresponding enthalpies of formation by using Eqs. (3.7.2) to (3.7.3), respectively, we get

$$\begin{aligned} \Delta_r H° &= \left[\Delta_f H°(SO_3, g) + H^\circ_m(S, \text{rhombic}) + \tfrac{3}{2} H^\circ_m(O_2, g)\right] \\ &\quad -[\Delta_f H°(SO_2, g) + H^\circ_m(S, \text{rhombic}) + H^\circ_m(O_2, g)] - \tfrac{1}{2} H^\circ_m(O_2, g) \\ &= \Delta_f H°(SO_3, g) - \Delta_f H°(SO_2, g) \end{aligned} \qquad (3.7.4)$$

From Eq. (3.7.4), if follows that the enthalpy of reaction can be computed from the data on enthalpies of formation of SO_2 and SO_3. This way of computing enthalpy of reaction can be extended to any reaction. Equation to be used is

$$\Delta_r H° = \Sigma \nu_B\, \Delta_f H°(B) \qquad (3.7.5)$$

Equation (3.7.5) holds good for any reaction as the same reference state is used for reactants and products (Fig. 3.7.1)

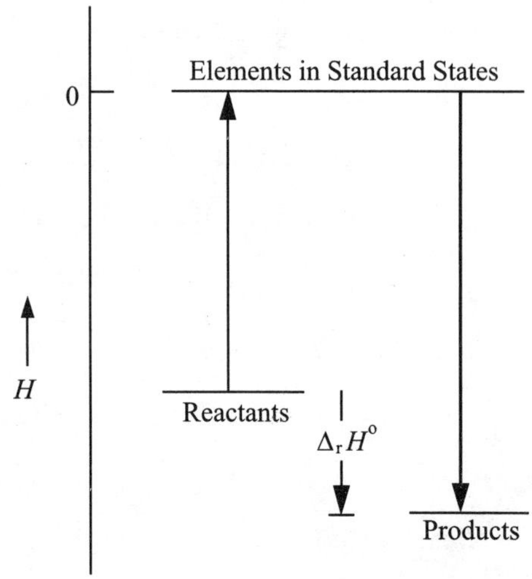

Fig. 3.7.1 Enthalpy of Reaction

The standard enthalpy of formation of a few substances have been tabulated in Appendix I.†

Example 3.7.1 Calculate the standard enthalpy of reaction

$$ZnO(s) + CO(g) \rightarrow Zn(s) + CO_2(g)$$

Given $\Delta_f H^\circ(ZnO, s) = -348.28\ kJ\ mol^{-1}$; $\Delta_f H^\circ(CO_2, g) = -393.51\ kJ\ mol^{-1}$

$\Delta_f H^\circ(CO, g) = -110.53\ kJ\ mol^{-1}$

Solution We have $\Delta_r H^\circ = \sum \nu_B\ \Delta_f H^\circ(B)$

$$\Delta_r H^\circ = \Delta_f H^\circ(Zn, s) + \Delta_f H^\circ(CO_2, g) - \Delta_f H^\circ(ZnO, s) - \Delta_f H^\circ(CO, g)$$

$$= \{0 + (-393.51) - (-348.28) - (-110.53)\}\ kJ\ mol^{-1} = 65.3\ kJ\ mol^{-1}$$

3.8 HESS'S LAW OF CONSTANT HEAT SUMMATION

Since the molar enthalpies of reactants and products involved in a chemical equation have definite values, it is obvious that the enthalpy change of a chemical equation would also have a definite value, irrespective of the way the reaction is carried out. Thus, if we transform a specified set of reactants to a specified set of products by more than one sequence of chemical equations, the total enthalpy change must be same for every sequence. This rule, which is a consequence of the first law of thermodynamics, is known as Hess's law of constant heat summation, which can be stated as follows:

The heat absorbed or evolved in a given chemical equation is the same whether the process occurs in one step or several steps.

---

† Prior to the standard-state pressure of 1 bar, the values of standard enthalpies of formation were tabulated for the standard-state pressure of 1 atm. The enthalpies of solids and liquids are not affected significantly by the small decrease in pressure from 1 atm to 1 bar. The standard enthalpies of formation of gases are also the same as the standard state is the ideal gas for which enthalpy is independent of pressure.

In support of Hess's law we cite below two different methods of synthesising sodium chloride from sodium and chlorine.

Method I

$$Na(s) + H_2O(1) \rightarrow NaOH(s) + \frac{1}{2}H_2(g) \qquad \Delta_r H^\circ = -140.87 \text{ kJ mol}^{-1}$$

$$\frac{1}{2}H_2(g) + \frac{1}{2}Cl_2(g) \rightarrow HCl(g) \qquad \Delta_r H^\circ = -92.30 \text{ kJ mol}^{-1}$$

$$HCl(g) + NaOH(s) \rightarrow NaCl(s) + H_2O(1) \qquad \Delta_r H^\circ = -177.83 \text{ kJ mol}^{-1}$$

Add

Net Change: $Na(s) + \frac{1}{2} Cl_2(g) \rightarrow NaCl(s) \qquad \Delta_r H^\circ = -411.00 \text{ kJ mol}^{-1}$

Method II

$$\frac{1}{2}H_2(g) + \frac{1}{2}Cl_2(g) \rightarrow HCl(g) \qquad \Delta_r H^\circ = -92.30 \text{ kJ mol}^{-1}$$

$$Na(s) + HCl(g) \rightarrow NaCl(s) + \frac{1}{2}H_2(g) \qquad \Delta_r H^\circ = -318.70 \text{ kJ mol}^{-1}$$

Add

Net Change: $Na(s) + \frac{1}{2} Cl_2(g) \rightarrow NaCl(s) \qquad \Delta_r H^\circ = -411.00 \text{ kJ mol}^{-1}$

Consequences of Hess's Law

The chemical equation can be treated as ordinary algebraic expressions and can be added or subtracted to yield the required equation. The corresponding enthalpies of reactions are also manipulated in the same way so as to give the enthalpy of reaction for the desired chemical equation.

Since $\Delta_r H$ stands for the change in enthalpy when reactants (substances on the left hand side of the arrow) are converted into products (substances on the right hand side of the arrow) at the same temperature and pressure, it follows that if the reaction is reversed (i.e. products are written on the left hand side and reactants on the right hand side), then the numerical value of $\Delta_r H$ remains the same, but its sign changes.[†]

Utility of Hess's Law

The utility of Hess's law is considerable. In almost all the thermochemical numericals, some way or the other, Hess's law is used. One of the important applications of Hess's law is to determine enthalpy of reaction which is difficult to determine experimentally. For example, the value $\Delta_r H$ for the reaction

$$C(\text{graphite}) + \frac{1}{2} O_2(g) \rightarrow CO(g)$$

which is difficult to determine experimentally, can be estimated from the following two reactions for which $\Delta_r H$ can be determined experimentally.

$$C(\text{graphite}) + O_2(g) \rightarrow CO_2(g) \qquad \Delta_r H_1$$

$$CO(g) + \frac{1}{2} O_2(g) \rightarrow CO_2(g) \qquad \Delta_r H_2$$

[†] This statement is known as Lavoisier and Laplace law.

Subtracting the latter from the former, we get

$$C(graphite) + \tfrac{1}{2}\, O_2\,(g) \rightarrow CO(g)$$

Consequently, $\Delta_r H = \Delta_r H_1 - \Delta_r H_2$.

Lattice Energy of a Crystal (Born-Haber Cycle)

The lattice energy is defined as the energy required to completely separate one mole of a solid ionic compound into gaseous ions.

The larger the lattice energy, the more stable the ionic compound and the more tightly the ions held.

Lattice energy cannot be measured directly. However, this can be determined from the Born-Haber cycle. Consider the following sequence of steps for the formation of NaCl crystals from Na(s) and Cl_2(g)

(i) Vaporization of Na(s)	$Na(s) \rightarrow Na(g)$	$\Delta_r H_1$
(ii) Ionization of Na(g)	$Na(g) \rightarrow Na^+(g) + e^-$	$\Delta_r H_2$
(iii) Dissociation of chlorine	$\tfrac{1}{2} Cl_2(g) \rightarrow Cl(g)$	$\Delta_r H_3$
(iv) Formation of Cl^-(g)	$Cl(g) + e^- \rightarrow Cl^-(g)$	$\Delta_r H_4$
(v) Condensation of Na^+(g) and Cl^-(g)	$Na^+(g) + Cl^-(g) \rightarrow NaCl(s)$	$\Delta_r H_5$
Net change:	$Na(s) + \tfrac{1}{2} Cl_2(g) \rightarrow NaCl(s)$	$\Delta_r H_6$

According to Hess's law, we can write

$$\Delta_r H_6 = \Delta_r H_1 + \Delta_r H_2 + \Delta_r H_3 + \Delta_r H_4 + \Delta_r H_5$$

Except $\Delta_r H_5$ all of these changes of enthalpy can be determined experimentally. Hence, $\Delta_r H_5$ can be determined from the above relation. The lattice energy is the negative of $\Delta_r H_5$ value.

Example 3.8.1

Set up a Born-Haber cycle to find the lattice energy of NaCl crystal. Given: $\Delta_f H°$(NaCl) = – 410.87 kJ mol^{-1}. Ionization enthalpy of Na = 495.80 kJ mol^{-1}, electron affinity of chlorine = 365.26 kJ mol^{-1}, sublimation enthalpy of Na = 317.57 kJ mol^{-1} and dissociation enthalpy of Cl_2(g) = 241.84 kJ mol^{-1}.

Solution

The following is the Born-Haber cycle for sodium chloride:

(i) Sublimation of Na	$Na(s) \rightarrow Na(g)$;	$\Delta_r H_1 = 317.57$ kJ mol^{-1}
(ii) Ionization of Na(g)	$Na(g) \rightarrow Na^+(g) + e^-$;	$\Delta_r H_2 = 495.80$ kJ mol^{-1}
(iii) Dissociation of half of chlorine	$\frac{1}{2} Cl_2(g) \rightarrow Cl(g)$;	$\Delta_r H_3 = \frac{1}{2} \times 241.84$ kJ mol^{-1}
(iv) Formation of Cl^-(g)	$Cl(g) + e \rightarrow Cl^-(g)$;	$\Delta_r H_4 = -\ 365.26$ kJ mol^{-1}
(v) Formation of NaCl(s)	$Na^+(g) + Cl^-(g) \rightarrow NaCl(s)$;	$\Delta_r H_5 = ?$
Net reaction	$Na(s) + \frac{1}{2} Cl_2(g) \rightarrow NaCl(s)$;	$\Delta_r H_6 = -\ 410.87$ kJ mol^{-1}

Now according to Hess's law

$$\Delta_r H_6 = \Delta_r H_1 + \Delta_r H_2 + \Delta_r H_3 + \Delta_r H_4 + \Delta_r H_5$$

Thus $\Delta_r H_5 = [-410.87 - 317.57 - 495.80 - 120.92 + 365.26]\ \text{kJ mol}^{-1}$
$= -979.9\ \text{kJ mol}^{-1}$

The lattice energy of NaCl(s) is negative of $\Delta_r H_5$, i.e. 979.9 kJ mol^{-1}.

Example 3.8.2 From the following thermochemical equations, calculate the enthalpy of formation of cane sugar ($C_{12}H_{22}O_{11}$):

(i) $C_{12}H_{22}O_{11}(s) + 12O_2(g) \rightarrow 12CO_2(g) + 11H_2O(l)$ $\Delta_r H = -5\ 644\ \text{kJ mol}^{-1}$

(ii) $C(s) + O_2(g) \rightarrow CO_2(g)$ $\Delta_r H = -393\ \text{kJ mol}^{-1}$

(iii) $H_2(g) + \frac{1}{2} O_2(g) \rightarrow H_2O(l)$ $\Delta_r H = -286\ \text{kJ mol}^{-1}$

Solution Multiplying Eq. (ii) by 12 and Eq. (iii) by 11 and adding them, we get

$$12 \times [C(s) + O_2(g) \rightarrow CO_2(g)] \qquad \Delta_r H = -12 \times 393\ \text{kJ mol}^{-1}$$

$$11 \times [H_2(g) + \frac{1}{2} O_2(g) \rightarrow H_2O(l)] \qquad \Delta_r H = -11 \times 286\ \text{kJ mol}^{-1}$$

$$12C(s) + 11H_2(g) + \frac{35}{2} O_2(g) \rightarrow 12CO_2(g) + 11H_2O(l); \quad \Delta_r H = -7\ 862\ \text{kJ mol}^{-1}$$

Subtracting Eq. (i) from the above resulting equation, we get

$$12C(s) + 11H_2(g) + \frac{35}{2} O_2(g) \rightarrow 12CO_2(g) + 11H_2O(l); \quad \Delta_r H = -7\ 862\ \text{kJ mol}^{-1}$$

$$-[C_{12}H_{22}O_{11}(s) + 12O_2(g) \rightarrow 12CO_2(g) + 11H_2O(l)] \qquad \Delta_r H = 5\ 644\ \text{kJ mol}^{-1}$$

$$12C(s) + 11H_2(g) + \frac{11}{2} O_2(g) \rightarrow C_{12}H_{22}O_{11}(s) \qquad \Delta_r H = -2\ 218\ \text{kJ mol}^{-1}$$

Example 3.8.3 From the data at 25 °C

(i) $Fe_2O_3(s) + 3C(\text{graphite}) \rightarrow 2Fe(s) + 3CO(g)$ $\Delta_r H = 490.78\ \text{kJ mol}^{-1}$

(ii) $FeO(s) + C(\text{graphite}) \rightarrow Fe(s) + CO(g)$ $\Delta_r H = 156.06\ \text{kJ mol}^{-1}$

(iii) $C(\text{graphite}) + O_2(g) \rightarrow CO_2(g)$ $\Delta_r H = -393.51\ \text{kJ mol}^{-1}$

(iv) $CO(g) + \frac{1}{2} O_2(g) \rightarrow CO_2(g)$ $\Delta_r H = -282.96\ \text{kJ mol}^{-1}$

Compute the enthalpy of formation of FeO(s) and of $Fe_2O_3(s)$, i.e. calculate $\Delta_r H$s corresponding to the following reactions.

(a) $Fe(s) + \frac{1}{2} O_2(g) \rightarrow FeO(s)$ and (b) $2Fe(s) + \frac{3}{2} O_2(g) \rightarrow Fe_2O_3(s)$

Solution **(a)** Multiplying each of Eq. (ii) and Eq. (iv) by –1 and adding the resultant equations in Eq. (iii), we get

$$Fe(s) + CO(g) \rightarrow FeO(s) + C(\text{graphite}) \qquad \Delta_r H = -156.06\ \text{kJ mol}^{-1}$$

$$C(\text{graphite}) + O_2(g) \rightarrow CO_2(g) \qquad \Delta_r H = -393.51\ \text{kJ mol}^{-1}$$

$$CO_2(g) \rightarrow CO(g) + \tfrac{1}{2} O_2(g) \qquad \Delta_r H = 282.96\ \text{kJ mol}^{-1}$$

$$Fe(s) + \tfrac{1}{2} O_2(g) \rightarrow FeO(s) \qquad \Delta_r H = -266.61\ \text{kJ mol}^{-1}$$

(b) Multiplying Eq. (i) by –1, Eq. (iii) by 3 and Eq. (iv) by –3, and adding the resultant equations, we get

$2Fe(s) + 3CO(g) \rightarrow Fe_2O_3(s) + 3C(\text{graphite});$	$\Delta_r H = -\,490.78 \text{ kJ mol}^{-1}$
$3C(\text{graphite}) + 3O_2(g) \rightarrow 3CO_2(g)$	$\Delta_r H = 3(-393.51 \text{ kJ mol}^{-1})$
$3CO_2(g) \rightarrow 3CO(g) + \frac{3}{2}O_2(g)$	$\Delta_r H = -3(-282.96 \text{ kJ mol}^{-1})$
$2Fe(s) + \frac{3}{2}O_2(g) \rightarrow Fe_2O_3(s)$	$\Delta_r H = -\,822.43 \text{ kJ mol}^{-1}$

3.9 VARIOUS TYPES OF ENTHALPIES OF REACTIONS

Enthalpy of Combustion

Enthalpy of combustion of a given compound is defined as follows:

It is the enthalpy change when one mole of this compound combines with the requisite amount of oxygen to give products in their stable forms.

The combustion of one mole of the compound implies that the substance appears as a reactant with stoichiometric number equal to one.

For example, the standard enthalpy of combustion of methane at 298.15 K is – 890.36 kJ mol^{-1}. This implies the following reaction:

$$CH_4(g) + 2O_2(g) \rightarrow CO_2(g) + 2H_2O(l) \qquad \Delta_c H^\circ = -\,890.36 \text{ kJ mol}^{-1}$$

The standard enthalpy of combustion of methane at 298.15 K may be written as

$$\Delta_c H^\circ(CH_4,\ g,\ 298.15\ K) = -\,890.36 \text{ kJ mol}^{-1}$$

Utility of Data on Enthalpy of Combustion

The data on the enthalpy of combustion can be determined experimentally. With the help of such data, we can determine the enthalpy of formation of a compound, which otherwise is difficult or impossible to determine experimentally. Consider, for example, the enthalpy of formation of $CH_4(g)$:

$$C(\text{graphite}) + 2H_2(g) \rightarrow CH_4(g)$$

First of all, the combination of carbon and hydrogen does not occur readily. Secondly, if the reaction is even completed, the end product would not be pure methane. Therefore, the enthalpy of formation of methane can be determined indirectly through the enthalpy of combustion of methane:

$$CH_4(g) + 2O_2(g) \rightarrow CO_2(g) + 2H_2O(l)$$

$$\Delta_c H^\circ(CH_4,\ g) = \Delta_f H^\circ(CO_2,\ g) + 2\ \Delta_f H^\circ(H_2O,\ l) - \Delta_f H^\circ(CH_4,\ g)$$

therefore

$$\Delta_f H^\circ(CH_4,\ g) = \Delta_f H^\circ(CO_2,\ g) + 2\ \Delta_f H^\circ(H_2O,\ l) - \Delta_c H^\circ(CH_4,\ g)$$

The enthalpies of formation of CO_2 and H_2O can be determined experimentally by the combustion of carbon (graphite) and hydrogen. Thus knowing the measured value of $\Delta_c H^\circ(CH_4, g)$, the enthalpy of formation of CH_4 can be calculated. The value is

$$\begin{aligned}\Delta_f H^\circ(CH_4,\ g) &= \Delta_f H^\circ(CO_2,\ g) + 2\ \Delta_f H^\circ(H_2O,\ l) - \Delta_c H^\circ(CH_4,\ g)\\ &= [-\,393.51 + 2\,(-\,285.83) - (-\,890.36)] \text{ kJ mol}^{-1}\\ &= -\,74.81 \text{ kJ mol}^{-1}\end{aligned}$$

Or, equivalently, we may add the following three chemical equations.

$$\begin{array}{ll} C(\text{graphite}) + O_2(g) \rightarrow CO_2(g) & \Delta_c H^\circ = -\ 393.51 \text{ kJ mol}^{-1} \\ 2[H_2(g) + \frac{1}{2}\ O_2(g) \rightarrow H_2O(l)] & \Delta_r H^\circ = 2\ (-\ 285.83) \text{ kJ mol}^{-1} \\ -\ [CH_4(g) + 2O_2(g) \rightarrow CO_2(g) + 2H_2O(l)] & \Delta_r H^\circ = -\ (-\ 890.36) \text{ kJ mol}^{-1} \\ \hline C(\text{graphite}) + 2H_2(g) \rightarrow CH_4(g) & \Delta_f H^\circ = -\ 74.81 \text{ kJ mol}^{-1} \end{array}$$

Example 3.9.1 Calculate the enthalpy change of the following reaction

$$3C_2H_2(g) \rightarrow C_6H_6(g)$$

Given: Enthalpy of combustion of $C_2H_2(g) = -1.30$ MJ mol^{-1} and that of $C_6H_6(g) = -3.302$ MJ mol^{-1}.

Solution The reactions corresponding to the combustion of C_2H_2 and C_6H_6 are

(i) $C_2H_2(g) + \frac{5}{2}\ O_2(g) \rightarrow 2CO_2(g) + H_2O(l)$

(ii) $C_6H_6(g) + \frac{15}{2}\ O_2(g) \rightarrow 6CO_2(g) + 3H_2O(l)$

Multiplying Eq. (i) by 3 and Eq. (ii) by –1, adding the resulting expressions, we get

$$\begin{array}{ll} 3 \times [C_2H_2(g) + \frac{5}{2}\ O_2(g) \rightarrow 2CO_2(g) + H_2O(1)] & \Delta_f H = -\ 3 \times 1.30 \text{ MJ mol}^{-1} \\ -1 \times [C_6H_6(g) + \frac{15}{2} O_2(g) \rightarrow 6CO_2 + 3H_2O(1)] & \Delta_f H = -\ (-\ 3.302 \text{ MJ mol}^{-1}) \\ \hline 3C_2H_2(g) \rightarrow C_6H_6(g) & \Delta_r H = -\ 0.598 \text{ MJ mol}^{-1} \end{array}$$

Example 3.9.2 Using the combustion data given below, compute the enthalpy of formation of isoprene(g) and its resonance energy. Given: $\Delta_f H$ (from bond enthalpies) = 103.31 kJ mol^{-1}.

Data:	Enthalpy of combustion of isoprene(g)	$\Delta_c H = -3\ 120$ kJ mol^{-1}
	Enthalpy of formation of $CO_2(g)$	$\Delta_f H = -393.71$ kJ mol^{-1}
	Enthalpy of formation of $H_2O(l)$	$\Delta_f H = -285.77$ kJ mol^{-1}

Solution Given that

$$\begin{array}{ll} C_5H_8(g) + 7O_2(g) \rightarrow 5CO_2(g) + 4H_2O(l) & \Delta_c H = -3\ 120 \text{ kJ mol}^{-1} \\ C(\text{graphite}) + O_2(g) \rightarrow CO_2(g) & \Delta_f H = -393.71 \text{ kJ mol}^{-1} \\ H_2(g) + \frac{1}{2}\ O_2(g) \rightarrow H_2O(l) & \Delta_f H = -285.77 \text{ kJ mol}^{-1} \end{array}$$

The reaction corresponding to the enthalpy of formation of $C_5H_8(g)$ can be obtained by the following manipulations:

$$\begin{array}{ll} -1 \times [C_5H_8(g) + 7O_2(g) \rightarrow 5CO_2(g) + 4H_2O(l)] & \Delta_r H = (-1) \times (-3\ 120.43 \text{ kJ mol}^{-1}) \\ 5 \times [C(\text{graphite}) + O_2(g) \rightarrow CO_2(g)] & \Delta_r H = -\ 5 \times 393.71 \text{ kJ mol}^{-1} \\ 4 \times [H_2(g) + \frac{1}{2}\ O_2\ (g) \rightarrow H_2O(1)] & \Delta_r H = -\ 4 \times 285.77 \text{ kJ mol} \end{array}$$

Adding these, we get

$$5C(\text{graphite}) + 4H_2(g) \rightarrow C_5H_8(g) \qquad \Delta_r H = 8.8 \text{ kJ mol}^{-1}$$

$$\text{Resonance energy} = \Delta_f H(\text{actual}) - \Delta_f H(\text{from bond enthalpies})$$
$$= 8.8 \text{ kJ mol}^{-1} - 103.31 \text{ kJ mol}^{-1}$$
$$= -94.51 \text{ kJ mol}^{-1}$$

Measurement of Enthalpy of Combustion

Enthalpies of combustion are usually measured by placing a known mass of the compound in a closed steel container (known as *bomb calorimetre*) which is filled with oxygen at about 30 bar pressure. The calorimetre is surrounded by a known mass of water. The entire apparatus is kept in an insulated jacket to prevent heat entering into or leaving from the container, as shown in Fig. 3.9.1. The sample is ignited electrically to bring about the combustion reaction. The heat evolved is used in raising the temperature of water and the calorimetre.

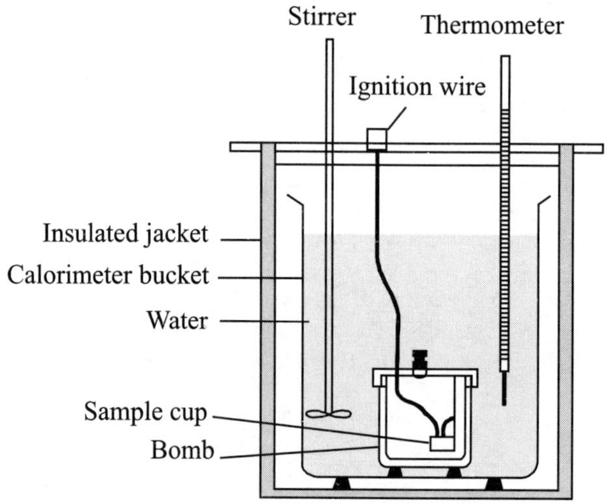

Fig. 3.9.1 Bomb calorimeter to determine enthalpy of combustion

Calculation of Enthalpy of Combustion

Since the heat released in combustion reaction is equal to the heat absorbed by water and bomb calorimetre, we will have

$$q_{\text{comb}} = -(q_{\text{water}} + q_{\text{bomb}})$$

The heat absorbed by water and bomb calorimeter are determined through their specific heat capacities. By definition, the specific heat capacity of a substance is the heat requried to raise the temperature of a unit mass of the substance by unit temperature. If q is heat required to raise the temperature of mass m of a substance by temperature ΔT, then

$$c = \frac{q}{m\Delta T} \qquad \text{or} \qquad q = mc\,\Delta T = C\,\Delta T$$

where $C(= mc)$ is known as heat capacity of the substance. With these, q_{comb} becomes

$$q_{\text{comb}} = -(m_{\text{water}}\, c_{\text{water}} + m_{\text{bomb}}\, c_{\text{bomb}})\,\Delta T$$
$$= -(m_{\text{water}}\, c_{\text{water}} + C_{\text{bomb}})\,\Delta T$$

Thus, knowing m_{water}, c_{water}, C_{bomb} and ΔT the value of q_{comb} can be determined. Since the experiment is carried out at a constant volume condition, the heat released per unit amount of substance will be equal to energy of combustion.

Hence $$\Delta_c U = \frac{q_{\text{comb}}}{n_{\text{subs}}} = \frac{q_{\text{comb}}}{m_{\text{subs}} / M_{\text{subs}}}$$

where m_{subs} and M_{subs} are the mass taken and the molar mass of the substance, respectively. Finally, the enthalpy of combustion will be given as

$$\Delta_c H = \Delta_c U + (\Delta v_g)\, RT$$

where Δv_g is the change in stoichiometric number of gaseous species in the balanced chemical equation representing the combustion process.†

Example 3.9.3 A 0.138 g sample of solid magnesium (molar mass = 24.30 g mol^{-1}) is burned in a constant volume bomb calorimetre that has a heat capacity of 1.77 kJ °C^{-1}. The calorimetre contains 300 mL of water (density = 1 g mL^{-1}) and its temperature is raised by 1.126 °C. Calculate the enthalpy of combustion of magnesium at 298 K.

Solution We have

$$\begin{aligned} q_{\text{water}} &= m_{\text{water}}\, c_{\text{water}}\, \Delta T \\ &= \{(300 \text{ mL})(1 \text{ g mL}^{-1})\}\,(4.184 \text{ J g}^{-1}\,°\text{C}^{-1})\,(1.126\ °\text{C}) \\ &= 1\,413 \text{ J} = 1.413 \text{ kJ} \end{aligned}$$

$$\begin{aligned} q_{\text{bomb}} &= (m_{\text{bomb}}\, c_{\text{bomb}})\, \Delta T \\ &= C_{\text{bomb}}\, \Delta T = (1.77 \text{ kJ }°\text{C}^{-1})\,(1.126\ °\text{C}) \\ &= 1.992 \text{ kJ} \end{aligned}$$

$$\begin{aligned} q_{\text{comb}} &= -(q_{\text{water}} + q_{\text{bomb}}) \\ &= -(1.413 + 1.992) \text{ kJ} = -3.405 \text{ kJ} \end{aligned}$$

$$\Delta_c U = \frac{q_{\text{comb}}}{m_{\text{subs}} / M_{\text{subs}}} = \frac{-3.405 \text{ kJ}}{(0.138 \text{ g})/(24.30 \text{ g mol}^{-1})} = -599.60 \text{ kJ mol}^{-1}$$

The combustion reaction is

$$\text{Mg(s)} + \frac{1}{2}\text{O}_2\text{(g)} \rightarrow \text{MgO(s)}; \qquad \Delta v_g = -\frac{1}{2}$$

$$\begin{aligned} \Delta_c H &= \Delta_c U + (\Delta v_g)\, RT \\ &= -599.60 \text{ kJ mol}^{-1} + \left(-\frac{1}{2}\right)(8.314 \times 10^{-3} \text{ kJ K}^{-1} \text{ mol}^{-1})\,(298 \text{ K}) \\ &= (-599.60 - 1.24) \text{ kJ mol}^{-1} = -600.89 \text{ kJ mol}^{-1} \end{aligned}$$

Integral Enthalpy of Solution

When a solute is dissolved in a solvent there is frequently an evolution or absorption of heat. The enthalpy change per unit amount of solute dissolved is not constant; it usually varies with concentration of the solution. Let ΔH be the enthalpy change when the amount n_2 of a solute is dissolved in a definite quantity of solvent, say 1 kg. We assume that the process of dissolution is an endothermic process. Now if ΔH is plotted against n_2, the resultant curve is as shown in Fig. 3.9.2.

Initially, ΔH increases almost linearly with n_2 but later the increase is not as fast as n_2. Finally, it reaches a constant value when the solution becomes saturated

† See Section 3.12

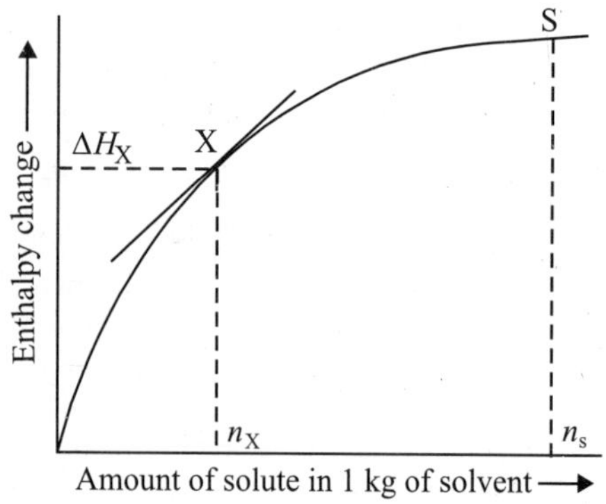

Fig. 3.9.2 Variation of enthalpy change when the amount n_2 of solute is dissolved in 1 kg solvent

with respect to solute. If the observed value of enthalpy change (ΔH) is divided by the amount of solute (n_2) that is dissolved to form a solution of a particular concentration, we get the quantity $\Delta H/n_2$, which is known as *integral enthalpy of solution at the given concentration*, Hence

The integral enthalpy of solution at the given concentration is the enthalpy change when one mole of the solute is dissolved in a definite quantity of solvent to produce a solution of a desired concentration.

While recording integral enthalpies of solution it is a general practice to state the amount of the solvent in which 1 mole of solute is dissolved. Thus

$$HCl(g) + 10\ H_2O(l) \rightarrow HCl(10\ H_2O) \qquad \Delta_r H_1 = -69.488\ \text{kJ mol}^{-1}$$

indicates that when 1 mol of hydrogen chloride gas is dissolved in 10 mol of water, there is an evolution of 69.488 kJ of heat. Other values are

(i) $HCl(g) + 25\ H_2O(l) \rightarrow HCl(25\ H_2O)$ $\qquad \Delta_r H_2 = -72.266\ \text{kJ mol}^{-1}$

(ii) $HCl(g) + 40\ H_2O(l) \rightarrow HCl(40\ H_2O)$ $\qquad \Delta_r H_3 = -73.023\ \text{kJ mol}^{-1}$

(iii) $HCl(g) + 200\ H_2O(l) \rightarrow HCl(200\ H_2O)$ $\qquad \Delta_r H_4 = -74.203\ \text{kJ mol}^{-1}$

(iv) $HCl(g) + aq \rightarrow HCl(aq)$ $\qquad \Delta_r H_5 = -75.145\ \text{kJ mol}^{-1}$

where $\Delta_r H_5$ represents the limit of enthalpy change when 1 mol of hydrogen chloride gas is dissolved in a very large quantity of water. The resultant solution is known as an *infinite diluted solution.*

Integral Enthalpy of Dilution

The fact that the enthalpy of solution of a solute varies with its concentration implies that there must be a change in enthalpy when a solution is diluted by adding more solvent.

The integral enthalpy of dilution is the change in enthalpy when a solution containing 1 mole of solute is diluted from one concentration to another.

According to Hess's law, it is equal to the difference between the integral enthalpies of solution at the two concentrations. For example, if to a solution of 1 mol of hydrogen chloride gas in 40 mol of water, enough water is added such that

$$HCl(40\ H_2O) + aq \rightarrow HCl(aq)$$

the associated enthalpy change can be obtained as follows:

$$HCl(g) + aq \rightarrow HCl(aq) \qquad \Delta_r H_5 = -75.145 \text{ kJ mol}^{-1}$$
$$HCl(g) + 40\ H_2O \rightarrow HCl(40\ H_2O) \qquad \Delta_r H_6 = -73.023 \text{ kJ mol}^{-1}$$

Subtracting, we have

$$HCl(40\ HCl) + aq \rightarrow HCl(aq) \qquad \Delta_r H = -2.122 \text{ kJ mol}^{-1}$$

Differential Enthalpy of Solution

The differential enthalpy of solution is defined as follows:

It is the enthalpy change when 1 mole of solute is dissolved in a very large volume of a solution of known concentration so that there occurs no appreciable change in concentration of the solution.

The value of differential enthalpy of solution for a given concentration of solution is given by the slope of the curve between ΔH and n_2 drawn at the point corresponding to the given concentration of solution. Mathematically, it may be expressed as $d(\Delta H)/dn_2$.

Alternatively, the differential enthalpy of solution may be defined as follows.

Let $d(\Delta H)$ be the change in value of ΔH when an infinitesimal amount dn_2 of solute is added to a solution of definite composition. By an infinitesimal amount dn_2 of solute, we mean that its addition does not cause any appreciable change in composition of solution. If we divide $d(\Delta H)$ by dn_2, we get the differential enthalpy of solution $d(\Delta H)/dn_2$. Thus, the differential enthalpy of solution may be defined as

It is the change in ΔH per unit amount of solute when an infinitesimal amount of the solute is added to a solution of definite concentration.

It may be seen from the shape of the curve of Fig. 3.9.1 that the value of the slope will depend upon the concentration of the solution. Thus, the differential enthalpy of a solution, besides depending on T and p, will also depend on the amount of solvent n_1 and solute n_2 present in the given solution, i.e. $d(\Delta H)/dn_2 = f(T, p, n_1, n_2)$. It is for this reason that the concentration of the solution is mentioned while defining the differential enthalpy of solution. The following conclusions may be drawn from the shape of the curve between ΔH and n_2 as shown in Fig. 3.9.1.

1. For smaller values of n_2, the curve is almost linear; thus its slope will have a constant value and will be equal to $\Delta H/n_2$. Since the latter represents the integral enthalpy of solution, if follows that the differential and integral enthalpies of solution are essentially equal for very dilute solutions.

2. For higher values of n_2, the curve is no more linear. In fact, ΔH does not increase as fast as n_2; thus the slope of the curve decreases as the value of n_2 increases. In other words, the differential enthalpy of solution decreases as the concentration of the solution increases, and becomes zero when the solution is saturated.

Differential Enthalpy of Dilution

The differential enthalpy of dilution may be defined as follows:

It is the enthalpy change when 1 mole of solvent is added to a large volume of the solution of known concentration so that there occurs no appreciable change in the concentration of the solution.

The value of differential enthalpy of dilution can be obtained by plotting ΔH of the solution at various concentrations against the amount of the solvent associated with a definite amount of solute, and finding the slope of the curve at a point corresponding to any particular concentration.

Comment on Differential Enthalpies

The differential enthalpies of solution and dilution are essentially the partial molar enthalpies of solution of solute and solvent, respectively, which are commonly used while dealing with solutions.† Consider a solution containing the amount n_1 of solvent and the amount n_2 of solute. In general,

$$\Delta H = f(T, p, n_1, n_2)$$

The differential of ΔH is given by

$$d(\Delta H) = \left(\frac{\partial(\Delta H)}{\partial T}\right)_{p,n_1,n_2} dT + \left(\frac{\partial(\Delta H)}{\partial p}\right)_{T,n_1,n_2} dp + \left(\frac{\partial(\Delta H)}{\partial n_1}\right)_{T,p,n_2} dn_1 + \left(\frac{\partial(\Delta H)}{\partial n_2}\right)_{T,p,n_1} dn_2$$

At constant temperature and pressure, this reduces to

$$d(\Delta H) = \left(\frac{\partial(\Delta H)}{\partial n_1}\right)_{T,p,n_2} dn_1 + \left(\frac{\partial(\Delta H)}{\partial n_2}\right)_{T,p,n_1} dn_2$$

or $$d(\Delta H) = (\Delta H_{1,\,pm})\, dn_1 + (\Delta H_{2,\,pm})\, dn_2$$

where $\Delta H_{1,\,pm}$ and $\Delta H_{2,\,pm}$ are known as partial molar enthalpies of solvent and solute in the solution, respectively. It can be seen that $\Delta H_{2,\,pm}$ is equal to the slope $d(\Delta H)/dn_2$ of the plot of ΔH versus n_2 for a series of solutions in which n_1 is kept constant. Comparison of this with the definition given at the beginning shows that the differential enthalpy of solution is really the partial molar enthalpy of solute in the solution, i.e. $\Delta H_{2,\,pm}$. Similarly, the differential enthalpy of dilution is the corresponding partial molar enthalpy of the solvent, i.e. $\Delta H_{1,\,pm}$.

Enthalpy of Hydration

Enthalpy of hydration of a given anhydrous or partially hydrated salt is the *enthalpy change when one mole of it combines with the requisite amount of water to form a new hydrated stable salt.* For example, the hydration of anhydrous cupric sulphate is represented by

$$CuSO_4(s) + 5H_2O(l) \rightarrow CuSO_4 \cdot 5H_2O(s)$$

There is almost invariably a liberation of heat in such reactions, i.e. the value of ΔH is negative. The value of enthalpy of hydration can be readily calculated from the integral enthalpies of solution of the hydrated and anhydrous salts. For example

$$CuSO_4(s) + 800\ H_2O(l) \rightarrow CuSO_4(800\ H_2O) \qquad \Delta_r H^\circ = -68.743\ \text{kJ mol}^{-1}$$

$$CuSO_4.5H_2O(s) + 795\ H_2O(l) \rightarrow CuSO_4(800\ H_2O) \qquad \Delta_r H^\circ = 10.125\ \text{kJ mol}^{-1}$$

† See Section 6.1.

Hence by subtraction, we get

$$CuSO_4(s) + 5H_2O(1) \rightarrow CuSO_4{\cdot}5H_2O(s) \qquad \Delta_r H^\circ = -78.868 \text{ kJ mol}^{-1}$$

Example 3.9.4

At 25 °C, 1 mole of $MgSO_4$ was dissolved in water. The heat evolved was found to be 91.211 kJ. One mol of $MgSO_4 \cdot 7H_2O$ on dissolution gives a solution of the same composition accompanied by an absorption of 13.807 kJ. Find the enthalpy of hydration, i.e. ΔH for the reaction.

$$MgSO_4(s) + 7H_2O(l) \rightarrow MgSO_4.7H_2O(s)$$

Solution

Given that

(i) $MgSO_4(s) + aq \rightarrow MgSO_4(aq)$ $\qquad \Delta_r H_1 = -\ 91.211 \text{ kJ mol}^{-1}$

(ii) $MgSO_4 \cdot 7H_2O(s) + aq \rightarrow MgSO_4(aq)$ $\qquad \Delta_r H_2 = +13.807 \text{ kJ mol}^{-1}$

Equation (i) can be considered to proceed through the following two steps:

$$MgSO_4(s) + 7H_2O(l) \rightarrow MgSO_4{\cdot}7H_2O(s) \qquad \Delta_{hyd}H = ?$$

$$MgSO_4{\cdot}7H_2O(s) + aq \rightarrow MgSO_4(aq) \qquad \Delta_r H_2 = +13.807 \text{ kJ mol}^{-1}$$

According to Hess's law, we get

$$\Delta_{hyd}H + \Delta_r H_2 = \Delta_r H_1$$

$$\Delta_{hyd}H = \Delta_r H_1 - \Delta_r H_2$$

$$= -\ 91.211 \text{ kJ mol}^{-1} - 13.807 \text{ kJ mol}^{-1} = -105.018 \text{ kJ mol}^{-1}$$

Enthalpy of Neutralization

Enthalpy of neutralization is defined as the *enthalpy change when one mole of* H^+ *in dilute solution combines with one mole of* OH^- *to give undissociated water, i.e.*

$$H^+(aq) + OH^-(aq) \rightarrow H_2O(l) \qquad \Delta_{neut}H = -\ 55.84 \text{ kJ mol}^{-1}$$

In this reaction, there is always a release of heat because of the bond formation H—OH. Whenever one mole of a strong monoprotic acid (HCl, HNO_3) is mixed with the one mole of a strong base (NaOH, KOH), the above neutralization reaction takes place, since these acids and bases are present in the completely dissociated from in dilute solutions. The corresponding enthalpy change is of the order of –55.8 kJ mol^{-1}.

In general, the enthalpy change of the reaction

$$H^+(nH_2O) + OH^-(nH_2O) \rightarrow H_2O(l)$$

depends on the value of n and may be visualized by mixing $HCl(nH_2O)$ and $NaOH(nH_2O)$. The reaction† is

$$HCl(nH_2O) + NaOH(nH_2O) \rightarrow NaCl(nH_2O) + H_2O(l)$$

The enthalpy change in the above reaction is

$$\Delta_r H = \Delta_f H(NaCl \cdot nH_2O) + \Delta_f H(H_2O, l)$$

$$-\Delta_f H(HCl \cdot nH_2O) - \Delta_f H(NaOH.nH_2O)$$

† Actually, we will get $NaCl((2n + 1)\ H_2O)$ instead of $NaCl(nH_2O) + H_2O(l)$. However, we may consider this hypothetical reaction just to show that ΔH of neutralization depends on the value of n.

For different values of n, the values are

$$n = 100; \quad \Delta_r H = [-407.07 - 285.83 - (-165.93) - (-469.65)] \text{ kJ mol}^{-1} = -57.32 \text{ kJ mol}^{-1}$$

$$n = 200; \quad \Delta_r H = [-406.92 - 285.83 - (-166.27) - (-469.61)] \text{ kJ mol}^{-1} = -56.87 \text{ kJ mol}^{-1}$$

$$n = \infty; \quad \Delta_r H = [-407.27 - 285.83 - (-167.16) - (-470.10)] \text{ kJ mol}^{-1} = -55.84 \text{ kJ mol}^{-1}$$

When $n = \infty$, the neutralization reaction may be written as

$$H^+(aq) + OH^-(aq) \rightarrow H_2O(l) \qquad \Delta_r H = -55.84 \text{ kJ mol}^{-1}$$

Enthalpy of Ionization

Whenever a weak acid (or base) reacts with a strong base (or acid), the release of heat is less than 55.84 kJ mol^{-1}. It is because of the fact that these acids or bases are not completely ionized in solution. Some of the heat is consumed in ionizing these acids and bases. This heat is known as *enthalpy of ionization.* Examples are

$$HCN + Na^+OH^- \rightarrow Na^+ + CN^- + H_2O \qquad \Delta_r H° = -12.13 \text{ kJ mol}^{-1}$$

$$CH_3COOH + Na^+OH^- \rightarrow Na^+ + CH_3COO^- + H_2O \qquad \Delta_r H° = -49.86 \text{ kJ mol}^{-1}$$

The enthalpy of ionization can be calculated as follows. The neutralization of a weak acid, say HCN, may be represented in two steps, namely,

(i) Ionization $\quad HCN \rightarrow H^+ + CN^- \quad \Delta_{ioniz} H° = ?$

(ii) Neutralization $\quad H^+ + OH^- \rightarrow H_2O \quad \Delta_{neut} H° = -55.84 \text{ kJ mol}^{-1}$

The complete reaction is obtained by adding the above two steps. Thus

$$HCN + OH^- \rightarrow H_2O + CN^- \qquad \Delta_r H° = -12.13 \text{ kJ mol}^{-1}$$

Obviously,

$$\Delta_r H° = \Delta_{ioniz} H° + \Delta_{neut} H°$$

or

$$\Delta_{ioniz} H° = \Delta_r H° - \Delta_{neut} H° = [-12.13 - (-55.84)] \text{ kJ mol}^{-1} = 43.71 \text{ kJ mol}^{-1}$$

Example 3.9.5

Enthalpy of neutralization of HCl by NaOH is –57.32 kJ mol^{-1} and by NH_4OH is –51.34 kJ mol^{-1}. Calculate the enthalpy of dissociation of NH_4OH.

Solution

Given that

$$H^+(aq) + NH_4OH(aq) \rightarrow NH_4^+(aq) + H_2O(l) \qquad \Delta_r H = -51.34 \text{ kJ mol}^{-1}$$

We may consider neutralization in two steps:

(i) Ionization $\quad NH_4OH(aq) \rightarrow NH_4^+(aq) + OH^-(aq) \quad \Delta_r H_1 = ?$

(ii) Neutralization $\quad H^+(aq) + OH^-(aq) \rightarrow H_2O(l) \quad \Delta_r H_2 = -57.32 \text{ kJ mol}^{-1}$

Thus, $\quad \Delta_r H = \Delta_r H_1 + \Delta_r H_2$

Therefore

$$\Delta_r H_1 = \Delta_r H - \Delta_r H_2 = -51.34 \text{ kJ mol}^{-1} + 57.32 \text{ kJ mol}^{-1} = 5.98 \text{ kJ mol}^{-1}$$

Enthalpy of Transition

Enthalpy of transition is the enthalpy change when one mole of one allotropic form changes to another. For example,

$$\text{C(graphite)} \rightarrow \text{C(diamond)} \qquad \Delta_{trs}H° = 1.90 \text{ kJ mol}^{-1}$$

The enthalpy of transition in the above example can be obtained from the enthalpies of combustion of C(graphite) and C(diamond).

$$\text{C(graphite)} + O_2(g) \rightarrow CO_2(g) \qquad \Delta_c H° = -393.51 \text{ kJ mol}^{-1}$$

$$\text{C(diamond)} + O_2(g) \rightarrow CO_2(g) \qquad \Delta_c H° = -395.41 \text{ kJ mol}^{-1}$$

Subtracting, we have

$$\text{C(graphite)} \rightarrow \text{C(diamond)} \qquad \Delta_{trs}H° = 1.90 \text{ kJ mol}^{-1}$$

Enthalpy of Precipitation

Enthalpy of precipitation is the enthalpy change when one mole of precipitate is formed. For example,

$$BaCl_2(aq) + Na_2SO_4(aq) \rightarrow BaSO_4(s) + 2NaCl(aq)$$

$$\Delta_r H° = -24.27 \text{ kJ mol}^{-1}$$

Enthalpy of Atomization

Enthalpy of atomization refers to a process in which a substance is separated into its constituent atoms in the gas phase. The corresponding reaction equation carries stoichiometric number of the substance equal to –1. Example is

$$H_2O(l) \rightarrow 2H(g) + O(g) \qquad \Delta_{at} H° = 69.94 \text{ kJ mol}^{-1}$$

Enthalpy of Formation of Ions

We have seen that

$$H^+(aq) + OH^-(aq) \rightarrow H_2O(l) \qquad \Delta_r H° = -55.84 \text{ kJ mol}^{-1}$$

For this reaction, we write

$$\Delta_r H° = \Delta_f H°(H_2O, l) - \{\Delta_f H°(H^+, aq) + \Delta_f H°(OH^-, aq)\}$$

Hence at 25 °C, we get

$$\Delta_f H°(H^+, aq) + \Delta_f H°(OH^-, aq) = \Delta_f H°(H_2O, l) - \Delta_r H°$$
$$= [-285.83 - (-55.84)] \text{ kJ mol}^{-1} = -229.99 \text{ kJ mol}^{-1}$$

By convention, *the standard enthalpy of formation of* H^+(aq) *is taken to be zero.*

Thus $\quad \Delta_f H°(OH^-, aq) = -\ 229.99 \text{ kJ mol}^{-1}$

Illustration

With the enthalpies of formation of these two ions, the enthalpy of formation of any other ion can be found from the enthalpy of formation and enthalpy of solution of its pure compound with H^+ or OH^-. For example, the enthalpy of formation of Na^+ can be calculated from the enthalpy of formation and enthalpy of infinite dilute solution of NaOH. The two values are

$\Delta_f H°(\text{NaOH, s}) = -\ 425.61 \text{ kJ mol}^{-1}$ and $\Delta_{aq} H°(\text{NaOH, s}) = -\ 44.50 \text{ kJ mol}^{-1}$

The chemical equation for the formation of infinite dilute solution of NaOH(s) is

$$NaOH(s) + nH_2O(l) \rightarrow Na^+(aq) + OH^-(aq)$$

$$\Delta_{aq} H°(\text{NaOH, s}) = -\ 44.50 \text{ kJ mol}^{-1}$$

Since there are equal amounts of water on both sides of the above equation, the two enthalpies give no net effect and thus

$$\Delta_{aq}H°(\text{NaOH, s}) = \Delta_f H°(\text{Na}^+, \text{aq}) + \Delta_f H°(\text{OH}^-, \text{aq}) - \Delta_f H°(\text{NaOH, s})$$

or $$\Delta_f H°(\text{Na}^+, \text{aq}) = \Delta_{aq}H°(\text{NaOH, s}) - \Delta_f H°(\text{OH}^-, \text{aq}) + \Delta_f H°(\text{NaOH, s})$$

$$= [-44.50 - (-229.99) + (-425.61)] \text{ kJ mol}^{-1}$$

$$= -240.12 \text{ kJ mol}^{-1}$$

Similarly, from NaCl(aq) or HCl(aq), the enthalpy of formation of Cl^-(aq) can be determined, and so on. These are recorded in Appendix I. The changes in enthalpy of any ionic reaction can then be found from these ionic enthalpies of formation and the usual enthalpies of formation of compounds.

Example 3.9.6 The enthalpy of formation of H_2O(l) is –285.83 kJ mol^{-1} and enthalpy of neutralization of a strong acid and a strong base is –55.84 kJ mol^{-1}. What is the enthalpy of formation of OH^- ions?

Solution Given that

$$\begin{array}{lccc} & \text{H}^+(\text{aq}) + \text{OH}^-(\text{aq}) & \rightarrow & \text{H}_2\text{O(l)} \\ \Delta_f H & 0 & & -285.83 \text{ kJ mol}^{-1} \end{array}$$

$$\Delta_{neut}H = \Delta_f H(\text{H}_2\text{O, l}) - \Delta_f H(\text{OH}^-, \text{aq})$$

Hence $$\Delta_f H(\text{OH}^-, \text{aq}) = \Delta_f H(\text{H}_2\text{O, l}) - \Delta_{neut}H$$

$$= [-285.83 - (-55.84)] \text{ kJ mol}^{-1}$$

$$= -229.99 \text{ kJ mol}^{-1}$$

Example 3.9.7 Calculate $\Delta_r H°$ for the reaction

$$\text{Ag}^+(\text{aq}) + \text{Cl}^-(\text{aq}) \rightarrow \text{AgCl(s)}$$

at 25 °C. Given: $\Delta_f H°(\text{Ag}^+, \text{aq}) = 105.58$ kJ mol^{-1}, $\Delta_f H°(\text{Cl}^-, \text{aq}) = -167.16$ kJ mol^{-1} and $\Delta_f H°(\text{AgCl, s}) = -127.07$ kJ mol^{-1}.

Solution For the reaction

$$\text{Ag}^+(\text{aq}) + \text{Cl}^-(\text{aq}) \rightarrow \text{AgCl(s)}$$

we have

$$\Delta_r H° = \Delta_f H°(\text{AgCl, s}) - \Delta_f H°(\text{Ag}^+, \text{aq}) - \Delta_f H°(\text{Cl}^-, \text{aq})$$

$$= [-127.07 - 105.58 - (-167.16)] \text{ kJ mol}^{-1} = -65.49 \text{ kJ mol}^{-1}$$

Example 3.9.8 Calculate the enthalpy change when one mole of HCl(g) is dissolved in a very large amount of water at 25 °C. The change in state is

$$\text{HCl(g)} + \text{aq} \rightarrow \text{H}^+(\text{aq}) + \text{Cl}^-(\text{aq})$$

Given: $\Delta_f H°(\text{HCl, g}) = -92.31$ kJ mol^{-1} and $\Delta_f H°(\text{Cl}^-, \text{aq}) = -167.16$ kJ mol^{-1}

Solution For the reaction

$$\text{HCl(g)} + \text{aq} \rightarrow \text{H}^+(\text{aq}) + \text{Cl}^-(\text{aq})$$

we have

$$\Delta_r H° = \Delta_f H°(\text{Cl}^-, \text{aq}) - \Delta_f H°(\text{HCl, g})$$

$$= [-167.16 - (-92.31)] \text{ kJ mol}^{-1}$$

$$= -74.85 \text{ kJ mol}^{-1}$$

3.10 BOND ENTHALPIES

Definition

Bond enthalpy of a given bond is defined as follows:

The bond enthalpy is the average of enthalpies required to dissociate the said bond present in different gaseous compounds into free atoms or radicals in the gaseous state.

The term bond enthalpy may be distinguished form the term bond dissociation enthalpy which is defined as follows:

The bond dissociation enthalpy is the enthalpy required to dissociate a given bond of some specific compound.

The distinction between these two terms may be more evident if described in terms of a simple example, say of the O—H bond. The enthalpy of dissociation of the O—H bond depends on the nature of molecular species from which the H atom is being separated. For example, in the water molecule

$$H_2O(g) \rightarrow H(g) + OH(g) \qquad \Delta_r H^\circ = 501.87 \text{ kJ mol}^{-1}$$

However, to break the O—H bond in the hydroxyl required a different quantity of heat

$$OH(g) \rightarrow O(g) + H(g) \qquad \Delta_r H^\circ = 423.38 \text{ kJ mol}^{-1}$$

The bond enthalpy, ε_{OH}, is defined as the average of these two values; that is

$$\varepsilon_{OH} = \frac{501.87 \text{ kJ mol}^{-1} + 423.38 \text{ kJ mol}^{-1}}{2} = 462.625 \text{ kJ mol}^{-1}$$

In the case of diatomic molecules, such as H_2, the bond enthalpy and bond dissociation enthalpy are identical because each refers to the reaction

$$H_2(g) \rightarrow 2H(g) \qquad \varepsilon_{H\text{–}H} = \Delta_r H^\circ = 435.93 \text{ kJ mol}^{-1}$$

Thus, the bond enthalpy given for any particular pair of atoms is the average value of the dissociation enthalpies of the bond for a number of molecules in which the pair of atoms appears. Appendix I records the recommended bond enthalpies of the various bonds.

Calculation of Bond Enthalpy

Bond enthalpies can be obtained from data on enthalpies of combustion and enthalpies of dissociation. Taking an example of the bond enthalpy of C—H, we have

$$CH_4(g) \rightarrow C(g) + 4H(g) \qquad \varepsilon_{C\text{–}H} = \Delta_r H/4$$

The value of $\Delta_r H$ and hence $\varepsilon_{C\text{–}H}$ for this reaction can be obtained from the summation of following equations:

$$\begin{array}{ll} CH_4(g) + 2O_2(g) \rightarrow CO_2(g) + 2H_2O(l) & \Delta_r H^\circ = -890.36 \text{ kJ mol}^{-1} \\ CO_2(g) \rightarrow C(\text{graphite}) + O_2(g) & \Delta_r H^\circ = 393.51 \text{ kJ mol}^{-1} \\ 2H_2O(l) \rightarrow 2H_2(g) + O_2(g) & \Delta_r H^\circ = 571.70 \text{ kJ mol}^{-1} \\ 2H_2(g) \rightarrow 4H(g) & \Delta_r H^\circ = 871.86 \text{ kJ mol}^{-1} \\ C(\text{graphite}) \rightarrow C(g) & \Delta_r H^\circ = 716.68 \text{ kJ mol}^{-1} \\ \hline CH_4(g) \rightarrow C(g) + 4H(g) & \Delta_r H^\circ = 1\,663.39 \text{ kJ mol}^{-1} \\ \hline \end{array}$$

Thus at 298.15 K, $$\varepsilon_{\text{C-H}} = \frac{1\,663.39 \text{ kJ mol}^{-1}}{4} = 415.85 \text{ kJ mol}^{-1}$$

Thus value of the C—H bond enthalpy does not correspond to the dissociation enthalpy of the carbon-hydrogen bond in methane, which is 426.77 kJ mol^{-1} and refers to the equation

$$CH_4(g) \rightarrow CH_3(g) + H(g)$$

Example 3.10.1 Find the bond enthalpy of S—S bond from the following data:

$C_2H_5—S—C_2H_5(g)$	$\Delta_f H^\circ = -147.23$ kJ mol^{-1}
$C_2H_5—S—S—C_2H_5(g)$	$\Delta_f H^\circ = -201.92$ kJ mol^{-1}
$S(g)$	$\Delta_f H^\circ = 222.80$ kJ mol^{-1}

Solution Given that

(i) $4C(s) + 5H_2(g) + S(s) \rightarrow C_2H_5—S—C_2H_5(g)$ $\quad \Delta_f H^\circ = -147.23$ kJ mol^{-1}

(ii) $4C(s) + 5H_2(g) + 2S(s) \rightarrow C_2H_5—S—S—C_2H_5(g)$ $\quad \Delta_f H^\circ = -201.92$ kJ mol^{-1}

Subtracting Eq. (i) from Eq. (ii), we get

$C_2H_5—S—C_2H_5(g) + S(s) \rightarrow C_2H_5—S—S—C_2H_5(g); \Delta_r H^\circ = -54.69$ kJ mol^{-1}

Adding to this, the following equation

$S(g) \rightarrow S(s)$ $\quad \Delta_r H^\circ = -222.80$ kJmol^{-1}

we get

$C_2H_5—S—C_2H_5(g) + S(g) \rightarrow C_2H_5—S—S—C_2H_5(g); \Delta_r H^\circ = -277.49$ kJ mol^{-1}

In the last equation 277.49 kJ of heat is evolved because of the S—S bond formation. Hence, the bond enthalpy of S—S is 277.49 kJ mol^{-1}

Diagrammatically, we may represent the above calculations as follows:

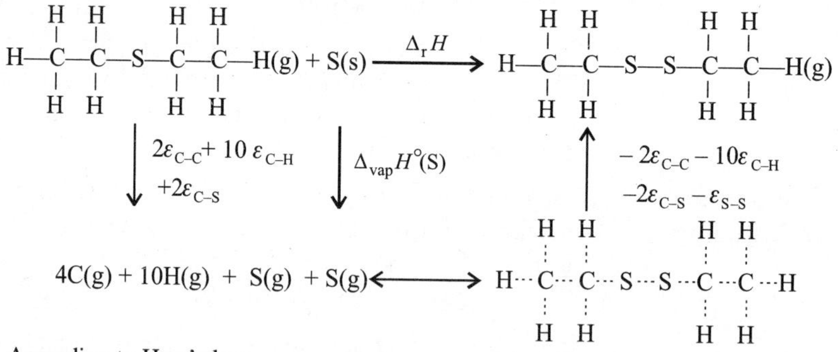

According to Hess's law

$$\begin{aligned}\Delta_r H^\circ &= \text{Enthalpy involved in bond breaking} - \text{Enthalpy involved in bond making}\\ &= [2\varepsilon_{\text{C-C}} + 10\varepsilon_{\text{C-H}} + 2\varepsilon_{\text{C-S}} + \Delta_{vap}H(S)] + [-2\varepsilon_{\text{C-C}} - 10\varepsilon_{\text{C-H}} - 2\varepsilon_{\text{C-S}} - \varepsilon_{\text{S-S}}]\\ &= \Delta_{vap}H^\circ(S) - \varepsilon_{\text{S-S}}\end{aligned}$$

or
$$\begin{aligned}\varepsilon_{\text{S-S}} &= \Delta_{vap}H^\circ(S) - \Delta_r H^\circ\\ &= \Delta_{vap}H^\circ(S) - [\Delta_f H^\circ(C_2H_5 - S - S - C_2H_5) - \Delta_f H^\circ(C_2H_5 - S - C_2H_5)]\\ &= [222.80 - \{-201.92 - (-147.23)\}] \text{ kJ mol}^{-1}\\ &= 277.49 \text{ kJ mol}^{-1}\end{aligned}$$

Example 3.10.2 Using the bond enthalpy data given below, calculate the enthalpy change for the reaction

$$C_2H_4(g) + H_2(g) \rightarrow C_2H_6(g)$$

Data:

Bond	*Bond enthalpy*
C—C	336.81 kJ mol^{-1}
C=C	606.68 kJ mol^{-1}
C—H	410.87 kJ mol^{-1}
H—H	431.79 kJ mol^{-1}

Solution Diagrammatically, we may represent the given reaction as follows:

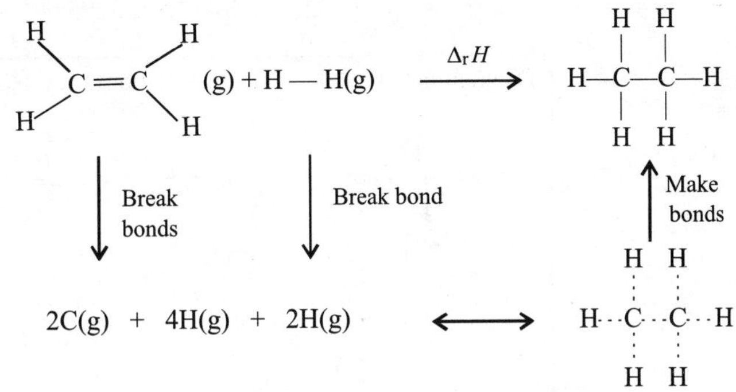

The heat required to dissociate $C_2H_4(g)$ and $H_2(g)$ into the gaseous atoms is

For breaking	1 C=C	606.68 kJ mol^{-1}
For breaking	4 C — H	4 × 410.87 kJ mol^{-1}
For breaking	1 H — H	431.79 kJ mol^{-1}
	Total	2 681.95 kJ mol^{-1}

If the same atoms now combine to give $C_2H_6(g)$, energy released will be

For making	1 C — C	– 336.81 kJ mol^{-1}
For making	6 C — H	– 6 × 410.87 kJ mol^{-1}
	Total	– 2 802.03 kJ mol^{-1}

Thus $\Delta_r H = -2\ 802.03 \text{ kJ mol}^{-1} + 2\ 681.95 \text{ kJ mol}^{-1} = -120.08 \text{ kJ mol}^{-1}$

Example 3.10.3 From the following data:

Enthalpy of formation of CH_3CN = 87.86 kJ mol^{-1}

Enthalpy of formation of C_2H_6 = – 83.68 kJ mol^{-1}

Enthalpy of sublimation of graphite = 719.65 kJ mol^{-1}

Enthalpy of dissociation of nitrogen = 945.58 kJ mol^{-1}

Enthalpy of dissociation of H_2 = 435.14 kJ mol^{-1}

C—H bond enthalpy = 414.22 kJ mol^{-1}

Calculate (i) ε_{C-C} and (ii) $\varepsilon_{C\equiv N}$.

Solution

(i) ε_{C-C} : We have

$$C_2H_6(g) \rightarrow 2C(\text{graphite}) + 3H_2(g) \qquad \Delta_r H_1 = 83.68 \text{ kJ mol}^{-1}$$

$$2C(\text{graphite}) \rightarrow 2C(g) \qquad \Delta_r H_2 = 2 \times 719.65 \text{ kJ mol}^{-1}$$

$$3H_2(g) \rightarrow 6H(g) \qquad \Delta_r H_3 = 3 \times 435.14 \text{ kJ mol}^{-1}$$

Adding, we get

$$C_2H_6(g) \rightarrow 2C(g) + 6H(g) \qquad \Delta_r H_4 = 2\,828.4 \text{ kJ mol}^{-1}$$

Now $\quad \Delta_r H_4 = \varepsilon_{C-C} + 6\,\varepsilon_{C-H}$

Therefore

$$\varepsilon_{C-C} = [2\,828.4 - 6 \times 414.22] \text{ kJ mol}^{-1} = 343.08 \text{ kJ mol}^{-1}$$

Diagrammatically, the above calculations may be represented as follows:

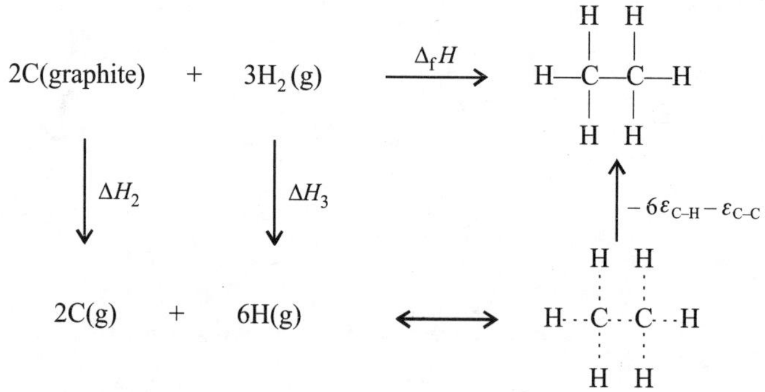

Applying Hess's law, we get

$$\Delta_f H = \Delta H_2 + \Delta H_3 - 6\,\varepsilon_{C-H} - \varepsilon_{C-C}$$

$$-83.68 \text{ kJ mol}^{-1} = (2 \times 719.65 + 3 \times 435.14 - 6 \times 414.22) \text{ kJ mol}^{-1} - \varepsilon_{C-C}$$

$$\varepsilon_{C-C} = (1\,439.3 + 1\,305.42 - 2\,485.32 + 83.68) \text{ kJ mol}^{-1}$$

$$= 343.08 \text{ kJ mol}^{-1}$$

(ii) $\varepsilon_{C\equiv N}$: We have

$$CH_3CN(g) \rightarrow 2C(\text{graphite}) + \tfrac{3}{2}H_2(g) + \tfrac{1}{2}N_2(g) \qquad \Delta_r H_1 = -87.86 \text{ kJ mol}^{-1}$$

$$2C(\text{graphite}) \rightarrow 2C(g) \qquad \Delta_r H_2 = 2 \times 719.65 \text{ kJ mol}^{-1}$$

$$\tfrac{3}{2}H_2(g) \rightarrow 3H(g) \qquad \Delta_r H_3 = \frac{3}{2} \times 435.14 \text{ kJ mol}^{-1}$$

$$\tfrac{1}{2}N_2(g) \rightarrow N(g) \qquad \Delta_r H_4 = \frac{1}{2} \times 945.58 \text{ kJ mol}^{-1}$$

Adding, we get

$$CH_3CN(g) \rightarrow 2C(g) + 3H(g) + N(g) \qquad \Delta_r H_5 = 2\,476.94 \text{ kJ mol}^{-1}$$

Now $\quad \Delta_r H_5 = \varepsilon_{C-C} + 3\varepsilon_{C-H} + \varepsilon_{C\equiv N}$

Therefore

$$\varepsilon_{C\equiv N} = [2\,476.94 - 3 \times 414.22 - 343.08] \text{ kJ mol}^{-1} = 891.2 \text{ kJ mol}^{-1}$$

Diagrammatically, the above calculations may be represented as follows:

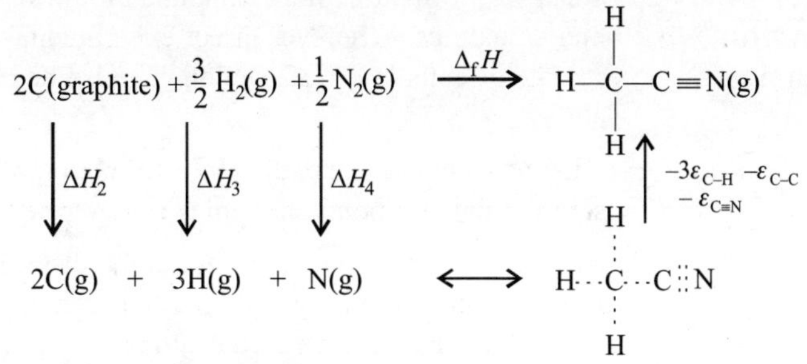

Applying Hess's law, we get

$$\Delta_f H = \Delta H_2 + \Delta H_3 + \Delta H_4 - (3\varepsilon_{C-H} + \varepsilon_{C-C} + \varepsilon_{C\equiv N})$$

$$87.86 = 2\times 719.65 + \frac{3}{2}\times 435.14 + \frac{1}{2}\times 945.58 - 3\times 414.22 - 343.08 - \varepsilon_{C\equiv N}/\text{kJ mol}^{-1}$$

$$\varepsilon_{C\equiv N} = (1\,439.30 + 652.71 + 472.79 - 1\,242.66 - 343.08 - 87.86)\ \text{kJ mol}^{-1}$$
$$= 891.2\ \text{kJ mol}^{-1}$$

Example 3.10.4 Calculate the bond enthalpy C—H from the following data at 298 K:

Enthalpy of combustion of methane	$\Delta_c H = -890.36$ kJ mol^{-1}
Enthalpy of combustion of C(graphite)	$\Delta_c H = -393.51$ kJ mol^{-1}
$H_2(g) + \frac{1}{2}O_2(g) \rightarrow H_2O(1)$	$\Delta_r H = -285.85$ kJ mol^{-1}
Enthalpy of dissociation of $H_2(g)$	$\Delta_{at} H = 435.93$ kJ mol^{-1}
Enthalpy of sublimation of C(graphite)	$\Delta_{sub} H = 716.68$ kJ mol^{-1}

Solution We have

(i)	$CH_4(g) + 2O_2(g) \rightarrow CO_2(g) + 2H_2O(1)$	$\Delta_c H_{298K} = -890.36$ kJ mol^{-1}
(ii)	$C(\text{graphite}) + O_2(g) \rightarrow CO_2(g)$	$\Delta_c H_{298K} = -393.51$ kJ mol^{-1}
(iii)	$H_2(g) + \frac{1}{2}O_2(g) \rightarrow H_2O(1)$	$\Delta_r H_{298K} = -285.85$ kJ mol^{-1}
(iv)	$H_2(g) \rightarrow 2H(g)$	$\Delta_{at} H_{298K} = 435.93$ kJ mol^{-1}
(v)	$C(\text{graphite}) \rightarrow C(g)$	$\Delta_{sub} H_{298K} = 716.68$ kJ mol^{-1}

If we do the following manipulations

Eq. (i) – Eq. (ii) – 2 Eq. (iii) + 2 Eq. (iv) + Eq. (v)

we get $CH_4(g) \rightarrow C(g) + 4H(g)$

The corresponding $\Delta_r H_{298K}$ is given by

$$\Delta_r H_{298K} = [-890.36 + 393.51 + 2\times 285.85 + 2\times 435.93 + 716.68]\ \text{kJ mol}^{-1}$$
$$= 1\,663.39\ \text{kJ mol}^{-1}$$

that is, 1 663.39 kJ is requried to dissociate one mole of $CH_4(g)$ into gaseous atoms in which 4 mol of C—H bonds are broken. Therefore

$$\varepsilon_{C-H} = \frac{1\,663.39\ \text{kJ mol}^{-1}}{4} = 415.85\ \text{kJ mol}^{-1}$$

Estimation of Enthalpy of Formation from Bond Enthalpies

The date on the bond enthalpies can be employed to calculate the approximate enthalpy of formation of a compound of known structure by adding the appropriate bond enthalpies. Wherever direct experimental data is not available, approximate estimate of enthalpy of a reaction can also be obtained through the bond enthalpy data.

Let us calculate the enthalpy of formation of $Se_2Cl_2(g)$. This can be carried out using the data on bond enthalpies as described in the following:

Since the bond enthalpy refers to the dissociation of gaseous Cl—Se—Se—Cl molecule into the gaseous atoms, the enthalpy change for the formation of this gaseous molecule from gaseous atoms is given by

$$2Se(g) + 2Cl(g) \rightarrow Se_2Cl_2(g)$$

$$\Delta H = -(\varepsilon_{Se-Se} + 2\varepsilon_{Se-Cl}) = -695.54 \text{ kJ mol}^{-1}$$

However, to estimate the enthalpy of formation it is necessary to add two reactions in the above equation since by definition the enthalpy of formation refers to the elements in their standard states. Therefore, we introduce the following enthalpy changes to convert the element from their standard states to the gaseous atoms at 298 K.

$$Cl_2(g) \rightarrow 2Cl(g) \qquad \Delta_r H_1 = 243.30 \text{ kJ mol}^{-1}$$

$$2Se(\text{hexagonal}) \rightarrow 2Se(g) \qquad \Delta_r H_2 = 2 \times 202.51 \text{ kJ mol}^{-1}$$

Thus adding these in the preceding expression, we get

$$2Se(\text{hexagonal}) + Cl_2(g) \rightarrow SeCl_2(g) \qquad \Delta_r H = -47.22 \text{ kJ mol}^{-1}$$

Diagrammatically the above calculations may be represented as follows:

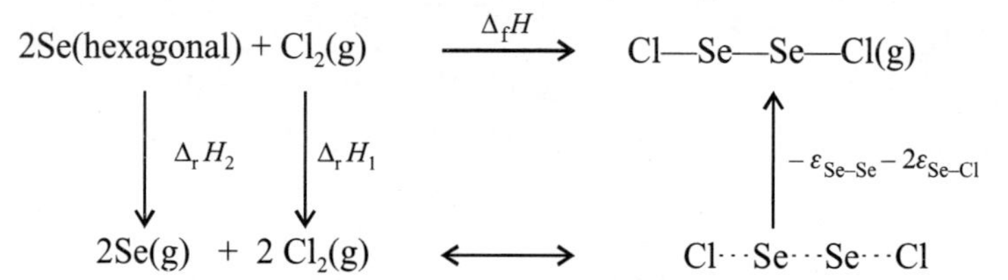

According to Hess's law,

$$\Delta_f H = \Delta_r H_1 + \Delta_r H_2 - (\varepsilon_{Se-Se} + 2\varepsilon_{Se-Cl})$$

$$= (243.30 + 2 \times 202.51 - 695.54) \text{ kJ mol}^{-1}$$

$$= -47.22 \text{ kJ mol}^{-1}$$

Example 3.10.5

Using the bond enthalpy date given below, estimate the enthalpy of formation of gaseous isoprene

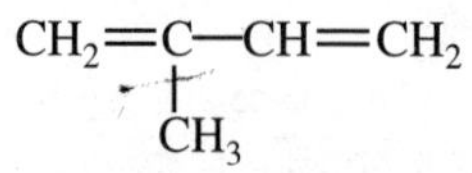

Data Bond enthalpy of C—H bond = 413.38 kJ mol^{-1}

Bond enthalpy of C—C bond = 347.69 kJ mol^{-1}

Bond enthalpy of C=C bond = 615.05 kJ mol^{-1}
Enthalpy of sublimation of carbon(graphite) = 718.39 kJ mol^{-1}
Enthalpy of dissociation of H_2(g) = 435.97 kJ mol^{-1}

Solution For isoprene, we have to form

2C—C bonds; 2C=C bonds and 8C—H bonds

For which energy released is

$$[2(-347.69) + 2(-615.05) + 8(-413.38)] \text{ kJ mol}^{-1} = -5\,232.52 \text{ kJ mol}^{-1}$$

that is, ΔH (from gaseous atoms) = – 5 232.52 kJ mol^{-1}

The reaction corresponding to this is

$$5C(g) + 8H(g) \rightarrow C_5H_8(g) \qquad \Delta_r H_1 = -5\,232.52 \text{ kJ mol}^{-1}$$

But we want $\Delta_f H$ corresponding to the following equation

$$5C(\text{graphite}) + 4H_2(g) \rightarrow C_5H_8(g) \qquad \Delta_f H = ?$$

This can be obtained by the following manipulations:

$$5C(g) + 8H(g) \rightarrow C_5H_6(g) \qquad \Delta_r H_2 = -5\,232.52 \text{ kJ mol}^{-1}$$
$$5C(\text{graphite}) \rightarrow 5C(g) \qquad \Delta_r H_3 = 5 \times 718.39 \text{ kJ mol}^{-1}$$
$$4H_2(g) \rightarrow 8H(g) \qquad \Delta_r H_4 = 4 \times 435.97 \text{ kJ mol}^{-1}$$

Adding, we get

$$5C(\text{graphite}) + 4H_2(g) \rightarrow C_5H_8(g) \qquad \Delta_f H = 103.31 \text{ kJ mol}^{-1}$$

Diagrammatically, the above calculations may be represented as follows:

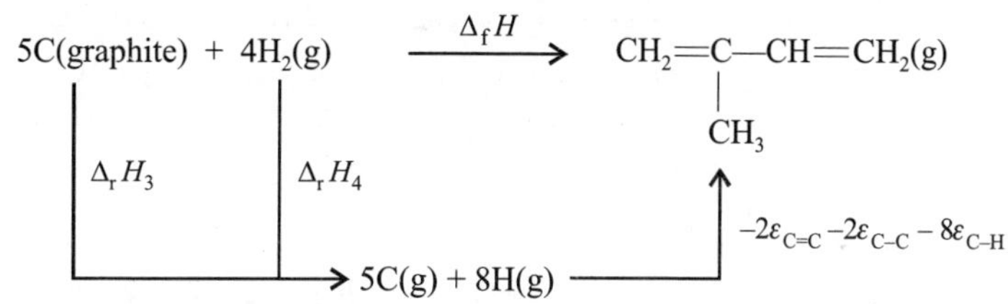

Applying Hess's law, we get

$$\Delta_f H = \Delta_r H_3 + \Delta_r H_4 - 2\varepsilon_{C=C} - 2\varepsilon_{C-C} - 8\varepsilon_{C-H}$$
$$= (5 \times 718.39 + 4 \times 435.97 - 2 \times 615.05 - 2 \times 347.69 - 8 \times 413.38) \text{ kJ mol}^{-1}$$
$$= 103.31 \text{ kJ mol}^{-1}$$

Estimation of Enthalpy of a Reaction from Bond Enthalpies

Let the enthalpy change for the reaction

$$C_2H_4(g) + HCl(g) \rightarrow C_2H_5Cl(g)$$

be required from the bond enthalpy data. This may be calculated as follows:

$$\Delta_r H = \begin{pmatrix}\text{Enthalpy required to}\\ \text{break reactant into}\\ \text{gaseous atoms}\end{pmatrix} + \begin{pmatrix}\text{Enthalpy released to from}\\ \text{products from the}\\ \text{gaseous atoms}\end{pmatrix}$$

$$= [4\varepsilon_{C-H} + \varepsilon_{C=C} + \varepsilon_{H-Cl}] + [-5\varepsilon_{C-H} - \varepsilon_{C-C} - \varepsilon_{C-Cl}]$$
$$= (\varepsilon_{C=C} + \varepsilon_{H-Cl}) - (\varepsilon_{C-H} + \varepsilon_{C-C} + \varepsilon_{C-Cl})$$

Substituting the bond enthalpy values, we have

$$\Delta_r H = [(615.05 + 431.79) - (413.38 + 347.69 + 328.44)] \text{ kJ mol}^{-1}$$
$$= -42.67 \text{ kJ mol}^{-1}$$

Bond Enthalpies and Resonance

Agreement between the calculated values of enthalpy of formation obtained by using the bond-enthalpy concept and any other method is usually good. For compounds involving alternate single and double bonds, however, large deviations are observed. For example, the reaction

$$\underset{\text{Benzene}}{C_6H_6(g)} \rightarrow 6C(g) + 6H(g)$$

will require enthalpy of 5 368.5 kJ mol^{-1} on the basis of bond enthalpies (3 $\varepsilon_{C–C}$ + 3 $\varepsilon_{C=C}$ + 6 $\varepsilon_{C–H}$) = (3 × 347.69 + 3 × 615.05 + 6 × 413.38) kJ mol^{-1} = 5 368.5 kJ mol^{-1}, whereas the experimental value is 5 535.1 kJ mol^{-1}. This amounts to the fact that benzene is more stable by 166.6 kJ mol^{-1}. This is due to resonance, i.e. there is no localization of single and double bonds in benzene, but the molecule is resonating hybrid.

Example 3.10.6

Calculate the resonance energy of benzene compared with one Kekule structure. Given the following data:

$\Delta_f H°$(methane, g) = –74.85 kJ mol^{-1}

$\Delta_f H°$(ethane, g) = – 84.68 kJ mol^{-1}

$\Delta_f H°$(ethylene, g) = 52.3 kJ mol^{-1}

$\Delta_f H°$(benzene, g) = 82.93 kJ mol^{-1}

Enthalpy of sublimation of carbon(graphite) = 718.39 kJ mol^{-1}

Dissociation enthalpy of H_2 = 435.89 kJ mol^{-1}

Solution

In order to calculate the resonance energy of benzene, we need to a compute $\Delta_f H°$ from the bond enthalpy data. For this, we need C—C, C=C and C—H bond enthalpies. These can be calculated as follows:

(i) *Bond enthalpy of* C—H *from* $\Delta_f H°$*(methane):* We have

$$CH_4(g) \rightarrow C(\text{graphite}) + 2H_2(g); \qquad \Delta_r H° = +\ 74.85 \text{ kJ mol}^{-1}$$

$$C(\text{graphite}) \rightarrow C(g); \qquad \Delta_r H° = +\ 718.39 \text{ kJ mol}^{-1}$$

$$2H_2(g) \rightarrow 4H(g); \qquad \Delta_r H° = 2 \times 435.89 \text{ kJ mol}^{-1}$$

Adding, we get

$$CH_4(g) \rightarrow C(g) + 4H(g); \qquad \Delta_r H_1^\circ = 1\ 665.02 \text{ kJ mol}^{-1}$$

Now $\quad \Delta_r H_1^\circ = 4\varepsilon_{C\text{–}H}, \quad$ therefore

$$\varepsilon_{C\text{–}H} = \frac{1\,665.02 \text{ kJ mol}^{-1}}{4} = 416.255 \text{ kJ mol}^{-1}$$

(ii) *Bond enthalpy of* C—C *from* $\Delta_f H°$*(ethane):* We have

$$C_2H_6(g) \rightarrow 2C(\text{graphite}) + 3H_2(g); \qquad \Delta_r H° = 84.68 \text{ kJ mol}^{-1}$$

$$2C(\text{graphite}) \rightarrow 2C(g); \qquad \Delta_r H° = 2 \times 718.39 \text{ kJ mol}^{-1}$$

$$3H_2(g) \rightarrow 6H(g); \qquad \Delta_r H° = 3 \times 435.89 \text{ kJ mol}^{-1}$$

Adding, we get

$$C_2H_6(g) \rightarrow 2C(g) + 6H(g); \qquad \Delta_r H_2^\circ = 2\ 829.13 \text{ kJ mol}^{-1}$$

Now $\quad \Delta_r H_2^\circ = \varepsilon_{C-C} + 6\varepsilon_{C-H}$

Thus $\quad \varepsilon_{C-C} = [2\,829.13 - 6 \times 416.225]$ kJ mol^{-1} = 331.60 kJ mol^{-1}

(iii) *Bond enthalpy of* C = C *from* $\Delta_f H^\circ$ *(ethylene):* We have

$$C_2H_4(g) \rightarrow 2C(\text{graphite}) + 2H_2(g); \qquad \Delta_r H^\circ = -52.3 \text{ kJ mol}^{-1}$$
$$2C(\text{graphite}) \rightarrow 2C(g); \qquad \Delta_r H^\circ = 2 \times 718.39 \text{ kJ mol}^{-1}$$
$$2H_2(g) \rightarrow 4H(g); \qquad \Delta_r H^\circ = 2 \times 435.89 \text{ kJ mol}^{-1}$$

Adding, we get

$$C_2H_4(g) \rightarrow 2C(g) + 4H(g); \qquad \Delta_r H_2^\circ = 2\,256.26 \text{ kJ mol}^{-1}$$

Now $\quad \Delta_r H_3^\circ = \varepsilon_{C=C} + 4\varepsilon_{C-H}$

Therefore

$$\varepsilon_{C=C} = (2\,256.26 - 4 \times 416.255) \text{ kJ mol}^{-1} = 591.24 \text{ kJ mol}^{-1}$$

(iv) $\Delta_f H^\circ$ *(benzene) from the bond enthalpy data:* We have

$$6C(g) + 6H(g) \rightarrow C_6H_6(g); \qquad \Delta_r H^\circ = -(3\varepsilon_{C-C} + 3\varepsilon_{C=C} + 6\varepsilon_{C-H}) = -5\,266.05 \text{ kJ mol}^{-1}$$
$$6C(\text{graphite}) \rightarrow 6C(g); \qquad \Delta_r H^\circ = 6 \times 718.39 \text{ kJ mol}^{-1}$$
$$3H_2(g) \rightarrow 6H(g); \qquad \Delta_r H^\circ = 3 \times 435.89 \text{ kJ mol}^{-1}$$

Adding, we get

$$6C(\text{graphite}) + 3H_2(g) \rightarrow C_6H_6(g); \qquad \Delta_r H^\circ = +315.96 \text{ kJ mol}^{-1}$$

(v) *Resonance energy of benzene*(g):

Actual value of $\Delta_f H^\circ$ = 82.93 kJ mol^{-1}

Calculated value of $\Delta_f H^\circ$ = 351.66 kJ mol^{-1}

Thus, benzene becomes more stable by +269.03 kJ mol^{-1}. Therefore, its resonance energy is 269.03 kJ mol^{-1}.

Example 3.10.7 The enthalpy of formation of ethane, ethylene and benzene from the gaseous atom are – 2 839.2, – 2 275.2 and – 5 536 kJ mol^{-1}, respectively. Calculate the resonance energy of benzene, compared with one Kekule structure. The bond enthalpy of C—H bond is given as equal to 410.87 kJ mol^{-1}.

Solution Bond enthalpy of C—C bond = Enthalpy required to break C_2H_6 into gaseous atoms – 6 × Bond enthalpy of C—H bond

= 2 839.2 kJ mol^{-1} – 6 × 410.87 kJ mol^{-1}

= 373.98 kJ mol^{-1}

Bond enthalpy of C═C bond = Enthalpy required to break C_2H_4 into gaseous atoms – 4 × Bond enthalpy of C—H bond

= 2 275.2 kJ mol^{-1} – 4 × 410.87 kJ mol^{-1}

= 631.72 kJ mol^{-1}

For the formation of benzene having Kekule structure, we have to from 3 C—C bond, 3 C═C bonds and 6 C—H bonds for which enthalpy released is

$$[3(-373.98 + 3(-631.72) + 6(-410.87) \text{ kJ mol}^{-1} = -5\,482.32 \text{ kJ mol}^{-1}$$

But the given value of $\Delta_f H$ is

$$\Delta_f H(\text{actual}) = -5\,536 \text{ kJ mol}^{-1}$$

Hence the resonance energy of benzene compared with one Kekule structure

$= \Delta_f H(\text{actual}) - \Delta_f H(\text{Kekule structure})$

$= (-5\,536 + 5\,482.32)\ \text{kJ mol}^{-1}$

$= -53.68\ \text{kJ mol}^{-1}$

3.11 VARIATION IN ENTHALPY OF A REACTION WITH TEMPERATURE (KIRCHHOFF'S RELATION)

Sometimes it is necessary to know the enthalpy of a reaction at a temperature different from that at which the value is available. Therefore, we consider below the procedure that can be used for this purpose.

Consider a general reaction

$$a\,\text{A} + b\,\text{B} + \cdots \rightarrow l\,\text{L} + m\,\text{M} + \cdots$$

we have $\quad \Delta_r H = (l\,H_{\text{m, L}} + m\,H_{\text{m, M}} + \cdots) - (a\,H_{\text{m, A}} + b\,H_{\text{m, B}} + \cdots)$

Its variation with temperature at constant pressure is given as

$$\left(\frac{\partial(\Delta_r H}{\partial T}\right)_p = \left\{ l\left(\frac{\partial H_{\text{m,L}}}{\partial T}\right)_p + m\left(\frac{\partial H_{\text{m,M}}}{\partial T}\right)_p + \cdots \right\}$$

$$- \left\{ a\left(\frac{\partial H_{\text{m,A}}}{\partial T}\right)_p + b\left(\frac{\partial H_{\text{m,B}}}{\partial T}\right)_p + \cdots \right\}$$

Now the variation of enthalpy of a substance with temperature at a given pressure is given by

$$\left(\frac{\partial H}{\partial T}\right)_p = C_p$$

With this, the previous equation becomes

$$\left(\frac{\partial(\Delta_r H)}{\partial T}\right)_p = \{(l\,C_{p,\text{m}}(\text{L}) + m C_{p,\text{m}}(\text{M}) + \cdots)\} - \{(a\ C_{p,\text{m}}(\text{A})$$

$$+ b\ C_{p,\text{m}}(\text{B}) + \cdots)\} = \Delta_r C_p \qquad (3.11.1)^\dagger$$

Rearranging this equation, we have

$$\text{d}(\Delta_r H) = \Delta_r C_p\ \text{d}T \qquad \text{(constant pressure)} \qquad (3.11.2)$$

which on integration at constant pressure gives

$$\int \text{d}(\Delta_r H) = \int (\Delta_r C_p)\text{d}T$$

---

†For a general reaction $0 = \sum_\text{B} \nu_\text{B}\text{B}$, we have $\quad \Delta_r C_p = \sum_\text{B} \nu_\text{B} C_{p,\text{m}}(\text{B})$

Thus $\quad \left(\frac{\partial(\Delta_r H)}{\partial T}\right)_p = \sum_\text{B} \nu_\text{B}\left(\frac{\partial H_\text{m}(\text{B})}{\partial T}\right)_p = \sum_\text{B} \nu_\text{B} C_{p,\text{m}}(\text{B}) = \Delta_r C_p$

or $$\Delta_r H = \int (\Delta_r C_p)\, dT + I \tag{3.11.3}$$

where I is the constant of integration. Equation (3.11.3) is known as *Kirchhoff's relation.* In order to carry out the above integration, it is necessary to know the temperature dependence of molar heat capacities of the reactants and products.

We may consider two possibilities as given below.

$\Delta_r C_p$ independent of temperature In this case, we have

$$\Delta_r H = (\Delta_r C_p)\Delta T + I \tag{3.11.4}$$

The value of integration constant I can be obtained if the value of $\Delta_r H$ at some temperature (say, 298 K) is available. Alternatively, integrating Eq. (3.11.2) within the limits gives

$$\Delta_r H_{T_2} - \Delta_r H_{T_1} = \Delta_r C_p (T_2 - T_1)$$

Example 3.11.1

Calculate $\Delta_r H^\circ_{373\text{K}}$ for the reaction

$$\tfrac{1}{2} N_2(g) + O_2(g) \rightarrow NO_2(g) \qquad \Delta_r H^\circ_{298\text{K}} = -33.18 \text{ kJ mol}^{-1}$$

Given: $C_{p,\text{m}}(NO_2, g) = 37.20 \text{ J K}^{-1} \text{ mol}^{-1}$; $C_{p,\text{m}}(O_2, g) = 29.36 \text{ J K}^{-1} \text{ mol}^{-1}$

$C_{p,\text{m}}(N_2, g) = 29.13 \text{ J K}^{-1} \text{ mol}^{-1}$

Solution

We have $\Delta_r C_p = \sum_B \nu_B C_{p,\text{m}}(B) = C_{p,\text{m}}(NO_2, g) - \dfrac{1}{2} C_{p,\text{m}}(N_2, g) - C_{p,\text{m}}(O_2, g)$

$$= (37.20 - \frac{1}{2} \times 29.13 - 29.36) \text{ J K}^{-1} \text{ mol}^{-1}$$

$$= -6.73 \text{ J K}^{-1} \text{ mol}^{-1} = -6.73 \times 10^{-3} \text{ kJ mol}^{-1}$$

Now $\Delta_r H^\circ_{373\text{ K}} = \Delta_r H^\circ_{298\text{ K}} + (\Delta_r C_p)(\Delta T)$

$$= [-33.18 + (-6.73 \times 10^{-3})(373 - 298)] \text{ kJ mol}^{-1}$$

$$= -33.68 \text{ kJ mol}^{-1}$$

$\Delta_r C_p$ dependent on temperature The variation of $C_{p,\text{m}}$ with temperature is usually given by the relation

$$C_{p,\text{m}} = a' + b'T + c'T^2 + \cdots \tag{3.11.5}$$

where a', b', c', are constants. Thus, the variation of $\Delta_r C_p$ of a chemical equation with temperature is given by

$$\Delta_r C_p = (\Delta_r a') + (\Delta_r b')\, T + (\Delta_r c')\, T^2 + \cdots \tag{3.11.6}$$

where $\Delta_r a' = (l\, a'_L + m\, a'_M \ldots) - (a\, a'_A + b\, a'_B \ldots)$ are so on. Substituting Eq. (3.11.6) in Eq. (3.11.3), we get

$$\Delta_r H = I + \int \{(\Delta_r a') + (\Delta_r b')T + (\Delta_r c')T^2 + \cdots\} dT$$

which on integration yields

$$\Delta_r H = I + (\Delta_r a')T + (\Delta_r b')\frac{T^2}{2} + (\Delta_r c')\frac{T^3}{3} + \cdots \tag{3.11.7}$$

The constant of integration I in Eq. (3.11.7) can be evaluated if the value of $\Delta_r H$ at some temperature (say, 298 K) is available.

Contribution of Phase Change

When a change of phase occurs in the course of change in temperature, the change in enthalpy for the change of state of aggregation must be included. For example, in the following reaction

$$\begin{array}{c} CH_4(g, 298\ K) + 2O_2(g, 298\ K) \rightarrow CO_2(g, 298\ K) + 2H_2O(1, 298\ K) \\ \downarrow \\ CH_4(g, 500\ K) + 2O_2(g, 500\ K) \rightarrow CO_2(g, 500\ K) + 2H_2O(g, 500\ K) \end{array}$$

a change in aggregation of water from liquid to gas occurs. Therefore, we must also add $\Delta H°$ corresponding to the following change

$$2H_2O(1,\ 298\ K) \rightarrow 2H_2O(g,\ 500\ K)$$

This change may be considered as the sum of three of changes of state, that is

$$2H_2O(1,\ 298\ K) \rightarrow 2H_2O(1,\ 373\ K)$$

$$2H_2O(1,\ 373\ K) \rightarrow 2H_2O(g,\ 373\ K)$$

$$2H_2O(g,\ 373\ K) \rightarrow 2H_2O(g,\ 500\ K)$$

The required $\Delta H°$ is thus the sum of the changes of enthalpy for all the three steps.

One of the most important things to remember while employing Kirchhoff's relation is that the units of $\Delta_r H°$ and $\Delta_r C_p$ should be the same. If $\Delta_r H°$ is expressed in kJ mol^{-1} (which is mostly the case), then $\Delta_r C_p$ should also be expressed in kJ K^{-1}mol^{-1}. The values of C_p are usually tabulated in J K^{-1} mol^{-1}. Therefore, these should be converted into kJ K^{-1} mol^{-1} while using the Kirchhoff's relation. Alternatively, $\Delta_r H°$ may be converted into joules if C_p are to be employed as such.

Example 3.11.2

Calculate $\Delta_f H^\circ_{373K}$ for the reaction

$$\frac{1}{2}N_2(g) + O_2(g) \rightarrow NO_2(g) \qquad \Delta_f H^\circ_{298K} = -33.18\ \text{kJ mol}^{-1}$$

Solution

Given: $C_{p,m}(NO_2, g)/\text{J K}^{-1}\text{ mol}^{-1} = 27.78 + 30.85 \times 10^{-3}\ (T/K)$

$C_{p,m}(N_2, g)/\text{J K}^{-1}\text{ mol}^{-1} = 28.46 + 2.26 \times 10^{-3}\ (T/K)$

$C_{p,m}(O_2, g)/\text{J K}^{-1}\text{ mol}^{-1} = 26.85 + 8.49 \times 10^{-3}\ (T/K)$

For the given reaction

$$\Delta_r C_p = \sum_B \nu_B C_{p,m}(B)$$

$$= \left[\left(27.78 - \frac{1}{2}\times 28.46 - 26.85\right) + \left(30.85 - \frac{1}{2}\times 2.26 - 8.49\right)10^{-3}(T/K)\right]\text{J K}^{-1}\text{ mol}^{-1}$$

$$= [-13.30 + 21.23 \times 10^{-3}\ (T/K)]\ \text{J K}^{-1}\text{ mol}^{-1}$$

$$\Delta_f H^\circ_{373K} = \Delta_f H^\circ_{298K} + \int_{T_1}^{T_2} (\Delta_r C_p) dT$$

$$= -33180\,\text{J mol}^{-1} + \int_{298K}^{373K} \left[\{-13.30 + 21.23 \times 10^{-3}(T/K)\}\,\text{J K}^{-1}\,\text{mol}^{-1}\right] dT$$

$$= -33180\,\text{J mol}^{-1} + \left[-13.30(373\,\text{K} - 298\,\text{K}) + 21.23 \times 10^{-3} \times \left\{\frac{(373\,\text{K})^2}{2\,\text{K}} - \frac{(298\,\text{K})^2}{2\,\text{K}}\right\}\right] \text{J K}^{-1}\,\text{mol}^{-1}$$

$$= (-33\,180 - 13.30 \times 75 + 21.23 \times 10^{-3} \times 25\,163)\ \text{J mol}^{-1}$$

$$= (-33\,180 - 997.5 + 534.21)\ \text{J mol}^{-1} = -33\,643\ \text{J mol}^{-1} = -33.64\ \text{kJ mol}^{-1}$$

3.12 RELATION BETWEEN ENERGY AND ENTHALPY OF A REACTION

For a chemical equation, the expression of $\Delta_r H$ is

$$\Delta_r H = \sum_B \nu_B H_m(B)$$

where ν_B is the stoichiometric number of B in the chemical equation (it is positive for products and negative for reactants).

The molar enthalpy of B is given as

$$H_m = U_m + pV_m$$

Substituting this in the previous expression, we get

$$\Delta_r H = \sum_B \nu_B [U_m(B) + (pV_m)_B]$$

$$= \sum_B \nu_B U_m(B) + \sum_B \nu_B (pV_m)_B$$

Using the fact that

$$pV_m \simeq 0 \text{ for one mole of solid or liquid}$$

$$pV_m = RT \text{ for one mole of gaseous species}$$

we get $$\Delta_r H = \sum_B \nu_B U_m(B) + \Big(\sum_{B(g)} \nu_B\Big) RT$$

$$= \Delta_r U + (\Delta \nu_g) RT \qquad (3.12.1)$$

where $\Delta\nu_g$ is the change in the stoichiometric number of gaseous species in going from reactants to products.

It should be noted that while computing $\Delta\nu_g$ of a reaction, only the stoichiometric numbers of gaseous species is counted and those of liquids and solids are completely ignored.

Example 3.12.1

For the reaction

$$\frac{1}{2} H_2(g) + \frac{1}{2} Br_2(g) \rightarrow HBr(g)$$

$\Delta_r H = -51.823$ kJ mol^{-1} at 373 K. Calculate the value of $\Delta_r U$ for this reaction at 373 K.

Solution In order to calculate $\Delta_r U_{373\,K}$, we employ the equation

$$\Delta_r U = \Delta_r H - (\Delta \nu_g) RT$$

Since $\Delta \nu_g = 0$ for this reaction, therefore

$$\Delta_r U = \Delta_r H = -\ 51.823 \text{ kJ mol}^{-1}$$

Example 3.12.2 For the reaction

$$2A_2(s) + 5B_2(g) \rightarrow 2A_2B_5(g) \qquad \Delta_r U_{298K} = 62.76 \text{ kJ mol}^{-1}$$

Calculate $\Delta_r H_{298K}$ for this reaction.

Solution We have

$$\Delta_r H = \Delta_r U + (\Delta \nu_g) RT$$

$\Delta \nu_g$ for this reaction $= 2 - 5 = -3$

Thus $$\begin{aligned} \Delta_r H &= 62.76 \text{ kJ mol}^{-1} + (-3)\ (8.314 \times 10^{-3} \text{ kJ K}^{-1} \text{ mol}^{-1})\ (298 \text{ K}) \\ &= 62.76 \text{ kJ mol}^{-1} - 7.43 \text{ kJ mol}^{-1} \\ &= 55.33 \text{ kJ mol}^{-1} \end{aligned}$$

3.13 ADIABATIC FLAME TEMPERATURE

At Constant Pressure

Adiabatic flame temperature is a temperature which the system attains if the changes in the system are carried out under adiabatic conditions. For example, if we are carrying out an exothermic reaction under adiabatic conditions, the heat involved would raise the temperature of the system. The rise in temperature can be calculated by considering the reaction to take place in the following two steps, both at constant pressure.

Step I: Reactants$(T_0, p) \rightarrow$ Products(T_0, p) $\qquad \Delta_r H_{T_0}$

Step II: Products$(T_0, p) \rightarrow$ Products(T_f, p) $\qquad \Delta_r H_2$

The heat involved in the second step is given by

$$\Delta_r H_2 = \int_{T_0}^{T_f} C_p(\text{products})\, dT$$

Since the overall reaction is the sum of the above two steps, i.e.

$$\text{Reactants}(T_0, p) \rightarrow \text{Products}(T_f, p)$$

the net heat change is given by

$$\Delta_r H = \Delta_r H_{T_0} + \Delta_r H_2$$

Since the reaction is carried out under adiabatic conditions, it is obvious that

$$\Delta_r H = 0.$$

Thus $$\Delta_r H_{T_0} + \Delta_r H_2 = 0$$

or $$\Delta_r H_2 = -\Delta_r H_{T_0}$$

or $$\int_{T_0}^{T_f} C_p(\text{products})\, dT = -\Delta_r H_{T_0}$$

If C_ps are considered independent of temperature, then

$$C_p(\text{products})\,(T_f - T_0) = -\Delta_r H_{T_0}$$

or
$$T_f = \frac{-\Delta_r H_{T_0}}{C_p(\text{products})} + T_0 \qquad (3.7.1)$$

However, the approximation that C_ps are independent of temperature is not correct since T_f is a very large quantity ($\Delta_r H_{T_0} \simeq$ kJ and $C_p \simeq$ J) of the order of thousands of kelvin.

Illustration

Taking the example of

$$CH_4(g) + 2O_2(g) \rightarrow CO_2(g) + 2H_2O(g)$$

we have

$$\begin{aligned}\Delta_r H^\circ_{298K} &= \Delta_f H^\circ(CO_2, g) + 2\Delta_f H^\circ(H_2O, g) - \Delta_f H^\circ(CH_4, g)\\ &= [-393.51 + 2\,(-241.82) - (-74.81)]\ \text{kJ mol}^{-1}\\ &= -802.34\ \text{kJ mol}^{-1}\end{aligned}$$

$$\begin{aligned}C_p(CO_2 + 2H_2O)_{298K} &= [37.11 + 2(33.58)]\ \text{J K}^{-1}\ \text{mol}^{-1}\\ &= 104.27\ \text{J K}^{-1}\ \text{mol}^{-1}\end{aligned}$$

Thus
$$T_f = \frac{802.34 \times 10^3\ \text{J mol}^{-1}}{104.27\ \text{J K}^{-1}\ \text{mol}^{-1}} + 298\ \text{K} = 7\,694.83\ \text{K} + 298\ \text{K}$$

$$= 7\,992.83\ \text{K} \simeq 8\,000\ \text{K}$$

Burning the Compound in Air

If the compound is burnt in air, each mole of oxygen is associated with four moles of nitrogen and hence these will also be raised to the final temperature. In the case of methane which requires two moles of oxygen, eight moles of nitrogen must be raised to the final temperature. In this case, the heat capacity of the products will be

$$\begin{aligned}C_p(CO_2 + 2H_2O + 8N_2)_{298K} &= [37.11 + 2(33.58) + 8(29.13)]\ \text{J K}^{-1}\ \text{mol}^{-1}\\ &= 337.31\ \text{J K}^{-1}\ \text{mol}^{-1}\end{aligned}$$

$$T_f = \frac{802.34 \times 10^3\ \text{J mol}^{-1}}{337.3\ \text{J K}^{-1}\ \text{mol}^{-1}} + 298\ \text{K} = 2\,378.71\ \text{K} + 298\ \text{K}$$

$$= 2\,676.71\ \text{K} = 2\,700\ \text{K}$$

At Constant Volume

If the reaction takes place in a closed vessel, i.e. under the condition of constant volume, then $\Delta_r U$ and C_V replace $\Delta_r H$ and C_p, respectively, i.e.

$$T_f = \frac{(-\Delta_r U_{T_0})}{C_V(\text{products})} + T_0$$

Example 3.13.1

(a) Calculate the adiabatic flame temperature for the following reaction carried out under constant volume condition:

$$CH_4(g) + 2O_2(g) \rightarrow CO_2(g) + 2H_2O(g) \qquad \Delta_r H_{298K} = -803.33\ \text{kJ mol}^{-1}$$

(b) If the above reaction is carried out using air, what would be the flame temperature (air to be taken as 80% N_2 and 20% O_2).

Given: $C_{p,\mathrm{m}}(\mathrm{CO_2, g}) = 36.29\ \mathrm{J\ K^{-1}\ mol^{-1}}$
$C_{p,\mathrm{m}}(\mathrm{H_2O, g}) = 33.58\ \mathrm{J\ K^{-1}\ mol^{-1}}$
$C_{p,\mathrm{m}}(\mathrm{N_2, g}) = 29.29\ \mathrm{J\ K^{-1}\ mol^{-1}}$

Solution **(a)** For the reaction under constant volume condition, we have

$$\Delta T_\mathrm{f} = T_\mathrm{f} - T_\mathrm{i} = \frac{-\Delta_\mathrm{r} U_{298\mathrm{K}}}{C_V(\text{products})}$$

Using $C_{V,\mathrm{m}} = C_{p,\mathrm{m}} - R$, we have

$$C_V\{\mathrm{CO_2 + 2H_2O}\} = [27.976 + 2 \times 25.266]\ \mathrm{J\ K^{-1}\ mol^{-1}}$$
$$= 78.508\ \mathrm{J\ K^{-1}\ mol^{-1}}$$

Since $\Delta \nu_\mathrm{g} = 0$, therefore $\Delta_\mathrm{r} U = \Delta_\mathrm{r} H$

Thus

$$T_\mathrm{f} = \frac{803\ 330\ \mathrm{J\ mol^{-1}}}{78.508\ \mathrm{J\ K^{-1}\ mol^{-1}}} + 298\ \mathrm{K}$$
$$= 10\ 232\ \mathrm{K} + 298\ \mathrm{K} = 10\ 530\ \mathrm{K}$$

(b) If the reaction is carried out using air, then we have

$$C_V = [27.976 + 2 \times 25.266 + 8 \times 20.976]\ \mathrm{J\ K^{-1}\ mol^{-1}}$$
$$= 246.32\ \mathrm{J\ K^{-1}\ mol^{-1}}$$

$$T_\mathrm{f} = \frac{803\ 330\ \mathrm{J\ mol^{-1}}}{246.32\ \mathrm{J\ K^{-1} mol^{-1}}} + 298\ \mathrm{K} = 3\ 559\ \mathrm{K}$$

REVISIONARY PROBLEMS

3.1 Explain with examples the following terms:
(a) Standard molar enthalpy
(b) Standard enthalpy of an element
(c) Standard molar enthalpy of formation
(d) Hess's Law of constant heat summation
(e) Enthalpy of formation
(f) Enthalpy of combustion
(g) Integral enthalpies of solution and dilution
(h) Differential enthalpies of solution and dilution
(i) Enthalpy of hydration
(j) Enthalpy of neutralization
(k) Enthalpy of ionization
(l) Enthalpy of transition
(m) Enthalpy of precipitation
(n) Enthalpy of formation of ions
(o) Bond enthalpy and bond dissociation enthalpy
(p) Adiabatic flame temperature

3.2 Derive the following relations:
(i) Kirchhoff's relation
(ii) $\Delta_\mathrm{r} H = \Delta_\mathrm{r} U + (\Delta \nu_\mathrm{g})\, RT$

TRY YOURSELF PROBLEMS

3.1 A student made the following erroneous statement in a laboratory record on bomb calorimeter: $\Delta_r H = \Delta_r U + p\Delta V$. Since the bomb calorimeter process is a constant volume one, $\Delta V = 0$ and hence $\Delta_r H = \Delta_r U$. Why is this argument incorrect?

3.2 For a reaction involving only condensed phase, $\Delta_r H = \Delta_r U$. Explain why.

3.3 It is often stated that the integral enthalpy of solution is identical to the differential enthalpy of solution for a very dilute solution. Explain why.

3.4 While stating the $\Delta_r H$ value of a chemical reaction, it is understood that the temperature and pressure of the products are identical with those of reactants? Explain why.

3.5 Show mathematically that the differential enthalpies of solution and dilution are essentially the partial molar enthalpies of solute and solvent, in solution respectively.

3.6 It is stated that the bond enthalpy and bond dissociation enthalpy of a diatomic molecule are identical. Explain why it should be so.

3.7 The expression of the adiabatic flame temperature is an approximate one:

$$T = T_0 + \frac{(-\Delta_r H°)}{\Sigma C_p(\text{products})}$$

State what assumptions were used in deriving the above expression.

3.8 Derive the expression of the adiabatic flame temperature if the reaction is carried out under constant volume condition.

3.9 Suppose the dissolution of a solute in water is an exothermic process. Draw a typical diagram illustrating the variation of integral enthalpy of solution with the amount of the solute dissolved in a fixed quantity (say, 1 kg) of water.

3.10. Comment upon the statement:

The differential enthalpy of solution (or dilution) at a given concentration may be regarded as the instantaneous rate of change of the integral enthalpy of solution (or dilution) with the solute (or solvent) concentration, i.e.

$$\Delta H_2 = \lim_{\delta m_2 \to 0} \frac{\delta(\Delta H)}{\delta m_2}; \qquad \Delta H_1 = \lim_{\delta m_1 \to 0} \frac{\delta(\Delta H)}{\delta m_1}$$

NUMERICAL PROBLEMS

Enthalpy of Reaction

3.1 From the following data at 25 °C

Reaction	$\Delta_r H°$/kJ mol^{-1}
$\frac{1}{2} H_2(g) + \frac{1}{2} O_2(g) \rightarrow OH(g)$	42.09
$H_2(g) + \frac{1}{2} O_2(g) \rightarrow H_2O(g)$	–241.84
$H_2(g) \rightarrow 2H(g)$	435.88
$O_2(g) \rightarrow 2O(g)$	495.04

calculate $\Delta_r H°$ for the following reactions.

(a) $OH(g) \rightarrow H(g) + O(g)$

(b) $H_2O(g) \rightarrow 2H(g) + O(g)$

(c) $H_2O(g) \rightarrow H(g) + OH(g)$

(*Ans.* 423.37, 925.24, 501.87 kJ mol^{-1})

3.2 The $\Delta_r H^\circ = -41.84$ kJ mol^{-1} for the neutralization reaction

$$HCO_3^-(aq) + OH^-(aq) \rightarrow H_2O(1) + CO_3^{2-}(aq)$$

Compute $\Delta_r H^\circ$ for the reaction

$$HCO_3^-(aq) \rightarrow H^+(aq) + CO_3^{2-}(aq)$$ (*Ans.* 14.0 kJ mol^{-1})

3.3 Calculate the enthalpy of formation of I_2O_5 from the following data:

(i) $I_2O_5(s) + H_2O(1) \rightarrow 2HIO_3(aq)$ $\quad \Delta_r H = 3.598$ kJ mol^{-1}

(ii) $KI(aq) + 3HClO(aq) \rightarrow HIO_3(aq) + 2HCl(aq) + KCl(aq)$

$\Delta_r H = -322.42$ kJ mol^{-1}

(iii) $NaOH(aq) + HClO(aq) \rightarrow NaOCl(aq) + H_2O(1)$

$\Delta_r H = -44.35$ kJ mol^{-1}

(iv) $NaOH(aq) + HCl(aq) \rightarrow NaCl(aq) + H_2O(1)$

$\Delta_r H = -57.32$ kJ mol^{-1}

(v) $2NaOH(aq) + Cl_2(g) \rightarrow NaOCl(aq) + NaCl(aq) + H_2O(1)$

$\Delta_r H = -99.79$ kJ mol^{-1}

(vi) $2KI(aq) + Cl_2(g) \rightarrow 2KCl(aq) + I_2(s)$ $\quad \Delta_r H = -223.85$ kJ mol^{-1}

(vii) $H_2(g) + \frac{1}{2}O_2(g) \rightarrow H_2O(l)$ $\quad \Delta_r H = -285.00$ kJ mol^{-1}

(viii) $\frac{1}{2}H_2(g) + \frac{1}{2}Cl_2(g) \rightarrow HCl(g)$ $\quad \Delta_r H = -92.30$ kJ mol^{-1}

(ix) $HCl(g) + nH_2O \rightarrow HCl(aq)$ $\quad \Delta_r H = -75.15$ kJ mol^{-1}

(*Ans.* – 163.81 kJ mol^{-1})

3.4 From the enthalpy of formation at 25 °C

Solution	$H_2SO_4.600$ aq	KOH.200 aq	$KHSO_4.800$ aq	$K_2SO_4 \cdot 1000$ aq
$\Delta_f H$/kJ mol^{-1}	– 888.47	– 480.41	– 1 147.67	– 1 408.96

calculate $\Delta_r H$ for the reactions:

(i) $H_2SO_4 \cdot 600\text{ aq} + KOH \cdot 200\text{ aq} \rightarrow KHSO_4 \cdot 800\text{ aq} + H_2O(1)$

(ii) $KHSO_4 \cdot 800\text{ aq} + KOH \cdot 200\text{ aq} \rightarrow K_2SO_4 \cdot 1000\text{ aq} + H_2O(1)$

Given that

$H_2(g) + \frac{1}{2}O_2(g) \rightarrow H_2O(1)$ $\quad \Delta_r H = -285.85$ kJ mol^{-1}

(*Ans.* – 64.64 kJ mol^{-1}, – 66.73 kJ mol^{-1})

3.5 On the basis of the following data, evaluate the standard enthalpy of formation of tungsten carbide WC(s).

(i) $W(s) + \frac{3}{2}O_2(g) \rightarrow WO_3(s)$ $\quad \Delta_r H^\circ_{298K} = -837.47$ kJ mol^{-1}

(ii) $WC(s) + \frac{5}{2}O_2(g) \rightarrow WO_3(s) + CO_2(g)$ $\quad \Delta_r H^\circ_{298K} = -1\,195.79$ kJ mol^{-1}

(iii) $C(\text{graphite}) + O_2(g) \rightarrow CO_2(g)$ $\quad \Delta_r H^\circ_{298K} = -393.51$ kJ mol^{-1}

(*Ans.* –35.19 kJ mol^{-1})

3.6 If the enthalpy of formation of HCl(g) and Cl^-(aq) are –92.30 kJ mol^{-1} and –167.44 kJ mol^{-1}, find the enthalpy of solution of hydrogen chloride gas.

(*Ans.* –75.14 kJ mol^{-1})

Enthalpy of Solution

3.7 At 25 °C, the enthalpy change for the reaction

$$H_2SO_4 + 5H_2O \rightarrow H_2SO_4 \cdot 5H_2O \text{ (all liquids)}$$

is –58.032 kJ mol^{-1}. Calculate the temperature change if 1 mol of H_2SO_4 is dropped into 5 mol of H_2O at 25 °C. Assume no heat loss to the surroundings and that the specific heat capacity of the solution is 4.184 J K^{-1} g^{-1}. (*Ans.* 73.78 °C)

3.8 Given the following information:

$$1\ H_2SO_4 + 1\ H_2O \rightarrow 1\ H_2SO_4 \cdot 1\ H_2O \qquad \Delta_r H = -28.075 \text{ kJ mol}^{-1} \text{ at } 25\ °C$$

$$1\ H_2SO_4 + 25\ H_2O \rightarrow 1\ H_2SO_4 \cdot 25\ H_2O \qquad \Delta_r H = -72.300 \text{ kJ mol}^{-1} \text{ at } 25\ °C$$

Calculate the enthalpy change at 25 °C for the following dilution

$$1\ H_2SO_4 \cdot 1\ H_2O + 24\ H_2O \rightarrow 1\ H_2SO_4 \cdot 25\ H_2O$$

Calculate the increase in temperature for this dilution if there is no heat loss to the surroundings. Assuming the specific heat capacity of the more dilute solution to be 4.184 J K^{-1} g^{-1}. (*Ans.* 44.225 kJ mol^{-1}, 19.3 K rise)

3.9 Using the bond enthalpy data given below, estimate the enthalpy of formation of acetic acid.

Bond	*Bond enthalpies*
C—H	413.38 kJ mol^{-1}
C—C	347.69 kJ mol^{-1}
C==O	728.02 kJ mol^{-1}
C—O	351.46 kJ mol^{-1}
O—H	462.75 kJ mol^{-1}

Enthalpy of formation	
C(g)	718.39 kJ mol^{-1}
H	217.96 kJ mol^{-1}
O	247.52 kJ mol^{-1}

The observed $\Delta_f H°$ for acetic acid is – 438.15 kJ mol^{-1}. Compute the resonance energy of acetic acid. (*Ans.* –326.40 kJ mol^{-1}, –111.75 kJ mol^{-1})

3.10 The bond dissociation enthalpies of H_2(g) and N_2(g) are +435.95 kJ mol^{-1} and +941.8 kJ mol^{-1} and the enthalpy of formation of NH_3(g) is – 46.024 kJ mol^{-1}.

(i) What is the enthalpy of atomization of NH_3(g)?

(ii) What is the average bond enthalpy of N—H bond?

(*Ans.* (i) 1.171 MJ mol^{-1}, (ii) 390.367 kJ mol^{-1})

3.11. (a) Value of $\Delta_r H^\circ_{298K}$ for the reaction

$$N_2H_4(g) \rightarrow N_2H_2(g) + H_2(g)$$

is found to be 108.76 kJ mol^{-1}. Assuming that the structure of N_2H_2 is HN = NH, calculate $\varepsilon_{N=N}$. Given the following bond enthalpies: ε_{N-N} = 163.18 kJ mol^{-1}, ε_{N-H} = 390.79 kJ mol^{-1} and ε_{H-H} = 435.89 kJ mol^{-1}.

(b) Given that $\Delta_f H$(298 K) of hydrazine N_2H_4 is 94.98 kJ mol^{-1}, calculate $\Delta_f H$(298 K) of N_2H_2(g).

(c) Calculate $\Delta_f H°$(298 K) of N_2H_4 and N_2H_2 from the bond enthalpy data and compare with those of part (b). Given: $\varepsilon_{N\equiv N}$ = 942.24 kJ mol^{-1}.

(d) Calculate $\Delta_r U_{298K}$ for the reaction of part (a). Compute the required heat capacities by using principle of equipartition of energy.

(*Ans.* (a) 400.1 kJ mol^{-1} (b) 203.76 kJ mol^{-1}
(c) 87.68 kJ mol^{-1} and 196.65 kJ mol^{-1}
(d) 108.38 kJ mol^{-1})

(e) The enthalpy of atomization of PH_3 is +953.95 kJ mol^{-1} and that of P_2H_4 is +1485.3 kJ mol^{-1}. What is the bond enthalpy of the P—P bond?

(*Ans.* 213.38 kJ mol^{-1})

Kirchhoff 's Relation

3.12 Calculate $\Delta_r H$ for each of the following change of state:

(a) $H_2O(1, 1 \text{ bar}, 300 \text{ K}) \rightarrow H_2O(1, 1 \text{ bar}, 350 \text{ K})$

(b) $3H_2O(s, 1 \text{ bar}, 263 \text{ K}) \rightarrow 3H_2O(g, 1 \text{ bar}, 500 \text{ K})$

(c) $H_2O(s, 1 \text{ bar}, 263 \text{ K}) \rightarrow H_2O(1, 1 \text{ bar}, 263 \text{ K})$

Data for water, all at 1 bar pressure, are

$$\Delta_{fus}H = 6.025 \text{ kJ mol}^{-1} \text{ at } 273 \text{ K}$$
$$\Delta_{vap}H = 40.585 \text{ kJ mol}^{-1} \text{ at } 373 \text{ K}$$
$$C_p(s) = 39.748 \text{ J K}^{-1} \text{ mol}^{-1}$$
$$C_p(1) = 75.312 \text{ J K}^{-1} \text{ mol}^{-1}$$
$$C_p(g) = 31.129 \text{ J K}^{-1} \text{ mol}^{-1} + (0.008\ 37 \text{ J K}^{-2} \text{ mol}^{-1})\ T$$

(*Ans.* (a) 3.77 kJ mol^{-1}, (b) 176.93 kJ mol^{-1}, (c) 5.67 kJ mol^{-1})

3.13 The $\Delta_r H$ of vaporization of water at 0 °C, i.e.

$$H_2O(1, 4.58 \text{ mmHg}, 273 \text{ K}) \rightarrow H_2O(g, 4.58 \text{ mmHg}, 273 \text{ K})$$

is given by +44.877 kJ mol^{-1}. Make the generally acceptable assumptions that the ΔH for compression of a liquid (or solid) is approximately zero and that $H_2O(g)$ is an ideal gas. Also assume that the heat capacity for liquid water C_p = 75.31 J K^{-1} mol^{-1} and that for gaseous water C_p = 33.30 J K^{-1} mol^{-1} are constant. Estimate $\Delta_r H°$ of vaporization at 1 bar and 100 °C. (*Ans.* $\Delta_r H°$ = 40.68 kJ mol^{-1})

3.14 For the reaction

$$CaO(s) + CO_2(g) \rightarrow CaCO_3(s)$$

$$\Delta_r H°/\text{J mol}^{-1} = -\ 177\ 820 - 2.761\ (T/\text{K}) + 9.016 \times 10^{-3}\ (T/\text{K})^2 + \frac{17.154 \times 10^3}{(T/\text{K})}$$

determine, $\Delta_r C_p$ and $\Delta_r C_V$ as functions of *T*.

3.15 From the value of $C_{p,\text{m}}$ as a function of temperature given as

$$C_{p,\text{m}}(H_2, g)/\text{J K}^{-1} \text{ mol}^{-1} = 29.066 - 0.836 \times 10^{-3}\ (T/\text{K})$$
$$C_{p,\text{m}}(Br_2, g)/\text{J K}^{-1} \text{ mol}^{-1} = 35.241 + 4.075 \times 10^{-3}(T/\text{K})$$
$$C_{p,\text{m}}(HBr, g)/\text{J K}^{-1} \text{ mol}^{-1} = 27.521 + 3.995 \times 10^{-3}\ (T/\text{K})$$

and from the date $\Delta_f H(HBr, g)$ = –36.23 kJ mol^{-1} and $\Delta_{vap}H(Br_2, 1)$ = 30.71 kJ mol^{-1}, calculate $\Delta_r H_{373K}$ for the reaction $\frac{1}{2} H_2(g) + \frac{1}{2} Br_2(g) \rightarrow HBr(g)$

(*Ans.* –51.823 kJ mol^{-1})

Relation between $\Delta_r H$ and $\Delta_r U$

3.16 The enthalpy changes of the following reactions at 25°C are

$$Na(s) + \tfrac{1}{2} Cl_2(g) \rightarrow NaCl(s); \qquad \Delta_r H = -\ 411.0 \text{ kJ mol}^{-1}$$

$H_2(g) + S(s) + 2O_2(g) \rightarrow H_2SO_4(1)$; $\Delta_r H = -811.3$ kJ mol^{-1}

$2Na(s) + S(s) + 2O_2(g) \rightarrow Na_2SO_4(s)$; $\Delta_r H = -1\,382.3$ kJ mol^{-1}

$\frac{1}{2} H_2(g) + \frac{1}{2} Cl_2(g) \rightarrow HCl(g)$; $\Delta_r H = -92.3$ kJ mol^{-1}

Form these data, find the heat change of reaction at constant volume at 25 °C for the process

$2NaCl(s) + H_2SO_4(1) \rightarrow Na_2SO_4(s) + 2HCl(g)$ (*Ans.* $\Delta_r U = 61.44$ kJ mol^{-1})

3.17 The combustion of 1 g of benzene in a bomb calorimeter evolves 41.746 kJ of heat at 25 °C,

(i) What is $\Delta U°$ for combustion of benzene?

(ii) Calculate $\Delta_f H°$(benzene) if

$\Delta_f H°(CO_2, g) = -393.129$ kJ mol^{-1} and $\Delta_f H°$ $(H_2O, 1) = -285.577$ kJ mol^{-1}.

(*Ans.* (i) –3 256.2 kJ mol^{-1}; (ii) –44.4 kJ mol^{-1})

Adiabatic Flame Temperature

3.18 Estimate the maximum possible temperature of a bunsen burner flame. Assume that the gas is pure methane and that it is premixed with sufficient air to permit complete combustion. Given that

$CH_4(g) + 2O_2(g) \rightarrow 2H_2O(1) + CO_2(g)$ $\Delta_r H_{298K} = -890.36$ kJ mol^{-1}

$H_2O(1) \rightarrow H_2O(g)$ $\Delta_r H_{298K} = 44.02$ kJ mol^{-1}

Calculate $C_{p,\,m}$ of $H_2O(g)$, $CO_2(g)$ and $N_2(g)$ from the equipartition principle (including vibrational contributions). (*Ans.* $T \approx 2\,000$ K)

What would be the final temperature if the methane is allowed to burn with the requisite amount of oxygen? (*Ans.* $T \approx 4\,800$ K)

3.19 Suppose 2.0 mol of $CH_4(g)$ is mixed with 5.0 mol of $O_2(g)$ in an adiabatic enclosure at 298 K. A spark is produced in the mixture and the CH_4 is completely burnt in the oxygen to CO_2 and H_2O. Assume ideal gas behaviour and compute the final temperature of the gas mixture. Given heat capacities data are

$O_2(g)$ $C_p/\text{J K}^{-1}\text{ mol}^{-1} = 29.96 + 4.184 \times 10^{-3}\ (T/\text{K})$

$CO_2(g)$ $C_p/\text{J K}^{-1}\text{ mol}^{-1} = 44.23 + 8.786 \times 10^{-3}\ (T/\text{K})$

$H_2O(g)$ $C_p/\text{J K}^{-1}\text{ mol}^{-1} = 30.54 + 10.293 \times 10^{-3}\ (T/\text{K})$

$CH_4(g)$ $C_p/\text{J K}^{-1}\text{ mol}^{-1} = 23.64 + 47.865 \times 10^{-3}\ (T/\text{K})$

Miscellaneous Numericals

3.20 Given the following data:

(i) Standard molar enthalpy of formation at 298 K and 1 bar pressure, are

$C_2H_5OH(1)$	–277.65 kJ mol^{-1}	$CO(g)$	–110.54 kJ mol^{-1}
$CO_2(g)$	–393.51 kJ mol^{-1}	$ZnO(s)$	–347.98 kJ mol^{-1}
$H_2O(1)$	–285.85 kJ mol^{-1}		

(ii) Molar changes of enthalpy on changes of state of aggregation:

Evaporation of H_2O at 373 K and 1 bar = 40.58 kJ mol^{-1}

Melting of Zn at 692 K and 1 bar = 6.569 kJ mol^{-1}

Evaporation of Zn at 1 180 K and 1 bar = 127.61 kJ mol^{-1}.

(iii) Molar heat capacity at constant pressure in J K^{-1} mol^{-1}.

$H_2O(1)$	75.312
$H_2O(g)$	$30.070 + 0.009\,92\ (T/\text{K}) - 0.870 \times 10^{-6}\ (T/\text{K})^2$
Zn(s)	25.104
Zn(1)	33.472
Zn(g)	20.92

$ZnO(s)$	41.84
$CO(g)$	27.196 + 0.004 18 (T/K)
$CO_2(g)$	33.472 + 0.016 7 (T/K)

(a) Calculate $\Delta_r H°$ for the reaction at 298 K and 1 bar

$C_2H_5OH(1) + 3O_2(g) \rightarrow 2CO_2(g) + 3H_2O(1)$ (*Ans.* –1.366 9 MJ mol^{-1})

(b) $\Delta_r H°$ for the reaction

$$C_6H_6(1) + \frac{15}{2} O_2(g) \rightarrow 6CO_2(g) + 3H_2O(1)$$

is –3.2677 MJ mol^{-1}. Calculate $\Delta_f H°$ for liquid benzene. (*Ans.* 49.09 kJ mol^{-1})

(c) Calculate $\Delta_r H°$ of the reaction at 298 K and 1 bar

$C_{12}H_{22}O_{11}(s) + H_2O(1) \rightarrow 4C_2H_5OH(1) + 4CO_2(g)$

Given $\Delta_r H°$ of the following reaction

$C_{12}H_{22}O_{11}(s) + 12O_2(g) \rightarrow 12CO_2(g) + 11H_2O(1)$; $\Delta H° = –5.646\ 7$ MJ mol^{-1}

(*Ans.* –179.04 kJ mol^{-1})

(d) Calculate $\Delta_f H°$ for water in the hypothetical state of $H_2O(g)$ at 298 K and 1 bar. (*Ans.* –242.11 kJ mol^{-1})

(e) Calculate $\Delta_f H$ for each of the following changes at 298 K and 1 bar.

$ZnO(s) + CO(g) \rightarrow Zn(s) + CO_2(g)$ (*Ans.* 65.02 kJ mol^{-1})

$ZnO(s) + CO(g) \rightarrow Zn(1) + CO_2(g)$ (*Ans.* 68.30 kJ mol^{-1})

$ZnO(s) + CO(g) \rightarrow Zn(g) + CO_2(g)$ (*Ans.* 206.94 kJ mol^{-1})

(f) Calculate $\Delta_r H$ for the following reaction at 1 300 K and 1 bar

$ZnO(s) + CO(g) \rightarrow Zn(g) + CO_2(g)$ (*Ans.* 346.1 kJ mol^{-1})

3.21 The enthalpies of formation at 298 K of gaseous CO_2, water vapour and liquid CH_3COOH are – 393.3 kJ mol^{-1}, – 241.84 kJ mol^{-1} and – 629.78 kJ mol^{-1}, respectively. The enthalpy of combustion of methane gas to CO_2 and water vapour is – 806.26 kJ mol^{-1}. The enthalpy of vaporization of water at 100 °C is 39.33 kJ mol^{-1}. Heat capacity values (C_p) in J K^{-1} mol^{-1} are

$CH_4(g)$	37.656	$H_2O(g)$	30.54
$CH_3CHO(g)$	52.3	$H_2O(1)$	75.31
$CO(g)$	31.38		

(a) Calculate the enthalpy of formation of liquid water at 298 K.

(b) Calculate the enthalpy change at 298 K for the reaction

$CH_3COOH(1) \rightarrow CH_4(g) + CO_2(g)$

(c) Calculate the temperature at which $\Delta_r H$ for the reaction

$CH_3CHO(g) \rightarrow CH_4(g) + CO(g)$

will be zero. $\Delta_r H_{298K}$ is –16.735 kJ mol^{-1}

(*Ans.* (a) –284.51 kJ mol^{-1}; (b) 165.76 kJ mol^{-1}; (c) 1000 K)

3.22 Calculate the enthalpy change for the following reaction.

$Hg(1) + I_2(g) \rightarrow HgI_2(1)$; $\Delta_r H$ = ? at 600 K, 1 bar

Given the enthalpy change for the reaction

$Hg(1) + I_2(s) \rightarrow HgI_2(\alpha)$; $\Delta_r H = –105.44$ kJ mol^{-1} at 298 K and 1 bar

and following thermal data

Hg(1) : $C_{p,\mathrm{m}} = 27.656\ \mathrm{J\ K^{-1}\ mol^{-1}}$

I_2(s) : $C_{p,\mathrm{m}} = 40.125\ \mathrm{J\ K^{-1}\ mol^{-1}} + (49.79 \times 10^{-3}\ \mathrm{J\ K^{-2}\ mol^{-1}})T$

: $\Delta_{\mathrm{fus}} H = 15.774\ \mathrm{kJ\ mol^{-1}}$ at 386.8 K

I_2(1) : $C_{p,\mathrm{m}} = 80.33\ \mathrm{J\ K^{-1}\ mol^{-1}}$; $\Delta_{\mathrm{vap}} H = 41.714\ \mathrm{kJ\ mol^{-1}}$ at 456 K

I_2(g) : $C_{p,\mathrm{m}} = 37.196\ \mathrm{J\ K^{-1}\ mol^{-1}}$

$HgI_2(\alpha)$: $C_{p,\mathrm{m}} = 77.404\ \mathrm{J\ K^{-1}\ mol^{-1}}$

$\alpha - \beta$ transition; $\Delta_{\mathrm{trs}} H = 2.720\ \mathrm{kJ\ mol^{-1}}$ at 403 K

$HgI_2(\beta)$: $C_{p,\mathrm{m}} = 84.517\ \mathrm{J\ K^{-1}\ mol^{-1}}$; $\Delta_{\mathrm{fus}} H = 18.828\ \mathrm{kJ\ mol^{-1}}$ at 523 K

HgI_2(l) : $C_{p,\mathrm{m}} = 104.6\ \mathrm{J\ K^{-1}\ mol^{-1}}$

(*Ans.* – 139.4 kJ $\mathrm{mol^{-1}}$)

4 Second Law of Thermodynamics

4.1 NECESSITY OF THE SECOND LAW

Limitation of First Law

Our primary interest in thermodynamics is to use it to establish a criterion for the feasibility of a given chemical or physical transformation under specified conditions. The first law of thermodynamics does not supply this information as it deals with only conservation of energy. This law has to be satisfied whether a reaction is spontaneous or otherwise. It cannot, however explain naturally occurring processes, for example, flow of heat from a warmer to a cooler body, expansion of gases into vacuum, interdiffusion of two gases, freezing of supercooled water, reaction of a mixture of H_2 and F_2, reaction between H_2 and O_2, etc.

The reverse of any one of the above processes, which will also be in accordance with the first law, does not however, occur. It would be possible to predict the nature of the process (natural or unnatural) if the system possessed one or more properties which always change in one direction when the system undergoes a spontaneous change, in the opposite direction if it undergoes a nonspontaneous change.

Criterion of Equilibrium for a Mechanical System

Everyday experience is that whenever there exists a difference in certain properties such as pressure, temperature, electric potential or concentration, the spontaneous change is always that which eliminates the difference. If such a type of difference exists in two systems, and they are put together, the system as a whole is in the nonequilibrium state and it will move towards the equilibrium state by removing the difference in the property, e.g., heat flows if temperature is not uniform, gas flows if pressure is not uniform, etc. When the property has attained a uniform value, it reaches the equilibrium state. The system will continue to be in this equilibrium state unless it disturbed by an external agent.

Extension of Criterion to a Chemical System

Analogous to the mechanical system, in which the system tries to have the minimum potential energy, it was suggested that for a chemical reaction, we should have the minimization of heat content of the system. This means that a reaction in which heat is evolved must be spontaneous whereas a reaction in which heat is absorbed must be nonspontaneous, i.e. all exothermic reactions must be spontaneous while all endothermic reactions must be nonspontaneous. But there exist a few exceptions in which ΔU is positive but nevertheless the reaction proceeds spontaneously, such as the vaporization of water at low pressure.

$$H_2O(1,\ 1.0\ \text{kPa},\ 25\ °C) \rightarrow H_2O(g,\ 1.0\ \text{kPa},\ 25\ °C); \quad \Delta_r U = 41.42\ \text{kJ mol}^{-1}$$

Hence $\Delta_r U$ cannot be used as the criterion of spontaneity. Similarly, $\Delta_r H$ cannot be used because there are many reactions that proceeds spontaneously with a positive $\Delta_r H$. One example is the polymorphic transformation of low temperature quartz to high temperature quartz at 848 K.

$$\text{SiO}_2(\text{low quartz}) \rightarrow \text{SiO}_2(\text{high quartz}) \qquad \Delta_r H = 0.88 \text{ kJ mol}^{-1}$$

Thus, it is seen that functions U and H derived from the first law of thermodynamics cannot predict the spontaneity of a chemical reaction.

Entropy Function

It is the second law of thermodynamics which identifies a new state function called the *entropy* (and the related functions) which provides a criterion for identifying this equilibrium state of a system. We will see later that the entropy of the universe (system + surroundings) increases for irreversible processes whereas it remains constant for reversible processes.

The function entropy was established from the work of Carnot, who was interested in the factors which limit the efficiency of a steam engine (a device which converts heat into work). We will, for convenience, first discuss Carnot cycle and then introduce the second law of thermodynamics. Then it will be shown how the work of Carnot introduces a new state function, entropy S, which can be used to predict the nature of a chemical reaction, whether it is spontaneous or nonspontaneous.

4.2 CARNOT CYCLE

Description of the Carnot Cycle

The Carnot cycle was named after Sadi Carnot, who was the first to describe this type of idealized reversible cycle. We consider the system to be any fluid, single-phase substance. The system is contained in a frictionless piston and cylinder arrangement. We also use two thermal reservoirs, one at higher temperature θ_2 and the other one at a lower temperature θ_1. The symbols θ_2 and θ_1 are used to indicate temperatures without reference to any specific temperature scale. The system is subjected consecutively to the following reversible changes in state (Fig. 4.2.1).

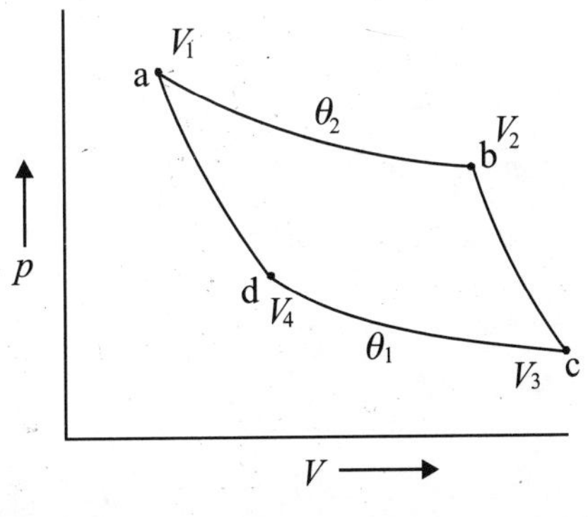

Fig. 4.2.1 **Carnot cycle**

Four Steps of a Carnot Cycle

Step 1 *The isothermal reversible expansion from volume V_1 to volume V_2 at the higher temperature θ_2*

Let w_2 be the work involved in this step. The numerical value of this work will be negative and the magnitude of this work will be represented by the area under the curve ab. (Actually here the work, equal to $| w_2 |$, is done by the system.) Moreover, a quantity of heat q_2 will be absorbed by the system from the thermal reservoir.†

Step 2 *The adiabatic reversible expansion from volume V_2 to volume V_3*

In this step the temperature of the system decreases to θ_1. Let w be the work involved in this step. The numerical value of this work will be negative and the magnitude of this work will be represented by the area under the curve bc. (Actually speaking, the work equal $| w |$ is done by the system.) The heat absorbed by the system is zero as the expansion takes place under adiabatic conditions.

Step 3 *The isothermal reversible compression from volume V_3 to volume V_4 at the temperature θ_1*

In this step, an amount of work w_1, represented by the area under the curve cd, will be done on the system; the numerical value of this work will be positive. Also a quantity of heat q_1, which is negative numerically, is involved in this process. The negative sign means that heat q_1 will be released by the system to the thermal reservoir at the temperature θ_1.

Step 4 *Adiabatic reversible compression from volume V_4 to volume V_1*

In this step, the temperature changes back to θ_2. The amount of work done on the system is w', represented by the area under the curve da. There is no heat absorbed by the system in this step as the compression process is adiabatic.

In the cyclic process described above, the total work involved is

$$w_{\text{total}} = w_2 + w + w_1 + w'$$

The magnitude of the total work is represented by the area within the closed cycle shown in Fig. 4.2.1.

Since the system has come back to its original state, therefore, $\Delta U = 0$. According to the first law of thermodynamics

$$(-w_{\text{total}}) = q_{\text{total}} \qquad (4.2.1)$$

where $q_{\text{total}} = q_2 + q_1$

and $w_{\text{total}} = w_2 + w + w_1 + w'$

Efficiency of a Carnot Cycle

The efficiency of the cycle is defined as the *ratio of the work done by the system to the amount of heat transferred to the system at the higher temperature.* Thus, efficiency

$$\eta = \frac{(-w_{\text{total}})}{q_2} = \frac{q_2 + q_1}{q_2} = 1 + \frac{q_1}{q_2} \qquad (4.2.2)$$

---

†**Note**: q and w carry the same subscript as that of temperature. The subscript 2 and 1 represent higher and lower temperatures, respectively.

4.3 EXPRESSION FOR THE EFFICIENCY OF A CARNOT CYCLE INVOLVING IDEAL GAS AS A WORKING SUBSTANCE

Let the system have one mole of an ideal gas and let the temperatures of the two reservoirs expressed in kelvin be T_2 and T_1, respectively, such that $T_2 > T_1$. We proceed as follows to determine the efficiency of the Carnot cycle.

Step 1 *Isothermal expansion at temperature T_2 from volume V_1 to V_2*
Since $\Delta T = 0$, therefore

$$\Delta U_1 = 0$$

According to the first law of thermodynamics, we have

$$q_2 = -w_2 = RT_2 \ln \frac{V_2}{V_1} \tag{4.3.1}$$

Step 2 *Adiabatic expansion from volume V_2 to V_3*

Temperature of the system after expansion is T_1. Since for adiabatic process $q = 0$, therefore, it follows that

$$w = \Delta U_2 = \int_{T_2}^{T_1} C_{V,\mathrm{m}}\, \mathrm{d}T = C_{V,\mathrm{m}}(T_1 - T_2) \tag{4.3.2}$$

Step 3 Isothermal compression at temperature T_1 from volume V_3 to V_4
Here, we will have

$$\Delta U_3 = 0$$

and
$$q_1 = -w_1 = RT_1 \ln \frac{V_4}{V_3} \tag{4.3.3}$$

(Note that q_1 has negative sign whereas w_1 has positive sign since $V_4 < V_3$.)

Step 4 Adiabatic compression form volume V_4 to V_1. Temperature of the system returns to T_2. Therefore

$$w' = \Delta U_4 = \int_{T_1}^{T_2} C_{V,\mathrm{m}}\, \mathrm{d}T = C_{V,\mathrm{m}}(T_2 - T_1) \tag{4.3.4}$$

Net work involved in the cyclic process is

$$w_{\mathrm{total}} = w_2 + w + w_1 + w'$$

The quantities w and w' cancel each other, being equal but opposite in sign for the ideal gas. Thus, the above expression reduces to

$$w_{\mathrm{total}} = w_2 + w_1$$

which according to Eqs (4.3.1) and (4.3.3) becomes

$$w_{\mathrm{total}} = -RT_2 \ln \frac{V_2}{V_1} - RT_1 \ln \frac{V_4}{V_3} \tag{4.3.5}$$

The ratios V_2/V_1 and V_4 / V_3 are not independent of each other. The relationship between them can be found through Steps 2 and 4.

For step 2

$$\int_{T_2}^{T_1} \frac{C_{V,\mathrm{m}}}{T}\,\mathrm{d}T = -R\ln\frac{V_3}{V_2} \qquad \text{(Eq. 2.11.6)}$$

and for the Step 4

$$\int_{T_1}^{T_2} \frac{C_{V,\mathrm{m}}}{T}\,\mathrm{d}T = -R\ln\frac{V_1}{V_4}$$

The integrals on the left hand side of these expressions are equal in magnitude but opposite in sign. Therefore

$$R\ln\frac{V_3}{V_2} = -R\ln\frac{V_1}{V_4}$$

or $$\frac{V_3}{V_2} = \frac{V_4}{V_1} \quad \text{i.e.} \quad \frac{V_2}{V_1} = \frac{V_3}{V_4}$$

With this relation, the expression for the total work involved in the cyclic process becomes

$$w_{\text{total}} = -\left(R\ln\frac{V_2}{V_1}\right)(T_2 - T_1)$$

and the efficiency is given by

$$\eta = \frac{(-w_{\text{total}})}{q_2} = \frac{\{R\ln(V_2/V_1)\}(T_2 - T_1)}{RT_2\ln(V_2 - V_1)} = \frac{T_2 - T_1}{T_2} \qquad (4.3.6)$$

Thus, the efficiency of a heat engine operating in a Carnot cycle depends only on the two temperatures and primarily, on the difference of the two temperatures between which the engine operates. The greater the difference, the greater the efficiency.

Two special cases of interest are:

(i) If T_1 equals T_2, then no work is done in the cycle. It is because of the fact that the work done by the system during reversible expansion at temperature T is equal to the work done on the system during reversible compression, and thus, the net work done by the cycle is zero. In other words, there cannot be any net conversion of heat into work in an isothermal cycle alone.

(ii) T_1 must be zero or T_2 must be infinity in order to obtain efficiency equal to one. However, this is not possible since neither of these two situations are physically realizable. Thus the efficiency is always less than one. Only a fraction of heat absorbed at a higher temperature can be converted into work.

4.4 TWO STATEMENTS OF SECOND LAW OF THERMODYNAMICS

Having investigated the cyclic operation of a heat engine, we can now state two forms of the second law which can be easily understood from the efficiency of the Carnot cycle.

Kelvin-Planck Statement

We have seen that the work done by a heat engine, operating in a cycle is zero when the two temperatures are the same. We can then state:

It is impossible for a system operating in a cycle and connected to a single heat reservoir to produce a positive amount of work in the surroundings.

Denial of Perpetual Motion of Second Kind

The *Kelvin-Planck statement* of the second law is the denial of what is known as a perpetual motion machine of the second kind. As an example, it is impossible to operate a heat engine solely by the removal of heat from a thermal reservoir such as an ocean or a river. The first law may be thought of as the denial of a perpetual motion machine of the first kind, which is a heat engine that performs work without absorption of an equivalent quantity of heat.

Clausius Statement

The second law states that a heat engine, operating in cycles can perform work only by absorbing heat from a reservoir at a higher temperature and by rejecting the difference between the heat absorbed and the work done by the engine to another thermal reservoir at a lower temperature. This is the second statement of the second law of thermodynamics as given by *Clausius*. It can be stated as follows.

It is impossible for a cyclic process to convert heat into work without the simultaneous transfer of heat from a body at a higher temperature to one at a lower temperature or vice versa, i.e. it is impossible for a cyclic process to transfer heat from a body at a lower temperature to one at higher temperature without the simultaneous conversion of work into heat.

Alternatively, *it is impossible for an engine operating in a cycle to have as its only effect the transfer of a quantity of heat from a reservoir at a lower temperature to a reservoir at a higher temperature.*

4.5 EFFICIENCY OF THE CARNOT CYCLE IS INDEPENDENT OF THE WORKING SUBSTANCE

One might suppose that the expression $(T_2 - T_1)/T_2$ for the efficiency of a Carnot cycle is applicable only when the working substance is an ideal gas. This is however not so, and the expression is applicable for any type of reversible cyclic process carried out between two temperatures T_1 and T_2. In other words, the efficiency of a Carnot cycle is independent of the make-up of the engine and also that of the working substance of the engine. This can be proved as follows.

Working of a Carnot Cycle in the Reverse Direction

First, let us note that the Carnot cycle can be operated equally will in the reverse direction, with the following steps:

Step 1 Adiabatic expansion from volume V_1 to V_4. Temperature of the system will change from T_2 to T_1.

Step 2 Isothermal expansion at temperature T_1 from volume V_4 to V_3.

Step 3 Adiabatic compression from volume V_3 to V_2. Temperature rises to T_2.

Step 4 Isothermal compression at T_2 from volume V_2 to V_1.

As a result of this reversed cycle, the amount q_1' of heat will be absorbed by the system at the lower temperature T_1 in Step 2 and an amount q_2' of heat will be transferred to the reservoir at higher temperature T_2 in Step 4. At the same time a net amount of work equal to $w_1' + w_2'$ will be done on the system (w_2' will have a positive value in Step 4 and w_1' will have a negative value in Step 2). In carrying out this cycle, the system acts as a refrigerator.† By the expenditure of work which is supplied by the surroundings since $w_2' > |w_1'|$, a certain amount of heat is pumped out from a lower temperature to a higher one.

Construction of a Composite Cycle

Let us have two systems, containing two different substances. We imagine that these systems are coupled together in such a way that the work produced in system I is done on system II. In other words, system I acts as an engine and system II acts as a refrigerator. The net work $w(\text{I}) + w(\text{II})$ involved in the coupled system is zero. We assume that both systems carry out Carnot cycles between the same two temperatures T_2 and T_1. This coupled system is shown in Fig. 4.5.1.

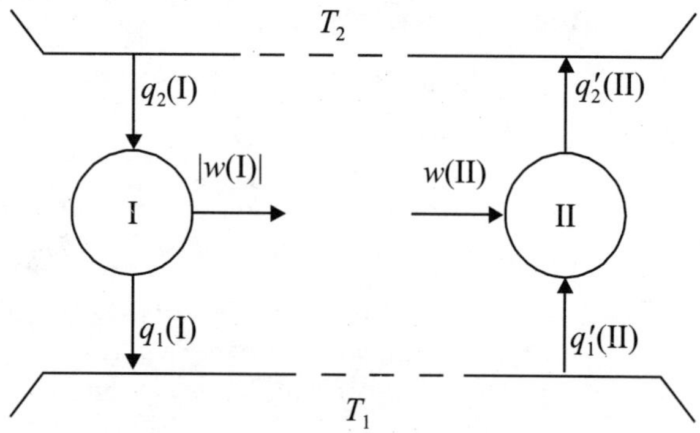

Fig. 4.5.1 Coupled system

Proof of Efficiency is Independent of Working Substance

Let us assume that the cyclic process for system I has a higher efficiency than that of system II, i.e.

---

† The coefficient of performance of a refrigerator is defined as the ratio of heat transferred from a lower temperature to a higher temperature to the work done on the machine to cause this removal, i.e.

$$\eta' = \frac{|q_c|}{w}$$

The less the work done, the more efficient the operation, and greater the coefficient of performance. Since $w_{irr} > w_{rev}$, the coefficient of performance of an irreversible machine is less than that of the reversible machine. Since the heat transferred to hot reservoir is sum of $|q_c|$ and w, we will have

$$\eta' = \frac{|q_c|}{|q_h| - |q_c|} = \frac{T_1}{T_2 - T_1}$$

Note as $T_1 \to 0$ K, $\eta' \to 0$. Since $\eta' = |q_c|/w$, the work done to bring about the transfer of heat from cold reservoir to hot reservoir approaches infinity. It follows that as the temperature of a system is lowered the amount of work required to lower the temperature further increases rapidly, and approaches infinity as the zero kelvin is attained. This fact illustrates that zero kelvin is unattainable.

$$\eta(\mathrm{I}) > \eta(\mathrm{II}) \quad \text{or} \quad \frac{|w(\mathrm{I})|}{q_2(\mathrm{I})} > \frac{w(\mathrm{II})}{|q_2'(\mathrm{II})|}$$

Since $|w(\mathrm{I})| = w(\mathrm{II})$, therefore,

$$q_2(\mathrm{I}) < |q_2'(\mathrm{II})| \tag{4.5.1}$$

that is, the heat absorbed by the system (I) from the reservoir at the higher temperature T_2 is smaller than the heat released by the system (II) to this reservoir. Therefore, there is a net transfer of heat by the coupled system to this reservoir, i.e. the reservoir at the higher temperature gains heat energy.

Now $w(\mathrm{I})$ and $w(\mathrm{II})$ are given by

$$|w(\mathrm{I})| = q_2(\mathrm{I}) + q_1(\mathrm{I})$$

$$w(\mathrm{II}) = |q_1'(\mathrm{II}) + q_2'(\mathrm{II})|$$

Since $|w(\mathrm{I})| = w(\mathrm{II})$, we have

$$q_2(\mathrm{I}) + q_1(\mathrm{I}) = |q_1'(\mathrm{II}) + q_2'(\mathrm{II})| \tag{4.5.2}$$

But $q_2(\mathrm{I}) < |q_2'(\mathrm{II})|$, therefore, in order to satisfy the above equality, we must have

$$|q_1(\mathrm{I})| < q_1'(\mathrm{II}) \tag{4.5.3}$$

that is, the heat released by the system (I) to the reservoir at temperature T_1 is smaller than the heat absorbed by the system (II) and, therefore, there is a net absorption of heat by the coupled system from this reservoir, i.e. the reservoir at the lower temperature loses heat energy.

Thus, the overall effect of the coupled system would be to transfer some heat from a lower to a higher temperature without any work being converted into heat. Such a process will not be permitted according to the second law. Therefore, our assumption that system (I) has a higher efficiency than system (II) must be invalid. On a similar line, it can be demonstrated that the assumption that system (I) having a lower efficiency than system (II) contradicts the second law. Thus, we may conclude that the efficiency of a Carnot cycle must be independent of the working substance.

Hence the efficiency, $(T_2 - T_1)/T_2$, which was derived earlier by taking an ideal gas as the working substance, is applicable to a reversible cyclic process involving any substance.

Example 4.5.1 A certain engine which operates in a Carnot cycle absorbs 3.347 kJ at 400 °C how much work is done on the engine per cycle and how much heat is evolved at 100 °C in each cycle?

Solution The efficiency of the Carnot cycle is given by

$$\eta = \frac{T_2 - T_1}{T_2} = \frac{q_2 + q_1}{q_2}$$

Thus $\quad -\dfrac{T_1}{T_2} = \dfrac{q_1}{q_2}$ and hence $q_1 = -\left(\dfrac{T_1}{T_2}\right) q_2$

Thus, the heat evolved in the present case is

$$q_1 = -\left(\frac{373.15\ \text{K}}{673.15\ \text{K}}\right)(3.347\ \text{kJ}) = -1.855\ \text{kJ}$$

and the work done on the engine is

$$w = -(q_2 + q_1) = -3.347\ \text{kJ} + 1.855\ \text{kJ} = -1.492\ \text{kJ}$$

(**Note:** The negative sign indicates that the work is actually done by the engine.)

Example 4.5.2 What percentage T_1 is of T_2 for a 10 per cent efficiency of a heat engine?

Solution Equating the efficiency equal to 0.1, we have

$$\eta = \frac{T_2 - T_1}{T_2} = 1 - \frac{T_1}{T_2} = 0.1$$

Therefore $\quad \dfrac{T_1}{T_2} \times 100 = (1 - 0.1) \times 100 = 0.9 \times 100 = 90$

or $\quad T_1 = 90$ per cent of T_2

Problem 4.5.1 Show that the expression for the efficiency of a Carnot cycle involving a van der Waals gas as the working substance is the same as that involving an ideal gas.

Solution The expressions of w and q involving in each of the four reversible steps of a Carnot cycle are described below.

(i) *Isothermal expansion form* V_1 to V_2 at T_2

$$w_2 = -nRT_2 \ln \frac{V_2 - nb}{V_1 - nb} - n^2 a\left(\frac{1}{V_2} - \frac{1}{V_1}\right) \qquad \text{(Eq. 2.13.2)}$$

$$q_2 = nRT_2 \ln \frac{V_2 - nb}{V_1 - nb} \qquad \text{(Eq. 2.13.5)}$$

(ii) *Adiabatic expansion from* V_2 *to* V_3 *Temperature changes from* T_2 *to* T_1.

$$w = nC_{V,\text{m}}(T_1 - T_2) - n^2 a\left(\frac{1}{V_3} - \frac{1}{V_2}\right) \qquad \text{(Eq. 2.13.2)}$$

$$q = 0$$

(iii) *Isothermal compression from* V_3 *to* V_4 *at* T_1

$$w = -nRT_1 \ln \frac{V_4 - nb}{V_3 - nb} - n^2 a\left(\frac{1}{V_4} - \frac{1}{V_3}\right)$$

$$q_1 = nRT_1 \ln \frac{V_4 - nb}{V_3 - nb}$$

(iv) *Adiabatic compression from* V_4 *to* V_1 *Temperature changes from* T_1 *to* T_2.

$$w' = nC_{V,\text{m}}(T_2 - T_1) - n^2 a\left(\frac{1}{V_1} - \frac{1}{V_4}\right)$$

$$q' = 0$$

The total work involved is given by

$$w_{\text{total}} = w_2 + w + w_1 + w'$$

$$= -\,nRT_2 \ln \frac{V_2 - nb}{V_1 - nb} - nRT_1 \ln \frac{V_4 - nb}{V_3 - nb} \tag{4.5.4}$$

The expression connecting various volumes is provided by Eq. (2.14.5), i.e.

$$T^{C_{V,\text{m}}/R}\,(V - nb) = \text{constant}$$

This relation is applicable for an adiabatic reversible expansion or compression. Hence

For step (ii) $\quad (T_2)^{C_{V,\text{m}}/R}\,(V_2 - nb) = (T_1)^{C_{V,\text{m}}/R}\,(V_3 - nb)$

For step (iv) $\quad (T_2)^{C_{V,\text{m}}/R}\,(V_1 - nb) = (T_1)^{C_{V,\text{m}}/R}\,(V_4 - nb)$

Dividing the two, we get

$$\frac{V_2 - nb}{V_1 - nb} = \frac{V_3 - nb}{V_4 - nb}$$

In view of the above expression, Eq. (4.5.4) becomes

$$w_{\text{total}} = -\left\{ nR \ln \frac{V_2 - nb}{V_1 - nb} \right\} (T_2 - T_1) \tag{4.5.5}$$

The efficiency of the Carnot cycle is given by

$$\eta = \frac{(-w_{\text{total}})}{q_2}$$

Substituting the expressions of w_{total} and q_2, we get

$$\eta = \frac{T_2 - T_1}{T_2}$$

which is the same as that of an ideal gas.

4.6 COMPARISON OF EFFICIENCIES OF REVERSIBLE AND IRREVERSIBLE CYCLIC PROCESSES

The efficiency of a reversible Carnot cycle is the theoretically possible maximum value which an engine can have. Since the various processes of this type of engine are to be carried out reversibly, therefore, such type of an engine does not have any realistic basis because reversible processes are idealized concepts which can never be realized. A real heat engine, which is irreversible in nature, will have efficiency smaller than the reversible heat engine. This can be proved as follows.

Let us have two cycles, one operating reversibly and the other irreversibly. Let both of them operate between the same two temperatures T_1 and T_2 and involve ideal gas as the working substance. These two cycles along with q values, are shown in Fig. 4.6.1.

Step 1 *Isothermal expansion form volume V_1 to V_2* The expressions for the work involved are

$$-\,w_2(\text{rev}) = q_2(\text{rev}) = RT_2 \ln \frac{V_2}{V_1}$$

and $\quad -\,w_2(\text{irr}) = q_2(\text{irr}) = p_{\text{ext}}(V_2 - V_1)$

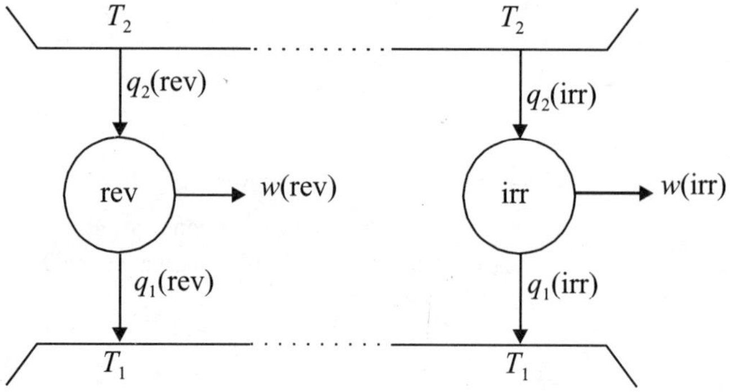

Fig. 4.6.1 Reversible and irreversible cycles

Since we known that $| w_2(\text{rev}) | > | w_2(\text{irr}) |$, therefore

$$q_2(\text{rev}) > q_2(\text{irr}) \qquad (4.6.1)$$

Step 3 *Isothermal compression from volume V_3 to V_4* The expressions for the work involved are

$$-w_1(\text{rev}) = q_1(\text{rev}) = RT_1 \ln \frac{V_4}{V_3}$$

$$-w_1(\text{irr}) = q_1(\text{irr}) = p'_{\text{ext}}(V_4 - V_3)$$

Now since in the irreversible process, more work is done as compared to that in the reversible process, we have

$$w_1(\text{irr}) > w_1(\text{rev})$$

It follows that

$$|q_1(\text{irr})| > |q_1(\text{rev})| \quad \text{i.e.} \quad |q_1(\text{rev})| < |q_1(\text{irr})| \qquad (4.6.2)$$

Now the efficiencies of the two cycles are

$$\eta(\text{rev}) = \frac{q_2(\text{rev}) + q_1(\text{rev})}{q_2(\text{rev})} = \frac{q_2(\text{rev}) - |q_1(\text{rev})|}{q_2(\text{rev})} = 1 - \frac{|q_1(\text{rev})|}{q_2(\text{rev})}$$

$$\eta(\text{irr}) = \frac{q_2(\text{irr}) + q_1(\text{irr})}{q_2(\text{irr})} = \frac{q_2(\text{irr}) - |q_1(\text{irr})|}{q_2(\text{irr})} = 1 - \frac{|q_1(\text{irr})|}{q_2(\text{irr})}$$

Now since $q_2(\text{rev}) > q_2(\text{irr})$ and $|q_1(\text{rev})| < |q_1(\text{irr})|$, therefore, it follows that

$$\frac{|q_1(\text{rev})|}{q_2(\text{rev})} < \frac{|q_1(\text{irr})|}{q_2(\text{irr})} \quad \text{or} \quad \left\{1 - \frac{|q_1(\text{rev})|}{q_2(\text{rev})}\right\} > \left\{1 - \frac{|q_1(\text{irr})|}{q_2(\text{irr})}\right\}$$

i.e. $$\eta(\text{rev}) > \eta(\text{irr}) \qquad (4.6.3)$$

4.7 THE THERMODYNAMIC OR KELVIN TEMPERATURE SCALE

The efficiency of a reversible Carnot cycle is given by

$$\eta = \frac{q_2 + q_1}{q_2} = 1 + \frac{q_1}{q_2} \tag{4.7.1}$$

and is independent of the working substance in the cycle. With this fact, Kelvin showed that it is possible to introduce a thermodynamics temperature scale which will be independent of the material used for the thermometric substance.

Consider the three Carnot cycles a b c d, a e f d and e b c f, each working between the same two temperatures (θ_3 and θ_1, such as shown in Fig. 4.7.1. The efficiencies of all these cycles will be equal. Thus

$$\left(\frac{q_1'}{q_3'}\right) = \left(\frac{q_1''}{q_3''}\right) = \left(\frac{q_1'''}{q_3'''}\right) \tag{4.7.2}$$ †

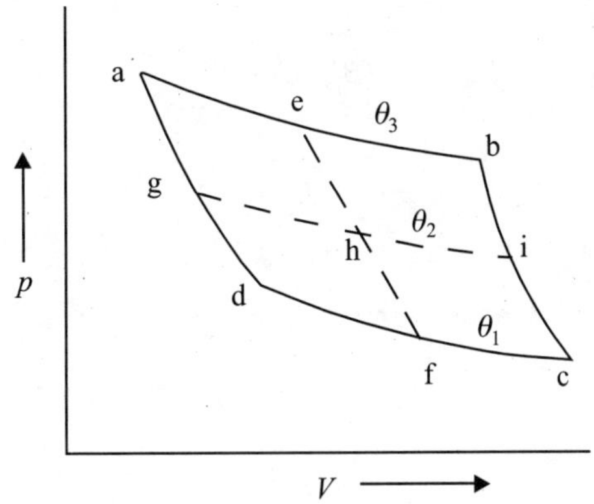

Fig. 4.7.1 Carnot cycles working between two temperatures

where (q_1'/q_3'), (q_1''/q_3'') and (q_1'''/q_3''') are the ratios of heats involved in the isothermal processes of the Carnot cycles a b c d, a e f d and e b c f, respectively. These ratios will be equal to one another provided the ratio q_1/q_3 is, in general, a function of the two temperatures θ_1 and θ_3, i.e.

$$\frac{|q_1|}{q_3} = f(\theta_1, \theta_3) \tag{4.7.3}$$

(Note that only the magnitude of q_1 is taken since we desire to make all temperatures positive.)

Now consider three cycles g i c d, a b i g and a b c d working between different temperatures θ_2 and θ_1, θ_3 and θ_2, and θ_3 and θ_1, respectively. We find that

† q carries the same subscript as that of T.

$$\left.\begin{aligned}\frac{|q_1|}{q_2} &= f(\theta_1, \theta_2) \quad \text{(i)}\\ \frac{|q_2|}{q_3} &= f(\theta_2, \theta_3) \quad \text{(ii)}\\ \text{and} \qquad \frac{|q_1|}{q_3} &= f(\theta_1, \theta_3) \quad \text{(iii)}\end{aligned}\right\} \tag{4.7.4}$$

Dividing (iii) by (ii), we get

$$\frac{|q_1|/q_3}{|q_2|/q_3} = \frac{|q_1|}{|q_2|} = \frac{f(\theta_1, \theta_3)}{f(\theta_2, \theta_3)} = f(\theta_1, \theta_2) \tag{4.7.5}$$

In other words, on dividing the ratio of two functions $f(\theta_1, \theta_3)$ and $f(\theta_2, \theta_3)$, we get a function $f(\theta_1, \theta_2)$ which is independent of temperature θ_3. This can be possible if the function $f(\theta_i, \theta_j)$ may be written as the ratio of two functions, each of which is dependent on only one temperature. Thus we may write

$$f(\theta_1, \theta_2) = \frac{\phi(\theta_1)}{\phi(\theta_2)}$$

$$f(\theta_2, \theta_3) = \frac{\phi(\theta_2)}{\phi(\theta_3)}$$

and $$f(\theta_1, \theta_3) = \frac{\phi(\theta_1)}{\phi(\theta_3)} \tag{4.7.6}$$

With these relations, Eq. (4.7.5) may be verified as follows.

$$\frac{|q_1|}{|q_2|} = \frac{|q_1|/q_3}{|q_2|/q_3} = \frac{f(\theta_1, \theta_3)}{f(\theta_2, \theta_3)} = \frac{\phi(\theta_1)/\phi(\theta_3)}{\phi(\theta_2)/\phi(\theta_3)} = \frac{\phi(\theta_1)}{\phi(\theta_2)} = f(\theta_1, \theta_2)$$

The relation

$$\frac{|q_1|}{|q_2|} = \frac{\phi(\theta_1)}{\phi(\theta_2)} \tag{4.7.7}$$

can be used to define a new temperature scale known as thermodynamic or Kelvin temperature scale.

The quantities q_1 and q_2 can be measured experimentally. The definition of the function $\phi(\theta)$ and of the required number of fixed points suffice to determine the scale. The simplest definition, and the one actually used, is to make $\phi(\theta)$ proportional to θ so that Eq. (4.7.7) becomes

$$\frac{|q_1|}{|q_2|} = \frac{\theta_1}{\theta_2} \tag{4.7.8}$$

Equation (4.7.8) relates the two unknown quantities θ_1 and θ_2. If q_1 is determined at the triple point of water and assign θ_1 to be 273.16 K, then the thermodynamic scale is completely defined, such that

$$\theta_2 = (273.16\ \text{K})\left(\frac{q_2}{q_{273.16\,\text{K}}}\right) \tag{4.7.9}$$

Thus, the temperature θ_2 can be determined from the heat that is exchanged at this temperature.

4.8 IDENTITY OF THERMODYNAMIC SCALE WITH IDEAL GAS TEMPERATURE SCALE

We have seen that the efficiency of a Carnot cycle involving an ideal gas is given by

$$\eta = \frac{T_2 - T_1}{T_2} = 1 - \frac{T_1}{T_2} \qquad (4.8.1)^*$$

and, in general, efficiency of a cycle is given by

$$\eta = \frac{q_2 + q_1}{q_2} = 1 - \frac{|q_1|}{q_2} \qquad (4.8.2)^{**}$$

Since efficiency of the cycle is independent of the wording substance, we must have

$$\frac{|q_1|}{q_2} = \frac{T_1}{T_2} \tag{4.8.3}$$

Comparing this with Eq. (4.7.8), we get

$$\frac{\theta_1}{\theta_2} = \frac{T_1}{T_2}$$

that is, the ratio of the two temperatures is the same on either scale. If θ_1 and T_1 are given a common value of 273.16 K at the triple point of water, the two temperature scales become identical.

4.9 DEFINITION OF THE ENTROPY FUNCTION

Basic Conclusion from Efficiency of a Carnot Cycle

For a reversible Carnot cycle operating between two temperatures T_2 and T_1, the efficiency is given as

$$\eta = \frac{q_2 + q_1}{q_2} = \frac{T_2 - T_1}{T_2}$$

where q_2 and q_1 are the heats exchanged with the thermal reservoirs at temperatures T_2 and T_1, respectively. Rewriting the above expression, we have

or $\quad 1 + \dfrac{q_1}{q_2} = 1 - \dfrac{T_1}{T_2} \quad$ or $\quad \dfrac{q_1}{q_2} = -\dfrac{T_1}{T_2}$

or $$\frac{q_2}{T_2} + \frac{q_1}{T_1} = 0 \tag{4.9.1}$$

---

*The subscripts 1 and 2 represent lower and higher temperatures, respectively.

** q carries the subscript of T at which it is being exchanged.

that is, the sum of the ratios of the heat involved and the corresponding temperature is zero for a Carnot cycle.

Replacement of a Bigger Carnot Cycle by Smaller Carnot Cycles

Now let us see what happens when a Carnot cycle operating between two temperatures T_2 and T_1 is replaced by a series of smaller Carnot cycles involving heat exchanges with a number of the heat reservoirs with temperatures between T_2 and T_1. Figure 4.9.1 shows a case where a large Carnot cycle is replaced by four smaller Carnot cycles.

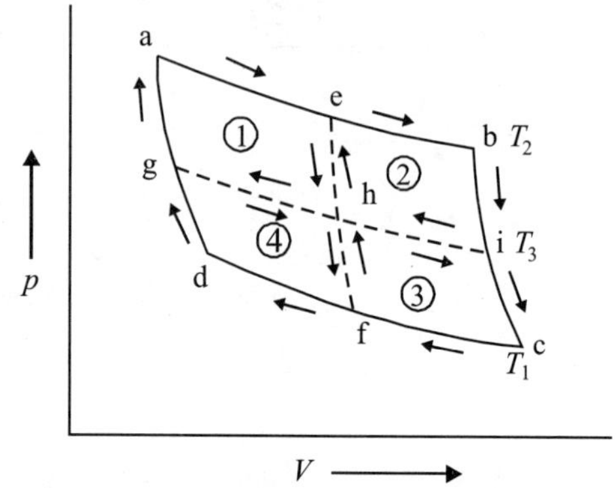

Fig. 4.9.1 Replacement of a Carnot cycle with four smaller cycles

Since the magnitude of the work involved in a reversible cyclic process is equal to the area enclosed by its graph in the p-V plane, it is follows from Fig. 4.9.1 that the sum of the areas of four smaller Carnot cycles is equal to the area of the single large Carnot cycle. In addition, it may be seen that when all the four small Carnot cycles are completed, each interior line is traversed twice, once in each direction. This means that the thermal reservoir at intermediate temperature absorbs and gives up the same amount of heat. Moreover, the total amount of heat absorbed from the reservoir at T_2 and given up to the reservoir at T_1 are the same as for the large Carnot cycle. Thus, we have

$$q_{3(1)} = -\, q_{3(4)} \qquad (4.9.2)^{\dagger}$$

$$q_{3(2)} = -\, q_{3(3)} \qquad (4.9.3)$$

$$q_2 = q_{2(1)} + q_{2(2)} \qquad (4.9.4)$$

$$q_1 = q_{1(3)} + q_{1(4)} \qquad (4.9.5)$$

For each of the four cycles, we can write expressions similar to that given by Eq. (4.9.1), i.e.,

$$\frac{q_{2(1)}}{T_2} + \frac{q_{3(1)}}{T_3} = 0; \qquad \frac{q_{2(2)}}{T_2} + \frac{q_{3(2)}}{T_3} = 0$$

$$\frac{q_{3(3)}}{T_3} + \frac{q_{1(3)}}{T_1} = 0; \qquad \frac{q_{3(4)}}{T_3} + \frac{q_{1(4)}}{T_1} = 0$$

---

†The number within the brackets represents the cycle number.

Adding these, we get

$$\frac{q_{2(1)}}{T_2}+\frac{q_{3(1)}}{T_3}+\frac{q_{2(2)}}{T_2}+\frac{q_{3(2)}}{T_3}+\frac{q_{3(3)}}{T_3}+\frac{q_{1(3)}}{T_1}+\frac{q_{3(4)}}{T_3}+\frac{q_{1(4)}}{T_1}=0$$

which, in view of Eqs (4.9.2) and (4.9.3), becomes

$$\frac{q_{2(1)}}{T_2}+\frac{q_{2(2)}}{T_2}+\frac{q_{1(3)}}{T_1}+\frac{q_{1(4)}}{T_1}=0$$

Making use of Eqs (4.9.4) and (4.9.5), the above expression becomes

$$\frac{q_2}{T_2}+\frac{q_1}{T_1}=0 \qquad (4.9.6)$$

which is an expression identical with Eq. (4.9.1). This means that the sum of efficiencies of the four Carnot cycles is equal to the efficiency of the large Carnot cycle. Thus, it may be concluded that a large Carnot cycle operating between temperatures T_2 and T_1 can be replaced by four small Carnot cycles (or in general, with any number of small Carnot cycles) involving heat exchanges with a number of reservoirs with temperatures between T_2 and T_1. The general form of Eq. (4.9.6), which is applicable to the heat transfers with any number of heat reservoirs, is

$$\sum_i \frac{q_i}{T_i}=0 \qquad (4.9.7)$$

where q_i is the heat transferred with the reservoir i at temperature T_i. The summation over i is to be carried over all thermal reservoirs.

Replacement of a Carnot Cycle by Infinite Smaller Carnot Cycles

In a similar way, it may be proved that any reversible cycle (need not be Carnot cycle) can be approximated by a sum of Carnot cycles, the smaller the Carnot cycles and larger the number of intermediate temperatures, the better the approximation (Fig. 4.9.2).

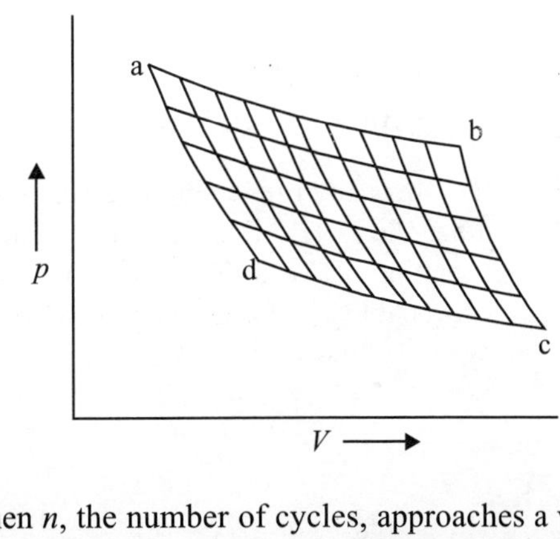

Fig. 4.9.2 Replacement of a reversible cycle with a number a smaller Carnot cycles

In the limit when n, the number of cycles, approaches a very large value, the heat transferred at each step in each cycle becomes infinitesimally small, and can be written as dq_i(rev); the symbol 'rev' represents that the heat is transferred reversibly. When n becomes very large, the summation in Eq. (4.9.7) can be replaced by the cyclic integration, i.e.

$$\oint \frac{dq_i(\text{rev})}{T_i} = 0 \tag{4.9.8}$$

The above expression can be written as

$$\oint d\left(\frac{q(\text{rev})}{T}\right) = 0 \tag{4.9.9}$$

Identification of Entropy Function

Since we know that if $\oint dZ = 0$, then dZ is an exact differential and Z is a state function. Similarly, we can conclude that $q(\text{rev})/T$ represents a state function because the value of its differential d(q(rev)/T) over a cyclic process is zero. This state function is given the name *entropy* of the system and is represented by the symbol S. Thus

$$dS = d\left(\frac{q(\text{rev})}{T}\right) \tag{4.9.10}$$

and $$S = \frac{q(\text{rev})}{T} \tag{4.9.11}$$

The entropy function is a *function of the independent variables which are used to define the state of a system. It is an extensive function.* The change in the value of the entropy function in going from one state to another, is independent of the path, and the cyclic integral of dS for a cyclic change of state is always zero. This function is defined, like the energy function, in terms of a differential, and consequently the absolute value of the entropy function for a system in any state may not be exactly known (*see* Section 4.21 for the third law entropy). It is well to emphasize that *whatever might be the path employed in going form one state to another, the determination of the corresponding entropy change is always calculated using a reversible path connecting the two states.*

The unit of entropy function is energy unit divided by temperature unit, e.g., J K^{-1}, cal K^{-1}. A unit of 1 cal K^{-1} (non-SI) is known as an entropy unit and is represented by the symbol eu.

4.10 THE VALUE OF d*q* (irr)/*T* FOR AN IRREVERSIBLE CYCLIC PROCESS

If there is any irreversibility at any stage of a cycle, the net work obtained $|w_{\text{net}}|$ in the cycle is less then the maximum work obtainable from the reversible cycle operating between the same two temperatures. Consequently, the efficiency of an irreversible cycle is always less than the efficiency of the corresponding reversible cycle. It follows that

$$\frac{|w_{\text{net}}|}{q_2} < \frac{T_2 - T_1}{T_2}$$

or $$\frac{q_2 + q_1}{q_2} < \frac{T_2 - T_1}{T_2}$$

or $$\frac{q_1}{q_2} < -\frac{T_1}{T_2}$$

or $$\frac{q_2}{T_2} + \frac{q_1}{T_1} < 0 \tag{4.10.1}$$

In the limiting case of infinitesimal heat transfer, this inequality becomes

$$\oint \frac{dq(\text{irr})}{T} < 0 \tag{4.10.2}$$

The fact that the sum of q_2/T_2 and q_1/T_1 is zero for a reversible Carnot cycle and is negative for a cycle involving irreversible isothermal expansion or/and irreversible isothermal compression may be rationalized as follows.

Since, in the isothermal reversible process of a Carnot cycle, the heat transferred is directly proportional to the temperature of the reservoir (Eq. 4.8.3), we have q_2/T_2 = constant and q_1/T_1 = – constant (since q_1 is negative). It follows that the sum of q_2/T_2 and q_1/T_1 is zero.

In an irreversible cycle operating between the same two temperatures T_1 and T_2, we have

(i) $q_2(\text{irr}) < q_2(\text{rev})$ and hence $q_2(\text{irr})/T_2$ will be less positive as compared to the value of $q_2(\text{rev})/T_2$.

(ii) $q_1(\text{irr}) < q_1(\text{rev})$ or $|q_{\text{irr}}| > |q_{\text{rev}}|$ and hence $q_1(\text{irr})/T_1$ will be more negative as compared to the value of $q_1(\text{rev})/T_1$.

Thus, in a cycle involving either of these two irreversible steps or involving both of these steps, the sum

$$\frac{q_2(\text{irr})}{T_2} + \frac{q_1(\text{irr})}{T_1}$$

will always be negative, because $q_1(\text{irr})/T_1$ is negative and its magnitude is greater than $q_2(\text{irr})/T_2$.

A General Proof for $\oint dq(\text{rev})/T = 0$ and $\oint dq(\text{irr})/T < 0$

We have seen that for a Carnot cycle

$$-w = \oint dq \qquad \text{and} \qquad \oint \frac{dq(\text{rev})}{T} = 0 \tag{4.10.3}$$

Now consider any other cycle, howsoever complicated (containing any number of temperature reservoirs and any working substance). For this engine, according to the first law, we have

$$-w' = \oint dq' \tag{4.10.4}$$

and let for this engine

$$\oint \frac{dq'}{T} > 0 \tag{4.10.5}$$

The two engines are coupled together to make a composite cyclic engine, represented by the symbol 'cyc'. The work involved in the composite cyclic engine is

$$w_{\text{cyc}} = w + w'$$

which, by Eqs (4.10.3) and (4.10.4), are equal to

$$-w_{\text{cyc}} = \oint dq + \oint dq' = \oint dq(\text{cyc}) \tag{4.10.6}$$

If Eqs (4.10.3) and (4.10.5) are added, we get

$$\oint\left(\frac{dq(\text{rev})}{T} + \frac{dq'}{T}\right) > 0$$

or

$$\oint \frac{dq(\text{cyc})}{T} > 0 \tag{4.10.7}$$

If it is adjusted that the composite cyclic engine does not produce any work, then $w_{\text{cyc}} = 0$. With this condition, Eq. (4.10.6) becomes

$$\oint dq(\text{cyc}) = 0 \tag{4.10.8}$$

Since each of the cyclic integral can be considered as a sum of terms, we may write Eqs (4.10.8) and (4.10.7) as

$$q_1 + q_2 + q_3 + \cdots = 0 \tag{4.10.9}$$

$$\frac{q_1}{T_1} + \frac{q_2}{T_2} + \frac{q_3}{T_3} + \cdots > 0 \tag{4.10.10}$$

The expression on the left hand side of Eq. (4.10.9) includes a number of terms, some positive and some negative. The positive ones just balance the negative ones and the sum is zero.

In Eq. (4.10.10) the net sum can be positive only if the positive terms are divided by small numbers and negative terms are divided by large numbers. This amounts to the fact that we are associating positive values of q with low temperature and negative values with high temperatures, i.e. we are extracting heat from reservoirs at low temperatures and rejecting heat to reservoirs at higher temperatures without spending any thing. This is against the second law of thermodynamics. Thus, our approximation that

$$\oint \frac{dq'}{T} > 0$$

must be wrong. It follows that for any engine, we must have

$$\oint \frac{dq'}{T} \leq 0$$

Now we will show that for a reversible cycle $\oint dq'/T = 0$ whereas $\oint dq'/T < 0$ for an irreversible cycle.

Case I *Reversible cycle* Let us assume for this cycle

$$\oint \frac{dq}{T} < 0$$

We can reverse this engine, which changes all the signs but not the magnitudes of the qs. Thus, we have

$$\oint \frac{dq}{T} > 0$$

which obviously is wrong as deduced earlier. This leads to the conclusion that for any system

$$\oint \frac{dq(\text{rev})}{T} = 0$$

for all reversible cycles.

Case II *Irreversible cycle* Since the heat and work associated with an irreversible cycle are different from those associated with the reversible cycles, therefore, the only possibility left is

$$\oint \frac{dq'}{T} < 0$$

for all irreversible cycles.

4.11 THE CLAUSIUS INEQUALITY

A Characteristic of Entropy Function

Since S is a state function, it follows that $\oint dS = 0$. If the cyclic process involves going from A to B and then coming back to A, then

$$\oint dS = \int_A^B dS + \int_B^A dS = 0$$

or $$\int_A^B dS = -\int_B^A dS \qquad (4.11.1)$$

that is, the entropy change in going B to A is equal but of opposite sign to that in going from A to B, for only then the addition of these two will be zero.

Entropy Changes in a Reversible Process

If we assume that the surroundings always transfer heat reversibly, then the entropy change in any process carried out reversibly will be the negative of the entropy change of the surroundings, i.e.

$$\Delta S(\text{system}) = -\Delta S(\text{surroundings})$$

or $$\Delta S(\text{total}) = \Delta S(\text{system}) + \Delta S(\text{surroundings}) = 0 \qquad (4.11.2)$$

Thus, the sum of entropy changes for the system and of the surroundings will always be zero. Hence, it may be concluded that in reversible processes entropy is merely transferred between the system and surroundings and that the total entropy change is zero.

Expression of Clausius Inequality

Now let us consider what happens when the process A → B is irreversible. No matter what the nature of this process might be, we can assume that the reverse process B → A is carried out reversibly. Then we have the following cycle.

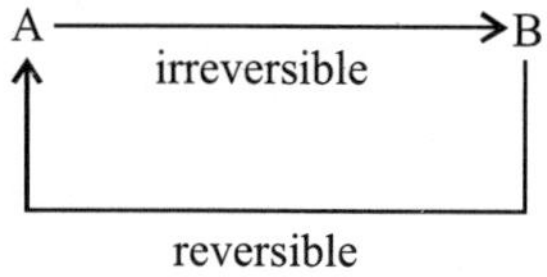

Now $$\oint \frac{dq(\text{irr})}{T} = \int_{A}^{B} \frac{dq(\text{irr})}{T} + \int_{B}^{A} \frac{dq(\text{rev})}{T}$$

Making use of Eq. (4.11.1), we have

$$\oint \frac{dq(\text{irr})}{T} = \int_{A}^{B} \frac{dq(\text{irr})}{T} - \int_{A}^{B} \frac{dq(\text{rev})}{T}$$

Now, according to Eq. (4.10.2), we have

$$\oint \frac{dq(\text{irr})}{T} < 0$$

Thus, it follows that

$$\int_{A}^{B} \frac{dq(\text{irr})}{T} < \int_{A}^{B} \frac{dq(\text{rev})}{T}$$

Using Eq. (4.9.11), this becomes

$$\int_{A}^{B} \frac{dq(\text{irr})}{T} < \int_{A}^{B} dS$$

or $$\int_{A}^{B} \frac{dq(\text{irr})}{T} < \Delta S_{AB} \qquad (4.11.3)$$

where ΔS_{AB} is the change of entropy of the system in going from A to B.

The expression of Eq. (4.11.3) is known as *Clausius inequality* which is a fundamental requirement for a real transformation. The inequality of Eq. (4.11.3) enables us to decide whether or not, some proposed transformation will occur in nature. We will not use this expression as such, but will manipulate it to express the inequality in terms of properties of the state of a system rather than in terms of path property such as $dq(\text{irr})$.

Clausius Inequality Applied to an Isolated System

For any change in an isolated system

$$dq(\text{irr}) = 0$$

and the inequality of Eq. (4.11.3) becomes

$$0 < \Delta S_{AB} \quad \text{or} \quad \Delta S_{AB} > 0 \qquad (4.11.4)$$

Thus, the requirement for a real transformation in an isolated system is that ΔS be positive, i.e. the entropy of the system increases whenever a natural change is occurring within an isolated system. The entropy continues to increase so long as changes occur in it. When the changes cease, the system is in equilibrium, and the entropy has reached a maximum value.

Thus the condition of equilibrium in an isolated system is that the entropy has a maximum value.

Clausius Inequality Applied to System and Surroundings

We can always imagine that system I and its surroundings form a larger system called system II, which is isolated from its surroundings. Consequently, no matter what the interaction between system I and its surroundings in the irreversible process might be, the entropy of system II always increases. We, therefore, have

$$\Delta S_{sys}(\text{II}) = \Delta S_{sys}(\text{I}) + \Delta S_{surr}(\text{I}) > 0 \qquad (4.11.5)$$

Clausius assumed that the entire universe could be considered as an isolated system, in which all naturally occurring processes are irreversible. This is the basis for his often quoted statement:

The energy of the universe is constant, the entropy of the universe always tends toward a maximum.

4.12 STATE FUNCTION ENTROPY FROM FIRST LAW OF THERMODYNAMICS

According to the first law of thermodynamics, we have

$$\mathrm{d}q = \mathrm{d}U - \mathrm{d}w \qquad (4.12.1)$$

If the work is of p-V type, we have

$$\mathrm{d}q = \mathrm{d}U + p_{ext}\,\mathrm{d}V \qquad (4.12.2)$$

Since $U = f(T, V)$, therefore, we have

$$\mathrm{d}U = \left(\frac{\partial U}{\partial T}\right)_V \mathrm{d}T + \left(\frac{\partial U}{\partial V}\right)_T \mathrm{d}V \qquad (4.12.3)$$

Substituting this in Eq. (4.12.2), we get

$$\mathrm{d}q = \left(\frac{\partial U}{\partial T}\right)_V \mathrm{d}T + \left(\frac{\partial U}{\partial V}\right)_T \mathrm{d}V + p_{ext}\,\mathrm{d}V \qquad (4.12.4)$$

Dividing this by T, we have

$$\frac{\mathrm{d}q}{T} = \frac{1}{T}\left(\frac{\partial U}{\partial T}\right)_V \mathrm{d}T + \frac{1}{T}\left\{\left(\frac{\partial U}{\partial V}\right)_T + p_{ext}\right\}\mathrm{d}V \qquad (4.12.5)$$

If the work is performed reversibly, then

$$p_{ext} = p \pm \mathrm{d}p \simeq p$$

and $\quad \mathrm{d}q = \mathrm{d}q_{rev}$

Now if $\mathrm{d}q_{rev}/T$ is to be a state function, then the Euler's condition of reciprocity has to be satisfied. According to this condition, we must have

$$\frac{\partial}{\partial V}\left\{\frac{1}{T}\left(\frac{\partial U}{\partial T}\right)_V\right\}_T = \frac{\partial}{\partial T}\left\{\frac{1}{T}\left(\frac{\partial U}{\partial V}\right)_T + \frac{p}{T}\right\}_V \tag{4.12.6}$$

or
$$\frac{1}{T}\frac{\partial^2 U}{\partial V\,\partial T} = -\frac{1}{T^2}\left(\frac{\partial U}{\partial V}\right)_T + \frac{1}{T}\frac{\partial^2 U}{\partial T\,\partial V} - \frac{p}{T^2} + \frac{1}{T}\left(\frac{\partial p}{\partial T}\right)_V \tag{4.12.7}$$

Since $\dfrac{\partial^2 U}{\partial V\,\partial T} = \dfrac{\partial^2 U}{\partial T\,\partial V}$, therefore, we must have

$$-\frac{1}{T^2}\left(\frac{\partial U}{\partial V}\right)_T - \frac{p}{T^2} + \frac{1}{T}\left(\frac{\partial p}{\partial T}\right)_V = 0$$

or
$$\left(\frac{\partial U}{\partial V}\right)_T + p = T\left(\frac{\partial p}{\partial T}\right)_V \tag{4.12.8}$$

This is the thermodynamic equation of state† which holds good for all types substances. Hence, dq_{rev}/T is a state function. Note that dq_{rev} alone is not a state function, but it becomes one after being divided by T. The factor $1/T$ is known as an *integrating factor*, since dq_{rev}/T can be directly integrated between the two state variables. Combining Eqs (4.12.5) and (4.12.8) and then integrating, we have

$$\int\frac{dq_{rev}}{T} = \int\frac{1}{T}\left(\frac{\partial U}{\partial T}\right)_V dT + \int\left(\frac{\partial p}{\partial T}\right)_V dV$$

$$= nC_{V,m}\int\frac{dT}{T} + \int\left(\frac{\partial p}{\partial T}\right)_V dV$$

For an ideal gas ($pV = nRT$), we have

$$\left(\frac{\partial p}{\partial T}\right)_V = \frac{nR}{V}$$

Hence
$$\Delta S = \int\frac{dq_{rev}}{T} = nC_{V,m}\ \ln\frac{T_2}{T_1} + nR\ln\frac{V_2}{V_1} \tag{4.12.9}$$

4.13 CHARACTERISTICS OF THE ENTROPY FUNCTION

The change in entropy of a system when the heat dq_{rev} is exchanged reversibly at temperature T is given by

$$dS = \frac{dq_{rev}}{T} \tag{4.13.1}$$

†This equation can be verified by applying it to an ideal gas. For such a gas

$(\partial U/\partial V)_T = 0$ and $(\partial p/\partial T)_V = nR/V$

Substituting these in Eq. (4.12.8), we have $p = nRT/V$ which obviously is true since for an ideal gas $pV = nRT$.

If the system undergoes only p-V work, then according to the first law of thermodynamics, dq_{rev} is given by

$$dq_{rev} = dU + p\,dV \tag{4.13.2}$$

Dividing both sides by T, we get

$$\frac{dq_{rev}}{T} = dS = \frac{1}{T}dU + \frac{p}{T}dV \tag{4.13.3}$$

Equation (4.13.3) represents the variation in entropy when the internal energy and volume of the system are changed. Since both $1/T$ and p/T are positive, we can conclude that dS increases when (i) dU increases at constant volume, and (ii) dV increases at constant U, i.e. this equation represents the characteristics of state function entropy.

The above characteristics can be expressed mathematically. From Eq. (4.13.3) we conclude that the $S = f(U, V)$. Its differential is given by

$$dS = \left(\frac{\partial S}{\partial U}\right)_V dU + \left(\frac{\partial S}{\partial V}\right)_U dV \tag{4.13.4}$$

Comparing Eqs (4.13.3) and (4.13.4), we get

$$\left(\frac{\partial S}{\partial U}\right)_V = \frac{1}{T} \quad \text{and} \quad \left(\frac{\partial S}{\partial V}\right)_U = \frac{p}{T} \tag{4.13.5}$$

Equation (4.13.3) can be rewritten (only for a reversible process) as

$$dU = T\,dS - p\,dV$$

Since dU is an exact differential, we can apply Euler's condition of reciprocity to this equation. This yields

$$\left(\frac{\partial T}{\partial V}\right)_S = -\left(\frac{\partial p}{\partial S}\right)_V \tag{4.13.6}$$

This constitutes one of the important relations in thermodynamics, named as *Maxwell relations.* The other such equations will be derived later (*see* Section 5.9).

4.14 ENTROPY AS A FUNCTION OF TEMPERATURE AND VOLUME

Expression for $(\partial S/\partial T)_V$ and $(\partial S/\partial V)_T$

We can write entropy as a function of temperature and volume of the system, such that

$$S = f(T, V) \tag{4.14.1}$$

Its differential is given by

$$dS = \left(\frac{\partial S}{\partial T}\right)_V dT + \left(\frac{\partial S}{\partial V}\right)_T dV \tag{4.14.2}$$

For a system involving p-V work, dS as given by Eq. (4.13.3) is

$$dS = \frac{1}{T}dU + \frac{p}{T}dV \tag{4.14.3}$$

The differential dU can be replaced in terms of dT and dV. This can be done by taking U as a function of temperature and volume, i.e.

$$U = f(T, V)$$

Writing its differential, we have

$$\mathrm{d}U = \left(\frac{\partial U}{\partial T}\right)_V \mathrm{d}T + \left(\frac{\partial U}{\partial V}\right)_T \mathrm{d}V \tag{4.14.4}$$

Substituting this in Eq. (4.14.3), we get

$$\mathrm{d}S = \frac{1}{T}\left(\frac{\partial U}{\partial T}\right)_V \mathrm{d}T + \frac{1}{T}\left[p + \left(\frac{\partial U}{\partial V}\right)_T\right]\mathrm{d}V \tag{4.14.5}$$

Comparing Eq. (4.14.5) with Eq. (4.14.2), we get

$$\left(\frac{\partial S}{\partial T}\right)_V = \frac{1}{T}\left(\frac{\partial U}{\partial T}\right)_V = \frac{nC_{V,\mathrm{m}}}{T} \tag{4.14.6}$$

and

$$\left(\frac{\partial S}{\partial V}\right)_T = \frac{1}{T}\left[p + \left(\frac{\partial U}{\partial T}\right)_V\right] \tag{4.14.7}$$

Entropy Change in an Isochoric Change in Temperature

Equation (4.14.6) gives variation in entropy of the system with temperature when its volume is held constant. Since $C_{V,\mathrm{m}}$ has positive value, it follows that $(\partial S/\partial T)_V$ has a positive value. This implies that the entropy of a system increases on increasing its temperature at constant volume. For a finite change of temperature at constant volume, we have

$$\Delta S = \int_{T_1}^{T_2} \frac{nC_{V,\mathrm{m}}}{T}\,\mathrm{d}T \tag{4.14.8}$$

Derivation of Thermodynamic Equation of State

Equation (4.14.7) gives the variation in entropy with volume of the system when its temperature is held constant, i.e. isothermal volume dependence of entropy. This expression can be simplified using the following procedure.

Differentiating Eq. (4.14.6) with respect to volume at constant temperature, we get

$$\frac{\partial}{\partial V}\left\{\left(\frac{\partial S}{\partial T}\right)_V\right\}_T = \frac{1}{T}\frac{\partial}{\partial V}\left\{\left(\frac{\partial U}{\partial T}\right)_V\right\}_T \tag{4.14.9}$$

Differentiation of Eq. (4.14.7) with temperature at constant volume gives

$$\frac{\partial}{\partial T}\left\{\left(\frac{\partial S}{\partial V}\right)_T\right\}_V = \frac{\partial}{\partial T}\left\{\frac{1}{T}\left[p + \left(\frac{\partial U}{\partial V}\right)_T\right]\right\}_V$$

$$= \frac{1}{T}\left[\left(\frac{\partial p}{\partial T}\right)_V + \frac{\partial^2 U}{\partial T\,\partial V}\right] - \frac{1}{T^2}\left[p + \left(\frac{\partial U}{\partial V}\right)_T\right] \tag{4.14.10}$$

Since S is a state function, we must have

$$\frac{\partial^2 S}{\partial V\,\partial T} = \frac{\partial^2 S}{\partial T\,\partial V}$$

Thus, from Eqs (4.14.9) and (4.14.10), we get

$$\frac{1}{T}\frac{\partial^2 U}{\partial V\,\partial T} = \frac{1}{T}\left(\frac{\partial p}{\partial T}\right)_V + \frac{1}{T}\frac{\partial^2 U}{\partial T\,\partial V} - \frac{1}{T^2}\left[p + \left(\frac{\partial U}{\partial V}\right)_T\right]$$

Since U is also a state function, we will have

$$\left(\frac{\partial U}{\partial V}\right)_T + p = T\left(\frac{\partial p}{\partial T}\right)_V \tag{4.14.11}$$

This is an important expression, known as the *thermodynamic equation of state,* as it relates various variables of the states.

Entropy Change in an Isothermal Change in Volume

Comparing Eq. (4.14.11) with Eq. (4.14.7), we get

$$\left(\frac{\partial S}{\partial V}\right)_T = \left(\frac{\partial p}{\partial T}\right)_V \tag{4.14.12}$$

For a finite change of volume at constant temperature, we have

$$\Delta S = \int_{V_1}^{V_2}\left(\frac{\partial p}{\partial T}\right)_V \mathrm{d}V$$

We have seen in Problem 1.3.13 that

$$\left(\frac{\partial p}{\partial T}\right)_V = \frac{\alpha}{\kappa_T}$$

where α and κ_T are cubic expansion coefficient and isothermal compressibility, respectively. Therefore, Eq. (4.14.12) becomes

$$\left(\frac{\partial S}{\partial V}\right)_T = \frac{\alpha}{\kappa_T} \tag{4.14.13}$$

Thus, the variation in entropy with volume at constant temperature depends on how the pressure of the system changes with temperature at constant volume or on the sign of α, since κ_T is always positive. For most cases, α is positive and hence entropy increases with increase in volume at constant temperature. Water between temperatures 0 °C and 4 °C has a negative value of α and thus has a negative value of $(\partial S/\partial V)_T$.

Complete Expression of d*S* in terms of d*T* and d*V*

Substituting Eqs (4.14.6) and (4.14.12) in Eq. (4.14.2), we get

$$\mathrm{d}S = \frac{nC_{V,\mathrm{m}}}{T}\,\mathrm{d}T + \left(\frac{\partial p}{\partial T}\right)_V \mathrm{d}V$$

or

$$\mathrm{d}S = \frac{nC_{V,\mathrm{m}}}{T}\,\mathrm{d}T + \frac{\alpha}{\kappa_T}\,\mathrm{d}V \tag{4.14.14}$$

Equation (4.14.14) describes the variation in S with temperature and volume of the system. It may be mentioned that for the condensed system, the value of (α/κ_T) is negligible. Therefore, the dependency of S on volume of the substance, except in the case of gases, can be ignored.

Example 4.14.1

For $C_6H_6(1)$, $\alpha = 1.24 \times 10^{-3}\ K^{-1}$, and $\kappa_T = 9.228 \times 10^{-4}\ MPa^{-1}$ at 298 K and 101.325 kPa pressure. Find the change in molar volume which will be required to produce an entropy change of 2.092 J K^{-1} mol^{-1} at 298 K. Assume α and κ_T to be constants.

Solution

We know that

$$\left(\frac{\partial S}{\partial V}\right)_T = \frac{\alpha}{\kappa_T} \qquad \text{(Eq. 4.14.13)}$$

Therefore, $$\Delta V = \left(\frac{\kappa_T}{\alpha}\right)\Delta S = \left(\frac{9.228\times 10^{-4}\ \text{MPa}^{-1}}{1.24\times 10^{-3}\ \text{K}^{-1}}\right)(2.092\ \text{J K}^{-1}\ \text{mol}^{-1})$$

$$= 1.56\ \text{cm}^3\ \text{mol}^{-1}$$

Problem 4.14.1

Show that

(i) For an ideal gas $(\partial U/\partial V)_T$ is zero.

(ii) $C_p - C_V = T\left(\frac{\partial p}{\partial T}\right)_V\left(\frac{\partial V}{\partial T}\right)_p$ and is equal to R for one mole of an ideal gas. Also show that $C_p = C_V$ for water at 4 °C. Given that the density of water is maximum at 4 °C.

(iii) $C_p - C_V = -\dfrac{T(\partial V/\partial T)_p^2}{(\partial V/\partial p)_T} = \dfrac{T\alpha^2 V}{\kappa_T}$

(iv) $(\partial C_V/\partial V)_T = T(\partial^2 p/\partial T^2)_V$

(v) C_V is independent of volume for ideal and van der Waals gases.

Solution

(i) The expression of $(\partial U/\partial V)_T$ in terms of easily evaluable derivative is given by the thermodynamic equation of state

$$\left(\frac{\partial U}{\partial V}\right)_T = T\left(\frac{\partial p}{\partial T}\right)_V - p$$

Since for an ideal gas $pV = nRT$, therefore

$$\left(\frac{\partial p}{\partial T}\right)_V = \frac{nR}{V} = \frac{p}{T}$$

so that $$\left(\frac{\partial U}{\partial V}\right)_T = T\frac{p}{T} - p = 0 \qquad (4.14.15)$$

(ii) The expression for $C_p - C_V$ is given by

$$C_p - C_V = \left\{\left(\frac{\partial U}{\partial V}\right)_T + p\right\}\left(\frac{\partial V}{\partial T}\right)_p \qquad \text{(Eq. 2.6.3)}$$

Substituting $(\partial U/\partial V)_T$ from the thermodynamic equation of state (Eq. 4.14.11), we get

$$C_p - C_V = T\left(\frac{\partial p}{\partial T}\right)_V \left(\frac{\partial V}{\partial T}\right)_p \tag{4.14.16}$$

For one mole of an ideal gas $pV_m = RT$, which gives

$$\left(\frac{\partial p}{\partial T}\right)_{V_m} = \frac{R}{V_m} \quad \text{and} \quad \left(\frac{\partial V_m}{\partial T}\right)_p = \frac{R}{p}$$

Substituting these expressions in Eq. (4.14.16), we get

$$C_{p,\,m} - C_{V,\,m} = T\left(\frac{R}{V_m}\right)\left(\frac{R}{p}\right) = R \tag{4.14.17}$$

Since the density of water is maximum at 4 °C, we must have

$$\frac{d\rho}{dT} = 0$$

(We write complete differential assuming that the density of a liquid is independent of pressure.)

Now since $\rho = m/V$, we have

$$\frac{d\rho}{dT} = \frac{d(m/V)}{dT} = -\frac{m}{V^2}\frac{dV}{dT}$$

In order that $d\rho/dT = 0$, we must have $dV/dT = 0$ as both m and V are positive quantities. Now, according to Eq. (4.14.16), we get

$$C_p - C_V = 0 \quad \text{or} \quad C_p = C_V$$

(iii) The cyclic rule for the relation $p = f(T, V)$ is

$$\left(\frac{\partial p}{\partial T}\right)_V \left(\frac{\partial T}{\partial V}\right)_p \left(\frac{\partial V}{\partial p}\right)_T + 1 = 0$$

This gives $$\left(\frac{\partial p}{\partial T}\right)_V = -\frac{1}{(\partial T/\partial V)_p(\partial V/\partial p)_T} = -\frac{(\partial V/\partial T)_p}{(\partial V/\partial p)_T}$$

Thus, Eq. (4.14.16) becomes

$$C_p - C_V = -\frac{T(\partial V/\partial T)_p^2}{(\partial V/\partial p)_T} = \frac{T\alpha^2 V}{\kappa_T} \tag{4.14.18}$$

(iv) By definition, we have

$$C_V = \left(\frac{\partial U}{\partial T}\right)_V$$

Thus $$\left(\frac{\partial C_V}{\partial V}\right)_T = \frac{\partial}{\partial V}\left\{\left(\frac{\partial U}{\partial T}\right)_V\right\}_T = \frac{\partial}{\partial T}\left\{\left(\frac{\partial U}{\partial V}\right)_T\right\}_V \quad \text{(as } U \text{ is a state function)}$$

Making use of the thermodynamic equation of state, we get

$$\left(\frac{\partial C_V}{\partial V}\right)_T = \frac{\partial}{\partial T}\left\{T\left(\frac{\partial p}{\partial T}\right)_V - p\right\}_V = \left(\frac{\partial p}{\partial T}\right)_V + T\left(\frac{\partial^2 p}{\partial T^2}\right)_V - \left(\frac{\partial p}{\partial T}\right)_V$$

$$= T\left(\frac{\partial^2 p}{\partial T^2}\right)_V \tag{4.14.19}$$

(v) For an ideal gas, $p = nRT/V$, therefore

$$\left(\frac{\partial p}{\partial T}\right)_V = \frac{nR}{V} \quad \text{and} \quad \left(\frac{\partial^2 p}{\partial T^2}\right)_V = 0$$

Thus $\left(\frac{\partial C_V}{\partial V}\right)_T = T\left(\frac{\partial^2 p}{\partial T^2}\right)_V = 0$

For a van der Waals gas, we have

$$\left(\frac{\partial U}{\partial V}\right)_T = \frac{n^2 a}{V^2} \qquad \text{(Eq. 2.13.3)}$$

Thus $\left(\frac{\partial C_V}{\partial V}\right)_T = \frac{\partial}{\partial V}\left\{\left(\frac{\partial U}{\partial T}\right)_V\right\}_T = \frac{\partial}{\partial T}\left(\frac{n^2 a}{V^2}\right) = 0$

Since $(\partial C_V/\partial V)_T$ is zero for ideal and van der Waals gases, it follows that C_V is independent of volume for these gases.

Problem 4.14.2

(a) Evaluate the expression $(\partial U/\partial V)_T$ for the amount n of a van der Waals gas (use van der Waals equation together with the thermodynamic equation of state) and compare it with that of an ideal gas.

(b) By integrating the total differential dU for a van der Waals gas, show that if C_V is a constant, then

$$U = nC_{V,\mathrm{m}}T - \frac{n^2 a}{V} + U'$$

where U' is a constant of integration.

Solution

(a) Thermodynamic equation of state is

$$\left(\frac{\partial U}{\partial V}\right)_T = T\left(\frac{\partial p}{\partial T}\right)_V - p$$

For the amount n of a van der Waals gas, we have

$$p = \frac{nRT}{V - nb} - \frac{n^2 a}{V^2}$$

Therefore $\left(\frac{\partial p}{\partial T}\right)_V = \frac{nR}{V - nb}$

Thus
$$\left(\frac{\partial U}{\partial V}\right)_T = T\frac{nR}{V - nb} - p = \frac{nRT}{V - nb} - \left(\frac{nRT}{V - nb} - \frac{n^2 a}{V^2}\right) = \frac{n^2 a}{V^2} \tag{4.14.20}$$

The internal energy of a van der Waals gas increases with increase in volume at constant temperature. This is attributed to the van der Waals constant a, i.e. molecular interactions. In case of an ideal gas, $(\partial U/\partial V)_T = 0$ indicating the absence of any molecular interactions.

(b) $$dU = \left(\frac{\partial U}{\partial T}\right)_V dT + \left(\frac{\partial U}{\partial V}\right)_T dV$$

$$= nC_{V,\text{m}}\, dT + \frac{n^2 a}{V^2}\, dV$$

On carrying out open inegration, we have

$$U = nC_{V,\text{m}}\, T - \frac{n^2 a}{V} + U' \tag{4.14.21}$$

4.15 ENTROPY AS A FUNCTION OF TEMPERATURE AND PRESSURE

Expression for $(\partial S/\partial T)_p$ and $(\partial S/\partial p)_T$

We can write the dependence of entropy on temperature and pressure as

$$S = f(T, p) \tag{4.15.1}$$

Its differential is given by

$$dS = \left(\frac{\partial S}{\partial T}\right)_p dT + \left(\frac{\partial S}{\partial p}\right)_T dp \tag{4.15.2}$$

In order to write the equation

$$dS = \frac{1}{T} dU + \frac{p}{T} dV \tag{4.15.3}$$

in the form of Eq. (4.15.2), let us consider the following relation:

$$U = H - pV$$

Therefore $dU = dH - p dV - V\, dp$

Taking $H = f(T, p)$, we may write

$$dU = \left\{\left(\frac{\partial H}{\partial T}\right)_p dT + \left(\frac{\partial H}{\partial p}\right)_T dp\right\} - p\, dV - V\, dp \tag{4.15.4}$$

Substituting Eq. (4.15.4) in Eq. (4.15.3), we get

$$dS = \frac{1}{T}\left\{\left(\frac{\partial H}{\partial T}\right)_p dT + \left(\frac{\partial H}{\partial p}\right)_T dp\right\} - \frac{p}{T} dV - \frac{V}{T} dp + \frac{p}{T} dV$$

or $$dS = \frac{1}{T}\left(\frac{\partial H}{\partial T}\right)_p dT + \frac{1}{T}\left\{\left(\frac{\partial H}{\partial p}\right)_T - V\right\} dp \tag{4.15.5}$$

Comparing this with Eq. (4.15.2), we get

$$\left(\frac{\partial S}{\partial T}\right)_p = \frac{1}{T}\left(\frac{\partial H}{\partial T}\right)_p = \frac{nC_{p,\text{m}}}{T} \tag{4.15.6}$$

and $$\left(\frac{\partial S}{\partial p}\right)_T = \frac{1}{T}\left[\left(\frac{\partial H}{\partial p}\right)_T - V\right] \tag{4.15.7}$$

Entropy Change in an Isobaric Variation in Temperature Equation (4.15.6) gives the dependence of entropy on temperature at constant pressure. Since both $C_{p,\,m}$ and T are positive, therefore, S increases with increase in temperature of the system when the pressure is held constant. For a finite change of temperature at constant pressure, we have

$$\Delta S = \int_{T_1}^{T_2} \frac{nC_{p,\,m}}{T}\,\mathrm{d}T \tag{4.15.8}$$

$C_{p,\,m}$ independent of temperature If $C_{p,\,m}$ is considered independent of temperature in the range T_1 to T_2, then

$$\Delta S = nC_{p,\,m} \ln \frac{T_2}{T_1} \tag{4.15.9}$$

$C_{p,\,m}$ dependent on temperature If temperature dependence of $C_{p,\,m}$ is available in the form of the analytic expression

$$C_{p,\,m} = a + bT + cT^2 + \cdots$$

then we have

$$\Delta S = \int_{T_1}^{T_2} n\frac{(a + bT + cT^2 + \cdots)}{T}\,\mathrm{d}T$$

$$= n\left[a \ln \frac{T_2}{T_1} + b\,(T_2 - T_1) + \frac{c}{2}(T_2^2 - T_1^2) + \cdots\right] \tag{4.15.10}$$

Example 4.15.1 Calculate ΔS for 2 mol of nitrogen heated at constant pressure from 298 K to 373 K. Given the temperature variation of $C_{p,\,m}$ of nitrogen as

$$C_{p,\,m}/\mathrm{J\,K^{-1}\,mol^{-1}} = 27.296 + 5.23 \times 10^{-3}\,(T/\mathrm{K}) - 0.042 \times 10^{-7}\,(T/\mathrm{K})^2$$

Solution We start with the relation for one mole of a gas

$$\left(\frac{\partial S}{\partial T}\right)_p = \frac{C_{p,\mathrm{m}}}{T}$$

Substituting the given expression of $C_{p,\,m}$, we get

$$\left(\frac{\partial S}{\partial T}\right)_p \bigg/ \mathrm{J\,K^{-1}\,mol^{-1}} = \frac{27.296}{T} + 5.23 \times 10^{-3}/\mathrm{K} - 0.042 \times 10^{-7}\,(T/\mathrm{K}^2)$$

or $$\mathrm{d}S / \mathrm{J\,K^{-1}\,mol^{-1}} = \left(\frac{27.296}{T} + 5.23 \times 10^{-3}/\mathrm{K} - 0.042 \times 10^{-7}\,(T/\mathrm{K}^2)\right)\mathrm{d}T$$

Integrating this, we have

$$\Delta S/\mathrm{J\,K^{-1}\,mol^{-1}} = \int_{298\,\mathrm{K}}^{373\,\mathrm{K}} \left(\frac{27.296}{T} + 5.23 \times 10^{-3}/\mathrm{K} - 0.042 \times 10^{-7}\,(T/\mathrm{K}^2)\right)\mathrm{d}T$$

$$= 27.296 \ln (T/\mathrm{K})\Big|_{298\,\mathrm{K}}^{373\,\mathrm{K}} + 5.23 \times 10^{-3}\left(\frac{T}{\mathrm{K}}\right)\Big|_{298\,\mathrm{K}}^{373\,\mathrm{K}}$$

$$- \frac{0.042 \times 10^{-7}}{2} \left(\frac{T^2}{K^2} \right) \Bigg|_{298\,K}^{373\,K}$$

$$= 27.296 \times 2.303 \times \log \left(\frac{373}{298} \right) + 5.23 \times 10^{-3} \times (373 - 298)$$

$$- 0.021 \times 10^{-7} (373^2 - 298^2)$$

$$\Delta S/\text{J K}^{-1}\ \text{mol}^{-1} = 6.129 + 0.392 - 1.057 \times 10^{-4} = 6.52$$

Thus ΔS for 2 mol = (2 mol) (6.52 J K^{-1} mol^{-1}) = 13.04 J K^{-1}

Derivation of Thermodynamic Equation of State

Equation Eq. (4.15.7) gives the variation in entropy with pressure of the system, when temperature is held constant, i.e. the isothermal pressure dependence of entropy. This expression can be simplified by using the following procedure:

Differentiating Eq. (4.15.6) with respect to pressure at constant temparature, we have

$$\frac{\partial}{\partial p}\left\{\left(\frac{\partial S}{\partial T}\right)_p\right\}_T = \frac{1}{T}\left\{\frac{\partial}{\partial p}\left(\frac{\partial H}{\partial T}\right)_p\right\}_T = \frac{1}{T}\frac{\partial^2 H}{\partial p\,\partial T} \tag{4.15.11}$$

Differentiating Eq. (4.15.7) with respect to temperature at constant pressure, we get

$$\frac{\partial}{\partial T}\left\{\left(\frac{\partial S}{\partial p}\right)_T\right\}_p = \frac{\partial}{\partial T}\left\{\frac{1}{T}\left[\left(\frac{\partial H}{\partial p}\right)_T - V\right]\right\}_p$$

$$= \frac{1}{T}\left[\frac{\partial^2 H}{\partial T\,\partial p} - \left(\frac{\partial V}{\partial T}\right)_p\right] - \frac{1}{T^2}\left[\left(\frac{\partial H}{\partial p}\right)_T - V\right] \tag{4.15.12}$$

Since S is a state function, we must have

$$\frac{\partial^2 S}{\partial p\,\partial T} = \frac{\partial^2 S}{\partial T\,\partial p}$$

Thus, from Eqs (4.15.11) and (4.15.12), we get

$$\frac{1}{T}\frac{\partial^2 H}{\partial p\,\partial T} = \frac{1}{T}\frac{\partial^2 H}{\partial T\,\partial p} - \frac{1}{T}\left(\frac{\partial V}{\partial T}\right)_p - \frac{1}{T^2}\left[\left(\frac{\partial H}{\partial p}\right)_T - V\right]$$

Since H is also a state function, the above expression reduces to

$$\left(\frac{\partial H}{\partial p}\right)_T - V = -T\left(\frac{\partial V}{\partial T}\right)_p \tag{4.15.13}$$

This is another expression of the thermodynamic equation of state as it gives variation of enthalpy with pressure at constant temperature in terms of the measurable derivative $(\partial V/\partial T)_p$ of the system.

Entropy Change in an Isothermal Variation in Pressure

Comparing Eq. (4.15.13) with Eq. (4.15.7), we get

$$\left(\frac{\partial S}{\partial p}\right)_T = -\left(\frac{\partial V}{\partial T}\right)_p = -V\alpha \tag{4.15.14}$$

For a finite change of pressure at constant temperature, ΔS is given by

$$\Delta S = -\int_{p_1}^{p_2}\left(\frac{\partial V}{\partial T}\right)_p \mathrm{d}p = -\int_{p_1}^{p_2} V\alpha\, \mathrm{d}p$$

The variation in entropy with pressure at constant temperature depends on how the volume of the system changes with temperature. For most substances this is positive. Because of negative sign, we can conclude that entropy decreases with increase in pressure at constant temperature.

Complete Expression of dS in terms of dT and dp

Substituting Eqs (4.15.6) and (4.15.14) in Eq. (4.15.2), we get

$$\mathrm{d}S = \frac{nC_{p,\mathrm{m}}}{T}\mathrm{d}T - \left(\frac{\partial V}{\partial T}\right)_p \mathrm{d}p$$

or

$$\mathrm{d}S = \frac{nC_{p,\mathrm{m}}}{T}\,\mathrm{d}T - V\,\alpha\,\mathrm{d}p \tag{4.15.15}$$

Equation (4.15.15) describes the variation in S with temperature and pressure of the system. For liquids and solids, α is negligibly small. Hence the variation in entropy with pressure at constant temperature for such substances is small, and can, therefore, be ignored.

Problem 4.15.1

Show that

(i) $\left(\frac{\partial H}{\partial p}\right)_T$ is zero for an ideal gas.

(ii) $\mu_{\mathrm{JT}} = \dfrac{T(\partial V/\partial T)_p - V}{C_p} = \dfrac{V(\alpha T - 1)}{C_p}$ and is zero for an ideal gas.

(iii) $\left(\frac{\partial C_p}{\partial p}\right)_T = -T\left(\frac{\partial^2 V}{\partial T^2}\right)_p = -TV\left[\alpha^2 + \left(\frac{\partial \alpha}{\partial T}\right)_p\right]$

(iv) $C_{p,\,\mathrm{m}}$ is independent of pressure for ideal gases whereas it depends both on temperature and pressure for van der Waals gases.

(v) Show that $C_{p,\,\mathrm{m}}$ has a maximum value at $p = RT/4b$ for a van der Waals gas.

(vi) Show that the effects of pressure on $C_{p,\,\mathrm{m}}$ of a van der Waals gas is the same as that of $C_{p,\,\mathrm{m}} - C_{V,\,\mathrm{m}}$.

Solution

(i) The expression of $(\partial H/\partial p)_T$ in terms of easily determinable derivative is given by the thermodynamic equation of state

$$\left(\frac{\partial H}{\partial p}\right)_T = -T\left(\frac{\partial V}{\partial T}\right)_p + V$$

For an ideal gas $pV = nRT$, thus

$$\left(\frac{\partial V}{\partial T}\right)_p = \frac{nR}{p}$$

Therefore $\left(\frac{\partial H}{\partial p}\right)_T = -T\frac{nR}{p} + V = -V + V = 0$ (4.15.16)

(ii) Joule-Thomson coefficient, by definition, is

$$\mu_{JT} = \left(\frac{\partial T}{\partial p}\right)_H = -\frac{1}{C_p}\left(\frac{\partial H}{\partial p}\right)_T \qquad \text{(Eq. 2.8.3)}$$

Making use of the thermodynamic equation of state, we have

$$\mu_{JT} = -\frac{1}{C_p}\left\{-T\left(\frac{\partial V}{\partial T}\right)_p + V\right\}$$

Since $\alpha = (1/V)\,(\partial V/\partial T)_p$, we have

$$\mu_{JT} = \frac{T(\partial V/\partial T)_p - V}{C_p} = \frac{V(\alpha T - 1)}{C_p} \qquad (4.15.17)$$

For an ideal gas $(\partial H/\partial p)_T = 0$, therefore $\mu_{JT} = 0$.

(iii) By definition, we have

$$C_p = \left(\frac{\partial H}{\partial T}\right)_p$$

Thus $\left(\frac{\partial C_p}{\partial p}\right)_T = \frac{\partial}{\partial p}\left\{\left(\frac{\partial H}{\partial T}\right)_p\right\}_T = \frac{\partial}{\partial T}\left\{\left(\frac{\partial H}{\partial p}\right)_T\right\}_p$ (as H is a state function)

Making use of the thermodynamic equation of state (Eq. 4.15.13), we get

$$\left(\frac{\partial C_p}{\partial p}\right)_T = \frac{\partial}{\partial T}\left\{-T\left(\frac{\partial V}{\partial T}\right)_p + V\right\}_p = -\left(\frac{\partial V}{\partial T}\right)_p - T\left(\frac{\partial^2 V}{\partial T^2}\right)_p + \left(\frac{\partial V}{\partial T}\right)_p$$

$$= -T\left(\frac{\partial^2 V}{\partial T^2}\right)_p \qquad (4.15.18)$$

Now since $\alpha = (1/V)\,(\partial V/\partial T)_p$, therefore

$$\left(\frac{\partial C_p}{\partial p}\right)_T = -T\left(\frac{\partial^2 V}{\partial T^2}\right)_p = -T\frac{\partial}{\partial T}\left\{\left(\frac{\partial V}{\partial T}\right)_p\right\}_p$$

$$= -T\left(\frac{\partial (V\alpha)}{\partial T}\right)_p = -T\left[V\left(\frac{\partial \alpha}{\partial T}\right)_p + \alpha\left(\frac{\partial V}{\partial T}\right)_p\right]$$

$$= -T\left[V\left(\frac{\partial \alpha}{\partial T}\right)_p + V\alpha^2\right] = -TV\left[\left(\frac{\partial \alpha}{\partial T}\right)_p + \alpha^2\right]$$

(iv) For one mole of an ideal gas, $V_m = RT/p$, therefore

$$\left(\frac{\partial V_m}{\partial T}\right)_p = \frac{R}{p} \quad \text{and} \quad \left(\frac{\partial^2 V_m}{\partial T^2}\right)_p = 0$$

Thus $$\left(\frac{\partial C_{p,m}}{\partial p}\right)_T = -T\left(\frac{\partial^2 V_m}{\partial T^2}\right) = 0$$

For one mole of a van der Waals gas, we have from Eq. (2.6.15)

$$\left(\frac{\partial V_m}{\partial T}\right)_p = \left(\frac{V_m - b}{T}\right)\left[1 + \frac{2ap}{R^2T^2} - \frac{4abp^2}{R^3T^3}\right] \simeq \frac{V_m - b}{T} + \frac{2a}{RT^2} - \frac{4abp}{R^2T^3}$$

(**Note:** In the second and third terms, the approximation $p\,(V_m - b) = RT$ has been used.)

Thus $$\left(\frac{\partial^2 V_m}{\partial T^2}\right)_p = \frac{1}{T}\left(\frac{\partial V_m}{\partial T}\right)_p - \frac{V_m - b}{T^2} - \frac{4a}{RT^3} + \frac{12\,abp}{R^2T^4}$$

$$= \frac{1}{T}\left(\frac{V_m - b}{T} + \frac{2a}{RT^2} - \frac{4abp}{R^2T^3}\right) - \frac{V_m - b}{T^2} - \frac{4a}{RT^3} + \frac{12\,abp}{R^2T^4}$$

$$= -\frac{2a}{RT^3} + \frac{8abp}{R^2T^4}$$

Hence $$\left(\frac{\partial C_{p,m}}{\partial p}\right)_T = -T\left(\frac{\partial^2 V_m}{\partial T^2}\right)_p = +\frac{2a}{RT^2} - \frac{8abp}{R^2T^3}$$

Hence $(\partial C_{p,m}/\partial p)_T$ depends both on T and p for van der Waals gases.
(v) For $C_{p,m}$ to have a maximum value, we set $(\partial C_{p,m}/\partial p)_T = 0$.
Thus, we get

$$\frac{2a}{RT^2} - \frac{8abp}{R^2T^3} = 0 \quad \text{or} \quad p = \frac{RT}{4b}$$

(vi) We have

$$\left(\frac{\partial(C_{p,m} - C_{V,m})}{\partial p}\right)_T = \left(\frac{\partial C_{p,m}}{\partial p}\right)_T - \left(\frac{\partial C_{V,m}}{\partial p}\right)_T$$

Now since C_V is independent of pressure for a van der Waals gas, we have

$$\left(\frac{\partial C_{V,m}}{\partial p}\right)_T = 0$$

Hence $$\left(\frac{\partial(C_{p,m} - C_{V,m})}{\partial p}\right)_T = \left(\frac{\partial C_{p,m}}{\partial p}\right)_T$$

Problem 4.15.2 Using $\mathrm{d}S = \dfrac{C_p}{T}\,\mathrm{d}T - V\alpha\,\mathrm{d}p$ or $\mathrm{d}S = \dfrac{C_V}{T}\,\mathrm{d}T + \dfrac{\alpha}{\kappa_T}\,\mathrm{d}V$, show that

(i) $$\left(\frac{\partial S}{\partial p}\right)_V = \frac{C_V}{T}\frac{\kappa_T}{\alpha}$$

(ii) $\left(\dfrac{\partial S}{\partial V}\right)_p = \dfrac{C_p}{TV\alpha}$

and (iii) $-\dfrac{1}{V}\left(\dfrac{\partial V}{\partial p}\right)_S = \dfrac{\kappa_T}{\gamma}$ where $\gamma = C_p / C_V$.

Solution **(i)** We have

$$dS = \frac{C_V}{T}\,dT + \frac{\alpha}{\kappa_T}\,dV \tag{4.15.19}$$

Since $T = f(p, V)$, therefore

$$dT = \left(\frac{\partial T}{\partial p}\right)_V dp + \left(\frac{\partial T}{\partial V}\right)_p dV$$

Substituting this in Eq. (4.15.19), we have

$$dS = \frac{C_V}{T}\left(\frac{\partial T}{\partial p}\right)_V dp + \left[\frac{C_V}{T}\left(\frac{\partial T}{\partial V}\right)_p + \frac{\alpha}{\kappa_T}\right]dV \tag{4.15.20}$$

Dividing by dp and introducing the condition of constant V, we get

$$\left(\frac{\partial S}{\partial p}\right)_V = \frac{C_V}{T}\left(\frac{\partial T}{\partial p}\right)_V = \frac{C_V}{T}\left(-\frac{(\partial V/\partial p)_T}{(\partial V/\partial T)_p}\right) = \frac{C_V}{T}\frac{\kappa_T}{\alpha} \tag{4.15.21}$$

(ii) We have

$$dS = \frac{C_p}{T}\,dT - V\alpha\,dp \tag{4.15.22}$$

or
$$dS = \frac{C_p}{T}\left\{\left(\frac{\partial T}{\partial p}\right)_V dp + \left(\frac{\partial T}{\partial V}\right)_p dV\right\} - V\alpha\,dp$$

or
$$dS = \frac{C_p}{T}\left(\frac{\partial T}{\partial V}\right)_V dV + \left\{\frac{C_p}{T}\left(\frac{\partial T}{\partial p}\right)_V - V\alpha\right\}dp \tag{4.15.23}$$

Dividing by dV and introducing the condition of constant p, we get

$$\left(\frac{\partial S}{\partial V}\right)_p = \frac{C_p}{T}\left(\frac{\partial T}{\partial V}\right)_p = \frac{C_p}{TV\alpha} \tag{4.15.24}$$

(iii) Dividing Eq. (4.15.20) by dp and introducing the condition of constant S, we get

$$0 = \frac{C_V}{T}\left(\frac{\partial T}{\partial p}\right)_V + \left\{\frac{C_V}{T}\left(\frac{\partial T}{\partial V}\right)_p + \frac{\alpha}{\kappa_T}\right\}\left(\frac{\partial V}{\partial p}\right)_S$$

or
$$\left(\frac{\partial V}{\partial p}\right)_S = \frac{-\dfrac{C_V}{T}\left(\dfrac{\partial T}{\partial p}\right)_V}{\dfrac{C_V}{T}\left(\dfrac{\partial T}{\partial V}\right)_p + \dfrac{\alpha}{\kappa_T}} = \frac{-\dfrac{C_V}{T}\dfrac{\kappa_T}{\alpha}}{\dfrac{C_V}{TV\alpha} + \dfrac{\alpha}{\kappa_T}} = -\frac{C_V V \kappa_T}{C_V + \dfrac{T\alpha^2 V}{\kappa_T}}$$

Now according to Eq. (4.14.18), we have

$$C_p - C_V = \frac{T\alpha^2 V}{\kappa_T}$$

With this the above expression modifies to

$$\left(\frac{\partial V}{\partial p}\right)_S = -\frac{C_V V \kappa_T}{C_V + C_p - C_V} = -V\frac{\kappa_T}{\gamma}$$

or $$-\frac{1}{V}\left(\frac{\partial V}{\partial p}\right)_S = \frac{\kappa_T}{\gamma} \qquad (4.15.25)$$

Problem 4.15.3

(a) Show that, if $(\partial U/\partial V)_T = 0$, the equation of state of the substance must be of the form $p = T\,f(V)$.

(b) Show that, if $(\partial H/\partial p)_T = 0$, the equation of state of the substance must be of the form $V = T\,f(p)$.

(c) If for a substance both $(\partial U/\partial V)_T = 0$ and $(\partial H/\partial p)_T = 0$, the equation of state is pV/T = constant.

Solution

(a) In the thermodynamic equation of state

$$\left(\frac{\partial U}{\partial V}\right)_T = T\left(\frac{\partial p}{\partial T}\right)_V - p$$

if we substitute $(\partial U/\partial V)_T = 0$, we get

$$\left(\frac{\partial p}{\partial T}\right)_V = \frac{p}{T} \qquad (4.15.26)$$

or $$\frac{\mathrm{d}p}{p} = \frac{\mathrm{d}T}{T} \quad ;\ (V\text{ constant}) \qquad (4.15.27)$$

Integrating this, we have

$$\ln\frac{p_2}{p_1} = \ln\frac{T_2}{T_1} \quad \text{or} \quad \frac{p_2}{p_1} = \frac{T_2}{T_1} \quad \text{or} \quad \frac{p_2}{T_2} = \frac{p_1}{T_1}$$

or $$p = TA \qquad (4.15.28)$$

where A is constant. Since V is held constant in Eq. (4.15.27), we conclude that the value of A will depend on V, i.e. $A = f(V)$. Hence, Eq. (4.15.28) may be written as

$$p = T\,f(V)$$

(b) In the thermodynamic equation of state

$$\left(\frac{\partial H}{\partial p}\right)_T = V - T\left(\frac{\partial V}{\partial T}\right)p$$

if we substitute $(\partial H/\partial p)_T = 0$, we get

$$\left(\frac{\partial V}{\partial T}\right)_p = \frac{V}{T} \qquad (4.15.29)$$

or $$\frac{\mathrm{d}V}{V} = \frac{\mathrm{d}T}{T} \quad ;\ (p\text{ constant}) \qquad (4.15.30)$$

Integrating this, we have

$$\ln \frac{V_2}{V_1} = \ln \frac{T_2}{T_1} \quad \text{or} \quad \frac{V_2}{V_1} = \frac{T_2}{T_1} \quad \text{or} \quad \frac{V_2}{T_2} = \frac{V_1}{T_1}$$

or $$V = TA \tag{4.15.31}$$

where A is constant. Since p is held constant in Eq. (4.15.30), we conclude that the value of A will depend on p, i.e., $A = f(p)$. Hence, Eq. (4.15.31) may be written as

$$V = T\, f(p)$$

(c) Making use of the cyclic rule

$$\left(\frac{\partial p}{\partial V}\right)_T \left(\frac{\partial V}{\partial T}\right)_p \left(\frac{\partial T}{\partial p}\right)_V + 1 = 0$$

we have

$$\left(\frac{\partial p}{\partial V}\right)_T = -\frac{1}{(\partial V/\partial T)_p (\partial T/\partial p)_V} = -\left(\frac{\partial T}{\partial V}\right)_p \left(\frac{\partial p}{\partial T}\right)_V$$

Substituting the expressions of $(\partial T/\partial V)_p$ and $(\partial p/\partial T)_V$ from Eqs (4.15.29) and (4.15.26), respectively, we get

$$\left(\frac{\partial p}{\partial V}\right)_T = -\frac{T}{V}\cdot\frac{p}{T} = -\frac{p}{V} \tag{4.15.32}$$

Form the given expression of equation of state, we conclude that $p = f(T, V)$. Writing the complete differential of p, we get

$$\mathrm{d}p = \left(\frac{\partial p}{\partial T}\right)_V \mathrm{d}T + \left(\frac{\partial p}{\partial V}\right)_T \mathrm{d}V$$

Substituting the expressions of $(\partial p/\partial T)_V$ and $(\partial p/\partial V)_T$ form Eqs (4.15.26) and (4.15.32), respectively, we get

$$\mathrm{d}p = \frac{p}{T}\mathrm{d}T - \frac{p}{V}\mathrm{d}V$$

or $$\frac{\mathrm{d}p}{p} = \frac{\mathrm{d}T}{T} - \frac{\mathrm{d}V}{V}$$

Integrating the above expression, we get

$$\ln \frac{p_2}{p_1} = \ln \frac{T_2}{T_1} - \ln \frac{V_2}{V_1} \quad \text{or} \quad \ln \frac{p_2 V_2}{p_1 V_1} = \ln \frac{T_2}{T_1} \quad \text{or} \quad \frac{p_2 V_2}{p_1 V_1} = \frac{T_2}{T_1}$$

or $$\frac{p_2 V_2}{T_2} = \frac{p_1 V_1}{T_1} \quad \text{i.e.} \quad \frac{pV}{T} = \text{constant} \tag{4.15.33}$$

4.16 ENTROPY CHANGES FOR AN IDEAL GAS

Entropy changes for an ideal gas due to variations in temperature and volume or temperature and pressure can be determined by using Eqs (4.14.14) and (4.15.15). But, for the sake of simplicity, we drive the necessary relations directly from the definition of entropy function and the first law of thermodynamics.

Temperature and Volume Variations

According to the first law of thermodynamics, we have

$$dq_{\text{rev}} = dU + p\ dV$$

Dividing this by T, we have

$$dS = \frac{dq_{\text{rev}}}{T} = \frac{dU}{T} + \frac{p}{T}\ dV$$

Taking U to be a function of T and V and writing its differential, we have

$$dU = \left(\frac{\partial U}{\partial T}\right)_V dT + \left(\frac{\partial U}{\partial V}\right)_T dV$$

Substituting this in the previous expression, we get

$$dS = \frac{1}{T}\left\{\left(\frac{\partial U}{\partial T}\right)_V dT + \left(\frac{\partial U}{\partial V}\right)_T dV\right\} + \frac{p}{T}\ dV$$

Now for an ideal gas

$$\left(\frac{\partial U}{\partial V}\right)_T = 0 \qquad \text{(Joule's law)}$$

and $\quad \dfrac{p}{T} = \dfrac{nR}{V}$

With these, alongwith the fact that $(\partial U/\partial T)_V = C_V$, the previous expression becomes

$$dS = \frac{C_V}{T}\ dT + \frac{nR}{V}\ dV \tag{4.16.1}$$

For a finite change, we have

$$\Delta S = \int_{T_1}^{T_2} \frac{C_V}{T}\ dT + \int_{V_1}^{V_2} \frac{nR}{V}\ dV$$

Considering C_V to be independent of temperature, we have

$$\Delta S = C_V \ln \frac{T_2}{T_1} + nR \ln \frac{V_2}{V_1} \tag{4.16.2}$$

or $$\Delta S = n\left[C_{V,\text{m}} \ln \frac{T_2}{T_1} + R \ln \frac{V_2}{V_1}\right] \tag{4.16.3}$$

This is the expression when both the volume and temperature of an ideal gas are changed. For an isothermal change in volume, the entropy change is given by the relation

$$\Delta S = nR \ln \frac{V_2}{V_1} \tag{4.16.4}$$

For the change of temperature at constant volume, the expression is

$$dS = \int_{T_1}^{T_2} \frac{nC_{V,\text{m}}}{T}\ dT \tag{4.16.5}$$

$C_{V,\,\mathrm{m}}$ independent of temperature From Eq. (4.16.5), we get

$$\Delta S = nC_{V,\,\mathrm{m}} \ln \frac{T_2}{T_1} \tag{4.16.6a}$$

$C_{V,\,\mathrm{m}}$ dependent on temperature If $C_{V,\,\mathrm{m}}$ depends upon temperature, we have to express $C_{V,\,\mathrm{m}}$ in terms of T before carrying out the integration in Eq. (4.16.5). For one mole of an ideal gas, we can write

$$C_{V,\,\mathrm{m}} = C_{p,\,\mathrm{m}} - R$$

Dependency of $C_{p,\,\mathrm{m}}$ on temperature is given by the analytic expression

$$C_{p,\,\mathrm{m}} = a + bT + cT^2 + \cdots$$

Thus, the expression of $C_{V,\,\mathrm{m}}$ for an ideal gas is given by

$$C_{V,\,\mathrm{m}} = a + bT + cT^2 + \cdots - R$$

Substituting this in Eq. (4.16.5), we get

$$\Delta S = n\int_{T_1}^{T_2} \frac{a + bT + cT^2 + \cdots - R}{T}\,\mathrm{d}T$$

$$= n\left[\int_{T_1}^{T_2} \frac{a - R}{T}\,\mathrm{d}T + \int_{T_1}^{T_2} (b + cT + \cdots)\,\mathrm{d}T\right]$$

Thus $$\Delta S = n\left[(a - R)\ln\frac{T_2}{T_1} + b(T_2 - T_1) + \frac{c}{2}(T_2^2 - T_1^2) + \cdots\right] \tag{4.16.6b}$$

Temperature and Pressure Variations

For an ideal gas, we have

$$\frac{V_2}{V_1} = \frac{p_1 T_2}{p_2 T_1}$$

Substituting this in Eq. (4.16.3), we get

$$\Delta S = n\left[C_{V,\mathrm{m}} \ln\frac{T_2}{T_1} + R\ln\frac{p_1 T_2}{p_2 T_1}\right]$$

or $$\Delta S = n\left[(C_{V,\mathrm{m}} + R)\ln\frac{T_2}{T_1} + R\ln\frac{p_1}{p_2}\right]$$

or $$\Delta S = n\left[C_{p,\,\mathrm{m}} \ln\frac{T_2}{T_1} + R\ln\frac{p_1}{p_2}\right] \tag{4.16.7}$$

This is the expression for the change in entropy of an ideal gas when both its pressure and temperature are changed. However, if the temperature is varied at constant pressure, then

$$\Delta S = nC_{p,\,\mathrm{m}} \ln\frac{T_2}{T_1} \tag{4.16.8}$$

In deriving the above expression, it has been assumed that $C_{p,\,m}$ is independent of temperature. If $C_{p,\,m}$ varies with temperature as

$$C_{p,\,m} = a + bT + cT^2 + \cdots$$

then $$\Delta S = n\left[a \ln \frac{T_2}{T_1} + b\,(T_2 - T_1) + \frac{c}{2}(T_2^2 - T_1^2) + \cdots\right] \qquad \text{(Eq. 4.15.10)}$$

For the isothermal change in pressure, we have

$$\Delta S = nR \ln \frac{p_1}{p_2} \qquad (4.16.9)$$

Example 4.16.1 For an ideal gas $C_{p,\,m} = (5/2)R$. Calculate the change in entropy suffered by 3 mol of the gas on being heated from 300 K to 600 K at (a) constant pressure, and (b) constant volume.

Solution **(a)** The change in entropy as a result of variation in temperature at constant pressure is given by

$$\Delta S_p = nC_{p,\,m} \ln \frac{T_2}{T_1}$$

Substituting the given values, we get

$$\Delta S_p = (3 \text{ mol})\left(\frac{5}{2} \times 8.314 \text{ J K}^{-1} \text{ mol}^{-1}\right) \times 2.303 \log\left(\frac{600 \text{ K}}{300 \text{ K}}\right)$$

$$= 43.22 \text{ J K}^{-1}$$

(b) The corresponding expression at constant volume is

$$\Delta S_V = nC_{V,\,m} \ln \frac{T_2}{T_1}$$

which on substituting the given values gives

$$\Delta S_V = (3 \text{ mol})\left(\frac{3}{2} \times 8.314 \text{ J K}^{-1} \text{ mol}^{-1}\right) \times 2.303 \log\left(\frac{600 \text{ K}}{300 \text{ K}}\right)$$

$$= 25.93 \text{ J K}^{-1}$$

Example 4.16.2 Calculate ΔS for 3 mol of a diatomic ideal gas which is heated and compressed from 298 K and 1 bar to 398 K and 5 bar. Given: $C_{p,\,m} = (7/2)R$.

Solution The entropy change as a result of variation in both temperature and pressure is given by

$$\Delta S = n\left[C_{p,\mathrm{m}} \ln \frac{T_2}{T_1} + R \ln \frac{p_1}{p_2}\right]$$

Substituting the given values, we get

$$\Delta S = (3 \text{ mol})\left[\left(\frac{7}{2} \times 8.314 \text{ J K}^{-1} \text{ mol}^{-1}\right) \times 2.303 \log\left(\frac{398 \text{ K}}{298 \text{ K}}\right)\right.$$

$$\left. + \left(8.314 \text{ J K}^{-1} \text{ mol}^{-1}\right) \times 2.303 \log\left(\frac{1 \text{ bar}}{5 \text{ bar}}\right)\right]$$

$$= 3(8.422 \text{ J K}^{-1} - 13.383 \text{ J K}^{-1})$$

$$= -\,14.883 \text{ J K}^{-1}$$

Example 4.16.3 By how much does the entropy of 3 mol of an ideal gas change in going from a pressure of 2 bar to a pressure of 1 bar, without any change in temperature. If the surroundings too are at 1 bar pressure and 300 K, and the expansion is against the constant external pressure of the surroundings, show that the process of expansion is irreversible.

Solution Entropy change of the gas

$$\Delta S_{gas} = nR \ln \frac{p_1}{p_2} = (3 \text{ mol}) (8.314 \text{ J K}^{-1} \text{ mol}^{-1}) \times 2.303 \log \left(\frac{2 \text{ bar}}{1 \text{ bar}}\right)$$

$$= 17.29 \text{ J K}^{-1}$$

Entropy change of the surroundings

Since $\Delta T = 0$, therefore $\Delta U = 0$ and according to the first law of thermodynamics

$$q = - w = p_{ext} (\Delta V) = p_{ext} (V_2 - V_1)$$

$$= p_{ext} \left(\frac{nRT}{p_2} - \frac{nRT}{p_1}\right) = p_{ext}\, nRT \left(\frac{1}{p_2} - \frac{1}{p_1}\right)$$

$$= nRT \left(1 - \frac{p_{ext}}{p_1}\right) \qquad \text{(since } p_{ext} = p_2\text{)}$$

Substituting the values, we get

$$q = (3 \text{ mol}) (8.314 \text{ J K}^{-1} \text{ mol}^{-1}) (T) \left(1 - \frac{1 \text{ bar}}{2 \text{ bar}}\right) = (12.47 \text{ J K}^{-1})T$$

$$\Delta S_{surr} = - \frac{q}{T} = - 12.47 \text{ J K}^{-1}$$

$$\Delta S_{total} = \Delta S_{sys} + \Delta S_{surr}$$

$$= (17.29 - 12.47) \text{ J K}^{-1}, \text{ which is positive.}$$

Since ΔS_{total} is positive, the process is irreversible.

4.17 A FEW DERIVATIONS INVOLVING A VAN DER WAALS GAS

(i) $$\left(\frac{\partial S}{\partial V}\right)_T = \frac{nR}{V - nb}$$

We know that

$$\left(\frac{\partial S}{\partial V}\right)_T = \left(\frac{\partial p}{\partial T}\right)_V \qquad \text{(Eq. 4.14.12)}$$

For the amount n of a van der Waals gas, we have

$$\left(p + \frac{n^2 a}{V^2}\right) (V - nb) = nRT$$

Differentiating this with respect to temperature keeping volume constant, we get

$$\left(\frac{\partial p}{\partial T}\right)_V (V - nb) = nR$$

or $$\left(\frac{\partial p}{\partial T}\right)_V = \frac{nR}{V - nb} \tag{4.17.1a}$$

Thus $$\left(\frac{\partial S}{\partial V}\right)_T = \frac{nR}{V - nb} \tag{4.17.1b}$$

(ii) $$\left(\frac{\partial S}{\partial p}\right)_T = \frac{V - nb}{\dfrac{2na}{V^3R}(V - nb)^2 - T}$$

We know that

$$\left(\frac{\partial S}{\partial p}\right)_T = -\left(\frac{\partial V}{\partial T}\right)_p \qquad \text{(Eq. 4.15.14)}$$

For the amount n of a van der Waals gas, we have

$$\left(p + \frac{n^2a}{V^2}\right)(V - nb) = nRT$$

Differentiating this with respect to temperature keeping pressure constant, we get

$$\left\{-\frac{2n^2a}{V^3}\left(\frac{\partial V}{\partial T}\right)_p\right\}(V - nb) + \left(p + \frac{n^2a}{V^2}\right)\left(\frac{\partial V}{\partial T}\right)_p = nR$$

or $$\left(\frac{\partial V}{\partial T}\right)_p \left\{-\frac{2n^2a}{V^3}(V - nb) + \left(p + \frac{n^2a}{V^2}\right)\right\} = nR$$

Multiplying by $(V - nb)$, we get

$$\left(\frac{\partial V}{\partial T}\right)_p \left\{-\frac{2n^2a}{V^3}(V - nb)^2 + \left(p + \frac{n^2a}{V^2}\right)(V - nb)\right\} = nR\,(V - nb)$$

$$\left(\frac{\partial V}{\partial T}\right)_p \left\{-\frac{2n^2a}{V^3}(V - nb)^2 + nRT\right\} = nR\,(V - nb)$$

Dividing by nR and rearranging, we get

$$\left(\frac{\partial V}{\partial T}\right)_p = -\frac{V - nb}{\dfrac{2na}{V^3R}(V - nb)^2 - T} \tag{4.17.2a}$$

Hence, it follows that

$$\left(\frac{\partial S}{\partial p}\right)_T = \frac{V - nb}{\dfrac{2na}{V^3R}(V - nb)^2 - T} \tag{4.17.2b}$$

(iii) $$\left(\frac{\partial H}{\partial p}\right)_T = V + \frac{V - nb}{\dfrac{2na}{V^3RT}(V - nb)^2 - 1}$$

The differential $(\partial H/\partial p)_T$ as given by thermodynamic equation of state is

$$\left(\frac{\partial H}{\partial p}\right)_T = V - T\left(\frac{\partial V}{\partial T}\right)_p$$

But $$\left(\frac{\partial V}{\partial T}\right)_p = -\frac{V - nb}{\dfrac{2na}{V^3R}(V - nb)^2 - T} \qquad \text{(Eq. 4.17.2a)}$$

Therefore $$\left(\frac{\partial H}{\partial p}\right)_T = V + \frac{(V - nb)T}{\dfrac{2na}{V^3R}(V - nb)^2 - T}$$

$$= V + \frac{(V - nb)}{\dfrac{2na}{V^3RT}(V - nb)^2 - 1} \qquad (4.17.3)$$

(iv) $$\left(\frac{\partial U}{\partial p}\right)_T = \frac{(V - nb)^2}{\dfrac{2}{V}(V - nb)^2 - \dfrac{V^2RT}{na}}$$

We can write

$$\left(\frac{\partial U}{\partial p}\right)_T = \left(\frac{\partial U}{\partial V}\right)_T \left(\frac{\partial V}{\partial p}\right)_T \qquad (4.17.4)$$

From the thermodynamic equation of state, we have

$$\left(\frac{\partial U}{\partial V}\right)_T = T\left(\frac{\partial p}{\partial T}\right)_T - p$$

Substituting $(\partial p/\partial T)_V$ from Eq. (4.17.1a), we get

$$\left(\frac{\partial U}{\partial V}\right)_T = T\frac{nR}{V - nb} - p = \left(p + \frac{n^2a}{V^2}\right) - p = \frac{n^2a}{V^2} \qquad (4.17.5)$$

Now making use of the cyclic

$$\left(\frac{\partial V}{\partial p}\right)_T \left(\frac{\partial p}{\partial T}\right)_V \left(\frac{\partial T}{\partial V}\right)_p + 1 = 0$$

we get $$\left(\frac{\partial V}{\partial p}\right)_T = -\frac{(\partial V/\partial T)_p}{(\partial p/\partial T)_V}$$

Substituting $(\partial p/\partial T)_V$ and $(\partial V/\partial T)_p$ from Eqs (4.17.1a) and (4.17.2a), we get

$$\left(\frac{\partial V}{\partial p}\right)_T = \left(\frac{V - nb}{\dfrac{2na}{V^3 R}(V - nb)^2 - T}\right)\left(\frac{V - nb}{nR}\right) = \frac{(V - nb)^2}{\dfrac{2n^2 a}{V^3}(V - nb)^2 - nRT} \qquad (4.17.6)$$

Substituting (4.17.5) and (4.17.6) in Eq. (4.17.4), we get

$$\left(\frac{\partial U}{\partial p}\right)_T = \left(\frac{n^2 a}{V^2}\right)\left(\frac{(V - nb)^2}{\dfrac{2n^2 a}{V^3}(V - nb)^2 - nRT}\right) = \frac{(V - nb)^2}{\dfrac{2}{V}(V - nb)^2 - \dfrac{V^2 RT}{na}} \qquad (4.17.7)$$

(v) $\left(\dfrac{\partial T}{\partial V}\right)_U = -\dfrac{na}{C_{V,\mathrm{m}} V^2}$

Taking U to be a function of T and V, and writing its differential, we get

$$\mathrm{d}U = \left(\frac{\partial U}{\partial T}\right)_V \mathrm{d}T + \left(\frac{\partial U}{\partial V}\right)_T \mathrm{d}V$$

Dividing by $\mathrm{d}V$ and introducing the condition of constant U, we get

$$0 = \left(\frac{\partial U}{\partial T}\right)_V \left(\frac{\partial T}{\partial V}\right)_U + \left(\frac{\partial U}{\partial V}\right)_T$$

or $$\left(\frac{\partial T}{\partial V}\right)_U = -\frac{(\partial U/\partial V)_T}{(\partial U/\partial T)_V}$$

Making use of Eqs (2.4.9) and (4.17.5), we get

$$\left(\frac{\partial T}{\partial V}\right)_U = -\frac{n^2 a/V^2}{nC_{V,\mathrm{m}}} = -\frac{na}{C_{V,\mathrm{m}} V^2} \qquad (4.17.8)$$

(vi) $\left(\dfrac{\partial V}{\partial T}\right)_S = -\dfrac{C_{V,\mathrm{m}}(V - nb)}{RT}$

Taking S to be a function of T and V, and writing its differential, we get

$$\mathrm{d}S = \left(\frac{\partial S}{\partial T}\right)_V \mathrm{d}T + \left(\frac{\partial S}{\partial V}\right)_T \mathrm{d}V$$

Dividing by $\mathrm{d}T$ an introducing the condition of constant S, we get

$$0 = \left(\frac{\partial S}{\partial T}\right)_V + \left(\frac{\partial S}{\partial V}\right)_T \left(\frac{\partial V}{\partial T}\right)_S \quad \text{or} \quad \left(\frac{\partial V}{\partial T}\right)_S = -\frac{(\partial S/\partial T)_V}{(\partial S/\partial V)_T}$$

Making use of Eqs (4.14.6) and (4.17.1b), we get

$$\left(\frac{\partial V}{\partial T}\right)_S = -\frac{nC_{V,\mathrm{m}}/T}{nR/(V - nb)} = -\frac{C_{V,\mathrm{m}}(V - nb)}{RT} \qquad (4.17.9)$$

Problem 4.17.1 A van der Waals gas changes its state from T_1, V_1 to T_2, V_2. Derive the expression for its entropy change.

Solution From Eq. (4.14.14), we have

$$dS = \frac{nC_{V,m}}{T}\,dT + \left(\frac{\partial p}{\partial T}\right)_V dT$$

For a van der Waals gas, we have

$$p = \frac{nRT}{V - nb} - \frac{n^2 a}{V^2}$$

Therefore

$$\left(\frac{\partial p}{\partial T}\right)_V = \frac{nR}{V - nb}$$

Hence, Eq. (4.14.14) becomes

$$dS = \frac{nC_{V,m}}{T}\,dT + \frac{nR}{V - nb}\,dV \qquad (4.17.10)$$

Integrating the above expression within the limits T_1, V_1 and T_2, V_2 we get

$$\Delta S = nC_{V,\,m}\ln\frac{T_2}{T_1} + nR\ln\frac{V_2 - nb}{V_1 - nb}$$

which is the required expression ΔS.

Example 4.17.1 One mole of a van der Waals gas undergoes a change from 298 K and 1 dm^3 to 373 K and 10 dm^3. What is the change in its entropy? Given $b = 0.06$ dm^3 mol^{-1}, $C_{V,\,m} = 29.0$ J K^{-1} mol^{-1}.

Solution Substituting the given data in the expression

$$\Delta S = C_{V,\,m}\ln\frac{T_2}{T_1} + R\ln\frac{V_2 - b}{V_1 - b}$$

we get $\Delta S = (29.0\ \text{J K}^{-1}\ \text{mol}^{-1})\ln\dfrac{373}{298} + (8.314\ \text{J K}^{-1}\ \text{mol}^{-1})\ln\left(\dfrac{10 - 0.06}{1 - 0.06}\right)$

$$= (6.51 + 19.61)\ \text{J K}^{-1}\ \text{mol}^{-1} = 26.12\ \text{J K}^{-1}\ \text{mol}^{-1}$$

Problem 4.17.2 For an adiabatic reversible expansion of a van der Waals gas, Eq. (4.17.10) will also be applicable. Show that ΔS_{sys} for such a process will be zero.

Solution From Eq. (2.14.4) which is applicable to an adiabatic reversible process, we have

$$C_{V,\,m}\,dT = -\frac{RT}{V - nb}\,dV$$

In the light of this expression, Eq. (4.17.10) will be zero. Hence for an adiabatic reversible expansion of a van der Waals gas, $\Delta S_{sys} = 0$.

4.18 STANDARD STATE FOR ENTROPY OF AN IDEAL GAS

The molar entropy change due to the isothermal change in pressure of an ideal gas is given by

$$\Delta S = R \ln \frac{p_1}{p_2} = - R \ln \frac{p_2}{p_1} \qquad \text{(Eq. 4.17.7)}$$

The above equation may be written as

$$S_{2,\,m} - S_{1,\,m} = - [R \ln (p_2/p^\circ) - R \ln (p_1/p^\circ)] \qquad (4.18.1)$$

where $S_{2,\,m}$ and $S_{1,\,m}$ are the molar entropies of an ideal gas at pressures p_2 and p_1, respectively, and p° is the standard-state unit pressure.

The standard molar entropy of an ideal gas at given temperature is defined as the entropy of one mole of the gas at pressure equal to 1 bar.† It is represented as S°. Substituting p_1 = 1 bar and replacing p_2 by the general term p in Eq. (4.18.1), we get

$$S_m - S_m^\circ = - R \ln (p/1 \text{ bar}) \qquad (4.18.2)$$

Equation (4.18.2) gives molar entropy of an ideal gas at pressure p relative to that at 1 bar pressure. A plot of $S_m - S_m^\circ$ for an ideal gas as a function of pressure is shown in Fig. 4.18.1 and as a function of ln (p/ bar) in Fig. 4.18.2. It is evident from Fig. 4.18.1 that the rate of decrease in entropy with pressure is rapid at low pressure and becomes less rapid at higher pressure.

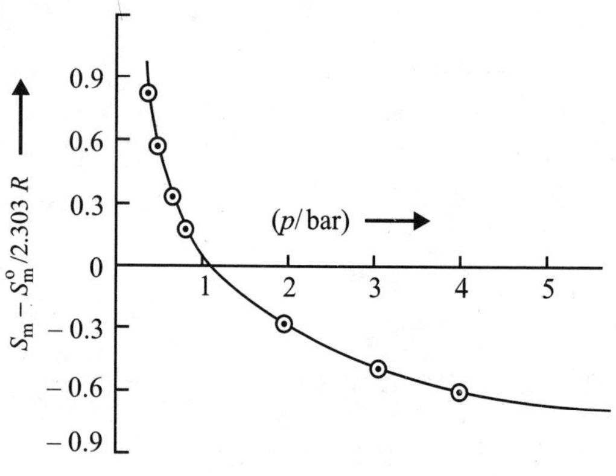

Fig 4.18.1 Plot of $S_m - S_m^\circ$ versus (p/ bar)

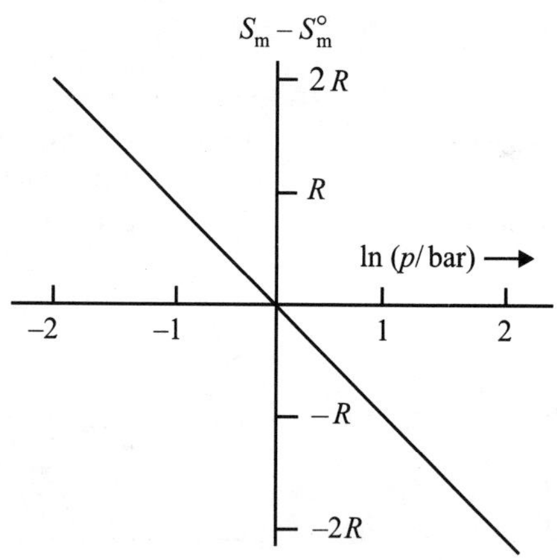

Fig. 4.18.2 $S_m - S_m^\circ$ versus ln (p/ bar)

† Prior to the recommendation of standard-state pressure of 1 bar, the value used was 1 atm (= 101.325 kPa). The change from 1 atm to 1 bar causes an increases of 0.109 J K^{-1} mol^{-1} (= R ln (101.325 kPa/100 kPa)} in the standard-state entropy of gaseous substances only.

4.19 ENTROPY AND DISORDERLINESS

Equations (4.16.1) to (4.16.9) describe the characteristics of entropy change for an ideal gas. These characteristics, in turn, reflect the basic nature of the entropy function. According to Eq. (4.16.6), we see that on increasing temperature of the gas at constant volume, there occurs an increase in the entropy of the system. The increase in temperature of the gas causes increase in the average kinetic energy of the molecules and thus the latter possess more energy and hence their motions become more random or disordered. Consequently, the increase in entropy of the system may be correlated with the increase in its disorderliness. The same conclusion will be drawn from other equations. For example, according to Eq. (4.16.4), the entropy of gas increases on increasing the volume of the system at a constant temperature. This increase in volume allows the molecules to move in a large space and consequently their motion become more random or disordered. Hence, the entropy may be considered to represent the disorderliness of the system—larger the disorderliness, larger the entropy of the system.

Take, for example, the three states of a substance, namely, the solid, liquid and gaseous states. In general, the molecules in the gaseous state are more disordered than those in the liquid state, while the molecules in the latter are more disordered than those in the solid state. Thus, the entropy of the substance in these three states of mater follows the order

$$S(\text{gaseous state}) >> S(\text{liquid state}) > S(\text{solid state})$$

The following reactions

$$Fe_2O_3(s) + 3H_2(g) \rightarrow 2Fe(s) + 3H_2O(l)$$

$$\tfrac{1}{2}N_2(g) + \tfrac{3}{2}H_2(g) \rightarrow NH_3(g)$$

will be attended to by a decrease in entropy since in converting reactants into products, the number of gaseous species decreases and thus the products are more ordered than the reactants. In the reaction

$$H_2(g) + Br_2(l) \rightarrow 2HBr(g)$$

entropy of the system increases as products contain larger number of gaseous molecules than the reactants and thus are more disordered than the latter.

4.20 ENTROPY CHANGE IN ISOTHERMAL EXPANSION OR COMPRESSION OF AN IDEAL GAS

In this section, we derive expressions for the change in total entropy (system + surroundings) when an ideal gas undergoes isothermal expansion or compression processes.

Entropy Change of the System

The change in entropy of one mole of an ideal gas as given by Eq. (4.16.4) is

$$\Delta S_{sys} = R \ln \frac{V_2}{V_1}$$

where V_2 is the final volume and V_1 is the initial volume.

Next we proceed to calculate the change in total entropy for the following categories.

Reversible Change If the expansion or compression is carried out reversibly, then

$$q_{\text{rev}} = -w = RT \ln \frac{V_2}{V_1} \quad \text{and} \quad \Delta S_{\text{sys}} = \frac{q_{\text{rev}}}{T} = R \ln \frac{V_2}{V_1}$$

If it is assumed that the heat q_{rev} is exchanged reversibly between the system and the surroundings at temperature T, it is obvious that

$$\Delta S_{\text{surr}} = -\frac{q_{\text{rev}}}{T}$$

Thus $\Delta S_{\text{total}} = \Delta S_{\text{sys}} + \Delta S_{\text{surr}} = 0$ (4.20.1)

Irreversible Change Two cases of expansion process may be considered:

Free expansion The gas expands into vacuum for this process, we have

$$w = 0 \quad \text{and} \quad q = 0$$

Since entropy is a state function, the entropy change of a system in going from volume V_1 to V_2 by any path will be same as that of a reversible change; therefore,

$$\Delta S_{\text{sys}} = R \ln \frac{V_2}{V_1}$$

Since no heat is supplied by the surroundings the entropy change of the latter would be zero, i.e.

$$\Delta S_{\text{surr}} = 0$$

$$\Delta S_{\text{total}} = \Delta S_{\text{sys}} + \Delta S_{\text{surr}} = R \ln \frac{V_2}{V_1} + 0 = R \ln \frac{V_2}{V_1} = \text{positive} \qquad (4.20.2)$$

Intermediate expansion Since ΔS_{sys} is the same as that of reversible change, therefore,

$$\Delta S_{\text{sys}} = R \ln \frac{V_2}{V_1} = \frac{q_{\text{rev}}}{T}$$

where q_{rev} is the amount of heat that the system would have absorbed had the process been carried out reversibly. In the present case the expansion is done against a constant pressure, thus

$$q_{\text{irr}} = -w = p_{\text{ext}}(V_2 - V_1)$$

The change in entropy of the surroundings will be given by

$$\Delta S_{\text{surr}} = -\frac{q_{\text{irr}}}{T} = -\frac{p_{\text{ext}}(V_2 - V_1)}{T}$$

Since the magnitude of work involved in the intermediate expansion is smaller than that involved in reversible expansion, it is obvious that $q_{\text{irr}} < q_{\text{rev}}$. With this, ΔS_{total} becomes

$$\Delta S_{\text{total}} = \Delta S_{\text{sys}} + \Delta S_{\text{surr}} = \frac{q_{\text{rev}}}{T} - \frac{q_{\text{irr}}}{T} = \text{positive} \qquad (4.20.3)$$

Example 4.20.1 One mole of an ideal gas is expanded isothermally at 298 K until its volume is tripled. Find the values of ΔS_{gas} and ΔS_{total} under the following conditions.

(i) Expansion is carried out reversibly.

(ii) Expansion is carried out irreversibly where 836.8 J of heat is less absorbed than in (i).

(iii) Expansion is free.

Solution The entropy change of the system ΔS_{sys} will be same in all the three processes as it is a state function. The expression of the change in entropy due to volume change at constant temperature is

$$\Delta S_{sys} = nR \ln \frac{V_2}{V_1} = \frac{q_{rev}}{T}$$

Substituting the given values, we get

$$\Delta S_{sys} = (1 \text{ mol})\ (8.314 \text{ J K}^{-1} \text{ mol}^{-1}) \times 2.303 \log \left(\frac{3}{1}\right) = 9.134 \text{ J K}^{-1}$$

The change in entropy of the surroundings ΔS_{surr}, however, will depend upon the process. We calculate below this change as well as the total change of entropy ($\Delta S_{sys} + \Delta S_{surr}$) for each of the three given processes.

(i) *Reversible expansion*

$$\Delta S_{gas} = -\Delta S_{surr} \qquad \text{and} \qquad \Delta S_{total} = 0$$

(ii) *Irreversible expansion* For the irreversible expansion, the quantity of heat absorbed by the system from surroundings would be

$$q_{irr} = q_{rev} - 836.8 \text{ J}$$

The quantity q_{irr} of heat will be supplied reversible by the surroundings at 298 K and hence

$$\Delta S_{surr} = -\frac{q_{irr}}{298 \text{ K}} = -\left(\frac{q_{rev}}{298 \text{ K}} - \frac{836.6 \text{ J}}{298 \text{ K}}\right) = -\left(\Delta S_{sys} - \frac{836.6 \text{ J}}{298 \text{ K}}\right)$$

and $$\Delta S_{total} = \Delta S_{sys} + \Delta S_{surr} = \Delta S_{sys} - \Delta S_{sys} + \left(\frac{836.6 \text{ J}}{298 \text{ K}}\right)$$

$$= 2.808 \text{ J K}^{-1}$$

(iii) *Free expansion* For this expansion, the system does not absorb any heat, thus $q = 0$. Hence

$$\Delta S_{surr} = 0 \qquad \text{and} \qquad \Delta S_{total} = \Delta S_{sys} = 9.134 \text{ J K}^{-1}$$

4.21 ENTROPY CHANGE IN ADIABATIC EXPANSION OR COMPRESSION OF AN IDEAL GAS

Entropy Change of the System Since in adiabatic processes $q = 0$, therefore

$$\Delta S_{surr} = 0$$

Since in an adiabatic process, both temperature and volume (or pressure) change, the expressions for the molar entropy change as given by Eq. (4.16.2) and (4,16.7) are

$$\Delta S_{sys} = C_{V,m} \ln \frac{T_2}{T_1} + R \ln \frac{V_2}{V_1}$$

$$\Delta S_{sys} = C_{p,\,m} \ln \frac{T_2}{T_1} + R \ln \frac{p_1}{p_2}$$

Now, we proceed to evaluate the change in total entropy for the following categories.

Reversible Change In this case

$$\Delta S_{sys} = 0$$

since for the adiabatic reversible process,

$$C_{V,\,m} \ln \frac{T_2}{T_1} = - R \ln \frac{V_2}{V_1} \qquad \text{(Eq. 2.11.6)}$$

and $$C_{p,\,m} \ln \frac{T_2}{T_1} = - R \ln \frac{p_1}{p_2}$$

Thus $$\Delta S_{total} = \Delta S_{sys} + \Delta S_{surr} = 0 + 0 = 0 \qquad (4.21.1)$$

In the present case of expansion (or compression), the increase (or decrease) in entropy due to the volume change just compensates the decrease (or increase) in entropy due to the decrease (or increase) in temperature.

Irreversible Change In the case,

$$\Delta S_{sys} = R \ln \frac{V_2}{V_1} + C_{V,\,m} \ln \frac{T_2'}{T_1} \qquad (4.21.2)$$

where T_2' is the temperature of the system in the final state. Making use of Eq. (2.11.6), we have

$$\Delta S_{sys} = - C_{V,\,m} \ln \frac{T_2}{T_1} + C_{V,\,m} \ln \frac{T_2'}{T_1} \qquad (4.21.3)$$

where T_2 is the temperature, if the process was reversible.

Since we know that

$$w_{irr} > w_{rev} \qquad \text{(including the sign of } w\text{)}$$

and moreover for adiabatic changes

$$\Delta U = w$$

it follows that

$$\Delta U_{irr} > \Delta U_{rev}$$

or $$C_{V,\,m} (T_2' - T_1) > C_{V,\,m} (T_2 - T_1)$$

Remembering that $T_2' < T_1$ in the expansion process and $T_2' > T_1$ in the compression process, we have

$$T_2' > T_2 \qquad (4.21.4)$$

that is, the decrease in temperature during the irreversible expansion will be lesser and the increase in temperature during the irreversible compression will be larger than the corresponding change in the reversible process. Thus, we have

$$C_{V,\,\mathrm{m}} \ln \frac{T_2'}{T_1} > C_{V,\,\mathrm{m}} \ln \frac{T_2}{T_1}$$

Substituting this relation in Eq. (4.21.3), we get

$$\Delta S_{\mathrm{sys}} = +\mathrm{ve}$$

and thus $\Delta S_{\mathrm{total}} = \Delta S_{\mathrm{sys}} + \Delta S_{\mathrm{surr}} = +\mathrm{ve}$

In the present case of expansion (or compression), the increase (or decrease) in entropy due to the volume change is larger (or smaller) than the decrease (or increase) in entropy due to the temperature change and hence ΔS_{sys} is positive.

Example 4.21.1

10 g of neon initially at a pressure of 506.625 kPa and temperature of 473 K expand adiabatically to a pressure of 202.65 kPa. Calculate entropy change of the system and total entropy change for the following ways of carrying out this expansion.

(i) Expansion is carried out reversibly.

(ii) Expansion occurs against a constant external pressure of 202.65 kPa.

(iii) Expansion is a free expansion.

Solution

(i) For an adiabatic reversible process

$$\Delta S_{\mathrm{sys}} = 0\,; \qquad \Delta S_{\mathrm{surr}} = 0 \qquad \text{and} \qquad \Delta S_{\mathrm{total}} = 0$$

(ii) First of all, we will have to calculate the temperature of the gas after it has undergone the said adiabatic irreversible expansion. This can be calculated as follows:

For an adiabatic process

$$\mathrm{d}q = 0$$

Therefore, according to the first law, we have

$$\mathrm{d}U = \mathrm{d}w$$

or $\qquad nC_{V,\,\mathrm{m}}\,\mathrm{d}T = -\,p_{\mathrm{ext}}\,\mathrm{d}V$

For an irreversible process against a constant pressure

$$nC_{V,\,\mathrm{m}}\,\Delta T = -\,p_{\mathrm{ext}}\,(V_2 - V_1)$$

or $\qquad nC_{V,\,\mathrm{m}}\,(T_2 - T_1) = p_{\mathrm{ext}}\,(V_1 - V_2)$

From the ideal gas equation, we have

$$V_1 = \frac{nRT_1}{p_1} = \frac{(0.5\ \mathrm{mol})\,(8.314\ \mathrm{kPa\ dm^3\ K^{-1}\ mol^{-1}})\,(473\mathrm{K})}{(506.625\ \mathrm{kPa})}$$

$$V_2 = \frac{nRT_2}{p_2} = \frac{(0.5\ \mathrm{mol})\,(8.314\ \mathrm{kPa\ dm^3\ K^{-1}\ mol^{-1}})T_2}{(202.65\ \mathrm{kPa})}$$

Therefore, we have

$$(0.5\ \mathrm{mol})\left(\frac{3}{2}\times 8.314\ \mathrm{J\ K^{-1}\ mol^{-1}}\right)(T_2 - 473\ \mathrm{K}) = (202.65\ \mathrm{kPa})$$

$$\times\left\{(0.5\ \mathrm{mol})(8.314\ \mathrm{J\ K^{-1}\ mol^{-1}})\left(\frac{473\ \mathrm{K}}{506.625\ \mathrm{kPa}} - \frac{T_2}{202.65\ \mathrm{kPa}}\right)\right\}$$

Cancelling the common terms, we have

$$3\left\{\left(\frac{T_2}{\text{K}}\right)-473\right\}=4\left(\frac{473}{5}-\frac{T_2/\text{K}}{2}\right)$$

or $$5(T_2/\text{K}) = \frac{1\,892}{5} + 1\,419$$

$$T_2 = 359.48 \text{ K}$$

Now employing the relation

$$\Delta S_{\text{sys}} = n\left[C_{p,\text{m}} \ln \frac{T_2}{T_1} + R \ln \frac{p_1}{p_2}\right]$$

for the entropy change of the system, we get

$$\Delta S_{\text{sys}} = (0.5 \text{ mol})\left[\left(\frac{5}{2}\times 8.314 \text{ J K}^{-1} \text{ mol}^{-1}\right)\times 2.303\times \log\left(\frac{359.48 \text{ K}}{473 \text{ K}}\right)\right.$$

$$\left. + (8.314 \text{ J K}^{-1} \text{ mol}^{-1}) \times 2.303 \times \log\left(\frac{506.625 \text{ kPa}}{202.65 \text{ kPa}}\right)\right]$$

$$= (0.5 \text{ mol}) (-5.704 \text{ J K}^{-1} \text{ mol}^{-1} + 7.169 \text{ J K}^{-1} \text{ mol}^{-1})$$

$$= 0.957 \text{ J K}^{-1}.$$

Since no heat is absorbed or given out to the surroundings

$$\Delta S_{\text{surr}} = 0$$

Thus $$\Delta S_{\text{total}} = \Delta S_{\text{sys}} = 0.957 \text{ J K}^{-1}$$

(iii) In a free expansion, we have

$$w = 0$$

Since the expansion is adiabatic, it follows that

$$q = 0$$

Thus, from the first law of thermodynamics, we have

$$\Delta U = 0$$

This implies that the temperature of the gas remains unchanged during the expansion. Thus, the entropy change will be there only because of pressure change. Employing the relation

$$\Delta S = nR \ln \frac{p_1}{p_2}$$

we get $$\Delta S_{\text{sys}} = \Delta S_{\text{total}}$$

$$= (0.5 \text{ mol}) (8.314 \text{ J K}^{-1} \text{ mol}^{-1}) \times 2.303 \log\left(\frac{506.625 \text{ kPa}}{202.65 \text{ kPa}}\right)$$

$$= 3.81 \text{ J K}^{-1}.$$

4.22 ENTROPY CHANGES IN A FEW TYPICAL CASES

In this section, we derive expressions for entropy change involved in a few typical cases given below.

Entropy Changes in a Reversible Phase Transformation

Since the reversible phase transformation takes place at a constant equilibrium temperature, it is obvious that

$$\Delta S = \frac{q_{rev}}{T}$$

where q_{rev} is the heat involved in the phase transformation. For the transformation of one mole of a substance at constant pressure, q_{rev} is equal to the molar enthalpy changes of the said transformation, i.e.

$$q_{rev} = \Delta H_m$$

Thus the entropy change per mole of phase transformation is given by

$$\Delta S_m = \frac{\Delta H_m}{T}$$

Trouton's Rule

According to Trouton's rule, the molar entropy of vaporization of most liquids which do not involve hydrogen bonding and also do not possess boiling point less then 150 K is about 10.5 R. Benzene is one of the examples for which we have

$$\Delta_{vap}H^\circ \text{ (benzene)} = 31.171 \text{ kJ mol}^{-1}$$

$$T_b = 353 \text{ K}$$

Thus $\quad \Delta_{vap}S^\circ \text{ (benzene)} = \dfrac{(31\,171 \text{ J mol}^{-1})}{(353 \text{ K})} = 88.3 \text{ J K}^{-1} \text{ mol}^{-1}$

For water which involves hydrogen bondings, the value of $\Delta_{vap}S^\circ$ is

$$\Delta_{vap}S^\circ\text{(water)} = \frac{(44\,012 \text{ J mol}^{-1})}{(373 \text{ K})} = 118.0 \text{ J K}^{-1} \text{ mol}^{-1}$$

which is much larger than 10.5 R, i.e. 87.3 J K^{-1} mol^{-1}.

No such rule exists for entropies of fusion at the melting point. For most substances

$$\Delta_{fus}S^\circ < \Delta_{vap}S^\circ$$

as the former involves only condensed phases whereas in the latter a condensed phase transforms into the chaotic gaseous phase.

Entropy Change in an Irreversible Phase Transition

As stated earlier, entropy change in an irreversible process can be calculated by transforming initial state to final state through the reversible paths. For example, the irreversible phase transformation of one mole of liquid water at –10 °C (T_1) to solid water at –10 °C can be calculated following the reversible paths given below.

(i) Heat the supercooled water reversibly (slowly) so that its temperature rises to 0 °C (T_2).

(ii) Convert the water at 0 °C reversibly to ice at 0 °C.

(iii) Cool the resulting ice reversibly (slowly) till it acquires a temperature of –10 °C.

The corresponding changes in entropies can be calculated as follows:

(i) $H_2O(1, 263.15\text{ K}) \rightarrow H_2O(1, 273.15\text{ K})$

$$\Delta S = \int_{T_1}^{T_2} \frac{dq}{T} = \int_{T_1}^{T_2} \frac{C_{p,\text{m}}(1)\, dT}{T} = C_{p,\text{m}}(1)\left(\ln \frac{T_2}{T_1}\right)$$

The heat capacity of liquid water is 75.312 J K^{-1} mol^{-1}. Substituting this value along with T_2 = 273.15 K and T_1 = 263.15 K, we get

$$\Delta S_1 = (75.312\text{ J K}^{-1}\text{ mol}^{-1}) \times 2.303 \times \log\left(\frac{273.15\text{ K}}{263.15\text{ K}}\right)$$
$$= 2.809\text{ J K}^{-1}\text{ mol}^{-1}$$

(ii) $H_2O(1, 273.15\text{ K}) \rightarrow H_2O(s, 273.15\text{ K})$

$$\Delta S_2 = \int \frac{dq}{T} = \frac{\Delta_{\text{freez}}H}{T_\text{m}}$$

Now $\Delta_{\text{freez}}H = -\ 6\ 008.2\text{ J mol}^{-1}$. Thus, we get

$$\Delta S_2 = \frac{-6\ 088.2\text{ J mol}^{-1}}{(273.15\text{ K})} = -\ 21.996\text{ J K}^{-1}\text{ mol}^{-1}$$

(iii) $H_2O(s, 273.15\text{ K}) \rightarrow H_2O(s, 263.15\text{ K})$

$$\Delta S_3 = \int_{T_2}^{T_1} \frac{dq}{T} = \int_{T_2}^{T_1} \frac{C_{p,\text{m}}(s)}{T} dT = C_{p,\text{m}}(s) \ln \frac{T_1}{T_2}$$

The heat capacity of water in the solid state is 36.401 J K^{-1} mol^{-1}. Thus, we get

$$\Delta S_3 = (36.401\text{ J K}^{-1}\text{ mol}^{-1}) \times 2.303 \times \log\left(\frac{263.15\text{ K}}{273.15\text{ K}}\right)$$
$$= -1.358\text{ J K}^{-1}\text{ mol}^{-1}$$

Thus, the change in entropy of the process

$$H_2O(1, 263.15\text{ K}) \rightarrow H_2O(s, 263.15\text{ K})$$

is given by

$$\Delta S_{\text{sys}} = \Delta S_1 + \Delta S_2 + \Delta S_3$$
$$= (2.809 - 21.996 - 1.358)\text{ J K}^{-1}\text{ mol}^{-1}$$
$$= -20.545\text{ J K}^{-1}\text{ mol}^{-1}$$

Thus, the transformation of liquid water at –10 °C to solid water is attended to by a decrease in entropy. Note that from this value alone, it is not possible to predict whether the said transformation is reversible or irreversible (or spontaneous). In order to predict this, we have to calculate the change in entropy of the surroundings and then determine the total change in entropy by adding entropy changes of the system and surroundings. White evaluating ΔS_{surr}, we assume that the surroundings receive heat equal to the heat of crystallization at 263.15 K in a reversible manner at temperature equal to 263.15 K. The value of $\Delta_{\text{freez}} H$ (263.15 K) can be obtained from that of $\Delta_{\text{freez}} H$ (273.15 K) by using the expression

$$\Delta_{\text{freez}} H(263.15 \text{ K}) - \Delta_{\text{freez}} H(273.15 \text{ K}) = (\Delta C_p)(\Delta T) \qquad \text{(Eq. 3.5.1)}$$
$$= (36.401 \text{ J K}^{-1} \text{ mol}^{-1} - 75.312 \text{ J K}^{-1} \text{ mol}^{-1})(-10 \text{ K})$$
$$= 389.1 \text{ J mol}^{-1}$$

Thus $\Delta_{\text{freez}}H(263.15 \text{ K}) = 389.1 \text{ J mol}^{-1} + \Delta_{\text{freez}}H(273.15 \text{ K})$

$$= 389.1 \text{ J mol}^{-1} - 6\,008.2 \text{ J mol}^{-1}$$
$$= -5\,619.1 \text{ J mol}^{-1}$$

Hence $\Delta S = \dfrac{-\Delta_{\text{freez}}H(263.15 \text{ K})}{T_1} = \dfrac{5\,619.1 \text{ J mol}^{-1}}{263.15 \text{ K}}$

$$= 21.353 \text{ J K}^{-1} \text{ mol}^{-1}$$

and $\Delta S_{\text{total}} = \Delta S_{\text{sys}} + \Delta S_{\text{surr}}$

$$= -20.545 \text{ J K}^{-1} \text{ mol}^{-1} + 21.353 \text{ J K}^{-1} \text{ mol}^{-1}$$
$$= 0.808 \text{ J K}^{-1} \text{ mol}^{-1}$$

Since ΔS_{total} = +ve, the said transformation will be irreversible (or spontaneous) in nature.

Example 4.22.1 Calculate the entropy change at 373 K for the transformation

$$H_2O(1, 1.013\ 25 \text{ bar}) = H_2O(g, 0.101\ 325 \text{ bar})$$

Given: $\Delta_{\text{vap}}H = 40.668 \text{ kJ mol}^{-1}$.

Solution We can calculate the change in entropy from the following steps.

Step I $H_2O(1, 1.013\ 25 \text{ bar}) \rightarrow H_2O(g, 1.013\ 25 \text{ bar})$

$$\Delta S_1 = \frac{40.668 \times 10^3 \text{ J mol}^{-1}}{373 \text{ K}} = 109.03 \text{ J K}^{-1} \text{ mol}^{-1}$$

Step II $H_2O(g, 1.013\ 25 \text{ bar}) \rightarrow H_2O(g, 0.101\ 325 \text{ bar})$

$$\Delta S_2 = R \ln \frac{p_1}{p_2} = (8.314 \text{ J K}^{-1} \text{ mol}^{-1}) \times 2.303 \times \log\left(\frac{1.013\ 25 \text{ bar}}{0.101\ 325 \text{ bar}}\right)$$
$$= 19.15 \text{ J K}^{-1} \text{ mol}^{-1}$$

Total change in entropy is given by

$$\Delta S = \Delta S_1 + \Delta S_2 = 109.03 \text{ J K}^{-1} \text{ mol}^{-1} + 19.15 \text{ J K}^{-1} \text{ mol}^{-1}$$
$$= 128.18 \text{ J K}^{-1} \text{ mol}^{-1}$$

Example 4.22.2 Calculate $\Delta_r U$, $\Delta_r H$ and $\Delta_r S$ for the process

$$1 \text{ mole } H_2O(1, 293 \text{ K}, 101.325 \text{ kPa}) \rightarrow 1 \text{ mol } H_2O(g, 523 \text{ K}, 101.325 \text{ kPa})$$

Given the following data:

$$C_{p,\text{m}}(1) = 75.312 \text{ J K}^{-1} \text{ mol}^{-1};\ C_{p,\text{m}}(g) = 35.982 \text{ J K}^{-1} \text{ mol}^{-1}$$
$$\Delta_{\text{vap}}H \text{ at } 373 \text{ K}, 101.325 \text{ kPa} = 40.668 \text{ kJ mol}^{-1}$$

Solution The changes in $\Delta_r U$, $\Delta_r H$ and $\Delta_r S$ can be calculated following the reversible paths given below.

Step I 1 mole $H_2O(1, 293 \text{ K}, 101.325 \text{ kPa}) \rightarrow$ 1 mole $H_2O(1, 373 \text{ K}, 101.325 \text{ kPa})$

$$q_p = \Delta_r H = C_{p,\text{m}}(\text{l})\,\Delta T = (75.312 \text{ J K}^{-1} \text{ mol}^{-1})(80 \text{ K}) = 6\,024.96 \text{ J mol}^{-1}$$

$$\Delta_r S = C_{p,\,m} \ln \frac{T_2}{T_1} = (75.312\ \text{J K}^{-1}\ \text{mol}^{-1}) \times 2.303 \times \log\left(\frac{373\ \text{K}}{293\ \text{K}}\right)$$

$$= 18.184\ \text{J K}^{-1}\ \text{mol}^{-1}$$

$$\Delta_r U = \Delta_r H - p\,\Delta_r V = \Delta_r H$$

Step II 1 mole H_2O(1, 373 K, 101.325 kPa) → 1 mole H_2O(g, 373 K, 101.325 kPa)

$$q_p = \Delta_{vap} H = 40.668\ \text{kJ mol}^{-1}$$

$$\Delta_r S = \frac{40\,668\ \text{J mol}^{-1}}{373\ \text{K}} = 109.03\ \text{J K}^{-1}\ \text{mol}^{-1}$$

$$\Delta_r U = \Delta_r H - p\,\Delta_r V$$

$$= 40\,668\ \text{J mol}^{-1} - (101.325\ \text{kPa})\left((22.414\ \text{dm}^3\ \text{mol}^{-1}) \times \frac{373\ \text{K}}{273\ \text{K}}\right)$$

$$= 40\,668\ \text{J mol}^{-1} - 3\,103\ \text{J mol}^{-1}$$

$$= 37\,565\ \text{J mol}^{-1}$$

Step III 1 mole H_2O (g, 373 K, 101.325 kPa) →1 mole H_2O(g, 523 K, 101.325 kPa)

$$\Delta_r H = C_{p,\,m}(\text{g})\,\Delta T = (35.982\ \text{J K}^{-1}\ \text{mol}^{-1})\ (150\ \text{K}) = 5\,397.3\ \text{J mol}^{-1}$$

$$\Delta_r S = C_{p,\,m}(\text{g}) \ln \frac{T_2}{T_1} = (35.982\ \text{J K}^{-1}\ \text{mol}^{-1}) \times 2.303 \times \log\left(\frac{523\ \text{K}}{373\ \text{K}}\right)$$

$$= (35\,982\ \text{J K}^{-1}\ \text{mol}^{-1}) \times 2.303 \times 0.146\,8$$

$$= 12.164\ \text{J K}^{-1}\ \text{mol}^{-1}$$

$$\Delta_r U = \Delta_r H - R(\Delta T)$$

$$= 5\,397.3\ \text{J mol}^{-1} - (8.314\ \text{J K}^{-1}\ \text{mol}^{-1})\ (150\ \text{K})$$

$$= 5\,397.3\ \text{J mol}^{-1} - 1\,247.1\ \text{J mol}^{-1} = 4\,150.2\ \text{J mol}^{-1}$$

Thus $\Delta U_{total} = (6\,024.96 + 37\,565 + 4\,150.2)\ \text{J mol}^{-1} = 47\,740.16\ \text{J mol}^{-1}$

$$\Delta H_{total} = (6\,024.96 + 40\,668 + 5\,397.3\ \text{J mol}^{-1} = 52\,090.26\ \text{J mol}^{-1}$$

$$\Delta S_{total} = (18.184 + 109.03 + 12.164)\ \text{J K}^{-1}\ \text{mol}^{-1} = 139.378\ \text{J K}^{-1}\ \text{mol}^{-1}$$

Example 4.22.3 10 g of ice is heated to become vapour at 373 K and 101.325 kPa. Calculate ΔS for the system. Given that

$\Delta_{fus}H$ of ice at 273 K = 334.72 J g^{-1}

$\Delta_{vap}H$ of water 373 K = 2 259.36 J g^{-1}

Average specific heat capacity of liquid water = 4.184 J K^{-1} g^{-1}

Solution The given transformation can be carried out by following the reversible paths given below.

10 g water(s, 273 K, 101.325 kPa) → 10 g water (1, 273 K, 101.325 kPa)

10 g water(1, 273 K, 101.325 kPa) → 10 g water (1, 373 K, 101.325 kPa)

10 g water(1, 373 K, 101.325 kPa) → 10 g water (g, 373 K, 101.325 kPa)

The entropy changes per gram of water in these processes are

$$\Delta S_1 = \frac{334.72\ \text{J g}^{-1}}{273\ \text{K}} = 1.226\ \text{J g}^{-1}\ \text{K}^{-1}$$

$$\Delta S_2 = \int \frac{C_p(1)}{T} dT = C_p(1) \ln \frac{T_2}{T_1} = (4.184 \text{ J K}^{-1} \text{ g}^{-1}) \ln \left(\frac{373 \text{ K}}{273 \text{ K}}\right)$$

$$= 1.305\ 8 \text{ J K}^{-1} \text{ g}^{-1}$$

$$\Delta S_3 = \frac{2\ 259.36 \text{ J g}^{-1}}{373 \text{ K}} = 6.057 \text{ J g}^{-1} \text{ K}^{-1}$$

Thus $\quad \Delta S_{\text{total}}/\text{g of water} = \Delta S_1 + \Delta S_2 + \Delta S_3 = 8.588 \text{ J K}^{-1} \text{ g}^{-1}$

$$\Delta S \text{ for } 10 \text{ g} = 10 \times 8.588\ 8 \text{ J K}^{-1} = 85.888 \text{ J K}^{-1}$$

Entropy Change when Two Solids at Different Temperatures are Brought Together

Consider two solid bodies of the same material each containing one mole of a substance. Let one be at higher temperature T_h and the other at lower temperature T_c. Let both of them be brought together. Heat will flow from the hotter body to the colder till the temperatures of both of them are the same. Let the equilibrium temperature be T, the value of which can be calculated as follows:

Heat lost by hot body = Heat gained by cold body

$$C_{p,\,\text{m}}\,(T_\text{h} - T) = C_{p,\,\text{m}}\,(T - T_\text{c})$$

or
$$T = \frac{T_\text{h} + T_\text{c}}{2}$$

Consequently, entropy changes of two bodies are

$$\Delta S_\text{h} = \int_{T_\text{h}}^{T} C_{p,\text{m}}\, dT = C_{p,\text{m}} \ln \frac{T}{T_\text{h}}$$

$$\Delta S_\text{c} = \int_{T_\text{c}}^{T} \frac{C_{p,\text{m}}}{T}\, dT = C_{p,\text{m}} \ln \frac{T}{T_\text{c}}$$

The total change in entropy is

$$\Delta S_\text{total} = \Delta S_\text{h} + \Delta S_\text{c}$$

$$= C_{p,\,\text{m}} \left(\ln \frac{T}{T_\text{h}} + \ln \frac{T}{T_\text{c}} \right) = C_{p,\text{m}} \ln \left(\frac{T^2}{T_\text{h} T_\text{c}} \right)$$

$$= C_{p,\,\text{m}} \ln \left\{ \frac{(T_\text{h} + T_\text{c})^2}{4 T_\text{h} T_\text{c}} \right\} = C_{p,\text{m}} \ln \left(\frac{T_\text{h}^2 + T_\text{c}^2 + 2 T_\text{h} T_\text{c}}{4 T_\text{h} T_\text{c}} \right)$$

$$= C_{p,\,\text{m}} \ln \left(\frac{T_\text{h}^2 + T_\text{c}^2 - 2 T_\text{h} T_\text{c}}{4 T_\text{h} T_\text{c}} + 1 \right)$$

$$= C_{p,\,\text{m}} \ln \left(\frac{(T_\text{h} - T_\text{c})^2}{4 T_\text{h} T_\text{c}} + 1 \right)$$

The quantity within brackets will always be greater than one. Hence, the logarithm of this quantity will always be greater than zero. Thus ΔS_total of this process is positive, indicating that it is an irreversible process.

Example 4.22.4 A 50 g mass of Cu at a temperature of 393 K is placed in contact with a 100 g mass of copper at a temperature of 303 K in a thermally insulated container. Calculate q and ΔS_{total} for the reversible process. Use a value of 0.418 4 J g^{-1} K^{-1} for the specific heat capacity of Cu.

Solution We have 50 g copper at 393 K and 100 g copper at 303 K. The temperature of the two bodies when they have come to thermal equilibrium can be calculated as follows.

$$\text{Heat gain} = \text{Heat loss}$$

$$m_1 C_p(\Delta T_1) = m_2 C_p(\Delta T_2) \quad \text{i.e.} \quad m_1 \Delta T_1 = m_2 \Delta T_2$$

Therefore

$$(100\ \text{g})\ (T - 303\ \text{K}) = (50\ \text{g})\ (393\ \text{K} - T)$$

$$2T - 606\ \text{K} = 393\ \text{K} - T$$

$$T = 333\ \text{K}$$

Thus $\quad \Delta S_{\text{h}} = nC_{p,\,\text{m}} \ln \dfrac{T}{T_{\text{h}}}$

$$= \left(\frac{50\ \text{g}}{63\ \text{g mol}^{-1}}\right) \{(0.418\ 4\ \text{J g}^{-1}\ \text{K}^{-1})\ (63\ \text{g mol}^{-1})\} \times 2.303 \log \left(\frac{333\ \text{K}}{393\ \text{K}}\right)$$

$$= -\ 3.466\ \text{J K}^{-1}$$

$$\Delta S_{\text{c}} = nC_{p,\,\text{m}} \ln \frac{T}{T_{\text{c}}}$$

$$= \left(\frac{100\ \text{g}}{63\ \text{g mol}^{-1}}\right) \{(0.418\ 4\ \text{J g}^{-1}\ \text{K}^{-1})(63\ \text{g mol}^{-1})\} \times 2.303 \log \left(\frac{333\ \text{K}}{303\ \text{K}}\right)$$

$$= 3.951\ \text{J K}^{-1}$$

and $\quad \Delta S_{\text{total}} = \Delta S_{\text{c}} + \Delta S_{\text{h}} = (3.951 - 3.466)\ \text{J K}^{-1} = 0.485\ \text{J K}^{-1}$

Example 4.22.5 5 g ice at 273 K is added to 30 g water at 323 K in a thermally insulated container. (a) What is the final temperature? (b) What is the total entropy change? Use the same physical constants of water as given in Example 4.22.3.

Solution Heat required to convert 5 g of ice at 273 K to 5 g water at 273 K

$$= (5\ \text{g})\ (334.72\ \text{J g}^{-1}) = 1\ 673.6\ \text{J}$$

Final temperature after mixing

$$\text{Heat gained by ice} = \text{Heat lost by hot water}$$

$$(5\ \text{g})\ (334.72\ \text{J g}^{-1}) + (5\ \text{g})\ (4.184\ \text{J g}^{-1}\ \text{K}^{-1})\ (T - 273\ \text{K})$$

$$= (30\ \text{g})\ (4.184\ \text{J K}^{-1}\ \text{g}^{-1})\ (323\ \text{K} - T)$$

$$1\ 673.6\ \text{J} + (20.92\ \text{J})\ (T/\text{K} - 273) = (125.52\ \text{J})\ (323 - T/\text{K})$$

Solving for T, we get

$$T = 304.43\ \text{K}$$

Entropy changes

$$C_{p,\,\text{m}}(1) = (4.184\ \text{J g}^{-1}\ \text{K}^{-1})\ (\ 18\ \text{g mol}^{-1}) = 75.312\ \text{J K}^{-1}\ \text{mol}^{-1}$$

$$\Delta S_{\text{ice}} = \frac{\Delta_{\text{fus}} H}{T_{\text{m}}} + nC_{p,\,\text{m}}(1) \ln \frac{T_2}{T_1}$$

$$= \frac{1673.6\ \text{J}}{273\ \text{K}} + \left(\frac{5\ \text{g}}{18\ \text{g mol}^{-1}}\right)(75.312\ \text{J K}^{-1}\ \text{mol}^{-1}) \times 2.303 \times \log\left(\frac{304.43\ \text{K}}{323\ \text{K}}\right)$$

$$= 6.130\ \text{J K}^{-1} + 2.280\ \text{J K}^{-1} = 8.410\ \text{J K}^{-1}$$

$$\Delta S_{\text{water}} = nC_{p,\,\text{m}}\,(1) \ln \frac{T_2}{T_1}$$

$$= \left(\frac{30\ \text{g}}{18\ \text{g mol}^{-1}}\right)(75.312\ \text{J K}^{-1}\ \text{mol}^{-1}) \times 2.303 \log\left(\frac{304.43\ \text{K}}{323\ \text{K}}\right)$$

$$= -7.434\ \text{K}^{-1}$$

$$\Delta S_{\text{total}} = \Delta S_{\text{ice}} + \Delta S_{\text{water}} = (8.410 - 7.434)\ \text{J K}^{-1}$$

$$= 0.976\ \text{J K}^{-1}$$

4.23 THE THIRD LAW OF THERMODYNAMICS

Planck's Statement Let one mole of a solid at a constant pressure be heated from 0 K to some temperature T below its melting point, i.e.

$$\text{Solid } (0\ \text{K}, p) \rightarrow \text{Solid } (T, p)$$

According to Eq. (4.16.8), the entropy change in the above process is given by

$$\Delta S = S_T - S_{0\text{K}} = \int_{0\text{K}}^{T} \frac{C_{p,\,\text{m}}}{T}\, \mathrm{d}T$$

or

$$S_T = S_{0\text{K}} + \int_{0\text{K}}^{T} \frac{C_{p,\,\text{m}}}{T}\, \mathrm{d}T \tag{4.23.1}$$

Now since the entropy function increases with increase in temperature, it may be expected that it has a minimum value at 0 K. In 1913, Max Planck suggested that this *minimum entropy may be assigned a zero value for a pure perfectly crystalline substance.* This suggestion is known as the *third law of thermodynamics.*

Third Law Entropy With this, Eq. (4.23.1) reduces to

$$S_T = \int_{0\text{K}}^{T} \frac{C_{p,\,\text{m}}}{T}\, \mathrm{d}T \tag{4.23.2}$$

where S_T is called the *third law entropy* or simply the entropy at temperature T and pressure p. If the latter has a value of 1 bar then S_T becomes the standard entropy S_T°.

Entropy of a Solid To evaluate the value of S_T for a solid, the heat capacity of the solid should be known over the range of temperature from absolute zero to the temperature of interest. Ordinarily, measurements of heat capacity of solid have been made up to a lower temperature that lies in the range of 10 to 15 K. Below this temperature, the heat capacity of the solid is determined from the Debye T-cubed law given as

$$C_{p,\,\text{m}} = a\,T^3 \tag{4.23.3}$$

where a is constant whose value is determined from the value of $C_{p,\,\text{m}}$ (or $C_{V,\,\text{m}}$ since $C_{p,\,\text{m}}$ and $C_{V,\,\text{m}}$ are indistinguishable at these temperatures) at some low temperature.

$$S_T = \int_{0\text{K}}^{T_{\text{min}}} \frac{C_{p,\text{m}}}{T}\,\text{d}T + \int_{T_{\text{min}}}^{T} \frac{C_{p,\text{m}}}{T}\,\text{d}T = \int_{0\text{K}}^{T_{\text{min}}} \frac{C_{p,\text{m}}}{T}\,\text{d}T + \int_{T_{\text{min}}}^{T} C_{p,\text{m}} \frac{\text{d}(T/\text{K})}{(T/\text{K})}$$

$$= \int_{0\text{K}}^{T_{\text{min}}} \frac{C_{p,\text{m}}}{T}\,\text{d}T + 2.303 \int_{T_{\text{min}}}^{T} C_{p,\text{m}}\,\text{d}\log(T/\text{K}) \qquad (4.23.4)$$

where T_{min} is the minimum temperature up to which the value of heat capacity of the solid is available. The first integral is evaluated using the Debye T-cubed law while the second one is usually evaluated following the graphical method where either $C_{p,\text{m}}/T$ versus T or $C_{p,\text{m}}$ versus log (T/K) is plotted (Figs. 4.23.1 and 4.23.2). The area under either of these curves gives the value of the integral.

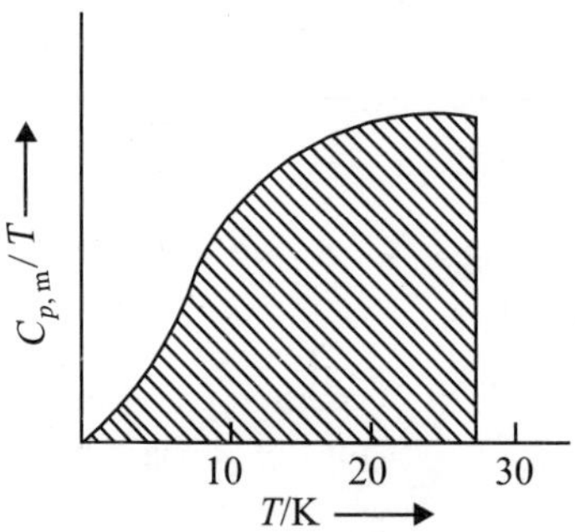

Fig. 4.23.1 Schematic plot of $C_{p,\text{m}}/T$ versus T

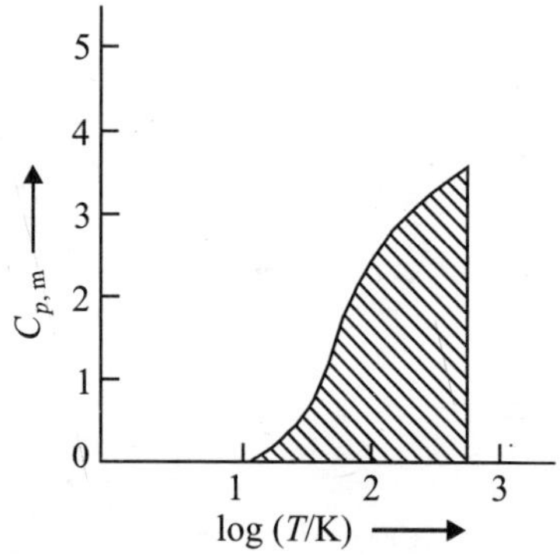

Fig. 4.23.2 Schematic plot of $C_{p,\text{m}}$ versus log (T/K)

Entropy of a Liquid The third law entropy for a liquid at temperature T can be determined by following the steps given below.

(1) Heat the solid substance from 0 K to its melting point T_{m}.
(2) Transfrom the substance from solid to liquid at its melting point T_{m}.
(3) Heat the liquid from T_{m} to the required temperature T.

The addition of entropy changes in the above three processes gives the third law entropy of the liquid at temperature T. Thus

$$S_T = \int_{0\text{K}}^{T_{\text{m}}} \frac{C_{p,\text{m}}(\text{s})}{T}\,\text{d}T + \frac{\Delta_{\text{fus}}H_{\text{m}}}{T_{\text{m}}} + \int_{T_{\text{m}}}^{T} \frac{C_{p,\text{m}}(\text{l})}{T}\,\text{d}T \qquad (4.23.5)$$

Entropy of a Gas Similarly, for a gas above its boiling point, the third law entropy is given as

$$S_T = \int_{0\text{K}}^{T_{\text{m}}} \frac{C_{p,\text{m}}(\text{s})}{T}\,\text{d}T + \frac{\Delta_{\text{fus}}H_{\text{m}}}{T_{\text{m}}} + \int_{T_{\text{m}}}^{T_{\text{b}}} \frac{C_{p,\text{m}}(\text{l})}{T}\,\text{d}T + \frac{\Delta_{\text{vap}}H}{T_{\text{b}}} + \int_{T_{\text{b}}}^{T} \frac{C_{p,\text{m}}(\text{g})}{T}\,\text{d}T$$

(4.23.6)

Figure 4.23 depicts the entropy variation as governed by Eq. (4.23.6)

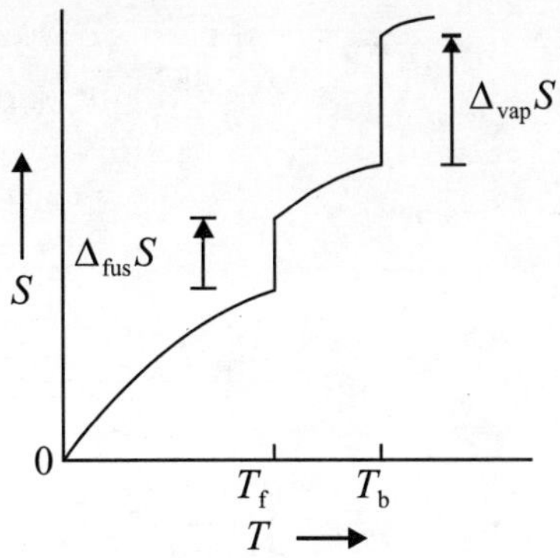

Fig. 4.23.3 Variation of entropy of a substance with temperature

If the solid under study undergoes any transition from one crystalline from to another, the entropy of transition

$$\Delta_{trs}S = \frac{\Delta_{trs}H}{T} \quad (4.23.7)$$

should also be added in the above entropy expressions.

Example 4.23.1

Calculate the third law entropy of a substance at 350 K using the following data.

(i) Heat capacity of solid from 0 K to normal melting point 200 K
$C_{p,\,m}(s) = [0.035\ (T/K) + 0.001\ 2\ (T/K)^2]$ J K^{-1} mol^{-1}

(ii) Enthalpy of fusion = 7.5 kJ mol^{-1}

(iii) Heat capacity of liquid from 200 K to normal boiling point 300 K
$C_{p,\,m}(1) = [60 + 0.016\ (T/K)]$ J K^{-1} mol^{-1}

(iv) Enthalpy of vaporization = 30 kJ mol^{-1}

(v) Heat capacity of gas from 300 K to 350 K at 1 atm
$C_{p,\,m}(g) = 50.0$ J K^{-1} mol^{-1}

Solution

Entropy changes for the given five steps are as follows.

(i) $$\Delta S_1 = \int_{0K}^{T_m} \frac{C_{p,m}(s)}{T}\, dT$$

$$\Delta S_1/\text{J K}^{-1}\text{ mol}^{-1} = \int_{0K}^{200K} \frac{0.035(T/K) + 0.001\,2\,(T/K)^2}{T}\, dT$$

$$= 0.035\ (200 - 0) + \frac{0.001\,2}{2}\ (200^2 - 0^2)$$

$$= 7.0 + 24 = 31.0$$

(ii) $$\Delta S_2 = \frac{\Delta_{fus}H}{T_m} = \frac{7.5\times 10^3}{200}\ \text{J K}^{-1}\text{ mol}^{-1} = 37.5\ \text{J K}^{-1}\text{ mol}^{-1}$$

(iii) $$\Delta S_3 = \int_{T_m}^{T_b} \frac{C_{p,m}(1)}{T}\, dT$$

$$\Delta S_3 / \text{J K}^{-1}\text{ mol}^{-1} = \int_{200\text{K}}^{300\text{K}} \frac{60 + 0.016\,(T/\text{K})}{T}\,\text{d}T$$

$$= 60 \ln \frac{300}{200} + 0.016\,(300 - 200)$$

$$= 24.33 + 1.60 = 25.93$$

(iv) $$\Delta S_4 = \frac{\Delta_{\text{vap}}H}{T_\text{b}} = \frac{30 \times 10^3}{300}\ \text{J K}^{-1}\text{ mol}^{-1} = 100\ \text{J K}^{-1}\text{ mol}^{-1}$$

(v) $$\Delta S_5 = \int_{T_\text{b}}^{T} \frac{C_{p,\text{m}}(\text{g})}{T}\,\text{d}T$$

$$\Delta S_5 / \text{J K}^{-1}\text{ mol}^{-1} = \int_{300\text{K}}^{350\text{K}} \frac{50.0}{T}\,\text{d}T = (50.0)\ln\frac{350}{300} = 7.71$$

Thus, the third law entropy of the substance at 350 K is

$$S_T = \Delta S_1 + \Delta S_2 + \Delta S_3 + \Delta S_4 + \Delta S_5$$

$$= (31.0 + 37.5 + 25.93 + 100 + 7.71)\ \text{J K}^{-1}\text{ mol}^{-1}$$

$$= 202.14\ \text{J K}^{-1}\text{ mol}^{-1}$$

4.24 ENTROPY OF REACTION AND ITS TEMPERATURE AND PRESSURE DEPENDENCE

Expression of Entropy of Reaction

Consider a general reaction

$$0 = \sum_\text{B} \nu_\text{B}\,\text{B}$$

The amount of species with the progress of reaction is given by expression

$$n_\text{B} = (n_\text{B})_0 + \nu_\text{B}\xi$$

where $(n_\text{B})_0$ is the amount of species B in the beginning of the reaction and n_B is the corresponding amount when the reaction has proceeded to the extent ξ.

For an infinitesimal change in extent of reaction, the change in the amount of species B is

$$\text{d}n_\text{B} = \nu_\text{B}\,\text{d}\xi$$

The corresponding entropy change of the reaction is

$$\text{d}S = \sum_\text{B} S_\text{m}(\text{B})\,\text{d}n_\text{B} = \sum_\text{B} S_\text{m}(\text{B})\,(\nu_\text{B}\,\text{d}\xi)$$

By definition, entropy of reaction is given as

$$\Delta_\text{r}S = \left(\frac{\partial S}{\partial \xi}\right)_{T,p} = \sum_\text{B} \nu_\text{B}\,S_\text{m}(\text{B}) \qquad (4.24.1)$$

that is, entropy change of a reaction is the rate of change of entropy with extent of reaction. It is equal to the entropy change per unit extent of reaction. The unit of $\Delta_\text{r}S$ is J K^{-1} mol^{-1}.

Computation of Entropy of Reaction

From Eq. (4.24.1), it follows that the standard entropy of reaction is computed from the tabulated data in much the same ways as the standard change in enthalpy. However, there is one important difference—the standard entropy of element in its stable state of aggregation at 1 bar at 25 °C is not assigned a conventional value of zero, but is determined through the third law of thermodynamics. As an example, in the reaction

$$Fe_2O_3(s) + 3H_2(g) \rightarrow 2\,Fe(s) + 3H_2O(l)$$

the standard entropy change is given by

$$\Delta_r S^\circ = \sum_B \nu_B S^\circ_m(B)$$

i.e. $\Delta_r S^\circ = S^\circ_{products} - S^\circ_{reactants}$

Since $S^\circ_{products} = 2S^\circ_m(Fe, s) + 3S^\circ_m(H_2O, l)$

and $S^\circ_{reactants} = S^\circ_m(Fe_2O_3, s) + 3S^\circ_m(H_2, g)$

therefore $\Delta_r S^\circ = [2S^\circ_m(Fe, s) + 3S^\circ_m(H_2O, l)] - [S^\circ_m(Fe_2O_3, s) + 3S^\circ_m(H_2, g)]$

The table of standard entropies (Appendix I) gives

$$\Delta_r S^\circ = \{2(27.28) + 3\,(69.91) - 87.40 - 3\,(130.68)\}\ \mathrm{J\,K^{-1}\,mol^{-1}}$$

$$= -\,215.15\ \mathrm{J\,K^{-1}\,mol^{-1}}$$

The entropy of reaction is more if there is a change in the value of $\Delta\nu_g$ (the change in the stoichiometric number of gaseous species), since entropy of gases is much larger than the entropy of condensed phases.

Effect of Temperature on Entropy of Reaction

The change in the value of $\Delta_r S^\circ$ for a reaction with temperature can be computed as follows:

Since $\Delta_r S^\circ = \sum_B \nu_B S^\circ_m(B)$

therefore
$$\left(\frac{\partial(\Delta_r S^\circ)}{\partial T}\right)_p = \sum_B \nu_B \left(\frac{\partial S^\circ_m(B)}{\partial T}\right)_p$$

$$= \sum_B \nu_B \frac{C_{p,m}(B)}{T} = \frac{\Delta_r C_p}{T} \qquad (4.24.2)$$

Thus $\mathrm{d}(\Delta_r S^\circ) = \dfrac{\Delta_r C_p}{T}\mathrm{d}T$ (p constant)

Integrating this within the limits T_0 and T, we obtain

$$\int_{T_0}^{T} \mathrm{d}(\Delta_r S^\circ) = \int_{T_0}^{T} \frac{\Delta_r C_p}{T}\mathrm{d}T$$

or
$$\Delta_r S^\circ_T - \Delta_r S^\circ_{T_0} = \int_{T_0}^{T} \frac{\Delta_r C_p}{T}\mathrm{d}T \qquad (4.24.3)$$

In order to carry out the above integration we need the temperature dependence of $\Delta_r C_p$. Two cases may be considered as given below.

$\Delta_r C_p$ independent of temperature In this case, we have

$$\Delta_r S_T^\circ - \Delta_r S_{T_0}^\circ = \Delta_r C_p \ln\left(\frac{T}{T_0}\right) \tag{4.24.4}$$

$\Delta_r C_p$ dependent on temperature As usual, the variation of C_p with temperature may be expressed as

$$C_p = a + b\,T + c\,T^2 + \cdots$$

where a, b, c, ... are constants. Thus, the variation of $\Delta_r C_p$ with temperature is given by

$$\Delta_r C_p = (\Delta_r a) + (\Delta_r b)\,T + (\Delta_r c)\,T^2 + \cdots$$

Substituting this in Eq. (4.24.3), we get

$$\Delta_r S_T^\circ - \Delta_r S_{T_0}^\circ = \int_{T_0}^{T}\left(\frac{\Delta_r a}{T} + \Delta_r b + (\Delta_r c)T + \cdots\right)\mathrm{d}T$$

which on integration gives

$$\Delta_r S_T^\circ - \Delta_r S_{T_0}^\circ = \Delta_r a \ln\left(\frac{T}{T_0}\right) + \Delta_r b\,(T - T_0) + \frac{\Delta_r c}{2}\,(T^2 - T_0^2) + \cdots \tag{4.24.5}$$

Example 4.24.1

Calculate $\Delta_r S_{348\,\mathrm{K}}^\circ$ for the reaction

$$\frac{1}{2}\,N_2(g) + O_2(g) \rightarrow NO_2(g)$$

Given: At 298 K, the standard entropies are

$S_m^\circ(N_2, g) = 191.61\ \mathrm{J\,K^{-1}\,mol^{-1}}$; $S_m^\circ(O_2, g) = 161.06\ \mathrm{J\,K^{-1}\,mol^{-1}}$

$S_m^\circ(NO_2, g) = 240.06\ \mathrm{J\,K^{-1}\,mol^{-1}}$

and heat capacities are

$C_p(N_2, g) = 29.13\ \mathrm{J\,K^{-1}\,mol^{-1}}$; $C_p(O_2, g) = 29.36\ \mathrm{J\,K^{-1}\,mol^{-1}}$

$C_p(NO_2, g) = 37.20\ \mathrm{J\,K^{-1}\,mol^{-1}}$

Solution

We have

$$\Delta_r S_{T_0}^\circ = S^\circ(NO_2, g) - \frac{1}{2}\,S^\circ(N_2, g) - S^\circ(O_2, g)$$

$$= \left(240.06 - \frac{1}{2}\times 191.61 - 161.06\right)\ \mathrm{J\,K^{-1}\,mol^{-1}} = -\,16.805\ \mathrm{J\,K^{-1}\,mol^{-1}}$$

$$\Delta_r C_p = C_{p,\,m}(NO_2, g) - \frac{1}{2}\,C_{p,\,m}(N_2, g) - C_{p,\,m}(O_2, g)$$

$$= \left(37.20 - \frac{1}{2}\times 29.13 - 29.36\right)\ \mathrm{J\,K^{-1}\,mol^{-1}} = -\,6.725\ \mathrm{J\,K^{-1}\,mol^{-1}}$$

Substituting the above data in the expression

$$\Delta S^\circ_T = \Delta_r S^\circ_{T_0} + \Delta_r C_p \ln \frac{T}{T_0}$$

we get $$\Delta S^\circ_{348\,\mathrm{K}} = \left[-16.805 + (-6.725)\ln\left(\frac{348}{298}\right)\right] \mathrm{J\,K^{-1}\,mol^{-1}}$$

$$= -\,17.85\ \mathrm{J\,K^{-1}\,mol^{-1}}$$

Example 4.24.2 Calculate $\Delta_r S^\circ_{348\,\mathrm{K}}$ = for the reaction

$$\frac{1}{2}\ N_2(g) + O_2(g) \rightarrow NO_2(g)$$

Given: At 298 K, the standard entropies are

$S^\circ_m(N_2, g) = 191.61\ \mathrm{J\,K^{-1}\,mol^{-1}}$; $S^\circ_m(O_2, g) = 161.06\ \mathrm{J\,K^{-1}\,mol^{-1}}$

$S^\circ_m(NO_2, g) = 240.06\ \mathrm{J\,K^{-1}\,mol^{-1}}$

and heat capacities are

$$C_p(N_2, g)/\mathrm{J\,K^{-1}\,mol^{-1}} = 28.46 + 2.26 \times 10^{-3}\ (T/\mathrm{K})$$

$$C_p(O_2, g)/\mathrm{J\,K^{-1}\,mol^{-1}} = 26.85 + 8.49 \times 10^{-3}\ (T/\mathrm{K})$$

$$C_p(NO_2, g)/\mathrm{J\,K^{-1}\,mol^{-1}} = 27.78 + 30.85 \times 10^{-3}\ (T/\mathrm{K})$$

Solution For the given reaction

$$\Delta_r S^\circ_{298\,\mathrm{K}} = \sum_{\mathrm{B}} \nu_{\mathrm{B}} S^\circ_m(\mathrm{B}) = S^\circ_m(NO_2) - \frac{1}{2}\ S^\circ_m(N_2) - S^\circ_m(O_2)$$

$$= \left(240.06 - \frac{1}{2} \times 191.61 - 161.06\right) \mathrm{J\,K^{-1}\,mol^{-1}} = -\,16.81\ \mathrm{J\,K^{-1}\,mol^{-1}}$$

Representing $C_p = a + bT$, we have

$$\Delta_r C_p = \Delta_r a + (\Delta_r b)T$$

$$= \left[\left(27.78 - \frac{1}{2} \times 28.46 - 26.85\right) + \left(30.85 - \frac{1}{2} \times 2.26 - 8.49\right) \times 10^{-3}\ (T/\mathrm{K})\right] \mathrm{J\,K^{-1}\,mol^{-1}}$$

$$= [-13.30 + 21.23 \times 10^{-3}\ (T/\mathrm{K})]\ \mathrm{J\,K^{-1}\,mol^{-1}}$$

Substituting this expression in the equation

$$\int_{T_0}^{T} \mathrm{d}(\Delta_r S^\circ) = \int_{T_0}^{T} \frac{\Delta_r C_p}{T}\, \mathrm{d}T$$

we get $$\Delta_r S^\circ_T - \Delta_r S^\circ_{T_0} = \left(\int_{T_0}^{T} -\frac{13.30}{T}\,\mathrm{d}T + \frac{21.23 \times 10^{-3}}{\mathrm{K}}\,\mathrm{d}T\right) \mathrm{J\,K^{-1}\,mol^{-1}}$$

$$= \left[-\,13.30 \ln \frac{T}{T_0} + \frac{21.23 \times 10^{-3}}{\mathrm{K}}(T - T_0)\right] \mathrm{J\,K^{-1}\,mol^{-1}}$$

Substituting $T = 348$ K and $T_0 = 298$ K along with $\Delta_r S^\circ_{298\,K} = -16.81$ J K^{-1} mol^{-1}, we get

$$\Delta_r S^\circ_{348\,K} = \left[-16.81 - 13.30 \ln \frac{348}{298} + 21.23\times10^{-3}(348 - 298)\right] \text{J K}^{-1} \text{mol}^{-1}$$

$$= (-16.81 - 2.06 + 1.06) \text{ J K}^{-1} \text{mol}^{-1}$$

$$= -17.81 \text{ J K}^{-1} \text{mol}^{-1}$$

Effect of Pressure on Entropy of Reaction

Since $\Delta_r S = \sum_B \nu_B S_m(B)$

therefore $\left(\frac{\partial(\Delta_r S)}{\partial p}\right)_T = \sum_B \nu_B \left(\frac{\partial S_{m,B}}{\partial p}\right)_T$

From Eq. (4.15.14), we get

$$\left(\frac{\partial S_m}{\partial p}\right)_T = -\left(\frac{\partial V_m}{\partial T}\right)_p$$

Substituting this in the previous expression, we get

$$\left(\frac{\partial(\Delta_r S)}{\partial p}\right)_T = -\sum_B \nu_B \left(\frac{\partial V_{m,B}}{\partial T}\right)_p = -\left(\frac{\partial(\Delta_r V)}{\partial T}\right)_p \qquad (4.24.6)$$

Since the variation in volume with temperature for condensed systems is usually very small, we include only gaseous constituents in the differential $(\partial\Delta_r V/\partial T)$. Assuming deal behaviour for gaseous constituents, we have

$$\Delta_r V = \sum_{B(g)} \nu_B V_{m,B} = \sum_{B(g)} \nu_B (RT/p) = \left(\sum_{B(g)} \nu_B\right)(RT/p) = (\Delta\nu_g)\, RT/p$$

Hence $\left(\frac{\partial(\Delta_r S)}{\partial p}\right)_T = -\left(\frac{\partial(\Delta_r V)}{\partial T}\right)_p = -\Delta\nu_g \frac{R}{p} \qquad (4.24.7)$

where $\Delta\nu_g = \sum_{B(g)} \nu_B(B) = \sum \nu_g(\text{product}) - \sum |\nu_g(\text{reactant})|$

The integrated form of Eq. (4.24.7) is

$$\Delta_r S_{p_2} - \Delta_r S_{p_1} = -\Delta\nu_g R \ln \frac{p_2}{p_1} \qquad (4.24.8)$$

From Eq. (4.24.7) or (4.24.8), it follows that $\Delta_r S$ of a reaction increases with increase in pressure if $\Delta\nu_g$ is negative and it decreases if $\Delta\nu_g$ is positive.

Example 4.24.3

Calculate $\Delta_r S$ for the reaction

$$\frac{1}{2} N_2(g) + O_2(g) \rightarrow NO_2(g) \qquad \Delta_r S^\circ_{298\,K} = -16.81 \text{ J K}^{-1} \text{mol}^{-1}$$

at 5 bar pressure and 298 K.

Solution

Substituting the given date in the expression

$$\Delta_r S_{p_2} - \Delta_r S^\circ_{p_1} = -(\Delta\nu_g) R \ln \frac{p_2}{p_1}$$

we get

$$\Delta_r S_{p_2} = \left[-16.81 - \left(-\frac{1}{2}\right)(8.314)\ln\left(\frac{5}{1}\right)\right] \text{ J K}^{-1}\text{ mol}^{-1}$$

$$= (-16.81 + 6.69) \text{ J K}^{-1}\text{ mol}^{-1}$$

$$= -10.12 \text{ J K}^{-1}\text{ mol}^{-1}$$

4.25 ENTROPY AND PROBABILITY

Introduction to Statistical Thermodynamics

A thermodynamic system is a macro system consisting of a large number of molecules. In establishing the various thermodynamic functions, the structural model of the system is nowhere required. In fact, all thermodynamic properties of a system can be derived without knowing whether the system is composed of atoms or molecules and how these are moving or interacting with one another. It is possible to develop a subject known as *statistical thermodynamics* where the properties of a system can be studied by treating the molecules on a statistical formalism. Such a study has shown that there exists a relationship between the entropy of the system and its probability. In establishing this relationship, we are not concerned with the motion of the particles but only with the number of ways of distributing the particles in the given energy levels or in a given volume. The various distributions are known as the various complexions or the microscopic states of the system.

Statistical Model of Tossing a Coin

The essential results of the statistical analysis can be easily obtained by considering simple, convenient mathematical models. For example, the number of ways of distributing the molecules in a given volume can be replaced by a model of distributing the given number of balls (equal to the number of molecules) in a given number of boxes. The latter may be assumed to be proportional to the given volume. Before considering this, we take another example of tossing a well-balanced coin repeatedly. We build the model step by step as given below.

Microstates Associated with the Coin Tossed Twice

Since the coin is well-balanced and is being played by an unbiased person, it is obvious that the outcome of the second toss is in no way related to the outcome of the first toss, i.e. the first toss can have no effect on the outcome of the second toss. If the first toss is head, the outcome of the second toss can be either head or tail and if the first toss is tail, the outcome of the second toss can again be either head or tail. This is predicted as follows:

First toss	*Second toss*	*Nature of combination*	*Nature and number of microstates*
H	H	H H	(i) 2H and 0T; HH; one
	T	H T	(ii) 1H and 1T; HT, TH; two
T	H	T H	
	T	T T	(iii) 0H and 2T; T T; one

We may regard all four results as equally probable. Each of these results is considered to be a distinct microstate. There are only three configurations which reflect the total number of heads and tails, irrespective of the order in which they are obtained. These are

2 H, 1 H and 1 T, and 2 T

It can be seen that the number of microstates associated with each of these configurations are one, two and one, respectively. Thus, it follows that the 1H and 1T possibility is twice as likely to materialize as either of the other two configurations, that is 2H and 0T and 0H and 2T.

Microstates Asso-ciated with the Coin Tossed Four Times

The results of the analysis are:

First toss	*Second toss*	*Third toss*	*Fourth toss*	*Nature of combination*		*Nature and number of microstates*
H	H	H	H	H H H H	(i) 4H and 0T;	H H H H; one
H	H	H	T	H H H T		
H	H	T	H	H H T H	(ii) 3H and 1T;	H H H T, H H T H, H T H H, T H H H; four
H	H	T	T	H H T T		
H	T	H	H	H T H H	(iii) 2H and 2T;	H H T T, H T H T, T H H T, H T T H, T H T H, T T H H; six
H	T	H	T	H T H T		
H	T	T	H	H T T H		
H	T	T	T	H T T T		
T	H	H	H	T H H H		
T	H	H	T	T H H T		
T	H	T	H	T H T H	(iv) 1H and 3T;	H T T T, T H T T, T T H T, T T T H; four
T	H	T	T	T H T T		
T	T	H	H	T T H H		
T	T	H	T	T T H T		
T	T	T	H	T T T H		
T	T	T	T	T T T T	(v) 0H and 4T;	T T T T; one

It can be seen from this analysis that the uniform distribution of 2H and 2T is larger than any of the nonuniform distribution of 4H and 0T, 3H and 1T, 1H and 3T, and 0H and 4T. Thus out of all these configurations, the 1 : 1 mixture of heads and tails is the predominant configuration.

Definition of *A*-ratios

We can measure all other configurations relative to the predominant configuration, by defining the ratio-term (*A*) as given below.

$$A_X = \frac{\text{Number of microstates associated with configuration X}}{\text{Number of microstates associated with 1:1 mixture}}$$

The A-ratios in the above case are

A_I	A_{II}	A_{III}	A_{IV}	A_V
0.166	0.666	1.000	0.666	0.166

Microstates Associated with the Coin Tossed *N* Times

In a similar way, we can analyse the results of playing the coin many times. The number of microstates (W) associated with each of the (N + 1) possible configurations can be calculated using the formula

$$W = \frac{N!}{H!T!} \tag{4.25.1}$$

where N is the number of times the coin is played and H and T are the respective number of heads and tails in a particular configuration. Knowing W, we can easily calculate the A-ratio, relative to the predominant (1 : 1) configuration. If smooth curves are drawn through the A-ratios plotted against the corresponding configuration index number H/N for increasing value of N, we get graphs as shown in Fig. 4.25.1.

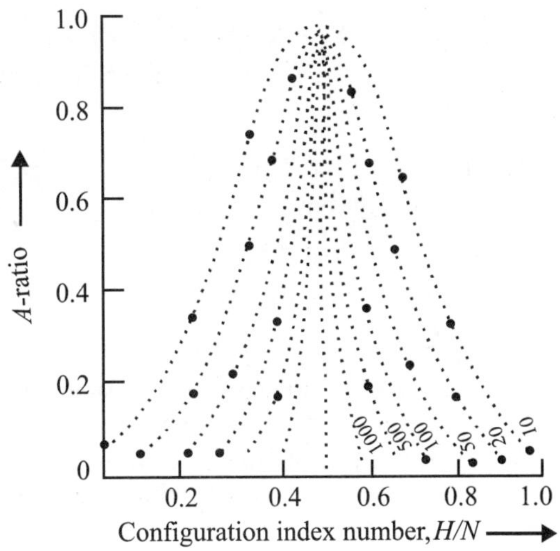

Fig. 4.25.1 Plots of *A*-ratio versus configuration index number for increasing value of *N*

It can be seen from the graph that as N increases the larger proportion of microstates are associated with the small set of configurations having H/N values falling in the range of 0.45 to 0.55. If N is extremely large, then it becomes highly probable that no actual set of N tosses will yield configurations having H/N appreciably different from 0.5. In other words, the probability of getting a uniform distribution if the coin is played unbiased becomes maximum whereas the probability of getting a nonuniform distribution becomes vanishingly lower. The same kind of analysis is applicable to other statistical analysis. Thus we conclude:

Out of the very large number of microstates that can be assumed by any large assembly, majority of them arise from one comparatively small set of configurations centred around, and very similar to the predominant configuration.

Relationship between Entropy and Maximum Number of Microstates

The macroscopic properties of a system depend upon the various microscopic states of the system. The fact that (i) majority of microstates of a system arise from the one comparatively small set of configurations centered around and very similar to the predominant configuration and that (ii) the methods used for the determination of a macroscopic property of a system are not very sensitive to distinguish one predominant configuration from another, result in a constant value of the macroscopic property. If the value of macroscopic properties do not change with time, we say that the system is in an equilibrium state. Thus, according to the statistical mechanics *an equilibrium state is characterized by the predominant configurations which are associated with a maximum number of microstates.* The classical thermodynamics tells us that if the system is at equilibrium then its entropy attains a maximum value. Consequently, it is logical to argue that there must exist some relationship which links the entropy maximum with the maximum number of microstates. Such a relationship can be derived if we assume that entropy S is a function of W only, i.e. $S = f(W)$, where W represents the maximum number of microstates that are associated with the predominant configuration.

Consider two separate samples A and B of the same material each in its equilibrium state at a given temperature and pressure. Thus, we have

$$S_A = f(W_A) \text{ and } S_B = f(W_B)$$

Let us unite the above two samples to form a single system AB. For this system we will have

$$S_{AB} = f(W_{AB})$$

Since entropy is an extensive quantity, it is obvious that

$$S_{AB} = S_A + S_B$$

which when expressed in terms of their functions becomes

$$f(W_{AB}) = f(W_A) + f(W_B)$$

Here W_{AB} is related in very simple way to W_A and W_B since any microstate of subsystem A can combine with any microstate of subsystem B to constitute a distinct microstate of the system AB. Take, for example, the case of a coin which is being played twice, so that the total number of microstates are four. If now the same coin is played twice again then the four old microstates can combine with any of the four new microstates, giving rise to a total of sixteen different microstates. Therefore, it follows that

$$W_{AB} = W_A \times W_B$$

Consequently,

$$f(W_A \times W_B) = f(W_A) + f(W_B)$$

The above relation will hold good provided the function f is a logarithmic function. Thus

$$S \propto \ln (W)$$

or

$$S = k \ln (W) \qquad (4.25.2)$$

where k is a constant of proportionality and has been identified with the Boltzmann's constant (R/N_A).

Entropy Expression for Volume Change in a Gaseous System

Let us now apply the relation to calculate the change in entropy of a gaseous system when its volume changes from V_1 to V_2 at constant temperature. As mentioned earlier, this problem can be replaced by a mathematical model in which various balls are being distributed over a given number of boxes. The former is equal to the number of molecules and the latter is assumed to be proportional to the volume of the system.

The number of ways of distributing the given number of balls in different boxes can be calculated as the follows.

We start with a small number of balls and the results of such analysis can be extended to larger number of balls. Let three balls be distributed in N different boxes. There will be N choices for placing the first ball since it can be placed in any of the boxes. Choices for the second ball will be $N - 1$ since it has to be placed in the $N - 1$ remaining boxes; and $N - 2$ choices for the third ball. Thus, the total possible arrangements are equal to $N(N - 1)(N - 2)$. Since we cannot distinguish various balls, therefore, many of the arrangements which differ only in the permutation of various balls between the boxes will give rise to only one arrangement. For example, if we have to distribute three balls in three boxes, x, y and z out N boxes then we can place them in the following six different ways (total of 3!).

Number of arrangement	*Position of balls*		
	First	*Second*	*Third*
I	x	y	z
II	x	z	y
III	y	x	z
IV	y	z	x
V	z	x	y
VI	z	y	x

If we cannot distinguish the balls, all of these distributions will be identical and give rise to one arrangement. Thus, due to the indistinguishability condition of the balls, many of the possible arrangements are being counted many times and hence in order to get different distinguishable arrangements, we have to divide the total possible arrangements by a factor 3!. Thus, we have

$$W = \frac{N(N-1)(N-2)}{3!}$$

If we have to distribute N_a balls to N different boxes, then the above expression generalises to

$$W = \frac{N(N-1)(N-2)\ldots(N-N_a+1)}{N_a!} \tag{4.25.3}$$

$$= \frac{N(N-1)(N-2)\ldots(N-N_a+1)}{N_a!}\frac{(N-N_a)!}{(N-N_a)!}$$

or $$W = \frac{N!}{N_a!(N - N_a)!} \qquad (4.25.4)^{\dagger}$$

If N is much larger than N_a, Eq. (4.25.3) can be approximated to

$$W = \frac{(N)^{N_a}}{N_a!} \qquad (4.25.5)$$

and the entropy of the system will be given by

$$S = k \ln W = k \ln N^{N_a} - k \ln (N_a!) \qquad (4.25.6)$$

The change in entropy in changing the number of boxes from N to N' can be easily calculated by using above equation. For N boxes.

$$S = k \ln (N^{N_a}) - k \ln (N_a!) \qquad (4.25.7)$$

while for N' boxes

$$S' = k \ln (N'^{N_a}) - k \ln (N_a!) \qquad (4.25.8)$$

Hence $$\Delta S = S' - S = N_a k \ln \left(\frac{N'}{N}\right) \qquad (4.25.9)$$

Since the number of boxes are proportional to the volume of the system we have $N'/N = V'/V$. Therefore, the change in entropy in changing the volume of the gas from V to V' is given by

$$\Delta S = N_a k \ln \left(\frac{V'}{V}\right), \text{ for } N_a \text{ molecules}$$

$$\Delta S = k \ln \left(\frac{V'}{V}\right), \text{ for one molecules}$$

$$\Delta S = R \ln \left(\frac{V'}{V}\right), \text{ for one mole of molecules} \qquad (4.25.10)$$

Equation (4.25.10) is identical to the expression obtained earlier for the change in entropy in the isothermal expansion of 1 mole of an ideal gas from volume V to V' (Eq. 4.16.4).

Derivation of Third Law of Thermodynamics

Now let us see how the relation $S = k \ln W$ helps understanding the third law of themrodynamics and the concept of residual entropy. The third law is applicable only to those substances which attain a completely ordered configuration at the absolute zero of temperature. In a pure crystal, for example, the atoms are located in an exact pattern of lattice sites. The number of different ways for arranging N atoms in N different lattice are $N!$. Since the atoms cannot be distinguished from one another, the different arrangements will differ only in the order of choosing the atoms and thus they are all indistinguishable. Thus

---

†Note that Eqs. (4.25.1) and (4.25.4) are identical where $N_a = H$ and $N - N_a = T$.

$$W = \frac{N!}{N!} = 1 \tag{4.25.11}$$

and $$S = k \ln (W) = 0 \tag{4.25.12}$$

This is the third law of thermodynamics.

Explanation of Residual Entropy

If we have to arrange different kinds of atoms a and b on the sites of the crystals, the number of complexions will be given by

$$W = \frac{N!}{N_a!N_b!} \tag{4.25.13}$$

where N is the total number of sites equal to $(N_a + N_b)$, and N_a and N_b are the numbers of atoms of a and b, respectively. Thus, the entropy of the mixed crystal which is given the name residual entropy is given by

$$S_{\text{residual}} = k \ln \left(\frac{N!}{N_a!N_b!} \right) \tag{4.25.14}$$

If N is very large, the evaluation of the above equation can be done by using Stirling approximation

$$\ln N! = N \ln N - N \tag{4.25.15}$$

Thus, Eq. (4.25.14) reduces to

$$S = k [(N \ln N - N) - (N_a \ln N_a - N_a) - (N_b \ln N_b - N_b)]$$

Since $N = N_a + N_b$, this becomes

$$S = -k [N_a \ln N_a + N_b \ln N_b - N \ln N]$$

Writing in terms of mole fractions, we have

$$\begin{aligned} S &= -k [N_a \ln N_a + N_b \ln N_b - (N_a + N_b) \ln N] \\ &= -k [N_a (\ln N_a - \ln N) + N_b (\ln N_b - \ln N)] \\ &= -k \left[N_a \ln \frac{N_a}{N} + N_b \ln \frac{N_b}{N} \right] \\ &= -N k \left[\frac{N_a}{N} \ln \frac{N_a}{N} + \frac{N_b}{N} \ln \frac{N_b}{N} \right] \\ &= -N k (x_a \ln x_a + x_b \ln x_b) \end{aligned} \tag{4.25.16}$$

Since the terms in parentheses are negative (the logarithm of a fraction is negative), the entropy of the mixed crystal is positive. If we imagine the mixed crystal to be formed from a pure crystal a and a pure crystal b, then for the mixing process

$$\text{pure a} + \text{pure b} \rightarrow \text{mixed crystal}$$

the entropy change is

$$\Delta S_{\text{mix}} = S(\text{mixed crystal}) - S(\text{pure a}) - S(\text{pure b})$$

Since the entropy of the pure crystal is zero, therefore

$$\Delta S_{\text{mix}} = -N k\,[x_{\text{a}} \ln x_{\text{a}} + x_{\text{b}} \ln x_{\text{b}}] \tag{4.25.17}$$

Examples of Residual Entropy

There are many crystals which do not follow the third law of thermodynamics. These include carbon monoxide crystals, nitric oxide crystals, crystalline ice, crystalline dihydrogen, etc. These substances are pure from the chemical standpoint but do not fulfil the requirement of perfectly ordered arrangement at the absolute zero of temperature.

In CO, perfect arrangement will be one in which all carbons are pointing in one direction and all oxygen in another directions, for example,

CO CO CO CO · · ·
CO CO CO CO · · ·
| |

In actual crystals, the two ends of the molecules are oriented randomly and the crystal behaves as if it were formed out of two different types of molecules of carbon monoxide CO and OC. The consequence of this is that the crystals of CO have the residual entropy which, in principle, can be calculated using Eq. (4.25.16), provided we know the number of the molecules having these two different orientations.

In case of nitric oxide we not only have random arrangements of the molecules but also the dimerization of molecules.

In ice, we have random arrangement of hydrogen bonding of water molecules in the crystal.

In case of dihydrogen, because of the total nuclear spin angular momentum, the rotational energy of *ortho*-dihydrogen at low temperatures does not approach zero as does that of *para*-dihydrogen, but achieves a finite value. *Ortho*-dihydorgen can be in any of nine states, all having the same energy, while *para*-dihydrogen exists in a single sta:e. Because of this randomness, dihydrogen has a residual entropy.

4.26 MISCELLANEOUS NUMERICALS

1. (i) One mole of an ideal monatomic gas at 298 K, occupying a volume of 3 dm^3 is expanded adiabatically and reversibly to a pressure of 101.325 kPa. It is then compressed isothermally reversibly at a new temperature until the volume is again 3 dm^3. Calculate q, w, ΔU, ΔH and ΔS.

(ii) The same gas in the same initial state is expanded isothermally reversibly to a pressure of 101.325 kPa. It is then compressed adiabatically and reversibly until the volume is again 3 dm^3. Calculate q, w, ΔU, ΔH and ΔS.

Solution (i) The two processes along with the given data are

$$\begin{matrix} n = 1\ \text{mol} & & n = 1\ \text{mol} & & n = 1\ \text{mol} \\ T = 298\ \text{K} & \xrightarrow[\text{reversible}]{\text{adiabatic}} & T_2 = ? & \xrightarrow[\text{reversible}]{\text{isothermal}} & T_3 = T_2 \\ V_1 = 3\ \text{dm}^3 & \text{expansion} & V_2 = ? & \text{compression} & V_3 = 3\ \text{dm}^3 \\ p_1 = ? & & p_2 = 101.325\ \text{kPa} & & \end{matrix}$$

For calculating the changes in the given quantities, we will have to calculate p_1, V_2 and T_2. These can be calculated as follows.

For p_1, we employ the ideal gas equations

$$p_1 = \frac{nRT}{V_1} = \frac{(1\ \text{mol})\ (8.314\ \text{J K}^{-1}\ \text{mol}^{-1})\ (298\ \text{K})}{(3\ \text{dm}^3)} = 825.857\ \text{kPa}$$

For T_2, we write the expression

$$T_1\, p_1^{-R/\,C_{p,\,\text{m}}} = T_2\, p_2^{-R/\,C_{p,\,\text{m}}}$$

as applicable to an adiabatic reversible change. Thus

$$T_2 = T_1 \left(\frac{p_2}{p_1}\right)^{R/C_{p,\text{m}}} = (298\ \text{K}) \left(\frac{101.325\ \text{kPa}}{825.857\ \text{kPa}}\right)^{R/(5/2)R}$$

$$= (298\ \text{K})\ (0.122\ 7)^{0.4} = 128.7\ \text{K}$$

For V_2, we again employ the ideal gas equation

$$V_2 = \frac{nRT_2}{p_2} = \frac{(1\ \text{mol})\ (8.314\ \text{kPa dm}^3\ \text{K}^{-1}\ \text{mol}^{-1})\ (128.7\ \text{K})}{(101.325\ \text{kPa})} = 10.56\ \text{dm}^3$$

Now we proceed to calculate the changes for the two processes.

***Process* 1** *Adiabatic reversible expansion* For this process

$$q = 0$$

$$\Delta U = q + w = w$$

and $$\Delta U = nC_{V,\,\text{m}}\,(T_2 - T_1)$$

$$= (1\ \text{mol}) \left(\frac{3}{2} \times 8.314\ \text{J K}^{-1}\ \text{mol}^{-1}\right) (128.7\ \text{K} - 298\ \text{K})$$

$$= -\,2\ 111.3\ \text{J}$$

$$w = \Delta U = -\,2\ 111.3\ \text{J}$$

$$\Delta H = \Delta U + \Delta(pV) = \Delta U + nR(\Delta T)$$

$$= -\,2\ 111.3\ \text{J} + (1\ \text{mol})\ (8.314\ \text{J K}^{-1}\ \text{mol}^{-1})\ (-\,169.3\ \text{K})$$

$$= -\,2\ 111.3\ \text{J} - 1\ 407.56\ \text{J} = -\,3\ 518.86\ \text{J}$$

$\Delta S = 0$ since expansion is adiabatic and reversible.

***Process* 2** *Isothermal reversible compression* For this process

$$\Delta T = 0$$

$$\Delta U = 0$$

and $$q = -\,w = nRT \ln \frac{V_3}{V_2}$$

$$= (1\text{mol})\ (8.314\ \text{J K}^{-1}\ \text{mol}^{-1})\ (128.7\ \text{K}) \times 2.303 \log \left(\frac{3\ \text{dm}^3}{10.56\ \text{dm}^3}\right)$$

$$= -\,1\ 346.8\ \text{J}$$

$$\Delta H = \Delta U + \Delta(pV) = \Delta U + nR(\Delta T) = 0 + 0 = 0$$

$$\Delta S = nR \ln \frac{V_3}{V_2}$$

$$= (1\text{mol})\ (8.314\ \text{J K}^{-1}\ \text{mol}^{-1}) \times 2.303 \log \left(\frac{3\ \text{dm}^3}{10.56\ \text{dm}^3}\right)$$

$$= -\,10.46\ \text{J K}^{-1}$$

Total changes in the required quantities are

$$q = 0 - 1\ 346.8\ \text{J} = -\ 1\ 346.8\ \text{J}$$
$$w = -\ 2\ 111.3\ \text{J} + 1\ 346.8\ \text{J} = -\ 764.5\ \text{J}$$
$$\Delta U = -\ 2\ 111.3\ \text{J} + 0 = -\ 2\ 111.3\ \text{J}$$
$$\Delta H = -\ 3\ 518.86 + 0 = -\ 3\ 518.86\ \text{J}$$
$$\Delta S = 0 - 10.46\ \text{J K}^{-1} = -\ 10.46\ \text{J K}^{-1}$$

(ii) The two processes with the given data are

$n = 1$ mol		$n = 1$ mol		$n = 1$ mol
$T_1 = 298$ K	$\xrightarrow[\text{reversible}]{\text{isothermal}}$	$T_2 = T_1$	$\xrightarrow[\text{reversible}]{\text{adiabatic}}$	$T_3 = ?$
$V_1 = 3\ \text{dm}^3$	expansion	$V_2 = ?$	compression	$V_3 = 3\ \text{dm}^3$
$p_1 = ?$		$p_2 = 101.325$ kPa		

Calculation of p_1, V_2 and T_3:

$$p_1 = \frac{nRT_1}{V_1} = \frac{(1\ \text{mol})\ (8.314\ \text{kPa dm}^3\ \text{K}^{-1}\ \text{mol}^{-1})\ (298\ \text{K})}{(3\ \text{dm}^3)} = 825.86\ \text{kPa}$$

$$V_2 = \frac{nRT_2}{p_2} = \frac{(1\ \text{mol})\ (8.314\ \text{kPa dm}^3\ \text{K}^{-1}\ \text{mol}^{-1})\ (298\ \text{K})}{(101.325\ \text{kPa})} = 24.45\ \text{dm}^3$$

For an adiabatic reversible change

$$T_2\, V_2^{R/C_{V,\text{m}}} = T_3\, V_3^{R/C_{V,\text{m}}}$$

or $$T_3 = T_2\left(\frac{V_2}{V_3}\right)^{R/C_{V,\text{m}}} = (298\ \text{K})\left(\frac{24.45\ \text{dm}^3}{3\ \text{dm}^3}\right)^{2/3} = 1\ 206.0\ \text{K}$$

***Process* 1** *Isothermal reversible expansion* For this process

$$\Delta T = 0$$
$$\Delta U = 0$$

and $$q = -\ w = nRT\ \ln\frac{V_2}{V_1}$$

$$= (1\ \text{mol})\ (8.314\ \text{J K}^{-1}\ \text{mol}^{-1}\ )\ (298\ \text{K}) \times 2.303\ \log\left(\frac{24.45\ \text{dm}^3}{3\ \text{dm}^3}\right)$$
$$= 5\ 198.9\ \text{J}$$

$$\Delta H = \Delta U + \Delta(pV) = \Delta U + nR(\Delta T) = 0$$

$$\Delta S = nR\ \ln\frac{V_2}{V_1} = (1\ \text{mol})\ (8.314\ \text{J K}^{-1}\ \text{mol}^{-1}) \times 2.303\ \log\left(\frac{24.45\ \text{dm}^3}{3\ \text{dm}^3}\right)$$

$$= 17.45\ \text{J K}^{-1}\ \left(\equiv \frac{q}{T} = \frac{5198.9\ \text{J}}{298\ \text{K}} = 17.45\ \text{J K}^{-1}\right)$$

***Process* 2** *Adiabatic reversible compression* For this process

$$q = 0$$
$$\Delta U = q + w = w$$

But $$\Delta U = nC_{V,\text{m}}\ (T_3 - T_2)$$

$$= (1\ \text{mol})\ \left(\frac{3}{2} \times 8.314\ \text{J K}^{-1}\ \text{mol}^{-1}\right)\ (1\ 206.6\ \text{K} - 298\ \text{K})$$
$$= 11\ 331\ \text{J}$$

$$w = \Delta U = 11\ 331\ \text{J}$$

$$\Delta H = \Delta U + nR(\Delta T)$$

$$= 11\ 331\ \text{J} + (1\ \text{mol})\ (8.314\ \text{J K}^{-1}\ \text{mol}^{-1})\ (908.6\ \text{K})$$

$$= 18\ 885\ \text{J}\ (\equiv C_{p,\text{m}}\ (T_3 - T_2))$$

$$\Delta S = 0$$

Total changes in the required quantities are

$$q = 5\ 198.9\ \text{J} + 0 = 5\ 198.9\ \text{J}$$

$$w = -\ 5\ 198.9 + 11\ 331\ \text{J} = 6\ 123.1\ \text{J}$$

$$\Delta U = 0 + 11\ 331\ \text{J} = 11\ 331\ \text{J}$$

$$\Delta H = 0 + 18\ 885\ \text{J} = 18\ 885\ \text{J}$$

$$\Delta S = 17.45\ \text{J K}^{-1} + 0 = 17.45\ \text{J K}^{-1}$$

---

2. One mole of helium is mixed with 2 mol of neon, both at the same temperature and pressure. Calculate ΔS for this process if the total volume remains constant.

Solution Both the gases before mixing are at the same temperature and pressure. Since the amount of neon is twice as that of helium. It is obvious that

$$V_{\text{Ne}} = 2\ V_{\text{He}}$$

Now the volume of the gas after mixing will be

$$V_{\text{total}} = V_{\text{He}} + V_{\text{Ne}} = V_{\text{He}} + 2\ V_{\text{He}} = 3\ V_{\text{He}}$$

Each of the two gases will suffer entropy changes due to the volume change. Thus

$$\Delta S_{\text{He}} = nR\ \ln\frac{V_2}{V_1} = (1\ \text{mol})\ (8.314\ \text{J K}^{-1}\ \text{mol}^{-1}) \times 2.303\ \log\left(\frac{3\ V_{\text{He}}}{V_{\text{He}}}\right)$$

$$= 9.136\ \text{J K}^{-1}$$

$$\Delta S_{\text{Ne}} = nR\ \ln\frac{V_2}{V_1} = (2\ \text{mol})\ (8.314\ \text{J K}^{-1}\ \text{mol}^{-1}) \times 2.303\ \log\left(\frac{3\ V_{\text{He}}}{2\ V_{\text{He}}}\right)$$

$$= 6.743\ \text{J K}^{-1}$$

Thus $\Delta S_{\text{mix}} = \Delta S_{\text{He}} + \Delta S_{\text{Ne}}$

$$= 9.136\ \text{J K}^{-1} + 6.743\ \text{J K}^{-1}$$

$$= 15.897\ \text{J K}^{-1}$$

---

3. Without doing explicit calculations, predict the signs of Δp, ΔV, ΔT, ΔU, ΔH and ΔS for one mole of an ideal gas taken through each of the following four steps of a Carnot cycle. Assume C_p and C_V constants.

Step (a) : Isothermal reversible expansion $p_1, V_1, T_1 \rightarrow p_2, V_2, T_1$
Step (b) : Adiabatic reversible expansion $p_2, V_2, T_1 \rightarrow p_3, V_3, T_2$
Step (c) : Isothermal reversible compression $p_3, V_3, T_2 \rightarrow p_4, V_4, T_2$
Step (d) : Adiabatic reversible compression $p_4, V_4, T_2 \rightarrow p_1, V_1, T_1$

Solution The changes in the given quantities in the above four processes are listed below.

Step (a):

$\Delta V = +\text{ve}$	as expansion takes place.
$\Delta T = 0$	as the process is isothermal.
$\Delta p = -\text{ve}$	as increase in volume at constant temperature will be attended to by decrease of pressure (Boyle's law).
$\Delta U = 0$	since for an ideal gas $U = f(T)$ and $\Delta T = 0$.

$\Delta H = 0$ since $\Delta H = \Delta U + \Delta(pV)$ and $\Delta U = 0$ and $\Delta(pV) = \Delta(RT) = 0$.

$\Delta S = +\text{ve}$ since $\Delta S = nR \ln (V_2/V_1)$ and $V_2 > V_1$

Step (b): Here we will have

$\Delta V = +\text{ve}$ as expansion takes place.

$\Delta S = 0$ as the adiabatic process is reversible.

$\Delta U = -\text{ve}$ since $\Delta U = w$ and w is negative (expansion process).

$\Delta T = -\text{ve}$ since $\Delta U = C_{V,\,\text{m}}\,(\Delta T)$

$\Delta H = -\text{ve}$ since $\Delta H = \Delta U + \Delta(pV) = \Delta U + R\,\Delta T$.
both ΔU and ΔT are negative.

$\Delta p = -\text{ve}$

Step (c): Here the sings of various quantities will be just opposite to those listed in step (a).

Step (d): Here the sings of various quantities will be just opposite to those listed in step (b).

---

4. Sketch diagrams for the reversible Cannot cycle of an ideal gas with constant C_p and C_V; p versus S; U versus p; T versus p; H versus p; V versus T; S versus V; V versus U; V versus H; U versus H; U versus T; H versus T; T versus S; U versus S and H versus S.

Solution The required graphs are shown in Fig. 4.26.1.

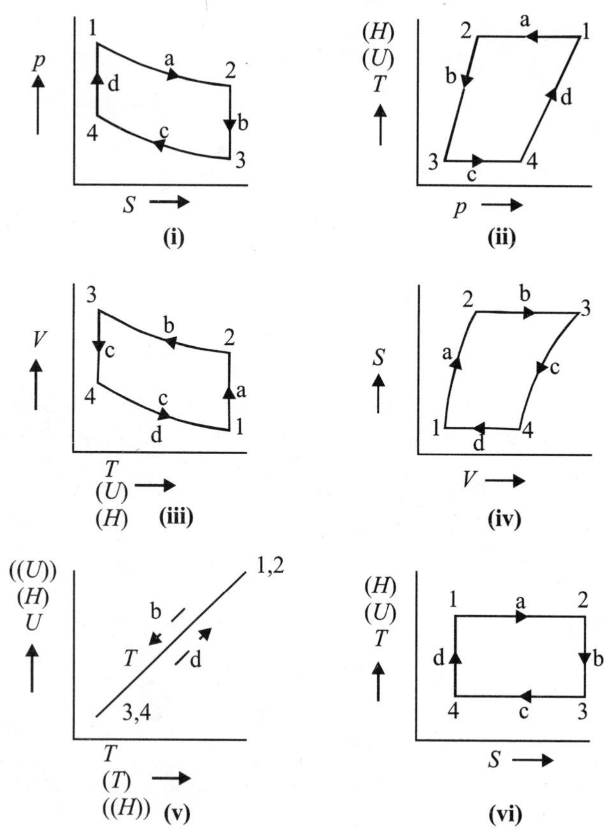

Fig. 4.26.1 (i) p v. S graph. (ii) T v. p graph. This graph also represents U v. p and H v. p graphs with a scale factor of C_V on the energy axis and C_p on the enthalpy axis, (iii) V v. T graph. This also represents V v. U and V v. H graphs. (iv) S v. V graph. (v) U v. T graph. This also represents H v. T and U v. H graphs. (vi) T v. S graph. This also represents U v. S and H v. S graphs

---

5. The data of a reversible Carot cycle is given in Question 3. Assuming $C_{V,\,\mathrm{m}}$ constant, derive expressions for Δp, ΔV, ΔT, ΔU, ΔH and ΔS for one mole of an ideal gas for each step of the above Carnot cycle.

Solution **Step (a)** *Isothermal expansion* Expressions for the required quantities are

$$\Delta T = 0; \qquad \Delta U = 0\,; \qquad \Delta H = 0\,; \qquad \Delta V = V_2 - V_1$$

$$\Delta p = RT_1\,\Delta\left(\frac{1}{V}\right) = RT_1\left[\frac{1}{V_2} - \frac{1}{V_1}\right]$$

$$\Delta S = \frac{q}{T_1} = \frac{(-w)}{T_1} = R\ln\frac{V_2}{V_1}$$

Step (b) *Adiabatic expansion* Here expressions are

$$\Delta T = T_2 - T_1$$

$$\Delta U = C_{V,\,\mathrm{m}}\,(T_2 - T_1)$$

$$\Delta H = \Delta U + \Delta(pV) = \Delta U + R(\Delta T)$$

$$= (C_{V,\,\mathrm{m}} + R)\,\Delta T = C_{p,\,\mathrm{m}}\,(T_2 - T_1)$$

$$\Delta S = 0 \qquad \text{(because the change is reversible and adiabatic)}$$

$$\Delta V = V_3 - V_2$$

where V_3 is the volume of the gas after it has undergone the adiabatic reversible expansion. V_3 can be substituted in terms of V_2 by making use of Eq. (2.11.8). Thus, we have

$$T_1\,V_2^{\gamma-1} = T_2\,V_3^{\gamma-1}$$

or

$$V_3 = \left(\frac{T_1}{T_2}\right)^{1/\gamma-1} V_2$$

Substituting this in the previous expression, we have

$$\Delta V = \left[\left(\frac{T_1}{T_2}\right)^{1/(\gamma-1)} - 1\right]V_2 = \left[\left(\frac{T_1}{T_2}\right)^{C_{V,\mathrm{m}}/R} - 1\right]V_2$$

$$\Delta p = p_3 - p_2$$

Since for an adiabatic reversible process pV^{γ} is constant, we have

$$\Delta p = p_2\left(\frac{V_2}{V_3}\right)^{\gamma} - p_2 = \frac{RT_1}{V_2}\left[\left(\frac{V_2}{V_3}\right)^{\gamma} - 1\right]$$

Making use of the relation $TV^{\gamma-1}$, we get

$$\Delta p = \frac{RT_1}{V_2}\left[\left(\frac{T_2}{T_1}\right)^{C_{p,\mathrm{m}}/R} - 1\right]$$

Step (c) *Isothermal compression* In this step, the various expressions are

$$\Delta T = 0\,; \qquad \Delta U = 0\,; \qquad \Delta H = 0$$

$$\Delta S = R\ln\frac{V_4}{V_3}$$

Since for steps (b) and (d), $\Delta S = 0$, it follows that the above entropy change will be just equal and opposite to that of step (a). Thus

$$\Delta S = R \ln \frac{V_4}{V_3} = -R \ln \frac{V_2}{V_1}$$

This gives us the relation

$$\frac{V_4}{V_3} = \frac{V_1}{V_2} \qquad \text{or} \qquad V_4 = \left(\frac{V_1 V_3}{V_2}\right)$$

$$\Delta V = V_4 - V_3 = \frac{V_1 V_3}{V_2} - V_3 = \frac{V_3}{V_2}(V_1 - V_2) = \left(\frac{T_1}{T_2}\right)^{C_{V,\text{m}}/R}(V_1 - V_2)$$

$$\Delta p = p_4 - p_3 = \frac{RT_2}{V_4} - \frac{RT_2}{V_3} = RT_2\left[\frac{1}{V_4} - \frac{1}{V_3}\right]$$

$$= RT_2\left[\frac{V_2}{V_1 V_3} - \frac{1}{V_3}\right] = \frac{RT_2}{V_3 V_1}(V_2 - V_1)$$

$$= \frac{RT_2}{V_2 V_1}\left(\frac{T_2}{T_1}\right)^{C_{V,\text{m}}/R}(V_1 - V_2)$$

Step (d) *Adiabatic compression* In this step, we will have

$$\Delta T = T_1 - T_2$$
$$\Delta U = C_{V,\text{m}}\,\Delta T = C_{V,\text{m}}(T_1 - T_2)$$
$$\Delta H = C_{p,\text{m}}\,\Delta T = C_{p,\text{m}}(T_1 - T_2)$$
$$\Delta S = 0$$

$$\Delta V = (V_1 - V_4) = V_1\left(1 - \frac{V_3}{V_2}\right) = V_1\left[1 - \left(\frac{T_1}{T_2}\right)^{C_{V,\text{m}}/R}\right]$$

Now $\quad \Delta p = p_1 - p_4$

and $\quad p_4 = p_1 + (\Delta p)_\text{a} + (\Delta p)_\text{b} + (\Delta p)_\text{c}$

where $(\Delta p)_\text{a}$, $(\Delta p)_\text{b}$, and $(\Delta p)_\text{c}$, are the pressure changes in steps (a), (b) and (c), respectively,

Thus $\quad \Delta p = -(\Delta p)_\text{a} - (\Delta p)_\text{b} - (\Delta p)_\text{c}$

$$= -RT_1\left(\frac{1}{V_2} - \frac{1}{V_1}\right) - \frac{RT_1}{V_2}\left[\left(\frac{T_2}{T_1}\right)^{C_{p,\text{m}}/R} - 1\right] - \frac{RT_2}{V_2 V_1}\left(\frac{T_2}{T_1}\right)^{C_{V,\text{m}}/R}(V_2 - V_1)$$

---

6. One mole of an ideal gas undergoes a reversible Carnot cycle with $V_1 = 20\ \text{dm}^3$, $V_2 = 40\ \text{dm}^3$ and $T_1 = 300$ K. Let $T_2 = 200$ K and $C_{V,\text{m}} = (3/2)R$. Using the expressions derived in Question 5, calculate Δp, ΔV, ΔT, ΔU, ΔH and ΔS involved in each of the four steps.

Solution **Step (a)**

$$\Delta T = 0\ ; \quad \Delta U = 0\ ; \quad \Delta H = 0$$

$$\Delta V = (V_2 - V_1) = 20\ \text{dm}^3$$

$$\Delta p = RT_1\left(\frac{1}{V_2}-\frac{1}{V_1}\right) = (8.314\ \text{kPa dm}^3\ \text{K}^{-1}\ \text{mol}^{-1})\ (300\ \text{K})$$

$$\times\left(\frac{1}{40\ \text{dm}^3\ \text{mol}^{-1}}-\frac{1}{20\ \text{dm}^3\ \text{mol}^{-1}}\right)$$

$$= -\ 62.355\ \text{kPa}$$

$$\Delta S = R\ \ln\left(\frac{V_2}{V_1}\right) = (8.314\ \text{J K}^{-1}\ \text{mol}^{-1}) \times 2.303\ \log\left(\frac{40\ \text{dm}^3\ \text{mol}^{-1}}{20\ \text{dm}^3\ \text{mol}^{-1}}\right)$$

$$= 5.763\ \text{J K}^{-1}\ \text{mol}^{-1}$$

Step (b)

$$\Delta T = T_2 - T_1 = -100\ \text{K}$$

$$\Delta U = C_{V,\,\text{m}}\ (T_2 - T_1) = \frac{3}{2}R\ (\Delta T)$$

$$= \frac{3}{2}\ (8.314\ \text{J K}^{-1}\ \text{mol}^{-1})\ (-100\ \text{K}) = -1\ 247.1\ \text{J mol}^{-1}$$

$$\Delta H = C_{p,\,\text{m}}(T_2 - T_1) = \frac{5}{2}\ R(\Delta T)$$

$$= \frac{5}{2}\ (8.314\ \text{J K}^{-1}\ \text{mol}^{-1})\ (-100\ \text{K}) = -\ 2\ 078.5\ \text{J mol}^{-1}$$

$$\Delta V = \left[\left(\frac{T_1}{T_2}\right)^{C_{V,\text{m}}/R} - 1\right] V_2 = \left[\left(\frac{300\ \text{K}}{200\ \text{K}}\right)^{1.5} - 1\right](40\ \text{dm}^3)$$

$$= 33.48\ \text{dm}^3$$

$$\Delta p = \frac{RT_1}{V_2}\left[\left(\frac{T_2}{T_1}\right)^{C_{p,\text{m}}/R} - 1\right]$$

$$= \frac{(8.314\ \text{kPa dm}^3\ \text{K}^{-1}\ \text{mol}^{-1})(300\ \text{K})}{(40\ \text{dm}^3\ \text{mol}^{-1})}\left[\left(\frac{200\ \text{K}}{300\ \text{K}}\right)^{5/2} - 1\right]$$

$$= -39.727\ \text{kPa}$$

$$\Delta S = 0$$

Step (c)

$$\Delta T = 0\ ;\quad \Delta U = 0;\quad \Delta H = 0$$

$$\Delta V = \left(\frac{T_1}{T_2}\right)^{C_{V,\text{m}}/R}\ (V_1 - V_2) = \left(\frac{300\ \text{K}}{200\ \text{K}}\right)^{3/2}\ (-\ 20\ \text{dm}^3\ \text{mol}^{-1}) = -\ 36.74\ \text{dm}^3\ \text{mol}^{-1}$$

$$\Delta p = \frac{RT_2}{V_2V_1}\left(\frac{T_2}{T_1}\right)^{C_{V,\mathrm{m}}/R}(V_2 - V_1)$$

$$= \frac{(8.314\ \mathrm{kPa\ dm^3\ K^{-1}\ mol^{-1}})(200\ \mathrm{K})}{(40\ \mathrm{dm^3\ mol^{-1}})(20\ \mathrm{dm^3\ mol^{-1}})}\left(\frac{200\ \mathrm{K}}{300\ \mathrm{K}}\right)^{1.5}\{(40-20)\ \mathrm{dm^3\ mol^{-1}}\}$$

$$= 22.63\ \mathrm{kPa}$$

$$\Delta S = R\ln\frac{V_4}{V_3} = R\ln\frac{V_1}{V_2} = (8.314\ \mathrm{J\ K^{-1}\ mol^{-1}})\ln\left(\frac{20}{40}\right) = -5.763\ \mathrm{K^{-1}\ mol^{-1}}$$

Step (d)

$$\Delta T = 100\ \mathrm{K};\quad \Delta U = 1\ 247.1\ \mathrm{J\ mol^{-1}};\quad \Delta H = 2\ 078.5\ \mathrm{J\ mol^{-1}}$$

$$\Delta V = V_1\left[1-\left(\frac{T_1}{T_2}\right)^{C_{V,\mathrm{m}}/R}\right] = (20\ \mathrm{dm^3\ mol^{-1}})\left[1-\left(\frac{300\ \mathrm{K}}{200\ \mathrm{K}}\right)^{1.5}\right]$$

$$= -16.74\ \mathrm{dm^3\ mol^{-1}}$$

$$\Delta p = -[(\Delta p)_a + (\Delta p)_b + (\Delta p)_c] = (+62.355 + 39.727 - 22.63)\ \mathrm{kPa}$$

$$= 79.452\ \mathrm{kPa}$$

$$\Delta S = 0$$

REVISIONARY PROBLEMS

4.1 Discuss what was the necessity of introducing second law of thermodynamics?

4.2 What is a Carnot cycle? Derive the expression of the efficiency of such a cycle, taking an ideal gas as a working substance. Is it possible to have a Carnot cycle of efficiency one?

4.3 Second law of thermodynamics is stated in any of the following forms:

(a) It is impossible for a system operating in a cycle and connected to a single heat reservoir to produce a positive amount of work in the surroundings (Kelvin).

(b) It is impossible for a cyclic process to convert heat into work without the simultaneous transfer of heat from a body at higher temperature to one at lower temperature or vice versa, i.e. it is impossible for a cyclic process to transfer heat from a body at a lower temperature to one at a higher temperature without the simultaneous conversion of work into heat (Clausius).

(c) It is impossible for an engine operative in a cycle to have as its only effect the transfer of a quantity of heat from a reservoir of lower temperature to a reservoir at higher temperature.

(d) The change of mechanical work into heat may be complete, but on the contrary, that of heat into work must need be incomplete, since whenever a certain quantity of heat is transformed into work, another quantity of heat must undergo a corresponding and compensating change (Planck).

Show that the efficiency of a Carnot cycle is a direct consequence of these laws.

4.4 Does that the efficiency of a Carnot cycle depends upon the make up of the engine or a working substance? If not, why? Hence show that all reversible engines operating between the same two temperatures have the same efficiency.

4.5 Show that the efficiency of an irreversible Carnot cycle is less than the efficiency of a reversible Carnot cycle.

4.6 What is Kelvin scale of temperature? Discuss, logically, how Kelvin introduced this scale. How would you establish the identity of Kelvin scale of temperature and the ideal gas temperature scale?

4.7 (a) Show how the following equation can be derived from a Carnot cycle operating between two temperatures T_1 and T_2

$$\frac{q_1}{T_1} + \frac{q_2}{T_2} = 0$$

(b) Does the above expression fails, if a bigger Carnot cycle is replaced by a number of smaller Carnot cycles involving heat exchange with number of heat reservoirs with temperatures between T_1 and T_2? If not, explain why it is so?

(c) Show that any cyclic process can be replaced by a larger number of Carnot cycles— larger the number of cycles better the replacement. Hence, show that for such a cyclic process

$$\oint \mathrm{d}\left(\frac{q_{\text{rev}}}{T}\right) = 0$$

(d) What conclusions do you draw from the above expression? Show that it leads to a state function S, called entropy.

4.8 Show that the value of integral $\oint \mathrm{d}(q_{\text{rev}}/T)$ is zero for any reversible engine whereas it is less than zero (i.e. negative) for an irreversible engine.

4.9 Show that the value of ΔS_{total} is zero if the process A $\rightarrow$ B is carried out reversibly and is greater than zero if it is carried out irreversibly.

4.10 What is Clausius inequality? Show that it leads to the following statement:

'The energy of the universe is constant, the entropy of the universe always tends towards a maximum.'

4.11 How would you derive the entropy function starting from the first law of thermodynamics alone? You can make use of the expression

$$\left(\frac{\partial U}{\partial V}\right)_T + p = T\left(\frac{\partial p}{\partial T}\right)_V$$

4.12 Derive the following relations:

(i) $(\partial S/\partial U)_V = 1/T$ and $(\partial S/\partial V)_U = p/T$ (ii) $(\partial T/\partial V)_S = -(\partial p/\partial S)_V$

(iii) $\mathrm{d}S = \dfrac{C_V}{T}\mathrm{d}T + \dfrac{\alpha}{\kappa_T}\,\mathrm{d}V$ (iv) $(\partial S/\partial T)_V = C_V/T$

(v) $(\partial S/\partial V)_T = \dfrac{1}{T}\,[p + (\partial U/\partial V)_T] = (\partial p/\partial T)_V$

(vi) $(\partial U/\partial V)_T = T\,(\partial p/\partial T)_V - p$

(vii) $C_p - C_V = T\,(\partial p/\partial T)_V\,(\partial V/\partial T)_p = -T\,(\partial V/\partial T)^2_p / (\partial V/\partial p)_T = \dfrac{T\alpha^2 V}{\kappa_T}$

4.13 (a) Show that

(i) $(\partial U/\partial V)_T$ is zero for an ideal gas

(ii) $C_p - C_V = R$ for one mole of an ideal gas

(b) Evaluate $(\partial U/\partial V)_T$ for a van der Waals gas. Use the van der Waals equation together with the thermodynamic equation of state.

(c) By integrating the total differential $\mathrm{d}U$ for a van der Waals gas, show that if $C_{V,\text{m}}$ is constant, then

$$U = C_{V,\mathrm{m}}\, T - \frac{a}{V} + U'$$

where U' is a constant of integration. How does U vary with (i) change in V at constant T and (ii) change of T at constant V.

4.14. Derive the following relations:

(i) $\mathrm{d}S = \frac{C_p}{T}\,\mathrm{d}T - V\,\alpha\,\mathrm{d}p$ (ii) $(\partial T/\partial p)_S = (\partial V/\partial S)_p$

(iii) $(\partial S/\partial T)_p = C_p/T$ (iv) $(\partial S/\partial p)_T = V - T(\partial V/\partial T)_p$

(v) $(\partial S/\partial p)_T = \frac{1}{T}\,[(\partial H/\partial p)_T - V] = -(\partial V/\partial T)_p = -V\,\alpha$

(vi) $\mu_{\mathrm{JT}} = \dfrac{T(\partial V/\partial T)_p - V}{C_p} = \dfrac{V(\alpha T - 1)}{C_p}$

4.15 (a) Show that for an ideal gas $(\partial H/\partial p)_T = 0$

(b) μ_{JT} is zero for an ideal gas

(c) Using $\mathrm{d}S = \frac{C_p}{T}\,\mathrm{d}T - V\alpha\,\mathrm{d}p$ or $\mathrm{d}S = \frac{C_V}{T}\,\mathrm{d}T + \frac{\alpha}{\kappa_T}\,\mathrm{d}V$, show that

(i) $\left(\dfrac{\partial S}{\partial p}\right)_V = \dfrac{\kappa_T C_V}{T\alpha}$ (ii) $\left(\dfrac{\partial S}{\partial V}\right)_p = \dfrac{C_p}{T\,V\alpha}$

and (iii) $-\dfrac{1}{V}\left(\dfrac{\partial V}{\partial p}\right)_S = \dfrac{\kappa_T}{\gamma}$ where $\gamma = \dfrac{C_p}{C_V}$

4.16 (a) Show that for one mole of an ideal gas

(i) $\left(\dfrac{\partial S}{\partial V}\right)_T = \dfrac{R}{V}$ (ii) $\left(\dfrac{\partial S}{\partial T}\right)_V = \dfrac{C_{V,\mathrm{m}}}{T}$

(b) The change in entropy function when the amount n of an ideal gas undergoes a change from T_1, V_1 to T_2, V_2 is given by the expression

$$\Delta S = n\left[C_{V,\mathrm{m}}\ \ln \frac{T_2}{T_1} + R\ \ln \frac{V_2}{V_1}\right]$$

and the corresponding expression for T_1, p_1 to T_2, p_2 is

$$\Delta S = n\left[C_{p,\mathrm{m}}\ \ln \frac{T_2}{T_1} + R\ \ln \frac{p_1}{p_2}\right]$$

4.17 Derive expressions for ΔS_{sys}, ΔS_{surr} and $\Delta S_{\mathrm{total}}$ for expansion (or compression) of an ideal gas under the following circumstances:

(a) Isothermal reversible change.

(b) Isothermal irreversible change involving (i) free expansion, and (ii) intermediate expansion.

(c) Adiabatic reversible change.

(d) Adiabatic irreversible change.

4.18 (a) Derive expressions of ΔS for the following transformations choosing a typical example for each.

(i) Isothermal reversible phase transformation.

(ii) Isothermal irreversible phase transformation.

(b) What is Trouton's rule?

4.19 Derive expression of ΔS for the entropy change when two solids at different temperatures are brought together.

4.20 Define the necessary condition for standard state for the entropy of an ideal gas (i.e. $S°$). How does the quantity $S - S°$ vary with pressure p and with ln (p/bar)?

4.21 What is the third law of thermodynamics? Evaluate theroretically the third law entropy of a gas above its boiling point. Write all the steps involved precisely.

4.22 According to the statistical thermodynamics, the number of microstates associated with any one of the configurations involving H number of heads and T number of tails, when a coin is played N times, is given by

$$W = \frac{N!}{H!T!}$$

(a) Calculate the values of W associated with following configurations.

$$\frac{H}{N} = 0.2,\ 0.4,\ 0.6,\ 0.8 \text{ and } 1.0$$

for the values of N = 20, 50, 100, 500 and 1000. With the values of W thus obtained, justify the statement that:

Of all configuration in a set, the 1 : 1: : H : T configuration is the predominant one.

(b) Calculate the relative ratio (known as A-ratio) of other configurations in a given set (i.e. with different values of H/N) with respect to 1 : 1 :: H : T configuration and then plot a graph between the A-ratio and the corresponding H/N ratio of the configuration. With the help of the graph, justify the following statement.

Out of very large number of microstates that can be assumed by a large assembly, majority of them arise from one comparatively small set of configurations centred on, and very similar to the predominant configuration.

(c) According to statistical mechanics, the equilibrium state of a system is characterized by the predominant configuration and the latter is always associated with the maximum number of microstates. On the other hand, according to the classical thermodynamics the equilibrium state is characterized by its maximum entropy. Thus, there exists a link between the maximum entropy and the maximum number of microstates when the system is in equilibrium. Show that this relation is given by

$$S = k \ln W$$

where k is Boltzmann's constant.

4.23 Using the equation $S = k \ln W$, how would calculate the change in entropy of a gas when its volume is changed from V_1 to V_2 at a constant temperature. Replace your gaseous system with a mathematical model of distribution of balls (equal to the number of molecules) over a given number of boxes (number being proportional to the volume of the gas).

4.24 What is the residual entropy? Show how the relation $S = k \ln W$ could explain the third law of thermodynamics and the concept of residual entropy.

4.25 (a) Show that if $(\partial U/\partial V)_T = 0$, the equation of state of the substance must be of the form $p = T f(V)$.

(b) Show that if $(\partial H/\partial p)_T = 0$, the equation of state of the substance must be of the form $V = T f(p)$.

(c) If for a substance both $(\partial U/\partial V)_T = 0$ and $(\partial H/\partial p)_T = 0$, then equation of state is pV/T = constant.

TRY YOURSELF PROBLEMS

4.1 (a) Which causes the greater increase in efficiency of a reversible engine—raising the source temperature by ΔT or lowering refrigerator temperature by ΔT?

(b) If the rate of heat received by a nonperfectly insulated body at a lower temperature (T_1) from its hotter surroundings (T_2) is proportional to $T_2 - T_1$ (Newton's law), then show that the rate at which work must be done to sustain the temperature difference $T_2 - T_1$ is proportional to $(T_2 - T_1)^2/T_1$.

[*Hint:* $\mathrm{d}w/\mathrm{d}t = \mathrm{d}\,(q_c/\eta')/\mathrm{d}t = (1/\eta')\,\mathrm{d}q_c/\mathrm{d}t =$
$\{(T_2 - T_1)/\,T_1\}\,\{A(T_2 - T_1)\} \propto (T_2 - T_1)^2/T_1$]

4.2 A system is taken reversibly around the cycle ABCD shown in the diagram. The temperature on the Kelvin scale is indicated at each corner. The pressure and volume of the system are as indicated on the scale of the diagram. The heat capacities of the system are $C_{p,\mathrm{m}} = 8.381\ \mathrm{J\ K^{-1}\ mol^{-1}}$ and $C_{V,\mathrm{m}} = 5.048\ \mathrm{J\ K^{-1}\ mol^{-1}}$.

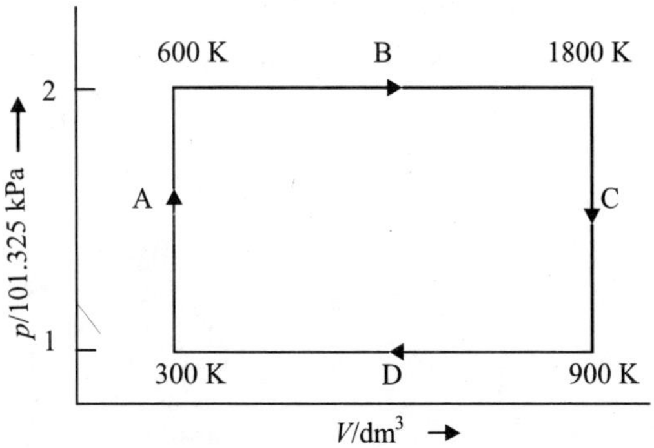

(a) Calculate the value of $\int \mathrm{d}q_{\mathrm{rev}}$ along each part of the cycle. The sum of these is $\oint \mathrm{d}q_{\mathrm{rev}}$. What is its significance according to the first law?

(b) Calculate the value of $\int \mathrm{d}(q_{\mathrm{rev}}/T)$ along each part of the cycle. The sum of these is $\oint \mathrm{d}(q_{\mathrm{rev}}/T)$. What is its significance according to the second law?

(c) If °C were converted to Kelvin by adding a conversion constant other than 273.15 K what would happen to the value of the integral $\oint \mathrm{d}(q_{\mathrm{rev}}/T)$? Discuss the significance of this according to the second law and the definition of an absolute temperature scale.

[*Ans.* (a) $q = +\ 1\,999.8$ J; $w = -\ 1\,999.8$ J, (b) zero, (c) nonzero]

4.3 State whether entropy of system increases, decreases or remains the same in the following changes of state.

(a) One gram of water evaporates into steam at the same p and T.

(b) One gram of water freezes to ice at the same p and T.

(c) One mole of a gas (not necessarily ideal) is compressed reversibly and adiabatically.

(d) One mole of gas (not necessarily ideal) is compressed reversibly and isothermally.

(e) One mole of a gas (not necessarily ideal) expands irreversibly and adiabatically into a vacuum chamber and thereby increases its volume.

4.4 Derive the relations

$$C_p = \alpha VT\left(\frac{\partial p}{\partial T}\right)_S; \qquad \left(\frac{\partial p}{\partial V}\right)_S = \frac{C_p}{C_V}\left(\frac{\partial p}{\partial V}\right)_T$$

$$\left(\frac{\partial S}{\partial p}\right)_T = -\alpha V; \qquad \left(\frac{\partial S}{\partial V}\right)_p = \alpha/\kappa_T$$

4.5 Given the equation of state $pV_m = RT + Bp$, where $B = f(T)$, show that

$$\left(\frac{\partial U}{\partial V_m}\right)_T = \frac{RT^2}{(V_m - B)^2}\frac{dB}{dT}$$

Also obtain expressions for $(\partial S/\partial V_m)_T$, $(\partial S/\partial p)_T$ and $(\partial H/\partial p)_T$.

4.6 A system is changed from a single initial equilibrium state to the same final equilibrium state by two different processes, one reversible and one irreversible. Which of the following is true, (ΔS refers to the system)?

$$\Delta S_{irr} = \Delta S_{rev}; \quad \Delta S_{irr} > \Delta S_{rev} \quad \text{and} \quad \Delta S_{irr} < \Delta S_{rev}$$

4.7 Show that the expression $(\partial p/\partial V)_T\,(\partial T/\partial p)_S\,(\partial S/\partial T)_p$ is equivalent to $-(\partial p/\partial T)_V$.

4.8 When an ideal gas is allowed to expand isothermally in a piston, $\Delta U = q + w = 0$. Thus, the work done by the system is equal to the heat transferred from the reservoir to the gas and the efficiency of turning heat into work is 100 per cent. Explain, why this is not a violation of the second law.

4.9 Derive the expression for thermal efficiency of a reversible Carnot engine using CO_2 (obeying van der Waals equation of state) as a working substance. Compare it with that of a Carnot engine involving ideal gas as the working substance. If the two are not equal, what will happen?

4.10 For an adiabatic process $dq = 0$. If one were to write

$$\Delta S = \int \frac{dq}{T} = \int \frac{0}{T} = 0$$

then every adiabatic process would be an isentropic process. Comment upon this statement.

4.11 A reversible cycle consists of the following stages:

(i) An expansion at constant pressure p_2
(ii) An adiabatic expansion to a lower pressure p_1
(iii) A compression at constant pressure p_1
(iv) An adiabatic compression which restores the system to its initial state.

Draw the indicator ($p - V$) diagram for the cycle and prove that with an ideal gas as the working substance, the efficiency is given by

$$\frac{(-w)}{q} = 1 - \left(\frac{p_1}{p_2}\right)^{(\gamma-1)/\gamma}$$

where $\gamma = C_p/C_V$ and q is the heat taken up in stage (i).

4.12 Prove for any homogeneous substance, that

$$\left(\frac{\partial U}{\partial p}\right)_T = -T\left(\frac{\partial V}{\partial T}\right)_p - p\left(\frac{\partial V}{\partial p}\right)_T$$

and that

$$\left(\frac{\partial H}{\partial V}\right)_T = T\left(\frac{\partial p}{\partial T}\right)_V + V\left(\frac{\partial p}{\partial T}\right)_T$$

4.13 Will ΔS be positive or negative in the following processes? Discuss qualitatively.

(a) $H_2O(s) \rightarrow H_2O(l)$; (b) $H_2O(l) \rightarrow H_2O(g)$

(b) $H_2(g) + Cl_2(g) \rightarrow 2HCl(g)$ (d) $2H_2(g) + N_2(g) \rightarrow N_2H_4(l)$

(c) $O_2(g) + N_2(g) \rightarrow$ Mixture of N_2 and O_2 (f) $Cl_2(g) \rightarrow 2Cl(g)$

(g) Dissolution of a solute in a solvent

(h) (1/2) $N_2(g)$ + (3/2) $H_2(g) \rightarrow NH_3(g)$

4.14 Rank the following in order of increasing entropy.

(a) 1 mole of $H_2O(l)$ at 298 K and 0.101 MPa

(b) 1 mole of $H_2O(s)$ at 273 K and 0.101 MPa

(c) 1 mole of $H_2O(g)$ at 373 K and 50.65 kPa

(d) 1 mole of $H_2O(g)$ at 373 K and 101.3 kPa

(e) 1 mole of $H_2O(l)$ at 273 K and 101.3 kPa

(f) 1 mole of $H_2O(l)$ at 373 K and 101.3 kPa

4.15 Consulting the table of standard entropy, calculate the entropy change for each of the following reactions:

$$N_2(g) + 3H_2(g) \rightarrow 2NH_3(g)$$

$$N_2O_4(g) \rightarrow 2NO_2(g)$$

$$H_2(g) + CO_2(g) \rightarrow H_2O(g) + CO(g)$$

4.16 Show that

$$\left(\frac{\partial U}{\partial V}\right)_T = -\left(\frac{\partial(p/T)}{\partial(1/T)}\right)_V; \qquad \left(\frac{\partial H}{\partial p}\right)_T = \left(\frac{\partial(V/T)}{\partial(1/T)}\right)_p$$

NUMERICAL PROBLEMS

4.1 (a) An ideal heat engine operates between – 40 °C and 40 °C. It absorbs 100 J of heat at higher temperature. Determine the work it delivers and the heat it rejects at lower temperature. (*Ans.* 25.55 J, 74.45 J)

(b) The machine of part (a) is operated as a refrigerator. How much power (in watts $\equiv$ J s^{-1}) must be supplied to the machine so as to maintain a cryostat bath at – 40 °C. The bath absorbs 418.4 J s^{-1} from the surroundings. How much heat it rejects at higher temperature? (*Ans.* 143.59 W, 561.99 J s^{-1})

4.2 Prove that a heat engine with an efficiency of 0.150 working between 25 °C and 100 °C is not undergoing a reversible work cycle.

4.3 Calculate the work obtained in 1000 cycles of a reversible engine operating between 1 000 K and 300 K if the heat absorbed at the higher temperature is 2.1 kJ cycle^{-1}. If this engine operates in the reverse direction, how much of work is needed to transfer 2.1 kJ of heat from the sink (300 K)? (*Ans.* 1.470 MJ, 4.9 MJ)

4.4 A heat engine absorbs 760 kJ of heat from a source at 380 K. It rejects (i) 650 kJ, (ii) 560 kJ and (iii) 504 kJ of heat to the sink at 280 K. State which of these represents a reversible, an irreversible and an impossible cycle.

(*Ans.* (i) irreversible, (iii) reversible (iii) impossible)

Entropy as a Function of Temperature and Volume

4.5 For C_6H_6, $\alpha = 1.24 \times 10^{-3}$ K^{-1} and $\kappa_T = 91.784 \times 10^{-5}$ MPa^{-1} at 20 °C and 0.101 3 MPa pressure. Assuming α and κ_T to be constant, find the change in molar volume which will be required to produce an entropy change of 2.092 J K^{-1} mol^{-1} at 20 °C. (*Ans.* 1.548 cm^3 mol^{-1})

(a) The temperature of an ideal monatomic gas is raised from 25 °C to 250 °C. What must be the accompanying volume change in order that the entropy of the gas may be unaffected by the complete process? (*Ans.* 4.08 fold decrease)

(b) Calculate the change in entropy suffered by 2 mol of a gas when its temperature changes from 300 K to 400 K at constant volume. (*Ans.* 7.18 J K^{-1} mol^{-1})

4.6 Calculate ΔS for 2 mol of NO_2 heated at constant volume from 298 K to 373 K. Given the temperature variation of $C_{p,\,m}$ of NO_2 as $C_{p,\,m}$/J K^{-1} mol^{-1} = 27.78 + 30.85 $\times 10^{-3}$ (T/K). (*Ans.* 13.37 J K^{-1})

4.7 Two moles of an ideal gas is expanded isothermally at 20 °C from a volume V_1 to 2.5 V_1. Determine the value of ΔS_{gas}, ΔS_{surr} and ΔS_{total} under the following conditions.

(i) Expansion is carried out reversibly.

(ii) Expansion is carried out irreversibly and 400 J mol^{-1} of heat is less absorbed than in step (i).

(iii) Expansion is free. (*Ans.* (i) 15.24 J K^{-1}, – 15.24 J K^{-1}, zero
(ii) 15.24 J K^{-1}, – 12.56 J K^{-1}, 2.68 J K^{-1}
(iii) 15.24 J K^{-1}, zero, 15.24 J K^{-1})

Entropy as a Function of Temperature and Pressure

4.8 (a) The temperature of an ideal monatomic gas is raised from 25 °C to 250 °C. What must be the accompanying pressure change in order that the entropy of the gas may be unaffected by the complete process? (*Ans.* 4.08 fold increase)

(b) Calculate the change in entropy suffered by two moles of a gas when its temperature changes from 300 K to 400 K at constant presure. (*Ans.* 11.96 J K^{-1})

4.9 Calculate ΔS for 2 mol of oxygen heated at constant pressure from 298 K to 373 K. Given the temperature variation of $C_{p,\,m}$ of oxygen as

$$C_{p,\,m}/\text{J K}^{-1}\,\text{mol}^{-1} = 26.85 + 8.49 \times 10^{-3}\,(T/\text{K})$$ (*Ans.* 13.33 J K^{-1})

4.10 The standard molar entropy of graphite at 298.15 K is 5.777 J K^{-1} mol^{-1}. Calculate its value at 1 500 K. Given: C_p /J K^{-1} mol^{-1} = –5.293 + 58.610 × 10^{-3} (T/K). (*Ans.* 67.67 J K^{-1} mol^{-1})

4.11 The heat capacity of solid iodine from 25 °C to the melting point (113.6 °C) at 101.325 kPa pressure is given by

$$C_p/\text{J K}^{-1}\,\text{mol}^{-1} = 54.685 + (13.43 \times 10^{-4}\,\text{K}^{-1})\,(T - 298\text{ K})$$

The enthalpy of fusion at the melting point is 15.648 kJ mol^{-1}. The heat capacity of liquid is approximately constant at 81.588 J K^{-1} mol^{-1} and the enthalpy of vaporization at the normal boiling point (184 °C) is 25.522 kJ mol^{-1}. Determine the increase in entropy accompanying the change from one mole of iodine from solid at 25 °C to vapour at 184 °C and 101.325 kPa pressure. (*Ans.* 124.23 J K^{-1} mol^{-1})

4.12 The moar heat capacity of water vapour is given by the equation

$$C_p/\text{J K}^{-1}\,\text{mol}^{-1} = 30.07 + 0.009\,6\,(T/\text{K})$$

and that of liquid water is 75.31 J K^{-1} mol^{-1}. The molar change in enthalpy on evaporation at 373 K and 101.325 kPa pressure is 40.58 kJ K^{-1} mol^{-1}. Calculate ΔS for the change of state

$$3H_2O(1,\ 298\text{ K},\ 101.325\text{ kPa}) \rightarrow 3H_2O\ (g,\ 500\text{ K},\ 101.325\text{ kPa})$$

(*Ans.* 407.25 J K^{-1} mol^{-1})

Entropy Changes in a Real Gas

4.13 Three moles of a gas with $pV = n(RT + Bp)$; where $B = 30\ cm^3\ mol^{-1}$ and for which $C_p\ /\ J\ K^{-1}\ mol^{-1} = 27.196 + 4.812 \times 10^{-3}\ (T\ /K)$
undergoes (in an irreversible process) the change in state

$$(600\ K,\ 1.013\ MPa) \rightarrow (300\ K,\ 0.506\ MPa).$$

Calculate ΔU, ΔH and ΔS. (*Ans.* – 18.93 kJ, – 26.43 kJ, – 43.03 J K^{-1})

4.14 A gas obeys the equation of state $p(V_m - b) = RT$
Calculate the entropy change when one mole of this gas undergoes the change from 1 dm^3, 27 °C to 2 dm^3, 127 °C. Given: $C_{V,m}/J\ K^{-1}\ mol^{-1} = 15.95 + 3.1 \times 10^{-3}(T/K)$ and $b = 0.039\ dm^3\ mol^{-1}$. (*Ans.* 10.83 J K^{-1} mol^{-1})

Reversible and Irreversible Processes

4.15 A system undergoes a process in which the entropy change is 10.08 J K^{-1}. During the process, 4.184 kJ of heat is added to the system from a thermostat at a constant temperature of 500 K. Is the process reversible or irreversible? (*Ans.* Irreversible)

4.16 Compute ΔS for the following process $H_2O(1,\ 253\ K) \rightarrow H_2O(s,\ 253\ K)$

Given: $C_p(H_2O,\ 1) = 75.31\ J\ K^{-1}\ mol^{-1}$; $C_p(H_2O,\ s) = 36.40\ J\ K^{-1}\ mol^{-1}$

$\Delta_{freez}H(273\ K) = -\ 6.008\ kJ\ mol^{-1}$

Is it possible to predict the nature of the process (spontaneous or nonspontaneous) from the obtained value of ΔS? (*Ans.* – 19.03 J K^{-1} mol^{-1}, No)

4.17 (a) At 100 °C, water vapour at 1 bar is in equilibrium with the liquid. Given:

$H_2O(l,\ 1\ bar,\ 373\ K) \rightarrow H_2O\ (g,\ 1\ bar,\ 373\ K)$; $\Delta_r H = 40.639\ kJ\ mol^{-1}$

$C_p(g) = 30.305\ J\ K^{-1}\ mol^{-1}$

$C_p(l) = 75.31\ J\ K^{-1}\ mol^{-1}$

Compute $\Delta_r S$ for the reaction $H_2O(l,\ 1\ bar,\ 273\ K) \rightarrow H_2O\ (g,\ 1\ bar,\ 273\ K)$ and also ΔS_{surr} and ΔS_{total}; hence demonstrate that the reaction is not spontaneous.

(*Ans.* $\Delta_r S_{sys} = 123\ J\ K^{-1}\ mol^{-1}$, $\Delta S_{surr} = -\ 165.35\ J\ K^{-1}\ mol^{-1}$, $\Delta S_{total} = -\ 42.35\ J\ K^{-1}\ mol^{-1}$)

(b) Now consider the reaction $H_2O(1,\ 6.13\ mbar,\ 273\ K) \rightarrow H_2O(g,\ 6.13\ mbar,\ 273\ K)$ Assuming that the entropy change and heat flow when liquid water is compressed is negligible and

$C_p(\text{vapour, 1 bar}) = C_p(\text{vapour, 6 bar})$

Comment on the nature of the reaction.

(*Ans.* $\Delta_r S_{sys} = 165.35\ J\ K^{-1}\ mol^{-1}$, $\Delta S_{surr} = -\ 165.35\ J\ K^{-1}\ mol^{-1}$, $\Delta S_{total} = 0$, reaction is at equilibrium)

4.18 Show that the change

$$2\ mol\ (g,\ 202.65\ kPa,\ 273\ K) \rightarrow 2\ mol\ (g,\ 101.325\ kPa,\ 273\ K)$$

and carried out irreversibly against an external pressure of 101.325 kPa is spontaneous.

(*Ans.* $\Delta S_{sys} = 11.527\ J\ K^{-1}$, $\Delta S_{surr} = -\ 8.314\ J\ K^{-1}$, $\Delta S_{total} = 3.213\ J\ K^{-1}$)

4.19 Take the specific heat capacity of water to be 4.18 J K^{-1} g^{-1} and the enthalpy of fusion of ice to be 333 J g^{-1} in the following problems.

(a) 100 g of water at 100 °C is placed in thermal contact with 1 kg of ice at 0 °C and the system is allowed to come into equilibrium. Calculate the entropy change for the original 100 g of water, for the original 1 kg of ice and for the total system. Verify that the inequality $\Delta S > \int dq/T$ holds for this irreversible process.

(*Ans.* –130.46 J K^{-1}, 153.11 J K^{-1})

(b) Calculate the entropy change for the change in state

$100\ g\ H_2O(373\ K) \rightarrow 100\ g\ H_2O(323\ K)$ (*Ans.* – 60.16 J K^{-1})

4.20 (a) At 100 °C water vapour at 101.325 kPa is in equilibrium with the liquid. Given:

$$H_2O(l, 101.325 \text{ kPa}, 373 \text{ K}) \rightarrow H_2O (g, 101.325 \text{ kPa}, 373 \text{ K})$$

$$\Delta_r H = 40.638 \text{ kJ mol}^{-1}$$

Compute ΔS for the system and the surroundings.

(*Ans.* 108.95 J K^{-1} mol^{-1} and –108.95 J K^{-1} mol^{-1})

(b) Compute ΔS_{sys} for the reaction

$$H_2O(1, 101.325 \text{ kPa}, 273 \text{ K}) \rightarrow H_2O(g, 101.325 \text{ kPa}, 273 \text{ K})$$

Also compute ΔS_{surr} and show that this reaction does not occur spontaneously. Given: C_p(water vapour) = 33.3 J K^{-1} mol^{-1}

C_p(liquid water) = 75.31 J K^{-1} mol^{-1}

(*Ans.* ΔS_{sys} = 122.09 J K^{-1}, ΔS_{surr} = –164.26 J K^{-1}, ΔS_{total} = – 42.17 J K^{-1})

(c) The reaction

$$H_2O(1, p, 273 \text{ K}) \rightarrow H_2O(g, p, 273 \text{ K})$$

represents an equilibrium reaction for some pressure p called the vapour pressure. Assume that the entropy change and heat flow when liquid water is compressed is negligible. Estimate the vapour pressure at 273 K. (*Ans.* p = 634 Pa)

4.21 The normal boiling point of methanol is 65 °C and the $\Delta H°$ for the vaporization, i.e.

$$CH_3OH(1, 101.325 \text{ kPa}, 338 \text{ K}) \rightarrow CH_3OH(g, 101.325 \text{ kPa}, 338 \text{ K}),$$

is 35.229 kJ mol^{-1}

(a) Calculate ΔS of vaporization.

(b) Given:

$C_p(CH_3OH, 1)$ = 100.42 J K^{-1} mol^{-1}

$C_p(CH_3OH, g)$ = 50.21 J K^{-1} mol^{-1}

estimate ΔS and ΔH for the given reaction at 357 K. Does the reaction occur spontaneously?

(c) Compute ΔS for the system and surroundings when the following occurs:

$$CH_3OH(1, 202.65 \text{ kPa}, 357 \text{ K}) \rightarrow CH_3OH(g, 202.65 \text{ kPa}, 357 \text{ K})$$

Are these values consistent with the fact that the vapour pressure of methanol at 357 K is about 202.65 kPa?

(*Ans.* (a) 104.23 J K^{-1}, (b) ΔS = 101.48 J K^{-1}, ΔH = 34.275 kJ, ΔS_{surr} = – 96.01 J K^{-1}, ΔS_{total} = +ve, spontaneous, ΔS_{sys} = 96.00 J K^{-1})

4.22 Given: C_p(ice) = 37.112 J K^{-1} mol^{-1}

$C_p(H_2O, 1)$ = 75.312 J K^{-1} mol^{-1}

$\Delta_r H$ for the reaction

$$H_2O(s, 613.28 \text{ Pa}, 273 \text{ K}) \rightarrow H_2O(1, 613.28 \text{ Pa}, 273 \text{ K})$$

is 6.025 kJ mol^{-1}. Compute the change in entropy when ice at 263 K is heated to form water at 283 K. (*Ans.* 26.16 J K^{-1} mol^{-1})

4.23 (a) A quantity of an ideal gas in an isolated system is expanded isothermally and reversibly at 400 K from a volume V_1 to V_2. During expansion the gas absorbs 836.8 J of heat from the reservoir in contact with it. Find (i) ΔS_{gas}, (ii) $\Delta S_{reservoir}$ and (iii) $\Delta S_{total\ system.}$ (*Ans.* (i) 2.092 J K^{-1}, (ii) –2.092 J K^{-1}, (iii) zero)

(b) If the above process is carried out irreversibly, then an absorption of 418.4 J of heat occurs. Calculate (i) ΔS_{gas}, (ii) $\Delta S_{reservoir}$ and (iii) $\Delta S_{total\ system}$.

(*Ans.* (i) 2.092 J K^{-1}, (ii) –1.046 J K^{-1}, (iii) 1.046 J K^{-1})

(c) Suppose now the isothermal expansion takes place freely (free expansion). What will be the (i) ΔS_{gas}, (ii) $\Delta S_{reservoir}$ and (iii) $\Delta S_{total\ system}$.

(*Ans.* (i) 2.092 J K^{-1}, (ii) zero, (iii) 2.092 J K^{-1})

Residual Entropy

4.24 Pure *ortho*-hydrogen can exist in any of the nine quantum states at the absolute zero. Calculate the entropy of this mixture of nine 'kinds' of *ortho*-hydrogen; each has a mole fraction of 1/9. (*Ans.* 18.268 J K^{-1} mol^{-1})

Third-law Entropy

4.25 One mole of Zn(s) is heated from 300 K to 1 500 K. Calculate the entropy changes from the following data: m. pt. of Zn = 692 K, b. pt. of Zn = 1 180 K, enthalpy of fusion = 7.53 kJ mol^{-1}, enthalpy of vaporization = 115.9 kJ mol^{-1}, specific heat capacity of Zn in solid, liquid and gaseous phases are 0.384, 0.512 and 0.312 J g^{-1} K^{-1} respectively, molar mass of Zn = 65.38 g mol^{-1}.

(*Ans.* 152.86 J K^{-1} mol^{-1})

MISCELLANEOUS NUMERICALS

4.26 A possible equation of state for a liquid is

$$V = V_0[1 + \alpha t - \kappa_T (p - 0.101\ 325\ \text{MPa})]$$

in which V_0 is the molar volume at 0 °C and 0.101 325 MPa, t is temperature in Celsius degree and α and κ_T are constants. Under these conditions, prove that

$$\left(\frac{\partial S}{\partial p}\right)_T = -V_0\,\alpha\,; \qquad \left(\frac{\partial S}{\partial V}\right)_T = \alpha/\kappa_T$$

$$\left(\frac{\partial H}{\partial p}\right)_T = V_0\,[1 - \alpha\,(273\ \text{K}) - \kappa_T\,(p - 0.101325\ \text{MPa})]$$

Calculate ΔS and ΔH when 1 mole of water is compressed at 20 °C from 0.101 3 MPa to 2.532 5 MPa. (For water near 20 °C and 101.325 kPa, $\alpha = 2.1 \times 10^{-4}\ K^{-1}$ and $\kappa_T = 48.36 \times 10^{-6}\ MPa^{-1}$.) (*Ans.* – 43.59 J K^{-1}, 41.25 J)

4.27 Compute the entropy change on transferring of 2 mol of liquid ammonia at a constant pressure of 101.325 kPa from – 40 °C to the gaseous state at 200 °C. Given:

$C_p(NH_3, 1) = 74.894$ J K^{-1} mol^{-1}

$C_p(NH_3, g)/\text{J K}^{-1}\text{ mol}^{-1} = 33.640 + 29.288 \times 10^{-4}\,(T/\text{K}) + 21.338 \times 10^{-5}\,(T/\text{K})^2$

$\Delta_{vap}H(NH_3, 1) = 23.263$ kJ mol^{-1} at 239.7 K (= normal boiling point)

(*Ans.* 280.9 J K^{-1})

4.28 The following data are available for liquid water at 25 °C and atmospheric pressure: $\alpha = 2.56 \times 10^{-4}\ K^{-1}$ and $(\partial\alpha/\partial T)_p = 9.6 \times 10^{-6}\ K^{-2}$, $V = 1\ 003\ cm^3\ g^{-1}$. Determine the value of $(\partial C_p/\partial p)_T$ for water at these conditions.

(*Ans.* – 2.889 × 10^{-3} J K^{-1} g^{-1} MPa^{-1})

4.29 Suppose the amount dq of heat is transferred from a thermal reservoir at T_1 to a thermal reservoir at T_2. Show that

$$dS = \frac{dq}{T_2} - \frac{dq}{T_1}$$

also show that

(i) $T_1 > T_2$ for a spontaneous flow of heat

(ii) $T_1 = T_2$ for the entire system to be at thermal equilibrium.

4.30 A sample of gas (with $\gamma = 4/3$) in a state A(22.4 dm^3, 1 bar, 300 K) is taken to a state C(33.6 dm^3, 2 bar, 900 K) by the following two routes

(i) State A $\rightarrow$ State B(22.4 dm^3, 600 K, 2 bar) $\rightarrow$ State C

(ii) State A $\rightarrow$ State D(33.6 dm^3, 450 K, 1 bar) $\rightarrow$ State C

Calculate the entropy changes in each step.

(*Ans.* (i) 15.525 J K^{-1}, 12.109 J K^{-1})
(ii) 12.109 J K^{-1}, 15.525 J K^{-1})

5 Equilibrium Criteria, *A* and *G* Functions

5.1 CRITERIA FOR EQUILIBRIUM UNDER DIFFERENT CONDITIONS

Criterion of Total Change in Entropy

We have already seen that for all reversible processes, the total change in entropy ($\Delta S_{\text{total}} = \Delta S_{\text{sys}} + \Delta S_{\text{surr}}$) is zero and for all irreversible processes or natural processes, the value of total entropy increases, i.e.

$$\Delta S_{\text{total}} = 0, \text{ for a reversible process}$$

$$\Delta S_{\text{total}} > 0, \text{ for an irreversible process}$$

Thus for an irreversible process, the total entropy goes on increasing and when the equilibrium is reached (where no further process takes place), the entropy attains a maximum value. No further change in entropy is possible at the equilibrium stage. Thus

$$\Delta S_{\text{total}}(\text{equilibrium}) = 0$$

Limitation of Criterion of ΔS_{total}

This criterion of equilibrium is very general and is not much useful because it requires knowledge not only of the possible processes in the system, but also of what the interaction with the surrounding might be. It will be desirable and also more useful to have a criterion which depends on the state of the system only. Several such criterion can be found by synthesizing the first and the second laws of thermodynamics.

Basic Expression to Establish other Criteria

From the first law, we have

$$\mathrm{d}q = \mathrm{d}U - \mathrm{d}w \tag{5.1.1}$$

Here dw represents all types of work (including type p-V work) that is involved in system undergoing a general process.

The value of dq is related to entropy via second law of thermodynamics, that is

$$\mathrm{d}q_{\text{rev}} = T\,\mathrm{d}S$$

Since $\mathrm{d}q_{\text{irr}} < \mathrm{d}q_{\text{rev}}$

we get $\mathrm{d}q_{\text{irr}} < T\,\mathrm{d}S$

Thus, in general, we have

$$\mathrm{d}q \leq T\,\mathrm{d}S$$

where the 'less than' sign is for an irreversible process and 'equality' sign is for a reversible process. Since we are interested in finding out the condition of equilibrium for a reaction that is in progress (i.e. spontaneous or irreversible reaction), we retain only a 'less than' sign. Thus for a spontaneous reaction, we have

$$dq < T\,dS \quad \text{or} \quad T\,dS > dq \tag{5.1.2}$$

Substituting Eq. (5.1.2) in Eq. (5.1.1), we get

$$T\,dS > dU - dw \tag{5.1.3}$$

Equation (5.1.3) forms the basis for deriving the various criteria of equilibrium for a process that is taking place under different conditions such as under constant temperature and volume, constant temperature and pressure, etc. We derive below a few such criteria of equilibrium.

Criterion when Internal Energy and Volume are Constants

If the system undergoes only *p-V* work, then Eq. (5.1.3) reduces to

$$T\,dS > dU + p_{ext}\,dV$$

or
$$dS > \frac{dU}{T} + \frac{p_{ext}}{T}\,dV \tag{5.1.4}$$

If we impose the conditions that the process under consideration is taking place at constant energy and volume, then according to Eq. (5.1.4), we get

$$dS_{U,V} > 0 \tag{5.1.5}$$

that is, under the conditions of constant U and V of the system, the entropy of the system will increase when a spontaneous process is taking place and its value goes on increasing as long as the reaction proceeds. When the reaction stops, i.e. when equilibrium is reached, it will attain a maximum value and thus $dS_{U,V} = 0$. Thus, *the criterion of equilibrium under the conditions of constant U and V is that S_{system} will attain a maximum value and $dS_{U,V} = 0$.*

A system at constant energy and volume is, however, an isolated system because *p-V* work and energy in any form including heat cannot be exchanged between the system and the surroundings. Though this criterion depends only on the system, yet it is not a practical one.

Criterion when Entropy and Volume are Constants

Equation (5.1.4) can be rewritten as

$$dU < T\,dS - p_{ext}\,dV \tag{5.1.6}$$

If the reaction is taking place under the condition of constant S and V, i.e. $dS = 0$ and $dV = 0$, we have

$$dU_{S,V} < 0 \tag{5.1.7}$$

In this case, a decrease in the energy of the system is observed during the spontaneous process. When the energy attains a minimum value (i.e. no further decrease can be observed), the process ceases and it attains the equilibrium position. Thus, *the criterion for a process to be at equilibrium under the conditions of constant S and V is that the energy function attains a minimum value and thus $dU_{S,V} = 0$.* Again this criterion is not a practical one.

Criterion when Temperature and Volume are Constants

Equation (5.1.3) can be rewritten as

$$dw > dU - T\,dS \tag{5.1.8}$$

The symbols d*w* represents the total work of all types that is involved in the system undergoing an infinitesimal process. If this process is carried out isothermally, we have

$$d(TS) = T\,dS \qquad \text{(constant } T)$$

and Eq. (5.1.8) becomes

$$dw > d(U - TS) \qquad \text{(constant } T) \qquad (5.1.9)$$

Since the expression $(U - TS)$ contains only state functions, it also represents a state function. It is represented by the symbol A (from the German *Arbeit* which means work). Thus

$$A = U - TS \qquad (5.1.10)$$

The function A being a state function, the change in its value while proceeding from state 1 to state 2 does not depend upon the path and is always determined following the reversible path (or paths) connecting the two states. For a reversible process, we will have

$$dw_{max} = dA \qquad \text{(constant } T)$$

or

$$w_{max} = \Delta A \qquad \text{(constant } T) \qquad (5.1.11)$$

In an irreversible process with the same initial and final states, though the work involved is different from w_{max}, yet ΔA has the same value in both cases.

The function A is called the *work function or the work content or the Helmholtz free energy* (or simply *Helmholtz energy* or *Helmholtz function*). The name free energy was given to this function by Helmholtz because the decrease in A (i.e. ΔA = –ve) represents the maximum amount of energy that is free or available for being converted into work (i.e. $w_{max} < 0$). Rewriting Eq. (5.1.9), we have

$$dw > dA \qquad \text{(constant } T)$$

or

$$dA - dw < 0 \qquad \text{(constant } T) \qquad (5.1.12)$$

If we assume that the system can do only p-V work and that it is subjected to constant volume condition, then it is obvious that

$$dw = 0$$

If the temperature of the system is also kept constant then from Eq. (5.1.12), it follows that

$$dA_{T,\,V} < 0 \qquad (5.1.13)$$

that is, *for a spontaneous process, occurring at constant T and V, the value of the function A for the system will decrease and it will continue to decrease as long as the process proceeds*. Since $A = U - TS$, therefore $dA = dU - T\,dS$. Thus, it follows that the Helmholtz free energy decreases as a result of a decrease in energy (i.e. dU = –ve) and increase in the entropy (i.e. dS = +ve) of the system. When equilibrium is reached, it will attain a minimum value. Thus, at equilibrium $dA_{T,V} = 0$ for any process.

Criterion when Temperature and Pressure are Constants

We now develop the criterion for equilibrium appropriate for the conditions which are important in chemistry, namely, constancy of temperature and pressure. In order to do this, we define the total work that can be obtained in a process as the sum of the net work (i.e. nonmechanical) and p-V work (i.e. mechanical):

$$\begin{aligned} dw_{total} &= dw_{net} + dw_{p\text{-}V} \\ &= dw_{net} - p_{ext}\,dV \end{aligned} \qquad (5.1.14)$$

Now from Eq. (5.1.8), we have

$$dw > dU - T\,dS$$

i.e. $$dw_{net} - p_{ext}\,dV > dU - T\,dS$$

If $p_{ext} = p_{sys} = p$, we will have

$$dw_{net} > dU + p\,dV - T\,dS$$

Since $H = U + pV$, for a process at constant pressure, we have

$$dH = dU + p\,dV \qquad \text{(constant } p\text{)}$$

Substituting this in the previous expression, we get

$$dw_{net} > dH - T\,dS \qquad \text{(constant } p\text{)}$$

and if the process is also carried out at constant temperature when $d(TS) = T\,dS$, then

$$dw_{net} > dH - d(TS) \qquad \text{(constant } T \text{ and } p\text{)}$$

or $$dw_{net} > d(H - TS) \qquad \text{(constant } T \text{ and } p\text{)} \qquad (5.1.15)$$

We once again define a new thermodynamic function by the equation

$$G = H - TS \qquad (5.1.16)$$

The function G is a state function because it is composed of functions of the state and therefore its differential is exact. This function is called *Gibbs free energy* (or simply *Gibbs energy* or *Gibbs function*). Thus, we have

$$dw_{net} > dG_{T,\,p} \qquad (5.1.17)$$

or $$dG_{T,\,p} < dw_{net}$$

If the spontaneous process is carried out in such a way that no net work is involved, $dw_{net} = 0$, then from the above expression, we get

$$\Delta G_{T,\,p} < 0$$

that is, *for a spontaneous process occurring at constant T and p, there occurs a decrease in free energy of the system and this decrease is continued till the equilibrium is reached. At this state, G attains a minimum value and hence* $dG_{T,\,p} = 0$.

Physical Interpretation of Gibbs Function

For a reversible process, we will have

$$dw_{net} = dG_{T,\,p}$$

that is, the decrease in Gibbs free energy (i.e. $dG_{T,\,p} = -\text{ve}$) for a reversible process is equal to the net amount of nonmechanical work that can be obtained (i.e. $dw_{net} = -\text{ve}$) from the system.

Stability of Equilibrium

If G is the free energy of a state neighbouring the equilibrium state and G_0 is that of the equilibrium state, then

$$dG = G - G_0 \qquad (5.1.18)$$

Regardless of the choice of neighbouring state, we have

If $dG > 0$ The equilibrium is stable

$dG = 0$ The equilibrium is neutral

$dG < 0$ The equilibrium is unstable

Also, for an infinitesimal displacement $\mathrm{d}\xi$ from the equilibrium state, the change in free energy is given by

$$\mathrm{d}G = \left(\frac{\partial G}{\partial \xi}\right)_{\xi = \xi_{\mathrm{eq}}} \mathrm{d}\xi \qquad (5.1.19)$$

Hence, regardless of the value of $\mathrm{d}\xi$, the system will be at equilibrium if $(\partial G/\partial \xi)_{\xi = \xi_{\mathrm{eq}}}$ has a zero value. But to examine the stability of the equilibrium, higher derivatives are needed as described in the following.

Let us assume the existence of a Taylor series for the free energy at the equilibrium state (assuming that there exists no discontinuity in the value of G or any of its derivatives at the quilibrium state). We will have

$$\begin{aligned}\mathrm{d}G &= G - G_0 \\ &= \left(\frac{\partial G}{\partial \xi}\right)_{\xi = \xi_{\mathrm{eq}}} \xi + \frac{1}{2}\left(\frac{\partial^2 G}{\partial \xi^2}\right)_{\xi = \xi_{eq}} \xi^2 + \cdots + \frac{1}{n!}\left(\frac{\partial^n G}{\partial \xi^n}\right)_{\xi = \xi_{\mathrm{eq}}} \xi^n + \cdots\end{aligned} \qquad (5.1.20)$$

At equilibrium $(\partial G/\partial \xi)_{\xi = \xi_{\mathrm{eq}}} = 0$. If $(\partial^2 G/\partial \xi^2)_{\xi = \xi_{\mathrm{eq}}} > 0$, the sign of $\mathrm{d}G$ will be positive, since ξ^2 is always positive. This leads to a stable equilibrium.

If $(\partial^2 G/\partial \xi^2)_{\xi = \xi_{\mathrm{eq}}} = 0$, then there is a need of examining the higher derivatives. A nonzero value of $(\partial^3 G/\partial \xi^3)_{\xi = \xi_{\mathrm{eq}}}$ may result in the negative value of $\mathrm{d}G$ as ξ^3 can have positive or negative value, and thus equilibrium will be unstable.

If $(\partial^3 G/\partial \xi^3)_{\xi = \xi_{\mathrm{eq}}} = 0$, then the equilibrium will be stable if $(\partial^4 G/\partial \xi^4)_{\xi = \xi_{\mathrm{eq}}} > 0$ since ξ^4 always has a positive value. If this derivative is zero, then higher derivatives have to be examined.

For a chemical reaction, $\mathrm{d}G$ is taken to be zero if $(\partial G/\partial \xi)_{\xi = \xi_{\mathrm{eq}}} = 0$ and the stability of equilibrium is rarely studied, because the equilibrium is attained spontaneously, it is necessarily stable. It is for this reason, $\partial G/\partial \xi$ is identified as a thermodynamic driving force toward the equilibrium state.

5.2 RELATION BETWEEN ΔG AND $\Delta S_{\mathrm{total}}$ FOR ISOTHERMAL AND ISOBARIC PROCESSES

The criterion of $\Delta G < 0$ for predicting the spontaneous nature of a process is completely equivalent with that of $\Delta S_{\mathrm{total}} > 0$. This equivalence can be shown as follows.

The free energy change ΔG_{sys} of an isothermal process is given by

$$\Delta G_{\mathrm{sys}} = \Delta H_{\mathrm{sys}} - T\,\Delta S_{\mathrm{sys}} \qquad (5.2.1)$$

where ΔH_{sys} and ΔS_{sys} are the enthalpy and entropy changes of the system, respectively. Under the condition of constant pressure, we have

$$\Delta H_{sys} = q_p \tag{5.2.2}$$

where q_p is the heat transferred between the system and the surrounding. If a given system gains heat than surroundings supply equivalent heat and vice versa. Thus, it is obvious that

$$q_{surr} = -q_p \tag{5.2.3}$$

If it be assumed that the above transfer of heat is done reversibly by surroundings at constant temperature T, then

$$\Delta S_{surr} = \frac{q_{surr}}{T} = -\frac{q_p}{T} = -\frac{\Delta H_{sys}}{T} \tag{5.2.4}$$

or $\quad \Delta H_{sys} = -T\,\Delta S_{surr}$

Substituting this in Eq. (5.2.1), we have

$$\begin{aligned}\Delta G_{sys} &= -T\,\Delta S_{surr} - T\,\Delta S_{sys}\\ &= -T\,(\Delta S_{surr} + \Delta S_{sys})\\ &= -T\,\Delta S_{total}\end{aligned} \tag{5.2.5}$$

Thus, if ΔS_{total} is positive, ΔG_{sys} in negative and vice versa.

Reversible and Irreversible Nature of Isothermal Expansion of an Ideal Gas

The prediction of the nature of the expansion process of an ideal gas from the sign of ΔS_{total} is, in fact, based on the principle of comparing entropy change of the surroundings under the given conditions with that under the reversible conditions. It the entropy fluctuations are the same, the actual process is said to be reversible. If the two values are different, the actual process is irreversible in nature. Let q_{actual} be the amount of heat absorbed (or released) by the system from (or to) its surroundings. It is assumed that the surroundings supply (or gain) this heat reversibly at temperature T, then

$$(\Delta S_{surr})_{actual} = -\frac{q_{actual}}{T} \tag{5.2.6}$$

Now if the same process were carried out reversibly, we would have

$$(\Delta S_{surr})_{rev} = -\Delta S_{sys} = -\frac{q_{rev}}{T} \tag{5.2.7}$$

While determining ΔS_{total}, we, in fact, compare $(\Delta S_{surr})_{actual}$ with $(\Delta S_{surr})_{rev}$ as shown below.

$$\begin{aligned}\Delta S_{total} &= \Delta S_{sys} + \Delta S_{surr}\\ &= -(\Delta S_{surr})_{rev} + (\Delta S_{surr})_{actual}\end{aligned} \tag{5.2.8}$$

$$= \frac{q_{rev}}{T} - \frac{q_{actual}}{T} \tag{5.2.9}$$

Now if the actual process is carried out reversibly, it is obvious that

$$q_{actual} = q_{rev}$$

or $\quad (\Delta S_{surr})_{actual} = (\Delta S_{surr})_{rev}$ (5.2.10)

and thus $\Delta S_{total} = 0$ (5.2.11)

In other words if $\Delta S_{total} = 0$ for an actual expansion process we say that the process is reversible in nature. Now if the actual process is carried out irreversibly (or spontaneously) then

$$q_{actual} < q_{rev} \quad \text{or} \quad -q_{actual} > -q_{rev}$$

or $\quad (\Delta S_{surr})_{actual} > (\Delta S_{surr})_{rev}$

and thus $\Delta S_{total} = \text{positive}$ (5.2.12)

In other words, if $\Delta S_{total} > 0$ for an actual process, we say that the process is irreversible in nature. Now since q_{irr} can never be greater than q_{rev} for any irreversible process, it is obvious that ΔS_{total} can never be negative. In other words, if we get ΔS_{total} negative for a process, we say that the process can never occur in actual practice. In fact, the process will have a tendency to proceed in the reverse direction.

Coming to the criterion of G, we write

$$\Delta G = \Delta H - T\,\Delta S \tag{5.2.13}$$

where ΔH and ΔS are respectively the enthalpy and entropy changes of the system. Now ΔS of the system is always to be calculated following the reversible path connecting the initial and final states of the system. Thus, for a process at constant T, we have

$$\Delta S_{sys} = -(\Delta S_{surr})_{rev} = \frac{q_{rev}}{T} \tag{5.2.14}$$

For an ideal gas

$$\Delta H_{sys} = 0 \tag{5.2.15}$$

Equation (5.2.13) therefore becomes

$$\Delta G = -T\,\Delta S_{sys} = T(\Delta S_{surr})_{rev} = -q_{rev} \tag{5.2.16}$$

Since q is positive for isothermal expansion of an ideal gas, the sign of ΔG is negative. This implies that the expansion of an ideal gas from a given volume V_1 (or pressure p_1) to a larger volume V_2 (or lesser pressure p_2) is a spontaneous process. Note that the said expansion may be carried out reversibly or irreversibly, but how exactly it is achieved, is not answerable unless we take into account of q actually involved in the expansion process.

5.3 GIBBS FREE-ENERGY CHANGE OF A CHEMICAL EQUATION

Consider a general chemical equation

$$0 = \sum_{B} \nu_B B$$

The amount of species with the progress of reaction is given by the expression

$$n_B = (n_B)_0 + \nu_B \xi$$

where $(n_B)_0$ is the amount of species B in the beginning of the reaction and n_B is the corresponding amount when the reaction has proceeded to the extent ξ.

For the infinitesimal change in the extent of reaction, the change in the amount of species B is

$$dn_B = \nu_B \, d\xi$$

The corresponding Gibbs free-energy change of the reaction is

$$dG = \sum_B G_m(B) dn_B = \sum_B G_m(B)(\nu_B \, d\xi)$$

By definition, Gibbs free energy change of a reaction is given as

$$\Delta_r G = \left(\frac{\partial G}{\partial \xi}\right)_{T,p} = \sum_B \nu_B G_m(B) \tag{5.3.1}$$

that is, *Gibbs free energy change of a reaction is the rate of change of Gibbs free energy with extent of reaction.* It is equal to the free energy change per unit extent of reaction. The unit of $\Delta_r G$ is J mol^{-1}.
The $\Delta_r G$ of a reaction is related to its $\Delta_r H$ and $\Delta_r S$ by the expression

$$\Delta_r G = \Delta_r H - T\,\Delta_r S$$

Exergonic and Endergonic Reactions

The reactions having $\Delta_r G$ negative are known as *exergonic reactions while those having $\Delta_r G$ positive are known as endergonic reactions.*

Driving Forces of a Reaction

For a reaction to be feasible, its $\Delta_r G$ should be negative, i.e.

$$\Delta_r G = \Delta_r H - T\,\Delta_r S < 0$$

This implies that decrease in $\Delta_r H$ and increase in $\Delta_r S$ favour a chemical reaction. The latter condition becomes more important at higher temperatures.

We can distinguish the following four cases depending upon the values of $\Delta_r H$ and $\Delta_r S$.

When $\Delta_r H < 0$ and $\Delta_r S > 0$

Reactions with these characteristics can be carried out at any temperature, since for this $\Delta_r G$ will be all the time negative. Example is:

$$2H_2O_2(g) \rightarrow 2H_2O(g) + O_2(g)$$

$\Delta_r H^\circ_{298K} = -211.29$ kJ mol^{-1}; $\quad T\,\Delta_r S^\circ_{298K} = 39.33$ kJ mol^{-1}

$\Delta_r G^\circ_{298K} = -250.62$ kJ mol^{-1}

When $\Delta_r H < 0$ and $\Delta_r S < 0$

Reactions can be carried only when $|T\,\Delta_r S| < |\Delta_r H|$, since only then $\Delta_r G$ will have a negative value. The reaction must be strongly exothermic to overcome the handicap of the entropy decrease; at very high temperature $-T\,\Delta_r S$ overcomes $\Delta_r H$ and then the reaction can no longer proceed. Examples are:

(a) $3H_2(g) + N_2(g) \rightarrow 2NH_3(g)$

$\Delta_r H^\circ_{298K} = -92.22$ kJ mol^{-1}; $\quad T\,\Delta_r S^\circ_{298K} = -59.23$ kJ mol^{-1}

$\Delta_r G^\circ_{298K} = -32.99$ kJ mol^{-1}

(b) $H_2(g) + \frac{1}{2}O_2(g) \rightarrow H_2O(l)$

$\Delta_r H^\circ_{298K} = -285.83$ kJ mol^{-1}; $\quad T\,\Delta_r S^\circ_{298K} = -48.68$ kJ mol^{-1}

$\Delta_r G^\circ_{298K} = -237.15$ kJ mol^{-1}

When $\Delta_r H > 0$ and $\Delta_r S > 0$

Reactions can proceed only if $T\,\Delta_r S > \Delta_r H$, i.e. temperature of the reaction should be such that entropy factor $T\,\Delta_r S$ predominates over the $\Delta_r H$ factor. Examples are

(a) $N_2O_4(g) \rightarrow 2NO_2(g)$

$\Delta_r H^\circ_{298K} = 57.20$ kJ mol^{-1}; $T\,\Delta_r S^\circ_{298K} = 52.40$ kJ mol^{-1}

$\Delta_r G^\circ_{298K} = 4.80$ kJ mol^{-1}

Since for a spontaneous reaction, $\Delta_r G$ should be negative, it is obvious that the decomposition of $N_2O_4(g)$ to $NO_2(g)$ will not be spontaneous. However, at a higher temperature $|T\,\Delta S|$ becomes larger then $\Delta_r H$ which makes $\Delta_r G$ negative and thus the reaction becomes spontaneous in nature.†

(b) $Ag(s) + \frac{1}{2}\ Hg_2Cl_2(s) \rightarrow AgCl(s) + Hg(1)$

$\Delta_r H^\circ_{298K} = 5.54$ kJ mol^{-1}; $T\,\Delta_r S^\circ_{298K} = 9.96$ kJ mol^{-1}

$\Delta_r G^\circ_{298K} = -4.42$ kJ mol^{-1}

Another examples of a reaction in which entropy dominates energy in determining the direction of reaction is the hard boiling of an egg. Egg albumen is protein whose structure is maintained in part by an immense number of hydrogen bonds. During the process of boiling, the hydrogen bonds are broken and much of the order of the protein molecules is lost. If the temperature is sufficiently high, the disordering of the system produced by the collapse of the structure of the protein molecules, and expressed as $T\,\Delta S$, is more than enough to counteract the large ΔH required to break all the hydrogen bonds. The result is that ΔG is negative and reaction at or above such a temperature is spontaneous.

When $\Delta_r H > 0$ and $\Delta_r S < 0$

Both the factors make $\Delta_r G$ more and more positive and, therefore, such type of reactions are prohibited at all temperatures. Examples include:

$N_2(g) + 2O_2(g) \rightarrow 2NO_2(g)$

$\Delta_r H^\circ_{298K} = 66.36$ kJ mol^{-1}; $T\,\Delta_r S^\circ_{298K} = -36.29$ kJ mol^{-1}

$\Delta_r G^\circ_{298K} = 102.65$ kJ mol^{-1}

---

† Note: The decomposition reaction of N_2O_4 is, in fact, an equilibrium reaction, i.e.

$N_2O_4 \rightleftharpoons 2NO_2$

The degree of dissociation of N_2O_4 at equilibrium is 0.19, i.e. if we start with pure N_2O_4 19% of it is decomposed into NO_2. The calculated value of $\Delta_r G^\circ_{298K} = 4.80$ kJ mol^{-1} predicts that one mole of N_2O_4 cannot be completely decomposed into NO_2. To determine whether N_2O_4 decomposes or not and if yes, to what extent, the change in the total free energy of the system with extent of reaction (i.e. $\partial G_{total}/\partial\xi$) must be evaluated at various stages of decomposition (*see* Section 7.13). If the quantity comes out to be negative, it implies that $d\xi$ amount of N_2O_4 can decompose into NO_2. A positive value implies that the decomposition of amount $d\xi$ of N_2O_4 to NO_2 is not possible. The decomposition reaction continues till $(\partial G_{total}/\partial\xi)$ becomes zero.

These four cases are tabulated below.

$\Delta_r H$	$\Delta_r S$	$\Delta_r G$	Comment
–	+	–	Always spontaneous
+	–	+	Never spontaneous
–	–	?	$\Delta_r G$ depends upon
+	+	?	conditions

A Few Comments Regarding Free Energy Change of a Reaction

1. It may be pointed out that thermodynamics cannot predict the rate of chemical reactions. What it can tell is whether a given reaction is feasible or not. If it is, and if its rate is very slow, than a suitable catalyst can be chosen which can accelerate the reaction. The value of $\Delta_r G$ for the reaction is independent of the catalyst used. No catalyst can be used for a thermodynamically prohibited reaction.

2. We know that at equilibrium $\Delta_r G = 0$, which gives rise to $\Delta_r H = T\,\Delta_r S$ and hence $T_{eq} = \Delta_r H/\Delta_r S$. If both $\Delta_r H$ and $\Delta_r S$ are positive, then the tendency towards greater randomness measured by $\Delta_r S$ and weighed by multiplication with T, is just sufficient of compensate for the endothermic character of the reaction. If both are negative, the exothermic character of the reaction just compensates for the decrease in randomness again weighted by T.

3. Since $\Delta_r G = \Delta_r H - T\,\Delta_r S$, it follows that at low temperatures enthalpy changes dominate the $\Delta_r G$ expression whereas at high temperature it is the entropy changes which dominate the value of $\Delta_r G$. Take, for example, the reaction

$$H_2O(l) \rightarrow H_2O(g)$$

A change in the value of $\Delta_r H$ with temperature of this reactions follows the equation

$$\Delta_r H_{T_2} - \Delta_r H_{T_1} = (\Delta_r C_p)\,\Delta T = \{C_{p,m}(g) - C_{p,m}(l)\}\{T_2 - T_1\}$$

Since $C_{p,m}(1) > C_{p,m}(g)$, it follows that $\Delta_r H$ of this reaction decreases with increase in temperature. The value of $T\,\Delta_r S$ will increase with increase in temperature since the reaction is attended to by an increase in entropy. Thus

At lower temperatures, $\Delta_{vap} H > T\,\Delta_{vap} S$ and $\Delta_{vap} G = +ve$

At higher temperatures, $\Delta_{vap} H < T\,\Delta_{vap} S$ and $\Delta_{vap} G = -ve$

Computation of Boiling Point of a Liquid In between when $\Delta_{vap} H = T\,\Delta_{vap} S$ and $\Delta_{vap} G = 0$, the system will be at equilibrium. The temperature at which this exist, is the boiling point of the liquid. It can be computed from the expression $T_b = \Delta_{vap} H/\Delta_{vap} S$. At this point, heat given to the system is utilized only in increasing the entropy of the system at constant temperature.

Example 5.3.1

Sulphur exists in more than one solid form. The stable form at room temperature is rhombic sulphur. But above room temperature the following reaction occurs:

$$S(\text{rhombic}) \rightarrow S(\text{monoclinic})$$

Thermodynamic measurements reveal that at 101.325 kPa and 298 K, $\Delta_r H = 276.144$ J mol^{-1} and $\Delta_r G = 75.312$ J mol^{-1}.

(a) Compute $\Delta_r S$ at 298 K.

(b) Assume the $\Delta_r H$ and $\Delta_r S$ do not vary significantly with temperature, compute T_{eq}, the temperature at which rhombic and monoclinic sulphur exist in equilibrium with each other.

Solution

(a) Since $\Delta_r G = \Delta_r H - T\,\Delta_r S$, therefore

$$\Delta_r S = \frac{\Delta_r H - \Delta_r G}{T} = \frac{276.144 \text{ J mol}^{-1} - 75.312 \text{ J mol}^{-1}}{298 \text{ K}}$$

$$= 0.674 \text{ J K}^{-1} \text{ mol}^{-1}$$

(b) When the rhombic sulphur is in equilibrium with monoclinic sulphur, we would have

$$\Delta G = 0 = \Delta_r H - T_{eq}\Delta_r S$$

Thus $$T_{eq} = \frac{\Delta_r H}{\Delta_r S} = \frac{276.144 \text{ J mol}^{-1}}{0.674 \text{ J K}^{-1} \text{mol}^{-1}} = 409.7 \text{ K}$$

---

4. Since G is a state function, therefore, ΔG value for any given process does not depend upon the path. Values of $\Delta_r G$ for different chemical reactions may be combined in the same way as the chemical equations of the reactions considered may be combined into the equation of a new reaction. The following two examples illustrate the procedure.

(i) Consider the reaction

$$S\text{(rhombic)} + \tfrac{3}{2}O_2\,(g) \rightarrow SO_3\,(g)$$

for which $\Delta_r G°$ is required. Its value can be determined from the following two reactions:

$$S\text{(rhombic)} + O_2(g) \rightarrow SO_2(g) \qquad \Delta_r G_1^\circ = -300.19 \text{ kJ mol}^{-1}$$

$$SO_2(g) + \tfrac{1}{2}O_2\,(g) \rightarrow SO_3(g) \qquad \Delta_r G_2^\circ = -70.87 \text{ kJ mol}^{-1}$$

Adding these two reactions, we get

$$S\text{(rhombic)} + \tfrac{3}{2}O_2\,(g) \rightarrow SO_3(g)$$

$$\Delta_r G° = \Delta_r G_1^\circ + \Delta_r G_2^\circ = -371.06 \text{ kJ mol}^{-1}$$

(ii) Suppose $\Delta_r G°$ for the following reaction is required.

$$8H^+ + MnO_4^- + 5Fe^{2+} \rightarrow 5Fe^{3+} + Mn^{2+} + 4H_2O(1)$$

This can be done by combining the following two reactions:

$$16H^+ + 2MnO_4^- + 10Cl^- \rightarrow 5Cl_2(g) + 2Mn^{2+} + 8H_2O(1)$$

$$\Delta_r G_1^\circ = -182.81 \text{ kJ mol}^{-1}$$

$$Cl_2(g) + 2Fe^{2+} \rightarrow 2Fe^{3+} + 2Cl^- \qquad \Delta_r G_2^\circ = -113.55 \text{ kJ mol}^{-1}$$

Multiplying the first reaction by 1/2 and the second reaction by 5/2 and then adding the two, we get the required equation, $\Delta_r G°$ for which will be:

$$\Delta_r G° = \frac{1}{2}\Delta_r G_1^\circ + \frac{5}{2}\Delta_r G_2^\circ = \left[\left(\frac{1}{2}\right)(-182.81) + \left(\frac{5}{2}\right)(-113.55)\right] \text{kJ mol}^{-1}$$

$$= -375.13 \text{ kJ mol}^{-1}$$

The above procedure of computing the free energy change of a coupled reaction finds its practical utility in the human body where endergonic synthesis of ATP (adenosine triphosphate) from ADP (adenosine diphosphate) and P_i (inorganic phosphate) is carried out by coupling it with the exergonic combustion of glucose.

$$C_6H_{12}O_6 + 6O_2 \rightarrow 6CO_2 + 6H_2O \qquad \Delta_r G = -2\,870 \text{ kJ mol}^{-1}$$

$$36ADP + 36P_i \rightarrow 36ATP + 36H_2O \qquad \Delta_r G = 1\,205.0 \text{ kJ mol}^{-1}$$

$$C_6H_{12}O_6 + 6O_2 + 36ADP + 36P_i \rightarrow 36ATP + 6CO_2 + 42H_2O \qquad \Delta_r G = -1\,665 \text{ kJ mol}^{-1}$$

The energy that is not trapped in ATP synthesis is used in maintaining normal body temperature.

Example 5.3.2 Calculate $\Delta_r G°$ for the reaction

$$C(\text{graphite}) + 2H_2(g) + N_2(g) + \frac{1}{2}O_2(g) \rightarrow CO(NH_2)_2$$

from the following data

(i) $CO_2(g) + 2NH_3(g) \rightarrow H_2O(g) + CO(NH_2)_2;\quad \Delta_r G_1^\circ = 1.91 \text{ kJ mol}^{-1}$

(ii) $H_2O(g) \rightarrow H_2\,(g) + \frac{1}{2}O_2\,(g);\quad \Delta_r G_2^\circ = 227.44 \text{ kJ mol}^{-1}$

(iii) $C(\text{graphite}) + O_2(g) \rightarrow CO_2(g);\quad \Delta_r G_3^\circ = -394.38 \text{ kJ mol}^{-1}$

(iv) $N_2(g) + 3H_2(g) \rightarrow 2NH_3(g);\quad \Delta_r G_4^\circ = -32.43 \text{ kJ mol}^{-1}$

Solution The given reaction is

$$C(\text{graphite}) + 2H_2(g) + N_2(g) + \frac{1}{2}O_2(g) \rightarrow CO(NH_2)_2$$

The above reaction can be obtained by adding the given reactions. Thus, we have

$$\Delta_r G° = \Delta_r G_1^\circ + \Delta_r G_2^\circ + \Delta_r G_3^\circ + \Delta_r G_4^\circ = -197.46 \text{ kJ mol}^{-1}$$

Computation of Free Energy Change of a Reaction

There are two methods for computing the free energy change of a reaction.

Using the Data on Enthalpies of Formation and Third-Law Entropies

Suppose the free energy change of a reaction

$$2N_2O_5(g) \rightarrow 4NO_2(g) + O_2(g)$$

is required. The required data for the computation are as follows.

	$2N_2O_5(g)$	$\rightarrow 4NO_2(g)$	$+ O_2(g)$
$\Delta_f H°/\text{kJ mol}^{-1}$	11.3	33.18	0
$S°/\text{J K}^{-1}\text{ mol}^{-1}$	355.7	240.06	205.14

Now the enthalpy of reaction and entropy of reaction are

$$\begin{aligned}\Delta_r H° &= \sum \nu_B\, \Delta_f H°(B)\\ &= 4\,\Delta_f H°(NO_2, g) + \Delta_f H°(O_2, g) - 2\,\Delta_f H°(N_2O_5, g)\\ &= [4 \times 33.18 + 0 - 2 \times 11.3] \text{ kJ mol}^{-1}\\ &= 110.12 \text{ kJ mol}^{-1}\end{aligned}$$

$$\Delta_r S^\circ = \sum \nu_B S^\circ_m(B)$$
$$= 4\, S^\circ_m(NO_2, g) + S^\circ_m(O_2, g) - 2\, S^\circ_m(N_2O_5, g)$$
$$= [4 \times 240.06 + 205.14 - 2 \times 355.7]\ \text{J K}^{-1}\ \text{mol}^{-1}$$
$$= 453.98\ \text{J K}^{-1}\ \text{mol}^{-1}$$

Finally, at 298 K, we have

$$\Delta_r G^\circ = \Delta_r H^\circ - T\,\Delta_r S^\circ$$
$$= 110.12\ \text{kJ mol}^{-1} - (298\ \text{K})\,(453.98 \times 10^{-3}\ \text{kJ K}^{-1}\ \text{mol}^{-1})$$
$$= -\,25.17\ \text{kJ mol}^{-1}$$

Example 5.3.3

It has been suggested that α-cyanopyridine might be prepared from cyanogen and butadiene by the reaction

$$C_4H_6(g) + C_2N_2(g) \rightarrow C_6H_4N_2(s) + H_2(g)$$

In view of the thermodynamic data given below, would you consider it worthwhile to attempt to work out this reaction?

Molecule	$\Delta_f H^\circ$(298 K) / kJ mol^{-1}	S°(298 K) / J K^{-1} mol^{-1}
Butadiene(g)	111.9	277.9
Cyanogen(g)	300.5	241.2
α-Cyanopyridine(s)	225.2	322.5
H_2(g)	0.0	103.6

Solution

We have

$$\Delta_r H^\circ = \Delta_f H^\circ(\alpha\text{-cyanopyridine}) - \Delta_f H^\circ(\text{butadiene}) - \Delta_f H^\circ(\text{cyanogen})$$
$$= (225.2 - 111.9 - 300.5)\ \text{kJ mol}^{-1}$$
$$= -187.2\ \text{kJ mol}^{-1}$$
$$\Delta_r S^\circ = S^\circ(\alpha\text{-cyanopyridine}) + S^\circ(H_2) - S^\circ(\text{butadiene}) - S^\circ(\text{cyanogen})$$
$$= (322.5 + 103.6 - 277.9 - 241.2)\ \text{J K}^{-1}\ \text{mol}^{-1}$$
$$= -93\ \text{J K}^{-1}\ \text{mol}^{-1}$$
$$\Delta_r G^\circ = \Delta_r H^\circ - T\,\Delta_r S^\circ$$
$$= -187.2\ \text{kJ mol}^{-1} - (298\ \text{K})\,(-93 \times 10^{-3}\ \text{kJ K}^{-1}\ \text{mol}^{-1})$$
$$= -159.486\ \text{kJ mol}^{-1}$$

Since ΔG° is negative, α-cyanopyridine can be prepared starting from cyanogen and butadiene.

Using the Data on Free Energy of Formation

Similar to the definition of enthalpy of formation and its use in the determination of enthalpy of reaction, one can define the free energy of formation of a substance and from these values the value of free energy of a reaction can be determined.

By definition, *the standard free energy of formation of a compound is the change in the standard free energy when one mole of the compound is formed starting from the requisite amounts of elements in their stable states of aggregation.*

By convention, the standard free energy of formation of every element in its stable state of aggregation at one bar pressure and at specified temperature is assigned a zero value. The specified temperature is usually taken as 25 °C.

The standard free energy of formation of a few substances are recorded in Appendix I. Using these values, one can calculate the value of $\Delta_r G°$ of a reaction using the expression

$$\Delta_r G° = \sum_B \nu_B \, \Delta_f G°(B)$$

i.e.,
$$\Delta_r G° = \sum_{\text{products}} \nu_B \, \Delta_f G°(B) - \sum_{\text{reactants}} |\nu_B| \, \Delta_f G°(B)$$

The following two examples illustrate the procedure.

(i) $4NH_3(g) + 5O_2(g) \rightarrow 4NO(g) + 6H_2O(1)$

The values of standard free energy of formation of the involved species are

	$4NH_3(g)$	$+\ 5O_2(g)$	$\rightarrow 4NO(g)$	$+\ 6H_2O(1)$
$\Delta_f G°/\text{kJ mol}^{-1}$	−16.45	0	86.57	−237.13

Thus,

$$\Delta_r G°_{298K} = \sum_B \nu_B \, \Delta_f G°(B)$$
$$= 4\Delta_f G°(NO, g) + 6\Delta_f G°(H_2O, 1) - 4\Delta_f G°(NH_3, g) - 5\Delta_f G°(O_2, g)$$
$$= \{4 \times 86.57 + 6\,(-237.13) - 4\,(-16.45) - 0\} \text{ kJ mol}^{-1}$$
$$= -1\,010.7 \text{ kJ mol}^{-1}$$

(ii) Oxidation of Cl^- by $MnO_2(s)$.

The reaction along with the appropriate values of $\Delta_f G°$ is

	$4H^+$	$+\ MnO_2(s)$	$+\ 2Cl^-$	$\rightarrow Mn^{2+}$	$+\ Cl_2(g)$	$+\ 2H_2O(l)$
$\Delta_f G°/\text{kJ mol}^{-1}$	0	− 464.84	− 131.17	− 223.43	0	− 237.19

Thus
$$\Delta_r G° = \Delta_f G°(Mn^{2+}) + 2\Delta_f G°(H_2O) - \Delta_f G°(MnO_2) - 2\Delta_f G°(Cl^-)$$
$$= [-223.43 + 2 \times (-237.19) - (-464.84) - 2(-131.17)] \text{ kJ mol}^{-1}$$
$$= 29.37 \text{ kJ mol}^{-1}$$

Example 5.3.4 Compute $\Delta_r H°$, $\Delta_r S°$, $T\Delta_r S°$, $\Delta_r G°$ and the equilibrium vapour pressure of the transformation

$$H_2O(l) \rightarrow H_2O(g)$$

at temperatures 323.15 K, 348.15 K, 373.15 K, 398.15 K and 423.15 K. Given:

$$\Delta_f H°(H_2O, 1, 298.15\text{ K}) = -285.83 \text{ kJ mol}^{-1}$$
$$\Delta_f H°(H_2O, g, 298.15\text{ K}) = -241.82 \text{ kJ mol}^{-1}$$
$$S°(H_2O, l, 298.15\text{ K}) = 69.91 \text{ J K}^{-1} \text{ mol}^{-1}$$
$$S°(H_2O, g, 298.15\text{ K}) = 188.83 \text{ J K}^{-1} \text{ mol}^{-1}$$
$$C_p(H_2O, l) = 75.29 \text{ J K}^{-1} \text{ mol}^{-1}$$
$$C_p(H_2O, g) = 33.58 \text{ J K}^{-1} \text{ mol}^{-1}$$

Solution Assume C_p's to be independent of temperature.

We have

$$\Delta_r H°_{298.15K} = \Delta_f H°(H_2O, g) - \Delta_f H°(H_2O, 1)$$
$$= [-241.82 - (-285.83)] \text{ kJ mol}^{-1} = 44.01 \text{ kJ mol}^{-1}$$
$$\Delta_r S°_{298.15K} = S°(H_2O, g) - S°(H_2O, 1)$$
$$= (188.83 - 69.91) \text{ J K}^{-1} \text{ mol}^{-1} = 118.92 \text{ J K}^{-1} \text{ mol}^{-1}$$

$$\Delta_r C_p = C_p(H_2O, g) - C_p(H_2O, 1)$$
$$= (33.58 - 75.29)\ J\ K^{-1}\ mol^{-1} = -41.71\ J\ K^{-1}\ mol^{-1}$$

The values of $\Delta_r H_T^\circ$, $\Delta_r S_T^\circ$ and $\Delta_r G_T^\circ$ at temperature T can be obtained by using the expressions

$$\Delta_r H_T^\circ = \Delta_r H_{298.15K}^\circ + \Delta_r C_p\ (T - 298.15\ K)$$

$$\Delta_r S_T^\circ = \Delta_r S_{298.15K}^\circ + \Delta_r C_p\ \ln\left(\frac{T}{298.15\ K}\right)$$

$$\Delta_r G_T^\circ = \Delta_r H_T^\circ - T\,\Delta_r S_T^\circ$$

For the transformation to be at equilibrium, we have

$$(\Delta_r G_T^\circ)_{eq} = 0$$

or $$(\Delta_r S_T^\circ)_{eq} = \frac{\Delta_r H_T^\circ}{T}$$

For the reaction to be at equilibrium, we must change $\Delta_r S_T^\circ$ to $(\Delta_r S_T^\circ)_{eq}$. If this change is carried out only by changing the vapour pressure of water, we will have

$$(\Delta_r S_T^\circ)_{eq} - \Delta_r S_T^\circ = -R\ln\left(\frac{p_{eq}}{1\ bar}\right) \qquad \text{(Eq. 4.24.8)}$$

where $(\Delta_r S_T^\circ)_{eq}$ and $(\Delta_r S_T^\circ)$ are the entropy changes of the transformation at pressure p_{eq} and 1 bar, respectively.

The calculated values of $\Delta_r H_T^\circ$, $\Delta_r S_T^\circ$, $T\Delta_r S_T^\circ$, $\Delta_r G_T^\circ$, $(\Delta_r S_T^\circ)_{eq}$ and p_{eq} are recorded in Table 5.3.1.

Table 5.3.1 Computing thermodynamic data for $H_2O(1) \rightleftharpoons H_2O(g)$ reaction

T/K	$\frac{\Delta_r H_T^\circ}{kJ\ mol^{-1}}$	$\frac{\Delta_r S_T^\circ}{J\ K^{-1}\ mol^{-1}}$	$\frac{T\,\Delta_r S_T^\circ}{kJ\ mol^{-1}}$	$\frac{\Delta_r G_T^\circ}{kJ\ mol^{-1}}$	$\frac{(\Delta_r G_T^\circ)_{eq}}{J\ K^{-1}\ mol^{-1}}$	$\frac{p_{eq}}{bar}$ †
323.15	42.97	115.56	37.34	5.63	132.97	0.123
348.15	41.93	112.45	39.15	2.78	120.44	0.383
373.15	40.88	109.56	40.88	0	109.56	1.0
398.15	39.84	106.86	42.55	–2.71	100.06	2.265
423.15	38.80	104.32	44.14	–5.34	91.69	4.56

† Thus at 1 bar external pressure, the transformation of liquid water to water vapour is spontaneous provided $T > 373.15$ K. At $T < 373.15$ K, the reverse of the process is spontaneous. In general, the above transformation will be spontaneous at temperature T provided the external pressure (or the pressure of water vapour in equilibrium with liquid water) is smaller than the corresponding equilibrium vapor pressure.

Alternatively, p_{eq} may be determined from the expression $p_{eq} = \exp(-\Delta_r G_T^\circ/RT)$. *See* Section 5.5.

Example 5.3.5 Taking the following typical example, justify the fact that the sulphide ores are generally roasted to oxides for the extraction of metals instead of being reduced directly.

Reaction	$\frac{\Delta_r G}{\text{kJ mol}^{-1}}$	$\frac{\Delta_r H}{\text{kJ mol}^{-1}}$	$\frac{\Delta_r S}{\text{J K}^{-1}\text{ mol}^{-1}}$
$2ZnS(s) + C(s) \rightarrow Zn(s) + CS_2(g)$	469.7	528.6	158.2
$2ZnS(s) + 3O_2(g) \rightarrow 2ZnO(s) + 2SO_2(g)$	−834.4	−879.0	−146.9
$2ZnO(s) + C(s) \rightarrow 2Zn(s) + CO_2(g)$	242.2	303.1	203.8

Solution The reduction of ZnS with C is an endergonic reaction ($\Delta_r G$ = +ve) with positive value of both $\Delta_r H$ and $\Delta_r S$. This reaction will be feasible only when $T\,\Delta_r S$ becomes greater than $\Delta_r H$. Assuming $\Delta_r H$ and $\Delta_r S$ to be independent of temperature, the temperature above which this reaction will be feasible is given as

$$T_{eq} = \frac{\Delta_r H}{\Delta_r S} = \frac{528.6 \times 10^3 \text{ J mol}^{-1}}{158.2 \text{ J K}^{-1} \text{ mol}^{-1}} = 3\,341 \text{ K}$$

Similarly, the reduction of ZnO with C is an endergonic reaction with positive values of $\Delta_r H$ and $\Delta_r S$. However, the temperature at which this reaction becomes feasible is much lower than 3 341 K.

$$T_{eq} = \frac{\Delta_r H}{\Delta_r S} = \frac{303.1 \times 10^3 \text{ J mol}^{-1}}{203.8 \text{ J K}^{-1} \text{ mol}^{-1}} = 1\,487 \text{ K}$$

However, the coupled reaction, i.e. roasting of ZnS and simultaneously reduction, is feasible at all temperature as $\Delta_r G$ is negative with negative value of $\Delta_r H$ and positive value of $\Delta_r S$ as may be seen from the following computation.

	Reaction	$\frac{\Delta_r G}{\text{kJ mol}^{-1}}$	$\frac{\Delta_r H}{\text{kJ mol}^{-1}}$	$\frac{\Delta_r S}{\text{J K}^{-1}\text{ mol}^{-1}}$
	$2ZnS(s) + 3O_2(g) \rightarrow 2ZnO(s) + 2SO_2(g)$	−834.4	−879.0	−146.9
	$2ZnO(s) + C(s) \rightarrow 2Zn(s) + CO_2(g)$	242.2	303.1	203.8
Add	$2ZnS(s) + 3O_2(g) + C(s) \rightarrow 2Zn(s) + CO_2(g) + 2SO_2(g)$	−592.2	−575.9	56.9

5.4 THERMODYNAMIC RELATIONS INVOLVING FUNCTIONS *A* AND *G*

Relations Involving the Function *A*

The Helmholtz free energy is given by

$$A = U - TS$$

Therefore $\mathrm{d}A = \mathrm{d}U - T\,\mathrm{d}S - S\,\mathrm{d}T$

For a system involving only reversible q and p–V work, we have

$$\mathrm{d}U = \mathrm{d}q_{rev} + \mathrm{d}w = T\,\mathrm{d}S - p\,\mathrm{d}V$$

Substituting this in the previous expression, we get

$$\mathrm{d}A = (T\,\mathrm{d}S - p\,\mathrm{d}V) - T\,\mathrm{d}S - S\,\mathrm{d}T$$

$$= -p\,\mathrm{d}V - S\,\mathrm{d}T \qquad (5.4.1)$$

It follows from this that

$$\left(\frac{\partial A}{\partial V}\right)_T = -p \tag{5.4.2}$$

$$\left(\frac{\partial A}{\partial T}\right)_V = -S \tag{5.4.3}$$

$$\left(\frac{\partial S}{\partial V}\right)_T = \left(\frac{\partial p}{\partial T}\right)_V \tag{5.4.4}$$

Equation (5.4.4) is one of the *Maxwell relations* which was derived earlier (Eq. 4.14.12) from the entropy dependence on T and V. Equations (5.4.2) and (5.4.3) describe how the function A varies with the change of V and T, respectively. The negative signs in these two equations imply that the function A decreases with the increase in volume at constant temperature and with the increase in temperature at constant volume.

Relations Involving the Function *G*

The Gibbs free energy is given by

$$G = H - TS$$

Therefore $\quad dG = dH - T\,dS - S\,dT \qquad (5.4.5)$

Since $\quad H = U + pV$

therefore $\quad dH = dU + p\,dV + V\,dp$

For a system involving only reversible q and p–V work, we have

$$dU = T\,dS - p\,dV$$

Therefore $\quad dH = (T\,dS - p\,dV) + p\,dV + V\,dp = T\,dS + V\,dp$

Substituting this in Eq. (5.4.5), we get

$$\begin{aligned} dG &= (T\,dS + V\,dp) - T\,dS - S\,dT \\ &= V\,dp - S\,dT \end{aligned} \tag{5.4.6}$$

It follows from this that

$$\left(\frac{\partial G}{\partial p}\right)_T = V \tag{5.4.7}$$

$$\left(\frac{\partial G}{\partial T}\right)_p = -S \tag{5.4.8}$$

$$\left(\frac{\partial S}{\partial p}\right)_T = -\left(\frac{\partial V}{\partial T}\right)_p \tag{5.4.9}$$

Equation (5.4.9) is another Maxwell relation which was derived earlier (Eq. 4.15.14) from the entropy dependence on T and p. Equations (5.4.7) and (5.4.8) describe how the function G varies with p and T, respectively.

Equation (5.4.7) implies that the increase in pressure increases the free energy at constant temperature. The larger the volume of the system, the greater the increase in free energy for a given increase in pressure. Thus, the free energy of a gas, which has a comparatively large volume, increases much more rapidly with pressure as compared to that of liquid or a solid (Fig. 5.4.1).

The negative sign in Eq. (5.4.8) indicates that the function free energy decreases with the increase in temperature at constant pressure. The rate will be greater for gases which have larger entropies than for liquids or solids (Fig. 5.4.2).

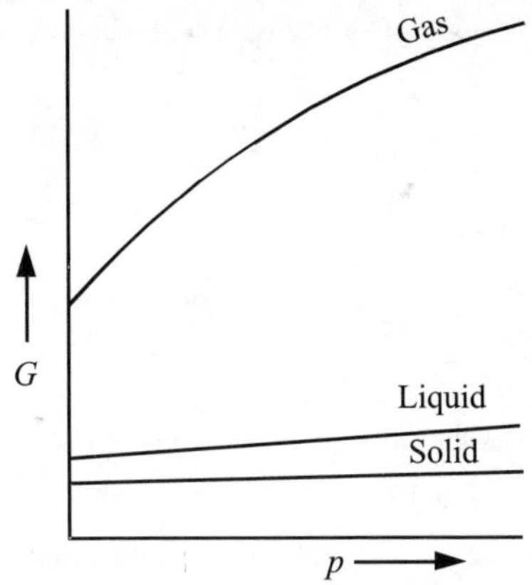

Fig. 5.4.1 Variation of *G* with *T*

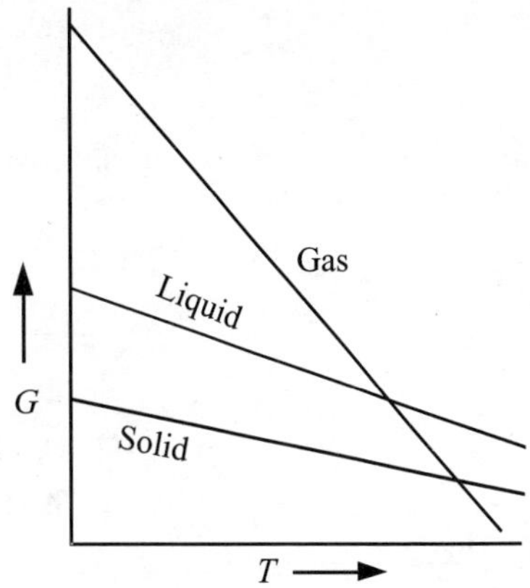

Fig. 5.4.2 Variation of *G* with *T*

Problem 5.4.1 Show that for an ideal gas undergoing isothermal reversible expansion $\Delta G = \Delta A$.

Solution The function G and A are given by

$$G = H - TS \quad \text{and} \quad A = U - TS$$

The function H is given by

$$H = U + pV$$

Substituting this in the function G, we get

$$G = (U + pV) - TS = (U - TS) + pV$$

or $\quad G = A + pV$

Thus, for a process

$$\Delta G = \Delta A + \Delta(pV)$$

Now for an isothermal expansion of an ideal gas, we have

$$\Delta(pV) = \Delta(nRT) = 0 \quad \text{as} \quad \Delta T = 0$$

so that $\quad \Delta G = \Delta A$

5.5 RELATIONSHIP BETWEEN $\Delta_r G^\circ$ AND $\Delta_r A^\circ$

For a chemical reaction

$$\Delta_r G^\circ = \sum_B \nu_B G^\circ_m(B) \quad \text{and} \quad \Delta_r A^\circ = \sum_B \nu_B A^\circ_m(B)$$

Since $\quad G = A + pV, \quad$ we have

$$G^\circ_m = A^\circ_m + pV_m$$

With this, $\Delta_r G^\circ$ becomes

$$\Delta_r G^\circ = \sum_B \nu_B G^\circ_m(B) = \sum_B \nu_B [A^\circ_m(B) + pV_m]$$

Taking $pV_m \simeq 0$ for condensed phase and $pV_m = RT$ for gasesous phases, we have

$$\Delta_r G^\circ = \sum_B \nu_B A^\circ_m(B) + \left(\sum_B \nu_B\right)_g RT$$

$$= \Delta_r A^\circ + (\Delta \nu_g) RT$$

where $\Delta\nu_g$ is the change in the stoichiometric numbers of gaseous species in going from reactants to products.

Example 5.5.1 $\Delta_r G^\circ$ at 298 K for the following reaction is – 3.202 MJ mol^{-1}.

$$C_6H_6(1) + 7\frac{1}{2}O_2(g) \rightarrow 6CO_2(g) + 3H_2O(1)$$

Find $\Delta_r A^\circ$ at 298 K

Solution For this reaction

$$\Delta\nu_g = 6 - 7\frac{1}{2} = -1\frac{1}{2}$$

Now $\quad \Delta_r G^\circ = \Delta_r A^\circ + \Delta_r(pV) = \Delta_r A^\circ + (\Delta\nu_g)RT$

or $\quad \Delta_r A^\circ = \Delta_r G^\circ - (\Delta\nu_g)RT$

Substituting the given values, we get

$$\begin{aligned}\Delta_r A^\circ &= -3.202 \times 10^6 \text{ J mol}^{-1} - \{(-1.5)(8.314 \text{ J K}^{-1} \text{ mol}^{-1})(298 \text{ K})\}\\ &= -3.202 \times 10^6 \text{ J mol}^{-1} + 3\,716.36 \text{ J mol}^{-1}\\ &= -3.198\,3 \times 10^6 \text{ J mol}^{-1} = -3.198\,3 \text{ MJ mol}^{-1}\end{aligned}$$

5.6 PRESSURE DEPENDENCE OF FREE ENERGY

Although the change in free energy is useful as a criterion of equilibrium for a system whose temperature and pressure are constant, we need to calculate the changes in the free energy as a function of these variables. Here we derive the relations corresponding to the pressure change at constant temperature.

The variation of free energy with pressure as given by Eq. (5.4.7) is

$$\left(\frac{\partial G}{\partial p}\right)_T = V$$

or $\quad dG = V\,dp \qquad (T \text{ constant}) \qquad (5.6.1)$

Integrating this within the limits p_1 and p_2, we get

$$\int_{p_1}^{p_2} dG = \int_{p_1}^{p_2} V\,dp \qquad (T \text{ constant}) \qquad (5.6.2)$$

For Condensed Phases

The volume occupied by condensed phases (liquids and solids) is nearly independent of pressure. Therefore, on integrating Eq. (5.6.2), we get

$$G_2 - G_1 = V(p_2 - p_1) \qquad (5.6.3)$$

where G_2 and G_1 are the free energies at pressure p_2 and p_1, respectively. Since the volume occupied by condensed phases is small, unless the pressure change is very large, the right hand side of Eq. (5.6.3) in negligibly small. Thus, the pressure dependence of free energy for condensed phases may be ignored and thus G may be regarded to be temperature dependent only, i.e.

$$G = f(T) \quad \text{or} \quad G(T, p) = G(T) \qquad (5.6.4)$$

For Ideal Gases

Here, the volume is related to the pressure through the ideal gas equation

$$V = \frac{nRT}{p}$$

Substituting this in Eq. (5.6.2) and integrating, we get

$$\Delta G = G_2 - G_1 = nRT \ln \frac{p_2}{p_1} \tag{5.6.5}$$

In terms of volume change, we have

$$\Delta G = nRT \ln \frac{V_1}{V_2} \tag{5.6.6}$$

Example 5.6.1

Calculate ΔG for the conversion of 3 mol of liquid benzene at 80 °C (normal boiling point) to vapour at the same temperature and a pressure of 0.66 bar. Consider the vapour as an ideal gas.

Solution

The transformation

Benzene(1, 80 °C, 1 bar) → Benzene(g, 80 °C, 0.66 bar)

may be brought about by following the two steps given below.

Step 1 3 mol of liquid benzene $\xrightarrow[\text{boiling point}]{\text{Reversibly at}}$ 3 mol of benzene vapour at 1 bar
80 °C and at 1 bar

$\Delta G = 0$

Step 2 3 mol of benzene vapour at 1 bar $\xrightarrow[\text{at 80 °C}]{}$ 3 mol of benzene vapour at 0.66 bar

$$\Delta G = nRT \ln \frac{p_2}{p_1} = (3.0 \text{ mol})\,(8.314 \text{ J K}^{-1} \text{ mol}^{-1})\,(353.15 \text{ K}) \ln \left(\frac{0.66}{1}\right)$$

$$= -\ 3\ 660.0 \text{ J mol}^{-1} = -3.66 \text{ kJ mol}^{-1}$$

Example 5.6.2

Calculate the free energy change in the freezing of 18 g of water at 263.15 K, given that the vapour pressure of water and ice at 263.15 K are 0.287 Pa and 0.260 Pa, respectively.

Solution

The crystallization process

H_2O(1, 263.15 K) → H_2O(s, 263.15 K)

can be replaced by the following three reversible steps.

Step 1 H_2O(1, 263.15 K, 0.287 Pa) $\xrightarrow[\text{evaporation at 263.15 K and 0.287 Pa}]{\text{Reversible}}$ H_2O(g, 263.15 K, 0.287 Pa)

For this process $\Delta G_1 = 0$ since the evaporation is reversible in nature.

Step 2 H_2O(g, 263.15 K, 0.287 Pa) → H_2O(g, 263.15 K, 0.260 Pa)

The free energy change in this process is

$$\Delta G_{\text{II}} = RT \ln \frac{p_2}{p_1}$$

$$= (8.314 \text{ J K}^{-1} \text{ mol}^{-1})\,(263.15 \text{ K})\,(2.303) \times \log \left(\frac{0.260 \text{ Pa}}{0.287 \text{ Pa}}\right)$$

$$= -216.198 \text{ J mol}^{-1}$$

Step 3 H_2O(g, 263.15 K, 0.260 Pa) $\xrightarrow[\text{263.15 K and 0.260 Pa}]{\text{Reversible condensation}}$ H_2O(s, 263.15 K, 0.260 Pa)

Again for this process $\Delta G_{III} = 0$,

The total change in ΔG is

$$\Delta G = \Delta G_I + \Delta G_{II} + \Delta G_{III} = \Delta G_{II} = -216.198 \text{ J mol}^{-1}$$

The Standard Free Energy

The standard free energy $G°$ of a substance at a given temperature is the free energy of the substance at 1 bar pressure.† Since the pressure has been stated, it may be noted that the standard free energy is a function of temperature only. The free energy of an ideal gas at pressure p relative to the standard free energy can be obtained from Eq. (5.6.5) by substituting p_1 = 1 bar. Thus, we have

$$G(T, p) = G°(T) + nRT \ln (p/1 \text{ bar}) \quad (5.6.7)$$

or
$$\frac{G(T, p)}{n} = \frac{G°(T)}{n} + RT \ln (p/1 \text{ bar})$$

or
$$\mu(T, p) = \mu°(T) + RT \ln (p/1 \text{ bar}) \quad (5.6.8)$$

where μ is the free energy per mole of the substance and is called the *chemical potential*. Figures (5.6.1) and (5.6.2) show the variations of $\mu - \mu°$ for an ideal gas with (p/1 bar) and ln (p/1 bar), respectively,

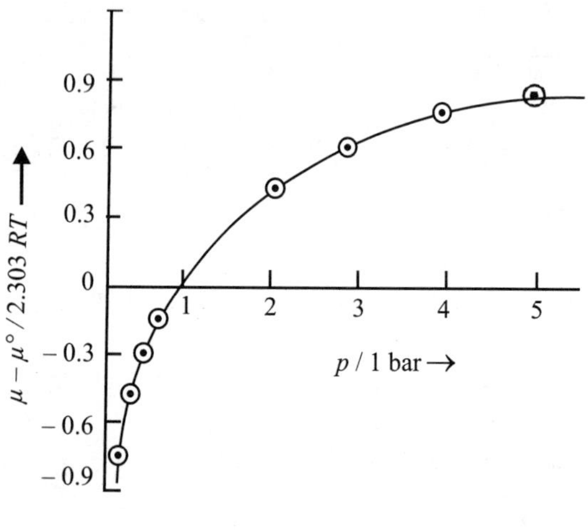

Fig. 5.6.1 Variation of $\mu - \mu°$ for an ideal gas with p/1 bar

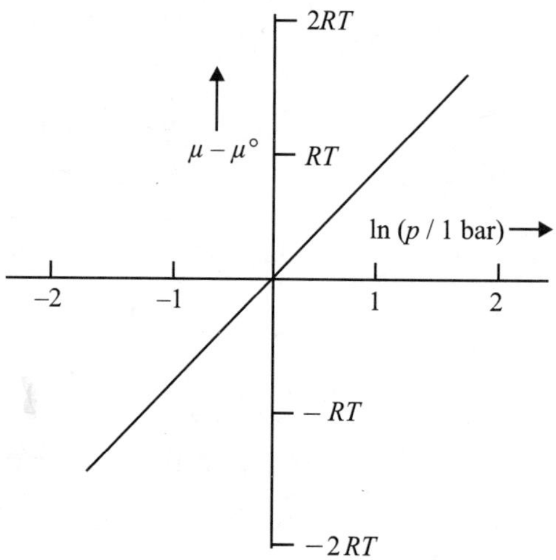

Fig. 5.6.2 Variation of $\mu - \mu°$ for an ideal gas with ln (p/1 bar)

† Prior to the recommendation of 1 bar as the standard-state pressure, the value used was 1 atm (= 101.325 kPa). This change causes the changes in Gibbs free energy by an amount – (0.109 J K^{-1}) $T \Delta \nu_g$, where $\Delta \nu_g$ is the change in stoichiometric number of gaseous species.

Problem 5.6.1 For an isothermal reversible expansion of a van der Waals gas, show that

$$\text{(i)}\quad \Delta A = -nRT\ \ln\frac{V_2 - nb}{V_1 - nb} - n^2 a\left(\frac{1}{V_2} - \frac{1}{V_1}\right)$$

$$\text{(ii)}\quad \Delta G = -nRT\ \ln\frac{V_2 - nb}{V_1 - nb} - 2n^2 a\left(\frac{1}{V_2} - \frac{1}{V_1}\right) + n^2 bRT\left(\frac{1}{V_2 - nb} - \frac{1}{V_1 - nb}\right)$$

Solution **(i)** From Eq. (5.4.2), we have

$$\left(\frac{\partial A}{\partial V}\right)_T = -p$$

or $\mathrm{d}A = -p\ \mathrm{d}V$ (T constant)

For the van der Waals equation, we have

$$\mathrm{d}A = -\left(\frac{nRT}{V - nb} - \frac{n^2 a}{V^2}\right)\mathrm{d}V$$

Hence $$\Delta A = -\int_{V_1}^{V_2}\frac{nRT}{V - nb}\,\mathrm{d}V + \int_{V_1}^{V_2}\frac{n^2 a}{V^2}\,\mathrm{d}V = -nRT\ \ln\frac{V_2 - nb}{V_1 - nb} - n^2 a\left(\frac{1}{V_2} - \frac{1}{V_1}\right)$$

(ii) From Eq. (5.4.7), we have

$$\left(\frac{\partial G}{\partial p}\right)_\mathrm{T} = V$$

or $\mathrm{d}G = V\ \mathrm{d}p$ (T constant)

or $\Delta G = \int V\ \mathrm{d}p$ (Eq. 5.6.2)

While evaluating the integral, it is convenient to replace dp in terms of dV. The van der Waals equation of state is

$$p = \frac{nRT}{V - nb} - \frac{n^2 a}{V^2}$$

Therefore $$\mathrm{d}p = -\frac{nRT}{(V - nb)^2}\mathrm{d}V + \frac{2n^2 a}{V^3}\mathrm{d}V$$

Substituting dp in the expression of ΔG, we get

$$\Delta G = -nRT\int_{V_1}^{V_2}\frac{V}{(V - nb)^2}\mathrm{d}V + \int_{V_1}^{V_2}\frac{2n^2 a}{V^2}\mathrm{d}V$$

$$= -nRT\int_{V_1}^{V_2}\frac{V - nb + nb}{(V - nb)^2}\,\mathrm{d}V + \int_{V_1}^{V_2}\frac{2n^2 a}{V^2}\,\mathrm{d}V$$

$$= -nRT\int_{V_1}^{V_2}\frac{1}{V - nb}\,\mathrm{d}V - nRT\int_{V_1}^{V_2}\frac{nb}{(V - nb)^2}\,\mathrm{d}V + \int_{V_1}^{V_2}\frac{2n^2 a}{V^2}\,\mathrm{d}V$$

$$= -nRT\ln\frac{V_2 - nb}{V_1 - nb} + n^2 bRT\left(\frac{1}{V_2 - nb} - \frac{1}{V_1 - nb}\right) - 2\,n^2 a\left(\frac{1}{V_2} - \frac{1}{V_1}\right)$$

5.7 FUGACITY FUNCTION AND ITS DETERMINATION FOR REAL GASES

In principle, the integration over pressure in Eq. (5.6.2) can be carried out if the equation of state of a real gas relating V with p is known. This procedure will, however, depend upon the nature of the gas. G. N. Lewis removed the above dependence by proposing an alternative procedure which does not depend upon the nature of the gas. In order to keep the form of the resultant expression as that of an ideal gas (Eq. 5.6.5), a function f, called that fugacity of the real gas, is defined such that

$$\Delta G = G_2 - G_1 = nRT \ln \frac{f_2}{f_1}$$

or $$\Delta \mu = \mu_2 - \mu_1 = RT \ln \frac{f_2}{f_1} \tag{5.7.1}$$

The fugacity f plays the role of pressure and need not be equal to the actual pressure of the real gas. Thus, it may be known as corrected pressure which applies to real gases.

The Fugacity Coefficient

In general, the fugacity of a real gas is related to its pressure by an equation

$$\frac{f}{p} = \gamma$$

where γ is known as the fugacity coefficient and is a measure of deviations of a real gas from the ideal gas behaviour. Since all gases approach ideality in the limit of zero pressure, it is obvious that

$$\lim_{p\to 0} \frac{f}{p} = \lim_{p\to 0} \gamma = 1 \tag{5.7.2}$$

The fugacity is expressed in the same unit as pressure and thus the fugacity coefficient is a pure number.

The Standard State of μ of Real Gases

The chemical potential of a real gas at fugacity f relative to that at standard-state fugacity f° as given by Eq. (5.7.1) is

$$\mu = \mu^\circ + RT \ln (f/f^\circ)$$

where μ and μ° are the chemical potentials at fugacities f and f°, respectively. Substituting f in terms of p in the above equation, we get

$$\mu = \mu^\circ + RT \ln (p/f^\circ) + RT \ln \gamma \tag{5.7.3}$$

Since the first two terms on the right hand side have the form of chemical potential of an ideal gas, it can be said that the third term measures the deviations of a real gas from ideality. If whole of the deviations are to be ascribed to γ, it is obvious that μ° in (5.7.3) must be identical with that appeared in Eq. (5.6.8) which is applicable to ideal gases, i.e. μ° of Eq. (5.7.3) must have the characteristics of an ideal gas. Thus, the standard state of a real gas is, in fact, some hypothetical state in which the gas behaves ideally at f° = 1 bar.

Evaluation of Fugacity

Equation (5.6.1) is applicable for all substances. Expressing it in terms of molar quantities and then using Eqs (5.6.5) and (5.7.1) separately, we get

$$\int_{p_1}^{p_2} V_{\text{m,ideal}}\,\text{d}p = \mu_{p_2,\text{ideal}} - \mu_{p_1,\text{ideal}} = RT\ln\frac{p_2}{p_1}$$

$$\int_{p_1}^{p_2} V_{\text{m,real}}\,\text{d}p = \mu_{p_2,\text{real}} - \mu_{p_1,\text{real}} = RT\ln\frac{f_2}{f_1}$$

Taking the difference, we have

$$\int_{p_1}^{p_2} (V_{\text{m,real}} - V_{\text{m,ideal}})\,\text{d}p = RT\ln\frac{f_2}{f_1} - RT\ln\frac{p_2}{p_1}$$

or
$$RT\ln\left(\frac{f_2}{f_1}\right)\left(\frac{p_1}{p_2}\right) = \int_{p_1}^{p_2} (V_{\text{m,real}} - V_{\text{m,ideal}})\,\text{d}p$$

Let p_1 be allowed to approach zero value, then from Eq. (5.7.2), we get

$$\lim_{p_1\to 0}\frac{f_1}{p_1} = 1$$

With this, the previous equation becomes

$$RT\ln\left(\frac{f_2}{p_2}\right) = \int_0^{p_2} (V_{\text{m,real}} - V_{\text{m,ideal}})\,\text{d}p \tag{5.7.4}$$

Now for a real gas

$$V_{\text{m,real}} = Z\frac{RT}{p} = Z\,V_{\text{m,ideal}} \quad \text{and} \quad V_{\text{m,ideal}} = \frac{RT}{p}$$

where Z is the *compression factor*. Introducing this in the previous expression, we get

$$\ln\left(\frac{f_2}{p_2}\right) = \int_0^{p_2}\frac{Z-1}{p}\,\text{d}p \tag{5.7.5}$$

The integral over p in the above expression may be performed graphically if the data on Z (or V_m) are available in numerical tables or analytically if the analytical expression for $Z(T, p)$ (or V_m in terms of p) is available.

Interpretation of the Value of Fugacity for a van der Waals Gas

In general, the value of f depends upon the value of Z. If Z is less than one, the right hand side of Eq. (5.7.5) will be negative and hence f_2 will be smaller than p_2. On the other hand, if Z is greater than one, the right hand side will be positive and hence f_2 will be greater than p_2. For most gases (except H_2 and He), Z is initially less than one but becomes greater then one at higher pressures. For such gases, f is initially less than p and becomes greater than p at higher pressures. These effects may be correlated with the van der Waals constants a and b. At lower pressures, where the attractive molecular forces predominate, the van der Waals equation for one mole of the gas takes the form

$$\left(p + \frac{a}{V_m^2}\right)(V_m) = RT$$

or $$Z = \frac{pV_m}{RT} = 1 - \frac{a}{V_m RT}$$

i.e. $$Z < 1$$

Hence, f less than p is due to the constant a (the molecules tend to stick together) At higher pressures, where the excluded volume plays the dominating role, the van der Waals equation takes the form

$$(p)\,(V_m - b) = RT$$

or $$Z = \frac{pV_m}{RT} = 1 + b\,\frac{p}{RT}$$

i.e. $$Z > 1$$

Hence, f greater than p is due to constant b (the repulsive part of the intermolecular potential dominates and the molecules tend to fly apart).

Thus, the neglect of the term b in van der Waals equation makes f less than p and that of a makes it greater than p.

Problem 5.7.1 Show that if equation of state for a gas is

$$p(V_m - b) = RT$$

where b is a constant, the fugacity of the gas is given by

$$\ln \frac{f}{p} = \frac{bp}{RT}$$

Also show that if bp/RT is small (applicable at low pressure) this equation reduces to

$$\frac{f}{p} = \frac{p}{p_{ideal}}$$

where p_{ideal} is the pressure the gas would exert if it were to be ideal but with the same molar volume as the real gas at pressure p.

Solution We have $p(V_m - b) = RT$

or $$Z = \frac{pV_m}{RT} = 1 + \frac{pb}{RT}$$

Substituting this in the relation

$$\ln \frac{f}{p} = \int_0^p \frac{Z-1}{p}\,dp$$

and integrating the resultant expression, we get

$$\ln \frac{f}{p} = \frac{bp}{RT} \quad \text{or} \quad \frac{f}{p} = \exp(bp/RT)$$

If bp/RT is small, we may write

$$\frac{f}{p} = 1 + \frac{bp}{RT} = \frac{RT + bp}{RT}$$

From the given equation of state, we have

$$RT + bp = pV_m$$

Hence $\quad \dfrac{f}{p} = p\left(\dfrac{V_m}{RT}\right) = \dfrac{p}{p_{ideal}}$

or $\quad p^2 = f\,p_{ideal}$

that is, pressure of the gas is the geometric mean of the ideal pressure and fugacity.

Problem 5.7.2

The fugacity of a certain gas is given by the expression

$$f = p + \alpha p^2$$

in which α is a function of temperature.

(a) Show that the gas must obey the following equation of state

$$\frac{pV_m}{RT} = 1 + \frac{\alpha p}{1+\alpha p}$$

(b) Also show that

$$\left(\frac{\partial H_m}{\partial p}\right)_T = -\frac{RT^2}{(1+\alpha p)^2}\left(\frac{d\alpha}{dT}\right)$$

Solution

(a) We have

$$f = p + \alpha p^2$$

or $\quad \dfrac{f}{p} = 1 + \alpha p \quad$ or $\quad \ln \dfrac{f}{p} = \ln(1+\alpha p)$

Since $\quad \ln \dfrac{f}{p} = \displaystyle\int_0^p \frac{Z-1}{p}\,dp \qquad$ (Eq. 5.5.13)

we have $\ln(1+\alpha p) \equiv \displaystyle\int_0^p \frac{\alpha}{1+\alpha p}\,dp = \int_0^p \frac{Z-1}{p}\,dp$

Thus $\quad \dfrac{\alpha}{1+\alpha p} = \dfrac{Z-1}{p} = \dfrac{(V_m/V_{m,ideal})-1}{p} = \dfrac{(V_m p/RT)-1}{p}$

Hence $\quad \dfrac{pV_m}{RT} = 1 + \dfrac{\alpha p}{1+\alpha p}$

(b) Substituting the given expression of f in the relation

$$\mu = \mu^\circ + RT\ln(f/f^\circ)$$

we get $\quad \mu = \mu^\circ + RT\ln\{p(1+\alpha p)/f^\circ\}$

or $\quad \dfrac{\mu}{T} = \dfrac{\mu^\circ}{T} + R\ln\{p(1+\alpha p)/f^\circ\}$

Differentiation of the above expression with respect to T at constant p gives

$$\left(\frac{\partial \mu/T}{\partial T}\right)_p = \left(\frac{\partial \mu^\circ/T}{\partial T}\right)_p + \frac{R}{1+\alpha p}\left\{p\left(\frac{d\alpha}{dT}\right)\right\}$$

Making use of Gibbs-Helmholtz equation (see, Problem 5.8.1), we get

$$-\frac{H_\text{m}}{T^2} = -\frac{H_\text{m}^\circ}{T^2} + \frac{Rp}{1+\alpha p}\left(\frac{\text{d}\alpha}{\text{d}T}\right)$$

or $$H_\text{m} = H_\text{m}^\circ - \frac{RT^2 p}{1+\alpha p}\left(\frac{\text{d}\alpha}{\text{d}T}\right)$$

Differentiation of the above expression with respect to p at constant T gives

$$\left(\frac{\partial H_\text{m}}{\partial p}\right)_T = -RT^2\left(\frac{\text{d}\alpha}{\text{d}T}\right)\left\{\frac{1}{1+\alpha p} - \frac{\alpha p}{(1+\alpha p)^2}\right\}$$

$$= -\frac{RT^2}{(1+\alpha p)^2}\left(\frac{\text{d}\alpha}{\text{d}T}\right)$$

Problem 5.7.3 A function $\alpha(T, p)$ is defined as

$$\alpha = V_\text{m, ideal} - V_\text{m, real}$$

where $V_\text{m, ideal}$ and $V_\text{m, real}$ are the molar volumes of an ideal gas and a real gas at the same T and p, respectively. Evaluate the function α for the general equation of state

$$\frac{pV_\text{m}}{RT} = 1 + Bp + Cp^2 + \cdots$$

Further show that the fugacity of the gas is given by the expression

$$\ln(f/p) = Bp + \frac{Cp^2}{2} + \cdots$$

Solution From the given equation of state, we have

$$V_\text{m, real} = \frac{RT}{p}(1 + Bp + Cp^2 + \cdots)$$

Substituting this in the equation

$$\alpha = V_\text{m, ideal} - V_\text{m, real} = \frac{RT}{p} - V_\text{m, real}$$

we get $$\alpha = \frac{RT}{p} - \frac{RT}{p}(1 + Bp + Cp^2 + \cdots)$$

or $$\alpha = -(BRT + CRTp + \cdots)$$

Form Eq. (5.7.5), we have

$$\ln\frac{f}{p} = \int_0^p \frac{Z-1}{p}\text{d}p = \int_0^p \frac{(V_\text{m,real}/V_\text{m,ideal}) - 1}{p}\text{d}p$$

$$= \int_0^p \frac{(V_\text{m,real} - V_\text{m,ideal})}{V_\text{m,ideal}\,p}\text{d}p = \int_0^p \frac{-\alpha}{(RT/p)p}\text{d}p = -\frac{1}{RT}\int_0^p \alpha\,\text{d}p$$

Substituting the expression of α and carrying out the integration, we get

$$\ln(f/p) = Bp + \frac{Cp^2}{2} + \cdots$$

Problem 5.7.4 Show that for a van der Waals gas:

(a) $$\ln\left(\frac{f}{p}\right) = \ln\frac{RT}{p(V_m - b)} + \frac{b}{V_m - b} - \frac{2a}{RTV_m}$$

(b) $$\ln\left(\frac{f}{p}\right) = -\frac{ap}{(RT)^2} + \frac{bp}{RT} + \frac{abp^2}{2(RT)^3}$$

Solution (a) We have

$$RT\ln\frac{f}{p} = \int_0^p (V_m - V_{m,\text{ideal}})\,dp \qquad \text{(Eq. 5.7.4)}$$

$$= \int_0^p V_m\,dp - \int_0^p \frac{RT}{p}\,dp \qquad (5.7.6)$$

While evaluating the first integral, it is convenient to replace dp in terms of dV_m. The van der Waals equation of state for one mole of the gas is

$$p = \frac{RT}{V_m - b} - \frac{a}{V_m^2}$$

Therefore $$dp = -\frac{RT}{(V_m - b)^2}dV_m + \frac{2a}{V_m^3}dV_m$$

Substituting dp in Eq. (5.7.6), we get

$$RT\ln\frac{f}{p} = -\int_\infty^{V_m}\frac{RT\,V_m}{(V_m - b)^2}dV_m + \int_\infty^{V_m}\frac{2a}{V_m^2}dV_m - RT\ln p\Big|_0^p$$

$$= -\int_\infty^{V_m}\frac{RT\,(V_m - b + b)}{(V_m - b)^2}dV_m + \int_\infty^{V_m}\frac{2a}{V_m^2}dV_m - RT\ln p\Big|_0^p$$

$$= -\int_\infty^{V_m}\frac{RT}{V_m - b}dV_m - \int_\infty^{V_m}\frac{RT\,b}{(V_m - b)^2}dV_m + \int_\infty^{V_m}\frac{2a}{V_m^2}dV_m - RT\ln p\Big|_0^p$$

$$= -RT\ln(V_m - b)\Big|_\infty^{V_m} + \frac{RTb}{(V_m - b)}\Big|_\infty^{V_m} - \frac{2a}{V_m}\Big|_\infty^{V_m} - RT\ln p\Big|_0^p$$

Combining first and fourth terms on the right hand side before inserting the limits, we have

$$RT\ln\frac{f}{p} = -RT\ln p(V_m - b)\Big|_\infty^{V_m} + \frac{RTb}{(V_m - b)}\Big|_\infty^{V_m} - \frac{2a}{V_m}\Big|_\infty^{V_m}$$

Remembering that

$$\lim_{V_m\to\infty} p(V_m - b) = pV_m = RT$$

we have

$$RT \ln \frac{f}{p} = -RT \ln \{p(V_m - b)/RT\} + \frac{RTb}{V_m - b} - \frac{2a}{V_m}$$

or
$$\ln \frac{f}{p} = -\ln \{p(V_m - b)/RT\} + \frac{b}{V_m - b} - \frac{2a}{V_m RT}$$

or
$$\ln \frac{f}{p} = \ln \left(\frac{RT}{p(V_m - b)} \right) + \frac{b}{V_m - b} - \frac{2a}{V_m RT} \tag{5.7.7}$$

(b) For one mole of the gas, the van der Waals equation is

$$\left(p + \frac{a}{V_m^2} \right)(V_m - b) = RT$$

or
$$pV_m - pb + \frac{a}{V_m} - \frac{ab}{V_m^2} = RT$$

To the first approximation V_m in the correction terms may be replaced by RT/p. Thus, we have

$$pV_m - pb + \frac{ap}{RT} - \frac{abp^2}{(RT)^2} = RT \quad \text{or} \quad V_m = \frac{RT}{p} + b - \frac{a}{RT} + \frac{abp}{(RT)^2}$$

Now
$$V_{m,\,ideal} - V_{m,\,real} = -\left(b - \frac{a}{RT} + \frac{apb}{(RT)^2} \right)$$

Since

$$\ln \frac{f}{p} = -\frac{1}{RT} \int_0^p (V_{m,\,ideal} - V_{m,\,real})\, dp$$

we get
$$\ln \frac{f}{p} = \frac{1}{RT} \int_0^p \left(b - \frac{a}{RT} + \frac{abp}{(RT)^2} \right) dp = \frac{1}{RT} \left(bp - \frac{ap}{RT} + \frac{abp^2}{2(RT)^2} \right) \tag{5.7.8}$$

Example 5.7.1 Calculate the fugacity of *n*-octance (a van der Waals gas) at 430 K and 2 bar. Given: $a = 3.8 \times 10^{-5}$ bar m^6 mol^{-2} and $b = 2.37 \times 10^{-4}$ m^3 mol^{-1}.

Solution Form the expression

$$\ln \frac{f}{p} = \frac{bp}{RT} - \frac{ap}{(RT)^2} + \frac{abp^2}{2(RT)^3}$$

we get

$$\frac{bp}{RT} = \frac{(2.37 \times 10^{-4} \text{m}^3 \text{ mol}^{-1})(2 \text{ bar})}{(8.314 \times 10^{-5} \text{ bar m}^3 \text{ K}^{-1} \text{ mol}^{-1})(430 \text{ K})} = 1.326 \times 10^{-2}$$

$$\frac{ap}{(RT)^2} = \frac{(3.8 \times 10^{-5} \text{ bar m}^6 \text{ mol}^{-2})(2 \text{ bar})}{\{(8.314 \times 10^{-5} \text{ bar m}^3 \text{ K}^{-1} \text{ mol}^{-1})(430 \text{ K})\}^2} = 5.946 \times 10^{-2}$$

$$\frac{abp^2}{2(RT)^3} = \frac{(3.8 \times 10^{-5} \text{ bar m}^6 \text{ mol}^{-2})(2.37 \times 10^{-4} \text{ m}^3 \text{ mol}^{-1})(2 \text{ bar})^2}{2\{(8.314 \times 10^{-5} \text{ bar m}^3 \text{ K}^{-1} \text{ mol}^{-1})(430 \text{ K})\}^3}$$

$$= 3.942 \times 10^{-4}$$

$$\ln (f/p) = 1.326 \times 10^{-2} - 5.946 \times 10^{-2} + 3.942 \times 10^{-4}$$
$$= -4.581 \times 10^{-2}$$

Hence $\dfrac{f}{p} = 0.955$ or $f = (0.955)(2\text{ bar}) = 1.91\text{ bar}$

Problem 5.7.5

The expression for the variation of fugacity of a pure gas with temperature may be obtained by differentiating the expression

$$\mu_2 - \mu_1 = RT \ln (f_2/p) - RT \ln (f_1/p) \qquad \text{(Eq. 5.7.1)}$$

(a) Show that the desired expression is

$$\left(\frac{\partial \ln (f/p)}{\partial T}\right)_p = \frac{H_m^0 - H_m}{RT^2}$$

where H_m^0 is the molar enthalpy of the gas at zero pressure.

(b) Evaluate $H_m^0 - H_m$ for one mole of a van der Waals gas.

(c) Evaluate $C_{p,m}^0 - C_{p,m}$ for one mole of a van der Waals gas.

Solution

(a) We have

$$\mu_2 - \mu_1 = RT \ln (f_2/p) - RT \ln (f_1/p)$$

$$\frac{\mu_2}{T} - \frac{\mu_1}{T} = R \ln (f_2/p) - R \ln (f_1/p)$$

Differentiating with respect to T at constant p, we get

$$\left(\frac{\partial(\mu_2/T)}{\partial T}\right)_p - \left(\frac{\partial(\mu_1/T)}{\partial T}\right)_p = R\left(\frac{\partial \ln (f_2/p)}{\partial T}\right)_p - R\left(\frac{\partial \ln (f_1/p)}{\partial T}\right)_p$$

$$-\frac{H_{2,m}}{T^2} + \frac{H_{1,m}}{T^2} = R\left(\frac{\partial \ln (f_2/p)}{\partial T}\right)_p - R\left(\frac{\partial \ln (f_1/p)}{\partial T}\right)_p$$

At zero pressure, $f_1 \rightarrow p$ and hence $\left(\dfrac{\partial \ln (f_1/p)}{\partial T}\right)_p = 0$

At zero pressure, we may write $H_{1,m}$ as H_m^0. Dropping the subscript 2 and rearranging, we get

$$\left(\frac{\partial \ln (f/p)}{\partial T}\right)_p = \frac{H_m^0 - H_m}{RT^2} \qquad (5.7.9)$$

(b) For a van der Waals gas, we have

$$\ln \frac{f}{p} = -\frac{ap}{(RT)^2} + \frac{bp}{RT} + \frac{abp^2}{2(RT)^3} \qquad \text{(Eq. 5.7.8)}$$

Thus $$\left(\frac{\partial \ln (f/p)}{\partial T}\right)_p = \frac{2ap}{R^2T^3} - \frac{bp}{RT^2} - \frac{3abp^2}{2R^3T^4}$$

Hence $$\frac{H_m^0 - H_m}{RT^2} = \frac{2ap}{R^2T^3} - \frac{bp}{RT^2} - \frac{3abp^2}{2R^3T^4}$$

or $$H_{\rm m}^{0} - H_{m} = \frac{2ap}{RT} - bp - \frac{3abp^2}{2R^2T^2} \tag{5.7.10}$$

(c) Differentiation of Eq. (5.7.10) with respect to T at constant p gives

$$C_{p,\rm m}^{0} - C_{p,\rm m} = -\frac{2ap}{RT^2} + \frac{3abp^2}{R^2T^3}$$

Problem 5.7.6

The pressure dependence of the quantity $H_{\rm m}^{0} - H_{\rm m}$ for a pure gas is

$$\left(\frac{\partial(H_{\rm m}^{0} - H_{\rm m})}{\partial p}\right)_T = -\left(\frac{\partial H_{\rm m}}{\partial p}\right)_T$$

because $(\partial H_{\rm m}^{0}/\partial p)_T$ is zero as $H_{\rm m}^{0}$ is the molar enthalpy at fixed (zero) pressure. Now

$$\left(\frac{\partial H_{\rm m}}{\partial p}\right)_T = -C_{p,\rm m}\,\mu_{\rm JT} \qquad \text{(Eq. 2.8.3)}$$

Thus $$\left(\frac{\partial(H_{\rm m}^{0} - H_{\rm m})}{\partial p}\right)_T = C_{p,\rm m}\,\mu_{\rm JT}$$

Making use of the above relation, drive the expression for $\mu_{\rm JT}$ for a van der Waals gas.

Solution

For one mole of a van der Waals gas, we have

$$H_{\rm m}^{0} - H_{\rm m} = \frac{2ap}{RT} - pb - \frac{3abp^2}{2R^2T^2}$$

Hence $$\left(\frac{\partial(H_{\rm m}^{0} - H_{\rm m})}{\partial p}\right)_T = \frac{2a}{RT} - b - \frac{3abp}{R^2T^2}$$

Thus $$\mu_{\rm JT} = \frac{1}{C_{p,\rm m}}\left(\frac{2a}{RT} - b - \frac{3abp}{R^2T^2}\right) \qquad \text{(Eq. 2.9.3)}$$

5.8 TEMPERATURE DEPENDENCE OF FREE ENERGY

The change in the free energy with temperature at constant pressure as given by Eq. (5.4.8) is

$$\left(\frac{\partial G}{\partial T}\right)_p = -S \tag{5.8.1}$$

From the definition of G (= $H - TS$), we have

$$-S = \frac{G - H}{T} \tag{5.8.2}$$

Substituting this in Eq. (5.8.1), we get

$$\left(\frac{\partial G}{\partial T}\right)_p = \frac{G - H}{T} \quad \text{or} \quad G = H + T\left(\frac{\partial G}{\partial T}\right)_p \tag{5.8.3}$$

Gibbs-Helmholtz Equation

In actual practice, one commonly employs the relations which give the variation in $\Delta_{\rm r}G$ with temperature. Such expressions can be conveniently written starting from Eq. (5.8.1) with the replacement of G with $\Delta_r G$ and that of S with $\Delta_r S$. Thus, we have

$$\left(\frac{\partial(\Delta_r G)}{\partial T}\right)_p = \Delta_r S = \frac{\Delta_r G - \Delta_r H}{T} \tag{5.8.4}$$

or

$$\Delta_r G = \Delta_r H + T\left(\frac{\partial(\Delta_r G)}{\partial T}\right)_p \tag{5.8.5}$$

This equation is known as the *Gibbs-Helmholtz equation.*

Temperature Dependence of $\Delta_r G/T$

To calculate the value of $\Delta_r G$ at one temperature from a known value at some other temperature, it is more convenient to make use of the temperature dependence of $\Delta_r G/T$, rather than the temperature dependence of $\Delta_r G$ itself. At a constant temperature, $\Delta_r G = \Delta_r H - T\,\Delta_r S$, so that

$$\frac{\Delta_r G}{T} = \frac{\Delta_r H - T\,\Delta_r S}{T} = \frac{\Delta_r H}{T} - \Delta_r S$$

Differentiating $\Delta_r G/T$ with respect to temperature at constant pressure, we get

$$\left(\frac{\partial(\Delta_r G/T)}{\partial T}\right)_p = \left(\frac{\partial(\Delta_r H/T)}{\partial T}\right)_p - \left(\frac{\partial(\Delta_r S)}{\partial T}\right)_p$$

$$= -\frac{1}{T^2}\Delta_r H + \frac{1}{T}\left(\frac{\partial(\Delta_r H)}{\partial T}\right)_p - \left(\frac{\partial(\Delta_r S)}{\partial T}\right)_p$$

Employing Eqs (3.11.1) and (4.24.2), we get

$$\left\{\frac{\partial(\Delta_r G/T)}{\partial T}\right\}_p = -\frac{\Delta_r H}{T^2} + \frac{\Delta_r C_p}{T} - \frac{\Delta_r C_p}{T} = -\frac{\Delta_r H}{T^2} \tag{5.8.6}$$

or

$$\left\{\frac{\partial(\Delta_r G/T)}{-(1/T^2)\partial T}\right\}_p = \Delta_r H$$

Now $\mathrm{d}\left(\frac{1}{T}\right) = -\frac{1}{T^2}\mathrm{d}T$. With this, the previous equatoin becomes

$$\left\{\frac{\partial(\Delta_r G/T)}{\partial(1/T)}\right\}_p = \Delta_r H \tag{5.8.7}$$

If $\Delta_r H$ is independent of temperature, the plot $\Delta_r G/T$ versus $1/T$ will be a straight line with slope equal to $\Delta_r H$. If $\Delta_r H$ depends on temperature, then the slope of the plot at $1/T$ gives $\Delta_r H$ at temperature T.

Computation of $\Delta_r G$ at Some Other Temperature

While using Eq. (5.8.7) to calculate the value of $\Delta_r G$ at one temperature from a known value at some other temperature, two cases may be distinguished.

$\Delta_r H$ independent of temperature In this case, we have

$$d\left(\frac{\Delta_r G}{T}\right) = -\frac{\Delta_r H}{T^2}dT \qquad (p \text{ constant})$$

Integrating this, we have

$$\frac{\Delta_r G}{T} = \frac{\Delta_r H}{T} + I$$

or $\Delta_r G = \Delta_r H + IT$

where I is the constant of integration whose value may be calculated from the value of $\Delta_r G_0$ at the given temperature T_0. Thus

$$I = \frac{1}{T_0}(\Delta_r G_0 - \Delta_r H)$$

Substituting this in the previous expression, we have

$$\Delta_r G = \Delta_r H + \frac{T}{T_0}(\Delta_r G_0 - \Delta_r H) \tag{5.8.8}$$

Example 5.8.1 For the reaction $\frac{1}{2}N_2(g) + \frac{3}{2}H_2(g) \rightarrow NH_3(g)$, $\Delta_r G^\circ_{298K} = -16.45$ kJ mol^{-1} and $\Delta_r H^\circ_{298K}$ $= -46.11$ kJ mol^{-1}. Estimate the value of $\Delta_r G^\circ_{500K}$ if $\Delta_r H^\circ$ is considered independent of temperature.

Solution From the expression

$$\Delta_r G^\circ_T = \Delta_r H^\circ + \frac{T}{T_0}(\Delta_r G^\circ_{T_0} - \Delta_r H^\circ)$$

we get $\Delta_r G^\circ_T = -46.11\text{ kJ mol}^{-1} + \left(\frac{500\text{ K}}{298\text{ K}}\right)(-16.45\text{ kJ mol}^{-1} + 46.11\text{ kJ mol}^{-1})$

$= 3.66$ kJ mol^{-1}

$\Delta_r H$ dependent on temperature The expression relating $\Delta_r H$ with temperature may be derived starting from Eq. (3.11.1), which gives

$$\left\{\frac{\partial(\Delta_r H)}{\partial T}\right\}_p = \Delta_r C_p$$

or $d(\Delta_r H) = (\Delta_r C_p)\, dT \qquad (p \text{ constant})$

The variation of $\Delta_r C_p$ with temperature is, as usual, given by the expression

$$\Delta_r C_p = (\Delta_r a) + (\Delta_r b)T + (\Delta_r c)T^2 + \cdots$$

where $(\Delta_r a)$, $(\Delta_r b)$ and $(\Delta_r c)$ are constants. Substituting this in the previous expression and the carrying out the integration, we get

$$\Delta_r H = (\Delta_r a)T + (\Delta_r b)\frac{T^2}{2} + (\Delta_r c)\frac{T^3}{3} + \cdots + K_1 \tag{5.8.9}$$

where K_1 is the constant of integration, whose value may be determined from the known value of $\Delta_r H$ at some temperature T. Let $\Delta_r H = \Delta_r H_0$ when $T = T_0$. Then

$$\Delta_r H_0 = (\Delta_r a)T_0 + (\Delta_r b)\frac{T_0^2}{2} + (\Delta_r c)\frac{T_0^3}{3} + \cdots + K_1$$

Thus $$K_1 = (\Delta_r H_0) - \left\{(\Delta_r a)T_0 + (\Delta_r b)\frac{T_0^2}{2} + (\Delta_r c)\frac{T_0^3}{3} + \cdots\right\}$$

Substituting Eq. (5.8.9) in Eq. (5.8.6), we get

$$\left\{\frac{\partial(\Delta_r G/T)}{\partial T}\right\}_p = -\frac{\Delta_r H}{T^2} = -\frac{(\Delta_r a)}{T} - \frac{(\Delta_r b)}{2} - \frac{(\Delta_r c)T}{3} - \cdots - \frac{K_1}{T^2}$$

which on integration gives

$$\frac{\Delta_r G}{T} = -\Delta_r a \ln\frac{T}{\mathrm{K}} - \frac{\Delta_r b}{2}T - \frac{\Delta_r c}{3}\frac{T^2}{2} - \cdots + \frac{K_1}{T} + K_2$$

or $$\Delta_r G = -\Delta_r a\, T \ln\frac{T}{\mathrm{K}} - \frac{\Delta_r b}{2}T^2 - \frac{\Delta_r c}{6}T^3 - \cdots + K_1 + K_2 T \qquad (5.8.10)$$

where K_2 is the constant of integration whose value may be determined from the known value of $\Delta_r G$ at a certain temperature T. Let $\Delta_r G = \Delta_r G_0$ when $T = T_0$

Then

$$K_2 = \frac{\Delta_r G_0}{T_0} + \Delta_r a \ln\frac{T_0}{\mathrm{K}} + \frac{\Delta_r b}{2}T_0 + \frac{\Delta_r c}{6}T_0^2 + \cdots - \frac{K_1}{T_0} \qquad (5.8.11)$$

Example 5.8.2 For $H_2O(1) \rightarrow H_2O(g)$

$\Delta_r G_{298\mathrm{K}} = 8.596$ kJ mol^{-1} and

$\Delta_r H/\mathrm{J\ mol^{-1}} = 57\ 086.5 - 45.327\ (T/\mathrm{K}) + 4.966 \times 10^{-3}\ (T/\mathrm{K})^2$

Compute $\Delta_r G_{380\mathrm{K}}$.

Solution Gibbs-Helmholtz equation is

$$\left\{\frac{\partial(\Delta_r G/T)}{\partial T}\right\}_p = -\frac{\Delta_r H}{T^2}$$

or $$\mathrm{d}(\Delta_r G/T) = -\frac{\Delta_r H}{T^2}\mathrm{d}T \qquad (p \text{ constant})$$

i.e. $$\mathrm{d}(\Delta_r G/T) = -\frac{1}{T^2}[(57\ 086.5\ \mathrm{J\ mol^{-1}}) - (45.327\ \mathrm{J\ mol^{-1}})(T/\mathrm{K}) + (4.966 \times 10^{-3}\ \mathrm{J\ mol^{-1}})(T/\mathrm{K})^2]\,\mathrm{d}T$$

Integrating this, we have

$$\frac{\Delta_r G}{T} = \frac{57\ 086.5\ \mathrm{J\ mol^{-1}}}{T} + (45.327\ \mathrm{J\ mol^{-1}\ K^{-1}}) \ln(T/\mathrm{K}) - (4.966 \times 10^{-3}\ \mathrm{J\ K^{-2}\ mol^{-1}})\,T + K$$

where K is the constant of integration, the value of which can be calculated from

$T = 298$ K $\qquad \Delta G = 8.596$ kJ mol^{-1}

Thus $$K = \frac{8.596\times10^{3}\ \mathrm{J\ mol^{-1}}}{298\ \mathrm{K}} - \frac{57\ 086.5\ \mathrm{J\ mol^{-1}}}{298\ \mathrm{K}} - (45.327\ \mathrm{J\ mol^{-1}\ K^{-1}})$$

$$\times\ 2.303 \times \log (298) + (4.966 \times 10^{-3}\ \mathrm{J\ K^{-2}\ mol^{-1}})\ (298\ \mathrm{K})$$

$$= (28.846 - 191.565 - 258.23 + 1.480)\ \mathrm{J\ K^{-1}\ mol^{-1}}$$

$$= -\ 419.47\ \mathrm{J\ K^{-1}\ mol^{-1}}$$

Therefore, the dependence of $\Delta_r G$ on temperature is given by

$$\Delta_r G/\mathrm{J\ mol^{-1}} = 57\ 086.5 + 45.327\ (T/\mathrm{K})\ \ln (T/\mathrm{K}) - 4.999 \times 10^{-3}\ (T/\mathrm{K})^2 - 419.47\ (T/\mathrm{K})$$

Thus at 380 K, we have

$$\Delta_r G_{380\mathrm{K}}/\ \mathrm{J\ mol^{-1}} = 57\ 086.5 + 45.327\ (380)\ (\ln 380) - (4.966 \times 10^{-3})\ (380)^2 - 419.47 \times 380$$

$$= 57\ 086.5 + 102.315.05 - 717.09 - 159\ 398.6 = -\ 714.14$$

$$\Delta_r G_{380\mathrm{K}} = -\ 714.14\ \mathrm{J\ mol^{-1}}$$

Example 5.8.3 For a chemical reaction $\Delta_r H^\circ_{298\mathrm{K}} = 195.267\ \mathrm{kJ\ mol^{-1}}$ and $\Delta_r G^\circ_{298\mathrm{K}} = 160.917\ \mathrm{kJ\ mol^{-1}}$ and $\Delta_r C_p$ is given by

$$\Delta_r C_p/\mathrm{J\ K^{-1}\ mol^{-1}} = -54.183 + (15.69 \times 10^{-3}\ \mathrm{K^{-1}})\ T - (2.105 \times 10^5\ \mathrm{K^2})T^{-2}$$

Derive the expression for $\Delta_r G^\circ$ as a function of T.

Solution We write the given expression of $\Delta_r C_p$ as

$$\Delta_r C_p = (\Delta_r a) + (\Delta_r b)\ T + (\Delta_r c)\ T^{-2}$$

where $(\Delta_r a) = -54.183\ \mathrm{J\ K^{-1}\ mol^{-1}}$; $(\Delta_r b) = 15.69 \times 10^{-3}\ \mathrm{J\ K^{-2}\ mol^{-1}}$

$(\Delta_r c) = -2.105 \times 10^5\ \mathrm{J\ K\ mol^{-1}}$

Now we have

$$\mathrm{d}(\Delta_r H) = (\Delta_r C_p)\ \mathrm{d}T \qquad (p \text{ constant})$$

$$= \{(\Delta_r a) + (\Delta_r b)\ T + (\Delta_r c)T^{-2}\}\ \mathrm{d}T$$

Thus $$\Delta_r H = (\Delta_r a)T + (\Delta_r b)\frac{T^2}{2} - (\Delta_r c)\frac{1}{T} + K_1 \tag{5.8.12a}$$

where K_1 is the constant of integration.

Now, from Gibbs-Helmholtz equation, we have

$$\mathrm{d}\left(\frac{\Delta_r G}{T}\right) = -\frac{\Delta_r H}{T^2}\mathrm{d}T = \left\{-\frac{(\Delta_r a)}{T} - \frac{(\Delta_r b)}{2} + (\Delta_r c)\frac{1}{T^3} - \frac{K_1}{T^2}\right\}\mathrm{d}T$$

Thus, $$\frac{\Delta_r G}{T} = -\ (\Delta_r a)\ \ln (T/\mathrm{K}) - \frac{(\Delta_r b)}{2}T - \frac{(\Delta_r c)}{2}\frac{1}{T^2} + \frac{K_1}{T} + K_2 \tag{5.8.12b}$$

where K_2 is the constant of integration. Hence

$$\Delta_r G = -(\Delta_r a)T\ \ln (T/\mathrm{K}) - \frac{(\Delta_r b)}{2}T^2 - \frac{(\Delta_r c)}{2}\frac{1}{T} + K_1 + K_2\,T$$

The expression of K_1 as given by Eq. (5.8.12a) is

$$K_1 = \Delta_r H - (\Delta_r a)T - \frac{(\Delta_r b)}{2}T^2 + (\Delta_r c)\frac{1}{T}$$

From the given value of $\Delta_r H$ of 298 K, we have

$$K_1 = 195\,267 \text{ J mol}^{-1} - (-54.183 \text{ J K}^{-1} \text{ mol}^{-1})(298 \text{ K})$$

$$-\frac{(15.69 \times 10^{-3} \text{ J K}^{-2} \text{ mol}^{-1})}{2}(298 \text{ K})^2 + (-2.105 \times 10^5 \text{ J K mol}^{-1})\frac{1}{(298 \text{ K})}$$

$$= [195\,267 + 16\,146.5 - 696.67 - 706.37] \text{ J mol}^{-1}$$

$$= 210\,010.46 \text{ J mol}^{-1}$$

The expression of K_2 as given by Eq. (5.8.12b) is

$$K_2 = \frac{1}{T}\left[\Delta_r G + (\Delta_r a)T \ln (T/\text{K}) + \frac{(\Delta_r b)}{2}T^2 + \frac{(\Delta_r c)}{2}\frac{1}{T} - K_1\right]$$

From the given value of $\Delta_r G$ at 298 K, we have

$$K_2 = \frac{1}{(298 \text{ K})}\left[160\,917 \text{ J} + (-54.183 \text{ J K}^{-1})(298 \text{ K}) \ln (298) + \frac{(15.69 \times 10^{-3} \text{ J K}^{-2})}{2}\right.$$

$$\left. \times (298 \text{ K})^2 + \frac{(-2.105 \times 10^5 \text{ J K})}{2} \times \frac{1}{(298 \text{ K})} - 210\,010.46 \text{ J}\right] \text{mol}^{-1}$$

$$= \frac{1}{(298 \text{ K})}(160\,917 - 91\,988.3 + 696.67 - 353.19 - 210\,010.46) \text{ J mol}^{-1}$$

$$= -472.28 \text{ J K}^{-1} \text{ mol}^{-1}$$

Hence, the expression of $\Delta_r G$ as a function of T is

$$\Delta_r G = (54.183 \text{ J K}^{-1} \text{ mol}^{-1})T \ln (T/\text{K}) - \frac{(15.69 \times 10^{-3} \text{ J K}^{-2} \text{ mol}^{-1})}{2}T^2$$

$$+ \frac{(2.105 \times 10^5 \text{ J K mol}^{-1})}{2}\frac{1}{T} + (210\,010.46 \text{ J mol}^{-1}) - (472.28 \text{ J K}^{-1} \text{ mol}^{-1})T$$

Problem 5.8.1 Show that

$$\left\{\frac{\partial(G/T)}{\partial(1/T)}\right\}_p = H$$

Solution Starting with the differentiation of G/T by T, we have

$$\left\{\frac{\partial(G/T)}{\partial T}\right\}_p = \frac{\partial}{\partial T}\left(\frac{H}{T} - S\right)_p$$

$$= -\frac{1}{T^2}H + \frac{1}{T}\left(\frac{\partial H}{\partial T}\right)_p - \left(\frac{\partial S}{\partial T}\right)_p = -\frac{1}{T^2}H + \frac{C_p}{T} - \frac{C_p}{T}$$

$$= -\frac{H}{T^2}$$

or
$$\left\{\frac{\partial(G/T)}{-(1/T^2)\partial T}\right\}_p = H$$

Since $\mathrm{d}(1/T) = -(1/T^2)\,\mathrm{d}T$ therefore, we have

$$\left\{\frac{\partial(G/T)}{-(1/T^2)\partial T}\right\}_p = \left\{\frac{\partial(G/T)}{\partial(1/T)}\right\}_p = H$$

Problem 5.8.2 Show that

(i) $A = U + T\left(\frac{\partial A}{\partial T}\right)_V$ (ii) $\Delta_r A = \Delta_r U + T\left(\frac{\partial(\Delta_r A)}{\partial T}\right)_V$

(iii) $\left(\frac{\partial(\Delta_r A/T)}{\partial T}\right)_V = -\frac{\Delta_r U}{T^2}$ (iv) $\left(\frac{\partial(\Delta_r A/T)}{\partial(1/T)}\right)_V = \Delta_r U$

Solution **(i)** We start with the expression

$$\left(\frac{\partial A}{\partial T}\right)_V = -S$$

From the definition of A (= $U - TS$), we have

$$-S = \frac{A - U}{T}$$

Substituting this in the previous expression and rearranging, we get

$$A = U + T\left(\frac{\partial A}{\partial T}\right)_V \tag{5.8.13}$$

(ii) Replacing A and U by $\Delta_r A$ and $\Delta_r U$ in Eq. (5.8.13), we get

$$\Delta_r A = \Delta_r H + T\left(\frac{\partial(\Delta_r A)}{\partial T}\right)_V \tag{5.8.14}$$

(iii) Dividing by T in the expression

$$\Delta_r A = \Delta_r U - T\,\Delta_r S$$

we get $\frac{\Delta_r A}{T} = \frac{\Delta_r U}{T} - \Delta_r S$

Differentiation of the above equation with respect to temperature at constant volume gives

$$\left\{\frac{\partial(\Delta_r A/T)}{\partial T}\right\}_V = \left\{\frac{\partial(\Delta_r U/T)}{\partial T}\right\}_V - \left\{\frac{\partial(\Delta_r S)}{\partial T}\right\}_V$$

$$= -\frac{1}{T^2}\,\Delta_r U + \frac{1}{T}\left\{\frac{\partial(\Delta_r U)}{\partial T}\right\}_V - \left\{\frac{\partial(\Delta_r S)}{\partial T}\right\}_V$$

Employing the appropriate forms of Eqs. (2.4.9) and (4.14.6), we get

$$\left\{\frac{\partial(\Delta_r A/T)}{\partial T}\right\}_V = -\frac{\Delta_r U}{T^2} + \frac{\Delta_r C_V}{T} - \frac{\Delta_r C_V}{T} = -\frac{\Delta_r U}{T^2} \tag{5.8.15}$$

(iv) Equation (5.8.15) may be written as

$$\left\{\frac{\partial(\Delta_r A/T)}{-(1/T^2)\partial T}\right\}_V = \Delta_r U$$

Since $\mathrm{d}\left(\frac{1}{T}\right) = -\frac{1}{T^2}\mathrm{d}T$, therefore, we have

$$\left\{\frac{\partial(\Delta_r A/T)}{\partial(1/T)}\right\}_V = \Delta_r U \tag{5.8.16}$$

Problem 5.8.3

Variation of $\Delta_r G$ of a reaction with temperature can be obtained by integrating the equation

$$\left\{\frac{\partial(\Delta_r G)}{\partial T}\right\}_p = -\Delta_r S \qquad (5.8.17)$$

where $\Delta_r S$ of the system can be obtained by integrating the equation

$$\left\{\frac{\partial(\Delta_r S)}{\partial T}\right\}_p = \frac{\Delta_r C_p}{T}$$

With the help of the above two expressions, derive the following expression.

$$\Delta_r G = I_2 - \Delta_r C_p\left(T \ln\frac{T}{\text{K}}\right) + (\Delta_r C_p - I_1)T$$

where I_2 and I_1 are the constants of integration.

Solution

Since $\left(\frac{\partial(\Delta_r S)}{\partial T}\right)_p = \frac{\Delta_r C_p}{T}$

we have $d(\Delta_r S) = \frac{\Delta_r C_p}{T} dT$ (p constant)

Considering $\Delta_r C_p$ to be independent to temperature, on integrating, we get

$$\Delta_r S = \Delta_r C_p \ln\left(\frac{T}{\text{K}}\right) + I_1 \qquad (5.8.18)$$

where I_1 is the constant of integration.

Substituting the above expression in Eq. (5.8.17), we get

$$d(\Delta_r G) = -\left\{\Delta_r C_p \ln\left(\frac{T}{\text{K}}\right) + I_1\right\} dT \qquad (p \text{ constant})$$

On integrating, we have

$$\Delta_r G = \left(-\Delta_r C_p T \ln\frac{T}{\text{K}} + \Delta_r C_p T\right) - I_1 T + I_2$$

or

$$\Delta_r G = I_2 - \Delta_r C_p T \ln\frac{T}{\text{K}} + (\Delta_r C_p - I_1)T \qquad (5.8.19)$$

where I_2 is the constant of integration.

Example 5.8.2

It is possible to supercool water without freezing. 18 g of water is supercooled to 263.15 K (–10 °C) in a thermostat held at this temperature, and then crystallization takes place. Calculate $\Delta_r G$ for this process. Given:

$C_p(H_2O, 1) = 75.312$ J K^{-1} mol^{-1} $\qquad$ $C_p(H_2O, s) = 36.400$ J K^{-1} mol^{-1}

$\Delta_{fus}H$ (at 0 °C) = 6.008 kJ mol^{-1}

Solution

The process of crystallization at 0 °C and at 101.325 kPa pressure is an equilibrium process, for which $\Delta G = 0$. The crystallization of supercooled water is a spontaneous phase transformation, for which ΔG must be less than zero. Its value for this process can be calculated as shown below.

Arithmetic Method

In this method, the given process

$$H_2O\,(1, -10\ °C) \rightarrow H_2O(s, -10\ °C)$$

is replaced by the following reversible steps.

(a) $H_2O(l, -10\ °C) \rightarrow H_2O(l, 0\ °C)$ (5.8.20)

$$\Delta_r H_1 = \int_{263.15K}^{273.15K} C_{p,m}(l)\, dT = (75.312\ J\ K^{-1}\ mol^{-1})\ (10\ K) = 753.12\ J\ mol^{-1}$$

$$\Delta_r S_1 = \int_{263.15K}^{273.15K} \frac{C_{p,m}(l)}{T}\, dT = (75.312\ J\ K^{-1}\ mol^{-1}) \times \ln\left(\frac{273.15\ K}{263.15\ K}\right)$$

$$= 2.809\ J\ K^{-1}\ mol^{-1}$$

(b) $H_2O(l, 0\ °C) \rightarrow H_2O(s, 0\ °C)$ (5.8.21)

$$\Delta_r H_2 = -\ 6.008\ kJ\ mol^{-1}$$

$$\Delta_r S_2 = -\frac{(6\ 008\ J\ mol^{-1})}{(273.15\ K)} = -\ 21.995\ J\ K^{-1}\ mol^{-1}$$

(c) $H_2O(s, 0\ °C) \rightarrow H_2O(s, -10\ °C)$ (5.8.22)

$$\Delta_r H_3 = \int_{273.15K}^{263.15K} C_{p,m}(s)\, dT = (36.400\ J\ K^{-1}\ mol^{-1})\ (-10\ K) = -364.0\ J\ mol^{-1}$$

$$\Delta_r S_3 = \int_{273.15K}^{263.15K} \frac{C_{p,m}(s)}{T}\, dT = (36.400\ J\ K^{-1}\ mol^{-1}) \times \ln\left(\frac{263.15\ K}{273.15\ K}\right)$$

$$= -1.358\ J\ K^{-1}\ mol^{-1}$$

The overall process is obtained by adding Eqs (5.8.20), (5.8.21) and (5.8.22), i.e.

$$H_2O(l, -10\ °C) \rightarrow H_2O(s, -10\ °C)$$

The total changes in $\Delta_r H$ and $\Delta_r S$ are given by

$$\Delta_r H = \Delta_r H_1 + \Delta_r H_2 + \Delta_r H_3$$

$$= (753.12 - 6\ 008 - 364.0)\ J\ mol^{-1} = -\ 5\ 618.88\ J\ mol^{-1}$$

$$\Delta_r S = \Delta_r S_1 + \Delta_r S_2 + \Delta_r S_3$$

$$= (2.809 - 21.995 - 1.358)\ J\ K^{-1}\ mol^{-1}$$

$$= -\ 20.544\ J\ K^{-1}\ mol^{-1}$$

Now $\Delta_r G$ of this process is given by

$$\Delta_r G = \Delta_r H - T\,\Delta_r S$$

$$= -\ 5\ 618.88\ J\ mol^{-1} - (263.15\ K)\ (-\ 20.544\ J\ K^{-1}\ mol^{-1})$$

$$= -\ 212.726\ J\ mol^{-1}$$

Analytical Method We start with Gibbs-Helmholtz equation

$$\left(\frac{\partial(\Delta_r G/T)}{\partial T}\right)_p = -\frac{\Delta_r H}{T^2} \tag{5.8.23}$$

Now $$\left(\frac{\partial(\Delta_r H)}{\partial T}\right)_p = \Delta_r C_p = C_{p,m}(\text{ice}) - C_{p,m}(\text{water})$$

$$= 36.400\ J\ K^{-1}\ mol^{-1} - 75.312\ J\ K^{-1}\ mol^{-1}$$

$$= -\ 38.912\ J\ K^{-1}\ mol^{-1}$$

Therefore

$$d(\Delta_r H) = -(38.912\ \text{J K}^{-1}\ \text{mol}^{-1})\ dT \qquad (p\ \text{constant})$$

Carrying out the integration, we get

$$\Delta_r H = -(38.912\ \text{J K}^{-1}\ \text{mol}^{-1})\ T + I \tag{5.8.24}$$

where I is the constant of integration, the value of which can be determined from the following data:

At 0 °C, $\Delta_r H = -6\ 008\ \text{J mol}^{-1}$. Therefore

$$I = \Delta_r H + (38.912\ \text{J K}^{-1}\ \text{mol}^{-1})T$$
$$= -6\ 008\ \text{J mol}^{-1} + (38.912\ \text{J K}^{-1}\ \text{mol}^{-1})\ (273.15\ \text{K})$$
$$= 4\ 620.81\ \text{J mol}^{-1}$$

Thus, the variation of $\Delta_r H$ with temperature is given by

$$\Delta_r H = (4\ 620.81\ \text{J mol}^{-1}) - (38.912\ \text{J K}^{-1}\ \text{mol}^{-1})T$$

Substituting Eq. (5.8.24) in Eq. (5.8.23), we get

$$d(\Delta_r G/T) = -\left(\frac{I - (38.921\ \text{J K}^{-1}\ \text{mol}^{-1})T}{T^2}\right)dT$$

Again, performing the integration, we get

$$\frac{\Delta_r G}{T} = \frac{I}{T} + (38.912\ \text{J K}^{-1}\ \text{mol}^{-1}) \times \ln\ (T\ /\ \text{K}) + K$$

where K is the constant of integration, the value of which can be determined from the fact that at $T = 273.15$ K, $\Delta_r G = 0$. Therefore

$$K = -\frac{I}{T} - (38.912\ \text{J K}^{-1}\ \text{mol}^{-1}) \times 2.303 \times \log\ (T\ /\text{K})$$
$$= -\left(\frac{4\ 620.81\ \text{J mol}^{-1}}{273.15\ \text{K}}\right) - (38.912\ \text{J K}^{-1}\ \text{mol}^{-1}) \times 2.303 \times \log\ (273.15)$$
$$= -16.917\ \text{J K}^{-1}\ \text{mol}^{-1} - 218.334\ \text{J K}^{-1}\ \text{mol}^{-1}$$
$$= -235.253\ \text{J K}^{-1}\ \text{mol}^{-1}$$

Therefore, the dependence of $\Delta_r G$ on temperature is given by

$$\Delta_r G = (4\ 620.81\ \text{J mol}^{-1}) + (38.912\ \text{J K}^{-1}\ \text{mol}^{-1})\ T \ln\ (T/\text{K}) - (235.253\ \text{J K}^{-1}\ \text{mol}^{-1})T$$

At $T = -10$ °C, i.e. 263.15 K, $\Delta_r G$ is given by

$$\Delta_r G_{263.15\text{K}} = (4\ 620.81\ \text{J mol}^{-1}) + (38.912\ \text{J K}^{-1}\ \text{mol}^{-1})\ (263.15\ \text{K}) \times (2.303)\ \log\ (263.15) - (235.253\ \text{J K}^{-1}\ \text{mol}^{-1})\ (263.15\ \text{K})$$
$$= 4\ 620.81\ \text{J mol}^{-1} + 57\ 073.27\ \text{J mol}^{-1} - 61\ 906.83\ \text{J mol}^{-1}$$
$$= -212.75\ \text{J mol}^{-1}$$

Alternatively, Eqs (5.8.18) and (5.8.19) may be employed to compute $\Delta_r G$ for the freezing of supercooled water. We have

$$\Delta_r S = \Delta_r C_p \ln\ (T/\text{K}) + I_1 \qquad \text{(Eq. 5.8.18)}$$

At the normal freezing point (273.15 K), $\Delta_r S$ is given by

$$\Delta_r S = \frac{\Delta_r H_{\text{freez}}}{T} = -\frac{(6\ 008\ \text{J mol}^{-1})}{(273.15\ \text{K})} = -21.995\ \text{J K}^{-1}\ \text{mol}^{-1}$$

Hence

$$I_1 = \Delta_r S - \Delta_r C_p \ln\ (T/\text{K})$$
$$= (-21.995\ \text{J K}^{-1}\ \text{mol}^{-1}) - (-38.912\ \text{J K}^{-1}\ \text{mol}^{-1})\ (2.303)\ (\log 273.15)$$
$$= 196.34\ \text{J K}^{-1}\ \text{mol}^{-1}$$

The constant of integration I_2 in Eq. (5.8.19) can be obtained by making use of the fact that $\Delta_r G$ is zero at 273.15 K. Thus from Eq. (5.8.19), we get

$$I_2 = \Delta_r C_p\, T \ln (T/\mathrm{K}) - (\Delta_r C_p - I_1)T$$

Substituting the data, we have

$$I_2 = (-38.912\ \mathrm{J\ K^{-1}\ mol^{-1}})(273.15\ \mathrm{K})(2.303)(\log 273.15)$$
$$- (-38.912\ \mathrm{J\ K^{-1}\ mol^{-1}} - 196.34\ \mathrm{J\ K^{-1}\ mol^{-1}})(273.15\ \mathrm{K})$$
$$= (-59\,638.60 + 64\,259.08)\ \mathrm{J\ mol^{-1}} = 4\,620.48\ \mathrm{J\ mol^{-1}}$$

Thus $\Delta_r G$ at 263.15 K is

$$\Delta_r G = I_2 - (\Delta_r C_p)T \ln (T/\mathrm{K}) + (\Delta_r C_p - I_1)T$$
$$= (4\,620.48\ \mathrm{J\ mol^{-1}}) + (38.912\ \mathrm{J\ K^{-1}\ mol^{-1}})(263.15\ \mathrm{K})$$
$$\times (2.303)(\log 263.15) + (-38.912\ \mathrm{J\ K^{-1}\ mol^{-1}}$$
$$- 196.34\ \mathrm{J\ K^{-1}\ mol^{-1}})(263.15\ \mathrm{K})$$
$$= (4\,620.48 + 57\,073.25 - 61\,906.56)\ \mathrm{J\ mol^{-1}}$$
$$= -212.83\ \mathrm{J\ mol^{-1}}$$

5.9 RESUME CONCERNING *U, H, S, A* AND *G*

We have so far established five functions U, H, S, A and G. These are the energy, the enthalpy, the entropy, the Helmholtz free energy and the Gibbs free energy. Of these five, H, A and G, were developed by suitable changes of variables to create functions with special useful characteristics. All the five are state functions and their differentials are exact. They are extensive variables.

From the first law, we have the relations

$$\mathrm{d}U = \mathrm{d}q - p\,\mathrm{d}V \tag{5.9.1}$$

and

$$\mathrm{d}H = \mathrm{d}q + V\,\mathrm{d}p \tag{5.9.2}$$

From the combination of the first and second laws, we have four equations:

$$\mathrm{d}U = T\,\mathrm{d}S - p\,\mathrm{d}V \tag{5.9.3}$$

$$\mathrm{d}H = T\,\mathrm{d}S + V\,\mathrm{d}p \tag{5.9.4}$$

$$\mathrm{d}A = -S\,\mathrm{d}T - p\,\mathrm{d}V \tag{5.9.5}$$

$$\mathrm{d}G = -S\,\mathrm{d}T + V\,\mathrm{d}p \tag{5.9.6}$$

Natural Independent Variables

We note that these equations give us the convenient independent variables (also known as *natural independent variables*) which may be used for each of the four functions. These are

$$U = f(S, V)$$
$$H = f(S, p)$$
$$A = f(T, V)$$

and

$$G = f(T, p)$$

Maxwell Relations and Other Partial Derivatives

Since $\mathrm{d}U$, $\mathrm{d}H$, $\mathrm{d}A$ and $\mathrm{d}G$ are all exact differentials, it follows that

$$\left(\frac{\partial U}{\partial S}\right)_V = T \quad \text{and} \quad \left(\frac{\partial U}{\partial V}\right)_S = -p \tag{5.9.7}$$

$$\left(\frac{\partial H}{\partial S}\right)_p = T \quad \text{and} \quad \left(\frac{\partial H}{\partial p}\right)_S = V \tag{5.9.8}$$

$$\left(\frac{\partial A}{\partial T}\right)_V = -S \quad \text{and} \quad \left(\frac{\partial A}{\partial V}\right)_T = -p \tag{5.9.9}$$

$$\left(\frac{\partial G}{\partial T}\right)_p = -S \quad \text{and} \quad \left(\frac{\partial G}{\partial p}\right)_T = V \tag{5.9.10}$$

Each of these equations tells us how one dependent variables changes with a change of one independent variable, keeping the appropriate variable constant.

Applying the Euler's reciprocity relations to Eqs. (5.9.3) to (5.9.6), we get the most important relations, known as the Maxwell relations:

$$\left(\frac{\partial T}{\partial V}\right)_S = -\left(\frac{\partial p}{\partial S}\right)_V \tag{5.9.11}$$

$$\left(\frac{\partial T}{\partial p}\right)_S = \left(\frac{\partial V}{\partial S}\right)_p \tag{5.9.12}$$

$$\left(\frac{\partial S}{\partial V}\right)_T = \left(\frac{\partial p}{\partial T}\right)_V \tag{5.9.13}$$

and

$$\left(\frac{\partial S}{\partial p}\right)_T = -\left(\frac{\partial V}{\partial T}\right)_p \tag{5.9.14}$$

All these relations are important because they equate the rate of change of a particular quantity, (which cannot be determined experimentally) with those experimentally observable quantities of pressure, volume and temperature.

The Thermodynamic Square

The thermodynamic square (Fig. 5.9.1) is a mnemonic diagram that can be used to obtain a number of useful thermodynamic relations. This was suggested by Max Born and F. O. Koeing and can be used for a system of fixed mass and composition involving only the work of expansion.

In this diagram, the thermodynamic functions U, H, G and A are placed on the sides of a square and the corners of the sides contain their respective independent variables; $U = f(S, V)$, $H = f(S, p)$, $G = f(T, p)$ and $A = f(T, V)$.

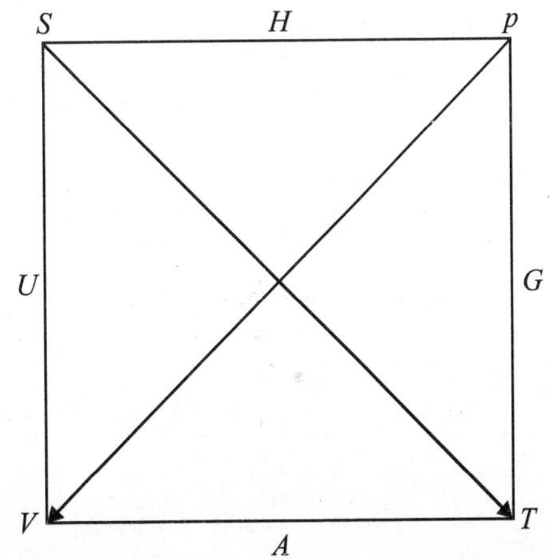

Fig. 5.9.1 The thermodynamic square to obtain a number of useful thermodynamic relations

To write the differential expression for any of these functions in terms of the differentials of its natural independent variables, we need only to note that the coefficient of the differential of the independent variable is placed diagonally opposite this variable on the square. An arrow pointing away from a natural variable indicates a positive sign. Thus, we have

$$dU = T\,dS - p\,dV$$
$$dH = T\,dS + V\,dp$$
$$dA = -S\,dT - p\,dV$$
$$dG = -S\,dT + V\,dp$$

The Maxwell relations can be read off form the above diagram using the corners of two opposite sides of the square with the signs determined by the placement of the arrows. If the arrows are symmetrically placed, the sign is positive, otherwise it is negative. Thus, we have

Sides U and G: $\left(\frac{\partial S}{\partial V}\right)_T = \left(\frac{\partial p}{\partial T}\right)_V$ $\qquad$ $\left(\frac{\partial V}{\partial S}\right)_p = \left(\frac{\partial T}{\partial p}\right)_S$

Sides H and A: $\left(\frac{\partial S}{\partial p}\right)_T = -\left(\frac{\partial V}{\partial T}\right)_p$ $\qquad$ $\left(\frac{\partial p}{\partial S}\right)_V = -\left(\frac{\partial T}{\partial V}\right)_S$

5.10 DERIVATIONS OF SOME THERMODYNAMIC RELATIONS

In this section, we derive a few thermodynamic relations concerning the functions S, A and G.

(i) $$\left(\frac{\partial T}{\partial V}\right)_S = -\frac{T(\partial p/\partial T)_V}{C_V} \tag{5.10.1}$$

We start with Eq. (4.14.14), such that

$$dS = \frac{C_V}{T}dT + \left(\frac{\partial p}{\partial T}\right)_V dV$$

For an isentropic change of state $dS = 0$. With this, the previous expression becomes

$$0 = \frac{C_V}{T}(\partial T)_S + \left(\frac{\partial p}{\partial T}\right)_V (\partial V)_S$$

or $$\left(\frac{\partial T}{\partial V}\right)_S = -\frac{T(\partial p/\partial T)_V}{C_V}$$

(ii) $$\left(\frac{\partial T}{\partial p}\right)_S = \frac{T(\partial V/\partial T)_p}{C_p} \tag{5.10.2}$$

We start with Eq. (4.15.15), such that

$$dS = \frac{C_p}{T}dT - \left(\frac{\partial V}{\partial T}\right)_p dp$$

For an isentropic change of state, we get

$$0 = \frac{C_p}{T}(\partial T)_S - \left(\frac{\partial V}{\partial T}\right)_p (\partial p)_S \quad \text{or} \quad \left(\frac{\partial T}{\partial p}\right)_S = \frac{T(\partial V/\partial T)_p}{C_p}$$

(iii) Equations of State for Adiabatic Reversible Processes

Equations (5.10.1) and (5.10.2) can be used to derive the equation of state for an adiabatic reversible process, involving an ideal gas since for this process, change in entropy during expansion or compression is zero. The required relations, which were derived earlier, can be rederived from these two relations as shown in the following.

Temperature and volume relation Equation (5.10.1) is

$$\left(\frac{\partial T}{\partial V}\right)_S = -\frac{T(\partial p/\partial T)_V}{C_V} \qquad \text{(Eq. 5.10.1)}$$

For an ideal gas $pV = nRT$, therefore

$$\left(\frac{\partial p}{\partial T}\right)_V = \frac{nR}{V}$$

With this Eq. (5.10.1) becomes

$$\left(\frac{\partial T}{\partial V}\right)_S = -\frac{nRT}{C_V V} = -\frac{RT}{C_{V,\mathrm{m}} V}$$

or $$\frac{\mathrm{d}T}{T} = -\left(\frac{R}{C_{V,\mathrm{m}}}\right)\frac{\mathrm{d}V}{V} \qquad (S \text{ constant})$$

Integrating this, we have

$$\ln\left(\frac{T_2}{T_1}\right) = -\left(\frac{R}{C_{V,\mathrm{m}}}\right)\ln\left(\frac{V_2}{V_1}\right) \quad \text{or} \quad \ln\left(\frac{T_2}{T_1}\right) = \ln\left(\frac{V_1}{V_2}\right)^{R/C_{V,\mathrm{m}}}$$

Thus $$\frac{T_2}{T_1} = \left(\frac{V_1}{V_2}\right)^{R/C_{V,\mathrm{m}}} \quad \text{or} \quad T_2 V_2^{R/C_{V,\mathrm{m}}} = T_1 V_1^{R/C_{V,\mathrm{m}}}$$

Hence $$T\,V^{R/C_{V,\mathrm{m}}} = \text{constant} \qquad (5.10.3)$$

Temperature and pressure relation We have from Eq. (5.10.2)

$$\left(\frac{\partial T}{\partial p}\right)_S = \frac{T(\partial V/\partial T)_p}{C_p} \qquad \text{(Eq. 5.10.2)}$$

For an ideal gas $pV = nRT$, therefore

$$\left(\frac{\partial V}{\partial T}\right)_p = \frac{nR}{p}$$

Using this relation, Eq. (5.10.2) becomes

$$\left(\frac{\partial T}{\partial p}\right)_S = \frac{nRT}{pC_p} = \frac{RT}{pC_{p,\mathrm{m}}}$$

Rearranging, we get

$$\frac{\mathrm{d}T}{T}=\frac{R}{C_{p,\mathrm{m}}}\frac{\mathrm{d}p}{p} \qquad (S \text{ constant})$$

On integration, we get

$$\ln\left(\frac{T_2}{T_1}\right)=\frac{R}{C_{p,\mathrm{m}}}\ln\left(\frac{p_2}{p_1}\right) \quad \text{or} \quad \ln\left(\frac{T_2}{T_1}\right)=\ln\left(\frac{p_1}{p_2}\right)^{-R/C_{p,\mathrm{m}}}$$

or $$\frac{T_2}{T_1}=\left(\frac{p_1}{p_2}\right)^{-R/C_{p,\mathrm{m}}} \quad \text{or} \quad T_2 p_2^{-R/C_{p,\mathrm{m}}}=T_1 p_1^{-R/C_{p,\mathrm{m}}}$$

or $$T\,p^{-R/C_{p,\mathrm{m}}}=\text{constant} \qquad (5.10.4)$$

Pressure and volume relation Pressure-volume relationship can be obtained starting from either Eq. (4.14.14) or Eq. (4.15.15).

Writing Eq. (4.14.14), we have

$$\mathrm{d}S=\frac{C_V}{T}\mathrm{d}T+\left(\frac{\partial p}{\partial T}\right)_V \mathrm{d}V$$

Taking $T = f(p, V)$, we have

$$\mathrm{d}T=\left(\frac{\partial T}{\partial p}\right)_V \mathrm{d}p+\left(\frac{\partial T}{\partial V}\right)_p \mathrm{d}V$$

Substituting the above expression of $\mathrm{d}T$ in the previous expression, we have

$$\mathrm{d}S=\frac{C_V}{T}\left\{\left(\frac{\partial T}{\partial p}\right)_V \mathrm{d}p+\left(\frac{\partial T}{\partial V}\right)_p \mathrm{d}V\right\}+\left(\frac{\partial p}{\partial T}\right)_V \mathrm{d}V$$

or $$\mathrm{d}S=\frac{C_V}{T}\left(\frac{\partial T}{\partial p}\right)_V \mathrm{d}p+\left\{\frac{C_V}{T}\left(\frac{\partial T}{\partial V}\right)_p+\left(\frac{\partial p}{\partial T}\right)_V\right\}\mathrm{d}V$$

For an ideal gas $pV = nRT$, therefore

$$\left(\frac{\partial T}{\partial p}\right)_V=\frac{V}{nR}; \quad \left(\frac{\partial T}{\partial V}\right)_p=\frac{p}{nR} \quad \text{and} \quad \left(\frac{\partial p}{\partial T}\right)_V=\frac{nR}{V}$$

With these, we obtain

$$\mathrm{d}S=\frac{C_{V,\mathrm{m}}}{T}\frac{V}{R}\mathrm{d}p+\left\{\frac{C_{V,\mathrm{m}}}{T}\frac{p}{R}+\frac{nR}{V}\right\}\mathrm{d}V$$

Dividing this equation by $\mathrm{d}V$ and introducing the condition of constant S (isentropic process), we have

$$0=\frac{C_{V,\mathrm{m}}}{T}\frac{V}{R}\left(\frac{\partial p}{\partial V}\right)_S+\left\{\frac{C_{V,\mathrm{m}}}{T}\frac{p}{R}+\frac{nR}{V}\right\}$$

Rearranging, we have

$$\left(\frac{\partial p}{\partial V}\right)_S = -\left(\frac{C_{V,\mathrm{m}}}{T}\frac{p}{R} + \frac{nR}{V}\right)\left(\frac{TR}{C_{V,\mathrm{m}}V}\right) = -\left\{\frac{p}{V} + \frac{nR^2T}{V^2C_{V,\mathrm{m}}}\right\}$$

$$= -\left(\frac{p}{V} + \frac{R^2}{V^2}\frac{pV}{RC_{V,\mathrm{m}}}\right) = -\frac{p}{V}\left\{1 + \frac{R}{C_{V,\mathrm{m}}}\right\}$$

$$= -\frac{p}{V}\left(1 + \frac{C_{p,\mathrm{m}} - C_{V,\mathrm{m}}}{C_{V,\mathrm{m}}}\right) = -\frac{p}{V}\left\{\frac{C_{p,\mathrm{m}}}{C_{V,\mathrm{m}}}\right\} = -\frac{p}{V}\gamma$$

Separating the variables, we get

$$\frac{\mathrm{d}p}{p} = -\gamma\frac{\mathrm{d}V}{V}$$

Integrating this, we have

$$\ln\left(\frac{p_2}{p_1}\right) = -\gamma\ \ln\left(\frac{V_2}{V_1}\right) \quad \text{or} \quad \ln\left(\frac{p_2}{p_1}\right) = \ln\left(\frac{V_1}{V_2}\right)^\gamma$$

or $$\frac{p_2}{p_1} = \left(\frac{V_1}{V_2}\right)^\gamma \quad \text{or} \quad p_2V_2^\gamma = p_1V_1^\gamma$$

or $$pV^\gamma = \text{constant} \tag{5.10.5}$$

(iv) Thermodynamic Equations of State The two thermodynamic equations of state

$$\left(\frac{\partial U}{\partial V}\right)_T = T\left(\frac{\partial p}{\partial T}\right)_V - p \tag{5.10.6}$$

$$\left(\frac{\partial H}{\partial p}\right)_T = V - T\left(\frac{\partial V}{\partial T}\right)_p \tag{5.10.7}$$

can be derived from Eqs (5.9.3) and (5.9.4). Writing Eq. (5.9.3), we get

$$\mathrm{d}U = T\,\mathrm{d}S - p\,\mathrm{d}V$$

Dividing this by $\mathrm{d}V$ and introducing the condition of constant temperature, we get

$$\left(\frac{\partial U}{\partial V}\right)_T = T\left(\frac{\partial S}{\partial V}\right)_T - p$$

But $$\left(\frac{\partial S}{\partial V}\right)_T = \left(\frac{\partial p}{\partial T}\right)_V \qquad \text{(Eq. 5.9.13)}$$

Therefore $$\left(\frac{\partial U}{\partial V}\right)_T = T\left(\frac{\partial p}{\partial T}\right)_V - p$$

Equation (5.9.4) is

$$\mathrm{d}H = T\,\mathrm{d}S + V\,\mathrm{d}p$$

Dividing this by dp and introducing the condition of constant temperature, we get

$$\left(\frac{\partial H}{\partial p}\right)_T = T\left(\frac{\partial S}{\partial p}\right)_T + V \tag{5.10.8}$$

But $\left(\frac{\partial S}{\partial p}\right)_T = -\left(\frac{\partial V}{\partial T}\right)_p$ (Eq. 5.9.14)

Therefore $\left(\frac{\partial H}{\partial p}\right)_T = V - T\left(\frac{\partial V}{\partial T}\right)_p$

(v) $$\left(\frac{\partial V}{\partial T}\right)_S = \frac{C_V}{T}\frac{(\partial V/\partial p)_T}{(\partial V/\partial T)_p} = \left[\frac{C_p}{T} + \frac{(\partial V/\partial T)_p^2}{(\partial V/\partial p)_T}\right]\frac{(\partial V/\partial p)_T}{(\partial V/\partial T)_p} \tag{5.10.9}$$

Since $S = f(T, V)$, therefore

$$\mathrm{d}S = \left(\frac{\partial S}{\partial T}\right)_V \mathrm{d}T + \left(\frac{\partial S}{\partial V}\right)_T \mathrm{d}V$$

Dividing by dT and introducing the condition of constant S, we get

$$0 = \left(\frac{\partial S}{\partial T}\right)_V + \left(\frac{\partial S}{\partial V}\right)_T\left(\frac{\partial V}{\partial T}\right)_S$$

Rearranging, we have

$$\left(\frac{\partial V}{\partial T}\right)_S = -\frac{(\partial S/\partial T)_V}{(\partial S/\partial V)_T}$$

Using Eqs (4.14.6) and (4.14.13), we have

$$\left(\frac{\partial V}{\partial T}\right)_S = -\frac{C_V}{T}\frac{\kappa_T}{\alpha} = \frac{C_V(\partial V/\partial p)_T}{T(\partial V/\partial T)_p}$$

Using Eq. (4.14.18), we have

$$\left(\frac{\partial V}{\partial T}\right)_S = \frac{(\partial V/\partial p)_T}{(\partial V/\partial T)_p}\left[\frac{C_p}{T} + \frac{(\partial V/\partial T)_p^2}{(\partial V/\partial p)_T}\right]$$

(vi) $$\left(\frac{\partial U}{\partial T}\right)_S = -p\frac{(\partial V/\partial p)_T}{(\partial V/\partial T)_p}\frac{C_V}{T} = -p\frac{(\partial V/\partial p)_T}{(\partial V/\partial T)_p}\left(\frac{C_p}{T} + \frac{(\partial V/\partial T)_p^2}{(\partial V/\partial p)_T}\right) \tag{5.10.10}$$

Since $U = f(T, V)$, therefore

$$\mathrm{d}U = \left(\frac{\partial U}{\partial T}\right)_V \mathrm{d}T + \left(\frac{\partial U}{\partial V}\right)_T \mathrm{d}V$$

Dividing by dT and introducing the condition of constant S, we have

$$\left(\frac{\partial U}{\partial T}\right)_S = \left(\frac{\partial U}{\partial T}\right)_V + \left(\frac{\partial U}{\partial V}\right)_T \left(\frac{\partial V}{\partial T}\right)_S$$

Using Eqs (5.10.6) and (5.9.11), we have

$$\left(\frac{\partial U}{\partial T}\right)_S = C_V + \left\{T\left(\frac{\partial p}{\partial T}\right)_V - p\right\}\left\{-\left(\frac{\partial S}{\partial p}\right)_V\right\}$$

Using Eq. (4.15.21), we have

$$\left(\frac{\partial U}{\partial T}\right)_S = C_V + \left\{T\left(\frac{\partial p}{\partial T}\right)_V - p\right\}\left\{-\frac{\kappa_T C_V}{T\alpha}\right\}$$

$$= C_V - T\left(\frac{\partial p}{\partial T}\right)_V \frac{\kappa_T C_V}{T\alpha} + \frac{p\kappa_T C_V}{T\alpha}$$

Since $\left(\frac{\partial p}{\partial T}\right)_V = \frac{\alpha}{\kappa_T}$ (Eq. 1.3.60a)

we have $\left(\frac{\partial U}{\partial T}\right)_S = \frac{pC_V}{T}\frac{\kappa_T}{\alpha} = \frac{C_V\, p}{T}\left(\frac{\partial T}{\partial p}\right)_V = -\frac{C_V\, p}{T}\frac{(\partial V/\partial p)_T}{(\partial V/\partial T)_p}$

where we have used the cyclic rule $(\partial T/\partial p)_V\,(\partial p/\partial V)_T\,(\partial V/\partial T)_p + 1 = 0$.
Using Eq. (4.14.18), we get

$$\left(\frac{\partial U}{\partial T}\right)_S = -p\frac{(\partial V/\partial p)_T}{(\partial V/\partial T)_p}\left\{\frac{C_p}{T} + \frac{(\partial V/\partial T)_p^2}{(\partial V/\partial p)_T}\right\}$$

(vii) $$\left(\frac{\partial H}{\partial T}\right)_S = \frac{VC_p}{(\partial V/\partial T)_p} \qquad (5.10.11)$$

Since $H = f(T, p)$, therefore

$$\mathrm{d}H = \left(\frac{\partial H}{\partial T}\right)_p \mathrm{d}T + \left(\frac{\partial H}{\partial p}\right)_T \mathrm{d}p$$

Dividing by dT and introducing the condition of constant S, we get

$$\left(\frac{\partial H}{\partial T}\right)_S = \left(\frac{\partial H}{\partial T}\right)_p + \left(\frac{\partial H}{\partial p}\right)_T \left(\frac{\partial p}{\partial T}\right)_S$$

Using Eqs (5.10.7) and (5.9.12), we have

$$\left(\frac{\partial H}{\partial T}\right)_S = C_p + \left\{V - T\left(\frac{\partial V}{\partial T}\right)_p\right\}\left(\frac{\partial S}{\partial V}\right)_p$$

which on using Eq. (4.15.24), becomes

$$\left(\frac{\partial H}{\partial T}\right)_S = C_p + \left\{V - T\left(\frac{\partial V}{\partial T}\right)_p\right\}\frac{C_p}{TV\alpha}$$

$$= C_p + \frac{C_p}{T\alpha} - C_p = \frac{C_p}{T\alpha} = \frac{VC_p}{T(\partial V/\partial T)_p}$$

(viii) $$\left(\frac{\partial A}{\partial T}\right)_S = -p\frac{(\partial V/\partial p)_T}{(\partial V/\partial T)_p}\frac{C_V}{T} - S = -p\frac{(\partial V/\partial p)_T}{(\partial V/\partial T)_p}\left[\frac{C_p}{T} + \frac{(\partial V/\partial T)_p^2}{(\partial V/\partial p)_T}\right] - S \tag{5.10.12}$$

We start with $A = U - TS$, which gives

$$dA = dU - T\,dS - S\,dT$$

Taking $U = f(T, V)$ and substituting its differential in the above equation, we have

$$dA = \left(\frac{\partial U}{\partial T}\right)_V dT + \left(\frac{\partial U}{\partial V}\right)_T dV - T\,dS - S\,dT$$

Dividing the above expression by dT and introducing the condition of constant S, we get

$$\left(\frac{\partial A}{\partial T}\right)_S = \left(\frac{\partial U}{\partial T}\right)_V + \left(\frac{\partial U}{\partial V}\right)_T\left(\frac{\partial V}{\partial T}\right)_S - S$$

Using Eqs (5.10.6) and (5.9.11), we have

$$\left(\frac{\partial A}{\partial T}\right)_S = C_V + \left\{T\left(\frac{\partial p}{\partial T}\right)_V - p\right\}\left\{\left(-\frac{\partial S}{\partial p}\right)_V\right\} - S$$

Using Eq. (4.15.21), we have

$$\left(\frac{\partial A}{\partial T}\right)_S = C_V - \left\{T\left(\frac{\partial p}{\partial T}\right)_V - p\right\}\left(\frac{\kappa_T C_V}{T\alpha}\right) - S$$

Since $$\frac{\alpha}{\kappa_T} = \left(\frac{\partial p}{\partial T}\right)_V \qquad \text{(Eq. 1.3.60a)}$$

therefore $$\left(\frac{\partial A}{\partial T}\right)_S = C_V - C_V + \frac{pC_V}{T}\left(\frac{\partial T}{\partial p}\right)_V - S = \frac{pC_V}{T}\left(\frac{\partial T}{\partial p}\right)_V - S$$

Using the cyclic rule $(\partial p/\partial T)_V\,(\partial T/\partial V)_p\,(\partial V/\partial p)_T + 1 = 0$, we have

$$\left(\frac{\partial A}{\partial T}\right)_S = -\frac{pC_V}{T}\left(\frac{\partial T}{\partial V}\right)_p\left(\frac{\partial V}{\partial p}\right)_T - S = -\frac{pC_V(\partial V/\partial p)_T}{T(\partial V/\partial T)_p} - S$$

Making use of Eq. (4.14.18), we get

$$\left(\frac{\partial A}{\partial T}\right)_S = -\frac{p(\partial V/\partial p)_T}{(\partial V/\partial T)_p}\left[\frac{C_p}{T}+\frac{(\partial V/\partial T)_p^2}{(\partial V/\partial p)_T}\right]-S$$

(ix) $$\left(\frac{\partial G}{\partial T}\right)_S = \frac{VC_p}{T(\partial V/\partial T)_p} - S \qquad (5.10.13)$$

We start with $G = H - TS$, therefore

$$\mathrm{d}G = \mathrm{d}H - T\,\mathrm{d}S - S\,\mathrm{d}T$$

Taking $H = f(T, p)$ and substituting its differential in the above equation, we have

$$\mathrm{d}G = \left(\frac{\partial H}{\partial T}\right)_p \mathrm{d}T + \left(\frac{\partial H}{\partial p}\right)_T \mathrm{d}p - T\,\mathrm{d}S - S\,\mathrm{d}T$$

Dividing this by dT and introducing the condition of constant S, we have

$$\left(\frac{\partial G}{\partial T}\right)_S = \left(\frac{\partial H}{\partial T}\right)_p + \left(\frac{\partial H}{\partial p}\right)_T\left(\frac{\partial p}{\partial T}\right)_S - S$$

Making use of Eqs (5.10.7) and (5.9.12), we get

$$\left(\frac{\partial G}{\partial T}\right)_S = C_p + \left\{V - T\left(\frac{\partial V}{\partial T}\right)_p\right\}\left(\frac{\partial S}{\partial V}\right)_p - S$$

Now using Eq. (4.15.24), we get

$$\left(\frac{\partial G}{\partial T}\right)_S = C_p + \left\{V - T\left(\frac{\partial V}{\partial T}\right)_p\right\}\left(\frac{C_p}{TV\alpha}\right) - S = C_p + \frac{C_p}{T\alpha} - C_p - S$$

$$= \frac{C_p}{T\alpha} - S = \frac{VC_p}{T(\partial V/\partial T)_p} - S$$

(x) $$\left(\frac{\partial p}{\partial T}\right)_G = \frac{S}{V} \qquad (5.10.14)$$

Since $G = f(T, p)$, we write

$$\mathrm{d}G = \left(\frac{\partial G}{\partial T}\right)_p \mathrm{d}T + \left(\frac{\partial G}{\partial p}\right)_T \mathrm{d}p$$

Dividing this by dT and introducing the condition of constant G, we get

$$\left(\frac{\partial p}{\partial T}\right)_G = \frac{-(\partial G/\partial T)_p}{(\partial G/\partial p)_T} = \frac{S}{V}$$

(xi) $$\left(\frac{\partial S}{\partial T}\right)_G = \frac{C_p}{T} - \frac{S}{V}\left(\frac{\partial V}{\partial T}\right)_p \qquad (5.10.15)$$

We start with $G = f(T, S)$. Its total differential is

$$\mathrm{d}G = \left(\frac{\partial G}{\partial T}\right)_S \mathrm{d}T + \left(\frac{\partial G}{\partial S}\right)_T \mathrm{d}S$$

Dividing by dT and introducing the condition of constant G, we have

$$\left(\frac{\partial S}{\partial T}\right)_G = -\frac{(\partial G/\partial T)_S}{(\partial G/\partial S)_T} \tag{5.10.15a}$$

Making use of Eq. (5.9.14), we have

$$\left(\frac{\partial G}{\partial S}\right)_T \equiv \left(\frac{\partial G}{\partial p}\right)_T \left(\frac{\partial p}{\partial S}\right)_T = -V\left(\frac{\partial T}{\partial V}\right)_p$$

and moreover, from Eq. (5.10.13)

$$\left(\frac{\partial G}{\partial T}\right)_S = \frac{VC_p}{T(\partial V/\partial T)_p} - S$$

Thus, Eq. (5.10.15a) reduces to

$$\left(\frac{\partial S}{\partial T}\right)_G = -\left\{\frac{VC_p}{T(\partial V/\partial T)_p} - S\right\}\left\{-\frac{1}{V(\partial T/\partial V)_p}\right\}$$

$$= \frac{C_p}{T} - \frac{S}{V}\left(\frac{\partial V}{\partial T}\right)_p$$

(xii) $$\frac{C_p}{C_V} = 1 + \frac{\alpha^2 VT}{\kappa_S C_p} \tag{5.10.16}$$

where α is coefficient of thermal expansion and κ_S is adiabatic compressibility. The latter is given as

$$\kappa_S = -\frac{1}{V}\left(\frac{\partial V}{\partial p}\right)_S$$

The given expression is

$$\frac{C_p}{C_V} = 1 + \frac{\alpha^2 VT}{\kappa_S C_p}$$

This rearranges to

$$C_p - C_V = \frac{\alpha^2 VTC_V}{\kappa_S C_p} = \frac{\alpha^2 VTC_V}{-(1/V)(\partial V/\partial p)_S C_p}$$

$$= \frac{\alpha^2 VTC_V}{-(1/V)(\partial V/\partial T)_S (\partial T/\partial p)_S C_p}$$

Making use of Eqs (5.10.9) and (5.10.2), we have

$$C_p - C_V = \frac{\alpha^2 VTC_V}{-\dfrac{1}{V}\dfrac{(\partial V/\partial p)_T}{(\partial V/\partial T)_p}\dfrac{C_V}{T}\dfrac{T(\partial V/\partial T)_p}{C_p}C_p} = \frac{\alpha^2 VT}{\kappa_T}$$

which is Eq. (4.14.18) and thus the given relation is obtained.

(xiii) $\dfrac{\kappa_T}{\kappa_S} = \dfrac{C_p}{C_V}$ (5.10.17)

Writing the expression for κ_T and κ_S, we get

$$\frac{\kappa_T}{\kappa_S} = \frac{-(1/V)(\partial V/\partial p)_T}{-(1/V)(\partial V/\partial p)_S}$$

$$= \frac{(\partial V/\partial p)_T}{(\partial V/\partial p)_S} \equiv \frac{(\partial V/\partial p)_T}{(\partial V/\partial T)_S(\partial T/\partial p)_S}$$

Making use of Eqs. (5.10.9) and (5.10.2), we get

$$\frac{\kappa_T}{\kappa_S} = \frac{(\partial V/\partial p)_T}{\dfrac{(\partial V/\partial p)_T}{(\partial V/\partial T)_p}\dfrac{C_V}{T}\dfrac{(\partial V/\partial T)_p}{C_p}} = \frac{C_p}{C_V}$$

which is the required relation.

(xiv) $\left(\dfrac{\partial A}{\partial T}\right)_p = -p\left(\dfrac{\partial V}{\partial T}\right)_p - S$

Since $\mathrm{d}A = -p\,\mathrm{d}V - S\mathrm{d}T$, dividing by $\mathrm{d}T$ and introducing the condition of constant p, we get

$$\left(\frac{\partial A}{\partial T}\right)_p = -p\left(\frac{\partial V}{\partial T}\right)_p - S$$

(xv) $\left(\dfrac{\partial A}{\partial p}\right)_T = -p\left(\dfrac{\partial V}{\partial p}\right)_T$ (5.10.18)

Since $\mathrm{d}A = -p\,\mathrm{d}V - S\,\mathrm{d}T$, dividing by $\mathrm{d}p$ and introducing the condition of constant T, we get

$$\left(\frac{\partial A}{\partial p}\right)_T = -p\left(\frac{\partial V}{\partial p}\right)_T$$

(xvi) $\kappa_T - \kappa_S = T\alpha^2 V/C_p$

where $\kappa_T = -(1/V)(\partial V/\partial p)_T$ and $\kappa_S = -(1/V)(\partial V/\partial p)_S$ and are known as isothermal compressibility and isentropic compressibility, respectively, and α is coefficient of thermal expansion

Taking $V = f(p, T)$, we write

$$\mathrm{d}V = \left(\frac{\partial V}{\partial p}\right)_T \mathrm{d}p + \left(\frac{\partial V}{\partial T}\right)_p \mathrm{d}T$$

Dividing by dp and introducing the condition of constant entropy, we get

$$\left(\frac{\partial V}{\partial p}\right)_S = \left(\frac{\partial V}{\partial p}\right)_T + \left(\frac{\partial V}{\partial T}\right)_p \left(\frac{\partial T}{\partial p}\right)_S$$

or $$\left(\frac{\partial V}{\partial p}\right)_T - \left(\frac{\partial V}{\partial p}\right)_S = -\left(\frac{\partial V}{\partial T}\right)_p \left(\frac{\partial T}{\partial p}\right)_S$$

From the cyclic rule

$$\left(\frac{\partial T}{\partial p}\right)_S \left(\frac{\partial p}{\partial S}\right)_T \left(\frac{\partial S}{\partial T}\right)_p + 1 = 0$$

we find that

$$\left(\frac{\partial T}{\partial p}\right)_S = -\frac{1}{(\partial p/\partial S)_T (\partial S/\partial T)_p} = -\frac{(\partial S/\partial p)_T}{(\partial S/\partial T)_p}$$

With this, the previous expression becomes

$$\left(\frac{\partial V}{\partial p}\right)_T - \left(\frac{\partial V}{\partial p}\right)_S = \left(\frac{\partial V}{\partial T}\right)_p \left\{\frac{(\partial S/\partial p)_T}{(\partial S/\partial T)_p}\right\}$$

From Maxwell relations, we find that

$$\left(\frac{\partial S}{\partial p}\right)_T = -\left(\frac{\partial V}{\partial T}\right)_p$$

Moreover, $\left(\frac{\partial S}{\partial T}\right)_p = \frac{C_p}{T}$

Hence $$\left(\frac{\partial V}{\partial p}\right)_T - \left(\frac{\partial V}{\partial p}\right)_S = \left(\frac{\partial V}{\partial T}\right)_p \left\{-\frac{(\partial V/\partial T)_p}{C_p/T}\right\}$$

i.e. $\kappa_T - \kappa_S = T\,\alpha^2 V/C_p$

5.11 BRIDGMAN FORMULAE TO WRITE THE EXPRESSIONS OF FIRST PARTIAL DERIVATIVES

Since each first partial derivative involves three variables (e.g., the derivative $(\partial G/\partial T)_p$ involves G, T, and p), one can write a total of 336 ($\equiv 8 \times 7 \times 6$) first partial derivatives out of eight common thermodyanmic variables p, V, T, U, H, S, A and G. Bridgman has devised a procedure which permits to write the expression of any of these first derivatives in terms of three experimentally obtainable derivatives $(\partial V/\partial T)_p$, $(\partial V/\partial p)_T$ and $(\partial H/\partial T)_p$. The procedure adopted is to write a given first derivative $(\partial x/\partial y)_z$ as $(\partial x)_z/(\partial y)_z$ and then to substitute the appropriate expressions of $(\partial x)_z$ and $(\partial y)_z$ from the total twenty-eight possible Bridgman formulae given in Table 5.11.1.

As an illustration, we derive the expression of $(\partial G/\partial T)_S$ which we write as $(\partial G)_S/(\partial T)_S$. Now consulting Bridgman formulae, we get

$$\left(\frac{\partial G}{\partial T}\right)_S = \frac{(\partial G)_S}{(\partial T)_S} = \frac{-VC_p/T + S(\partial V/\partial T)_p}{-(\partial V/\partial T)_p} = \frac{VC_p}{T(\partial V/\partial T)_p} - S$$

which we have derived earlier (Eq. 5.10.13).

Table 5.11.1 Bridgman Formulae

$(\partial T)_p = -(\partial p)_T = 1$

$(\partial V)_p = -(\partial p)_V = (\partial V/\partial T)_p$

$(\partial S)_p = -(\partial p)_S = C_p/T$

$(\partial U)_p = -(\partial p)_U = C_p - p(\partial V/\partial T)_p$

$(\partial H)_p = -(\partial p)_H = C_p$

$(\partial G)_p = -(\partial p)_G = -S$

$(\partial A)_p = -(\partial p)_A = -S - p(\partial V/\partial T)_p$

$(\partial V)_T = -(\partial T)_V = -(\partial V/\partial p)_T$

$(\partial S)_T = -(\partial T)_S = (\partial V/\partial T)_p$

$(\partial U)_T = -(\partial T)_U = T(\partial V/\partial T)_p + p(\partial V/\partial p)_T$

$(\partial H)_T = -(\partial T)_H = -V + T(\partial V/\partial T)_p$

$(\partial G)_T = -(\partial T)_G = -V$

$(\partial A)_T = -(\partial T)_A = p(\partial V/\partial p)_T$

$(\partial S)_V = -(\partial V)_S = C_p(\partial V/\partial T)_p/T + (\partial V/\partial T)_p^2$

$(\partial U)_V = -(\partial V)_U = C_p(\partial V/\partial p)_T + T(\partial V/\partial T)_p^2$

$(\partial H)_V = -(\partial V)_H = C_p(\partial V/\partial p)_T + T(\partial V/\partial T)_p^2 - V(\partial V/\partial T)_p$

$(\partial G)_V = -(\partial V)_G = -V(\partial V/\partial T)_p - S(\partial V/\partial p)_T$

$(\partial A)_V = -(\partial V)_A = -S(\partial V/\partial p)_T$

$(\partial U)_S = -(\partial S)_U = p\,C_p(\partial V/\partial p)_T/T + p(\partial V/\partial T)_p^2$

$(\partial H)_S = -(\partial S)_H = -V\,C_p/T$

$(\partial G)_S = -(\partial S)_G = -V\,C_p/T + S(\partial V/\partial T)_p$

$(\partial A)_S = -(\partial S)_A = p\,C_p(\partial V/\partial p)_T/T + p(\partial V/\partial T)_p^2 + S(\partial V/\partial T)_p$

$(\partial H)_U = -(\partial U)_H = -V[C_p - p(\partial V/\partial T)_p] - p[C_p(\partial V/\partial p)_T + T(\partial V/\partial T)_p^2]$

$(\partial G)_U = -(\partial U)_G = -V[C_p - p(\partial V/\partial T)_p] + S[T(\partial V/\partial T)_p + p(\partial V/\partial p)_T]$

$(\partial A)_U = -(\partial U)_A = p[C_p(\partial V/\partial p)_T + T(\partial V/\partial T)_p^2]$

$(\partial G)_H = -(\partial H)_G = -V(C_p + S) + TS(\partial V/\partial T)_p$

$(\partial A)_H = -(\partial H)_A = -[S + p(\partial V/\partial T)_p][V - T(\partial V/\partial T)_p] + p(\partial V/\partial p)_T$

$(\partial A)_G = -(\partial G)_A = -S[V + p(\partial V/\partial p)_T] - pV(\partial V/\partial T)_p$

5.12 MISCELLANEOUS NUMERICALS

1. One mole of steam is condensed reversibly to liquid water at 373 K and 101.325 kPa pressure. The heat of vaporization of water is 2.256 8 kJ g^{-1}. Assuming that the steam behaves as an ideal gas, calculate w, q, $\Delta_r U$, $\Delta_r S$, $\Delta_r A$ and $\Delta_r G$ for the condensation process.

Solution The process is

$$H_2O(g, 101.325 \text{ kPa}, 373 \text{ K}) \rightarrow H_2O(1, 101.325 \text{ kPa}, 373 \text{ K})$$

and $\Delta_{vap}H = (2.256\ 8 \text{ kJ g}^{-1})(18 \text{ g mol}^{-1}) = 40.624 \text{ kJ mol}^{-1}$

Since the process takes place at constant pressure, therefore

$$q_p = -\Delta_{vap}H = -40.624 \text{ kJ mol}^{-1}$$

$$w = -p\,\Delta V = -p(V_{m,1} - V_{m,g})$$

$$\simeq p(V_{m,g})$$

where $V_{m,g}$ is the molar volume of the gas at 373 K. Hence

$$w = (101.325 \text{ kPa})\{(22.414 \text{ dm}^3 \text{ mol}^{-1})(373 \text{ K}/273\text{K})\}$$

$$= 3\ 103.0 \text{ kPa dm}^3 \text{ mol}^{-1} = 3\ 103 \text{ J mol}^{-1}$$

$$\Delta_r H = q_p = -40.622\ 4 \text{ kJ mol}^{-1}$$

$$\Delta_r S = -\frac{40\ 622.4 \text{ J mol}^{-1}}{373 \text{ K}} = -108.907 \text{ J K}^{-1} \text{ mol}^{-1}$$

$$\Delta_r G = 0$$

Since $\Delta_r G = \Delta_r A + (\Delta v_g)RT$, we have

$$\Delta_r A = -(\Delta v_g)RT = -(-1)(8.314 \text{ J K}^{-1} \text{ mol}^{-1})(373 \text{ K}) = 3\ 101 \text{ J mol}^{-1}$$

$$\Delta_r U = q + w = -40.622\ 4 \text{ kJ mol}^{-1} + 3.101 \text{ kJ mol}^{-1} = -37.52 \text{ kJ mol}^{-1}$$

---

2. Show that the change

2 mol of an ideal gas (2 bar, 273 K) → 2 mol of gas (1 bar, 273 K) carried out irreversibly against an external pressure of 1 bar is spontaneous.

Solution ΔG for the process can be calculated using the expression

$$\Delta G = nRT \ln \frac{p_2}{p_1}$$

Substituting the given values, we get

$$\Delta G = (2.0 \text{ mol})(8.314 \text{ J K}^{-1} \text{ mol}^{-1})(273 \text{ K})(2.303) \log\left(\frac{1 \text{ bar}}{2 \text{ bar}}\right)$$

$$= -3\ 147.06 \text{ J}$$

Since there occurs a decrease in the value of ΔG, it must therefore, be a spontaneous change.

---

3. Compute $\Delta_r G$ for the following processes.

(a) H_2O (1, 1 atm, 373 K) → H_2O(g, 1 atm, 373 K)

(b) H_2O (1, 2 atm, 373 K) → H_2O(g, 2 atm, 373 K)

(c) H_2O (1, 1 atm, 300 K) → H_2O(g, 0.80 atm, 300 K)

where atm stands for atmospheric pressure (= 101.325 kPa). Given that

Normal boiling point of water = 373 K

$C_{p,m}(H_2O, l) = 75.312\ J\ K^{-1}\ mol^{-1}$

$C_{p,m}(H_2O, g) = 33.054\ J\ K^{-1}\ mol^{-1}$

$\Delta_{vap}H(1\ atm, 373\ K) = 40.639\ kJ\ mol^{-1}$

Solution **(a)** $H_2O(l, 1\ atm, 373\ K) \rightarrow H_2O(g, 1\ atm, 373\ K)$

373 K is the normal boiling point of water at 1 atm pressure. Therefore, the above process takes place at equilibrium conditions. Thus

$$\Delta_r G = 0$$

(b) $H_2O(l, 2\ atm, 373\ K) \rightarrow H_2O(g, 2\ atm, 373\ K)$

This process can be replaced by the following three reversible processes.

(i) $H_2O(l, 2\ atm, 373\ K) \rightarrow H_2O(l, 1\ atm, 373\ K)$

(ii) $H_2O(l, 1\ atm, 373\ K) \rightarrow H_2O(g, 1\ atm, 373\ K)$

(iii) $H_2O(g, 1\ atm, 373\ K) \rightarrow H_2O(g, 2\ atm, 373\ K)$

***For step* (i)** $\Delta_r G_1 = 0$ assuming no variation in $\Delta_r G$ for condensed phase due to the pressure variation alone.

***For step* (ii)** Process takes place at equilibrium conditions; that is

$$\Delta_r G_2 = 0$$

***For step* (iii)**

$$\Delta_r G_3 = RT \ln \frac{p_2}{p_1} = (8.314\ J\ K^{-1} mol^{-1})(373\ K) \times 2.303 \times \log\left(\frac{2\ atm}{1\ atm}\right)$$

$$= 2\ 149.92\ J\ mol^{-1}$$

For the overall process

$$\Delta_r G = \Delta_r G_1 + \Delta_r G_2 + \Delta_r G_3$$

$$= \Delta_r G_3 = 2\ 149.92\ J\ mol^{-1}$$

(c) $H_2O\ (l, 1\ atm\ 300\ K) \rightarrow H_2O(g, 0.08\ atm, 300\ K)$

This process may be replaced by the following four steps.

(i) $H_2O(l, 1\ atm, 300\ K) \rightarrow H_2O(l, 1\ atm, 373\ K)$

(ii) $H_2O(l, 1\ atm, 373\ K) \rightarrow H_2O(g, 1\ atm, 373\ K)$

(iii) $H_2O(g, 1\ atm, 373\ K) \rightarrow H_2O(g, 1\ atm, 300\ K)$

(iv) $H_2O(g, 1\ atm, 300\ K) \rightarrow H_2O(g, 0.08\ atm, 300\ K)$

The changes in enthalpy and entropy in the first three steps can be calculated as follows.

Step (i)

$$\Delta_r H_1 = \int_{300K}^{373K} C_p(l)\,dT = (75.312\ J\ K^{-1} mol^{-1})(73\ K) = 5\ 497.78\ J\ mol^{-1}$$

$$\Delta_r S_1 = \int_{300K}^{373K} \frac{C_p(l)}{T}\,dT = C_p(l) \ln \frac{T_2}{T_1}$$

$$= (75.312\ J\ K^{-1}\ mol^{-1})(2.303) \times \log\left(\frac{373\ K}{300\ K}\right)$$

$$= 16.406\ J\ K^{-1}\ mol^{-1}$$

Step (ii)

$$\Delta_r H_2 = 40\ 639\ J\ mol^{-1}$$

$$\Delta_r S_2 = \frac{(40\,639\ \text{J mol}^{-1})}{(373\ \text{K})} = 108.95\ \text{J K}^{-1}\ \text{mol}^{-1}$$

Step (iii)

$$\Delta_r H_3 = \int_{373\text{K}}^{300\text{K}} C_{p,\text{m}}(\text{g})\,\text{d}T = (33.054\ \text{J K}^{-1}\ \text{mol}^{-1})(-73\ \text{K}) = -2\,412.9\ \text{J mol}^{-1}$$

$$\Delta_r S_3 = \int_{373\text{K}}^{300\text{K}} \frac{C_{p,\text{m}}(\text{g})}{T}\,\text{d}T = (33.054\ \text{J K}^{-1}\ \text{mol}^{-1})(2.303)\log\left(\frac{300\ \text{K}}{373\ \text{K}}\right)$$

$$= -\ 7.200\ \text{J K}^{-1}\ \text{mol}^{-1}$$

The total changes of enthalpy and entropy in the above three steps are:

Reaction	$\dfrac{\Delta_r H}{\text{J mol}^{-1}}$	$\dfrac{\Delta_r S}{\text{J K}^{-1}\ \text{mol}^{-1}}$
H_2O(l, 1 atm 300 K) → H_2O(l, 1 atm, 373 K);	5 497.78	16.406
H_2O(l, 1 atm, 373 K) → H_2O(g, 1 atm, 373 K);	40 639.0	108.95
H_2O(g, 1 atm, 373 K) → H_2O(g, 1 atm, 300 K);	–2 412.9	–7.200
H_2O(l, 1 atm, 300 K) → H_2O(g, 1 atm, 300 K);	43 723.88	118.156

The change in $\Delta_r G$ is given by

$$\Delta_r G = \Delta_r H - T\,\Delta_r S$$

$$= (43\,723.88\ \text{J mol}^{-1}) - (300\ \text{K})\ (118.156\ \text{J K}^{-1}\ \text{mol}^{-1})$$

$$= (43\,723.88\ \text{J mol}^{-1}) - (35\,446.8\ \text{J mol}^{-1})$$

$$\Delta_r G = 8\,277.08\ \text{J mol}^{-1}$$

For the fourth step, $\Delta_r G$ is given by

$$\Delta_r G = RT\ln\frac{p_2}{p_1} = (8.314\ \text{K}^{-1}\ \text{mol}^{-1})\ (300\ \text{K})\ (2.303)\ \log\left(\frac{0.08\ \text{atm}}{1\ \text{atm}}\right)$$

$$= -\ 6\,300.8\ \text{J mol}^{-1}$$

Adding this contribution, we get

	$\Delta_r G$/J mol^{-1}
H_2O(l, 1 atm, 300 K) → H_2O(g, 1 atm, 300 K)	8 277.08
H_2O(g, 1 atm, 300 K) → H_2O(g, 0.08 atm, 300 K)	– 6 300.80
H_2O(l, 1 atm, 300 K) → H_2O(g, 0.08 atm, 300 K)	1 976.28

---

4. The reaction

$$H_2O(\text{l}, p, 273\ \text{K}) \rightarrow H_2O(\text{g}, p, 273\ \text{K})$$

represents an equilibrium reaction for some pressure p, called the vapour pressure. Assuming that the entropy change and heat flow when liquid water is compressed is negligible, estimate the vapour pressure at 273 K. Given that

$$C_{p,\,\text{m}}(H_2O,\ \text{g}) = 33.054\ \text{J K}^{-1}\ \text{mol}^{-1}$$
$$C_{p,\,\text{m}}(H_2O,\ \text{l}) = 75.312\ \text{J K}^{-1}\ \text{mol}^{-1}$$
$$\Delta_{\text{vap}}H(H_2O,\ 373\ \text{K}) = 40.639\ \text{kJ mol}^{-1}$$

Solution

The above process can be replaced by the following five equations.

(i) $H_2O(1, p, 273\ K) \rightarrow H_2O(1, 101.325\ kPa, 273\ K)$

(ii) $H_2O(1, 101.325\ kPa, 273\ K) \rightarrow H_2O(1, 101.325\ kPa, 373\ K)$

(iii) $H_2O(1, 101.325\ kPa, 373\ K) \rightarrow H_2O(g, 101.325\ kPa, 373\ K)$

(iv) $H_2O(g, 101.325\ kPa, 373\ K) \rightarrow H_2O(g, 101.325\ kPa, 273\ K)$

(v) $H_2O(g, 101.325\ kPa, 273\ K) \rightarrow H_2O(g, p, 273\ K)$

The corresponding changes in $\Delta_r H$ and $\Delta_r S$ are given below:

	$\Delta_r H$/J mol^{-1}	$\Delta_r S$/J K^{-1} mol^{-1}
(i)	0	0
(ii)	$75.312 \times 100 = 7\ 531.2$	$75.312 \times \ln \dfrac{373}{273} = 23.505$
(iii)	40 639	$\dfrac{40\ 639}{373} = 108.952$
(iv)	$33.054(-100) = -3\ 305.4$	$33.054 \ln \dfrac{273}{373} = -\ 10.316$
(v)	0	$8.314 \ln \left(\dfrac{101.325\ \text{kPa}}{p}\right)$
Total change	44 864.8	$122.141 + 8.314 \ln \left(\dfrac{101.325\ \text{kPa}}{p}\right)$

For this equilibrium process

$$\Delta_r G = 0 = \Delta_r H - T\,\Delta_r S$$

$$= (44\,864.8\ \text{J mol}^{-1}) - (273\ \text{K})\left\{(122.141\ \text{J K}^{-1}\,\text{mol}^{-1}) + (8.314\ \text{J K}^{-1}\,\text{mol}^{-1}) \ln \left(\frac{101.325}{p/\text{kPa}}\right)\right\}$$

Thus $\quad p = 0.632\ \text{kPa}$

5. The vapour pressure of liquid mercury at 433 K is 4.19 mmHg. Calculate the free energy change accompanying the expansion of one mole of mercury vapour in equilibrium with liquid at 433 K to a pressure of 1 atm at the same temperature assuming the vapour behaves like an ideal monatomic gas.

Solution

The transformation of Hg(l) to Hg(g) at temperature 433 K and at the corresponding equilibrium pressure 4.19 mmHg will be reversible in nature. Thus, ΔG for this process will be equal to zero. But there will occur a change of free energy when the pressure of Hg(g) is changed form 4.19 mmHg to 760 mmHg which can be calculated using the expression

$$\Delta G = nRT \ln \frac{p_2}{p_1}$$

Substituting $p_2 = 760$ mmHg and $p_1 = 4.19$ mmHg, we get

$$\Delta G = (1 \text{ mol})\,(8.314 \text{ J K}^{-1} \text{ mol}^{-1})\,(433 \text{ K})\,(2.303) \log \left(\frac{760 \text{ mmHg}}{4.19 \text{ mmHg}}\right)$$

$$= 18\,725.4 \text{ J mol}^{-1} = 18.725 \text{ kJ mol}^{-1}$$

---

6. Compute $\Delta_r G$ for the change H_2(g, 298 K, 1 atm) $\rightarrow$ H_2(g, 323 K, 2 atm); given that H_2(g) is ideal, that its $C_{p,\,m}$ is constant at 28.87 J K^{-1} mol^{-1} over the temperature range and that S_{298K} for H_2(g) is 130.58 J K^{-1} mol^{-1}.

Solution The change of free energy for the transformation

$$H_2(g,\ T_1 = 298 \text{ K},\ p_1 = 1 \text{ atm}) \rightarrow H_2(g,\ T_2 = 323 \text{ K},\ p_2 = 2 \text{ atm})$$

may be calculated as follows.

$$\Delta_r G = G(p_2, T_2) - G(p_1, T_1) = (H_2 - T_2 S_2) - (H_1 - T_1 S_1)$$

$$= (H_2 - H_1) - T_2 S_2 + T_1 S_1$$

$$\Delta_r G = (H_2 - H_1) - T_2\left[S_1 + \int_{T_1}^{T_2} \frac{C_{p,\mathrm{m}}}{T}\,\mathrm{d}T - \int_{p_1}^{p_2} R\frac{\mathrm{d}p}{p}\right] + T_1 S_1$$

$$= \int_{T_1}^{T_2} C_{p,\mathrm{m}}\,\mathrm{d}T - S_1(T_2 - T_1) - T_2\left[\int_{T_1}^{T_2} \frac{C_{p,\mathrm{m}}}{T}\,\mathrm{d}T - \int_{p_1}^{p_2} R\frac{\mathrm{d}p}{p}\right]$$

$$= C_{p,\mathrm{m}}(\Delta T) - S_1(\Delta T) - T_2\left[C_{p,\mathrm{m}} \ln\frac{T_2}{T_1} + R\ln\frac{p_1}{p_2}\right]$$

Substituting the given values, we get

$$\Delta_r G = (28.87 \text{ J K}^{-1} \text{ mol}^{-1})\,(25 \text{ K}) - (130.58 \text{ J K}^{-1} \text{ mol}^{-1})\,(25 \text{ K})$$

$$-(323 \text{ K})\left[(28.87 \text{ J K}^{-1}\text{mol}^{-1}) \ln\left(\frac{323 \text{ K}}{298 \text{ K}}\right) + (8.314 \text{ J K}^{-1}\text{mol}^{-1}) \ln(1 \text{ atm}/2 \text{ atm})\right]$$

$$= 721.75 \text{ J mol}^{-1} - 3\,264.5 \text{ J mol}^{-1} - (323)\,[2.326 \text{ J mol}^{-1} - 5.763 \text{ J mol}^{-1}]$$

$$= -\,1\,432.6 \text{ J mol}^{-1}$$

Alternative Solution

$$\Delta S = C_{p,\,\mathrm{m}} \ln\frac{T_2}{T_1} + R\ln\frac{p_1}{p_2}$$

$$= (28.87 \text{ J K}^{-1} \text{ mol}^{-1}) \ln\left(\frac{323 \text{ K}}{298 \text{ K}}\right) + (8.314 \text{ J K}^{-1}\text{mol}^{-1}) \ln\left(\frac{1 \text{ atm}}{2 \text{ atm}}\right)$$

$$= 2.326 \text{ J K}^{-1} \text{ mol}^{-1} - 5.764 \text{ J K}^{-1} \text{ mol}^{-1}$$

$$= -\,3.438 \text{ J K}^{-1} \text{ mol}^{-1}$$

$$S_{323K} = S_{298K} + \Delta S = 130.58 \text{ J K}^{-1} \text{ mol}^{-1} - 3.348 \text{ J K}^{-1} \text{ mol}^{-1}$$

$$= 127.142 \text{ J K}^{-1} \text{ mol}^{-1}$$

Now $\quad H_{298K} = 0$

$$H_{323K} = C_{p,\,\mathrm{m}}\,\Delta T = (28.87 \text{ J K}^{-1} \text{ mol}^{-1})\,(25 \text{ K}) = 721.75 \text{ J mol}^{-1}$$

Thus $\quad G_{323K} = H_{323K} - TS_{323K}$

$$= 721.75 \text{ J mol}^{-1} - (323 \text{ K})\,(127.42 \text{ J K}^{-1} \text{ mol}^{-1})$$

$$= -40\,345 \text{ J mol}^{-1} = -40.\,345 \text{ kJ mol}^{-1}$$

$$G_{298\text{K}} = H_{298\text{K}} - TS_{298\text{K}} = 0 - (298 \text{ K})\,(130.58 \text{ J K}^{-1} \text{ mol}^{-1})$$

$$= -38\,912.8 \text{ J mol}^{-1} = -38.913 \text{ kJ mol}^{-1}$$

Thus $\Delta G = G_{323\text{K}} - G_{298\text{K}} = -40.345 \text{ kJ mol}^{-1} + 38.913 \text{ kJ mol}^{-1}$

$$= -1.432 \text{ kJ mol}^{-1}$$

REVISIONARY PROBLEMS

5.1 Show that the criterion of thermodynamic equilibrium under the different conditions are as follows:

(a) *At constant energy and volume*: Entropy attains a maximum value and $dS_{U,\,V} = 0$

(b) *At constant entropy and volume*: Internal energy attains a minimum value and $dU_{S,\,V} = 0$

(c) *At constant entropy and pressure*: Enthalpy attains a minimum value and $dH_{S,\,p} = 0$

(d) *At constant temperature and volume*: Helmholtz free energy attains a minimum value and $dA_{T,\,V} = 0$

(e) *At constant temperature and pressure*: Gibbs free energy attains a minimum value and $dG_{T,\,p} = 0$

5.2 Show that for a system

(a) Decrease in the value of Helmholtz free energy at constant temperature represents the maximum amount of the work that can be obtained from the system.

(b) Decrease in the value of Gibbs free energy at constant temperature and pressure is equal to the net available nonmechanical work.

5.3 State which one of the following four conditions will always lead to a spontaneous change and which one will never lead to a spontaneous change:

(i) $\Delta H = -\text{ve}$; $\Delta S = +\text{ve}$

(ii) $\Delta H = +\text{ve}$; $\Delta S = -\text{ve}$

(iii) $\Delta H = -\text{ve}$; $\Delta S = -\text{ve}$

(iv) $\Delta H = +\text{ve}$; $\Delta S = +\text{ve}$

5.4 Justify the following statements:

(a) The laws of thermodynamics cannot predict the rate of a chemical reaction.

(b) A catalyst cannot change the values of $\Delta_r H$ and $\Delta_r G$ of given chemical reaction.

(c) At low temperature, enthalpy change dominates the $\Delta_r G$ expression whereas at high temperature, it is the entropy change which dominates the value of $\Delta_r G$.

(d) For an isothermal expansion of an ideal gas $\Delta_r G = \Delta_r A$.

5.4 Show that

$$(\partial A/\partial V)_T = p \quad \text{and} \quad (\partial A/\partial T)_V = -S$$

$$\text{and} \quad (\partial G/\partial p)_T = V \quad \text{and} \quad (\partial G/\partial T)_p = -S$$

5.6 (a) Define the necessary condition for the standard state for the free energy of a substance. Show that

(i) For condensed system $G \simeq G^\circ(T)$, i.e. the pressure dependence of G is ignored.

(ii) For an ideal gas

$$\mu = \mu^\circ + RT \ln (p/1 \text{ bar})$$

(b) Draw a graph showing the variation of $\mu - \mu^\circ$ with pressure of an ideal gas.

5.7 Show that for an isothermal change in an ideal gas

$$\Delta G = nRT \ln \frac{p_2}{p_1} = nRT \ln \frac{V_1}{V_2}$$

5.8 (a) Derive the following forms of Gibbs–Helmholtz equation:

$$G = H + T(\partial G/\partial T)_p \qquad \Delta_r G = \partial_r H + T\{\partial(\Delta_r G)/\partial T\}_p$$

$$\left(\frac{\partial(G/T)}{\partial T}\right)_p = -\frac{H}{T^2} \qquad \left(\frac{\partial(\Delta_r G/T)}{\partial T}\right)_p = -\frac{\Delta_r H}{T^2}$$

$$\left(\frac{\partial(G/T)}{\partial(1/T)}\right)_p = H \qquad \left(\frac{\partial(\Delta_r G/T)}{\partial(1/T)}\right)_p = \Delta_r H$$

(b) Derive the corresponding relations for Helmholtz free energy.

5.9 (a) Starting from the fundamental expressions of first and second laws of thermodynamics derive the following relations:

$$\mathrm{d}U = T\,\mathrm{d}S - p\,\mathrm{d}V \qquad \mathrm{d}H = T\,\mathrm{d}S + V\,\mathrm{d}p$$

$$\mathrm{d}A = -S\,\mathrm{d}T - p\,\mathrm{d}V \qquad \mathrm{d}G = -S\,\mathrm{d}T + V\,\mathrm{d}p$$

(b) What are natural independent variables? State these for the functions, *U*, *H*, *A* and *G*.

(c) Starting from the relations given in part (a), derive the following Maxwell relation and discuss their utility.

$$(\partial T/\partial V)_S = -(\partial p/\partial S)_V \qquad (\partial T/\partial p)_S = (\partial V/\partial S)_p$$

$$(\partial S/\partial V)_T = (\partial p/\partial T)_V \qquad (\partial S/\partial p)_T = -(\partial V/\partial T)_p$$

(d) Starting from the relations given in part (a), prove that:

$(\partial U/\partial S)_V = T$ and $(\partial U/\partial V)_S = -p$

$(\partial H/\partial S)_p = T$ and $(\partial H/\partial p)_S = V$

$(\partial A/\partial T)_V = -S$ and $(\partial A/\partial V)_T = -p$

and $(\partial G/\partial T)_p = -S$ and $(\partial G/\partial p)_T = V$

Hence show that

$$(\partial U/\partial S)_V = (\partial H/\partial S)_p \qquad (\partial U/\partial V)_S = (\partial A/\partial V)_T$$

$$(\partial H/\partial p)_S = (\partial G/\partial p)_T \qquad (\partial A/\partial T)_V = (\partial G/\partial T)_p$$

(e) Starting from the relation given in part (a), derive the relations

$$(\partial U/\partial V)_T = T(\partial p/\partial T)_V - p \quad \text{and} \quad (\partial H/\partial p)_T = V - T(\partial V/\partial T)_p$$

What are these called? Evaluate their values for an ideal gas. What do you conclude about the dependence of *U* on *V* and *H* on *p* at a constant temperature for an ideal gas?

5.10 (a) Derive the following relations:

$$(\partial T/\partial V)_S = -\frac{T(\partial p/\partial T)_V}{C_V} \qquad (\partial T/\partial p)_S = \frac{T(\partial V/\partial T)_p}{C_p}$$

(b) An adiabatic reversible process is an isentropic process. Equations given in part (a) express the variation of *V* versus *T* and *p* versus *T* for such types of processes. The required relations can be derived by integrating these relations. For an ideal gas carry out the integrations and show that it leads to

$$T\,V^{R/C_{V,\mathrm{m}}} = \text{constant} \quad \text{and} \quad T\,p^{-R/C_{p,\mathrm{m}}} = \text{constant}$$

(c) Derive the relation involving *p* and *V* for an isentropic process involving an ideal gas.

5.11 Derive the following thermodynamic relations:

(i) $(\partial p/\partial T)_S = \dfrac{C_p}{T}\dfrac{1}{(\partial V/\partial T)_p}$

(ii) $(\partial V/\partial T)_S = \dfrac{C_V}{T}\dfrac{(\partial V/\partial p)_T}{(\partial V/\partial T)_p} = \left[\dfrac{C_p}{T} + \dfrac{(\partial V/\partial T)_p^2}{(\partial V/\partial p)_T}\right]\dfrac{(\partial V/\partial p)_T}{(\partial V/\partial T)_p}$

(iii) $(\partial U/\partial T)_S = -p\dfrac{(\partial V/\partial p)_T}{(\partial V/\partial T)_p}\dfrac{C_V}{T} = -p\dfrac{(\partial V/\partial p)_T}{(\partial V/\partial T)_p}\left[\dfrac{C_p}{T} + \dfrac{(\partial V/\partial T)_p^2}{(\partial V/\partial p)_T}\right]$

(iv) $(\partial H/\partial T)_S = \dfrac{VC_p}{T(\partial V/\partial T)_p}$

(v) $\left(\dfrac{\partial A}{\partial T}\right)_S = -p\dfrac{(\partial V/\partial p)_T}{(\partial V/\partial T)_p}\dfrac{C_V}{T} - S = -p\dfrac{(\partial V/\partial p)_T}{(\partial V/\partial T)_p}\left[\dfrac{C_p}{T} + \dfrac{(\partial V/\partial T)_p^2}{(\partial V/\partial p)_T}\right] - S$

(vi) $\left(\dfrac{\partial G}{\partial T}\right)_S = \dfrac{VC_p}{T(\partial V/\partial T)_p} - S$ (vii) $\left(\dfrac{\partial p}{\partial T}\right)_G = \dfrac{S}{V}$

(viii) $\left(\dfrac{\partial S}{\partial T}\right)_G = \dfrac{C_p}{T} - \dfrac{S}{V}\left(\dfrac{\partial V}{\partial T}\right)_p$

(ix) $\dfrac{C_p}{C_V} = 1 + \dfrac{\alpha^2 VT}{\kappa_S C_p}$, where κ_S is adiabatic compressibility and is given as

$$\kappa_S = -\frac{1}{V}\left(\frac{\partial V}{\partial p}\right)_S$$

(x) $\dfrac{\kappa_T}{\kappa_S} = \dfrac{C_p}{C_V}$

(xi) $\left(\dfrac{\partial A}{\partial T}\right)_p = -p\left(\dfrac{\partial V}{\partial T}\right)_p - S$ (xii) $\left(\dfrac{\partial A}{\partial p}\right)_T = -p\left(\dfrac{\partial V}{\partial p}\right)_T$

5.12 (a) Define the terms fugacity and fugacity coefficient. Show that
- (i) $f = p$ for an ideal gas.
- (ii) $f > p$ for a van der Waals gas at higher pressures.
- (iii) $f < p$ for a van der Waals gas at lower pressures.

(b) In the lower pressure range, the pressure of a real gas is the geometric mean of the ideal pressure and fugacity.

(c) Derive the expression of fugacity for each of the following equation of state.

(i) $p(V_m - b) = RT$ (ii) $(p + a/V_m^2)V_m = RT$

(iii) $(p + a/V_m^2)(V_m - b) = RT$ (iv) $Z = 1 + Bp + Cp^2/2 + \cdots$

(d) Derive the following expressions:

(i) $\left(\dfrac{\partial \ln (f/p)}{\partial T}\right)_p = \dfrac{H_m^0 - H_m}{RT^2}$ (ii) $\left(\dfrac{\partial (H_m^0 - H_m)}{\partial p}\right)_p = C_{p,m}\,\mu_{JT}$

(e) Show that for a van der Waals gas

$$H^0_{\rm m} - H_{\rm m} = \frac{2ap}{RT} - bp - \frac{3abp^2}{2R^2T^2}; \qquad \mu_{\rm JT} = \frac{1}{C_{p,\rm m}}\left(\frac{2a}{RT} - b - \frac{3abp}{R^2T^2}\right)$$

$$C_{p,\rm m} = C^0_{p,\rm m} + \frac{2a}{RT^2}\,p - \frac{3abp^2}{R^2T^3}$$

TRY YOURSELF PROBLEMS

5.1 Derive the following relations:

$$p = -\left(\frac{\partial A}{\partial V}\right)_T; \qquad S = -\left(\frac{\partial A}{\partial T}\right)_V$$

$$U = A + TS = A - T\left(\frac{\partial A}{\partial T}\right)_V; \qquad C_V = \left(\frac{\partial U}{\partial T}\right)_V = -T\left(\frac{\partial^2 A}{\partial T^2}\right)_V$$

$$\left(\frac{\partial U}{\partial V}\right)_T = -p - T\left\{\frac{\partial}{\partial V}\left(\frac{\partial A}{\partial T}\right)_V\right\}_T; \qquad \left(\frac{\partial p}{\partial T}\right)_V = -\left\{\frac{\partial}{\partial T}\left(\frac{\partial A}{\partial V}\right)_T\right\}_V$$

$$\left(\frac{\partial V}{\partial T}\right)_p = -\frac{(\partial p/\partial T)_V}{(\partial p/\partial V)_T} = \frac{\left[\frac{\partial}{\partial T}\left(\frac{\partial A}{\partial V}\right)_T\right]_V}{(\partial^2 A/\partial V^2)_T}$$

$$C_p = C_V + \left[p + \left(\frac{\partial U}{\partial V}\right)_T\right]\left(\frac{\partial V}{\partial T}\right)_p = \frac{T\left[\frac{\partial}{\partial T}\left(\frac{\partial A}{\partial V}\right)_T\right]_V^2}{(\partial^2 A/\partial V^2)_T} - T\left(\frac{\partial^2 A}{\partial T^2}\right)_V$$

$$\alpha = \frac{1}{V}\left(\frac{\partial V}{\partial T}\right)_p = -\frac{\left[\frac{\partial}{\partial T}\left(\frac{\partial A}{\partial V}\right)_T\right]_V}{V\left(\frac{\partial^2 A}{\partial V^2}\right)_T}; \qquad \kappa_T = -\frac{1}{V}\left(\frac{\partial V}{\partial p}\right)_T = \frac{1}{V(\partial^2 A/\partial V^2)_T}$$

5.2 The Helmoholtz function of one mole of certain gas is given by

$$A = -\left(\frac{a}{V_{\rm m}}\right) - RT\ln(V_{\rm m} - b) + f(T)$$

where a and b are constants and $f(T)$ is a function of temperature only. Derive an expression for the pressure of the gas. [*Ans.* $p = RT/(V_{\rm m} - b) - a/V_{\rm m}^2$]

5.3 The Gibbs function of one mole of a certain gas is given by

$$G = RT\ln p + A + Bp + \frac{1}{2}Cp^2 + \frac{1}{3}Dp^3$$

where A, B, C and D are functions of temperature only. Find the equation of state of the gas. [*Ans.* $V = (RT/p) + B + Cp + Dp^2$]

5.4 Suppose that G is known as a function of p and T for a system. Derive the expressions for the other thermodynamic functions of the systems (V, S, H, U, A, C_p, C_V, α and κ_T) in terms of G and its derivatives with respect to p and T only.

5.5 Show that

$$\frac{1}{S^2}\left(\frac{\partial S}{\partial p}\right)_G = \frac{1}{S}\left(\frac{\partial V}{\partial G}\right)_p - \frac{V}{S^2}\left(\frac{\partial S}{\partial G}\right)_p$$

$$\left(\frac{\partial p}{\partial V}\right)_S = \left(\frac{\partial p}{\partial T}\right)_S\left(\frac{\partial T}{\partial V}\right)_S = \frac{C_p}{C_V}\left(\frac{\partial p}{\partial V}\right)_T$$

5.6 Show that the criterion of thermodynamic equilibrium under the condition of constant entropy and pressure is $dH_{S,p} = 0$.

5.7 What is meant by a system's being in: (a) thermal equilibrium, (b) chemical equilibrium, (c) mechanical equilibrium and (d) thermodynamic equilibrium.

5.8 (a) Show that the graphs of H versus T and G versus T will always have opposite slopes.
[*Hint:* $(\partial H/\partial T)_p = C_p > 0$ and $(\partial G/\partial T)_p = -S < 0$]

(b) What is the separation between these two plots at any temperature? How does this separation vary with temperature? [*Ans.* TS]

(c) Show that the two plots will approach each other as the temperature is lowered and ultimately will meet as $T \to 0$. Draw a schematic diagram showing these two curves.

(d) Given the experimental fact of T.W. Richards that ΔG and ΔH of any system approach each other as the temperature is lowered and that as $T \to 0$, they become identical, i.e. $\lim_{T\to 0} \Delta G = \Delta H$

Show that if graphs of ΔH versus T and ΔG versus T are plotted, the two graphs will have opposite slopes.
[*Hint:* $\Delta G = \Delta H - T\,\Delta S$ or $T\,\Delta S = \Delta H - \Delta G$]
For a finite temperature, we can have either

$$\Delta S = +\text{ve} \Rightarrow \Delta H > \Delta G$$

or

$$\Delta S = -\text{ve} \Rightarrow \Delta G > \Delta H$$

as $T \to 0$, $\Delta H = \Delta G$. Thus, the variation of ΔH and ΔG as the temperature is lowered will have opposite trends.]

(e) Nernst on examining the Richards plots observed that not only the two curves in any plot approach each other but their slopes seemed to approach zero, i.e.

$$\left(\frac{\partial(\Delta G)}{\partial T}\right)_p = 0 \text{ and } \left(\frac{\partial(\Delta H)}{\partial T}\right)_p = 0$$

These results are known as *Nernst heat theorem*. With this information, draw a schematic diagram illustrating the behaviour of ΔH versus T and ΔG versus T curves. (f) Show that as $T \to 0$, we must have $\Delta S = 0$ and $\Delta C_p = 0$.

$$\left[\textit{Hint:} \left(\frac{\partial(\Delta G)}{\partial T}\right)_p = -\Delta S \text{ and } \left(\frac{\partial(\Delta H)}{\partial T}\right)_p = \Delta C_p.\right]$$

(g) What is Planck's contribution about the absolute value of entropy of a substance at 0 K. Is it in agreement with the Nernst heat theorem?

(h) According to the third law of thermodynamics

$$\lim_{T \to 0} S = 0$$

This implies that in the limit of absolute zero of temperature the entropy of a perfect crystalline substance must be independent of changes in pressure or volume, such that

$$\lim_{T \to 0} \left(\frac{\partial S}{\partial p}\right)_T = 0 \quad ; \quad \lim_{T \to 0} \left(\frac{\partial S}{\partial V}\right)_T = 0$$

Using the appropriate Maxwell equation, show that

$$\lim_{T \to 0} \left(\frac{\partial V}{\partial T}\right)_p = 0 \quad ; \quad \lim_{T \to 0} \left(\frac{\partial p}{\partial T}\right)_V = 0$$

5.9 (a) A gas consisting of the amount n obeys the relation

$$pV = n(RT + Bp)$$

where B is constant, is being compressed from a volume V to a volume $V/2$ isothermally and reversibly. Derive the expressions for q, w, ΔU, ΔH, ΔS_{sys}, ΔS_{surr}, and ΔG.

$$\left[\textit{Ans.}\; q = -w = -nRT \ln\left(\frac{V - nB}{(V/2) - nB}\right); \qquad \Delta U = 0 \right.$$

$$\Delta S_{\text{sys}} = nR \ln\left(\frac{(V/2) - nB}{V - nB}\right) \quad ; \quad \Delta S_{\text{surr}} = -nR \ln\left(\frac{(V/2) - nB}{V - nB}\right)$$

$$\Delta H = n^2 B\, RT\left(\frac{1}{(V/2) - nB} - \frac{1}{V - nB}\right) \quad ; \quad \Delta A = -T\, \Delta S_{\text{surr}}$$

$$\left. \Delta G = \Delta H - T\, S_{\text{surr}} = nB(p_2 - p_1) + nRT \ln \frac{p_2}{p_1} \right]$$

(b) Suppose now the gas is allowed to return to its initial state by expanding isothermally into vacuum from a volume $V/2$ to a volume V. Calculate q, w, ΔU, ΔH, ΔS_{sys}, ΔS_{surr} and ΔG.

[*Ans.* Values of ΔU, ΔH, ΔS and ΔG are the same as those in part (a) but of opposite signs, $q = w = 0$]

(c) Suppose now the gas is compressed reversibly and adiabatically from the state p_1, T_1 to a final state with pressure p_2. Derive the expressions for T_2, q, w, ΔU, ΔH, ΔS_{sys} and ΔG.

Given that $C_{V,\text{m}}$ of the gas varies with temperature according to the expression

$$C_{V,\text{m}} = a + bT$$

where a and b are constants.

$$\left[\textit{Ans.}\; T_2 ;\, a \ln \frac{T_2}{T_1} + b(T_2 - T_1) = -R \ln \frac{V_2 - nB}{V_1 - nB} = R \ln \frac{p_2 T_1}{p_1 T_2} \right.$$

$$q = 0$$

$$\Delta U = n\left\{a(T_2 - T_1) + \frac{b}{2}(T_2^2 - T_1^2)\right\}$$

$$w = \Delta U$$

$$\Delta H = \Delta U = +n[R(T_2 - T_1) + B(p_2 - p_1)]$$

$$\left. \Delta G = \Delta H - T\, \Delta S \right]$$

[*Note:* ΔG cannot be calculated unless the value of ΔS is provided.]

(d) Now suppose gas is allowed to expand adiabatically into vacuum to a final pressure p_1. What will be the final temperature? Derive the expression for ΔU, ΔH, ΔS and ΔG.

$$\left[\textit{Ans.} \text{ Final temperature will also be } T_2;\ q = -w = \Delta U = 0 \right.$$

$$\left. \Delta H = nB(p_1 - p_2);\ \Delta S = nR \ln \frac{p_2}{p_1};\ \Delta G = nRT \ln \frac{p_1}{p_2} + nB(p_1 - p_2) \right]$$

5.10 Suppose one mole of the gas (assumed to be ideal) is transported reversibly and isothermally from its equilibrium state at the surface of the earth where pressure is p_0 to its equilibrium state at a height h from the earth where pressure is p(with $p < p_0$). One of the procedures to carry out the above process is to transport the gas by an infinitesimal vertical amount and then integrating the resultant expression from $h = 0$ to $h = h$. The only work done in this process (and will have a negative sign) is the gravitational work and is equal to $-Mg\,dh$ where M is the mass of 1 mole of gas and dh is the infinitesimal vertical movement. This work is the work other than p–V and must be equal to infinitesimal change in the value of Gibbs free energy dG such that

$$dG = dw = -Mg\,dh$$

But $dG = V\,dp - S\,dT$, therefore, the above expression becomes

$$V\,dp - S\,dT = -Mg\,dh$$

Since the transport takes place at constant temperature, therefore, we have

$$V\,dp = -Mg\,dh$$

Carry out the integration of the above equation from $h = 0, p = p_0$ to $h = h, p = p$ and show that it leads to the barometric formula.

5.11 A chemist reports that he has found the entropy of a particular system decreases during a specific spontaneous endothermic changes in state. Evaluate the statement.

5.12 (a) Show that V, H and S are related to f and its derivatives by the following expressions.

$$V_{\mathrm{m}} = V_{\mathrm{m}}^{\circ} + RT\left(\frac{\partial \ln (f/f^{\circ})}{\partial p}\right)_T; \quad H_{\mathrm{m}} = H_{\mathrm{m}}^{\circ} - RT^2\left(\frac{\partial \ln (f/f^{\circ})}{\partial T}\right)_p$$

$$S_{\mathrm{m}} = S_{\mathrm{m}}^{\circ} - RT\left(\frac{\partial \ln (f/f^{\circ})}{\partial T}\right)_p - R\ln (f/f^{\circ})$$

(b) For a gas obeying equation of state

$$pV = RT + Bp + Cp^2 + \cdots$$

where B, C, …, etc. are temperature dependent. Prove the following relations.

$$H = H^{\circ} + Bp - T\left(\frac{\partial B}{\partial T}\right)_p p + \cdots; \qquad S = S^{\circ} - R\ln p - p\left(\frac{\partial B}{\partial T}\right)_p + \cdots$$

$$C_p = C_p^{\circ} - T\left(\frac{\partial^2 B}{\partial T^2}\right)_p + \cdots; \qquad \mu_{\mathrm{JT}}\, C_p = T\left(\frac{\partial B}{\partial T}\right)_p - B + \cdots$$

5.13 A certain gas obeys the equations of state

$$pV = RT + \frac{B}{V} + \frac{C}{V^2} + \frac{D}{V^3} + \cdots$$

Show that the fugacity of the gas is given by

$$\ln\frac{f}{p} = \ln\frac{RT}{V} + \frac{2B}{RTV} + \frac{3C}{2RTV^2} + \cdots$$

5.14 (a) Show that

(i) $$\mu_{JT} = \frac{1}{C_p}\left[T\left(\frac{\partial V}{\partial T}\right)_p - V\right] = -\frac{RT^2}{C_p p}\left(\frac{\partial Z}{\partial T}\right)_p$$

(ii) $$\frac{\partial^2 f}{\partial p\,\partial T} = \frac{1}{RT^2}\left(\frac{\partial(H_m^0 - H_m)}{\partial p}\right)_T = \frac{1}{RT^2}C_p\mu_{JT} = -\frac{1}{p}\left(\frac{\partial Z}{\partial T}\right)_V$$

where Z is the compression factor of the system of a real gas.

(b) Derive the expressions of μ_{JT} and $(\partial^2 f/\partial p\,\partial T)$ for one mole of a van der Waals gas.

5.15 The formation of the oxide M_xO_y per mole of O_2 consumed is written as

$$\frac{2x}{y}M(s) + O_2(g) \rightarrow \frac{2}{y}M_xO_y$$

The free energy change of the reaction is given by $\Delta_r G = \Delta_r H - T\Delta_r S$

If the enthalpy change $\Delta_r H$ and entropy change $\Delta_r S$ are assumed to be independent of temperature, the plot of $\Delta_r G$ versus T is linear such as shown in Fig. 1, known as Ellingham diagram. Figure 1 is widely used to discuss the principles involved in the extraction of metal from its oxide.

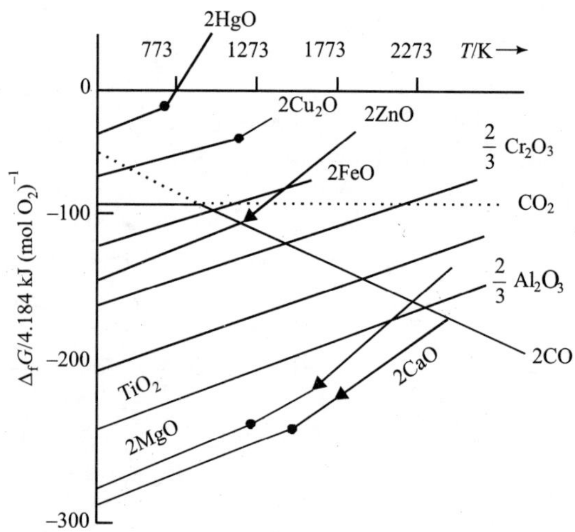

Fig. 1 Ellingham diagram

With the help of this figure, answer the following.

(i) The slope of linear plot of $\Delta_r G$ versus T is

(a) positive for M_xO_y with the exception when M is carbon,

(b) nearly zero but positive for CO_2 and

(c) negative for CO.

[Ans. (a) $\Delta_r S$ of the oxide formation is negative as $\Delta\nu_g = -1$ (decrease in gaseous species),

(b) $\Delta_r S$ is slightly positive as $\Delta\nu_g = 0$ (no change in the gaseous species),

(c) $\Delta_r S$ is positive as $\Delta\nu_g = +1$ (increase in gaseous species).]

(ii) The plots of $\Delta_r G$ versus T for the various oxides formed from M(s) or M(l) are more or less parallel to one another.

(*Ans.* $\Delta_r S$ of the reactions are, more or less, the same.)

(iii) The slope of $\Delta_r G$ versus T plot becomes more positive when the oxide is formed from M(g).

(*Ans.* Larger decrease in the gaseous species.)

(iv) At $T < 700$ °C, the stable oxide of C is CO_2 whereas at $T > 700$ °C, the stable oxide is CO. Given that at higher temperature the equilibrium reaction

$$C + CO_2 \rightleftharpoons 2CO$$

sets in with $\Delta_r H = 172.8$ kJ mol^{-1} and $\Delta_r S = 176.15$ J K^{-1} mol^{-1}.

(*Ans.* The temperature at which CO_2 reacts with C to give CO is equal to $\Delta_r H/\Delta_r S$ as $\Delta_r G = 0$ which gives $T = 981$ K = 708 °C)

(v) Mercury(II) oxide decomposes spontaneously to its element by heating alone.

(*Ans.* $\Delta_r G$ of oxide formation becomes positive at higher temperature. It becomes unstable and thus decomposes.)

(vi) The reduction of M_xO_y by carbon at $T > 700$ °C may be represented as

$$\frac{2}{y}M_xO_y + 2C \rightarrow \frac{2x}{y}M + 2CO$$

The free energy of this reaction can be obtained from the following two reactions.

$$\frac{2x}{y}M + O_2 \rightarrow \frac{2}{y}M_xO_y \qquad \Delta_r G_1$$

$$2C + O_2 \rightarrow 2CO \qquad \Delta_r G_2$$

Obviously,

$$\Delta_r G = \Delta_r G_2 - \Delta_r G_1$$

For the reduction to be spontaneous, we must have $\Delta_r G_1 > \Delta_r G_2$, that is, the line representing the oxide formation of the metal M should lie above that of C. For each oxide, the temperature at which this condition is achieved can be determined from Fig. 1. Show that the ascending order of temperature for various oxides of M is as follows.

$$T_{Fe} < T_{Zn} < T_{Cr} < T_{Ti} < T_{Mg} < T_{Al} < T_{Ca}$$

(vii) Show that reduction of Cr_2O_3 by Al is possible at all temperatures.

NUMERICAL PROBLEMS

Pressure Dependence of *G*

5.1 (a) One mole of an ideal monatomic gas expands isothermally at 300 K into vacuum from an initial volume of 10 dm^3 to a final volume of 20 dm^3.

(b) One mole of the same gas expands isothermally and reversibly at 300 K from 10 dm^3 to 20 dm^3.

For each of the above two processes, calculate q, w, ΔU, ΔH, ΔA and ΔG.

(c) Describe a process by which the gas could be restored to its initial state after each of the two processes. Show how surroundings could also be restored to their initial state in one of the above processes and for the second it cannot be restored.

(*Ans.* (a) $q = w = 0$; (b) $q = -w = 1.728$ kJ; For both processes $\Delta U = \Delta H = 0$; $\Delta A = \Delta G = -1.728$ kJ)

5.2 Three moles of a gas with $pV = n(RT + Bp)$ where $B = 30$ cm^3 mol^{-1} and for which C_p/J K^{-1} mol^{-1} = 27.196 + 46.65 × 10^{-3} (*T*/K) undergoes an irreversible change in state given by

600 K, 10 bar → 300 K, 5 bar

Calculate ΔU, ΔH and ΔS. Can you estimate the values of ΔA and ΔG from the given data? How do you reconcile the sign of ΔS calculated in this problem with the help of second law? (*Ans.* – 35.89 kJ, – 43.37 kJ, – 81.25 J K^{-1}, no)

5.3 An ideal gas is allowed to expand reversibly and isothermally at 298 K from a pressure of 1 bar to 0.05 bar. (a) What is the change in molar Gibbs energy? (b) What would be the change if the process occurs irreversibly?

(*Ans.* (a) –7 422 J mol^{-1}, (b) –7 422 J mol^{-1})

Relation Between $\Delta_r G$ and $\Delta_r A$

5.4 Calculate the difference between $\Delta_r G$ and $\Delta_r A$ at 20 °C for the reaction

$$H_2(g, 1\,atm) + \frac{1}{2} O_2\,(g,\ 1\,atm) = H_2O(1)$$ (*Ans.* –3.716 kJ mol^{-1})

Fugacity

5.5 The density of NH_3(g) at 200 °C and 5.066 MPa is 24.30 g dm^{-3}. Estimate its fugacity. [*Hint:* Make use of the expression $f = p^2_{real}/p_{ideal}$) (*Ans.* 4.569 MPa)

5.6 Calculate the fugacity of NH_3 at 200 °C and 10.1325 MPa from the following data.

p/MPa	2.026 5	6.079 5	10.132 5	15.198 8
V/cm^3	1 866	570.8	310.9	176.6

[*Hint:* Follow the graphical method, Eq. (5.7.5).] (*Ans.* 8.329 MPa)

5.7 For a van der Waals gas

$$Z = 1 + \left[b - \left(\frac{a}{RT}\right)\right]\frac{p}{RT}$$

calculate its fugacity at 100 bar and 298 K. Given: $a = 0.247\,6$ dm^6 bar mol^{-2} and $b = 0.026\,61$ dm^3 mol^{-1}. (*Ans.* 74.38 bar)

Temperature Dependence of *G*

5.8 In the formation of silver chloride from its elements under normal conditions, $\Delta_r G$ is –110.04 kJ mol^{-1} and $\Delta_r H$ = –126.72 kJ mol^{-1} at 18 °C. What is the corresponding entropy change? Calculate the change in free energy for a 10 °C rise in temperature, assuming the rate of change of $\Delta_r G$ with temperature, i.e. $\partial(\Delta_r G)/\partial T$ to remain constant.

(*Ans.* –57.32 J K^{-1} mol^{-1}; 573.2 J mol^{-1})

5.9 For a certain process, ΔG = –50.208 kJ and ΔH = –73.198 kJ at 400 K. Find for the process at this temperature ΔS, $[\partial(\Delta G)/\partial T]_p$ and $[\partial(\Delta A)/\partial T]_V$.

(*Ans.* –57.475 J K^{-1}, 57.475 J K^{-1}, 57.475 J K^{-1})

5.10 For a certain reaction

$$\Delta_r G/\text{J mol}^{-1} = 56\,818.7 + 67.36\,(T/\text{K}) \log (T/\text{K}) - 303.72\,(T/\text{K})$$

find $\Delta_r S$ and $\Delta_r H$ of the reaction at 25 °C.

(*Ans.* $\Delta_r S$ = 107.70 J K^{-1} mol^{-1}, $\Delta_r H$ = 48.074 kJ mol^{-1})

5.11 For the sublimation Au(s) → Au(g)

$\Delta_r H^\circ_{298K}$ = 378.65 kJ mol^{-1} and $\Delta_r G^\circ_{298K}$ = 33.89 kJ mol^{-1}

Further

Au(g): $C_p = 20.92$ J K^{-1} mol^{-1}

Au(s): C_p/J K^{-1} mol^{-1} = 23.47 + 6.025 × 10^{-3} (*T*/K)

From these data, find an expression for $\Delta_r G°$ as a function of temperature.

(*Ans.* $\Delta_r G°$/J mol^{-1} = 379 656.2 + 5.857 6 (*T*/K) log (*T*/K) –151.59 (*T*/K) + 3.013 × 10^{-3} (*T*/K)^2)

5.12 For the reaction

$FeCO_3(s) = FeO(g) + CO_2(g)$

$\Delta_r G°$/J mol^{-1} = 78 073.4 – 60.33 (*T*/K) log (*T*/K) – 25.397 (*T* /K) + 34.476 × 10^{-3} (*T*/K)^2

find $\Delta_r H°$ and $\Delta_r S°$ for the reaction at 25 °C

(*Ans.* $\Delta_r H°$ = 82.801 kJ mol^{-1}; $\Delta_r S°$ = 180.33 J K^{-1} mol^{-1})

5.13 For the reaction

$A(s) + B_2(g) = AB_2(g)$

$$\Delta_r G°/\text{J mol}^{-1} = 21\ 087.4 - (32.09\ \text{K}^{-1})T \log (T/\text{K}) + (6.32 \times 10^{-3}\ \text{K}^{-2})\ T^2 + \frac{(4.63\times 10^5\ \text{K})}{T} + 90.29(T/\text{K})$$

Further

$$C_p(\text{A})/\text{J K}^{-1}\ \text{mol}^{-1} = 11.185 + (10.950 \times 10^{-3}\ \text{K}^{-1})\ T - \frac{(4.89\times 10^5\ \text{K}^2)}{T^2}$$

$$C_p(\text{AB}_2)/\text{J K}^{-1}\ \text{mol}^{-1} = 57.53 + (2.050 \times 10^{-3}\ \text{K}^{-1})\ T - \frac{(14.142\times 10^5\ \text{K}^2)}{T^2}$$

Form these data, find the expression of C_p of $B_2(g)$ as a function of *T*.

(*Ans.* $C_p(B_2)$/J K^{-1} mol^{-1} = 32.412 + (3.74 × 10^{-3} K^{-1})*T* + (0.008 × 10^5 K^2)/T^2)

5.14 (a) Compute $\Delta_r G$ for the reaction

H_2O(1, 1 atm, 323 K) → H_2O(g, 1 atm, 323 K)

Given that

$\Delta_{vap}H$ of H_2O at 373 K = 40.639 kJ mol^{-1}

$C_p(H_2O$, 1) = 75.312 J K^{-1} mol^{-1}

$C_p(H_2O$, g) = 33.305 J K^{-1} mol^{-1} (*Ans.* $\Delta_r G$ = 5.59 kJ mol^{-1})

(b) If *p* is the vapour pressure of water at 323 K, then the following reaction will be at equilibrium:

H_2O(1, *p*, 323 K) → H_2O (g, *p*, 323 K)

and $\Delta_r G$ for this reaction will be zero. Compute the value of *p* at 323 K.

(*Ans.* 12.63 kPa)

Miscellaneous Numerical

5.15 One mole of steam is condensed reversibly to liquid water at 100 °C and 101.325 kPa pressure. The enthalpy of vaporization of water is 2.257 kJ g^{-1}. Assume that the steam behaves as an ideal gas, calculate *w*, *q*, ΔU, ΔH, ΔS, ΔA and ΔG for the condensation process.

6 Systems of Variable Composition

6.1 PARTIAL MOLAR QUANTITIES

Expression of Partial Molar Quantity

Thermodynamic relations derived earlier are applicable to closed systems. In a system where not only the work and heat but also several kinds of matter are being exchanged, a multicomponent open system has to be considered. Here, the amounts of the various substances are treated as variables like any other thermodynamic variables. For example, the Gibbs free energy of a system is a function not only of temperature and pressure, but also of the amount of each substance in the system, such that

$$G = f(T, p, n_1, n_2, ..., n_k)$$

where $n_1, n_2, ..., n_k$ represent the amounts of each of the k components in the system. For simplicity, let a system contain only two components. The total differential of G is

$$\mathrm{d}G = \left(\frac{\partial G}{\partial T}\right)_{p,n_1,n_2} \mathrm{d}T + \left(\frac{\partial G}{\partial p}\right)_{T,n_1,n_2} \mathrm{d}p + \left(\frac{\partial G}{\partial n_1}\right)_{T,p,n_2} \mathrm{d}n_1 + \left(\frac{\partial G}{\partial n_2}\right)_{T,p,n_1} \mathrm{d}n_2 \tag{6.1.1}$$

In this equation, the partial derivatives $(\partial G/\partial n_1)_{T,\,p,\,n_2}$ and $(\partial G/\partial n_2)_{T,\,p,\,n_1}$ are known as partial molar free energies of components 1 and 2, respectively. In general, *the partial derivative of a thermodynamic function Y with respect to the amount of component i of a mixture when, T, p and amounts of other constituents are kept constant, is known as the partial molar quantity of the ith component and is represented as* $Y_{i,\mathrm{pm}}$ (or simply as Y_i). Thus

$$Y_{i,\,\mathrm{pm}} = \left(\frac{\partial Y}{\partial n_i}\right)_{T,p,n_j\mathrm{s}\ (j\neq i)} \tag{6.1.2}$$

Definition of Partial Molar Quantity

The partial molar quantity may be defined in either of the following two ways:

1. *It is the change in Y when 1 mole of component i is added to a system which is so large that this addition has a negligible effect on the composition of the system.*

2. Let dY be the change in value of Y when an infinitesimal amount dn of component i is added to a system of definite composition. By an infinitesimal

amount dn_i we mean that its addition does not cause any appreciable change in the composition of the system. If we divide dY by dn_i, we get the partial molar quantity $(\partial Y/\partial n_i)$. Thus, the partial molar quantity of component i may be defined as *the change in Y per mole of component i when an infinitesimal amount of this component is added to a system of definite composition.*

Physical Significance of Partial Molar Quantity

The quantity $Y_{i,\ \mathrm{pm}}$ represents the actual value of Y per mole of the ith species in a system of known composition. This value may or may not be equal to the corresponding molar value Y^* of the species in the pure state. In fact, in only ideal systems where either there exists no interaction between the constituents (ideal gases) or all interactions are of the same type (ideal solutions) that the two values, viz., the partial molar value and the molar value, are the same. In nonideal solutions, because of interactions between the constituents, the molar value of Y_i^* usually gets modified to a new value $Y_{i,\ \mathrm{pm}}$. Since the extent of interactions depend upon the relative amounts of the components, it is obvious that the modified value of Y may not be the same at all compositions. Hence, its value will also depend upon the composition of the solution. Thus, although the partial molar quantities are meant for the individual components of the system, their values are not only dependent on the nature of the particular substance in question but also on the nature and amounts of the other components present in the system.

Specific Example of Partial Molar Volumes

To be specific, let us take the case of partial molar volume. The partial molar volume of the ith species in an ideal solution will be equal to its molar volume in the pure state. In a nonideal solution, the partial molar volume is the molar volume actually occupied by the substance in a solution of known composition. For example, at 50 mass percentage solution of ethanol in water at 25 °C, the partial molar volume are found to be

$$V_{\mathrm{pm}}(C_2H_5OH) = 56.76\ \mathrm{cm^3\ mol^{-1}}$$

$$V_{\mathrm{pm}}(H_2O) = 17.37\ \mathrm{cm^3\ mol^{-1}}$$

The corresponding values for the pure components are

$$V^*(C_2H_5OH) = 58.65\ \mathrm{cm^3\ mol^{-1}}$$

$$V^*(H_2O) = 18.07\ \mathrm{cm^3\ mol^{-1}}$$

Suppose 54 g of water (3 mol) is mixed with 54 g of ethanol (1.173 8 mol), we will then have

$$\text{Volume of water before mixing} = (3\ \mathrm{mol})\ (18.07\ \mathrm{cm^3\ mol^{-1}}) = 54.21\ \mathrm{cm^3}$$

$$\text{Volume of ethanol before mixing} = (1.173\ 8\ \mathrm{mol})\ (58.65\ \mathrm{cm^3\ mol^{-1}}) = 68.85\ \mathrm{cm^3}$$

The total volume on mixing is not equal to the sum of the individual volumes, i.e. $54.21\ \mathrm{cm^3} + 68.85\ \mathrm{cm^3} = 123.06\ \mathrm{cm^3}$ but will have a value as calculated below.

Since the partial molar volumes are the actual molar volumes present in a solution of known composition, it is obvious, therefore, that the total volume of the solution of the given composition will be given as†

$$V_{\mathrm{total}} = n_1 V_{1,\ \mathrm{pm}} + n_2 V_{2,\ \mathrm{pm}}$$

† See Section 6.10.

Thus in the above example of 50 mass percentage ethanol-water solution, we have

$$\begin{aligned} V_{\text{total}} &= n(\text{H}_2\text{O})\, V_{\text{pm}}(\text{H}_2\text{O}) + n(\text{C}_2\text{H}_5\text{OH})\, V_{\text{pm}}(\text{C}_2\text{H}_5\text{OH}) \\ &= (3\text{ mol})\,(17.37\text{ cm}^3\text{ mol}^{-1}) + (1.173\,8\text{ mol})\,(56.76\text{ cm}^3\text{ mol}^{-1}) \\ &= 52.11\text{ cm}^3 + 66.63\text{ cm}^3 \\ &= 118.74\text{ cm}^3 \end{aligned}$$

Example 6.1.1 The partial molar volumes of acetone and chloroform in a mixture in which the mole fraction of acetone is 0.53 are 74.17 $cm^3\ mol^{-1}$ and 80.24 $cm^3\ mol^{-1}$, respectively. What is the density of a solution? What will be the volume of solution weighing 1.0 kg?

Solution Let us consider a solution containing 0.53 mol of acetone and 0.47 mol of chloroform. The volume of this solution will be

$$\begin{aligned} V &= n_1 V_{1,\,\text{pm}} + n_2 V_{2,\,\text{pm}} \\ &= (0.53\text{ mol})\,(74.17\text{ cm}^3\text{ mol}^{-1}) + (0.47\text{ mol})\,(80.24\text{ cm}^3\text{ mol}^{-1}) \\ &= 39.31\text{ cm}^3 + 37.71\text{ cm}^3 \\ &= 77.02\text{ cm}^3 \end{aligned}$$

The mass of the solution will be

$$\begin{aligned} m &= n_1 M_1 + n_2 M_2 \\ &= (0.53\text{ mol})\,(58\text{ g mol}^{-1}) + (0.47\text{ mol})\,(119.5\text{ g mol}^{-1}) \\ &= 30.74\text{ g} + 56.17\text{ g} = 86.91\text{ g} \end{aligned}$$

The density of solution will be

$$\rho = \frac{m}{V} = \frac{86.91\text{ g}}{77.02\text{ cm}^3} = 1.128\text{ g cm}^{-3}$$

The volume of 1 kg of solution will be

$$V_1 = \left(\frac{1000\text{ g}}{86.91\text{ g}}\right)(77.02\text{ cm}^3) = 886.2\text{ cm}^3$$

Example 6.1.2 The partial molar volume of methanol in a methanol-water solution at $x_{\text{methanol}} = 0.39$ is 39.2 $cm^3\ mol^{-1}$. The density of solution is 0.91 g cm^{-3}. Calculate partial molar volume of water in the solution.

Solution Let we have 0.39 mol of methanol and 0.61 mol of water so that the total amount of these two substances is one mole and thus we have a solution of given composition. Now,

$$\begin{aligned} \text{Mass of solution, } m &= n_1 M_1 + n_2 M_2 \\ &= (0.39\text{ mol})\,(32\text{ g mol}^{-1}) + (0.61\text{ mol})\,(18\text{ g mol}^{-1}) \\ &= 23.46\text{ g} \end{aligned}$$

$$\text{Volume of solution, } V = \frac{m}{\rho} = \frac{23.46\text{ g}}{0.91\text{ g cm}^{-3}} = 25.78\text{ cm}^3$$

Now from the expression

$$V_{\text{total}} = n_1 V_{1,\,\text{pm}} + n_2 V_{2,\,\text{pm}}$$

we get

$$\begin{aligned} V_{2,\,\text{pm}} &= (V_{\text{total}} - n_1 V_{1,\,\text{pm}})/n_2 \\ &= \{(25.78\text{ cm}^3) - (0.39\text{ mol})\,(39.2\text{ cm}^3\text{ mol}^{-1})\} / 0.61\text{ mol} \\ &= 17.20\text{ cm}^3\text{ mol}^{-1} \end{aligned}$$

Example 6.1.3 It is required to prepare 10^3 cm^3 of methanol-water solution at 25 °C with mole fraction of methanol equal to 0.78. Determine the volumes of methanol and water to be mixed at 25 °C. Given: Partial molar volumes of water and methanol are 15.69 cm^3 mol^{-1} and 40.49 cm^3 mol^{-1}, respectively. The density of methanol at 25 °C is 0.79 g cm^{-3}.

Solution Let n_1 and n_2 be the required amounts of methanol and water, respectivly, to prepare 10^3 cm^3 of the required solution. We will have

$$V = n_1 V_{1,\,\text{pm}} + n_2 V_{2,\,\text{pm}}$$

and
$$x_1 = \frac{n_1}{n_1 + n_2}$$

From the given data, we have

$$10^3\ \text{cm}^3 = n_1\ (40.49\ \text{cm}^3\ \text{mol}^{-1}) + n_2\ (15.69\ \text{cm}^3\ \text{mol}^{-1})$$

$$0.78 = \frac{n_1}{n_1 + n_2} \quad \Rightarrow \quad 0.78\ n_2 = 0.22\ n_1$$

Substituting $n_1 = (0.78/0.22)\ n_2$ in the first expression, we get

$$10^3\ \text{cm}^3 = \left(\frac{0.78}{0.22} n_2\right)(40.49\ \text{cm}^3\ \text{mol}^{-1}) + n_2\ (15.69\ \text{cm}^3\ \text{mol}^{-1})$$

or
$$10^3 = (143.56\ n_2 + 15.69\ n_2)\ \text{mol}^{-1} = 159.25\ n_2\ \text{mol}^{-1}$$

or
$$n_2 = \frac{10^3}{159.25}\ \text{mol} = 6.28\ \text{mol}$$

Thus
$$n_1 = \frac{0.78}{0.22} n_2 = \frac{0.78}{0.22} \times 6.28\ \text{mol} = 22.27\ \text{mol}$$

Now
$$V_2 = \frac{n_2 M_2}{\rho_2} = \frac{(6.28\ \text{mol})\ (18\ \text{g mol}^{-1})}{(1.00\ \text{g cm}^{-3})} = 113.0\ \text{cm}^3$$

$$V_1 = \frac{n_1 M_1}{\rho_1} = \frac{(22.27\ \text{mol})\ (32\ \text{g mol}^{-1})}{(0.79\ \text{g cm}^{-3})} = 902.1\ \text{cm}^3$$

Example 6.1.4 The molar enthalpy of a binary liquid solution at constant T and p is given by the expression H_m /kJ mol^{-1} = $0.5x_\text{A} + x_\text{B} + (0.05\ x_\text{A} + 0.04\ x_\text{B})\ x_\text{A}\ x_\text{B}$. Determine the expressions of H_pm(A) and H_pm(B) in terms of x_A. Also determine the value of H^*_m(A), H^*_m(B), $\Delta_\text{soln}H$(0.5 mol A + 0.5 mol B).

Solution Since $H_\text{m} = H/(n_\text{A} + n_\text{B})$, we get

$$H/\text{kJ mol}^{-1} = 0.5\ n_\text{A} + n_\text{B} + (0.05\ n_\text{A} + 0.04\ n_\text{B}) \left(\frac{n_\text{A}}{n_\text{A} + n_\text{B}}\right)\left(\frac{n_\text{B}}{n_\text{A} + n_\text{B}}\right)$$

Since $H_\text{pm}(\text{A}) = (\partial H/\partial n_\text{A})_{T,\,p,\,n_\text{B}}$, we have

$$H_\text{pm}(\text{A})/\ \text{kJ mol}^{-1} = 0.5 + (0.05)\left(\frac{n_\text{A}}{n_\text{A} + n_\text{B}}\right)\left(\frac{n_\text{B}}{n_\text{A} + n_\text{B}}\right) + (0.05 n_\text{A} + 0.04 n_\text{B})$$
$$\times \left[\frac{n_\text{B}}{(n_\text{A} + n_\text{B})^2} - \frac{2 n_\text{A}\ n_\text{B}}{(n_\text{A} + n_\text{B})^3}\right]$$

$$= 0.5 + 0.05\ x_\text{A} x_\text{B} + (0.05\ x_\text{A} + 0.04\ x_\text{B})\ (x_\text{B} - 2\ x_\text{A} x_\text{B})$$

Substituting $x_B = 1 - x_A$ and rearranging, we get

$$H_{pm}(A)/kJ\ mol^{-1} = 0.54 - 0.06\ x_A + 0.02\ x_A^3$$

Similarly working out $H_{pm}(B) = (\partial H/\partial n_B)_{T,\,p,\,n_A}$, we get

$$H_{pm}(B)/kJ\ mol^{-1} = 1.0 + (0.04)\left(\frac{n_A}{n_A + n_B}\right)\left(\frac{n_B}{n_A + n_B}\right) + (0.05\ n_A + 0.04\ n_B) \times \left[\frac{n_A}{(n_A + n_B)^2} - \frac{2n_A n_B}{(n_A + n_B)^3}\right]$$

$$= 1.0 + 0.04\ x_A x_B + (0.05\ x_A + 0.04\ x_B)(x_A - 2x_A x_B)$$

Substituting $x_B = 1 - x_A$ and rearranging, we get

$$H_{pm}(B)/kJ\ mol^{-1} = 1.0 + 0.03\ x_A^2 + 0.02\ x_A^3$$

Now $$H_m^*(A) = \lim_{x_A \to 1} H_{pm}(A) = (0.54 - 0.06 + 0.02)\ kJ\ mol^{-1} = 0.50\ kJ\ mol^{-1}$$

$$H_m^*(B) = \lim_{x_A \to 0} H_{pm}(B) = 1.0\ kJ\ mol^{-1}$$

The enthalpy of solution for

$$0.5\ mol\ A + 0.5\ mol\ B \rightarrow soln\ (0.5\ mol\ A + 0.5\ mol\ B)$$

is given as

$$\begin{aligned}\Delta_{soln}H &= [(0.5\ mol)\ H_{pm}(A) + (0.5\ mol)\ H_{pm}(B)] - [(0.5\ mol)\ H_m^*(A) + (0.5\ mol)\ H_m^*(B)] \\ &= [\{(0.5)\ (0.54 - 0.06 \times 0.5 + 0.02 \times 0.5^3) + (0.5)\ (1.0 + 0.03 \times 0.5^2 + 0.02 \times 0.5^3)\} - [(0.5)\ (0.50) + (0.5)\ (1.0)]\ kJ \\ &= [(0.5)\ (0.512\ 5) + (0.5)\ (1.01) - (0.25 + 0.5)]\ kJ \\ &= (0.761\ 25 - 0.75)\ kJ = 0.011\ 25\ kJ \\ &= 11.25\ J\end{aligned}$$

Problem 6.1.1 The apparent molar volume of a solute in a solution of volume V is defined as

$$V_{2,\,app} = \frac{\text{Total volume of solution} - \text{Volume of pure solvent}}{\text{Amount of solute}} = \frac{V - n_1 V_1^*}{n_2}$$

Show that $$V_{2,\,pm} = n_2\left(\frac{\partial V_{2,\,app}}{\partial n_2}\right)_{n_1} + V_{2,\,app} \quad \text{and} \quad V_{1,pm} = n_2\left(\frac{\partial V_{2,\,app}}{\partial n_1}\right)_{n_2} + V_1^*$$

Solution We have $$V_{2,\,app} = \frac{V - n_1 V_1^*}{n_2}$$

Rearranging this, we get

$$V = n_2\ (V_{2,\,app}) + n_1 V_1^* \tag{6.1.3}$$

Differentiating Eq. (6.1.3) with respect to n_2 keeping n_1 constant, we get

$$V_{2,pm} = \left(\frac{\partial V}{\partial n_2}\right)_{n_1} = V_{2,app} + n_2\left(\frac{\partial V_{2,\,app}}{\partial n_2}\right)_{n_1} \tag{6.1.4}$$

(Note that the differentiation of second term in Eq. (6.1.3) is zero as both n_1 and V_1^* are independent of n_2.)

Similarly, differentiation of Eq. (6.1.3) with respect to n_1 gives

$$V_{1,\,\mathrm{pm}} = \left(\frac{\partial V}{\partial n_1}\right)_{n_2} = n_2\left(\frac{\partial V_{2,\mathrm{app}}}{\partial n_1}\right)_{n_2} + V_1^* \tag{6.1.5}$$

Problem 6.1.2

The apparent molar volume of a solute in a solution is given by the following analytic expression

$$V_{2,\,\mathrm{app}} = a + bm + cm^2$$

where a, b and c are constants and m is the molality of the solution. Show that

$$V_{2,\,\mathrm{pm}} = a + 2bm + 3\,cm^2$$

$$V_{1,\mathrm{pm}} = V_1^* - bM_1\,m^2 - 2c\,M_1 m^3$$

Solution

From Eq. (6.1.4), we have

$$V_{2,\,\mathrm{pm}} = V_{2,\,\mathrm{app}} + n_2\left(\frac{\partial V_{2,\mathrm{app}}}{\partial n_2}\right)_{n_1}$$

Let m_1 be the mass of solvent. Since the molality of the solution is $m = n_2/m_1$, we have

$$n_2 = mm_1 \qquad \text{and} \qquad \mathrm{d}n_2 = m_1\,\mathrm{d}m$$

With these, the previous expression becomes

$$V_{2,\,\mathrm{pm}} = V_{2,\mathrm{app}} + m\left(\frac{\partial V_{2,\,\mathrm{app}}}{\partial m}\right)_{n_1}$$

From the given expression of $V_{2,\,\mathrm{app}}$, we have

$$V_{2,\,\mathrm{pm}} = (a + bm + cm^2) + m(b + 2cm)$$
$$= a + 2bm + 3cm^2$$

Now $\qquad V = n_2(V_{2,\mathrm{app}}) + n_1 V_1^*$ (Eq. 6.1.3)

and also $\qquad V = n_1\,V_{1,\mathrm{pm}} + n_2\,V_{2,\mathrm{pm}}$

Equating these two equations, we have

$$n_1 V_{1,\,\mathrm{pm}} + n_2 V_{2,\,\mathrm{pm}} = n_2(V_{2,\,\mathrm{app}}) + n_1 V_1^* \tag{6.1.6}$$

For the present case, we have

$$n_2 = mm_1 \qquad \text{and} \qquad n_1 = m_1/M_1$$

Substituting these along with the expressions of $V_{2,\,\mathrm{pm}}$ and $V_{2,\,\mathrm{app}}$ in Eq. (6.1.6), we get

$$\frac{m_1}{M_1}V_{1,\mathrm{pm}} + m\,m_1(a + 2\,bm + 3\,cm^2\,) = m\,m_1(a + bm + cm^2) + \frac{m_1}{M_1}V_1^*$$

Rearranging, we get

$$V_{1,\,\mathrm{pm}} = V_1^* - b\,M_1 m^2 - 2c\,M_1 m^3$$

Example 6.1.5 The apparent molar volume of KI in CH_3OH may be expressed by the equation:

$$V_{2,\,app}/\text{cm}^3\ \text{mol}^{-1} = 21.45 + 11.5\sqrt{m/\text{mol kg}^{-1}}$$

The density of pure methanol is 0.786 5 g cm^{-3}. Find $\Delta_r V$ for the process

$CH_3OH(1) \rightarrow CH_3OH(\text{soln, [KI]} = 1\ \text{mol kg}^{-1})$.

Solution For 1 mol kg^{-1} solution, we will have:

$$n_1 = (1\ 000\ \text{g})/M_1 = (1\ 000\ \text{g})/(32\ \text{g mol}^{-1}) = 31.25\ \text{mol}$$

$$n_2 = m\ (1\ \text{kg}) = 1\ \text{mol}$$

$$V_1^* = \frac{M_1}{\rho_1} = \frac{(32\ \text{g mol}^{-1})}{(0.786\,5\ \text{g cm}^{-3})} = 40.69\ \text{cm}^3\ \text{mol}^{-1}$$

Hence for m = 1 mol kg^{-1}, we have

$$V_{2,\,app} = 21.45\ \text{cm}^3\ \text{mol}^{-1} + 11.5\ \text{cm}^3\ \text{mol}^{-1} = 32.95\ \text{cm}^3\ \text{mol}^{-1}$$

Total volume of the solution will be

$$V = n_2(V_{2,\,app}) + n_1V_1^* \qquad \text{(Eq. 6.1.3)}$$

$$= (1\ \text{mol})\ (32.95\ \text{cm}^3\ \text{mol}^{-1}) + (31.25\ \text{mol})\ (40.69\ \text{cm}^3\ \text{mol}^{-1})$$

$$= 32.95\ \text{cm}^3 + 1\ 271.56\ \text{cm}^3 = 1\ 304.51\ \text{cm}^3$$

Now since $$V_{2,\,pm} = V_{2,app} + n_2\left(\frac{\partial V_{2,app}}{\partial n_2}\right)_{n_1} \qquad \text{(Eq. 6.1.4)}$$

$$= V_{2,app} + m\left(\frac{\partial V_{2,app}}{\partial m}\right)_{n_1}$$

we get $$V_{2,\,pm} = \left\{21.45\ \text{cm}^3\ \text{mol}^{-1} + 11.5\ \text{cm}^3\ \text{mol}^{-1}\sqrt{m/\text{mol kg}^{-1}}\right\} + (m/\text{mol kg}^{-1})\left(\frac{11.5\ \text{cm}^3\ \text{mol}^{-1}}{2\sqrt{(m/\text{mol kg}^{-1})}}\right)$$

For m = 1 mol kg^{-1}, we have

$$V_{2,\,pm} = 21.45\ \text{cm}^3\ \text{mol}^{-1} + 11.5\ \text{cm}^3\ \text{mol}^{-1} + 5.75\ \text{cm}^3\ \text{mol}^{-1}$$

$$= 38.7\ \text{cm}^3\ \text{mol}^{-1}$$

Finally, we have

$$V = n_1V_{1,\,pm} + n_2V_{2,\,pm}$$

Hence $$V_{1,\,pm} = \frac{1}{n_1}\ (V - n_2V_{2,\,pm})$$

$$= \frac{1}{(31.25\ \text{mol})}\{1\ 304.51\ \text{cm}^3 - (1\ \text{mol})\ (38.7\ \text{cm}^3\ \text{mol}^{-1})\}$$

$$= 40.50\ \text{cm}^3\ \text{mol}^{-1}$$

Thus for the given process, we have

$$\Delta_r V = V_{1,pm} - V_1^*$$

$$= 40.50\ \text{cm}^3\ \text{mol}^{-1} - 40.69\ \text{cm}^3\ \text{mol}^{-1}$$

$$= -\ 0.19\ \text{cm}^3\ \text{mol}^{-1}$$

Problem 6.1.3 Suppose that the density of a solution is known as a function of the molar concentration c of solute, that is $\rho = f(c)$.

(a) If $c = n_2/V$, where n_2 is the amount of solute and V is the volume of solution, show that the partial molar volume of the solute is given by the expression

$$V_{2,\,\text{pm}} = \frac{M_2 - (\partial\rho/\partial c)_{T,p,n_1}}{\rho - c(\partial\rho/\partial c)_{T,p,n_1}}$$

(b) Also show that the partial molar volume of the solvent is given by the expression

$$V_{1,\,\text{pm}} = \frac{M_1}{\rho - c(\partial\rho/\partial c)_{T,\,p,\,n_2}}$$

(c) If the density of the solution is a linear function of the molar concentration of solute, show that the partial molar volume of solvent is the same as the molar volume of the pure solvent and that the partial molar volume of the solute is independent of concentration.

Solution **(a)** Density of the solution can be written as

$$\rho = \frac{n_1M_1 + n_2M_2}{V} \tag{1}$$

where V is the volume of the solution. Rewriting this, we have

$$V\rho = n_1M_1 + n_2M_2$$

or $$V\rho = n_1M_1 + cVM_2 \qquad \text{(Since } c = n_2/V\text{)} \tag{2}$$

Differentiating this with respect to n_2 keeping T, p and n_1 constant, we have

$$\left(\frac{\partial V}{\partial n_2}\right)_{T,p,n_1}\rho + V\left(\frac{\partial \rho}{\partial n_2}\right)_{T,p,n_1} = cM_2\left(\frac{\partial V}{\partial n_2}\right)_{T,p,n_1} + VM_2\left(\frac{\partial c}{\partial n_2}\right)_{T,p,n_1}$$

or $$\left(\frac{\partial V}{\partial n_2}\right)_{T,p,n_1}\rho + V\left(\frac{\partial \rho}{\partial c}\right)_{T,p,n_1}\left(\frac{\partial c}{\partial n_2}\right)_{T,p,n_1} = cM_2\left(\frac{\partial V}{\partial n_2}\right)_{T,p,n_1} + VM_2\left(\frac{\partial c}{\partial n_2}\right)_{T,p,n_1}$$

Now, since $c = n_2/V$, therefore

$$\left(\frac{\partial c}{\partial n_2}\right)_{T,p,n_1} = \frac{1}{V} - \frac{n_2}{V^2}\left(\frac{\partial V}{\partial n_2}\right)_{T,p,n_1}$$

Introducing this in the previous expression, we have

$$\left(\frac{\partial V}{\partial n_2}\right)_{T,p,n_1}\rho + V\left(\frac{\partial \rho}{\partial c}\right)_{T,p,n_1}\left[\frac{1}{V} - \frac{n_2}{V^2}\left(\frac{\partial V}{\partial n_2}\right)_{T,p,n_1}\right] = cM_2\left(\frac{\partial V}{\partial n_2}\right)_{T,p,n_1}$$
$$+ VM_2\left[\frac{1}{V} - \frac{n_2}{V^2}\left(\frac{\partial V}{\partial n_2}\right)_{T,p,n_1}\right]$$

Rearranging, we have

$$\left(\frac{\partial V}{\partial n_2}\right)_{T,p,n_1}\left[\rho - \frac{n_2}{V}\left(\frac{\partial \rho}{\partial c}\right)_{T,p,n_1} - cM_2 + \frac{n_2M_2}{V}\right] = M_2 - \left(\frac{\partial \rho}{\partial c}\right)_{T,p,n_1}$$

or $$\left(\frac{\partial V}{\partial n_2}\right)_{T,p,n_1}\left[\rho - c\left(\frac{\partial \rho}{\partial c}\right)_{T,p,n_1} - cM_2 + cM_2\right] = M_2 - \left(\frac{\partial \rho}{\partial c}\right)_{T,p,n_1}$$

or $$V_{2,\,\mathrm{pm}} = \left(\frac{\partial V}{\partial n_2}\right)_{T,p,n_1} = \frac{M_2 - (\partial\rho/\partial c)_{T,p,n_1}}{\rho - c(\partial\rho/\partial c)_{T,p,n_1}} \qquad (4)$$

(b) Differentiating Eq. (2) with respect to n_1 keeping T, p and n_2 constant, we get

$$\left(\frac{\partial V}{\partial n_1}\right)_{T,p,n_2}\rho + V\left(\frac{\partial \rho}{\partial n_1}\right)_{T,p,n_2} = M_1 + cM_2\left(\frac{\partial V}{\partial n_1}\right)_{T,p,n_2} + VM_2\left(\frac{\partial c}{\partial n_1}\right)_{T,p,n_2}$$

$$\left(\frac{\partial V}{\partial n_1}\right)_{T,p,n_2}\rho + V\left(\frac{\partial \rho}{\partial c}\right)_{T,p,n_2}\left(\frac{\partial c}{\partial n_1}\right)_{T,p,n_2} = M_1 + cM_2\left(\frac{\partial V}{\partial n_1}\right)_{T,p,n_2} + VM_2\left(\frac{\partial c}{\partial n_1}\right)_{T,p,n_2}$$

Now $c = n_2/V$, therefore

$$\left(\frac{\partial c}{\partial n_1}\right)_{T,p,n_2} = -\frac{n_2}{V^2}\left(\frac{\partial V}{\partial n_1}\right)_{T,p,n_2}$$

Substituting this in the previous expression, we have

$$\left(\frac{\partial V}{\partial n_1}\right)_{T,p,n_2}\rho + V\left(\frac{\partial \rho}{\partial c}\right)_{T,p,n_2}\left[-\frac{n_2}{V^2}\left(\frac{\partial V}{\partial n_1}\right)_{T,p,n_2}\right]$$

$$= M_1 + cM_2\left(\frac{\partial V}{\partial n_1}\right)_{T,p,n_2} + VM_2\left[-\frac{n_2}{V^2}\left(\frac{\partial V}{\partial n_1}\right)_{T,p,n_2}\right]$$

Rearranging, we have

$$\left(\frac{\partial V}{\partial n_1}\right)_{T,p,n_2}\left[\rho - \frac{n_2}{V}\left(\frac{\partial \rho}{\partial c}\right)_{T,p,n_2} - cM_2 + M_2\frac{n_2}{V}\right] = M_1$$

or $$\left(\frac{\partial V}{\partial n_1}\right)_{T,p,n_2}\left[\rho - c\left(\frac{\partial \rho}{\partial c}\right)_{T,p,n_2} - cM_2 + cM_2\right] = M_1$$

or $$V_{1,\mathrm{pm}} = \left(\frac{\partial V}{\partial n_1}\right)_{T,p,n_2} = \frac{M_1}{\rho - c(\partial\rho/\partial c)_{T,p,n_2}}$$

(c) Let the density of the solution be related to its molar concentration by the relation

$$\rho = a + bc$$

where a and b are constants. The constant a represents the density of pure solvent; substituting $c = 0$ in the above expression, we get $a = \rho_1$. Thus, we have

$$\rho = \rho_1 + bc$$

Therefore

$$\frac{\mathrm{d}\rho}{\mathrm{d}c} = b$$

Now $$V_{2,\,\mathrm{pm}} = \frac{M_2 - \mathrm{d}\rho/\mathrm{d}c}{(\rho - c\,\mathrm{d}\rho/\mathrm{d}c)} = \frac{M_2 - b}{(\rho_1 + bc - bc)} = \frac{M_2 - b}{\rho_1}$$

In the expression, c does not appear. Therefore, it is obvious that the partial molar volume of the solute is independent of concentration.

Now $$V_{1,\,m} = \frac{M_1}{(\rho - c\,d\rho/dc)} = \frac{M_1}{(\rho_1 + bc - bc)} = \frac{M_1}{\rho_1}$$

$$= \text{Molar volume of pure solvent}$$

6.2 EXPERIMENTAL DETERMINATION OF PARTIAL MOLAR VOLUMES

A straightforward method for determining $V_{1,\,pm}$ of component 1 in a binary solution of known composition is to measure volumes of solutions obtained by adding varying amount of component 1 in a fixed amount of component 2. A graph between volume of solution and amount of component 1 is plotted. The slope of the line at the given composition of the solution gives the partial molar volume of the component 1.

Method of Intercept A considerably more precise method for the determination of partial molar volumes (or in general any partial molar quantity) is the *method of intercepts.* The principle underlying this method is described below.

Let $V_{m,\,mix}$ be the volume of mixture containing a total of one mole of the two components. Obviously, it will be given as

$$V_{m,\,mix} = \frac{V}{n_1 + n_2} \tag{6.2.1}$$

where V is the volume of the mixture containing the amounts n_1 and n_2 of components 1 and 2, respectively. Since partial molar volume of component 1 is defined as $(\partial V/\partial n_1)_{T,\,p,\,n_2}$, (written hereafter as $(\partial V/\partial n_1)_{n_2}$ since T and p remain constant during the experiment), it follows that

$$V_{1,\,pm} = \left(\frac{\partial V}{\partial n_1}\right)_{n_2} = \left(\frac{\partial(n_1 + n_2)\,V_{m,\,mix}}{\partial n_1}\right)_{n_2}$$

$$= V_{m,\,mix} + (n_1 + n_2)\left(\frac{\partial V_{m,\,mix}}{\partial n_1}\right)_{n_2} \tag{6.2.2}$$

Since $x_1 + x_2 = 1$, the total molar volume $V_{m,\,mix}$ at a given temperature and pressure will depend only on one of the mole fraction terms. Writing it as

$$V_{m,\,mix} = f(x_2) \tag{6.2.3}$$

we get $$dV_{m,\,mix} = \left(\frac{dV_{m,mix}}{dx_2}\right)dx_2 \tag{6.2.4}$$

Division of the above equation by dn_1, and restriction to constant n_2 gives

$$\left(\frac{\partial V_{m,mix}}{\partial n_1}\right)_{n_2} = \left(\frac{dV_{m,mix}}{dx_2}\right)\left(\frac{\partial x_2}{\partial n_1}\right)_{n_2} \tag{6.2.5}$$

Now since $$x_2 = \frac{n_2}{n_1 + n_2}$$

we have $$\left(\frac{\partial x_2}{\partial n_1}\right)_{n_2} = -\frac{n_2}{(n_1+n_2)^2} = -\frac{x_2}{(n_1+n_2)} \qquad (6.2.6)$$

Hence, Eq. (6.2.5) becomes

$$\left(\frac{\partial V_{\mathrm{m,mix}}}{\partial n_1}\right)_{n_2} = \left(\frac{\mathrm{d}V_{\mathrm{m,mix}}}{\mathrm{d}x_2}\right)\left(-\frac{x_2}{n_1+n_2}\right)$$

Substituting the above equation in Eq. (6.2.2), we have

$$V_{1,\,\mathrm{pm}} = V_{\mathrm{m,\,mix}} - x_2\left(\frac{\partial V_{\mathrm{m,mix}}}{\mathrm{d}x_2}\right) \quad \text{i.e.} \quad V_{\mathrm{m,mix}} = V_{1,\,\mathrm{pm}} + x_2\left(\frac{\mathrm{d}V_{\mathrm{m,mix}}}{\mathrm{d}x_2}\right) \qquad (6.2.7)$$

A similar treatment for $V_{2,\,\mathrm{pm}}$ yields

$$V_{\mathrm{m,\,mix}} = V_{2,\,\mathrm{pm}} + x_1\left(\frac{\mathrm{d}V_{\mathrm{m,mix}}}{\mathrm{d}x_1}\right) \qquad (6.2.8)$$

Geometrical interpretation of Eq. (6.2.7) (or 6.2.8) is straightforward. This represents the tangent line drawn to the plot of $V_{\mathrm{m,\,mix}}$ versus x_2 (or $V_{\mathrm{m,\,mix}}$ versus x_1) with intercept equal to $V_{1,\,\mathrm{pm}}$ (or $V_{2,\,\mathrm{pm}}$) and slope equal to $\mathrm{d}V_{\mathrm{m,\,mix}}/\mathrm{d}x_2$ (or $\mathrm{d}V_{\mathrm{m,\,mix}}/\mathrm{d}x_1$). Both the plots of $V_{\mathrm{m,\,mix}}$ versus x_2 and $V_{\mathrm{m,\,mix}}$ versus x_1 represent a single plot as $x_1 + x_2$ is always equal to 1. This is illustrated in Fig. 6.2.1 where the intercepts are $V_{1,\,\mathrm{pm}}$ at $x_2 = 0$ (or $x_1 = 1$) and $V_{2,\,\mathrm{pm}}$ at $x_2 = 1$ (or $x_1 = 0$). In fact, Eq. (6.2.8) may be derived from Fig. 6.2.1. The slope of $V_{\mathrm{m,\,mix}}$ versus x_2 at P is given by expression

$$\frac{\mathrm{d}V_{\mathrm{m,\,mix}}}{\mathrm{d}x_2} = \frac{V_{2,\,\mathrm{pm}} - V_{\mathrm{m,\,mix}}}{1 - x_2}$$

On rearranging this expression, we get

$$V_{\mathrm{m,\,mix}} = V_{2,\,\mathrm{pm}} - (1 - x_2)\,\frac{\mathrm{d}V_{\mathrm{m,\,mix}}}{\mathrm{d}x_2}$$

Since $x_1 + x_2 = 1$, it follows that $\mathrm{d}x_1 = -\,\mathrm{d}x_2$. Hence, the above expression becomes

$$V_{\mathrm{m,\,mix}} = V_{2,\,\mathrm{pm}} + x_1\,\frac{\mathrm{d}V_{\mathrm{m,\,mix}}}{\mathrm{d}x_1}$$

Equations (6.2.7) and (6.2.8) are valid for any extensive property Y (say, V, U, H, S, A and G) of the mixture. Hence, these equations may be generalized to

$$Y_{\mathrm{m,\,mix}} = Y_{1,\,\mathrm{pm}} + x_2\left(\frac{\mathrm{d}Y_{\mathrm{m,mix}}}{\mathrm{d}x_2}\right)$$

$$Y_{\mathrm{m,\,mix}} = Y_{2,\,\mathrm{pm}} + x_1\left(\frac{\mathrm{d}Y_{\mathrm{m,mix}}}{\mathrm{d}x_1}\right)$$

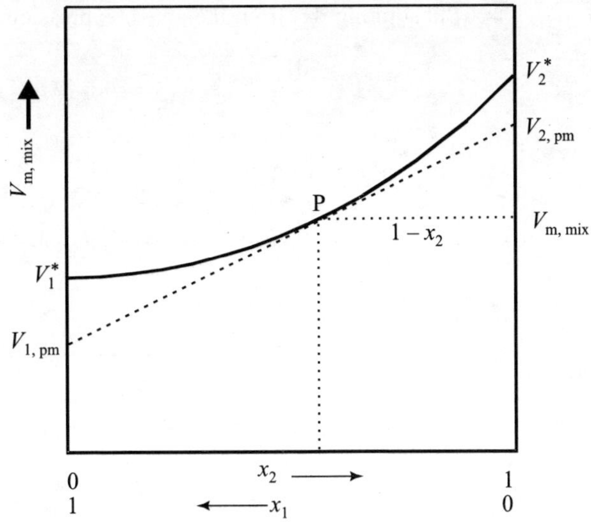

Fig. 6.2.1 The method of intercepts to determine the partial molar volumes of components of a binary solution

Alternative Method In older literatures, partial molar volumes have been determined by plotting a graph between the reciprocal of the density of the solution versus the mass per cent of component 1 (or component 2), rather than $V_{m,\,mix}$ versus x. Equation (6.2.7) can be rewritten in terms of the reciprocal of density of the mixture ($1/\rho_{mix}$) and mass per cent of component 2 (w_2), we have

$$V_{m,\,mix} = \frac{x_1M_1 + x_2M_2}{\rho_{mix}} = \frac{M_1 + x_2(M_2 - M_1)}{\rho_{mix}} \tag{6.2.9}$$

Therefore

$$\frac{dV_{m,mix}}{dx_2} = \frac{M_2 - M_1}{\rho_{mix}} + \{M_1 + x_2(M_2 - M_1)\}\left(\frac{d(1/\rho_{mix})}{dx_2}\right) \tag{6.2.10}$$

The dx will be related to dw as shown below.

$$w_2 = \frac{m_2}{m_1 + m_2} \times 100 = \frac{n_2M_2}{n_1M_1 + n_2M_2} \times 100 = \frac{x_2M_2}{x_1M_1 + x_2M_2} \times 100$$

or

$$w_2 = \frac{x_2M_2}{M_1 + x_2(M_2 - M_1)} \times 100 \tag{6.2.11}$$

Taking inverse of the above expression, we have

$$\frac{1}{w_2} = \frac{M_1}{x_2M_2 \times 100} + \frac{M_2 - M_1}{M_2 \times 100}$$

Differentiating the above expression, we have

$$-\frac{1}{w_2^2}\,dw_2 = -\frac{M_1}{x_2^2M_2 \times 100}\,dx_2 \quad \text{or} \quad dx_2 = \frac{x_2^2M_2 \times 100}{M_1w_2^2}(dw_2)$$

Substituting dx_2 from the above expression in Eq. (6.2.10), we get

$$\frac{dV_{m,\,mix}}{dx_2} = \frac{M_2 - M_1}{\rho_{mix}} + \{M_1 + x_2(M_2 - M_1)\}\frac{M_1 w_2^2}{x_2^2 M_2 \times 100}\frac{d(1/\rho_{mix})}{dw_2}$$

Making use of Eq. (6.2.11), we get

$$\frac{dV_{m,\,mix}}{dx_2} = \frac{M_2 - M_1}{\rho_{mix}} + \frac{x_2 M_2 \times 100}{w_2}\frac{M_1 w_2^2}{x_2^2 M_2 \times 100}\frac{d(1/\rho_{mix})}{dw_2}$$

$$= \frac{M_2 - M_1}{\rho_{mix}} + \frac{M_1 w_2}{x_2}\frac{d(1/\rho_{mix})}{dw_2} \qquad (6.2.12)$$

Substituting Eqs. (6.2.12) and (6.2.9) in Eq. (6.2.7), we get

$$\frac{M_1 + x_2(M_2 - M_1)}{\rho_{mix}} = \frac{M_1}{\rho_{1,pm}} + x_2\left[\frac{M_2 - M_1}{\rho_{mix}} + \frac{M_1 w_2}{x_2}\frac{d(1/\rho_{mix})}{dw_2}\right]$$

On simplifying, we get

$$\frac{1}{\rho_{mix}} = \frac{1}{\rho_{1,\,pm}} + w_2\frac{d(1/\rho_{mix})}{dw_2} \qquad (6.2.13)$$

Similarly, Eq. (6.2.8) will become

$$\frac{1}{\rho_{mix}} = \frac{1}{\rho_{2,\,pm}} + w_1\frac{d(1/\rho_{mix})}{dw_1} \qquad (6.2.14)$$

Hence, a tangent to the curve $1/\rho_{mix}$ versus w_2 at any particular w_2 has an intercept on $1/\rho_{mix}$ axis of $1/\rho_{1,\,pm}$, whereas a tangent to the curve $1/\rho_{mix}$ versus w_1 at any particular w_1 has an intercept on $1/\rho_{mix}$ of $1/\rho_{2,\,pm}$. Both the curves can be represented by a single plot as $w_1 + w_2$ is always equal to 100. The partial molar volumes are obtained by multiplying the two intercepts with the respective molar masses.

Partial Molar Volumes in a Solution Containing More than Two Components

If a solution contains more than two components (say, *m*), Eqs (6.2.2) to (6.2.6) will, respectively, take the form given below.

$$V_{i,\,pm} = V_{m,\,mix} + n_{total}\left(\frac{\partial V_{m,\,mix}}{\partial n_i}\right)_{\substack{n_j\\ j\neq i}} \qquad (6.2.15)$$

$$V_{m,\,mix} = f(x_1, x_2, \ldots, x_{i-1}, x_{i+1}, \ldots, x_m) \qquad (6.2.16)$$

$$dV_{m,\,mix} = \sum_{k\neq i}\left(\frac{\partial V_{m,mix}}{\partial x_k}\right)_{\substack{x_j\\ j\neq k,\,i}} dx_k \qquad (6.2.17)$$

$$\left(\frac{\partial V_{m,mix}}{\partial n_i}\right)_{\substack{n_j\\ j\neq i}} = \sum_{k\neq i}\left(\frac{\partial V_{m,\,mix}}{\partial x_k}\right)_{\substack{x_j\\ j\neq k,\,i}}\left(\frac{\partial x_k}{\partial n_i}\right)_{\substack{n_j\\ j\neq i}} \qquad (6.2.18)$$

$$\left(\frac{\partial x_k}{\partial n_i}\right)_{\substack{n_j\\ j\neq i}} = -\frac{x_k}{n_{\text{total}}} \tag{6.2.19}$$

Substituting Eqs. (6.2.18) and (6.2.19) in Eq. (6.2.15), we get

$$V_{i,\,\text{pm}} = V_{\text{m, mix}} - \sum_{k\neq i} x_k\left(\frac{\partial V_{\text{m, mix}}}{\partial x_k}\right)_{\substack{x_j\\ j\neq k,i}} \tag{6.2.20}$$

Equation (6.2.20) is the required expression for the partial molar volume in a solution containing more than two components.

Problem 6.2.1 Show that

$$V_{1,\,\text{pm}} = V_{\text{m, mix}} - x_2\frac{\mathrm{d}V_{\text{m, mix}}}{\mathrm{d}x_2}$$

can be written as

$$V_{1,\,\text{pm}} = \frac{M_1}{\rho} + \frac{mM_1(1+mM_2)}{\rho^2}\frac{\mathrm{d}\rho}{\mathrm{d}m}$$

where m is the molality of solute in the mixture and ρ is its density.

Solution By definitions,

$$x_2 = \frac{n_2}{n_1+n_2} \quad \text{and} \quad m = \frac{n_2}{m_1} = \frac{n_2}{n_1M_1}$$

Thus $$x_2 = \frac{n_2}{(n_2/mM_1)+n_2} = \frac{mM_1}{1+mM_1}$$

Hence $$\mathrm{d}x_2 = \frac{M_1}{1+mM_1}\mathrm{d}m - \frac{mM_1^2}{(1+mM_1)^2}\mathrm{d}m = \frac{M_1}{(1+mM_1)^2}\mathrm{d}m$$

The molar volume is

$$V_{\text{m, mix}} = \frac{x_1M_1+x_2M_2}{\rho} = \frac{M_1+x_2(M_2-M_1)}{\rho}$$

Hence, $$\frac{\mathrm{d}V_{\text{m, mix}}}{\mathrm{d}x_2} = \frac{M_2-M_1}{\rho} - \frac{M_1+x_2(M_2-M_1)}{\rho^2}\frac{\mathrm{d}\rho}{\mathrm{d}x_2}$$

Replacing x_2 and $\mathrm{d}x_2$ in terms of molality, we get

$$\frac{\mathrm{d}V_{\text{m, mix}}}{\mathrm{d}x_2} = \frac{M_2-M_1}{\rho} - \frac{M_1+\{mM_1/(1+mM_1)\}(M_2-M_1)}{\rho^2}\times\frac{(1+mM_1)^2}{M_1}\frac{\mathrm{d}\rho}{\mathrm{d}m}$$

$$= \frac{M_2-M_1}{\rho} - \frac{(1+mM_1)(1+mM_2)}{\rho^2}\frac{\mathrm{d}\rho}{\mathrm{d}m}$$

Thus $$x_2\frac{\mathrm{d}V_{\text{m, mix}}}{\mathrm{d}x_2} = \left(\frac{mM_1}{1+mM_1}\right)\left[\frac{M_2-M_1}{\rho} - \frac{(1+mM_1)(1+mM_2)}{\rho^2}\frac{\mathrm{d}\rho}{\mathrm{d}m}\right]$$

$$= \left(\frac{mM_1}{1+mM_1}\right)\left(\frac{M_2-M_1}{\rho}\right) - \frac{mM_1(1+mM_2)}{\rho^2}\frac{\mathrm{d}\rho}{\mathrm{d}m}$$

The molar volume of the mixture in terms of molality is

$$V_{\text{m, mix}} = \frac{M_1 + \{mM_1/(1+mM_1)\}(M_2 - M_1)}{\rho}$$

$$= \frac{M_1}{\rho} + \left(\frac{mM_1}{1+mM_1}\right)\left(\frac{M_2 - M_1}{\rho}\right)$$

Hence, $$V_{1,\text{ pm}} = V_{\text{m, mix}} - x_2 \frac{\mathrm{d}V_{\text{m,mix}}}{\mathrm{d}x_2}$$

$$= \frac{M_1}{\rho} + \frac{mM_1(1+mM_2)}{\rho^2}\frac{\mathrm{d}\rho}{\mathrm{d}m}$$

The corresponding expression of $V_{2,\text{ pm}}$ ($= V_{\text{m, mix}} - x_1\, \mathrm{d}V_{\text{m, mix}}/\mathrm{d}x_1$) is

$$V_{2,\text{ pm}} = \frac{M_2}{\rho} - \frac{1+mM_2}{\rho^2}\frac{\mathrm{d}\rho}{\mathrm{d}m}$$

Example 6.2.1

The density ρ of an aqueous solution of a solute (Molar mass = 58.5 g mol^{-1}) is given by

$$\rho/\text{g cm}^{-3} = 0.997\,1 + 0.047\,2\,(m/m^\circ) - 0.003\,1\,(m/m^\circ)^2 + 0.000\,3\,(m/m^\circ)^3$$

where m is the molality of solute in solution and $m^\circ = 1$ mol kg^{-1}. Determine partial molar volume of solvent in a solution of molality 0.5 mol kg^{-1}.

Solution

From the given expression of density, we get

$$\frac{\mathrm{d}(\rho/\text{g cm}^{-3})}{\mathrm{d}(m/m^\circ)} = 0.047\,2 - 2 \times 0.0031\,(m/m^\circ) + 3 \times 0.000\,3\,(m/m^\circ)^2$$

For 0.5 mol kg^{-1} solution, we get

$$\rho/\text{g cm}^{-3} = 0.997\,1 + 0.047\,2 \times 0.5 - 0.003\,1 \times 0.5^2 + 0.000\,3 \times 0.5^3$$

$$= 1.0200$$

Hence, $$\rho = 1.020\,0 \text{ g cm}^{-3}$$

$$\frac{\mathrm{d}(\rho/\text{g cm}^{-3})}{\mathrm{d}(m/m^\circ)} = 0.047\,2 - 2 \times 0.0031 \times 0.5 + 3 \times 0.000\,3 \times 0.5^2$$

$$= 0.044\,3$$

Hence, $$\frac{\mathrm{d}\rho}{\mathrm{d}m} = 0.044\,3\ (\text{g cm}^{-3}/\text{mol kg}^{-1})$$

$$= 0.044\,3 \times 10^3 \text{ g}^2 \text{ cm}^{-3} \text{ mol}^{-1}$$

Now $$mM_1 = (0.5 \text{ mol kg}^{-1})\,(0.018 \text{ kg mol}^{-1}) = 0.009$$

$$mM_2 = (0.5 \text{ mol kg}^{-1})\,(58.5 \times 10^{-3} \text{ kg mol}^{-1}) = 0.029\,25$$

Finally,

$$V_{1,\text{ pm}} = \frac{M_1}{\rho} + \frac{mM_1(1+mM_2)}{\rho^2}\frac{\mathrm{d}\rho}{\mathrm{d}m}$$

$$= \frac{18 \text{ g mol}^{-1}}{1.0200 \text{ g cm}^{-3}} + \frac{0.009(1+0.029\,25)}{(1.0200 \text{ g cm}^{-3})^2}(0.044\,3 \times 10^3 \text{ g}^2 \text{ cm}^{-3} \text{ mol}^{-1})$$

$$= (17.647 + 0.394) \text{ cm}^3 \text{ mol}^{-1}$$

$$= 18.041 \text{ cm}^3 \text{ mol}^{-1}$$

6.3 CHEMICAL POTENTIAL

The most useful partial molar quantity is the partial molar free energy $G_{i,\,\mathrm{pm}}$. It is so useful that it is given the name of chemical potential and a separate symbol μ_i to emphasis its generality:

$$\mu_i = \left(\frac{\partial G}{\partial n_i}\right)_{T,\,p,\,n_j\mathrm{s} \atop j\neq i} \tag{6.3.1}$$

Chemical potential is an intensive property because it is a molar quantity.

Problem 6.3.1 Using the criterion

$$f(\lambda x, \lambda y, \lambda z, \ldots) = \lambda^n f(x, y, z, \ldots)$$

for the homogeneity of a function of degree n, show that the partial molar quantities are intensive properties.

Solution Since the thermodynamic properties are homogeneous functions of degree 1 with respect to the amounts of substances, we may write

$$Y(\lambda n_1, \lambda n_2 \ldots, \lambda n_k) = \lambda^1\, Y(n_1\, n_2, \ldots, n_k) \tag{6.3.2}$$

Differentiating both sides of Eq. (6.3.2) with respect to n_i keeping the amounts of other substances constant, we get

$$\frac{\partial Y(\lambda n_1, \lambda n_2, \ldots, \lambda n_k)}{\partial n_i} = \lambda^1 \frac{\partial Y(n_1, n_2, \ldots, n_k)}{\partial n_i}$$

Dividing both sides by λ, we get

$$\frac{\partial Y(\lambda n_1, \lambda n_2, \ldots, \lambda n_k)}{\partial (\lambda n_i)} = \lambda^0 \frac{\partial Y(n_1, n_2, \ldots, n_k)}{\partial n_i}$$

Rewriting the above expression in the common alternative form, we have

$$Y_{i,\,\mathrm{pm}}(\lambda n_1, \lambda n_2, \ldots, \lambda n_k) = \lambda^0\, Y_{i,\,\mathrm{pm}}(n_1, n_2, \ldots n_k) \tag{6.3.3}$$

Equation (6.3.3) states that the partial molar quantity Y_i is a homogeneous function of zero degree with respect to amounts of substances, i.e. it is an intensive property.

Form Eq. (6.3.3), it is obvious that although $Y_{i,\,\mathrm{pm}}$, depends on the composition of the system (i.e. known values on n_1, n_2, etc.), it is not dependent on the total size of the system (i.e. whether the amounts of substances are simply $n_1, n_2, \ldots$ or $\lambda n_1, \lambda n_2, \ldots$. In other words, it may be stated that $Y_{i,\,\mathrm{pm}}$ is a function of mole fractions of the substances as these remain the same whether the amounts are $n_1, n_2, \ldots$ or $\lambda n_1, \lambda n_2, \ldots$

6.4 EXPRESSIONS OF d*U*, d*H*, d*A* AND d*G* FOR MULTICOMPONENT OPEN SYSTEM

If a system contains a total of k components, then

$$G = f(T, p, n_1, n_2, \ldots, n_k)$$

Its differential is given by

$$\mathrm{d}G = \left(\frac{\partial G}{\partial T}\right)_{p,n_j\mathrm{s}} \mathrm{d}T + \left(\frac{\partial G}{\partial p}\right)_{T,n_j\mathrm{s}} \mathrm{d}p + \sum_{i=1}^{k} \left(\frac{\partial G}{\partial n_i}\right)_{T,p,\,n_j\mathrm{s} \atop j\neq i} \mathrm{d}n_i$$

Substituting Eq. (6.3.1) in the above equation, we get

$$\mathrm{d}G = \left(\frac{\partial G}{\partial T}\right)_{p,n_j\mathrm{s}} \mathrm{d}T + \left(\frac{\partial G}{\partial p}\right)_{T,n_j\mathrm{s}} \mathrm{d}p + \sum_{i=1}^{k} \mu_i \,\mathrm{d}n_i$$

Making use of Eqs. (5.4.7) and (5.4.8), we get

$$\mathrm{d}G = -S\,\mathrm{d}T + V\,\mathrm{d}p + \sum_{i=1}^{k} \mu_i \,\mathrm{d}n_i \tag{6.4.1}$$

Now since $A = G - pV$, therefore

$$\mathrm{d}A = \mathrm{d}G - p\,\mathrm{d}V - V\,\mathrm{d}p$$

Substituting for dG from Eq. (6.4.1), we get

$$\mathrm{d}A = \left(-S\,\mathrm{d}T + V\,\mathrm{d}p + \sum_{i=1}^{k} \mu_i \,\mathrm{d}n_i\right) - p\,\mathrm{d}V - V\,\mathrm{d}p$$

$$= -S\,\mathrm{d}T - p\,\mathrm{d}V + \sum_{i=1}^{k} \mu_i \,\mathrm{d}n_i \tag{6.4.2}$$

Similarly, for enthalpy

$$H = G + TS$$

$$\mathrm{d}H = \mathrm{d}G + T\,\mathrm{d}S + \mathrm{S}\,\mathrm{d}T$$

$$= \left(-S\,\mathrm{d}T + V\,\mathrm{d}p + \sum_{i=1}^{k} \mu_i \,\mathrm{d}n_i\right) + T\mathrm{d}S + S\mathrm{d}T$$

$$= T\,\mathrm{d}S + V\,\mathrm{d}p + \sum_{i=1}^{k} \mu_i \,\mathrm{d}n_i \tag{6.4.3}$$

and for energy $U = H - pV$

$$\mathrm{d}U = \mathrm{d}H - p\,\mathrm{d}V - V\,\mathrm{d}p$$

$$= \left(T\,\mathrm{d}S + V\,\mathrm{d}p + \sum_{i=1}^{k} \mu_i \,\mathrm{d}n_i\right) - p\,\mathrm{d}V - V\,\mathrm{d}p$$

$$= T\,\mathrm{d}S - p\,\mathrm{d}V + \sum_{i}^{k} \mu_i \,\mathrm{d}n_i \tag{6.4.4}$$

If the amounts of all components are held constant, then Eqs. (6.4.1) to (6.4.4) are reduced to

$$\mathrm{d}G = -S\,\mathrm{d}T + V\,\mathrm{d}p \tag{6.4.5}$$

$$\mathrm{d}A = -S\,\mathrm{d}T - p\,\mathrm{d}V \tag{6.4.6}$$

$$\mathrm{d}H = T\,\mathrm{d}S + V\,\mathrm{d}p \tag{6.4.7}$$

$$\mathrm{d}U = T\,\mathrm{d}S - p\,\mathrm{d}V \tag{6.4.8}$$

Equations (6.4.5) to (6.4.8) are applicable to closed systems.

Different Ways of Defining Chemical Potential

From Eq. (6.4.2), it follows that

$$\mu_i = \left(\frac{\partial A}{\partial n_i}\right)_{T,V,n_j\mathrm{s}\atop j\neq i} \tag{6.4.9}$$

Similarly from Eqs. (6.4.3) and (6.4.4), we have

$$\mu_i = \left(\frac{\partial H}{\partial n_i}\right)_{S,p,n_j\text{s} \atop j\neq i} \tag{6.4.10}$$

$$\mu_i = \left(\frac{\partial U}{\partial n_i}\right)_{S,V,n_j\text{s} \atop j\neq i} \tag{6.4.11}$$

Hence $$\mu_i = \left(\frac{\partial G}{\partial n_i}\right)_{T,p,n_j\text{s} \atop j\neq i} = \left(\frac{\partial A}{\partial n_i}\right)_{T,V,n_j\text{s} \atop j\neq i} = \left(\frac{\partial H}{\partial n_i}\right)_{S,p,n_j\text{s} \atop j\neq i} = \left(\frac{\partial U}{\partial n_i}\right)_{S,V,n_j\text{s} \atop j\neq i} \tag{6.4.12}$$

From Eq. (6.4.12), it is obvious that the chemical potential plays a different role in each of Eqs (6.4.1) to (6.4.4), even though it is the same quantity in all cases.

6.5 THERMODYNAMIC RELATIONS INVOLVING PARTIAL MOLAR QUANTITIES

By definition, the partial molar quantities for functions *V, U, H, S, A* and *G* are given by

$$V_{i,\,\text{pm}} = \left(\frac{\partial V}{\partial n_i}\right)_{T,p,n_j\text{s};\ j\neq i} \qquad U_{i,\,\text{pm}} = \left(\frac{\partial U}{\partial n_i}\right)_{T,p,n_j\text{s};\ j\neq i}$$

$$H_{i,\,\text{pm}} = \left(\frac{\partial H}{\partial n_i}\right)_{T,p,n_j\text{s};\ j\neq i} \qquad S_{i,\,\text{pm}} = \left(\frac{\partial S}{\partial n_i}\right)_{T,p,n_j\text{s};\ j\neq i}$$

$$A_{i,\,\text{pm}} = \left(\frac{\partial A}{\partial n_i}\right)_{T,p,n_j\text{s};\ j\neq i} \qquad G_{i,\,\text{pm}} = \left(\frac{\partial G}{\partial n_i}\right)_{T,p,n_j\text{s};\ j\neq i} \tag{6.5.1}$$

It is important to see that while the chemical potential can be defined by any one expression of Eq. (6.4.12), but only the first one is a partial molar quantity.[†]

Now $\qquad H = U + pV$

Differentiating the above equation with respect to n_i, keeping *T, p* and all other n_js constant, we get

---

[†] In Eq. (6.4.12), the chemical potential has been defined in terms of $\partial G/\partial n_i$, $\partial A/\partial n_i$, $\partial H/\partial n_i$ and $\partial U/\partial n_i$. In each of these expressions, the variables kept constant are the corresponding natural independent variables and the amounts of all the constituents of the mixture except the one under study. The natural independent variables as defined earlier are *G*(*T, p*), *A*(*T, V*), *H*(*S, p*) and *U*(*S, V*). The expression of various partial molar quantities (Eq. 6.5.1) are very similar to those of chemical potential with the notable exception that in all the partial molar quantities, the variables *T* and *p* are kept constant whereas in the expressions of chemical potential, the variables kept constant are the corresponding natural independent variables.

$$\left(\frac{\partial H}{\partial n_i}\right)_{\substack{T,p,n_j\text{s}\\ j\neq i}} = \left(\frac{\partial U}{\partial n_i}\right)_{\substack{T,p,n_j\text{s}\\ j\neq i}} + p\left(\frac{\partial V}{\partial n_i}\right)_{\substack{T,p,n_j\text{s}\\ j\neq i}}$$

or $$H_{i,\,\text{pm}} = U_{i,\,\text{pm}} + pV_{i,\,\text{pm}} \tag{6.5.2}$$

Similarly for the functions A and G, we get

$$A_{i,\,\text{pm}} = U_{i,\,\text{pm}} - TS_{i,\,\text{pm}} \tag{6.5.3}$$

$$G_{i,\,\text{pm}} = \mu_i = H_{i,\,\text{pm}} - TS_{i,\,\text{pm}} \tag{6.5.4}$$

Equations (6.5.2) to (6.5.4) show that *the partial molar quantities are interrelated in the same way as the total quantities.* Other expressions involving G, V, H, U, and so on have their respective analogues with μ_i, $V_{i,\,\text{pm}}$, $H_{i,\,\text{pm}}$, $U_{i,\,\text{pm}}$, and so on. For example, the relations

$$\left(\frac{\partial G}{\partial T}\right)_p = -S \quad \text{and} \quad \left(\frac{\partial G}{\partial p}\right)_T = V$$

which are applicable to a single-component system become

$$\left(\frac{\partial \mu_i}{\partial T}\right)_{p,\,n_j\text{s}} = -S_{i,\,\text{pm}} \quad \text{and} \quad \left(\frac{\partial \mu_i}{\partial p}\right)_{T,\,n_j\text{s}} = V_{i,\,\text{pm}}$$

for a multicomponent open system. These equalities may be proved as follows:

$$\left(\frac{\partial \mu_i}{\partial T}\right)_{p,\,n_j\text{s}} = \left\{\frac{\partial}{\partial T}\left(\frac{\partial G}{\partial n_i}\right)_{\substack{T,\,p,\,n_j\text{s}\\ j\neq i}}\right\}_{p,n_j\text{s}}$$

$$= \left\{\frac{\partial}{\partial n_i}\left(\frac{\partial G}{\partial T}\right)_{p,\,n_j\text{s}}\right\}_{\substack{T,p,n_j\text{s}\\ j\neq i}}$$

(Since the order of differentiation does not matter as G is a state function.)

$$= \left\{\frac{\partial}{\partial n_i}(-S)\right\}_{\substack{T,p,n_j\text{s}\\ j\neq i}} = -S_{i,\,\text{pm}} \tag{6.5.5}$$

Similarly $$\left(\frac{\partial \mu_i}{\partial p}\right)_{T,\,n_j\text{s}} = \left\{\frac{\partial}{\partial p}\left(\frac{\partial G}{\partial n_i}\right)_{\substack{T,\,p,\,n_j\text{s}\\ j\neq i}}\right\}_{T,\,n_j\,\text{s}} = \left\{\frac{\partial}{\partial n_i}\left(\frac{\partial G}{\partial p}\right)_{T,\,n_j\,\text{s}}\right\}_{\substack{T,\,p,\,n_j\text{s}\\ j\neq i}}$$

$$= \left\{\frac{\partial}{\partial n_i}(V)\right\}_{\substack{T,\,p,\,n_j\text{s}\\ j\neq i}} = V_{i,\,\text{pm}} \tag{6.5.6}$$

Problem 6.5.1 Show that

$$\left(\frac{\partial(\Delta\mu_i/T)}{\partial T}\right)_{p,n_j\text{s}} = -\frac{\Delta H_{i,\text{pm}}}{T^2}$$

Solution We prove the above equation as follows:

$$\left(\frac{\partial(\Delta\mu_i/T)}{\partial T}\right)_{p,n_j\text{s}} = -\frac{1}{T^2}\Delta\mu_i + \frac{1}{T}\left(\frac{\partial}{\partial T}\Delta\mu_i\right)_{p,n_j\text{s}} = -\frac{1}{T^2}\Delta\mu_i + \frac{1}{T}(-\Delta S_{i,\text{pm}})$$

$$= -\frac{1}{T^2}(\Delta H_{i,\text{pm}} - T\,\Delta S_{i,\text{pm}}) + \frac{1}{T}(-\Delta S_{i,\text{pm}}) = -\frac{\Delta H_{i,\text{pm}}}{T^2}$$

6.6 THE ESCAPING TENDENCY

Suppose a substance is present in two different regions α and β of a system. Let $\mu^{(\alpha)}$ and $\mu^{(\beta)}$ be the values of chemical potentials of the substance in these two regions of the system. Suppose that we transfer the amount dn of the substance from the region α to region β. Then, according to Eq. (6.3.1), we have

$$\mathrm{d}G^{(\alpha)} = \mu^{(\alpha)}\,(-\mathrm{d}n) \tag{6.6.1}$$

and

$$\mathrm{d}G^{(\beta)} = \mu^{(\beta)}\,(\mathrm{d}n) \tag{6.6.2}$$

The total change in free energy of the substance is given as

$$\mathrm{d}G = \mathrm{d}G^{(\alpha)} + \mathrm{d}G^{(\beta)}$$

or

$$\mathrm{d}G = (-\,\mu^{(\alpha)} + \mu^{(\beta)})\,\mathrm{d}n \tag{6.6.3}$$

Now if $\mu^{(\alpha)} > \mu^{(\beta)}$, then d$G$ is negative and hence the above transfer of the substance from the region α to region β will occur spontaneously. Thus, the substance flows spontaneously from a region of higher chemical potential to a region where the chemical potential is low. This flow will continue till the substance in both the regions have the same value of chemical potential, since at this stage dG will be equal to zero and hence the system will be at equilibrium.

Based on the above characteristic that the matter flows spontaneously from a region of high chemical potential to a region of low chemical potential, G.N. Lewis proposed the term *escaping tendency* for the chemical potential. The chemical potential of a component in a system is directly proportional to the escaping tendency of that component.

6.7 CHEMICAL POTENTIAL OF A GAS

Chemical Potential of an Ideal Gas The chemical potential μ of an ideal gas at a given temperature is related to its pressure p through Eq. (5.6.8), i.e.

$$\mu = \mu^\circ + RT\ln(p/p^\circ) \tag{6.7.1}$$

where μ° is the standard chemical potential when the pressure of the gas is p°, (i,e. standard-state pressure of 1 bar). It may be noted that μ depends on both T and p whereas μ° depends only on T.

Equation (6.7.1) suggests that at a given temperature, the pressure of the gas is a measure of its chemical potential. If inequalities in pressure exist in a gas container, the gas flows spontaneously from the high pressure region (high chemical potential) to the lower pressure region (lower chemical potential) until the pressure is equalized throughout the vessel. In the later stage, the gas has the same value of chemical potential throughout the container.

Chemical Potential of a Pure Gas

The chemical potential of a pure gas (ideal or real) at temperature T and pressure p may be expressed in terms of chemical potential of an ideal gas (abbreviation; ig) by the following identity.

$$\mu_{g,T,p} \equiv \mu^{\circ}_{ig,T} + \{\mu_{ig,T,p} - \mu^{\circ}_{ig,T}\} - \{\mu_{ig,T,p} - \mu_{ig,T,p\to 0}\} + \{\mu_{g,T,p} - \mu_{g,T,p\to 0}\} \qquad (6.7.2)$$

where $\mu_{ig,T,p\to 0} = \mu_{g,T,p\to 0}$ as all gases behave as an ideal gas at $p = 0$. The above identity is equivalent to

$$\mu_{g,T,p} = \mu^{\circ}_{ig,T} + RT\ln\frac{p}{p^{\circ}} - \int_0^p \frac{RT}{p}\,dp + \int_0^p V_m\,dp$$

$$= \mu^{\circ}_{ig,T} + RT\ln\frac{p}{p^{\circ}} + \int_0^p \left(V_m - \frac{RT}{p}\right)dp \qquad (6.7.3)$$

Since for an ideal gas $V_m = RT/p$, the above expression for ideal gas is reduced to

$$\mu_{ig,T,p} = \mu^{\circ}_{ig,T} + RT\ln\frac{p}{p^{\circ}} \qquad \text{(Eq. 5.6.8)}$$

Equation (6.7.3) can be expressed in terms of T and V_m as independent variables by using the expression

$$d(pV_m) = p\,dV_m + V_m\,dp$$

i.e.

$$dp = \frac{1}{V_m}[d(pV_m) - p\,dV_m]$$

with the limits of integration are

$$RT \text{ to } pV_m \text{ for } d(pV_m)$$

and

$$\infty \text{ to } V_m \text{ for } dV_m$$

as the corresponding limits for dp are 0 to p. This follows from the fact that at

$$p = 0, \quad \lim_{p\to 0} pV_m = RT \quad \text{and} \quad \lim_{p\to 0} V_m = \infty$$

With these, Eq. (6.7.3) becomes

$$V_{g,T,p} = \mu^{\circ}_{ig,T} + RT\ln\frac{p}{p^{\circ}} + \int_{RT}^{pV_m}\left(V_m - \frac{RT}{p}\right)\frac{1}{V_m}\,d(pV_m) - \int_{\infty}^{V_m}\left(V_m - \frac{RT}{p}\right)\frac{p}{V_m}\,dV_m$$

$$= \mu^\circ_{\text{ig},T} + RT\ln\frac{p}{p^\circ} + \int_{RT}^{pV_\text{m}}\left(1-\frac{RT}{pV_\text{m}}\right)\text{d}(pV_\text{m}) - \int_\infty^{V_m}\left(p-\frac{RT}{V_\text{m}}\right)\text{d}V_\text{m}$$

$$= \mu^\circ_{\text{ig},T} + RT\ln\frac{p}{p^\circ} + (pV_\text{m}-RT) - RT\ln\frac{pV_\text{m}}{RT} - \int_\infty^{V_m}\left(p-\frac{RT}{V_\text{m}}\right)\text{d}V_\text{m}$$

$$= \mu^\circ_{\text{ig},T} + (pV_\text{m}-RT) - RT\ln\frac{p^\circ V_\text{m}}{RT} - \int_\infty^{V_m}\left(p-\frac{RT}{V_\text{m}}\right)\text{d}V_\text{m}$$

For a van der Waals gas, we have

$$p = \frac{RT}{V_\text{m}-b} - \frac{a}{V_\text{m}^2}$$

Hence,

$$\mu_{\text{g},T,p} = \mu^\circ_{\text{ig},T} + \left[\left(\frac{RT}{V_\text{m}-b} - \frac{a}{V_\text{m}^2}\right)V_\text{m} - RT\right] - RT\ln\frac{p^\circ V_\text{m}}{RT}$$

$$-\int_\infty^{V_\text{m}}\left(\frac{RT}{V_\text{m}-b} - \frac{a}{V_\text{m}^2} - \frac{RT}{V_\text{m}}\right)\text{d}V_\text{m}$$

$$= \mu^\circ_{\text{ig},T} + \frac{RT\,V_\text{m}}{V_\text{m}-b} - \frac{2a}{V_\text{m}} - RT - RT\ln\frac{p^\circ V_\text{m}}{RT} + RT\ln\frac{V_\text{m}}{V_\text{m}-b} \tag{6.7.4}$$

Equation (6.7.4) reduces to Eq. (5.6.8) for an ideal gas (for which $a = 0$, $b = 0$ and $V_\text{m}/RT = 1/p$).

6.8 CHEMICAL POTENTIAL OF A GAS IN A MIXTURE OF IDEAL GASES

The expression for the chemical potential of an ideal gas in a mixture of ideal gases can be obtained by replacing p in Eq. (6.7.1) by the partial pressure of the gas in the gaseous mixture. Thus

$$\mu_{i(\text{mix})} = \mu_i^\circ + RT\ln(p_i/p^\circ) \tag{6.8.1}$$

where $p^\circ = 1$ bar. Since $p_i = x_i p$, therefore

$$\mu_{i(\text{mix})} = \mu_i^\circ + RT\ln(p/p^\circ) + RT\ln x_i \tag{6.8.2}$$

$$= \mu_i^* + RT\ln x_i \tag{6.8.3}$$

where μ_i^* represents the chemical potential of the pure ideal gas at temperature T and pressure p and is given as

$$\mu_i^* = \mu_i^\circ + RT\ln(p/p^\circ) \tag{6.8.4}$$

Since $x_i < 1$, the term $RT\ln x_i$ in Eq. (6.8.3) has a negative value. It, therefore, follows that

$$\mu_{i(\text{mix})} < \mu_i^* \tag{6.8.5}$$

that is, the chemical potential of an ideal gas in a mixture of ideal gases with a total pressure p is less than the corresponding value for the pure ideal gas at the same total pressure p.

Phenomenon of Diffusion

Consider a system in which a pure gas A is separated from a mixture of gases, containing the gas A as one of the components, by a suitable membrane permeable to the pure gas as shown in Fig. (6.8.1). Let the pressure of the pure gas be the same as the total pressure of the mixture of gases.

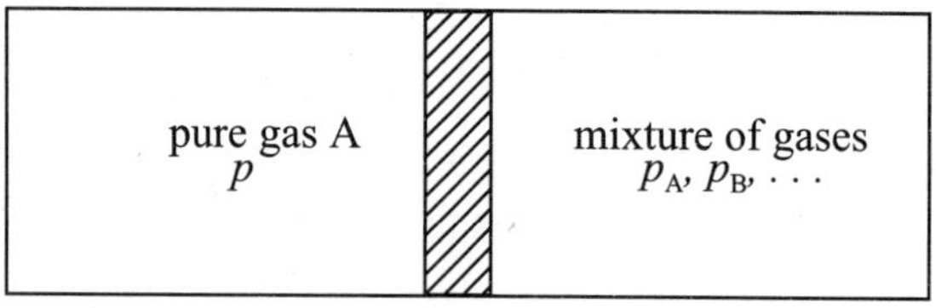

Fig. 6.8.1 Process of diffusion

Making use of Eq. (6.8.5) and the fact that the substance flows from a higher chemical potential region to a lower chemical potential region, we can conclude that gas A will diffuse from the region of the pure gas to that of the mixture of gases, i.e. from the left hand side to the right hand side of Fig. 6.8.1.

6.9 PARTIAL MOLAR QUANTITIES OF A GAS IN A MIXTURE OF IDEAL GASES

Expressions for the various partial molar quantities of an ideal gas in a mixture of ideal gases can be obtained from Eq. (6.8.3), which gives

$$\mu_{i\,(\mathrm{mix})} = \mu_i^* + RT \ln x_i \tag{6.9.1}$$

Partial Molar Entropy

Differentiating Eq. (6.9.1) with respect to temperature, keeping p and n_is constant, we get

$$\left(\frac{\partial \mu_{i(\mathrm{mix})}}{\partial T}\right)_{p,\,n_i\mathrm{s}} = \left(\frac{\partial \mu_i^*}{\partial T}\right)_p + R \ln x_i$$

or
$$S_{i,\mathrm{pm}} = S_{i,\mathrm{m}}^* - R \ln x_i \tag{6.9.2}$$

where $S_{i,\mathrm{m}}^*$ can be derived from Eq. (6.8.4) and is given by

$$S_{i,\mathrm{m}}^* = S_{i,\mathrm{m}}^\circ - R \ln (p/p^\circ)$$

Partial Molar Volume

Similarly, differentiating Eq. (6.9.1) with respect to p keeping T and n_is constant, we get

$$\left(\frac{\partial \mu_{i(\mathrm{mix})}}{\partial p}\right)_{T,n_i\mathrm{s}} = \left(\frac{\partial \mu_i^*}{\partial p}\right)_T$$

Deriving $(\partial \mu_i^* / \partial p)_T$ from Eq. (6.8.4), and then substituting in the above expression, we get

$$\left(\frac{\partial \mu_{i(\text{mix})}}{\partial p}\right)_{T,n_i\text{s}} = \frac{RT}{p} \quad \text{or} \quad V_{i,\text{pm}} = \frac{RT}{p} = V^*_{i,\text{pm}} \tag{6.9.3}$$

that is, the partial molar volume of an ideal gas in a mixture of ideal gases is the same as that of the molar volume of the pure ideal gas.

Comment on Partial Molar Volume

For an ideal gas mixture, we have

$$V = \frac{nRT}{p}$$

where n is the total amount of all the gases in the mixture. Comparing this with Eq. (6.9.3), we get

$$V_{i,\text{pm}} = \frac{V}{n} \tag{6.9.4}$$

that is, the partial molar volume is simple the average molar volume and thus the partial molar volume of all gases in a mixture has the same value.

6.10 ADDITIVITY RULES

Additivity Rule for Free Energy

In order to express the free energy of a system in terms of chemical potential and the amount of the components, we will require the integrated from of the equation

$$\mathrm{d}G = \sum_i \mu_i \,\mathrm{d}n_i \tag{6.10.1}$$

This can be done by considering a system consisting of a very large quantity of the mixture of uniform composition. This system will be in equilibrium and, therefore, will have the same values of μs throughout. Consider a small volume of this mixture. We are interested in finding how the free energy changes if this volume is increased so as to enclose the greater quantity of the mixture. This can be done by integrating Eq. (6.10.1). If n_i' and n_i are the amounts of ith component before and after the expansion and G' and G are the corresponding free energies, then we have

$$\int_{G'}^{G} \mathrm{d}G = \sum_i \int_{n_i'}^{n_i} \mu_i \,\mathrm{d}n_i$$

or

$$G - G' = \sum_i \mu_i (n_i - n_i') \tag{6.10.2}$$

since μ_i remains constant. Initially if the volume is zero, then $n_i' = 0$ and $G' = 0$. Thus, the above equation reduces to

$$G = \sum_i n_i \mu_i \tag{6.10.3}$$

This is the additivity rule for the Gibbs free energy. Knowing the chemical potential and the amount of each constituent of a mixture, we can compute the total free energy G of the mixture at the specified temperature and pressure. If the system contains only one component, then Eq. (6.10.3) reduces to

$$G = n\mu \qquad \text{or} \qquad \mu = \frac{G}{n} \tag{6.10.4}$$

Thus, the chemical potential of a pure substance is simply the molar free energy. In the mixture, μ_i stands for the partial molar free energy of the substance i.

The additivity rule is applicable to all the partial molar quantities. These can be derived as follows.

Additive Rule for Entropy

Differentiating Eq. (6.10.3) with respect to temperature keeping p and all n_is constant, we have

$$\left(\frac{\partial G}{\partial T}\right)_{p,n_i\text{s}} = \sum n_i \left(\frac{\partial \mu_i}{\partial T}\right)_{p,n_i\text{s}} \tag{6.10.5}$$

But we know that

$$\left(\frac{\partial G}{\partial T}\right)_{p,n_i\text{s}} = -S \quad \text{and} \quad \left(\frac{\partial \mu_i}{\partial T}\right)_{p,n_i\text{s}} = -S_{i,\text{pm}} \qquad \text{(Eq. 6.4.5)}$$

Thus $$S = \sum_i n_i\, S_{i,\text{pm}} \tag{6.10.6}$$

Additive Rule for Volume

Differentiating Eq. (6.10.3) with respect to pressure keeping T and all n_is constant, we have

$$\left(\frac{\partial G}{\partial p}\right)_{T,n_i\text{s}} = \sum n_i \left(\frac{\partial \mu_i}{\partial p}\right)_{T,n_i\text{s}}$$

But $$\left(\frac{\partial G}{\partial p}\right)_{T,n_i\text{s}} = V \quad \text{and} \quad \left(\frac{\partial \mu_i}{\partial p}\right)_{T,n_i\text{s}} = V_{i,\text{pm}} \qquad \text{(Eq. 6.4.5)}$$

Thus $$V = \sum_i n_i V_{i,\text{pm}} \tag{6.10.7}$$

Additive Rule for Enthalpy

The chemical potential of the ith component is given as

$$\mu_i = H_{i,\text{ pm}} - TS_{i,\text{ pm}} \qquad \text{(Eq. 6.5.4)}$$

Multiplying this equation by n_i and summing, we have

$$\sum_i n_i \mu_i = \sum_i n_i H_{i,\text{ pm}} - T\sum_i n_i S_{i,\text{ pm}}$$

which reduces to

$$G = \sum_i n_i H_{i,\text{ pm}} - TS$$

But by definition $G = H - TS$, therefore

$$H = \sum_i n_i H_{i,\text{pm}} \tag{6.10.8}$$

Additive Rule for Internal Energy

Multiplying the expression $H_{i,\text{ pm}} = U_{i,\text{ pm}} + pV_{i,\text{ pm}}$ with n_i and adding, we have

$$\sum_i n_i H_{i,\text{ pm}} = \sum_i n_i U_{i,\text{ pm}} + p\left(\sum_i n_i V_{i,\text{ pm}}\right)$$

which reduces to

$$H = \sum_i n_i U_{i,\text{pm}} + pV$$

But $H = U + pV$; therefore

$$U = \sum_i n_i U_{i,\text{pm}} \tag{6.10.9}$$

Additive Rule for Helmholtz Free Energy

Starting with the equation

$$A_{i,\text{m}} = U_{i,\text{pm}} - TS_{i,\text{pm}} \qquad \text{(Eq. 6.5.3)}$$

we have $\sum_i n_i A_{i,\text{pm}} = \sum_i n_i U_{i,\text{pm}} - T\sum_i n_i S_{i,\text{pm}}$

which reduces to

$$\sum_i n_i A_{i,\text{pm}} = U - TS$$

But $U - TS = A$, thus

$$A = \sum_i n_i A_{i,\text{pm}} \tag{6.10.10}$$

Hence, additivity rule is applicable to all extensive properties.

Derivation of Additive Rule using Euler's Theorem

The above additivity rules can also be derived with the help of Euler's theorem (Eq. 1.3.65) as applicable to homogeneous functions of first degree with respect to the amounts of the substances present in the system, and thus at constant T and p may be written as

$$Y = f(n_1, n_2, \ldots n_k)$$

where Y stands for any one of the above functions. Applying Euler's theorem, we get

$$Y = n_1\frac{\partial Y}{\partial n_1} + n_2\frac{\partial Y}{\partial n_2} + \cdots + n_k\frac{\partial Y}{\partial n_k}$$

$$= \sum_i n_i\left(\frac{\partial Y}{\partial n_i}\right)_{T,p,n_j\text{s}\atop j\neq i} = \sum_i n_i Y_{i,\text{pm}} \tag{6.10.11}$$

Additive Rule for the Amount and Mass of a Substance

The additivity rule is also applicable to the total amount or the total mass, so that

$$n = \sum_i (1)\, n_i \tag{6.10.12}$$

$$m = \sum_i M_i\, n_i \tag{6.10.13}$$

Thus, the partial molar amounts are all equal to unity and the partial molar mass of a substance is its molar mass.

Not that the Euler's theorem has been applied to the function Y written as

$$Y = f(n_1, n_2, \ldots, n_k)$$

and not as

$$Y = f(T, p, n_1, n_2, \ldots, n_k) \tag{6.10.4}$$

since the function Y may not be a homogeneous function of first degree with respect to either T or p. For instance, the function G is not a homogeneous function of either T or p (*see* Problem 6.10.1). Hence, Eq. (6.10.11) holds good for any given values of temperature and pressure.

Problem 6.10.1

Show that G is not a homogeneous function of first degree with respect to T and p.

Solution

If it be assumed that G is a homogeneous function of first degree with respect to T and p, according to Euler's theorem, we would have

$$G = T\left(\frac{\partial G}{\partial T}\right)_p + p\left(\frac{\partial G}{\partial p}\right)_T$$

$$= -TS + pV$$

But by definition

$$G = H - TS = U + pV - TS$$

Since the two equations of G are not identical, our assumption that G is a homogeneous function in T and p must be wrong.

Example 6.10.1

The volume of an aqueous solution of NaCl at 25 °C is expressed as a function of the amount m in 1 kg of solvent in the following form:

$$V/\text{cm}^3 = 1\,000.94 + 16.4\,(m/\text{mol}) + 2.14\,(m/\text{mol})^{3/2} - 0.002\,7\,(m/\text{mol})^{5/2}$$

(a) Find the partial molar volume of NaCl in water in 1 mol kg^{-1} solution.

(b) Given that the molar volumes of NaCl(s) and liquid water are 27.00 and 18.07 cm^3 mol^{-1}, respectively, find $\Delta_r V_m$ for

(i) $\frac{1000}{18} H_2O + NaCl(s) \rightarrow$ solution (1 mol kg^{-1})

(ii) $NaCl(s) \rightarrow NaCl$(1 mol kg^{-1} solution)

(iii) $H_2O(1) \rightarrow H_2O$(1 mol kg^{-1} solution)

Solution

(a) The partial molar volume of NaCl in the solution can be determined by evaluating the expression $\partial V/\partial m$. From the given expression of V, we get

$$\frac{\partial V}{\partial m} = (16.4\ \text{cm}^3\ \text{mol}^{-1}) + \frac{3}{2}(2.14\ \text{cm}^3\ \text{mol}^{-1})\,(m/\text{mol})^{1/2}$$

$$- \frac{5}{2}(0.0027\ \text{cm}\ \text{mol}^{-1})\,(m/\text{mol})^{3/2}$$

Substituting $m = 1$ mol, we get

$$V_{2,\,\text{pm}} = (16.4\ \text{cm}^3\ \text{mol}^{-1}) + \frac{3}{2}(2.14\ \text{cm}^3\ \text{mol}^{-1}) - \frac{5}{2}\,(0.002\,7\ \text{cm}^3\ \text{mol}^{-1})$$

$$= 19.60\ \text{cm}^3\ \text{mol}^{-1}$$

(b) From the given expression of total volume, we get for 1 mol kg^{-1} solution

$$V/\text{cm}^3 = 1\,000.94 + 16.4 + 2.14 - 0.002\,7$$

i.e.

$$V = 1\,019.48\ \text{cm}^3$$

In 1 mol kg^{-1} solution, we will have

$$n_1 = \frac{1\,000\ \text{g}}{18\ \text{g mol}^{-1}} = \frac{1\,000}{18}\text{mol} \quad \text{and} \quad n_2 = 1\ \text{mol}$$

Hence, following the additivity rule, we write

$$n_1 V_{1,\,\mathrm{pm}} + n_2 V_{2,\,\mathrm{pm}} = V_{\mathrm{total}}$$

$$\left(\frac{1000}{18}\,\mathrm{mol}\right) V_{1,\,\mathrm{pm}} + (1\ \mathrm{mol})\ (19.60\ \mathrm{cm^3\ mol^{-1}}) = 1\,019.48\ \mathrm{cm^3}$$

Thus $\quad V_{1,\,\mathrm{pm}} = 18.00\ \mathrm{cm^3\ mol^{-1}}$

Now, we have

(i) $\dfrac{1000}{18}\,H_2O + NaCl(s) \rightarrow$ solution (1 mol kg^{-1})

$$\Delta V = V_{\mathrm{soln.}} - \left(\frac{1000}{18}\,\mathrm{mol}\right) V_{\mathrm{m}}^{*}(H_2O) - (1\ \mathrm{mol})\ V_{\mathrm{m}}^{*}\,(NaCl,\ s)$$

$$= 1\,019.48\ \mathrm{cm^3} - \left(\frac{1000}{18}\,\mathrm{mol}\right) (18.07\ \mathrm{cm^3\ mol^{-1}}) - (1\ \mathrm{mol})\ (27.00\ \mathrm{cm^3\ mol^{-1}})$$

$$= -11.41\ \mathrm{cm^3}$$

(ii) $NaCl(s) \rightarrow NaCl$(1 mol kg^{-1} solution)

$$\Delta V = V_{2,\,\mathrm{pm}} - V_{2,\,\mathrm{m}}^{*}$$

$$= 19.60\ \mathrm{cm^3\ mol^{-1}} - 27.00\ \mathrm{cm^3\ mol^{-1}} = -7.4\ \mathrm{cm^3\ mol^{-1}}$$

(iii) $H_2O(1) \rightarrow H_2O$(1 mol kg^{-1} solution)

$$\Delta V = V_{1,\,\mathrm{pm}} - V_{1,\,\mathrm{m}}^{*}$$

$$= 18.00\ \mathrm{cm^3\ mol^{-1}} - 18\ 07\ \mathrm{cm^3\ mol^{-1}} = -\ 0.07\ \mathrm{cm^3\ mol^{-1}}$$

Example 6.10.2

What volume (in dm^3) of water should be added to 2 dm^3 of laboratory alcohol (96 mass per cent ethanol and 4 mass per cent water) so that it is converted into vodka (44 mass per cent ethanol and 56 mass per cent water)? How much volume in dm^3 of vodka is formed? Given the following information:

	96% ethanol	*Vodka*
$V_{\mathrm{pm}}(H_2O)$	0.916 cm^3 g^{-1}	0.953 cm^3 g^{-1}
V_{pm}(ethanol)	1.273 cm^3 g^{-1}	1.243 cm^3 g^{-1}

Specific volume of water = 1.003 cm^3 g^{-1}.

Solution

Let m_1 and m_2 be the respective masses of ethanol and water in the given 2 dm^3 of the laboratory alcohol and let m_3 be the mass of water added to convert laboratory alcohol into vodka. Following the additivity rule, we write

$$m_1\ (1.273\ \mathrm{cm^3\ g^{-1}}) + m_2\ (0.916\ \mathrm{cm^3\ g^{-1}}) = 2\,000\ \mathrm{cm^3} \tag{1}$$

$$m_1\ (1.243\ \mathrm{cm^3\ g^{-1}}) + (m_2 + m_3)\ (0.953\ \mathrm{cm^3\ g^{-1}}) = V \tag{2}$$

where V is total volume of vodka formed. From the given percentage of laboratory alcohol and vodka, we write

$$\frac{m_1}{m_2} = \frac{96}{4} = 24 \tag{3}$$

$$\frac{m_1}{m_2 + m_3} = \frac{44}{56} \tag{4}$$

From Eqs (3) and (4), we get

$$m_1 = 24\ m_2$$

$$m_2 + m_3 = \frac{56}{44} m_1 = \frac{56}{44} (24\ m_2) = 30.545\ 4\ m_2$$

i.e. $m_3 = 29.545\ 4\ m_2$

Substituting m_1 in terms of m_2 in Eq. (1), we get

$$(24\ m_2)\ (1.273\ \text{cm}^3\ \text{g}^{-1}) + m_2\ (0.916\ \text{cm}^3\ \text{g}^{-1}) = 2\ 000\ \text{cm}^3$$

which on solving for m_2 gives

$$m_2 = 63.556\ 6\ \text{g}$$

Hence $m_1 = 24\ m_2 = 1\ 525.358\ 4\ \text{g}$

$$m_3 = 29.545\ 4\ m_2 = (29.545\ 4)(63.556\ 6\ \text{g}) = 1\ 877.805\ \text{g}$$

Substituting m_1, m_2 and m_3 in Eq. (2), we get

$$(1\ 525.358\ 4\ \text{g})\ (1.243\ \text{cm}^3\ \text{g}^{-1}) + (63.556\ 6\ \text{g} + 1\ 877.805\ \text{g})\ (0.953\ \text{cm}^3\ \text{g}^{-1}) = V$$

which gives

$$V = 3\ 746.14\ \text{cm}^3$$

Finally $V_3 = (m_3)$ (specific volume)

$$= (1\ 877.805\ \text{g})\ (1.003\ \text{cm}^3\ \text{g}^{-1}) = 1\ 883.44\ \text{cm}^3 \simeq 1.883\ \text{dm}^3$$

Example 6.10.3

The volumetric data for liquid mixtures of A and B are represented by the simple quadratic expression

$$V/\text{cm}^3\ (\text{mol of mixture})^{-1} = 109.4 - 16.8\ x_B - 2.64\ x_B^2$$

where x_B is the mole fraction of B. Derive the expressions of $V_{A,\,pm}$, $V_{B,\,pm}$ and ΔV.

Solution

We are given that

$$V/\text{cm}^3\ (\text{mol of mixture})^{-1} = 109.4 - 16.8\ x_B - 2.64\ x_B^2$$

Let the mixture contain the amounts n_A and n_B of A and B, respectively. From the above expression, we have

$$\frac{V}{n_A + n_B} \Big/ \text{cm}^3 = 109.4 - 16.8 \frac{n_B}{n_A + n_B} - 2.64 \frac{n_B^2}{(n_A + n_B)^2}$$

or
$$V/\text{cm}^3 = 109.4\ (n_A + n_B) - 16.8\ n_B - 2.64\ \frac{n_B^2}{(n_A + n_B)} \qquad (1)$$

Now
$$V_{A,\,pm} = \left(\frac{\partial V}{\partial n_A}\right)_{n_B} = (109.4\ \text{cm}^3) + (2.64\ \text{cm}^3)\frac{n_B^2}{(n_A + n_B)^2}$$

$$= 109.4\ \text{cm}^3 + (2.64\ \text{cm}^3) x_B^2 \qquad (2)$$

$$V_{B,\,pm} = \left(\frac{\partial V}{\partial n_B}\right)_{n_A} = 109.4\ \text{cm}^3 - 16.8\ \text{cm}^3 - (2.64\ \text{cm}^3)\frac{2n_B}{(n_A + n_B)}$$

$$+ (2.64\ \text{cm}^3)\frac{n_B^2}{(n_A + n_B)^2}$$

$$= 92.6\ \text{cm}^3 - (5.28\ \text{cm}^3)\ x_B + (2.64\ \text{cm}^3)\ x_B^2 \qquad (3)$$

For finding ΔV, we determine $n_A V^*_{A,m}$ and $n_B V^*_{B,m}$ from Eq. (1) by substituting $n_B = 0$ and $n_A = 0$, respectively. Thus

$$n_A V^*_{A,m} = 109.4\ n_A$$

$$n_B V^*_{B,m} = 109.4\ n_B - 16.8\ n_B - 2.64\ n_B = 89.96\ n_B$$

Hence $\quad \Delta V = V_{(\text{mixture})} - V_{(\text{pure})}$

$$= \left\{109.4(n_A + n_B) - 16.8 n_B - 2.64\frac{n_B^2}{(n_A + n_B)}\right\} - \{109.4\ n_A + 89.96\ n_B\}$$

$$= 2.64\ n_B - 2.64\ \frac{n_B^2}{(n_A + n_B)} = 2.64 n_B\left(1 - \frac{n_B}{n_A + n_B}\right) = 2.64\ n_B(1 - x_B)$$

$$= 2.64\ n_B\ x_A$$

6.11 GIBBS-DUHEM EQUATION

Gibbs-Duhem Equation Involving Chemical Potential

For a multicomponent open system, we have

$$G = \sum_i n_i\ \mu_i$$

Differentiating the above equation, we get

$$dG = \sum_i \mu_i\ dn_i + \sum_i (d\mu_i)\ n_i$$

But, by the fundamental equation

$$dG = -S\ dT + V\ dp + \sum_i \mu_i\ dn_i \qquad \text{(Eq. 6.4.1)}$$

Equating the two equations, we get

$$\sum_i n_i\ d\mu_i = -S\ d\,T + V\ dp \qquad (6.11.1)$$

which is the Gibbs-Duhem equation.

If temperature and pressure of the system are held constant, we have

$$\sum_i n_i\ d\mu_i = 0 \qquad (6.11.2)$$

Physical Interpretation of Gibbs-Duhem Equation

Equation (6.11.2) shows that if the composition varies, the chemical potential do not change independently but in a related way. Taking, for example, a system of two constituents, Eq. (6.11.2) becomes

$$n_1\ d\mu_1 + n_2\ d\mu_2 = 0 \qquad (6.11.3)$$

Rearranging, we have

$$d\mu_2 = -\left(\frac{n_1}{n_2}\right) d\mu_1 \qquad (6.11.4)$$

that is, if a given variation in composition produces a change $d\mu_1$ in the chemical potential of the first component, the corresponding change in the chemical potential $d\mu_2$ of the second component is given by the above equation. It may be noted that if $d\mu_1$ is positive (i.e. an increase in the value of μ_1) then $d\mu_2$ is negative (i.e. a decrease in the value of μ_2) and vice vissa.

General Expression of Gibbs-Duhem Equation

Equation (6.11.4) has been derived for the changes in chemical potential caused by the variation in composition of a solution containing two constituents. Similar relations can be derived for any of the partial quantities. In general, we can write (when T and p are constant)

$$\sum_i n_i \,(\mathrm{d}Y_{i,\mathrm{pm}}) = 0 \tag{6.11.5}$$

where $Y_{i,\,\mathrm{pm},}$ is any partial molar quantity.

Gibbs-Duhem Equation for Volume

Taking, for example, the Gibbs-Duhem equation involving partial molar volume, we have

$$V = \sum_i n_i V_{i,\,\mathrm{pm}} \tag{6.11.6}$$

Therefore $$\mathrm{d}V = \sum_i \mathrm{d}n_i \, V_{i,\,\mathrm{pm}} + \sum_i n_i \, \mathrm{d}V_{i,\mathrm{pm}} \tag{6.11.7}$$

Taking $V = f(T, p, n_i)$, we have

$$\mathrm{d}V = \left(\frac{\partial V}{\partial T}\right)_{p,n_j\mathrm{s}} \mathrm{d}T + \left(\frac{\partial V}{\partial p}\right)_{T,n_j\mathrm{s}} \mathrm{d}p + \sum_i \left(\frac{\partial V}{\partial n_i}\right)_{\substack{T,p,n_j\mathrm{s} \\ j\neq i}} \mathrm{d}n_i$$

At constant T and p, we have

$$\mathrm{d}V = \sum_i \left(\frac{\partial V}{\partial n_i}\right)_{\substack{T,p,n_j\mathrm{s} \\ j\neq i}} \mathrm{d}n_i \quad \text{or} \quad \mathrm{d}V = \sum_i V_{i,\,\mathrm{pm}} \, \mathrm{d}n_i \tag{6.11.8}$$

Subtracting Eq. (6.11.8) from Eq. (6.11.7), we get

$$\sum_i n_i \, \mathrm{d}V_{i,\,\mathrm{pm}} = 0 \tag{6.11.9}$$

For a binary solution, we have

$$n_1 \, \mathrm{d}V_{1,\,\mathrm{pm}} + n_2 \, \mathrm{d}V_{2,\,\mathrm{pm}} = 0$$

or $$\mathrm{d}V_{1,\,\mathrm{pm}} = -\frac{n_2}{n_1} \, \mathrm{d}V_{2,\,\mathrm{pm}} \tag{6.11.10}$$

From Eq. (6.11.10), it follows that the changes in partial molar volumes of the two components vary in the opposite direction, i.e. if partial molar volume of one component increases then there occurs a decrease in the partial molar volume of the second component and vice versa.

6.12 FREE ENERGY AND ENTROPY OF MIXING OF IDEAL GASES

Let a number of gases at the same temperature and pressure be mixed as shown in Fig. 6.12.1. Since the total volume after mixing is the sum of the individual volumes, the temperature and pressure of the system after mixing will be the same as that of unmixed gases. The free energy of mixing of the above process can be computed as follows.

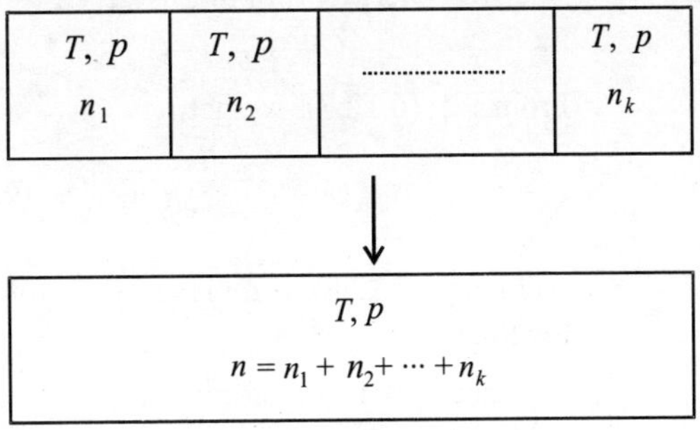

Fig. 6.12.1 Mixing of gases at the same temperature and pressure

Free Energy of Unmixed Gases

Since each gas is present at the same T and p, its chemical potential is given by the expression

$$\mu_i = \mu_i^\circ + RT \ln \left(\frac{p}{p^\circ} \right) \tag{6.12.1}$$

Hence $$G_{\text{initial}} = \sum_{i=1}^{k} n_i \mu_i \tag{6.12.2}$$

Free Energy of Mixed Gases

The pressure of the entire system is still p, but the individual gases have partial pressure given as

$$p_i = \frac{n_i}{\sum_i n_i} p = \frac{n_i}{n} p = x_i p \tag{6.12.3}$$

where x_i is the amount fraction of ith gas. The chemical potential of individual gases will be given by the expression

$$\mu_i' = \mu_i^\circ + RT \ln \left(\frac{p_i}{p^\circ} \right) \tag{6.12.4}$$

which in view of Eq. (6.12.3) becomes

$$\mu_i' = \mu_i^\circ + RT \ln \left(\frac{x_i p}{p^\circ} \right) = \left[\mu_i^\circ + RT \ln \left(\frac{p}{p^\circ} \right) + RT \ln x_i \right]$$

Using Eq. (6.12.1), the above expression becomes

$$\mu_i' = \mu_i + RT \ln x_i \tag{6.12.5}$$

Hence $$G_{\text{final}} = \sum_{i=1}^{k} n_i \mu_i' = \sum_{i=1}^{k} n_i (\mu_i + RT \ln x_i)$$

$$= \sum_{i=1}^{k} n_i \mu_i + RT \sum_{i=1}^{k} n_i \ln x_i \tag{6.12.6}$$

Using Eq. (6.12.2), we have

$$G_{\text{final}} = G_{\text{initial}} + RT \sum_{i=1}^{k} n_i \ln x_i \tag{6.12.7}$$

Free Energy of Mixing

The free energy of mixing is given as

$$\Delta_{mix}G = G_{final} - G_{initial}$$

From Eq. (6.12.7), we get

$$\Delta_{mix}G = RT \sum_{i=1}^{k} n_i \ln x_i = nRT \sum_{i=1}^{k} \left(\frac{n_i}{n}\right) \ln x_i = nRT \sum_{i=1}^{k} x_i \ln x_i \quad (6.12.8)$$

In Eq. (6.12.8), n represents the total amount of gases which are being mixed together.

Since $x_i < 1$, therefore, every term on the right side is negative and so the sum is always negative, i.e. a decrease in free energy occurs. It follows from this that the formation of a mixture from the pure constituents always occurs spontaneously.

If there are only two substances in the mixture (binary system), then we have

$$x_1 = x \quad \text{and} \quad x_2 = 1 - x$$

With these, Eq. (6.12.8) becomes

$$\Delta_{mix}G = nRT\,[x \ln x + (1 - x) \ln (1 - x)] \quad (6.12.9)$$

A plot of $\Delta_{mix}G/nRT$ versus x is shown in Fig. 6.11.2. The curve exhibits minimum at $x = 1/2$ and is symmetrical about this value of x.†

The greater decrease in free energy of mixing is associated with the formation of the mixture having equal amounts of the two constituents.

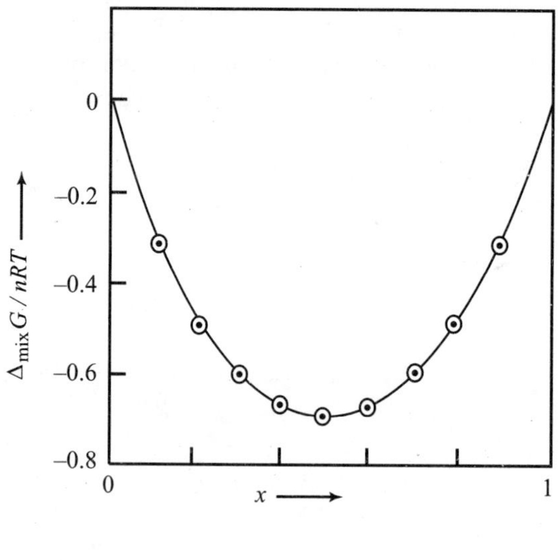

Fig. 6.12.2 $\Delta_{mix}G/nRT$ **versus** x **for a binary ideal mixture**

† The value of x can be obtained by differentiating Eq. (6.12.9) with respect to x and equating the resultant expression to zero.

$$\frac{\partial(\Delta_{mix}G/nRT)}{\partial x} = \ln x + x\frac{1}{x} + (-1)\ln(1-x) + (1-x)\left(\frac{-1}{1-x}\right) = 0$$

or $$\ln x - \ln(1 - x) = 0 \quad \text{or} \quad \ln\frac{x}{1-x} = 0 \quad \text{or} \quad \frac{x}{1-x} = 1 \quad \text{or} \quad x = \frac{1}{2}$$

Example 6.12.1 Calculate $\Delta_{mix}G$ for mixing the constituents of air (80 mole per cent nitrogen and 20 mole per cent oxygen) at 298 K so as to get 1 mol of air.

Solution Substituting $x_{N_2} = x_1 = 0.8$ and $x_{O_2} = x_2 = 0.2$ in the expression

$$\Delta_{mix}G = nRT\,[x_1 \ln x_1 + x_2 \ln x_2]$$

we get

$$\begin{aligned}\Delta_{mix}G &= (1\text{ mol})\,(8.314\text{ J K}^{-1}\text{ mol}^{-1})\,(298\text{ K})\,(0.8 \ln 0.8 + 0.2 \ln 0.2)\\ &= (2\,447.6\text{ J})\,(-\,0.178\,5 - 0.321\,9)\\ &= -1\,224.8\text{ J} \simeq -1.225\text{ kJ}\end{aligned}$$

Problem 6.12.1 Show that the free energy of mixing of three gases has a minimum value with respect to each gas only when $x_1 = x_2 = x_3 = 1/3$.

Solution The expression for free energy of mixing of three gases is

$$\Delta_{mix}G = nRT\,[x_1 \ln x_1 + x_2 \ln x_2 + x_3 \ln x_3]$$

Since $x_1 + x_2 + x_3 = 1$, we get

$$\Delta_{mix}G = nRT\,[x_1 \ln x_1 + x_2 \ln x_2 + (1 - x_1 - x_2) \ln (1 - x_1 - x_2)]$$

For a minimum value of $\Delta_{mix}G$ with respect to each gas, we set $(\partial\Delta_{mix}G/\partial x_1)_{x_2}$ and $(\partial\Delta_{mix}G/\partial x_2)_{x_1}$ separately equal to zero, i.e.

$$\begin{aligned}\frac{1}{nRT}\left(\frac{\partial\Delta_{mix}G}{\partial x_1}\right)_{x_2} &= 1 + \ln x_1 - 1 - \ln (1 - x_1 - x_2)\\ &= \ln x_1 - \ln x_3 = 0 \quad \Rightarrow \quad x_1 = x_3\end{aligned}$$

$$\begin{aligned}\frac{1}{nRT}\left(\frac{\partial\Delta_{mix}G}{\partial x_2}\right)_{x_1} &= 1 + \ln x_2 - 1 - \ln (1 - x_1 - x_2)\\ &= \ln x_2 - \ln x_3 = 0 \quad \Rightarrow \quad x_2 = x_3\end{aligned}$$

From these, it follows that

$$x_1 = x_2 = x_3$$

Now, since $x_1 + x_2 + x_3 = 1$, we will have

$$x_1 = x_2 = x_3 = 1/3$$

Comment: The above procedure may be extended to a mixture of n gases. It may be shown that, $\Delta_{mix}G$ for mixing n gases has a minimum value provided $x_1 = x_2 = \cdots = x_n = 1/n$.

Entropy of Mixing Since $(\partial G/\partial T)_p = -S$, therefore, differentiation with respect to temperature yields:

$$\begin{aligned}\left\{\frac{\partial(\Delta_{mix}G)}{\partial T}\right\}_{p,n_i} &= \left(\frac{\partial G_{final}}{\partial T}\right)_{p,n_i} - \left(\frac{\partial G_{initial}}{\partial T}\right)_{p,n_i}\\ &= -\,(S_{final} - S_{initial}) = -\,\Delta_{mix}S\end{aligned}$$

Thus, differentiating Eq. (6.12.8), we get

$$\left\{\frac{\partial(\Delta_{mix}G)}{\partial T}\right\}_p = nR\sum_i x_i \ln x_i$$

Therefore $$\Delta_{\text{mix}}S = -nR\sum_i x_i \ln x_i \tag{6.12.10}$$

The negative sign means that $\Delta_{\text{mix}}S$ is always positive. This corresponds to the increase in randomness which occurs in mixing the molecules of several kinds. A plot of $\Delta_{\text{mix}}S/nR$ versus x of a binary mixture according to the equation

$$\Delta_{\text{mix}}S = -nR\,[\,x \ln x + (1-x)\ln(1-x)] \tag{6.12.11}$$

is shown in Fig. 6.12.3. The entropy of mixing has a maximum value when $x = \frac{1}{2}$.†

This value can be obtained by differentiating Eq. (6.12.11) with respect to x and equating the resultant expression to zero. The numerical value of which per mole of mixture is

$$\frac{\Delta_{\text{mix}}S}{n} = -(8.314\ \text{J K}^{-1}\ \text{mol}^{-1})\left(\frac{1}{2}\ln\frac{1}{2}+\frac{1}{2}\ln\frac{1}{2}\right)$$

$$= -(8.314\ \text{J K}^{-1}\ \text{mol}^{-1})\ln\frac{1}{2} = +5.763\ \text{J K}^{-1}\ \text{mol}^{-1}$$

Thus, in a mixture containing only two substance the entropy of mixing per mole of the mixture varies between 0 and 5.763 J K^{-1}, depending on the composition of the mixture.

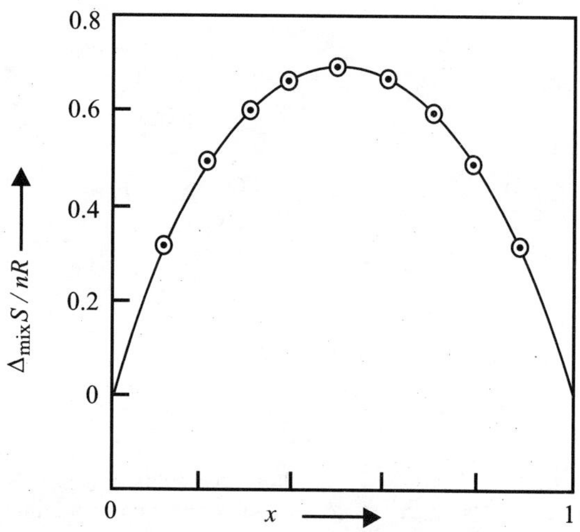

Fig. 6.12.3 Plot $\Delta_{\text{mix}}S/nR$ versus x for a binary ideal mixture

Enthalpy of Mixing That fact that

$$\Delta_{\text{mix}}G = -T\,\Delta_{\text{mix}}S \tag{6.12.12}$$

indicates that for the mixing of ideal gases $\Delta_{\text{mix}}H = 0$ since $\Delta G = \Delta H - T\,\Delta S$. This is to be expected because there exist no forces of attraction between the molecules of ideal gases (i.e. they move independent of each other). Equation (6.12.12) shows that the driving force, $\Delta_{\text{mix}}G$ which produces the mixing is entirely an entropy effect. The mixed state is a more random state, and is therefore, a more probable state.

† In general, the entropy of mixing of n gases has a maximum value provided $x_1 = x_2 = \cdots = x_n = 1/n$.

Volume of Mixing

The fact that the volume of the mixture is equal to the sum of the individual volumes can be verified by evaluating the value of $\Delta_{mix}V$. The latter can be obtained by differentiating the $\Delta_{mix}G$ with respect to pressure, temperature and composition being constant, that is

$$\Delta_{mix}V = \left\{\frac{\partial(\Delta_{mix}G)}{\partial p}\right\}_{T,n_i\text{s}} \tag{6.12.13}$$

Since $\Delta_{mix}G$ is independent of pressure (Eq. 6.12.8), the above derivative has a zero value. Hence

$$\Delta_{mix}V = 0$$

Generalization of the Expression of Entropy of Mixing

For the binary mixture, Eq. (6.12.10) may be written as

$$\Delta_{mix}S = -\,nR\,(x_1 \ln x_1 + x_2 \ln x_2) \tag{6.12.14}$$

From the Dalton's law of partial pressure, we have

$$x_1 = \frac{p_1}{p} \quad \text{and} \quad x_2 = \frac{p_2}{p} \tag{6.12.15}$$

where p_1 and p_2 are the partial pressures of the two gases in the mixture and p is the total pressure. As mentioned earlier, the total pressure p of the mixture will be the same as that of unmixed gases since the total volume of the gases is the sum of the volumes of the individual gases. Substituting Eq. (6.12.15) in Eq. (6.12.14), we get

$$\Delta_{mix}S = -\,nR\left(\frac{n_1}{n}\ln\frac{p_1}{p} + \frac{n_2}{n}\ln\frac{p_2}{p}\right)$$

or

$$\Delta_{mix}S = n_1 R \ln\frac{p}{p_1} + n_2 R \ln\frac{p}{p_2} \tag{6.12.16}$$

Since the change in entropy of a gas in an isothermal expansion from pressure p_1 to pressure p_2 is given by

$$\Delta S = nR \ln\frac{p_1}{p_2}$$

it follows immediately from Eq. (6.12.16) that the entropy of mixing is simply a sum of the entropy changes for each gas undergoing an expansion from its pressure p to its partial pressure p_i in the mixture, i.e. the entropy change in the mixing process is simply due to the pressure changes for each gas as given by Eq. (6.12.16).†

---

† *Alternatively,* the entropy change during the isothermal mixing process may be attributed to the change in volume of each gas from its inital volume to the same final volume. The expression for such a process is given by

$$\Delta S = nR \ln\frac{V_{final}}{V_{initial}}$$

Hence, for a binary mixture, we have

$$\Delta S = n_1 R \ln\frac{V_{final}}{V_1} + n_2 R \ln\frac{V_{final}}{V_2} \tag{6.12.17}$$

Equation (6.12.16) is applicable only when the initial pressures and temperatures of the gases before mixing are identical. The expression for the entropy of mixing in a case where this condition is not fulfilled can easily be derived. What we have to do is to find out the final temperature and the partial pressure of each gas and then calculate the change in entropy of each gas using the expression

$$\Delta S = n\left(C_{p,\mathrm{m}} \ln\frac{T_2}{T_1} + R\ln\frac{p_1}{p_2}\right) \tag{6.12.18}$$

and, finally adding all these changes. Alternatively, in terms of volume change, for each gas we may employ the expression

$$\Delta S = n\left(C_{V,\mathrm{m}} \ln\frac{T_2}{T_1} + R\ln\frac{V_2}{V_1}\right) \tag{6.12.19}$$

and, finally adding all these changes.

Example 6.12.2 Consider two pure gases A and B each at 298 K and 1 bar pressure. Calculate the free energy change, the entropy change and the enthalpy change relative to the unmixed gases of (i) a mixture of 10 mol of A and 10 mol of B, (ii) a mixture of 10 mol of A and 20 mol of B and also (iii) when 10 mol of B is added to the mixture of 10 mol of A and 10 mol of B.

Solution **(i)** The mole fractions of A and B in the given mixture are

$$x_\mathrm{A} = \frac{n_\mathrm{A}}{n_\mathrm{A}+n_\mathrm{B}} = \frac{(10\ \mathrm{mol})}{(10\ \mathrm{mol})+(10\ \mathrm{mol})} = 0.5$$

Substituting these in the relation

$$\Delta_\mathrm{mix}G = nRT\,[x_\mathrm{A}\ln x_\mathrm{A} + x_\mathrm{B}\ln x_\mathrm{B}]$$

we get

$$\Delta_\mathrm{mix}G = (20\ \mathrm{mol})\,(8.314\ \mathrm{J\ K^{-1}\ mol^{-1}})\,(298\ \mathrm{K})\,[0.5\ln 0.5 + 0.5\ln 0.5]$$

$$= -34\,352.6\ \mathrm{J} \simeq -34.353\ \mathrm{kJ}$$

Now

$$\Delta_\mathrm{mix}S = -\frac{\Delta_\mathrm{mix}G}{T} = \frac{34\,352.6\ \mathrm{J}}{298\ \mathrm{K}} = 115.28\ \mathrm{J\ K^{-1}}$$

$$\Delta_\mathrm{mix}H = 0$$

(ii) Here the mole fractions of A and B are

$$x_\mathrm{A} = \frac{n_\mathrm{A}}{n_\mathrm{A}+n_\mathrm{B}} = \frac{(10\ \mathrm{mol})}{(10\ \mathrm{mol})+(20\ \mathrm{mol})} = 0.333$$

$$x_\mathrm{B} = 1 - x_\mathrm{A} = 0.667$$

Substituting these in the relation

$$\Delta_\mathrm{mix}G = nRT\,[x_\mathrm{A}\ln x_\mathrm{A} + x_\mathrm{B}\ln x_\mathrm{B}]$$

we get

$$\Delta_\mathrm{mix}G = (30\ \mathrm{mol})\,(8.314\ \mathrm{J\ K^{-1}\ mol^{-1}})\,(298\ \mathrm{K}) \times (0.333\ln 0.333 + 0.667\ln 0.667)$$

$$= -47\,321\ \mathrm{J} = -47.321\ \mathrm{kJ}$$

Now

$$\Delta_\mathrm{mix}S = -\frac{\Delta G}{T} = \frac{47\,321\ \mathrm{J}}{298\ \mathrm{K}} = 158.80\ \mathrm{J\ K^{-1}}$$

$$\Delta_\mathrm{mix}H = 0$$

(iii) From parts (i) and (ii), we have

(a) 10 mol A + 10 mol B → Mixture of 10 mol A and 10 mol B

$$\begin{cases}\Delta_{\text{mix}} G_1 = -34.353 \text{ kJ}; \\ \Delta_{\text{mix}} S_1 = 115.28 \text{ J K}^{-1}\end{cases}$$

(b) 10 mol A + 20 mol B → Mixture of 10 mol A and 20 mol B

$$\begin{cases}\Delta_{\text{mix}} G_2 = -47.321 \text{ kJ}; \\ \Delta_{\text{mix}} S_2 = 158.80 \text{ J K}^{-1}\end{cases}$$

Subtracting the expression (a) from the expression (b), we get

10 mol B + Mixture of 10 mol A and 10 mol B → Mixture of 10 mol A and 20 mol B

Obviously, for this mixing we will have

$$\Delta_{\text{mix}} G = \Delta_{\text{mix}} G_2 - \Delta_{\text{mix}} G_1 = -47.321 \text{ kJ} - (-34.353 \text{ kJ})$$
$$= -12.968 \text{ kJ}$$
$$\Delta_{\text{mix}} S = \Delta_{\text{mix}} S_2 - \Delta_{\text{mix}} S_1 = 158.80 \text{ J K}^{-1} - 115.28 \text{ J K}^{-1}$$
$$= 43.52 \text{ J K}^{-1}$$
$$\Delta_{\text{mix}} H = 0$$

Example 6.12.3 (a) One dm^3 bulb containing nitrogen at 1 bar pressure and 25 °C is connected by a tube with a stop-cock to a 3 dm^3 bulb containing carbon dioxide at 2 bar pressure at 25 °C. The stop-cock is opened and the gases are allowed to mix until equilibrium is reached. Assuming that the gases are ideal, what is ΔS for this spontaneous change?

(b) What is the total change in entropy if the 3 dm^3 bulb contains oxygen at 2 bar instead of carbon dioxide? Explain the difference, if any, between this value and that obtained in part (a).

Solution **(a)**

$$\text{Amount of N}_2 = \frac{pV}{RT} = \frac{(100 \text{ kPa})(1 \text{ dm}^3)}{(8.314 \text{ kPa dm}^3 \text{ K}^{-1} \text{ mol}^{-1})(298 \text{ K})}$$
$$= 0.040\,4 \text{ mol}$$
$$\text{Amount of CO}_2 = \frac{(200 \text{ kPa})(3 \text{ dm}^3)}{(8.314 \text{ kPa dm}^3 \text{ K}^{-1} \text{ mol}^{-1})(298 \text{ K})}$$
$$= 0.242\,2 \text{ mol}$$

Now, employing the relation

$$\Delta_{\text{mix}} S = n_A R \ln \frac{V_{\text{final}}}{V_A} + n_B R \ln \frac{V_{\text{final}}}{V_B}$$

we get

$$\Delta_{\text{mix}} S = (0.040\,4 \text{ mol})(8.314 \text{ J K}^{-1} \text{ mol}^{-1}) \times \ln\left(\frac{4 \text{ dm}^3}{1 \text{ dm}^3}\right)$$
$$+ (0.242\,2 \text{ mol})(8.314 \text{ J K}^{-1} \text{ mol}^{-1}) \times \ln\left(\frac{4 \text{ dm}^3}{3 \text{ dm}^3}\right)$$
$$= 0.465\,6 \text{ J K}^{-1} + 0.579\,3 \text{ J K}^{-1} = 1.044\,9 \text{ J K}^{-1}$$

(b) $\Delta_{\text{mix}} S$ will be the same, since the two gases are different, assumed to be ideal and they do not interact with each other.

Example 6.12.4 Two dm^3 of methane under 4 bar and 298 K and 4 dm^3 of oxygen under 20 bar and 298 K are forced into a 3 dm^3 evacuated reaction vessel, the temperature being maintained at 298 K. Calculate the change in entropy of gases assuming that they are ideal.

Solution

$$\text{Amount of } CH_4 = \frac{pV}{RT} = \frac{(400\ \text{kPa})(2\ \text{dm}^3)}{(8.314\ \text{kPa dm}^3\ \text{K}^{-1}\ \text{mol}^{-1})(298\ \text{K})}$$

$$= 0.322\ 9\ \text{mol}$$

$$\text{Amount of } O_2 = \frac{(2\ 000\ \text{kPa})(4\ \text{dm}^3)}{(8.314\ \text{kPa dm}^3\ \text{K}^{-1}\ \text{mol}^{-1})(298\ \text{K})} = 3.229\ \text{mol}$$

Now, employing the relation

$$\Delta_{\text{mix}}S = n_A R \ln \frac{V_{\text{final}}}{V_A} + n_B R \ln \frac{V_{\text{final}}}{V_B}$$

we get $$\Delta_{\text{mix}}S = (0.322\ 9\ \text{mol})\ (8.314\ \text{J K}^{-1}\ \text{mol}^{-1}) \ln \left(\frac{3\ \text{dm}^3}{2\ \text{dm}^3}\right)$$

$$+ (3.22\ 9)\ (8.314\ \text{J K}^{-1}\ \text{mol}^{-1}) \ln \left(\frac{3\ \text{dm}^3}{4\ \text{dm}^3}\right)$$

$$= 1.088\ 5\ \text{J K}^{-1} - 7.723\ 1\ \text{J K}^{-1} = -\ 6.634\ 6\ \text{J K}^{-1}$$

Example 6.12.5 A thermally insulated vessel is divided into two compartments by a partition of insulating material. One compartment contains 0.4 mol of He at 293 K and 101.325 kPa. The other contains 0.6 mol of N_2 at 373 K and 202.65 kPa. The partition is removed so that the two gases mix. Calculate the resultant entropy change. Given that $C_V(\text{He}) = (3/2)\ R$, $C_V(N_2) = (5/2)R$ and may be assumed to be independent of temperature.

Solution *Temperature of the mixture* This may be calculated as follows.

$$\text{Heat gained by He} = nC_{V,\text{m}}\ dT$$
$$= (0.4\ \text{mol})\ \{(3/2)\ (8.314\ \text{J K}^{-1}\ \text{mol}^{-1})\}(T_f - 293\ \text{K})$$
$$= (4.988\ 4\ \text{J K}^{-1})\ (T_f - 293\ \text{K})$$

$$\text{Heat lost by } N_2 = nC_{V,\text{m}}\ dT$$
$$= (0.6\ \text{mol})\ \{(5/2)\ (8.314\ \text{J K}^{-1}\ \text{mol}^{-1})\}\ (373\ \text{K} - T_f)$$
$$= (12.471\ \text{J K}^{-1})\ (373\ \text{K} - T_f)$$

Heat gained = Heat lost

$$(4.988\ 4\ \text{J K}^{-1})\ (T_f - 293\ \text{K}) = (12.471\ \text{J K}^{-1})\ (373\ \text{K} - T_f)$$

$$T_f\ (4.988\ 4\ \text{J K}^{-1} + 12.471\ \text{J K}^{-1}) = 12.471 \times 373\ \text{J} + 4.988\ 4 \times 293\ \text{J}$$

$$T_f = \frac{6\ 113.28\ \text{J}}{17.459\ 4\ \text{J K}^{-1}} = 350.14\ \text{K}$$

Volume of the container containing He

$$= \frac{nRT}{p} = \frac{(0.4\ \text{mol})(8.314\ \text{kPa dm}^3\ \text{K}^{-1}\ \text{mol}^{-1})\ (293\ \text{K})}{(101.325\ \text{kPa})}$$

$$= 9.617\ \text{dm}^3$$

Volume of the container containing N_2

$$= \frac{nRT}{p} = \frac{(0.6\ \text{mol})(8.314\ \text{kPa dm}^3\ \text{K}^{-1}\ \text{mol}^{-1})(373\ \text{K})}{(202.65\ \text{kPa})}$$

$$= 9.182\ \text{dm}^3$$

Total volume after mixing = 9.617 dm^3 + 9.182 dm^3 = 18.799 dm^3

Now, employing the relation

$$\Delta S = n\left(C_{V,\text{m}} \ln \frac{T_\text{f}}{T_\text{i}} + R \ln \frac{V_\text{f}}{V_\text{i}}\right)$$

For each of the two gases, we get

$$\Delta S_{\text{He}} = (0.4\ \text{mol}) \left[\left(\frac{3}{2}\right)(8.314\ \text{J K}^{-1}\ \text{mol}^{-1}) \ln\left(\frac{350.14\ \text{K}}{293\ \text{K}}\right) + (8.314\ \text{J K}^{-1}\ \text{mol}^{-1}) \ln\left(\frac{18.799\ \text{dm}^3}{9.617\ \text{dm}^3}\right)\right]$$

$$= (0.4\ \text{mol})\ (2.221\ 8\ \text{J K}^{-1}\ \text{mol}^{-1} + 5.572\ 6\ \text{J K}^{-1}\ \text{mol}^{-1})$$

$$= 3.117\ 8\ \text{J K}^{-1}$$

$$\Delta S_{\text{N}_2} = (0.6\ \text{mol}) \left[\left(\frac{5}{2}\right)(8.314\ \text{J K}^{-1}\ \text{mol}^{-1}) \ln\left(\frac{350.14\ \text{K}}{373\ \text{K}}\right) + (8.314\ \text{J K}^{-1}\ \text{mol}^{-1}) \ln\left(\frac{18.799\ \text{dm}^3}{9.182\ \text{dm}^3}\right)\right]$$

$$= (0.6\ \text{mol})\ (-1.314\ 6\ \text{J K}^{-1}\ \text{mol}^{-1} + 5.957\ 5\ \text{J K}^{-1}\ \text{mol}^{-1})$$

$$= 2.785\ 7\ \text{J K}^{-1}$$

and

$$\Delta_{\text{mix}} S = \Delta S_{\text{He}} + S_{\text{N}_2}$$

$$= 3.117\ 8\ \text{J K}^{-1} + 2.785\ \text{J K}^{-1}$$

$$= 5.903\ 5\ \text{J K}^{-1}$$

Example 6.12.6 One mole of helium at 298 K is mixed with two moles of neon at the same pressure but at a temperature of 398 K in such a way that the total volume remains constant and there is no heat exchange with the surroundings. Calculate ΔS_{total} for this process.

Solution The changes in p and T of the two gases in the mixing process may be depicted as follows.

Before mixing — After mixing

Helium

$$\begin{cases} n_1 = 1\ \text{mol} \\ T_i = 298\ \text{K} \\ p \\ V = V_1 \end{cases} \rightarrow \begin{cases} n = 1\ \text{mol} \\ T_\text{f} = ? \\ p_{\text{He}} = ? \\ V = V_1 + V_2 \end{cases}$$

Neon

$$\begin{cases} n = 2\ \text{mol} \\ T_2 = 398\ \text{K} \\ p \\ V = V_2 \end{cases} \rightarrow \begin{cases} n = 2\ \text{mol} \\ T_\text{f} = ? \\ p_{\text{Ne}} = ? \\ V = V_1 + V_2 \end{cases}$$

Thus, we calculate

Final temperature after mixing

$$n_{\rm He}\, C_{V,\,\rm m}\,(T_{\rm f} - T_i) = n_{\rm Ne}\, C_{V,\,\rm m}\,(T_2 - T_1)$$

$$(1\ \text{mol})\,(T_{\rm f} - 298\ \text{K}) = (2\ \text{mol})\,(398\ \text{K} - T_{\rm f})$$

$$T_{\rm f} = \frac{1\,094\ \text{K}}{3} = 364.67\ \text{K}$$

Now, employing the relation

$$\Delta S = n\left[C_{\rm V,m}\ln\frac{T_{\rm f}}{T_{\rm i}} + R\ln\frac{V_{\rm f}}{V_{\rm i}}\right]$$

we have $$\Delta S_{\rm He} = (1\ \text{mol})\left[1.5\,R\ln\left(\frac{364.67\ \text{K}}{298\ \text{K}}\right) + R\ln\frac{V_1+V_2}{V_1}\right] \quad (1)$$

$$\Delta S_{\rm Ne} = (2\ \text{mol})\left[1.5\,R\ln\left(\frac{364.67\ \text{K}}{398\ \text{K}}\right) + R\ln\frac{V_1+V_2}{V_2}\right] \quad (2)$$

Assuming ideal behaviour, we will have

$$V_1 = \frac{n_1RT_1}{p} \quad \text{and} \quad V_2 = \frac{n_2RT_2}{p}$$

Hence $$\frac{V_1}{V_2} = \frac{n_1T_1}{n_2T_2} = \frac{298}{2(398)}$$

or $$\frac{V_1+V_2}{V_2} = \frac{298+2(398)}{2(398)} = \frac{1094}{2(398)} = 1.374\,; \qquad \ln(1.374) = 0.318\,0$$

$$\frac{V_2+V_1}{V_1} = \frac{2(398)+298}{298} = \frac{1094}{298} = 3.671\,; \qquad \ln(3.671) = 1.300\,5$$

Hence, Eqs (1) and (2) becomes

$$\begin{aligned}\Delta S_{\rm He} &= (1\ \text{mol})\,(1.5\,R\,(0.201\,9) + 1.300\,5\,R)\\ &= (1\ \text{mol})\,(0.303\,0 + 1.300\,5)\,R = (1\ \text{mol})\,(1.603\,5\,R)\end{aligned}$$

$$\begin{aligned}\Delta S_{\rm Ne} &= (2\ \text{mol})\,[1.5\,R\,(-\,0.087\,5) + 0.318\,0\,R]\\ &= (2\ \text{mol})\,(-\,0.131\,1 + 0.318\,0)\,R = (2\ \text{mol})\,(0.186\,9\,R)\end{aligned}$$

$$\Delta_{\rm mix}S = \Delta S_{\rm He} + \Delta S_{\rm Ne} = (1.977\,3\ \text{mol})\,R = 16.439\ \text{J K}^{-1}$$

Alternative Solution *Alternatively,* we may find out the partial pressures of helium and neon in the mixture and then make use of Eq. (6.12.17) to determine $\Delta_{\rm mix}S$. This method is also illustrated here. We have

$$p_{\rm initial}({\rm He}) = \frac{(1\ \text{mol})\,(R)\,(298\ \text{K})}{V_1}\,; \qquad p_{\rm initial}({\rm Ne}) = \frac{(2\ \text{mol})\,(R)\,(398\ \text{K})}{V_2}$$

$$p_{\rm final}({\rm He}) = \frac{(1\ \text{mol})\,(R)\,(364.67\ \text{K})}{V_1+V_2}\,; \qquad p_{\rm final}({\rm Ne}) = \frac{(2\ \text{mol})\,(R)\,(364.67\ \text{K})}{V_1+V_2}$$

Now, employing expression

$$\Delta S = n\left[C_{p,\rm m}\ln\frac{T_{\rm f}}{T_{\rm i}} + R\ln\frac{p_{\rm initial}}{p_{\rm final}}\right]$$

we get $$\Delta S_{He} = (1\text{ mol})\left[\left(\frac{5}{2}\right)R\ln\left(\frac{364.67\text{ K}}{298\text{ K}}\right) + R\ln\left(\frac{298}{364.67}\frac{V_1+V_2}{V_1}\right)\right]$$

$$= (1\text{ mol})\left[\left(\frac{5}{2}\right)R(0.202\,0) + R\ln\left(\frac{298}{364.67}\frac{1\,094}{298}\right)\right]$$

$$= (1\text{ mol})\,[0.505\,0\,R + 1.098\,6\,R]$$

$$= (1.603\,6\text{ mol})\,R = 13.332\text{ J K}^{-1}$$

$$\Delta S_{Ne} = (2\text{ mol})\left[\left(\frac{5}{2}\right)R\ln\left(\frac{364.67\text{ K}}{398\text{ K}}\right) + R\ln\left(\frac{398}{364.67}\frac{V_1+V_2}{V_2}\right)\right]$$

$$= (2\text{ mol})\left[\left(\frac{5}{2}\right)R\,(-0.0875) + R\ln\left(\frac{398}{364.67}\frac{1\,094}{2(398)}\right)\right]$$

$$= (2\text{ mol})\,[(-\,0.218\,6)\,R + R\ln(1.5)]$$

$$= (2\text{ mol})\,[(-\,0.218\,6)\,R + R\,(0.405\,5)]$$

$$= (0.373\,8\text{ mol})\,R = 3.108\text{ J K}^{-1}$$

Hence $$\Delta_{mix}S = \Delta S_{He} + \Delta S_{Ne}$$

$$= 13.332\text{ J K}^{-1} + 3.108\text{ J K}^{-1}$$

$$= 16.440\text{ J K}^{-1}$$

REVISIONARY PROBLEMS

6.1 The chemical potential of component in a homogeneous mixture is defined as

$$\mu_i = (\partial G/\partial n_i)_{T,p,n_j\text{s}\ (j\neq i)}$$

Show that it may be expressed in any of the following equivalent forms:

$$\mu_i = \left(\frac{\partial A}{\partial n_i}\right)_{T,V,n_j\text{s}\ (j\neq i)} = \left(\frac{\partial H}{\partial n_i}\right)_{S,p,n_j\text{s}\ (j\neq i)} = \left(\frac{\partial A}{\partial n_i}\right)_{S,V,n_j\text{s}\ (j\neq i)}$$

Is the chemical potential an intensive or extensive property?

6.2 (a) The partial molar derivative

$$\left(\frac{\partial Y}{\partial n_i}\right)_{T,p,n_j\text{s}\ (j\neq i)}$$

is known as partial molar quantity and is represented by the symbol Y_i. Depending upon the nature of variable Y, the partial derivative may be called partial molar volume, entropy, free energy, and so on. The chemical potential can be defined by any one of the expressions given in Q. 6.1 but only one of them is the partial molar quantity. Identify this from Q. 6.1.

(b) The derivative $(\partial Y/\partial n_i)_{T,p,n_j\text{s}\ (j\neq i)}$ may be interpreted by either of the following statements:

(i) This represents the change in Y per mole of component i when an infinitesimal amount of this component is added to a system of definite composition.

(ii) This represents the change in Y when one mole of component i is added to a system which is large enough so as to cause no change in the composition.

Justify these statements from the derivative $(\partial Y/\partial n_i)_{T,p,n_j\text{s}\ (j\neq i)}$

6.3 How would you show that the partial molar quantities are related to each other exactly by the same expressions as those of total quantities for a closed system, i.e. the expressions relating partial molar quantities (applicable to an open system) can be obtained by replacing the total quantities in the relations applicable to a closed system by the corresponding partial molar quantities?

6.4 How would you prove that 'a substance will flow spontaneously from a region of high chemical potential to a region of low chemical potential'? Thus justify how far it is true to identify the chemical potential with the escaping tendency of the substance.

6.5 Show that

(a) The chemical potential of a pure ideal gas is given by

$$\mu = \mu^\circ + RT \ln (p/1 \text{ bar})$$

On what variables will μ and μ° depend?

(b) Pressure of a gas is a measure of its chemical potential.

(c) The chemical potential of an ideal gas in a mixture of ideal gases is given as:

$$\mu_i(p_i, T) = \mu_i^*(p, T) + RT \ln x_i$$

(d) In a mixture of ideal gases, the partial molar volume is simply the average molar volume and that it has the same value for each gas of the mixture.

(e) The important condition of equilibrium is that μ_i should have uniform value throughout the system.

6.6 Show that partial molar quantities are additive and hence derive the following additivity rules.

$$U = \sum_i n_i U_i; \qquad S = \sum_i n_i S_i; \qquad H = \sum_i n_i H_i;$$

$$n = \sum_i (1) n_i; \qquad A = \sum_i n_i A_i; \qquad m = \sum_i n_i M_i;$$

$$G = \sum_i n_i G_i = \sum_i n_i \mu_i$$

6.7 (a) Derive the Gibbs-Duhem equation in the form

$$\sum_i n_i (\mathrm{d}Y_i) = 0 \text{ at constant } T \text{ and } p$$

where Y_i is a partial molar quantity. It may be μ_i, V_i, U_i, H_i and so on.

(b) For a binary system show that

$$\mathrm{d}V_1 = -\frac{n_2}{n_1}\mathrm{d}V_2$$

What is the physical significance of this expression?

6.8 (a) Derive the following expressions for the free energy of mixing and entropy of mixing involving ideal gases.

$$\Delta_{\text{mix}}G = nRT \sum_i x_i \ln x_i$$

$$\Delta_{\text{mix}}S = -nR \sum_i x_i \ln x_i$$

(b) Show that for the mixing of ideal gases $\Delta_{\text{mix}}V = 0$ and $\Delta_{\text{mix}}H = 0$

(c) Show mathematically that $\Delta_{\text{mix}}G$ will have a minimum value and $\Delta_{\text{mix}}S$ will have a maximum value at $x = 1/2$ when two gases are mixed together.

TRY YOURSELF PROBLEMS

6.1 Show that for an ideal gas

$$\left(\frac{\partial \mu_i}{\partial V}\right)_{T,n_1,n_2} = V_i\left(\frac{\partial p}{\partial V}\right)_{T,n_1,n_2}; \quad \left(\frac{\partial U}{\partial n_i}\right)_{T,V} = H_i - TV_i\left(\frac{\partial p}{\partial T}\right)_{V,n_i}$$

$$\left(\frac{\partial S}{\partial n_i}\right)_{T,V} = S_i - V_i\left(\frac{\partial p}{\partial T}\right)_{V,n_i}$$

where V_i, H_i and S_i are partial molar quantities.

6.2 (a) Suppose that V as a function of m can be expressed by the relation

$$V = a + bm + cm^2$$

where m is the molality and a, b and c and constants at a given T and p. Show that

$$V_2 = (b + 2\,c\,m)/n_1 m_1 \qquad \text{(where } n_1 m_1 = 1 \text{ kg)}$$

$$V_1 = \frac{a - cm^2}{n_1}$$

where V_2 and V_1 are the partial molar volumes of solute and solvent, respectively.
(b) For a certain binary solution at constant T and p

$$V_2 = a_2 + 2a_3 m + 3a_4 m^4$$

where m is molality and a_2, a_3 and a_4 are constants. Determine the expression for the total volume V of solution as a function of m.

6.3 Suppose that the density ρ of a solution is known as a function of the mass percentage (w) of solute. Show that the partial molar volumes of solute V_2 and solvent V_1, are given by the expressions

$$V_2 = \frac{M_2}{\rho}\left\{1 - \frac{(100-w)}{\rho}\frac{\partial \rho}{\partial w}\right\}; \qquad V_1 = \frac{M_1}{\rho}\left\{1 + \frac{w}{\rho}\frac{\partial \rho}{\partial w}\right\}$$

If ρ is a linear function of w and if $w\partial\rho/\partial w$ is small compared with ρ, show that to a good approximation V_1 and V_2 are constants.
(*Hint:* Make use of Eqs (6.2.13) and (6.2.14).)

6.4 (a) If V is the volume of a two-component solution containing the amounts n_1 and n_2 of solvent and solute, respectively, show that

$$V_2 = \frac{M_2 - V(\partial \rho / \partial n_2)}{\rho}$$

in which M_2 is the molar mass of the solute and ρ is the density of the solution.
(b) Show that

$$V_1 = \frac{\left(M_1 + \frac{n_2 V}{n_1}\frac{\partial \rho}{\partial n_2}\right)}{\rho}$$

(*Hint:* Make use of Eqs (6.2.13) and (6.2.14).)

6.5 Derive the following relations.

$$U = TS - pV + \sum_i n_i\,\mu_i$$

$$H = TS + \sum_i n_i\,\mu_i$$

$$A = -pV + \sum_i n_i\,\mu_i$$

6.6 Which of the following derivatives are equal to the chemical potential and which are equal to the partial molar quantity?

$$\left(\frac{\partial A}{\partial n_i}\right)_{T,V,n_j};\quad \left(\frac{\partial G}{\partial n_i}\right)_{T,p,n_j};\quad \left(\frac{\partial U}{\partial n_i}\right)_{T,p,n_j};\quad \left(\frac{\partial H}{\partial n_i}\right)_{S,p,n_j};$$

$$\left(\frac{\partial U}{\partial n_i}\right)_{S,V,n_j};\quad \left(\frac{\partial H}{\partial n_i}\right)_{T,p,n_j};\quad \left(\frac{\partial V}{\partial n_i}\right)_{T,p,n_j};\quad \left(\frac{\partial S}{\partial n_i}\right)_{T,p,n_j}$$

6.7 Show that

(i) $$\frac{\mathrm{d}(V_{\mathrm{m,\,mix}}/x_2)}{\mathrm{d}(1/x_2)} = V_1 \quad \text{and} \quad \frac{\mathrm{d}(V_{\mathrm{m,mix}}/x_1)}{\mathrm{d}(1/x_1)} = V_2$$

(ii) $$\frac{\mathrm{d}\left(\dfrac{1/\rho}{w_2}\right)}{\mathrm{d}(1/w_2)} = \frac{1}{\rho_1} \quad \text{and} \quad \frac{\mathrm{d}\left(\dfrac{1/\rho}{w_1}\right)}{\mathrm{d}(1/w_2)} = \frac{1}{\rho_2}$$

where $V_{\mathrm{m,\,mix}}$ and ρ are the molar volume and density of a binary solution containing a mole fraction x_2 (or mass per cent w_2) of component 2, respectively. ρ_1 and ρ_2 are the partial molar densities of components 1 and 2, respectively.

NUMERICAL PROBLEMS

Partial Molar Quantities

6.1 The volume of an aqueous solution of sodium chloride at 25 °C was found to obey the relation

$$V/\mathrm{cm^3\ kg^{-1}} = 1\ 003 + 16.62\ (m/\mathrm{mol\ kg^{-1}}) + 1.77\ (m/\mathrm{mol\ kg^{-1}})^{3/2} + 0.12\ (m/\mathrm{mol\ kg^{-1}})^2$$

Find the partial molar volume of the components at m = 0.1 mol kg^{-1} by explicit differentiation. (*Ans.* 17.483 cm^3 mol^{-1})

6.2 For iodine in methanol solution, the apparent molar volume of the solute is independent of the concentration of the solution and is equal to 62.3 cm^3 mol^{-1} at 25 °C.

The densities of solid I_2 and pure methanol at 25 °C are 4.93 g cm^{-3} and 0.786 5 g cm^{-3}, respectively. Find ΔV for the following.

$I_2(s) = I_2(m = 1\ \mathrm{mol\ kg^{-1}})$

$CH_3OH(l) \rightarrow CH_3OH(1\ \mathrm{mol\ kg^{-1}}\ I_2\ \text{solution})$

$I_2(s) + 32.2\ CH_3OH \rightarrow \text{Solution}\ (m = 1\ \mathrm{mol\ kg^{-1}})$

(*Ans.* 10.8 cm^3, 0 cm^3 10.8 cm^3)

6.3 At 25 °C, the density of succinic acid in water is given by the expression (valid up to 5.5% acid).

$$\rho/\mathrm{g\ cm^{-3}} = 0.997 + 0.003\ 04\ w$$

where w is the mass per cent of solute. Calculate V_1 and V_2 for 0, 2 and 4% solution of succinic acid. (*Ans.* 18.05, 90.70; 18.05, 82.70; 18.00, 82.98 cm^3 mol^{-1})

6.4 The partial molar volumes of water and glycerol in a solution of x_{H_2O} = 0.5 at 15.6 °C are found to be V_{H_2O} = 17.3 cm^3 mol^{-1} and V_{gly} = 72.68 cm^3 mol^{-1}. The corresponding values for the pure components are $V^*_{H_2O}$ = 18.0 cm^3 mol^{-1} and V^*_{gly} = 72.8 cm^3 mol^{-1}. Calculate the volume of the solution when 5 mol of water are mixed with 5 mol of glycerol. Compare this value with that obtained by adding the volumes of the pure components. (*Ans.* 450.5 cm^3, 454 cm^3)

6.5 The partial molar volume of ethanol in a 60 mole per cent ethanol-water solution is 57.5 cm^3 mol^{-1}. The density of solution is 0.8508 g cm^{-3}. Calculate the partial molar volume of water in the solution. (*Ans.* 16 cm^3 mol^{-1})

6.6 It is required to prepare 3 m^3 of a 60 mole per cent water solution containing ethanol and water. Determine the volumes of ethanol (ρ = 0.7945 g cm^{-3}) and water to be mixed in order to prepare the required solution. The partial molar volume of ethanol and water in 60 mole percent water solution are

$$V_{pm}(C_2H_5OH) = 57.5 \text{ cm}^3 \text{ mol}^{-1} \text{ and } V_{pm}(H_2O) = 16 \text{ cm}^3 \text{ mol}^{-1}$$

(*Ans.* 2.548 m^3 water and 0.528 m^3 ethanol)

Free Energy and Entropy of Mixing

6.7 Three moles of N_2(g), originally at 101.325 kPa pressure, are mixed isothermally with 5 mol of H_2(g), also at 101.325 kPa, to yield a mixture whose total pressure is 101.325 kPa. Assuming gas to be ideal, calculate the total entropy of mixing. (*Ans.* 44.016 J K^{-1})

6.8 Repeat Problem 6.7 on the basis that the total pressure of mixture is 1.013 MPa.

6.9 One mole of an ideal gas at 273 K and 101.325 kPa pressure is mixed adiabatically with one mole of a different gas at 373 K at 101.325 kPa pressure to yield a mixture whose pressure is also 101.325 kPa. If $C_{p,\,m}$ for each gas is $5R/2$, what is ΔS for the process? (*Ans.* 12.008 J K^{-1})

6.10 Calculate the entropy of unmixing of 100 g air considered as 21% vol O_2, 78% vol N_2 and 1% vol Ar. (*Ans.* 16.44 J K^{-1})

6.11 Calculate the entropy change on mixing a solution containing 5 mol of benzene (B) and 7 mol of toluene (T) with a second solution at the same temperature containing 3 mol of benzene, 2 mol of toluene and 4 mol of *p*-xylene (X). Assume the solutions are ideal in nature.

(*Hint:* Calculate ΔS_1 for 5B + 7T → mix (5B + 7 B))
ΔS_2 for 3B + 2T + 4X → mix (3B + 2T + 4X)
ΔS_3 for 8B + 9T + 4X → mix (8B + 9T + 4X)

and find ΔS for mix (5B + 7T) + mix (3B + 2T + 4X) → mix (8B + 9T + 4X) using the expression

$$\Delta S = \Delta S_3 - \Delta S_1 - \Delta S_2.$$ (*Ans.* 35.59 J K^{-1})

6.12 One dm^3 of O_2 and 4 dm^3 of N_2, each at a pressure 101.325 kPa at 300 K are mixed to form an ideal gas mixture of 3 dm^3 at the same temperature. Calculate the entropy change for the process. (*Ans.* – 0.0167 J K^{-1})

6.13 A 1 dm^3 vessel containing helium at 373 K and 101.325 kPa is connected to a 2 dm^3 vessel containing neon at 273 K and 40.53 kPa. The gases are allowed to mix with each other and reach thermal equilibrium adiabatically. What is the ΔS for the process? (*Ans.* 0.429 3 J K^{-1})

6.14 A vessel is divided into two compartments of equal volumes by a partition. Each compartment contains 1.0 mole of the same absolutely identical gases at the same temperature and pressure. The partition is removed and the gases are allowed to mix. Calculate the resultant entropy change. (*Ans.* $\Delta_{mix}S = 0$)

(*Note:* One may expect to get a value of $2R$ ln 2 on the basis of Eq. (6.12.10), but the actual value is zero. This contradiction is known as *Gibbs paradox*. The basic reason for this contradiction is that Eq. (6.12.10) is applicable for two different gases (may be two similar gases differing in some respects so that their molecules are distinguishable from each other, e.g., two gases comprising of different isotopes) and not to the two absolutely identical gases. The molecules in a mixture formed from the latter gases cannot be distinguished and, therefore, one cannot apply Dalton's law of partial pressure ($p_i = x_i\, p_{total}$) which has been used in Eq. (6.12.2) to derive the final expression (Eq. 6.12.10). Hence, in such a case, Eq. (6.12.10) will not valid.)

Experimental Determination of Partial Molar Volumes

6.15 (a) When chloroform is added to acetone at 25 °C, the volume of mixture varies with composition as follows.

$x_{\text{chloroform}}$	0	0.194	0.385	0.559	0.788	0.889	1.000
V_m/cm^3 mol^{-1}	73.99	75.29	76.5	77.55	79.08	79.82	80.67

Determine the partial molar volumes of the two components at $x_{\text{chloroform}} = 0.5$.

(b) What will be the volume of a solution of mass 1 kg at $x_{\text{chloroform}} = 0.5$? What will be the total volume of the unmixed components?

6.16 Given below are the densities of aqueous solutions of sulphuric acid at various mass per cent of H_2SO_4. Determine the partial volumes of H_2O and H_2SO_4 at 50 mass per cent of H_2SO_4.

mass %	10	20	30	40	50	60	70	80	90	100
density/g cm^{-3}	1.066	1.139	1.219	1.303	1.395	1.498	1.611	1.727	1.814	1.831

(*Ans.* 1.036 cm^3, 2.151 cm^3)

7 Thermodynamics of Chemical Reactions

7.1 DESCRIPTION OF A REACTION IN PROGRESS

Representation of a Chemical Equation

A chemical reaction, in general, may be written as

$$0 = \sum_{B} \nu_B B \tag{7.1.1}$$

where Bs represent the various constituents of the reaction and ν_Bs are the corresopnding stoichiometric coefficients (or numbers). The latter have positive values for products and negative for reactants.

Extent of Reaction

The change in the amount of a reactant or a product during the progress of a reaction may be represented in terms of the extent of reaction (ξ) defined as

$$n(B) = n_0(B) + \nu_B \xi \tag{7.1.2}$$

where $n_0(B)$ is the amount of B at $t = 0$. The unit of ξ is that of the amount of a substance, i.e. mol.

From Eq. (7.1.2), we may write

$$\frac{dn(B)}{d\xi} = \nu_B \tag{7.1.3}$$

that is, the change in the amount of a reactant or a product with the change in the extent of reaction is equal to its stoichiometric coefficient that appeared in the chemical reaction.

Illustration

To be specific, we take an example of a reaction containing only two reactants and two products:

$$|\nu_1|\, A_1 + |\nu_2|\, A_2 \rightarrow \nu_3 A_3 + \nu_4 A_4 \tag{7.1.4}$$

If the reaction is started with $n_0(A_1)$ and $n_0(A_2)$ as the amounts of A_1 and A_2, respectively, and if ξ is the extent of reaction, we will have

$$n(A_1) = n_0(A_1) + \nu_1 \xi$$

$$n(A_2) = n_0(A_2) + \nu_2 \xi$$

$$n(A_3) = \nu_3 \xi$$

$$n(A_4) = \nu_4 \xi$$

Since ν s are negative for reactants and positive for products, the amounts of A_1 and A_2 will decrease whereas those of A_3 and A_4 will increase with the increase in the extent of reaction.

7.2 THERMODYNAMICS OF CHEMICAL REACTIONS (REACTION POTENTIAL)

Definition of Reaction Potential and Affinity of a Chemical Reaction

The change in Gibbs function due to a chemical reaction or variation in temperature and pressure is given by

$$dG = -S\,dT + V\,dp + \sum_B \mu_B\,dn_B \tag{7.2.1}$$

Substituting dn_B in terms of $d\xi$ from Eq. (7.1.3), we get

$$dG = -S\,dT + V\,dp + \sum_B (\nu_B\mu_B)\,d\xi \tag{7.2.2}$$

The term $\sum_B (\nu_B\mu_B)$ is given the special name of the *reaction potential* and is expressed as

$$\sum_B \nu_B\mu_B = \Delta_r\tilde{G} \tag{7.2.3}$$

If the amounts of different species in a reaction change only because of the chemical reaction whose progress is measured by ξ, then from Eq. (7.2.2) if follows that

$$\Delta_r\tilde{G} = \left(\frac{\partial G}{\partial \xi}\right)_{T,p} \tag{7.2.4}$$

That is, *the reaction potential is equal of the rate of change of total free energy with extent of reaction at constant temperature and pressure.*

The reaction potential is identical to the free energy change of a reaction (Eq. 7.2.3), the value of which can be calculated from values of free energy of formation tabulated in Appendix II. Here after, we represent the reaction potential simply as $\Delta_r G$.

The expression $A = -\Delta_r\tilde{G} = -(\partial G/\partial\xi)_{T,\,p}$ is known as *affinity of a chemical reaction.*

Nature of Chemical Reaction

The nature of a reaction, that is, whether it is spontaneous or not, depends on the value of dG. Under constant temperature and pressure conditions, we have

$$\left.\begin{array}{lll} \text{(i)} & dG = \Delta_r G\,d\xi < 0 & \text{Spontaneous} \\ \text{(ii)} & dG = \Delta_r G\,d\xi > 0 & \text{Not possible} \\ \text{(iii)} & dG = \Delta_r G\,d\xi = 0 & \text{Equilibrium} \end{array}\right\} \tag{7.2.5}$$

Since $d\xi$ is positive for the forward reaction, the above conditions, therefore, lead to

Criterion for a Spontaneous Reaction

(i) $\Delta_r G < 0$, Spontaneous in the forward direction.

For the reaction given in Eq. (7.1.4), $\Delta_r G$ is given as

$$\Delta_r G = \nu_3\mu_3 + \nu_4\mu_4 - (|\nu_1|\,\mu_1 + |\nu_2|\,\mu_2)$$

Thus in the present case, we have

$$|\nu_1|\,\mu_1 + |\nu_2|\,\mu_2 > \nu_3\mu_3 + \nu_4\mu_4 \tag{7.2.6}$$

Criterion for a Nonspontaneous Reaction

(ii) $\Delta_r G > 0$, Not spontaneous in the forward direction. In fact, the spontaneous direction will be back-wards, i.e. the products will be converted to reactants.

For $\Delta_r G > 0$, we will have

$$\nu_3 \mu_3 + \nu_4 \mu_4 > |\nu_1| \mu_1 + |\nu_2| \mu_2 \tag{7.2.7}$$

Criterion of Reaction at Equilibrium

(iii) $\Delta_r G = 0$ The reaction will be at equilibrium since an infinitesimal change in ξ produces $dG = 0$.

This will happen when

$$|\nu_1| \mu_1 + |\nu_2| \mu_2 = \nu_3 \mu_3 + \nu_4 \mu_4 \tag{7.2.8}$$

Equation (7.2.8) is the requirement for the reaction at equilibrium. It should be noted that this equation contains only intensive variables and will not depend upon the relative amounts of different species involved in a chemical reaction.

7.3 HOMOGENEOUS IDEAL GAS REACTION

Reaction Potential of a Chemical Equation

The chemical potential of an ideal gas is related to the standard chemical potential and its partial pressure by the relation

$$\mu_B = \mu_B^\circ + RT \ln (p_B/p^\circ) \tag{7.3.1}$$

where p° is the standard-state pressure and has a value of 1 bar. With this, the reaction potential is given by

$$\Delta_r G = \sum_B \nu_B \mu_B = \Delta_r G^\circ + RT \sum_B \nu_B \ln (p_B/p^\circ)$$

or

$$\Delta_r G = \Delta_r G^\circ + RT \ln \left(\prod_B (p_B/p^\circ)^{\nu_B} \right) \tag{7.3.2}$$

where $\Delta_r G^\circ = \sum_B \nu_B \mu_B^\circ$

The symbol $\prod_B$ stands for the multiplication symbol.

Illustration

As an example, let us apply the above expression for the following gas phase reaction:

$$2HCl(g) + \frac{1}{2} O_2(g) \rightarrow H_2O(g) + Cl_2(g)$$

$\nu(HCl) = -2$; $\nu(O_2) = -1/2$; $\nu(H_2O) = 1$ and $\nu(Cl_2) = 1$

$$\Delta_r G = \Delta_r G^\circ + RT \ln \frac{(p_{H_2O}/p^\circ)(p_{Cl_2}/p^\circ)}{(p_{HCl}/p^\circ)^2 (p_{O_2}/p^\circ)^{1/2}}$$

Here $\Delta_r G$ is change in Gibbs function per unit extent of reaction at a temperature T and at a constant pressure p; the gases comprising reactants and products being at specified partial pressures. $\Delta_r G^\circ$ is the standard reaction potential when each of the gaseous reactants and products has partial pressure of 1 bar at temperature T.

Reaction Potential of a Reaction Started with Reactants only

The relative magnitudes of the $\Delta_r G^\circ$ term and the logarithm term in Eq. (7.3.2) determine the thermodynamic progress of the reaction. Starting with all reactants and no products, we have

$$\Delta_r G = \Delta_r G^\circ + RT \ln \frac{\prod\limits_{\substack{\text{B} \\ \text{(products)}}} (p_\text{B}/p^\circ)^{\nu_\text{B}}}{\prod\limits_{\substack{\text{B} \\ \text{(reactants)}}} (p_\text{B}/p^\circ)^{|\nu_\text{B}|}}$$

$$= \Delta_r G^\circ + RT \ln \frac{\text{zero}}{\prod\limits_{\substack{\text{B} \\ \text{(reactants)}}} (p_\text{B}/p^\circ)^{|\nu_\text{B}|}} = -\infty$$

Thus $\Delta_r G$ is negative and infinitely large, no matter what value $\Delta_r G^\circ$ might have. Thus, the reaction will proceed spontaneously to the right.

Chemical Potential of a Reaction Started with Products only

On the other hand, when all products and no reactants are present, we have

$$\Delta_r G = \Delta_r G^\circ + RT \ln \frac{\prod\limits_{\substack{\text{B} \\ \text{(products)}}} (p_\text{B}/p^\circ)^{\nu_\text{B}}}{\text{zero}} = +\infty$$

Thus, $\Delta_r G$ is positive and infinitely large. The formation of reactants is, therefore, inevitable.

Essential Requirement for a Reaction at Equilibrium

At some intermediate mixture of reactants and products, the two terms in Eq. (7.3.2) will cancel each other making $\Delta_r G$ equal to zero. At this stages, the reaction will be at equilibrium.

The above analysis for the reaction given in Eq. (7.1.4) is shown in Fig. 7.3.1 where G has been plotted against ξ.

Reaction Quotient and Equilibrium Constant

The term $\prod\limits_\text{B} (p_\text{B}/p^\circ)^{\nu_\text{B}}$ in Eq. (7.3.2) is known as the standard *reaction quotient* Q_p°. *At equilibrium condition,* Q_p° is given the name of standard equilibrium constant K_p°, such that

$$K_p^\circ = (Q_p^\circ)_\text{eq} = \prod_\text{B} \left\{(p_\text{B}/p^\circ)^{\nu_\text{B}}\right\}_\text{eq} \qquad (7.3.3)^*$$

Now, at equilibrium, we have

(i) $\Delta_r G = 0$

and (ii) $\prod\limits_\text{B} (p_\text{B}/p^\circ)^{\nu_\text{B}} = K_p^\circ$

With these, Eq. (7.3.2) becomes

$$\Delta_r G^\circ = -RT \ln K_p^\circ = -2.303\, RT \log K_p^\circ$$

or

$$K_p^\circ = (10)^{-\Delta_r G^\circ/(2303\, RT)} \qquad (7.3.4)$$

* Prior to the recommendation of standard-state pressure of 1 bar, a value of 1 atm (= 101.325 kPa = 1.013 25 bar) was used. The change from 1 atm to 1 bar changes the value of standard equilibrium constant by a factor of $(1.013\ 25)^{\Delta\nu_g}$.

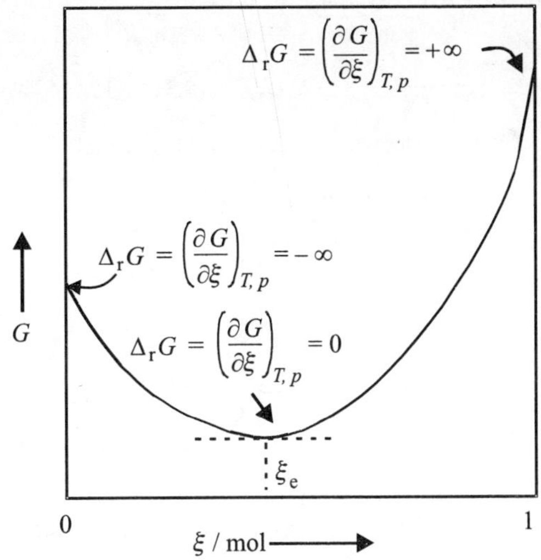

Fig. 7.3.1 Graph of free energy of a reacting system against extent of reaction at constant temperature and pressure. At $\xi = 0$, only reactants ($|\nu_1|$ mol of A_1 and $|\nu_2|$ mol of A_2) are present. At $\xi = 1$ mol, only products (ν_3 mol of A_3 and ν_4 mol of A_4) are present. At $\xi = \xi_e$ the reaction is at equilibrium

Factor Affecting Equilibrium Constant

Since $\Delta_r G^\circ$ is a function of temperature only and is independent of the partial pressures of the components of the reaction, it follows from Eq. (7.3.4) that the equilibrium constant K_p° also depends only on the temperature and is independent of the pressure of the system.

A Convenient Way of Expressing Reaction Potential

A convenient way of expressing reaction potential is provided by Eqs (7.3.2) and (7.3.4). Thus, we have

$$\Delta_r G = -RT \ln K_p^\circ + RT \ln Q_p^\circ = RT \ln \frac{Q_p^\circ}{K_p^\circ} \tag{7.3.5}$$

If for a particular reaction $Q_p^\circ < K_p^\circ$, then $\Delta_r G$ is negative and dG will also be negative. This leads to a spontaneous reaction. During the reaction, Q_p° increases and eventually it becomes equal to K_p°, $\Delta_r G$ becomes equal to zero. At this stage, the reaction stops resulting in equilibrium, i.e. no further change in the reaction is observed. If for a given reaction $Q_p^\circ > K_p^\circ$, then $\Delta_r G$ is positive and dG will also be positive. Thus, the given reaction will not proceed from reactants to products. In fact, it would have a tendency to proceed in the reverse direction.

Example 7.3.1

At 1 105 K, the value of K_p° for the reaction $SO_2(g) + \frac{1}{2} O_2 \rightleftharpoons SO_3(g)$ is 0.63. Calculate; (a) the standard free energy change for this reaction at 1 105 K; (b) the free energy change at 1 105 K for the reaction

$$SO_2(1 \text{ bar}) + \tfrac{1}{2} O_2(25 \text{ bar}) = SO_3(2 \text{ bar})$$

What is the significance of the signs of the answers in parts (a) and (b)?

Solution

(a) Employing the relation

$$\Delta_r G^\circ = -RT \ln K_p^\circ$$

we get $\Delta_r G^\circ = -(8.314 \text{ J K}^{-1} \text{ mol}^{-1})\ (1\,105 \text{ K})\ (2.303) \log (0.63)$

$$= 4\,245.5 \text{ J mol}^{-1}$$

(b) Employing the relation

$$\Delta_r G = \Delta_r G^\circ + RT \ln Q_p^\circ = \Delta_r G^\circ + RT \ln \frac{(p_{SO_3}/p^\circ)}{(p_{SO_2}/p^\circ)(p_{O_2}/p^\circ)^{1/2}}$$

we get $\Delta_r G = (4\,245.5 \text{ J mol}^{-1}) + (8.314 \text{ J K}^{-1} \text{ mol}^{-1})(1\,105 \text{ K}) \ln \left(\dfrac{2}{1 \times (25)^{1/2}}\right)$

$$= (4\,245.5 - 8\,419.5) \text{ J mol}^{-1} = -\,4\,174 \text{ J mol}^{-1}$$

In (a) $\Delta_r G^\circ$ positive indicates that the reaction is not possible, and in (b) $\Delta_r G$ becomes negative, thus the reaction is possible.

Example 7.3.2

The reaction

$$N_2O_3(g) \rightleftharpoons NO_2(g) + NO(g)$$

was studied at 25 °C with initial amount of N_2O_3 equal to 1.0 mol. At equilibrium, the extent of reaction is found to be 0.3 mol for a total pressure of 1 bar. Calculate K_p° and $\Delta_r G^\circ$ for the reaction.

Solution

The given reaction is

	$N_2O_3(g)$	$\rightleftharpoons$ NO_2	+ $NO(g)$
$t = 0$	1.0 mol	0	0
t_{eq}	1.0 mol − 0.3 mol	0.3 mol	0.3 mol

Total amount of substances at equilibrium = 1.3 mol

Now $p_{N_2O_3} = \dfrac{n_{N_2O_3}}{n_{total}} p = \dfrac{0.7}{1.3}(1 \text{ bar}) = 0.538 \text{ bar}$

$$p_{NO_2} = \frac{n_{NO_2}}{n_{total}} p = \frac{0.3}{1.3}(1 \text{ bar}) = 0.231 \text{ bar}$$

$$p_{NO} = p_{NO_2} = 0.231 \text{ bar}$$

$$K_p^\circ = \prod_B (p_B/p^\circ)^{\nu_B} = (p_{NO_2}/p^\circ)(p_{NO}/p^\circ)(p_{N_2O_3}/p^\circ)^{-1}$$

$$= (0.231 \text{ bar}/1 \text{ bar})\ (0.231 \text{ bar}/1 \text{ bar})\ (0.538 \text{ bar}/\ 1 \text{ bar})^{-1}$$

$$= 0.099$$

$$\Delta_r G^\circ = -RT \ln K_p^\circ = -\ (8.314 \text{ J K}^{-1} \text{ mol}^{-1})\ (298 \text{ K}) \ln (0.099)$$

$$= 5\,729.7 \text{ J mol}^{-1}$$

7.4 EXPRESSION OF K_p° FOR A REACTION INVOLVING HETEROGENEOUS PHASES

We have so far dealt with the equilibrium involving only ideal gases. Such a type of equilibrium is called homogeneous. If the equilibrium contains substances in more than one phase, the equilibrium is said to be heterogeneous. In this section, we develop the expression of the equilibrium constant for a heterogeneous equilibrium.

Illustration

We start with a specific reaction given below.

$$C(s) + \frac{1}{2} O_2(g) \rightleftharpoons CO(g) \tag{7.4.1}$$

The reaction potential of this reaction is given by an expression of the same form as for the homogeneous system. The general form of this expression is

$$\Delta_r G = \sum_B \nu_B \mu_B \tag{7.4.2}$$

For the given reaction, we have

$$\nu(CO) = +1; \quad \nu(C) = -1 \quad \text{and} \quad \nu(O_2) = -1/2$$

Thus $$\Delta_r G = \mu(CO, g) - \mu(C, s) - \frac{1}{2}\mu(O_2, g) \tag{7.4.3}$$

Now for an ideal gas, we write

$$\mu_g(T, p) = \mu_g^\circ(T) + RT \ln (p/p^\circ) \tag{7.4.4}$$

where μ_g° is the standard chemical potential of the gas and p is its pressure. The corresponding expression for a condensed phase is given by an expression analogous to Eq. (5.6.4), such that

$$\mu_{cond}(T, p) = \mu_{cond}^\circ(T) \tag{7.4.5}^\dagger$$

Substituting Eqs (7.4.4) and (7.4.5) in Eq. (7.4.3), we get

$$\Delta_r G = \{\mu_{CO(g)}^\circ + RT \ln (p_{CO}/p^\circ)\} - \mu_{C(s)}^\circ - \frac{1}{2}\{\mu_{O_2(g)}^\circ + RT \ln (p_{O_2}/p^\circ)\}$$

or $$\Delta_r G = \left(\mu_{CO(g)}^\circ - \mu_{C(s)}^\circ - \frac{1}{2}\mu_{O_2(g)}^\circ\right) + RT \ln\left(\frac{(p_{CO}/p^\circ)}{(p_{O_2}/p^\circ)^{1/2}}\right)$$

or $$\Delta_r G = \Delta_r G^\circ + RT \ln\left(\frac{(p_{CO}/p^\circ)}{(p_{O_2}/p^\circ)^{1/2}}\right)$$

† Equation 7.4.5 follows from the fact that V_m for condensed phases has a small value and thus $\mu = \mu^\circ + V_m\,dp \simeq \mu^\circ$.

At equilibrium, we have

$$\Delta_r G = 0$$

$$\Delta_r G^\circ = -RT \ln \left(\frac{(p_{CO}/p^\circ)_{eq}}{(p_{O_2}/p^\circ)_{eq}^{1/2}} \right) = -RT \ln K_p^\circ$$

Conclusion

Thus, the expression of K_p° for a reaction involving gases and condensed phases can be written as usual considering only the partial pressures of gases, and ignoring condensed phases. However, $\Delta_r G^\circ$ contains the standard free energies of all the reactants and products.

One of the important examples of heterogeneous equilibrium is an equilibrium between liquid (or solid) and its vapour. Taking, for example, the case of water, we have

$$H_2O(1) \rightleftharpoons H_2O(g)$$

Let p be the equilibrium vapour pressure, Then

$$K_p^\circ = (p/p^\circ)$$

Example 7.4.1

For the reaction

$$NH_4HS(s) \rightleftharpoons NH_3(g) + H_2S(g)$$

the equilibrium pressure at 298 K was found to be 0.67 bar. Calculate K_p° and $\Delta_r G^\circ$ of the reaction.

Solution

The given reaction is

$$NH_4HS(s) \rightleftharpoons NH_3(g) + H_2S(g)$$

At equilibrium

$$p_{NH_3} = p_{H_3S}$$

Since $p_{NH_3} + p_{H_2S} = 0.67$ bar, we have

$$p_{NH_3} = p_{H_2S} = \frac{0.67 \text{ bar}}{2} = 0.335 \text{ bar}$$

$$K_p^\circ = \prod_B (p_B / p^\circ)^{\nu_B} = (p_{NH_3} / p^\circ)(p_{H_2S} / p^\circ) = (0.335 \text{ bar} / 1 \text{ bar})^2$$

$$= 0.112$$

$$\Delta_r G^\circ = -RT \ln K_p^\circ = -(8.314 \text{ J K}^{-1} \text{ mol}^{-1})(298 \text{ K}) \ln (0.112)$$

$$= 5\,424.0 \text{ J mol}^{-1}$$

7.5 DYNAMIC EQUILIBRIUM (LAW OF MASS ACTION)

The existence of an equilibrium constant and its relation to the composition of the system were actually found from the study of the rates of some reversible reactions before the thermodynamic approach was developed. C.M. Guldberg and P. Waage expressed the dependence of concentrations on the rate of reaction in a very general form, known as the *law of mass action,* and applied the results to the problem of chemical equilibrium. They recognized *that chemical equilibrium is a dynamic and not a static equilibrium.* It is characterized not by the cessation of all reactions but by the fact the rates of forward and reverse reactions become the same.

Accodring to the law of mass action,

The rate of a reaction is proportional to the product of effective concentrations of the reacting species, each raised to a power which is equal to the corresponding stoichiometric number of the substance in the balanced chemical equation.

Illustration Let us apply the law of mass action to a general reaction

$$|\nu_1|\,A_1 + |\nu_2|\,A_2 \rightleftharpoons \nu_3A_3 + \nu_4A_4$$

Now the rate of the forward reaction is

$$-\frac{1}{|\nu_1|}\frac{d[A_1]}{dt} = k_f[A_1]^{|\nu_1|}[A_2]^{|\nu_2|} \tag{7.5.1}$$

where k_f is the proportionality constant.

Similarly for the backward reaction

$$-\frac{1}{\nu_3}\frac{d[A_3]}{dt} = k_b[A_3]^{\nu_3}[A_4]^{\nu_4} \tag{7.5.2}$$

At equilibrium, both the rates are equal; thus

$$k_f[A_1]^{|\nu_1|}[A_2]^{|\nu_2|} = k_b[A_3]^{\nu_3}\,[A_4]^{\nu_4}$$

or

$$\frac{[A_3]^{\nu_3}[A_4]^{\nu_4}}{[A_1]^{|\nu_1|}\,[A_2]^{|\nu_2|}} = \frac{k_f}{k_b} = K_{eq} \tag{7.5.3}$$

or

$$K_{eq} = \prod_B (c_B)^{\nu_B} \tag{7.5.4}$$

While dealing with the reactions involving ideal gases, the concentrations may be replaced in terms of partial pressures as the latter are directly proportional to the former. Thus

$$K_{eq} = \prod_B (p_B)^{\nu_B} \tag{7.5.5}$$

For reactions involving condensed phases in addition to gases, the effective concentrations of solids and liquids remain constant and thus are merged in the constant K_{eq}, with the result that the form K_{eq} is still given by Eq. (7.5.5).

Note K_{eq} of Eq. (7.5.4) is represented as K_c while that of Eq. (7.5.5) is reprsented as K_p.

7.6 GENERAL RULES TO WRITE Q_p° AND K_p° FOR ANY REACTION

Expressions of Q_p° and K_p° for any reaction can be derived following the rules given below.

- Products are written in the numerator and reactants in the denominator.
- For gases, partial pressures expressed in the units of standard-state pressure are used.
- Solids and liquids—these do not appear in Q_p° and K_p° because their chemical potentials remain constant and are merged in Δ_rG°.

- If a reaction is multiplied by some constant, say *n*, then the value of $K°$ for the new reaction is

$$K_p^\circ(\text{new reaction}) = (K_p^\circ)^n(\text{old reaction}) \tag{7.6.1}$$

- If a reaction can be obtained by the addition or subtraction of two or more separate reactions, then the K_p° value of the new reaction can be obtained from the K_p° values of the reactions being added or subtracted. The following rules (can be derived through $\Delta_r G^\circ$ considerations) will be helpful in calculating the value of K_p° in such a situation.
  - (i) If a reaction is multiplied by some constant then its K_p° value raises exponentially by the same constant (Eq. 7.6.1).
  - (ii) If a reaction is being added to some other reaction then the K_p° of the former gets multiplied with that of the latter.
  - (iii) If a reaction is being subtracted from some other reaction then the K_p° of the latter gets divided with that of the former.

  The following problem illustrates the procedure.

Example 7.6.1 Given

(i) $H_2(g) + \frac{1}{2} S_2(g) \rightleftharpoons H_2S(g)$; K_{p1}°

(ii) $2H_2(g) + S_2(g) \rightleftharpoons 2H_2S(g)$; K_{p2}°

Show that $K_{p2}^\circ = (K_{p1}^\circ)^2$

Solution For the first reaction $\Delta_r G_1^\circ = -RT \ln K_{p1}^\circ$

and for the second reaction $\Delta_r G_2^\circ = -RT \ln K_{p2}^\circ$

Since $\Delta_r G_2^\circ = 2\Delta_r G_1^\circ$, therefore, it follows that

$$-RT \ln K_{p2}^\circ = -2RT \ln K_{p1}^\circ$$

or

$$K_{p2}^\circ = (K_{p1}^\circ)^2$$

Example 7.6.2 If K_p° for $H_2(g) + \frac{1}{2} S_2(g) \rightleftharpoons H_2S(g)$ is 0.80 determine the value of K_p° for $2H_2(g) + S_2(g) \rightleftharpoons 2H_2S(g)$.

Solution For the reaction

$$H_2(g) + \tfrac{1}{2} S_2(g) = H_2S(g) \tag{7.6.2}$$

$$K_{p1}^\circ = \frac{(p_{H_2S}/p^\circ)}{(p_{H_2}/p^\circ)(p_{S_2}/p^\circ)^{1/2}} = 0.80$$

If the reaction is multiplied by 2, we have

i.e. $$2 \times \left[H_2(g) + \tfrac{1}{2}S_2(g) = H_2S(g) \right] \quad \text{i.e.} \quad 2H_2(g) + S_2(g) \rightleftharpoons 2H_2S(g)$$

K_p° for this reaction will be

$$K_{p2}^\circ = \frac{(p_{H_2S}/p^\circ)^2}{(p_{H_2}/p^\circ)^2\,(p_{S_2}/p^\circ)} = (K_{p1}^\circ)^2 = (0.\ 80)^2 = 0.64$$

Example 7.6.3 If $K_{p1}^\circ = 0.80$ for $H_2(g) + \frac{1}{2}S_2(g) \rightleftharpoons H_2S(g)$, determine the value of K_p° for

$$H_2S(g) \rightleftharpoons H_2(g) + \tfrac{1}{2}S_2(g)$$

Solution If the reaction is multiplied by –1, the reaction gets inverted, i.e. reactants become products and products become reactants. For example, if the given first reaction is multiplied by –1, we get the given second reaction, i.e.

$$-1 \times \left[H_2(g) + \tfrac{1}{2}S_2(g) = H_2S(g) \right] \quad \text{or} \quad H_2S(g) = H_2(g) + \tfrac{1}{2}S_2(g)$$

The equilibrium constant of this reaction will be

$$K_{p3}^\circ = \frac{(p_{H_2(g)}/p^\circ)\,(p_{S_2(g)}/p^\circ)^{1/2}}{(p_{H_2S(g)}/p^\circ)} = \frac{1}{K_{p_1}^\circ} = \frac{1}{0.80} = 1.25$$

Example 7.6.4 Given the following standard equilibrium constants at 1 362 K:

(i) $H_2(g) + \frac{1}{2}S_2(g) \rightleftharpoons H_2S(g)$ $\qquad K_{p_1}^\circ = 0.80$

(ii) $3H_2(g) + SO_2(g) \rightleftharpoons H_2S(g) + 2H_2O(g)$ $\qquad K_{p_2}^\circ = 1.8 \times 10^4$

Find the value of K_p° for the following reaction at 1 362 K.

$$4H_2(g) + 2SO_2(g) \rightleftharpoons S_2(g) + 4H_2O(g)$$

Solution The reaction can be obtained by multiplying reaction (ii) by 2 and the reaction (i) by 2 and then subtracting the latter from the former. Thus, the K_p° value of the resultant equation is given by

$$K_p^\circ = \frac{(K_{p_2}^\circ)^2}{(K_{p_1}^\circ)^2} = \left(\frac{1.8 \times 10^4}{0.8} \right)^2 = 5.06 \times 10^8$$

7.7 STANDARD EQUILIBRIUM CONSTANT IN UNITS OTHER THAN PARTIAL PRESSURES

The standard equilibrium constant K_p° used so far is defined in terms of partial pressures. In some cases, it is convenient to express compositions of various species in terms of other units such as molar concentrations, mole fractions, amounts of species, etc. In this section, we develop expressions for the equilibrium constant expressed in units other than partial pressures. We consider three such cases as given below.

Standard Equilibrium Constant K_c° We write the standard equilibrium constant expressed in terms of molar concentrations as K_c°. By definition, we have

$$K_p^\circ = \prod_B (p_B / p^\circ)^{\nu_B} \tag{7.7.1}$$

and

$$K_c^\circ = \prod_B (c_B / c^\circ)^{\nu_B} \tag{7.7.2}$$

where p° and c° are the standard units of pressure and concentration, respectively. The ideal gas equation, $pV = nRT$, may be written as

$$p = \frac{n}{V} RT = cRT \tag{7.7.3}$$

where c is the concentration of the gas expressed as amount of gas per unit volume. Substituting Eq. (7.7.3) in Eq. (7.7.1), we get

$$K_p^\circ = \prod_B (c_B RT/p^\circ)^{\nu_B}$$

$$= \prod_B (c_B / c^\circ)(c^\circ RT/p^\circ)^{\nu_B}$$

$$= K_c^\circ (c^\circ RT/p^\circ)^{\Delta \nu_g} \tag{7.7.4}$$

where $\Delta \nu_g$ is the difference in the stoichiometric numbers of gaseous species between products and reactants, that is,

$$\Delta \nu_g = \sum_B \nu_g(B) = \sum \nu_g(\text{product}) - \sum |\nu_g(\text{reactant})| \tag{7.7.5}$$

Example 7.7.1 For the reaction

$$CH_4(g) + 2H_2S(g) \rightleftharpoons CS_2(g) + 4H_2(g)$$

$K_p^\circ = 2.05 \times 10^9$ at 25 °C. Calculate (i) K_p° and (ii) K_c° at this temperature for the reaction

$$2H_2(g) + \tfrac{1}{2} CS_2(g) \rightleftharpoons H_2S(g) + \tfrac{1}{2} CH_4(g)$$

Solution For the reaction

$$CH_4(g) + 2H_2S(g) \rightleftharpoons CS_2(g) + 4H_2(g)$$

K_p° is given by

$$K_p^\circ = \frac{(p_{CS_2} / p^\circ)(p_{H_2} / p^\circ)^4}{(p_{CH_4} / p^\circ)(p_{H_2S} / p^\circ)^2} = 2.05 \times 10^9$$

and for the reaction

$$2H_2(g) + \tfrac{1}{2} CS_2(g) \rightleftharpoons H_2S(g) + \tfrac{1}{2} CH_4(g)$$

$$K_p^{\circ\prime} = \frac{(p_{H_2S} / p^\circ)(p_{CH_4} / p^\circ)^{1/2}}{(p_{H_2} / p^\circ)^2 (p_{CS_2} / p^\circ)^{1/2}} = \frac{1}{\sqrt{K_p^\circ}}$$

$$= \frac{1}{\sqrt{2.05 \times 10^9}} = 2.21 \times 10^{-5}$$

Now $$K_c^{\circ\prime} = K_p^{\circ\prime}\left(\frac{c^\circ RT}{p^\circ}\right)^{-\Delta\nu_g}$$

For the given reaction, we have

$$\Delta\nu_g = -1$$

Thus $$K_c^{\circ\prime} = (2.21\times10^{-5})\left[\frac{(1\ \text{mol dm}^{-3})(8.314\ \text{kPa dm}^3\ \text{K}^{-1}\ \text{mol}^{-1})(298\ \text{K})}{(101.325\ \text{kPa})}\right]$$

$$= 5.40\times10^{-4}$$

Equilibrium Constant K_x

The equilibrium constant expressed in terms of mole fractions is defined as

$$K_x = \prod_B (x_B)^{\nu_B} \tag{7.7.6}$$

The mole fraction of a gas in a gaseous mixture is its partial pressure divided by the total pressure of the system. Thus

$$x_B = p_B/p_{\text{total}} \quad \text{or} \quad p_B = x_B\, p_{\text{total}} \tag{7.7.7}$$

Substituting this in Eq. (7.7.1), we get

$$K_p^\circ = \prod_B (p_B/p^\circ)^{\nu_B}$$

$$= \prod_B \{(x_B)^{\nu_B}(p_{\text{total}}/p^\circ)^{\nu_B}\}$$

$$= K_x(p_{\text{total}}/p^\circ)^{\Delta\nu_g}$$

or $$K_x = K_p^\circ(p_{\text{totoal}}/p^\circ)^{-\Delta\nu_g} \tag{7.7.8}$$

Equilibrium Constant K_n

The equilibrium constant expressed in terms of amounts of substances is defined as

$$K_n = \prod_B (n_B)^{\nu_B}$$

Now $$p_B = x_B p_{\text{total}} = \left(\frac{n_B}{n_{\text{total}}}\right)p_{\text{total}} \tag{7.7.9}$$

where n_B is the amount of constitutent B and n_{total} is the total amount of gases in the system. Substituting this in Eq. (7.7.1), we get

$$K_p^\circ = \prod_B (p_B/p^\circ)^{\nu_B}$$

$$= \prod_B \left\{\left(\frac{n_B}{n_{\text{total}}}p_{\text{total}}\right)\Big/p^\circ\right\}^{\nu_B}$$

$$= \prod_B \left\{(n_B/n^\circ)^{\nu_B}(n^\circ/n_{\text{total}})^{\nu_B}(p_{\text{total}}/p^\circ)^{\nu_B}\right\}$$

$$= K_n^\circ(n^\circ/n_{\text{total}})^{\Delta\nu_g}(p_{\text{total}}/p^\circ)^{\Delta\nu_g} \tag{7.7.10}$$

where $n^\circ = 1$ mol. We have already seen that K_p° depends only on temperature and is independent of the total pressure of the system. Like K_p°, K_c° is also independent of pressure and depends only on temperature. K_x and K_n depend both on pressure and temperature.

7.8 PRINCIPLE OF LE CHATELIER AND BRAUN

Statement

This rule predicts qualitatively the effects on the system at equilibrium when some of the variables such as temperature, pressure and concentration are changed. It states.

If a system at equilibrium is subjected to a change, the system will react in such a way so as to oppose or reduce the change if this is possible, i.e. the system tends to balance or counteract the effects of any imposed stress.

Illustration

Examples of this principle are:

(i) Pressure increase	Reaction will shift in a direction where the number of gaseous molecules is reduced; thus lowering the pressure p.
(ii) Heat added or temperature raised	The equilibrium will shift in the endothermic direction, i.e. it is shifted to high enthalpy side.
(iii) One of the components of the system is added	Reaction proceeds in a direction so as to reduce the amount of this component.

Thermodynamic Proof

The thermodynamic proof of Le Chatelier-Braun principle as applicable to the temperature and pressure variations of chemical reactions is given below.

The reaction potential $\Delta_r G$ which is defined as

$$\Delta_r G = \left(\frac{\partial G}{\partial \xi}\right)_{T,p}$$

is a function of T, p and ξ, i.e.

$$\Delta_r G = f(T, p, \xi)$$

Writing its total differential, we have

$$d(\Delta_r G) = \left(\frac{\partial}{\partial T}\Delta_r G\right)_{p,\xi} dT + \left(\frac{\partial}{\partial p}\Delta_r G\right)_{T,\xi} dp + \left(\frac{\partial}{\partial \xi}\Delta_r G\right)_{T,p} d\xi \quad (7.8.1)$$

Following Eqs (5.4.7) and (5.4.8), we write the above expression as

$$d(\Delta_r G) = -\Delta_r S\, dT + \Delta_r V\, dp + G''\, d\xi \quad (7.8.2)$$

where

$$G'' = \left(\frac{\partial}{\partial \xi}\Delta_r G\right)_{T,p} = \left\{\frac{\partial}{\partial \xi}\left(\frac{\partial G}{\partial \xi}\right)_{T,p}\right\}_{T,p} \quad (7.8.3)$$

If we insist that the variations in temperature, pressure and the extent of reaction occur all the while keeping the reaction at equilibrium, then $(\Delta_r G) = 0$ and hence $d(\Delta_r G) = 0$. At equilibrium $\Delta_r S = \Delta_r H/T$, so that Eq. (7.8.2) becomes

$$0 = -\left(\frac{\Delta_r H}{T}\right)(\partial T)_{eq} + \Delta_r V(\partial p)_{eq} + G''_{eq}(\partial \xi)_{eq}$$

where G''_{eq} is the value of G'' at equilibrium. Since in the latter, G has a minimum value, therefore G''_{eq} must be positive. At constant pressure, $dp = 0$ and the above equation gives

$$\left(\frac{\partial \xi_{eq}}{\partial T}\right)_p = \frac{\Delta_r H}{TG''_{eq}} \tag{7.8.4}$$

and at constant temperature, $dT = 0$, the above equation gives

$$\left(\frac{\partial \xi_{eq}}{\partial p}\right)_T = -\frac{\Delta_r V}{G''_{eq}} \tag{7.8.5a}$$

For gases behaving ideally, we have

$$\left(\frac{\partial \xi_{eq}}{\partial p}\right)_T = -\frac{(\Delta \nu_g)RT}{pG''_{eq}} \tag{7.8.5b}$$

Equations (7.8.4) and (7.8.5) describe the dependence of the extent of reaction at equilibrium on temperature and pressure, respectively.

Temperature Dependence

Since G''_{eq} is positive, the sign of $(d\xi_{eq}/dT)_p$ in Eq. (7.8.4) depends on the sign of $\Delta_r H$.

- If $\Delta_r H$ is positive, an endothermic reaction, then $(d\xi /dT)_p$ is also positive, i.e. an increase in temperature increases the extent of reaction at equilibrium.
- For an exothermic reaction, $\Delta_r H$ is negative and thus $(d\xi_{eq}/dT)_p$ is negative, i.e. increase in temperature decreases the equilibrium extent of the reaction.

The above analysis amounts to the fact that the equilibrium is shifted in the forward direction for the endothermic reaction and in the backward direction for the exothermic reaction as the temperature of the system is increased. These conclusions are listed in Table 7.8.1.

Table 7.8.1 Effect of temperature on a reaction at equilibrium

Nature of reaction	*Sign of $\Delta_r H$*	*Sign of dT*	*Sign of $d\xi_{eq}$*	*Shift in the equilibrium*
Endothermic reaction	+ve	+ve	+ve	Forward direction or endothermic direction or high enthalpy side
		–ve	–ve	Backward direction or exothermic direction or low enthalpy side
Exothermic reaction	–ve	+ve	–ve	Backward direction or endothermic direction or high enthalpy side
		–ve	+ve	Forward direction or exothermic direction or low enthalpy side.

Highlighting Endothermic and Exothermic Directions

The designations of endothermic or exothermic direction and low or high enthalpy side may be understood from the following analysis.

We write the reaction as

Reactants → Products

with

$$\Delta_r H = \sum_B \nu_B H_m(B)$$

$$= \sum_{\substack{B \\ \text{(products)}}} \nu_B H_m(B) - \sum_{\substack{B \\ \text{(reactants)}}} |\nu_B| H_m(B)$$

We may consider two possibilities.

(1) $\Delta_r H$ *positive* In this case, we have

$$\sum_{\substack{B \\ \text{(products)}}} \nu_B H_m(B) > \sum_{\substack{B \\ \text{(reactants)}}} |\nu_B|\, H_m(B)$$

The reaction may be written as

Low enthalpy side → High enthalpy side

If the reaction proceeds in forward direction (or from low enthalpy side to high enthalpy side), heat is absorbed and it is for this reason, this direction is known as the endothermic direction. On the other hand, if the reaction proceeds in the backward direction (or from high enthalpy to low enthalpy side), heat is released and thus this direction is known as the exothermic direction.

(2) $\Delta_r H$ *negative* In this case, we have

$$\sum_{\substack{B \\ \text{(products)}}} \nu_B H_m(B) < \sum_{\substack{B \\ \text{(reactants)}}} |\nu_B|\, H_m(B)$$

and thus the reaction may be written as

High enthalpy side → Low enthalpy side

Based on the arguments given above, we can show that the forward direction (i.e. from high enthalpy to low enthalpy side) is the exothermic direction and the reverse one is the endothermic direction.

Following Table 7.8.1, the temperature effects may be stated in a general form which is applicable to both the exothermic and endothermic reactions. The statement is:

On increasing the temperature, equilibrium is shifted in the endothermic direction or towards high enthalpy side and on decreasing temperature, it is shifted in the exothermic direction or towards low enthalpy side.

Pressure Dependence

The sign of $(\partial\xi_{eq}/\partial p)_T$ in Eq. (7.8.5b) depends on that of $\Delta\nu_g$. If $\Delta\nu_g$ *is negative, that is, the sum of stoichiometric numbers of gaseous products is less than that of gaseous reactants, the derivative* $(\partial\xi_{eq}/\partial p)_T$ *has a positive value. Thus, increase in pressure increases the extent of reaction.*

If $\Delta\nu_g$ *is positive, then* $(\partial\xi_{eq}/\partial p)_T$ *is negative and thus increase in pressure decreases the extent of reaction.*

Hence, the net effect of increasing pressure is to shift the reaction in a direction where the sum of stoichiometric numbers of gaseous molecules is lowered, thus lowering the *p*. In other words, increase in pressure shifts the equilibrium to the low volume side of the reaction whereas a decrease in pressure shifts it to the high volume side.

7.9 TEMPERATURE DEPENDENCE OF STANDARD EQUILIBRIUM CONSTANT K_p°

Qualitative Predictions

We have seen that the standard equilibrium constant of a reaction depends only on temperature. Thus, it is of interest to study the effect of temperature on the standard equilibrium constant. The qualitative dependence of the standard equilibrium constant on temperature can easily be derived from the variation of extent of reaction with temperature as given in Table 7.8.1. Before deriving the dependence, we need to consider the relation of shift in the extent of reaction with that of the equilibrium constant. If the extent of reaction is increased, it means the equilibrium is shifted in the forward direction and thus more of the reactants are converted into products. In other words, the concentrations of products are increased and those of reactants are decreased. This, in turn, increases the value of the equilibrium constant. Hence, the increase in the extent of reaction leads to an increase in the value of equilibrium constant. Similarly, it can be shown that a decrease in the extent of reaction decreases the value of equilibrium constant. With these correlations, the results of Table 7.8.1 can easily be written in terms of variation in the value of equilibrium constant. These are given in Table 7.9.1.

Table 7.9.1 Effect of temperature on a reaction at Equilibrium Constant

Nature of reaction	*Sign of* $\Delta_r H$	*Sign of* $\mathrm{d}T$	*Sign of* $\mathrm{d}\xi_{eq}$	*Variation in the value of equilibrium constant*
Endothermic reaction	+ve	+ve	+ve	Increase
Exothermic reaction	–ve	+ve	–ve	Decrease

Quantitative Relationship Between ln K_p° and T

The quantitative relation connecting equilibrium constant and temperature can be derived thermodynamically starting from the Gibbs-Helmholtz equation (5.8.6).

van't Hoff Equation

We have[†]

$$\frac{\mathrm{d}}{\mathrm{d}T}\left\{\frac{\Delta_r G^\circ}{T}\right\} = -\frac{\Delta_r H^\circ}{T^2}$$

Since $\quad \Delta_r G^\circ = -RT \ln K_p^\circ$

therefore
$$\frac{\mathrm{d}}{\mathrm{d}T}\left\{-\frac{RT \ln K_p^\circ}{T}\right\} = -\frac{\Delta_r H^\circ}{T^2}$$

i.e.
$$\frac{\mathrm{d} \ln K_p^\circ}{\mathrm{d}T} = \frac{\Delta_r H^\circ}{RT^2} \tag{7.9.1}$$

[†] Note that the direct differential rather than the partial differential are used since both K_p° and $\Delta_r G^\circ$ are independent of pressure.

Equation (7.9.1) is known as van't Hoff equation.

The following conclusions may be drawn from Eq. (7.9.1).

- For an endothermic reaction, $\Delta_r H^\circ > 0$ and thus the right hand side of Eq. (7.9.1) is positive. This amounts to the fact that ln K_p° (and so also K_p°) increases with increase in temperature.
- For an exothermic reaction, $\Delta_r H^\circ < 0$ and thus the right hand side of the above equation is negative, which leads to the fact that ln K_p° (and so also K_p°) decreases with increase in temperature.

These results agree with those listed in Table 7.9.1.

Alternative Form of van't Hoff Equation

Equation (7.9.1) may be written as

$$\left\{\frac{\mathrm{d}\ln K_p^\circ}{\mathrm{d}(1/T)}\right\} = -\frac{\Delta_r H^\circ}{R} \tag{7.9.2}$$

Equation (7.9.2) is also known as *van't Hoff equation.*

Integrated Form of van't Hoff Equation

We can write Eq. (7.9.1) as

$$\mathrm{d}\ln K_p^\circ = \frac{\Delta_r H^\circ}{RT^2}\mathrm{d}T \tag{7.9.3}$$

Integrating the above expression, we have

$$\ln K_p^\circ = -\frac{\Delta_r H^\circ}{RT} + I \tag{7.9.4}$$

where I is a constant of integration whose value can be determined from a known value of K_p° at a given temperature. According to Eq. (7.9.4), ln K_p° varies linearly with $1/T$ with a slope equal to $-\Delta_r H^\circ/R$ (Figs 7.9.1a and 7.9.1b).

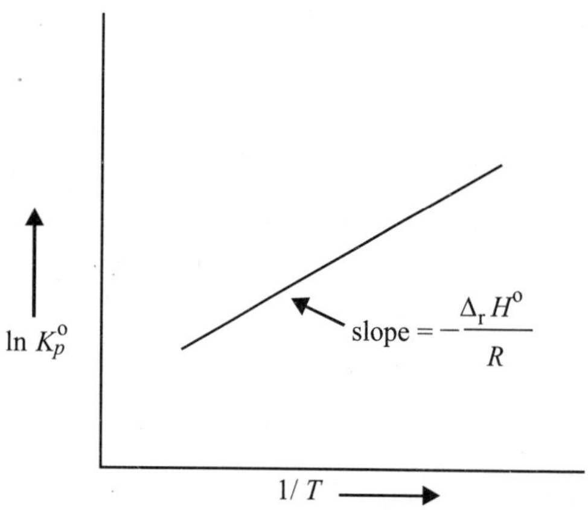

Fig. 7.9.1a Plot of ln K_p° versus $1/T$ where $\Delta_r H^\circ$ is negative. Example includes formation of ammonia from N_2 and H_2

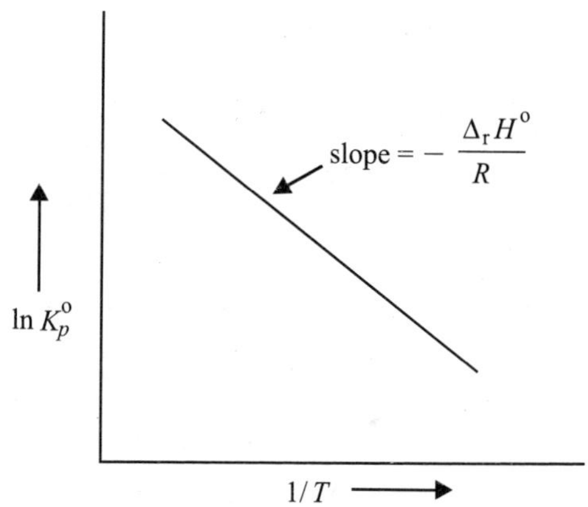

Fig. 7.9.1b Plot of ln K_p° versus $1/T$ where $\Delta_r H^\circ$ is positive. Example includes the dissociation of HI into H_2 and I_2

If we integrate Eq. (7.9.3) within the limits of T_1 and T_2, we get

$$\ln \frac{K^\circ_{p(T_2)}}{K^\circ_{p(T_1)}} = -\frac{\Delta_r H^\circ}{R}\left(\frac{1}{T_2} - \frac{1}{T_1}\right) \tag{7.9.5a}$$

i.e.

$$\ln \frac{K^\circ_{p(T_2)}}{K^\circ_{p(T_1)}} = \frac{\Delta_r H^\circ}{R}\left(\frac{T_2 - T_1}{T_2 T_1}\right) \tag{7.5.5b}$$

From this, it also follows that if the temperature is raised ($T_2 > T_1$), then

(i) $K^\circ_{p(T_2)} > K^\circ_{p(T_1)}$, if $\Delta_r H^\circ$ is positive

(ii) $K_{p(T_2)} < K^\circ_{p(T_1)}$, if $\Delta_r H^\circ$ is negative

---

Example 7.9.1 For the equilibrium

$$H_2O(l) \rightleftharpoons H_2O(g)$$

(a) show that

$$\ln\left(\frac{p_2}{p_1}\right) = -\frac{\Delta_{vap} H^\circ}{R}\left(\frac{1}{T_2} - \frac{1}{T_1}\right)$$

where p_2 and p_1 are the vapour pressure of liquid water at temperatures T_2 and T_1, respectively.

(b) The vapour pressure of liquid water at 373 K is 1 bar. Find its value at 323 K. Given $\Delta_{vap} H^\circ = 44.01$ kJ mol^{-1}

Solution **(a)** For the reaction

$$H_2O(l) \rightleftharpoons H_2O(g)$$

the standard equilibrium constant is

$$K^\circ_p = \frac{p_{H_2O}}{p^\circ}$$

Substituting this in the expression

$$\frac{d \ln K^\circ_p}{dT} = \frac{\Delta_r H^\circ}{RT^2} \quad \text{(Eq. 7.9.1)}$$

we get

$$\frac{d \ln (p_{H_2O}/p^\circ)}{dT} = \frac{\Delta_{vap} H^\circ}{RT^2} \quad \text{or} \quad d \ln(p_{H_2O}/p^\circ) = \frac{\Delta_{vap} H^\circ}{RT^2} dT$$

Integrating this expression, we have

$$\int_{p_1}^{p_2} d \ln (p/p^\circ) = \frac{\Delta_{vap} H^\circ}{R}\int_{T_1}^{T_2} \frac{dT}{T^2}$$

or

$$\ln\left(\frac{p_2}{p_1}\right) = -\frac{\Delta_{vap} H^\circ}{R}\left(\frac{1}{T_2} - \frac{1}{T_1}\right)$$

(b)

$$\log\left(\frac{p_2}{p_1}\right) = -\frac{44.01 \times 10^3 \text{ J mol}^{-1}}{(2.303)(8.314 \text{ J K}^{-1} \text{ mol}^{-1})}\left[\frac{1}{373 \text{ K}} - \frac{1}{323 \text{ K}}\right] = 0.954$$

Hence

$$\frac{p_2}{p_1} = 8.98 \quad \text{or} \quad p_1 = \frac{p_2}{8.98} = \frac{1 \text{ bar}}{8.98} = 0.111 \text{ bar.}$$

Expression for K_p° when $\Delta_r H^\circ$ is Dependent on Temperature

Equation (7.9.3) to (7.9.5) are applicable only when $\Delta_r H^\circ$ is independent of temperature. If $\Delta_r H^\circ$ depends on temperature, then the expression relating K_p° and T can be derived as follows.

The temperature dependence of $\Delta_r H^\circ$ as given by Eq. (3.11.2) is

$$d(\Delta_r H^\circ) = \Delta_r C_p \, dT \tag{7.9.6}$$

where $\Delta_r C_p = \sum_B \nu_B C_p(B)$

that is $\Delta_r C_p = \sum_{B \text{ (products)}} \nu_B C_p(B) - \sum_{B \text{ (reactants)}} |\nu_B| C_p(B)$

If the dependence of C_p is given by

$$C_p = a + bT + cT^2 + \cdots$$

then $\Delta_r C_p = \Delta_r a + (\Delta_r b)T + (\Delta_r c)T^2 + \cdots$ (7.9.7)

Substituting this in Eq. (7.9.6) and integrating, we have

$$\Delta_r H^\circ = I_1 + (\Delta_r a)T + \frac{\Delta_r b}{2}T^2 + \frac{\Delta_r c}{3}T^3 + \cdots \tag{7.9.8}$$

where I_1 is the constant of integration, the value of which is to be determined from a known value of $\Delta_r H^\circ$ at a given temperature. Substituting the above expression of $\Delta_r H^\circ$ in Eq. (7.9.1), we get

$$d \ln K_p^\circ = \left(\frac{I_1}{RT^2} + \frac{\Delta_r a}{RT} + \frac{\Delta_r b}{2R} + \frac{\Delta_r c}{3R}T + \cdots\right) dT \tag{7.9.9}$$

Integrating this, we have

$$\ln K_p^\circ = -\frac{I_1}{RT} + \frac{\Delta_r a}{R}\ln\left(\frac{T}{K}\right) + \frac{\Delta_r b}{2R}T + \frac{\Delta_r c}{6R}T^2 + \cdots + I_2 \tag{7.9.10}$$

where I_2 is the constant of integration, which is to be evaluated with the help of $\Delta_r G^\circ$ or K_p° at some temperature.

Example 7.9.2

For the reaction

$$\tfrac{1}{2} N_2(g) + O_2(g) \rightleftharpoons NO_2(g) \qquad \Delta_r H^\circ_{298K} = 33.18 \text{ kJ mol}^{-1}$$

$$\Delta_r G^\circ_{298K} = 51.31 \text{ kJ mol}^{-1}$$

obtain a general expression relating K° with temperature and compute the value of K_p° at 343 K. Given

$$C_p(N_2)/\text{J K}^{-1}\text{ mol}^{-1} = 28.46 + 2.32 \times 10^{-3}(T/\text{K})$$

$$C_p(O_2)/\text{J K}^{-1}\text{ mol}^{-1} = 26.85 + 8.49 \times 10^{-3}(T/\text{K})$$

$$C_p(NO_2)/\text{J K}^{-1}\text{ mol}^{-1} = 27.78 + 30.85 \times 10^{-3}(T/\text{K})$$

Solution

For the given reaction

$$\Delta_r C_p = \sum_B \nu_B C_p(B) = C_{p,m}(NO_2) - \tfrac{1}{2} C_{p,m}(N_2) - C_{p,m}(O_2)$$

$$= [-13.30 + 21.2 \times 10^{-3}(T/\text{K})] \text{ J K}^{-1}\text{ mol}^{-1}$$

From the expression

$$\frac{d(\Delta_r H^\circ)}{dT} = \Delta_r C_p$$

we get $\quad d(\Delta_r H^\circ) = [\{-13.30 + 21.2 \times 10^{-3}(T/K)\}\ \text{J K}^{-1}\ \text{mol}^{-1}]\ dT$

$$\Delta_r H^\circ = \left[-13.30\,T + 21.2\times10^{-3}\left(\frac{T^2}{2\text{K}}\right)\right]\text{J K}^{-1}\ \text{mol}^{-1} + I_1$$

where I_1 is the constant of integration whose value may be obtained from the fact that at 298 K, $\Delta_r H^\circ = 33.18\ \text{kJ mol}^{-1}$. Hence

$$I_1 = \Delta_r H^\circ - \left[-13.30\,T + 21.2\times10^{-3}\left(\frac{T^2}{2\text{K}}\right)\right]\text{J K}^{-1}\ \text{mol}^{-1}$$

$$= \left[33.18\times10^3 - \left\{-13.30\times298 + \frac{21.2\times10^{-3}}{2}(298)^2\right\}\right]\text{J mol}^{-1}$$

$$= [33.18 \times 10^3 - (-\,3\,963.4 + 941.32)]\ \text{J mol}^{-1}$$

$$= 36.202 \times 10^3\ \text{J mol}^{-1}$$

Hence, $\quad \Delta_r H^\circ = \left[-13.30T + 10.6\times10^{-3}\left(\frac{T^2}{\text{K}}\right)\right]\text{J K}^{-1}\ \text{mol}^{-1} + 36.202\times10^3\ \text{J mol}^{-1}$

Substituting this in the expression

$$\frac{d\ln K_p^\circ}{dT} = \frac{\Delta_r H^\circ}{RT^2}$$

we get $\quad \dfrac{d\ln K_p^\circ}{dT} = \left[-\dfrac{13.30}{RT} + \dfrac{10.6\times10^{-3}/\text{K}}{R}\right]\text{J K}^{-1}\ \text{mol}^{-1} + \dfrac{36.202\times10^3}{RT^2}\ \text{J mol}^{-1}$

Integrating this expression, we get

$$\ln K_p^\circ = \left[-\frac{13.30}{R}\ln(T/\text{K}) + \frac{10.6\times10^{-3}}{R}\left(\frac{T}{\text{K}}\right)\right]\text{J K}^{-1}\ \text{mol}^{-1}$$

$$-\frac{36.202\times10^3}{RT}\ \text{J mol}^{-1} + I_2$$

where I_2 is the constant of integration. Its value can be obtained from the fact that at 298 K,

$$\ln K_p^\circ = -\frac{\Delta_r G^\circ}{RT} = -\frac{51.31\times10^3\ \text{J mol}^{-1}}{(8.314\ \text{J K}^{-1}\ \text{mol}^{-1})(298\ \text{K})} = -20.71$$

Thus $\quad I_2 = \ln K_p^\circ - \left[-\dfrac{13.30}{R}\ln\left(\dfrac{T}{\text{K}}\right) + \dfrac{10.6\times10^{-3}}{R}\left(\dfrac{T}{\text{K}}\right)\right]\text{J K}^{-1}\text{mol}^{-1}$

$$+\frac{36.202\times10^3}{RT}\ \text{J mol}^{-1}$$

$$= -20.71 - \left[-\frac{13.30}{8.314}\ln(298) + \frac{10.6\times10^{-3}}{8.314}(298)\right] + \frac{36.202\times10^3}{(8.314)(298)}$$

$$= -\,20.71 + 9.114 - 0.380 + 14.612$$

$$= 2.64$$

Hence $$\ln K_p^\circ = \left[-\frac{13.30}{R}\ln\left(\frac{T}{\text{K}}\right)+\frac{10.6\times10^{-3}}{R}\left(\frac{T}{\text{K}}\right)\right]\text{J K}^{-1}\text{ mol}^{-1}$$

$$-\frac{36.202\times10^3}{RT}\text{J mol}^{-1}+2.64$$

At 343 K, we get

$$\ln K_p^\circ = -\frac{13.30}{8.314}\ln(343)+\frac{10.6\times10^{-3}}{8.314}(343)-\frac{36.202\times10^3}{(8.314)(343)}+2.64$$

$$= -9.34 + 0.44 - 12.69 + 2.64 = -18.95$$

$$K_p^\circ = 5.89\times10^{-9}$$

Example 7.9.3 Equilibrium constant for the relation

$$\tfrac{1}{2}\text{I}_2(\text{g})+\tfrac{1}{2}\text{Br}_2(\text{g}) \rightleftharpoons \text{IBr(g)}$$

can be represented by

$$\log K_p^\circ = \frac{(277.4\text{ K})}{T}+0.3811$$

Find $\Delta_r G^\circ_{298\text{K}}$, $\Delta_r H^\circ_{298\text{K}}$ and $\Delta_r S^\circ_{298\text{K}}$ for the reaction.

Solution The expression relating $\Delta_r G^\circ$ and K_p° is

$$\Delta_r G^\circ = -RT\ln K_p^\circ = -2.303\,RT\log K_p^\circ$$

Substituting the given expression of $\log K^\circ$, we get

$$\Delta_r G^\circ = -2.303\,RT\left(\frac{277.4\text{ K}}{T}+0.3811\right)$$

Thus at T = 298 K, we have

$$\Delta_r G^\circ = -2.303\,(8.314\text{ J K}^{-1}\text{ mol}^{-1})\,(298\text{ K})\left(\frac{277.4\text{ K}}{298\text{ K}}+0.3811\right)$$

$$= -7\,486\text{ J mol}^{-1}$$

The expression for $\Delta_r H^\circ$ is

$$\Delta_r H^\circ = RT^2\frac{\text{d}\ln K_p^\circ}{\text{d}T}$$

For the given expression of $\log K_p^\circ$, this becomes

$$\Delta_r H^\circ = RT^2\frac{\text{d}}{\text{d}T}\left\{2.303\left(\frac{277.4\text{ K}}{T}+0.3811\right)\right\}$$

$$= RT^2\left(-\frac{2.303\times277.4\text{ K}}{T^2}\right) = -2.303\,R(277.4\text{ K})$$

Hence $$\Delta_r H^\circ = -2.303(8.314\text{ J K}^{-1}\text{ mol}^{-1})\,(277.4\text{ K}) = -5\,311.4\text{ J mol}^{-1}$$

Now $$\Delta_r S^\circ = \frac{\Delta_r H^\circ-\Delta_r G^\circ}{T} = \frac{-5\,311.4\text{ J mol}^{-1}+7\,486\text{ J mol}^{-1}}{298\text{ K}}$$

$$= 7.30\text{ J K}^{-1}\text{ mol}^{-1}$$

Problem 7.9.1 Show that for gaseous equilibria

$$\frac{\mathrm{d}\ln K_c^\circ}{\mathrm{d}T} = \frac{\Delta_\mathrm{r}U^\circ}{RT^2}$$

where $K_p^\circ = K_c^\circ(c^\circ RT/p^\circ)^{\Delta\nu_\mathrm{g}}$.

Solution We start with the expression

$$\frac{\mathrm{d}\ln K_p^\circ}{\mathrm{d}T} = \frac{\Delta_\mathrm{r}H^\circ}{RT^2}$$

Substituting K_p° in terms of K_c° and $\Delta_\mathrm{r}H^\circ$ in terms of $\Delta_\mathrm{r}U^\circ$, we have

$$\frac{\mathrm{d}\ln\{K_c^\circ(c^\circ RT/p^\circ)^{\Delta\nu_\mathrm{g}}\}}{\mathrm{d}T} = \frac{\Delta_\mathrm{r}U^\circ + (\Delta\nu_\mathrm{g})RT}{RT^2}$$

or

$$\frac{\mathrm{d}\ln K_c^\circ}{\mathrm{d}T} + \Delta\nu_\mathrm{g}\frac{\mathrm{d}\ln(c^\circ RT/p^\circ)}{\mathrm{d}T} = \frac{\Delta_\mathrm{r}U^\circ}{RT^2} + \frac{(\Delta\nu_\mathrm{g})}{T}$$

or

$$\frac{\mathrm{d}\ln K_c^\circ}{\mathrm{d}T} + \Delta\nu_\mathrm{g}\frac{(c^\circ R/p^\circ)}{(c^\circ RT/p^\circ)} = \frac{\Delta_\mathrm{r}U^\circ}{RT^2} + \frac{(\Delta\nu_\mathrm{g})}{T}$$

Cancelling the term $\Delta\nu_\mathrm{g}/T$, we get

or

$$\frac{\mathrm{d}\ln K_c^\circ}{\mathrm{d}T} = \frac{\Delta_\mathrm{r}U^\circ}{RT^2}$$

7.10 PRESSURE DEPENDENCE OF EQUILIBRIUM CONSTANTS

In this section, we describe the pressure dependence of quilibrium constants K_p°, K_c° and K_x.

Pressure Dependence of K_p°

The equilibrium constant K_p° is related to the standard free energy change of the reaction by the relation

$$\Delta_\mathrm{r}G^\circ = -RT\ln K_p^\circ$$

Since $\Delta_\mathrm{r}G^\circ$ represents the free energy of the reaction where all its reactants and products are at the standard state of 1 bar pressure, its value depends only on temperature. It is thus obvious that K_p° also shows only temperature dependence and is independent of the pressure of the system.

Pressure Dependence of K_c°

Since $K_c^\circ = K_p^\circ(c^\circ RT/p^\circ)^{-\Delta\nu_\mathrm{g}}$

it is obvious that K_c° like K_p° depends only on temperature and is independent of pressure of the system.

Pressure Dependence of K_x

Since $K_x = K_p^\circ(p_\mathrm{total}/p^\circ)^{-\Delta\nu_\mathrm{g}}$, we have

$$\ln K_x = \ln K_p^\circ - \Delta\nu_\mathrm{g}\ln(p_\mathrm{total}/p^\circ)$$

Differentiating the above expression with respect to the total pressure, we get

$$\left(\frac{\partial \ln K_x}{\partial p_{\text{total}}}\right)_T = -\frac{1}{p_{\text{total}}}\Delta \nu_{\text{g}} = -\frac{\Delta_{\text{r}} V_{\text{g}}}{RT} \tag{7.10.1}$$

Pressure dependence of K_x can be described with the help of Eq. (7.10.1). We may consider two cases given below.

(i) $\Delta \nu_{\text{g}} = 0$, i.e. there is no change in the stoichiometric number of gaseous molecules in the system, then

$$K_x = K_p^{\circ}$$

In such a case, the position of equilibrium does not depend on the total pressure.

(ii) $\Delta_{\text{r}} \nu_{\text{g}} \neq 0$, the effect of pressure as predicted from Eq. (7.10.1) are:

(a) When $\Delta_{\text{r}} \nu_{\text{g}} < 0$, thus there is a decrease in the gaseous species, K_x increases with increase in pressure.

(b) When $\Delta_{\text{r}} \nu_{\text{g}} > 0$, thus there is an increase in the gaseous species, K_x decreases with increase in pressure.

7.11 EFFECT OF AN INERT GAS ON EQUILIBRIUM

Quantitative Predictions

The quantitative effect of adding an inert gas on the position of equilibrium of a reaction can be derived as follows.

Consider the general reaction

$$\nu_1 A_1 + \nu_2 A_2 \rightleftharpoons \nu_3 A_3 + \nu_4 A_4 \tag{7.11.1}$$ [†]

Let the reaction be carried out starting only from the reactants with n_1 and n_2 as the initial amounts of A_1 and A_2, respectively. Let the extent of reaction at equilibrium be represented as ξ_{eq}. Thus, we have

	$\nu_1 A_1$ +	$\nu_2 A_2$	$\rightleftharpoons$	$\nu_3 A_3$ +	$\nu_4 A_4$
$\xi = 0$	n_1	n_2		0	0
$\xi = \xi_{\text{eq}}$	$n_1 - \nu_1 \xi_{\text{eq}}$	$n_2 - \nu_2 \xi_{\text{eq}}$		$\nu_3 \xi_{\text{eq}}$	$\nu_4 \xi_{\text{eq}}$

Total amount of gases at equilibrium

$$n_{\text{total}} = (n_1 - \nu_1\, \xi_{\text{eq}}) + (n_2 - \nu_2\, \xi_{\text{eq}}) + \nu_3\, \xi_{\text{eq}} + \nu_4\, \xi_{\text{eq}}$$

The partial pressures of various constituents are

$$p_{A_1} = \frac{n_1 - \nu_1 \xi_{\text{eq}}}{n_{\text{total}}} p_{\text{total}}\,; \qquad p_{A_2} = \frac{n_2 - \nu_2 \xi_{\text{eq}}}{n_{\text{total}}} p_{\text{total}}$$

$$p_{A_3} = \frac{\nu_3 \xi_{\text{eq}}}{n_{\text{total}}} p_{\text{total}}\,; \qquad p_{A_4} = \frac{\nu_4 \xi_{\text{eq}}}{n_{\text{total}}} p_{\text{total}}$$

† In Eq. (7.11.1), all stoichiometric numbers are considered as positive numbers.

$$K_p = \frac{(p_{A_3})^{\nu_3}(p_{A_4})^{\nu_4}}{(p_{A_1})^{\nu_1}(p_{A_2})^{\nu_2}} = \frac{\left(\dfrac{\nu_3\xi_{eq}}{n_{total}}p_{total}\right)^{\nu_3}\left(\dfrac{\nu_4\xi_{eq}}{n_{total}}p_{total}\right)^{\nu_4}}{\left(\dfrac{n_1-\nu_1\xi_{eq}}{n_{total}}p_{total}\right)^{\nu_1}\left(\dfrac{n_2-\nu_2\xi_{eq}}{n_{total}}p_{total}\right)^{\nu_2}}$$

$$= \frac{(\nu_3\xi_{eq})^{\nu_3}(\nu_4\xi_{eq})^{\nu_4}}{(n_1-\nu_1\xi_{eq})^{\nu_1}(n_2-\nu_2\xi_{eq})^{\nu_2}}\left(\frac{p_{total}}{n_{total}}\right)^{\Delta\nu_g} \tag{7.11.2}$$

Now let the amount n of an inert gas be added at equilibrium and let the extent of reaction at equilibrium be ξ'_{eq}. The total amount of gases at equilibrium is

$$n'_{total} = n + (n_1 - \nu_1\xi'_{eq}) + (n_2 - \nu_2\xi'_{eq}) + \nu_3\xi'_{eq} + \nu_4\xi'_{eq}$$

Since $n'_{total} > n_{total}$, it follows that the mole fractions of various constituents will decrease as a result of adding inert gas. Deriving the expressions of partial pressures of various constituents and then substitutng in the expression for K_p, we get

$$K_p = \frac{(\nu_3\xi'_{eq})^{\nu_3}(\nu_4\xi'_{eq})^{\nu_4}}{(n_1-\nu_1\xi'_{eq})^{\nu_1}(n_2-\nu_2\xi'_{eq})^{\nu_2}}\left(\frac{p'_{total}}{n'_{total}}\right)^{\Delta\nu_g} \tag{7.11.3}$$

With Eqs (7.11.2) and (7.11.3), we can derive the effects produced as a result of addition of inert gases. Three cases may be distinguished depending on the value of $\Delta\nu_g$. These are:

Reaction Involving No Change in Number of Gaseous Species

Here $\Delta\nu_g = 0$. For this case, Eqs (7.11.2) and (7.11.3) become

$$K_p = \frac{(\nu_3\xi_{eq})^{\nu_3}(\nu_4\xi_{eq})^{\nu_4}}{(n_1-\nu_1\xi_{eq})^{\nu_1}(n_2-\nu_1\xi_{eq})^{\nu_2}} \tag{7.11.4}$$

$$K_p = \frac{(\nu_3\xi'_{eq})^{\nu_3}(\nu_4\xi'_{eq})^{\nu_4}}{(n_1-\nu_1\xi'_{eq})^{\nu_1}(n_2-\nu_1\xi'_{eq})^{\nu_2}} \tag{7.11.5}$$

Since K remains unaffected, it is obvious that

$$\xi_{eq} = \xi'_{eq}$$

that is, *there is no effect on the equilibrium position by the addition of an inert gas.*

Reaction with an Increase in Number of Gaseous Species

Here $\Delta\nu_g$ is positive. Two subcases may be considered.

(a) *Inert gas added keeping pressure of the system constant* In this case, the denominator of Eq. (7.11.3) will be greater than that of Eq. (7.11.2). Now, in order to keep the overall expression of Eq. (7.11.3) to a constant value of K_p, its numerator must also increase. The latter is possible provided the extent of reaction is increased, i.e.

$$\xi'_{eq} > \xi_{eq}$$

Thus, the reaction is shifted to the right side, i.e. more products are formed.

(b) *Inert gas added keeping volume of the system constant* Following the ideal gas equation

$$pV = nRT$$

we get $$\frac{p}{n} = \frac{RT}{V}$$

or $$\left(\frac{p}{n}\right)^{\Delta\nu_g} = \left(\frac{RT}{V}\right)^{\Delta\nu_g}$$

Therefore in Eq. (7.11.2) and (7.11.3), we have

$$\left(\frac{p_{\text{total}}}{n_{\text{total}}}\right)^{\Delta\nu_g} = \left(\frac{p'_{\text{total}}}{n'_{\text{total}}}\right)^{\Delta\nu_g} = \left(\frac{RT}{V}\right)^{\Delta\nu_g}$$

Substituting this in Eqs (7.11.2) and (7.11.3), we get

$$K_p = \frac{(\nu_3\xi_{\text{eq}})^{\nu_3}(\nu_4\xi_{\text{eq}})^{\nu_4}}{\{n_1 - \nu_1\xi_{\text{eq}}\}^{\nu_1}\{n_2 - n_2\xi_{\text{eq}}\}^{\nu_2}}\left(\frac{RT}{V}\right)^{\Delta\nu_g}$$

$$= \frac{(\nu_3\xi'_{\text{eq}})^{\nu_3}(\nu_4\xi'_{\text{eq}})^{\nu_4}}{\{n_1 - \nu_1\xi'_{\text{eq}}\}^{\nu_1}\{n_2 - n_2\xi'_{\text{eq}}\}^{\nu_2}}\left(\frac{RT}{V}\right)^{\Delta\nu_g}$$

Thus at constant V, this gives

$$\xi_{\text{eq}} = \xi'_{\text{eq}}$$

that is, there is no effect on the equilibrium position of the reaction.

Reaction with a Decrease in Number of Gaseous Species

Here $\Delta\nu_g$ is Negative. Again, we may consider two subcases.

(a) *Inert gas added keeping pressure of the system constant* Following the arguments given above, we can show that in the present case.

$$\xi'_{\text{eq}} < \xi_{\text{eq}}$$

Thus, the equilibrium is shifted to the left side i.e. less reactants are consumed.

(b) *Inert gas added keeping volume of the system constant* Here again, we get

$$\xi_{\text{eq}} = \xi'_{\text{eq}}$$

that is, there is no effect on the equilibrium position of the reaction.

Qualitative Predictions

The above effect on the value of ξ_{eq} caused by the addition of an inert gas may also be understood qualitatively as follows.

In general, the equilibrium constant may be written as

$$K_p = \frac{\prod_{\text{p}}(x_{\text{p}}\, p_{\text{total}})^{\nu_{\text{p}}}}{\prod_{\text{r}}(x_{\text{r}}\, p_{\text{total}})^{|\nu_{\text{r}}|}} = \left\{\frac{\prod_{\text{p}}(n_{\text{p}})^{\nu_{\text{p}}}}{\prod_{\text{r}}(n_{\text{r}})^{|\nu_{\text{r}}|}}\right\}\left\{\frac{p_{\text{total}}}{n_{\text{total}}}\right\}^{\Delta\nu_g} \tag{7.11.6}$$

where p and r represent products and reactants, respectively. Three cases discussed above are:

No Change in Number of Gaseous Species

Here $\Delta\nu = 0$; In this case, the expression of K_p becomes independent of p_{total} and n_{total}, hence

$$\xi'_{\text{eq}} = \xi_{\text{eq}}$$

Increase in Number of Gaseous Species

Here $\Delta\nu_{\text{g}} = +\text{ve}$. In this case, we have

(a) *p constant* The addition of an inert gas increases n_{total} and hence the second term of Eq. (7.11.6) is decreased. Now to keep K_p constant, the first term of Eq. (7.11.6) will increase and hence

$$\xi'_{\text{eq}} > \xi_{\text{eq}}$$

(b) *V constant* The addition of an inert gas increases not only n_{total} but also p_{total} proportionately. Thus, the second term of Eq. (7.11.6) remains constant and so also the first term and hence

$$\xi'_{\text{eq}} = \xi_{\text{eq}}$$

Decrease in Number of Gaseous Species

Here $\Delta\nu_{\text{g}} = -\text{ve}$: In this case, we have

(a) *p constant* The addition of an inert gas increases n_{total} and hence the second term of Eq. (7.11.6) is increased as $\Delta\nu_{\text{g}}$ is negative. Now to keep K_p constant, the first term of Eq. (7.11.6) will decrease and hence

$$\xi'_{\text{eq}} < \xi_{\text{eq}}$$

(b) *V constant* The addition of an inert gas increases not only n_{total} but also p_{total} proportionately. Thus, the second term of Eq. (7.11.6) remains constant and so also the first term and hence

$$\xi'_{\text{eq}} = \xi_{\text{eq}}$$

Effects Caused by Increase in Pressure without Adding Inert Gas

From Eqs (7.11.2) and (7.11.3), we can also derive the effects produced on the equilibrium position as a result of increasing pressure without the addition of an inert gas. Three cases may be considered.

(i) $\Delta\nu_{\text{g}}$ *is zero* Under this condition, Eqs (7.11.2) and (7.11.3) become independent of pressure and thus are identical. This gives us

$$\xi_{\text{eq}} = \xi'_{\text{eq}}$$

(ii) $\Delta\nu_{\text{g}}$ *is positive* In this case, $(p'_{\text{total}})^{\Delta\nu_{\text{g}}} > (p_{\text{total}})^{\Delta\nu_{\text{g}}}$. Thus, we must have

$$\xi'_{\text{eq}} < \xi_{\text{eq}}$$

(iii) $\Delta\nu_{\text{g}}$ *is negative* In this case $(p'_{\text{total}})^{\Delta\nu_{\text{g}}} < (p_{\text{total}})^{\Delta\nu_{\text{g}}}$ and thus, we must have

$$\xi'_{\text{eq}} > \xi_{\text{eq}}$$

Problem 7.11.1

Discuss what effects are produced on the equilibrium position of the reaction

$$SO_2(g) + \tfrac{1}{2}O_2(g) \rightleftharpoons SO_3(g)$$

when (i) pressure of the system is increased without addition of any inert gas, (ii) inert gas added keeping volume of the system constant, and (iii) inert gas added keeping pressure of the system constant.

Solution The given reaction is

$$SO_2(g) + \tfrac{1}{2}O_2(g) \rightleftharpoons SO_3(g)$$

Let at equilibrium, the amounts of various species be n_{SO_2} , n_{O_2} and n_{SO_3}, respectively. Let n_{total} be the amount of the gas including the inert gas and p be the total pressure. Then the partial pressures of various constituents are given by

$$p_{SO_2} = \frac{n_{SO_2}}{n_{total}}p; \qquad p_{O_2} = \frac{n_{O_2}}{n_{total}}p; \qquad p_{SO_3} = \frac{n_{SO_3}}{n_{total}}p$$

The equilibrium constant K_p is given by

$$K_p = \frac{p_{SO_3}}{p_{SO_2}\sqrt{p_{O_2}}} = \frac{(n_{SO_3}/n_{total})p}{(n_{SO_2}/n_{total})p\sqrt{(n_{O_2}/n_{total})p}}$$

Rearranging this, we get

$$\frac{n_{SO_3}}{(n_{O_2})^{1/2}n_{SO_2}} = \left(\frac{p}{n_{total}}\right)^{1/2} K_p \qquad (7.11.7)$$

Thus, we have three cases:

(i) If the pressure is increased by compressing the system without addition of gas from outside, the right hand side of Eq. (7.11.7) increases as K_p is a constant. Consequently, more of SO_3 is produced from the combination of SO_2 and O_2.

(ii) If an inert gas is added at constant volume both n_{total} and p increase in the same ratio. Thus, the right hand side of Eq. (7.11.7) does not change and hence the equilibrium remains unaffected.

(iii) If an inert gas is added at constant pressure, n_{total} is increased while p remains constant. Thus, the right hand side of Eq. (7.11.7) decreases. Consequently, the dilution of the mixture with the inert gas decreases the extent of conversion of SO_2 and O_2 to SO_3.

7.12 GENERAL TREATMENT OF A REACTION IN PROGRESS

Expression of Gibbs Function

A chemical reaction, in general, may be written as

$$0 = \sum_{B} \nu_B B \qquad (7.12.1)$$

where B represents constituent of the reaction and ν_B is the corresponding stoichiometric coefficient (assumed to be positive for a product and negative for a reactant) in the balanced chemical equation of the reaction.

The amount of a reactant (or product) during the progress of a reaction is given as

$$n_B = (n_B)_0 + \nu_B \xi \qquad (7.12.2)$$

where $(n_B)_0$ is the initial amount of the reactant (or product) and ξ is the extent of reaction (unit: mol).

The Gibbs function of a reaction in progress is given by

$$G = \sum_{B} n_B \mu_B \qquad (7.12.3)$$

We know that:

1. The chemical potential of a species in a condensed phase (solid or liquid; abbrevation: cd) is given as

$$\mu_{\mathrm{B(cd)}} = \mu^{\circ}_{\mathrm{B(cd)}} \tag{7.12.4}$$

2. The chemical potential of a species in gaseous phase (assumed to be ideal) is given as

$$\mu_{\mathrm{B(g)}} = \mu^{\circ}_{\mathrm{B(g)}} + RT \ln \frac{p_{\mathrm{B}}}{p^{\circ}} \tag{7.12.5}$$

where p_{B} is the partial pressure of the gas B in a mixture of gases and p° is the standard-state pressure (= 1 bar).

Substituting Eqs (7.12.4)–(7.12.5) in Eq. (7.12.3), we get

$$G = \sum_{\mathrm{B(cd)}} n_{\mathrm{B}}\, \mu^{\circ}_{\mathrm{B}} + \sum_{\mathrm{B(g)}} n_{\mathrm{B}} \left(\mu^{\circ}_{\mathrm{B}} + RT \ln \frac{p_{\mathrm{B}}}{p^{\circ}} \right)$$

$$= \sum_{\mathrm{B}} n_{\mathrm{B}}\, \mu^{\circ}_{\mathrm{B}} + \sum_{\mathrm{B(g)}} n_{\mathrm{B}}\, RT \ln \frac{p_{\mathrm{B}}}{p^{\circ}} \tag{7.12.6}$$

where the summations $\sum_{\mathrm{B(cd)}}$, $\sum_{\mathrm{B(g)}}$ and $\sum_{\mathrm{B}}$ are over condensed phases, gases and all species, respectively, in a chemical reaction.

The partial pressure, p_{B}, is related to the extent of reaction through the expression

$$p_{\mathrm{B}} = x_{\mathrm{B}}\, p$$

$$= \frac{n_{\mathrm{B}}}{n_{\mathrm{total}}}\, p = \frac{(n_{\mathrm{B}})_0 + \nu_{\mathrm{B}}\, \xi}{\sum_{\mathrm{B(g)}} \{(n_{\mathrm{B}})_0 + \nu_{\mathrm{B}}\, \xi\}}\, p \tag{7.12.7}$$

Substituting Eq. (7.12.7) in Eq. (7.12.6), we get

$$G = \sum_{\mathrm{B}} n_{\mathrm{B}}\, \mu^{\circ}_{\mathrm{B}} + RT \sum_{\mathrm{B(g)}} \left(n_{\mathrm{B}} \ln \frac{x_{\mathrm{B}}\, p}{p^{\circ}} \right)$$

$$= \sum_{\mathrm{B}} n_{\mathrm{B}}\, \mu^{\circ}_{\mathrm{B}} + RT \left(\sum_{\mathrm{B(g)}} n_{\mathrm{B}} \right) \ln \frac{p}{p^{\circ}} + RT \sum_{\mathrm{B(g)}} n_{\mathrm{B}} \ln x_{\mathrm{B}}$$

$$= \sum_{\mathrm{B}} n_{\mathrm{B}}\, \mu^{\circ}_{\mathrm{B}} + n_{\mathrm{total}}\, RT \ln \frac{p}{p^{\circ}} + RT \sum_{\mathrm{B(g)}} n_{\mathrm{B}} \ln x_{\mathrm{B}} \tag{7.12.8}$$

Equation (7.12.8) is usually written as

$$G = G_{\mathrm{pure}} + \Delta_{\mathrm{mix}} G \tag{7.12.9}$$

where

$$G_{\mathrm{pure}} = \sum_{\mathrm{B}} n_{\mathrm{B}}\, \mu^{\circ}_{\mathrm{B}} + n_{\mathrm{total}}\, RT \ln \frac{p}{p^{\circ}} \tag{7.12.10}$$

$$\Delta_{\mathrm{mix}} G = RT \sum_{\mathrm{B(g)}} n_{\mathrm{B}} \ln x_{\mathrm{B}} \tag{7.12.11}$$

Expression of Reaction Potential

The change of Gibbs function (Eq. 7.12.8) with extent of reaction is

$$\left(\frac{\partial G}{\partial \xi}\right)_{T,p} = \sum_{\mathrm{B}} \frac{\mathrm{d}n_{\mathrm{B}}}{\mathrm{d}\xi}\mu_{\mathrm{B}}^{\mathrm{o}} + \frac{\mathrm{d}n_{\mathrm{total}}}{\mathrm{d}\xi} RT \ln \frac{p}{p^{\circ}}$$

$$+ RT \sum_{\mathrm{B(g)}} \left(\frac{\mathrm{d}n_{\mathrm{B}}}{\mathrm{d}\xi} \ln x_{\mathrm{B}} + n_{\mathrm{B}} \frac{\mathrm{d} \ln x_{\mathrm{B}}}{\mathrm{d}\xi}\right) \qquad (7.12.12)$$

Now since

1. $n_{\mathrm{B}} = (n_{\mathrm{B}})_0 + \nu_{\mathrm{B}}\xi \quad \Rightarrow \quad \dfrac{\mathrm{d}n_{\mathrm{B}}}{\mathrm{d}\xi} = \nu_{\mathrm{B}}$

2. $n_{\mathrm{total}} = \sum\limits_{\mathrm{B(g)}} n_{\mathrm{B}} = \sum\limits_{\mathrm{B(g)}} \{(n_{\mathrm{B}})_0 + \nu_{\mathrm{B}}\,\xi\} \quad \Rightarrow \quad \dfrac{\mathrm{d}n_{\mathrm{total}}}{\mathrm{d}\xi} = \sum\limits_{\mathrm{B(g)}} \nu_{\mathrm{B}}$

3. $\ln x_{\mathrm{B}} = \ln \dfrac{n_{\mathrm{B}}}{n_{\mathrm{total}}} = \ln \dfrac{n_{\mathrm{B}}}{\sum n_{\mathrm{B}}} = \ln n_{\mathrm{B}} - \ln (\sum n_{\mathrm{B}})$

$$\Rightarrow \quad \frac{\mathrm{d} \ln x_{\mathrm{B}}}{\mathrm{d}\xi} = \frac{\mathrm{d} \ln n_{\mathrm{B}}}{\mathrm{d}\xi} - \frac{\mathrm{d} \ln \sum n_{\mathrm{B}}}{\mathrm{d}\xi} = \frac{1}{n_{\mathrm{B}}}\frac{\mathrm{d}n_{\mathrm{B}}}{\mathrm{d}\xi} - \frac{1}{\Sigma n_{\mathrm{B}}}\frac{\mathrm{d}\Sigma n_{\mathrm{B}}}{\mathrm{d}\xi}$$

$$= \frac{\nu_{\mathrm{B}}}{n_{\mathrm{B}}} - \frac{\Sigma \nu_{\mathrm{B}}}{\Sigma n_{\mathrm{B}}}$$

4. $\Sigma n_{\mathrm{B}} \dfrac{\mathrm{d} \ln x_{\mathrm{B}}}{\mathrm{d}\xi} = \Sigma\, n_{\mathrm{B}} \left(\dfrac{\nu_{\mathrm{B}}}{n_{\mathrm{B}}} - \dfrac{\Sigma \nu_{\mathrm{B}}}{\Sigma n_{\mathrm{B}}}\right) = \Sigma\left(\nu_{\mathrm{B}} - \dfrac{n_{\mathrm{B}}\Sigma \nu_{\mathrm{B}}}{\Sigma n_{\mathrm{B}}}\right)$

$$= \Sigma \nu_{\mathrm{B}} - \frac{\Sigma n_{\mathrm{B}} \Sigma \nu_{\mathrm{B}}}{\Sigma n_{\mathrm{B}}} = \Sigma \nu_{\mathrm{B}} - \Sigma \nu_{\mathrm{B}} = 0$$

In the above expressions, summation is over gaseous species, i.e. B(g). With these, Eq. (7.12.12) reduces to

$$\left(\frac{\partial G}{\partial \xi}\right)_{T,p} = \sum_{\mathrm{B}} \nu_{\mathrm{B}}\, \mu_{\mathrm{B}}^{\mathrm{o}} + \left(\sum_{\mathrm{B(g)}} \nu_{\mathrm{B}}\right) RT \ln \frac{p}{p^{\circ}} + RT \sum_{\mathrm{B(g)}} \nu_{\mathrm{B}} \ln x_{\mathrm{B}} \qquad (7.12.13)$$

Equation (7.12.13) may be written as

$$\left(\frac{\partial G}{\partial \xi}\right)_{T,p} = \sum_{\mathrm{B}} \nu_{\mathrm{B}}\, \mu_{\mathrm{B}}^{\mathrm{o}} + RT \sum_{\mathrm{B(g)}} \nu_{\mathrm{B}} \ln \left(\frac{x_{\mathrm{B}}\, p}{p^{\circ}}\right)$$

$$= \sum_{\mathrm{B}} \nu_{\mathrm{B}} \mu_{\mathrm{B}}^{\mathrm{o}} + RT \sum_{\mathrm{B(g)}} \nu_{\mathrm{B}} \ln \left(\frac{p_{\mathrm{B}}}{p^{\circ}}\right)$$

$$= \sum_{\mathrm{B}} \nu_{\mathrm{B}} \mu_{\mathrm{B}}^{\mathrm{o}} + RT \ln \prod_{\mathrm{B(g)}} \left(\frac{p_{\mathrm{B}}}{p^{\circ}}\right)^{\nu_{\mathrm{B}}}$$

$$= \Delta G^\circ + RT \ln Q_p^\circ \qquad (7.12.14)$$

where Q_p°, the standard reaction quotient, is given as

$$Q_p^\circ = \prod_{\mathrm{B(g)}} \left(\frac{p_\mathrm{B}}{p^\circ}\right)^{\nu_\mathrm{B}} \qquad (7.12.15)$$

(**Note:** The symbol $\prod_{\mathrm{B(g)}}$ stands for multiplication of $(p_\mathrm{B}/p^\circ)^{\nu_\mathrm{B}}$ for all gaseous species in a chemical equation.)

The derivative $(\partial G/\partial \xi)_{T,\,p}$ is known as reaction potential and is represented by the symbol $\Delta \tilde{G}$. It is equal to the change in Gibbs function for a unit change of extent of reaction.

Reaction at Equilibrium

For a reaction to be at equilibrium,

$$\mathrm{d}G = \Delta \tilde{G}\, \mathrm{d}\xi = 0 \quad \Rightarrow \quad \Delta \tilde{G} = 0$$

and $\quad Q_p^\circ = K_p^\circ$

With these, Eq. (7.12.14) reduces to

$$\Delta G^\circ = -RT \ln K_p^\circ \qquad (7.12.16)$$

Illustration

We derive the expressions of G $(= G_\mathrm{pure} + \Delta_\mathrm{mix} G)$ and $(\Delta G/\partial \xi)_{T,p}$ by using Eqs (7.12.10), (7.12.11) and (7.12.13) for the reaction

$$\mathrm{N_2O_4(g)} \rightleftharpoons \mathrm{2NO_2(g)}$$

which is initiated with reactant having amount equal to the corresponding stoichiometric coefficient expressed in moles. Thus, for the given reaction we have

$$(n_{\mathrm{N_2O_4}})_0 = 1\ \mathrm{mol}; \qquad (n_{\mathrm{NO_2}})_0 = 0$$

$$\nu_{\mathrm{N_2O_4}} = -1; \qquad \nu_{\mathrm{NO_2}} = 2; \qquad \sum_{\mathrm{B(g)}} \nu_\mathrm{B} = 2 - 1 = 1$$

$$n_\mathrm{total} = \sum_{\mathrm{B(g)}} n_\mathrm{B} = \sum_{\mathrm{B(g)}} \{(n_\mathrm{B})_0 + \nu_\mathrm{B}\, \xi\}$$

$$= \{(n_{\mathrm{N_2O_4}})_0 + \nu_{\mathrm{N_2O_4}}\xi\} + \{(n_{\mathrm{NO_2}})_0 + \nu_{\mathrm{NO_2}}\xi\}$$

$$= (1\ \mathrm{mol} - \xi) + (0 + 2\xi) = 1\ \mathrm{mol} + \xi \qquad (7.12.17)$$

Expression of G_pure

With these, we get

$$G_\mathrm{pure} = \sum_\mathrm{B} n_\mathrm{B}\, \mu_\mathrm{B}^\circ + n_\mathrm{total}\, RT \ln \frac{p}{p^\circ} \qquad \text{(Eq. 7.12.10)}$$

$$= \sum_\mathrm{B} \{(n_\mathrm{B})_0 + \nu_\mathrm{B}\, \xi\}\, \mu_\mathrm{B}^\circ + RT\, n_\mathrm{total} \ln \frac{p}{p^\circ}$$

$$= (1\ \mathrm{mol})\, \mu_{\mathrm{N_2O_4}}^\circ + \xi\,(-\mu_{\mathrm{N_2O_4}}^\circ + 2\,\mu_{\mathrm{NO_2}}^\circ) + RT\,(1\ \mathrm{mol} + \xi) \ln \frac{p}{p^\circ}$$

$$(7.12.18)$$

Expression of $\Delta_{mix}G$

$$\Delta_{mix}G = RT \sum_{B(g)} n_B \,\mathrm{In}\, x_B = RT \sum_{B(g)} n_B \ln \frac{n_B}{n_{total}} \qquad \text{(Eq 7.12.11)}$$

$$= RT \sum_{B(g)} \{(n_B)_0 + \nu_B\, \xi\} \ln \frac{\{(n_B)_0 + \nu_B\, \xi\}}{n_{total}}$$

$$= RT\left[(1\text{ mol} - \xi)\ln\left(\frac{1\text{ mol} - \xi}{1\text{ mol} + \xi}\right) + 2\xi \ln\left(\frac{2\xi}{1\text{ mol} + \xi}\right)\right] \qquad (7.12.19)$$

Expression of $\Delta_r\tilde{G}$

$$\left(\frac{\partial G}{\partial \xi}\right)_{T,p} = \sum_B \nu_B\, \mu_B^\circ + \left(\sum_{B(g)} \nu_B\right) RT \ln \frac{p}{p^\circ} + RT \sum_{B(g)} \nu_B \ln x_B \qquad \text{(Eq 7.12.13)}$$

$$= \sum_B \nu_B\, \mu_B^\circ + \left(\sum_{B(g)} \nu_B\right) RT \ln \frac{p}{p^\circ} + RT \sum_{B(g)} \nu_B \ln \frac{(n_B)_0 + \nu_B\, \xi}{n_{total}}$$

$$= \left(-\mu^\circ_{N_2O_4} + 2\,\mu^\circ_{NO_2}\right) + RT \ln \frac{p}{p^\circ}$$

$$+ RT\left[-\ln\left(\frac{1\text{ mol} - \xi}{1\text{ mol} + \xi}\right) + 2\ln\left(\frac{2\xi}{1\text{ mol} + \xi}\right)\right]$$

$$= \left(2\,\mu^\circ_{NO_2} - \mu^\circ_{N_2O_4}\right) + RT \ln \frac{p}{p^\circ} + RT \ln\left(\frac{4\xi^2}{1\text{ mol}^2 - \xi^2}\right) \qquad (7.12.20)$$

Computation of G_{pure}, $\Delta_{mix}G$ and Reaction Potential

Writing the above expressions in terms of dimensionless parameter $\xi' = \xi/\text{mol}$, we get

$$\frac{G_{pure}}{\text{mol}} = \mu^\circ_{N_2O_4} + \xi'\left(2\,\mu^\circ_{N_2} - \mu^\circ_{N_2O_4}\right) + RT(1+\xi') \ln \frac{p}{p^\circ} \qquad (7.12.21)$$

$$\frac{\Delta_{mix}G}{\text{mol}} = RT\left[(1-\xi')\ln\left(\frac{1-\xi'}{1+\xi'}\right) + 2\xi' \ln\left(\frac{2\xi'}{1+\xi'}\right)\right] \qquad (7.12.22)$$

$$\left(\frac{\partial G}{\partial \xi}\right)_{T,p} = \left(2\mu^\circ_{NO_2} - \mu^\circ_{N_2O_4}\right) + RT \ln \frac{p}{p^\circ} + RT \ln\left(\frac{4\xi'}{1-\xi'^2}\right) \qquad (7.12.23)$$

From the table of standard free energy of formation we find that

$$\mu^\circ_{NO_2(g)} = 51.31 \text{ kJ mol}^{-1} \qquad \text{and} \qquad \mu^\circ_{N_2O_4(g)} = 97.89 \text{ kJ mol}^{-1}$$

With these, the values of G_{pure}, $\Delta_{mix}G$, G_{total}, and $(\partial G/\partial \xi)_{T,\,p}$ at 298 K and 1 bar pressure for different values of ξ' are recorded in Table 7.12.1.

Table 7.12.1 Values of G_{pure}, $\Delta_{mix}G$, G_{total} and $(\partial G/\partial \xi)_{T,p}$ for the reaction $N_2O_4(g) \rightleftharpoons 2NO_2(g)$

ξ'	0.0	0.1	0.2	0.3	0.4	0.5	0.6	0.7	0.8	0.9	1.0
G_{pure}/kJ	97.89	98.36	98.84	99.31	99.78	100.26	100.73	101.21	101.67	102.15	102.62
$\Delta_{mix}G$/kJ	0	−1.29	−1.89	−2.22	−2.37	−2.37	−2.23	−1.96	−1.56	−0.97	0
G_{total}/kJ	97.89	97.07	96.95	97.09	97.41	97.89	98.50	99.25	100.18	101.18	102.62
$\dfrac{(\partial G/\partial \xi)_{T,p}}{\text{kJ mol}^{-1}}$	$-\infty$	−3.22	0.29	2.43	4.06	5.44	6.74	8.07	9.59	11.76	∞
	... spontaneous....		↑ ....nonspontaneous........ equilibrium $\xi'_{eq} = 0.19$								

Graphical Representations

When the values recorded in Table 7.12.1 are plotted against ξ', we get three curves as shown in Fig. 7.12.1. The curve of G_{pure} versus ξ' shows minimum at $\xi = 0$ (pure dimer) and that $\Delta_{mix}G$ versus ξ' shows at $\xi' = 0.5$. Thus, the intrinsic molar free energy of unmixed components favours pure dimer whereas the free energy of mixing favours at 50–50 mixture. The true equilibrium point represents a balance between these two trends and lies at 19.0% dissociation of N_2O_4 as is shown by the minimum of G_{total} versus ξ curve.

Summary of Fig. 7.12.1

Figure 7.12.1 represents, in fact, a typical variation of G_{total} with ξ of a reaction. This graphical representation is referred to as the *Gibbs valley*. In general, ξ varies from ξ_{min} to ξ_{max}.

- At ξ_{min}, one (or more) of the products has been exhausted while at ξ_{max} one or more) of the reactants has been exhausted.
- At some intermediate values, ξ_{eq}, G_{total} passes through a minimum. The values of ξ_{eq} is known as the equilibrium value of extent of reaction.
- To the left of the minimum the slope $\partial G_{total}/\partial \xi$ of the curve is negative, indicating spontaneity in the forward direction, while to right of the minimum, the slope of the curve is positive, indicating spontaneity in the reverse direction.
- Before the equilibrium, increase in the value of G_{pure} with increase in ξ is smaller than the corresponding decrease in $\Delta_{mix}G$, thus making G_{total} smaller and smaller.
- After the equilibrium, increase in G_{pure} is larger than the corresponding decrease in the value of $\Delta_{mix}G$, making G_{total} larger and larger.
- At equilibrium the increase in the value of G_{pure} is just equal to the decrease in $\Delta_{mix}G$: This can be expressed mathematically by differentiating Eq. (7.12.9) by dξ at constant T and p. This gives

$$\left(\frac{\partial G_{total}}{\partial \xi}\right)_{T,p} = \left(\frac{\partial G_{pure}}{\partial \xi}\right)_{T,p} + \left(\frac{\partial(\Delta_{mix}G)}{\partial \xi}\right)_{T,p}$$

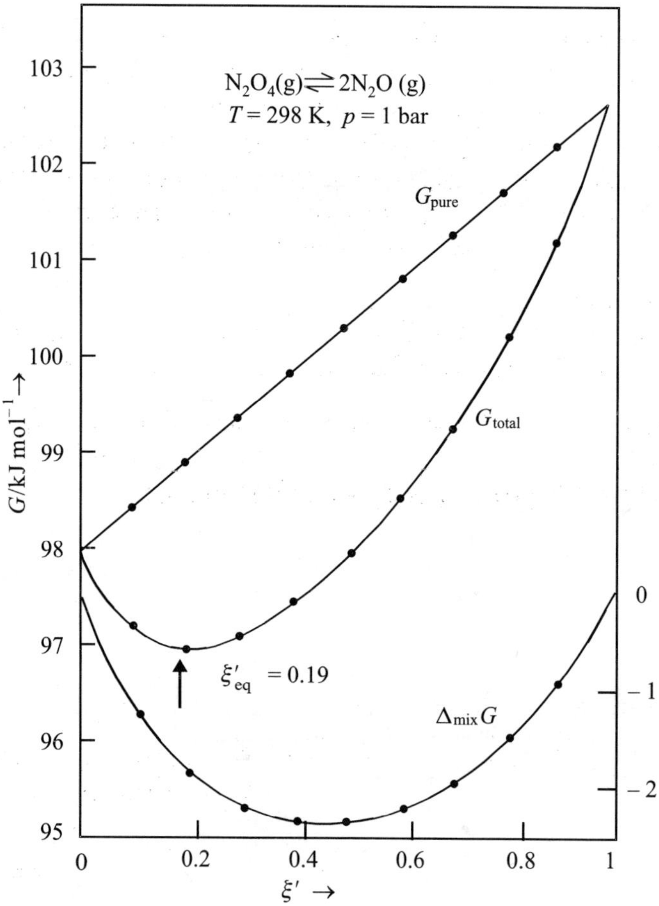

Fig. 7.12.1 Plot of G_{pure}, $\Delta_{mix}G$ and G_{total} versus ξ (Gibbs Valley)

At equilibrium G_{total} is minimum, thus $(\partial G_{total}/\partial \xi)_{T,p} = 0$, therefore, it follows that

$$\left(\frac{\partial G_{pure}}{\partial \xi}\right)_{T,p} = -\left(\frac{\partial(\Delta_{mix} G)}{\partial \xi}\right)_{T,p} \tag{7.12.24}$$

that is, the minimum in G_{total} occurs at the point where $\Delta_{mix}G$ decreases as rapidly as G_{pure} increases. Before the equilibrium

$$\left(\frac{\partial G_{pure}}{\partial \xi}\right)_{T,p} < -\left(\frac{\partial(\Delta_{mix} G)}{\partial \xi}\right)_{T,p} \tag{7.12.25}$$

and beyond the equilibrium

$$\left(\frac{\partial G_{pure}}{\partial \xi}\right)_{T,p} > -\left(\frac{\partial(\Delta_{mix} G)}{\partial \xi}\right)_{T,p} \tag{7.12.26}$$

The above facts have been illustrated in Table 7.12.1.

Computation of Equilibrium Constant

Equation (7.12.23), in fact, is

$$\left(\frac{\partial G}{\partial \xi}\right)_{T,p} = \Delta_r G^\circ + RT \ln Q_p^\circ$$

where $$Q_p^\circ = \left(\frac{4\xi'^2}{1-\xi'^2}\right)\left(\frac{p}{p^\circ}\right)$$

The value of equilibrium constant can be computed from $\Delta_r G^\circ$ or by substituting $\xi'_{eq} = 0.19$ in the expression of Q_p°. Thus,

$$K_p^\circ = \frac{4(0.19)^2}{1-(0.19)^2} = 0.150 \qquad \text{(since } p = p^\circ\text{)}$$

Alternatively, we have

$$K_p^\circ = \exp(-\Delta_r G^\circ/RT) = \exp(-4.73/(8.314 \times 10^{-3} \times 298))$$
$$= 0.148$$

Effect of Pressure on Equilibrium

Equations (7.12.21) – (7.12.23) may also be used to study the effect of pressure on the equilibrium value of extent of reaction (ξ'_{eq}). Figure 7.12.2 displays a few plots of G_{pure} and G_{total} along the value of ξ'_{eq} at different pressures. The value of ξ'_{eq} increases with decrease in pressure. This fact is in agreement with Le Chatelier's principle.

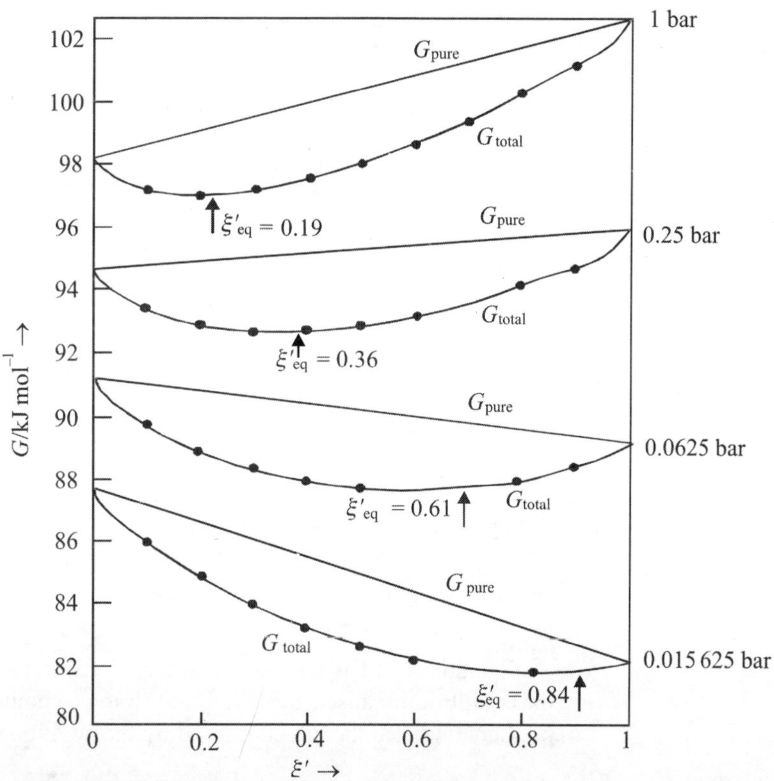

Fig. 7.12.2 Plots of G_{pure} and G_{total} at different pressures

Effect of Temperature on Equilibrium

Equations (7.12.21)–(7.12.23) may also be used to study the effect of temperature on the equilibrium value of extent of reaction and the equilibrium constant of the reaction.

The values of $\mu^\circ_{N_2O_4}$ and $\mu^\circ_{NO_2}$ at different temperatures may be computed using Gibbs-Helmholtz equation. The required data are given in Table 7.12.2.

Table 7.12.2 Data to compute $\mu^\circ_{N_2O_4}$ and $\mu^\circ_{N_2}$ at temperatures other than 298 K

Substance	C_p/J K^{-1} mol^{-1} = $a + b(T/K)^2 + c\,(T/K)^2$			$\Delta_f H^\circ_{298K}$	$\Delta_f G^\circ_{298K}$
	a	$b \times 10^3$	$c \times 10^6$	kJ mol^{-1}	kJ mol^{-1}
$N_2(g)$	29.295	–2.2634	5.6649	0	0
$O_2(g)$	26.648	9.55	–1.328	0	0
$NO_2(g)$	25.036	45.550	–18.420	33.18	51.31
$N_2O_4(g)$	36.208	160.98	–77.984	9.16	97.89

Equations to be used are

$$\Delta_f G^\circ_T = -(\Delta_r a)T \ln\left(\frac{T}{K}\right) - \frac{\Delta_r b}{2}T^2 - \frac{\Delta_r c}{6}T^3 + K_1 + K_2 T \quad \text{(Eq. 5.8.10)}$$

where

$$K_1 = \Delta_r H^\circ_{T_0} - \ln\left[(\Delta_r a)T_0 + (\Delta_r b)\frac{T_0^2}{2} + (\Delta_r c)\frac{T_0^3}{3}\right]$$

$$K_2 = \frac{\Delta_r G^\circ_{T_0}}{T_0} + (\Delta_r a)\ln\left(\frac{T_0}{K}\right) + \frac{\Delta_r b}{2}T_0 + \frac{\Delta_r c}{6}T_0^2 - \frac{K_1}{T_0}$$

$$T_0 = 298 \text{ K}$$

The computed values of $\mu^\circ_{NO_2}$ and $\mu^\circ_{N_2O_4}$ are as follows.

T/K	$\mu^\circ_{NO_2}$/kJ mol^{-1}	$\mu^\circ_{N_2O_4}$/kJ mol^{-1}
318	52.53	103.85
338	53.76	109.82
358	54.99	115.81

With the above data, the obtained values of G_{pure} and G_{total} are plotted in Fig. 7.12.3 at 1 bar pressure. The extent of reaction at equilibrium increases with increase in temperature. This fact is also in agreement with Le Chatelier's principle as the reaction is endothermic.

Equations (7.12.10), (7.12.11) and (7.12.13) can also be used to study the effect on equilibrium caused by taking any initial amounts of N_2O_4 and NO_2 at any pressure and also the effects caused by the addition of inert gas.

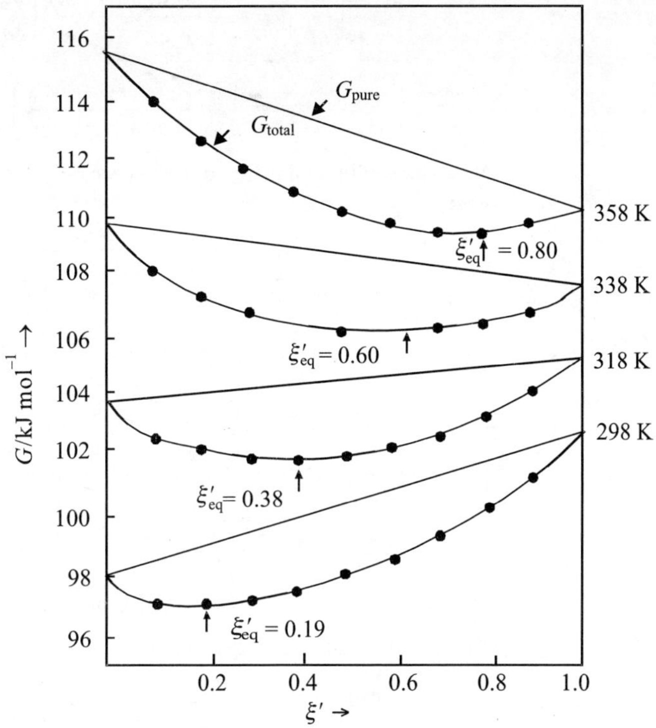

Fig. 7.12.3 Plot of G_{pure} and G_{total} at different temperatures

Problem 7.12.1

(a) The extent of reaction at equilibrium (ξ_{eq}) for the decomposition of 1 mol of N_2O_4, according to the reaction

$$N_2O_4 \rightleftharpoons 2NO_2$$

is a function of the pressure. Show that if the mixture remains in equilibrium as the pressure is changed, the apparent isothermal compressibility (κ_T) is given by

$$\kappa_T = -\frac{1}{V}\left(\frac{\partial V}{\partial p}\right)_T = (1/p)\left[1+\frac{1}{2}\xi_{eq}\ (1\ \text{mol} - \xi_{eq})\right]$$

(b) Show that κ_T has a maximum value at $\xi_{eq} = 0.5$ mol and also show that at this maximum $p = (3/4)\ K_p$.

Solution

(a) We have

$$\underset{1\ \text{mol}-\xi_{eq}}{N_2O_4} \rightleftharpoons \underset{2\xi_{eq}}{2NO_2}$$

Total amount of gases at equilibrium = 1 mol + ξ_{eq}
Now, from ideal gas equation, we have

$$pV = (1\ \text{mol} + \xi_{eq})\ RT$$

Hence $$\left(\frac{\partial V}{\partial p}\right)_T = -\frac{(1\ \text{mol} + \xi_{eq})RT}{p^2} + \frac{RT}{p}\left(\frac{\partial \xi_{eq}}{\partial p}\right)_T$$

Thus
$$-\frac{1}{V}\left(\frac{\partial V}{\partial p}\right)_T = \frac{(1\,\text{mol}+\xi_{\text{eq}})RT}{p^2V} - \frac{RT}{pV}\left(\frac{\partial \xi_{\text{eq}}}{\partial p}\right)_T$$
$$= \frac{1}{p} - \frac{1}{(1\,\text{mol}+\xi_{\text{eq}})}\left(\frac{\partial \xi_{\text{eq}}}{\partial p}\right)_T \qquad (7.12.27)$$

Now, the equilibrium constant of the decomposition reaction is given by

$$K_p = \frac{(p_{\text{NO}_2})^2}{p_{\text{N}_2\text{O}_4}} = \frac{\{2\xi_{\text{eq}}/(1\,\text{mol}+\xi_{\text{eq}})\}^2\,p^2}{\{(1\,\text{mol}-\xi_{\text{eq}})/(1\,\text{mol}+\xi_{\text{eq}})\}\,p}$$
$$= \frac{4\xi_{\text{eq}}^2}{1\,\text{mol}^2-\xi_{\text{eq}}^2}\,p \qquad (7.12.28)$$

or $\qquad K_p\,(1\,\text{mol}^2 - \xi^2_{\text{eq}}) = 4\,\xi^2_{\text{eq}}p$

Differentiating the above expression with respect to p at constant T, we get

$$K_p\left\{-2\xi_{\text{eq}}\left(\frac{\partial \xi_{\text{eq}}}{\partial p}\right)_T\right\} = 4\xi_{\text{eq}}^2 + 8\xi_{\text{eq}}p\left(\frac{\partial \xi_{\text{eq}}}{\partial p}\right)_T \qquad (7.12.29)$$

[The term $(\partial K_p/\partial p)_T = 0$ as K_p is independent of p.]

Rearranging Eq. (7.12.29), we get

$$\left(\frac{\partial \xi_{\text{eq}}}{\partial p}\right)_T = -\frac{4\xi_{\text{eq}}^2}{2\xi_{\text{eq}}\,K_p + 8\xi_{\text{eq}}\,p}$$

Substituting the expression of K_p from Eq. (7.12.28), we get

$$\left(\frac{\partial \xi_{\text{eq}}}{\partial p}\right)_T = -\frac{\xi_{\text{eq}}(1\,\text{mol}^2-\xi_{\text{eq}}^2)}{2p} \qquad (7.12.30)$$

Substituting Eq. (7.12.30) in Eq. (7.12.27), we get

$$-\frac{1}{V}\left(\frac{\partial V}{\partial p}\right)_T = \frac{1}{p} + \frac{1}{(1\,\text{mol}+\xi_{\text{eq}})}\frac{\xi_{\text{eq}}(1\,\text{mol}^2-\xi_{\text{eq}}^2)}{2\,p}$$
$$\kappa_T = \frac{1}{p}\left[1 + \frac{\xi_{\text{eq}}}{2}(1\,\text{mol}-\xi_{\text{eq}})\right]$$

(b) To show that κ_T has a maximum value at $\xi_{\text{eq}} = 0.5$, we set $(\partial\kappa_T/\partial\xi_{\text{eq}})_p = 0$. Thus, we have

$$\left(\frac{\partial \kappa_T}{\partial \xi_{\text{eq}}}\right)_p = \frac{1}{p}\left[\frac{1}{2}(1\,\text{mol}-\xi_{\text{eq}}) + \frac{\xi_{\text{eq}}}{2}(-1)\right] = 0$$

which gives $\xi_{\text{eq}} = (1/2)$ mol. Substituting this value in the expression of K_p, we get

$$K_p = \frac{4\xi_{\text{eq}}^2}{1\,\text{mol}^2-\xi_{\text{eq}}^2}\,p = \frac{4}{3}p \quad\Rightarrow\quad p = \frac{3}{4}K_p$$

Problem 7.12.2 Consider the reaction $H_2(g) + I_2(g) \rightleftharpoons 2HI(g)$.

(a) If there are one mole of H_2, one mole of I_2 and zero mole of HI present before the reaction advanced, express the free energy change for the reaction in terms of extent of reaction.

(b) What form would the expression G have if iodine were present as solid?

Solution **(a)** The reaction is

	$H_2(g)$	+	$I_2(g)$	$\rightleftharpoons$	$2HI(g)$	
At $t = 0$	1 mol		1 mol		0 mol	
At t_{eq}	1 mol $- \xi'$		1 mol $- \xi'$		$2\,\xi'$	Total amount = 2 mol

If we represent ξ'/mol by ξ, then the total value of G of the system is given by

$$\begin{aligned}
G/\text{mol} &= (1-\xi)\,\mu_{H_2(g)} + (1-\xi)\,\mu_{I_2(g)} + 2\xi\mu_{HI(g)} \\
&= (1-\xi)\,\{\mu^\circ_{H_2(g)} + RT\ln(p_{H_2}/p^\circ)\} + (1-\xi)\,\{\mu^\circ_{I_2(g)} + RT\ln(p_{I_2}/p^\circ)\} \\
&\qquad + 2\xi\,\{\mu^\circ_{HI(g)} + RT\ln(p_{HI}/p^\circ)\} \\
&= (1-\xi)\,\{\mu^\circ_{H_2(g)} + RT\ln x_{H_2} + RT\ln(p/p^\circ)\} \\
&\qquad + (1-\xi)\,\{\mu^\circ_{I_2(g)} + RT\ln x_{I_2} + RT\ln(p/p^\circ)\} \\
&\qquad + (2\xi)\,\{\mu^\circ_{HI(g)} + RT\ln x_{HI} + RT\ln(p/p^\circ)\} \\
&= \mu^\circ_{H_2(g)} + \mu^\circ_{I_2(g)} + \xi\,(2\mu^\circ_{HI(g)} - \mu^\circ_{H_2(g)} - \mu^\circ_{I_2(g)}) + 2RT\ln(p/p^\circ) \\
&\qquad + RT\,[(1-\xi)\ln x_{H_2} + (1-\xi)\ln x_{I_2} + 2\xi\ln x_{HI}] \\
&= \mu^\circ_{H_2(g)} + \mu^\circ_{I_2(g)} + \xi\,\Delta_r\tilde{G}^\circ + 2RT\ln(p/p^\circ) \\
&\qquad + RT\left[(1-\xi)\ln\frac{1}{2}(1-\xi) + (1-\xi)\ln\frac{1}{2}(1-\xi) + 2\xi\ln\xi\right] \\
&= \mu^\circ_{H_2(g)} + \mu^\circ_{I_2(g)} + \xi\,\Delta_r G^\circ + 2RT\ln(p/p^\circ) \\
&\qquad + 2RT\left[(1-\xi)\ln\frac{1}{2}(1-\xi) + \xi\ln\xi\right]
\end{aligned}$$

(b)

$$\begin{aligned}
G/\text{mol} &= (1-\xi)\,\mu_{H_2(g)} + (1-\xi)\mu_{I_2(s)} + 2\xi\,\mu^\circ_{HI(g)} \\
&= (1-\xi)\,\{\mu^\circ_{H_2(g)} + RT\ln(p_{H_2}/p^\circ)\} + (1-\xi)\,\mu^\circ_{I_2(s)} \\
&\qquad + 2\xi\,\{(\mu^\circ_{HI(g)} + RT\ln(p_{HI(g)}/p^\circ)\} \\
&= \mu^\circ_{H_2(g)} + \mu^\circ_{I_2(s)} + \xi\,(2\mu^\circ_{HI(g)} - \mu^\circ_{H_2(g)} - \mu^\circ_{I_2(s)}) \\
&\qquad + (1-\xi)\,RT\ln(p_{H_2}/p^\circ) + 2\xi\,RT\ln(p_{HI(g)}/p^\circ) \\
&= \mu^\circ_{H_2(g)} + \mu^\circ_{I_2(s)} + \xi\,\Delta_r G^\circ + RT\,(1-\xi)\,[\ln x_{H_2} + \ln(p/p^\circ)] \\
&\qquad + 2\,\xi\,RT\,[\ln x_{HI} + \ln(p/p^\circ)] \\
&= \mu^\circ_{H_2(g)} + \mu^\circ_{I_2(s)} + \xi\,\Delta_r G^\circ + (1+\xi)\,RT\ln(p/p^\circ) \\
&\qquad + RT\left[(1-\xi)\ln\frac{(1-\xi)}{(1+\xi)} + 2\xi\ln\left(\frac{2\xi}{1+\xi}\right)\right] \\
&= \mu^\circ_{H_2(g)} + \mu^\circ_{I_2(s)} + \xi\,\Delta_r G^\circ + (1+\xi)RT\ln(p/p^\circ) \\
&\qquad + RT\,[(1-\xi)\ln(1-\xi) + 2\xi\ln(2\xi) - (1+\xi)\ln(1+\xi)]
\end{aligned}$$

7.13 CHARACTERISTICS OF HOMOGENEOUS GASEOUS REACTIONS

In this section, we consider a few typical characteristics of homogeneous gaseous reactions having zero, positive and negative values of Δv_g.

REACTIONS WITH NO CHANGE IN GASEOUS SPECIES

Consider the reaction

$$A(g) + B(g) \rightleftharpoons C(g) + D(g) \tag{7.13.1}$$

Thus
$$K_p = \frac{p_C \times p_D}{p_A \times p_B} \tag{7.13.2}$$

Knowing the value of K_p and the initial amounts of A and B, the amounts of various species at equilibrium can be calculated as follows:

	A +	B ⇌	C +	D	Total amounts
At $t = 0$	a	b	0	0	$a + b$
At equilibrium	$a - \xi$	$b - \xi$	ξ	ξ	$a + b$
Partial pressures	$\frac{a-\xi}{a+b}p$	$\frac{b-\xi}{a+b}p$	$\frac{\xi}{a+b}p$	$\frac{\xi}{a+b}p$	

We have
$$K_p = \frac{\{\xi p/(a+b)\}^2}{\left(\frac{a-\xi}{a+b}p\right)\left(\frac{b-\xi}{a+b}p\right)} = \frac{\xi^2}{(a-\xi)(b-\xi)} \tag{7.13.3}$$

This quadratic equation can be solved for ξ. One of the values of ξ will determine the equilibrium composition of the reaction and the other will have an unreasonable value, which must be ignored.

Some Conclusions From Eqs. (7.13.2) and (7.13.3), the following conclusions can be drawn:

- Since $\Delta v_g = 0$, therefore
$$K_p = K_c = K_x = K_n \tag{7.13.4}$$
- If we have a set of equilibrium partial pressures, multiplication of each of the partial pressures by the same factor does not change the value of K_p, i.e. the relative amounts of components present at equilibrium are independent of the total pressure or volume of the system.
- The relative amounts of the components present at equilibrium remain unaltered by the addition of an inert gas.
- K_p is independent of the units in which pressures or concentrations are expressed.
- Addition of any extra gas A, B, C or D, when the reaction is at equilibrium, will change the relative amounts of various species without changing the value of K_p. Such changes can be calculated as follows:

Let the amount n of the gas A be introduced at equilibrium. The system will readjust its equilibrium, and let the amount ξ' of gas A be consumed at equilibrium. The amount of various species at equilibrium will be

$$\begin{matrix} A & + & B & \rightleftharpoons & C & + & D \\ a+n-\xi' & & b-\xi' & & \xi' & & \xi' \end{matrix}$$

Thus $$K_p = \frac{\xi'^2}{(a+n-\xi')(b-\xi')} \tag{7.13.5}$$

Since now the denominator is greater than in the previous case, ξ' will be greater than ξ in order to keep the value of K_p same. Thus, more of C and D will be formed. This is also in agreement with the Le Chatelier principle. If at equilibrium, the gas C or D is introduced, then ξ' will decrease, i.e. lesser of A and B will combine to give C and D.

Temperature Effect Effect of temperature on the equilibrium will depend upon the nature of the reaction—whether it is exothermic or endothermic:

Exothermic reaction Equilibrium will be shifted to the left, i.e. lesser amounts of products will be produced.

Endothermic reaction Equilibrium will be shifted to the right, i.e. larger amounts of products will be produced.

Example $H_2(g) + CO_2(g) \rightleftharpoons H_2O(g) + CO(g)$

REACTIONS WITH INCREASE IN GASEOUS SPECIES

Consider the relation

$$A \rightleftharpoons B + C \tag{7.13.6}$$

Thus, $$K_p = \frac{p_B \times p_C}{p_A} \tag{7.13.7}$$

Knowing the value of K_p and the initial amount of A, the amounts of A, B and C corresponding to the equilibrium at a given pressure can be calculated as follows:

	A ⇌	B +	C	Total amount
At $t = 0$	a	0	0	a
At equilibrium	$a - \xi$	ξ	ξ	$a + \xi$
Partial pressures	$\frac{a-\xi}{a+\xi}p$	$\frac{\xi}{a+\xi}p$	$\frac{\xi}{a+\xi}p$	

We have $$K_p = \frac{\left(\frac{\xi}{a+\xi}p\right)^2}{\left(\frac{a-\xi}{a+\xi}p\right)} = \frac{\xi^2 p}{(a-\xi)(a+\xi)} = \frac{\xi^2}{a^2-\xi^2}p \tag{7.13.8}$$

Some Conclusions Although K_p is independent of p for reactions of ideal gases, p appears on the right hand side of the above equation. This fact has the following consequences.

- At a given temperature, the composition of the system at equilibrium depends on the value of p. Since K_p has to be independent of pressure, it follows that if p increases ξ has to decrease, i.e. lesser of A will be dissociated. This also follows from the Le Chatelier principle.

- The numerical value of K_p depends on the units of pressure but the standard equilibrium constant K_p° is fundamentally dimensionless. This follows from the fact that in the expression

$$\mu_i = \mu_i^\circ + RT \ln (p_i/p^\circ)$$

the ratio of partial pressure to the standard-state pressure is involved. By convention, K_p is given the units of p.

- Since $\Delta\nu_g \neq 0$, therefore

$$K_p \neq K_c \neq K_x \neq K_n \qquad (7.13.9)$$

- On the basis of Le Chatelier principle, it can be predicted that the dissociation of a compound is partially suppressed, when one of the dissociation products is initially present. Let the amount b of B be already present; then the amount of A dissociated can be calculated as follows:

	A	$\rightleftharpoons$	B	+	C	Total amounts
At $t = 0$	a		b		0	$a + b$
At equilibrium	$a - \xi'$		$b + \xi'$		ξ'	$a + b + \xi'$
Partial pressures	$\frac{a-\xi'}{a+b+\xi'}p$		$\frac{b+\xi'}{a+b+\xi'}p$		$\frac{\xi'}{a+b+\xi'}p$	

Thus

$$K_p = \frac{(b+\xi')(\xi')}{(a-\xi')(a+b+\xi')}p \qquad (7.13.10)$$

Solving this quadratic equation in ξ' will give the amount of A dissociated.

Equilibrium Constant in terms of Degree of Dissociation

The equilibrium constant of this type of reactions can also be written in terms of degree of dissociation. The latter gives the fraction of the reactant dissociated, i.e. out of 1 mol of the reactant, how much of the reactant is present in the dissociated form. Thus, if we have 1 mol of A to start with and at equilibrium if α is the degree of dissociation, then

	A	$\rightleftharpoons$	B	+	C
Fraction in the beginning	1		0		0
Fraction at equilibrium	$1 - \alpha$		α		α

If a is the original amount of A, we will have

	A	$\rightleftharpoons$	B	+	C
In the beginning	a		0		0
At equilibrium	$a(1 - \alpha)$		$a\alpha$		$a\alpha$

Total amount at equilibrium = $a(1 - \alpha) + a\alpha + a\alpha = a\,(1 + \alpha)$

Deriving the expressions of partial pressure and then substituting in the expression of K_p would give

$$K_p = \frac{p_B\, p_C}{p_A} = \frac{\left\{\frac{a\alpha}{a(1+\alpha)}p\right\}\left\{\frac{a\alpha}{a(1+\alpha)}p\right\}}{\left\{\frac{a(1-\alpha)}{a(1+\alpha)}p\right\}} = \frac{\alpha^2}{1-\alpha^2}p \qquad (7.13.11)$$

Knowing K_p and p, we can solve Eq. (7.13.11) for α.

Relation Between Extent of Reaction and Degree of Dissociation

For the reaction

	A	$\rightleftharpoons$	B	+	C
Amount involving extent of reaction	$a - \xi$		ξ		ξ
Degree of dissociation	$a(1 - \alpha)$		$a\alpha$		$a\alpha$

On comparing the amount, it follows that

$$\xi = a\alpha$$

Example Decomposition of NH_3

$$2NH_3(g) \rightleftharpoons N_2(g) + 3H_2(g)$$
$$\Delta\nu_g = +1$$

REACTIONS WITH DECREASE IN NUMBER OF GASEOUS SPECIES

Consider the reaction of Eq. (7.13.6) written in the reverse direction

$$B + C \rightleftharpoons A$$

For this reaction, we will have

$$K'_p = \frac{p_A}{p_B p_C} = \frac{1}{K_p} \qquad (7.13.12)$$

where K_p is defined in Eq. (7.13.7).

Thus K'_p has the characteristics which are inverse of characteristics of K_p. For example, if p is increased then the amount of A produced from B and C is increased.

Example Formation of $NH_3(g)$ from $N_2(g)$ and $H_2(g)$

$$(1/2)\, N_2(g) + (3/2)\, H_2(g) \rightleftharpoons NH_3(g)$$
$$\Delta\nu_g = -1$$

Illustrations

The following three examples illustrate the kind of effects discussed above.

Example 7.13.1

Two moles of PCl_5 are heated at 502 K till equilibrium is reached at a total pressure of 101.325 kPa. Calculate the composition at equilibrium and also the percentage of PCl_5 decomposed ($K_p^\circ = 0.466\,1$).

Solution

PCl_5 decomposes on heating according to the reaction

$$PCl_5(g) \rightleftharpoons PCl_3(g) + Cl_2(g)$$

The expression of K_p°, as given by Eq. (7.13.8) is

$$K_p^\circ = \frac{\xi^2}{a^2 - \xi^2}(p/1\,\text{bar})$$

Substituting the given value of p and K_p°, we get

$$\frac{\xi^2}{4\,\text{mol}^2 - \xi^2} = \frac{0.466\,1}{1.013\,25} = 0.46$$

Solving the quadratic equation for ξ, we have

$$\xi = 1.12\ \text{mol}$$

Amount of PCl_5 = 2.0 mol – 1.12 mol = 0.88 mol

Amount of PCl_3 or Cl_2 = 1.12 mol

$$\text{Percentage of } PCl_5 \text{ decomposed} = \frac{1.12\ \text{mol}}{2\ \text{mol}} \times 100 = 56$$

Example 7.13.2 If pressure p of the system given in Example 7.13.1 is raised to 1 013.25 kPa at the same temperature, calculate the percentage of PCl_5 decomposed.

Solution Substituting p = 1 013.25 kPa = 10.132 5 bar in the expression of K_p° we get

$$K_p^\circ = \frac{\xi^2}{4\ \text{mol}^2 - \xi^2} 10.132\,5 = 0.466\,1$$

Solving the quadratic expression for ξ, we get

$$\xi = 0.42\ \text{mol}$$

Thus Percentage of PCl_5 decomposed $= \dfrac{0.42\ \text{mol}}{2\ \text{mol}} \times 100 = 21$

Example 7.13.3 If the vessel of the system given in Example 7.13.1 already contains 1 mol of Cl_2, calculate per cent of PCl_5 decomposed. The equilibrium pressure is 101.325 kPa.

Solution In the present case, the amounts of various species in the beginning and at equilibrium will be

	$PCl_5(g)$	$\rightleftharpoons$	$PCl_3(g)$	+	$Cl_2(g)$
To start with	2 mol		0		1 mol
At equilibrium	2 mol – ξ		ξ		1 mol + ξ

Total amount of gases at equilibrium = 3 mol + ξ

The partial pressures are

$$p_{PCl_5} = \frac{2\ \text{mol} - \xi}{3\ \text{mol} + \xi}p; \quad p_{PCl_3} = \frac{\xi}{3\ \text{mol} + \xi}p; \quad p_{Cl_2} = \frac{1\ \text{mol} + \xi}{3\ \text{mol} + \xi}p$$

Substituting these in the standard equilibrium constant expression

$$K_p^\circ = \frac{(p_{PCl_3}/p^\circ)(p_{Cl_2}/p^\circ)}{(p_{PCl_5}/p^\circ)}$$

we get
$$K_p^\circ = \frac{\xi(1\ \text{mol} + \xi)}{(2\ \text{mol} - \xi)(3\ \text{mol} + \xi)}(p/p^\circ)$$

Since p = 101.325 kPa, therefore

$$\frac{\xi(1\ \text{mol} + \xi)}{(2\ \text{mol} - \xi)(3\ \text{mol} + \xi)} = \frac{0.466\,1}{1.013\,25} = 0.46$$

Solving for ξ, we get

$$\xi = 0.96\ \text{mol}$$

Thus Percentage of PCl_5 decomposed $= \dfrac{0.96\ \text{mol}}{2\ \text{mol}} \times 100 = 48$

Example 7.13.4 0.10 mole each of SO_2 and SO_3 are mixed in a 2.0 dm^3 flask at 300 K. Equilibrium is attained as

$$2SO_2(g) + O_2(g) \rightleftharpoons 2SO_3(g)$$

The equilibrium pressure is 281.68 kPa. Calculate (a) the mole fraction O_2 at equilibrium, (b) K_p° and K_c° and (c) the percentage dissociation of SO_3 if initially the flask contained 0.1 mol of SO_3 and none of O_2 or SO_2.

Solution **(a)** Let ξ be the amount of O_2 that is formed at equilibrium. We will have

	$2SO_2$	+ O_2	$\rightleftharpoons$ $2SO_3$	
$t = 0$	0.1 mol	0	0.1 mol	Total amount at equilibrium
t_{eq}	0.1 mol + 2ξ	ξ	0.1 mol − 2ξ	0.2 mol + ξ

$V = 2\ dm^3$

Total amount of gases at equilibrium can be calculated by using ideal gas equation

$$pV = nRT$$

where $T = 300\ K,\quad p = 281.68\ kPa,\quad$ and $\quad V = 2\ dm^3$

Thus
$$n = \frac{pV}{RT} = \frac{(281.68\ kPa)\ (2\ dm^3)}{(8.314\ kPa\ dm^3\ mol^{-1}\ K^{-1})\ (300\ K)} = 0.226\ mol$$

Hence
$$0.2\ mol + \xi = 0.226\ mol$$
$$\xi = 0.026\ mol$$
$$\text{Mole fraction of oxygen} = \frac{0.026\ mol}{0.226\ mol} = 0.1151$$

(b) The expression of K_p° for the given reaction is

$$K_p^\circ = \frac{(p_{SO_3}/p^\circ)^2}{(p_{SO_2}/p^\circ)^2\ (p_{O_2}/p^\circ)} = \frac{(x_{SO_3}p/p^\circ)^2}{(x_{SO_2}p/p^\circ)^2(x_{O_2}p/p^\circ)} = \frac{(x_{SO_3})^2}{(x_{SO_2})^2(x_{O_2})}\left(\frac{p^\circ}{p}\right)$$

where *x*s are the mole fractions. Evaluating these *x*s and substituting in the above expression, we get

$$K_p^\circ = \frac{(0.048/0.226)^2}{(0.152/0.226)^2(0.026/0.226)}\left(\frac{100\ kPa}{281.68\ kPa}\right)$$

$$= \frac{(0.048)^2(0.226)}{(0.152)^2\ (0.026)}\left(\frac{100}{281.68}\right)$$

$$= 0.308$$

Now
$$K_p^\circ = K_c^\circ\left(\frac{c^\circ RT}{p^\circ}\right)^{\Delta\nu_g} \quad \text{or} \quad K_c^\circ = K_p^\circ\left(\frac{c^\circ RT}{p^\circ}\right)^{-\Delta\nu_g}$$

$\Delta\nu_g$ for the reaction is −1. Thus

$$K_c^\circ = 0.308 \times \frac{(1\ mol\ dm^{-3})\ (8.314\ kPa\ dm^3\ K^{-1}\ mol^{-1})\ (300\ K)}{(100\ kPa)}$$

$$= 7.68$$

(c) We are given with the following data.

$$2SO_2 + O_2 \rightleftharpoons 2SO_3$$

At $t = 0$ 0.1 mol

Let x be the amount of O_2 formed at equilibrium, then the amounts of various species are

Amount of SO_3 = 0.1 mol − $2x$

Amount of SO_2 = $2x$

Amount of O_2 = x

The amount of gases = $2x + x + 0.1$ mol $-2x$ = 0.1 mol + x

Evaluating the mole fractions of SO_2, O_2 and SO_3 and then substituting in the expression of K_p°, we get

$$K_p^\circ = \frac{(0.1\,\text{mol} - 2x)^2\,(0.1\,\text{mol} + x)}{(2x)^2 x}\left(\frac{100\,\text{kPa}}{p}\right)$$

The pressure p of the flask can be replaced in terms of total amount of gases with the help of ideal gas equation

$$pV = nRT$$

Thus for $n = (0.1\text{ mol} + x)$, $V = 2\text{ dm}^3$, $T = 300\text{ K}$, we get

$$p = \frac{(0.1\,\text{mol} + x)\,(8.314\,\text{kPa dm}^3\,\text{mol}^{-1}\,\text{K}^{-1})\,(300\,\text{K})}{(2\,\text{dm}^3)}$$

$$= (1\,247\,\text{kPa mol}^{-1})\,(0.1\,\text{mol} + x)$$

With this K_p° becomes

$$K_p^\circ = \frac{(0.1\,\text{mol} - 2x)^2}{4x^3}\left(\frac{100\,\text{kPa}}{1\,247\,\text{kPa mol}^{-1}}\right)$$

Solving the cubic equation, by approximation method, we get

$$x = 0.036\,4\text{ mol}$$

$$\text{Per cent dissociation of } SO_3 = \frac{2 \times 0.036\,4}{0.1} \times 100 = 72.8$$

7.14 STUDY OF A FEW IMPORTANT HOMOGENEOUS GASEOUS REACTIONS

DISSOCIATION OF WATER

Consider a system consisting of water vapour in a glass bulb at 298 K with a total pressure equal to its normal vapour pressure of 27 mmHg. We will be interested in knowing how much of water has been dissociated to hydrogen and oxygen gas at equilibrium.

Expression of K_p° in Terms of α

Let α be the degree of dissociation of water vapour which, according to its definition, is the fraction of the substance dissociated, i.e. out of one mole how much of the substance is present in the dissociated form.

The following table gives the requisite data for calculation:

	$H_2O(g)$	$\rightleftharpoons$	$H_2(g)$	+	$\frac{1}{2}O_2(g)$	*Total amount*
Initial amount	n		0		0	n
Amount at equilibrium	$n(1-\alpha)$		$n\alpha$		$\frac{1}{2}n\alpha$	$n\left(1+\frac{1}{2}\alpha\right)$
Mole fraction	$\frac{1-\alpha}{1+\alpha/2}$		$\frac{\alpha}{1+\alpha/2}$		$\frac{\alpha/2}{1+\alpha/2}$	
Partial pressure	$\frac{1-\alpha}{1+\alpha/2}p$		$\frac{\alpha}{1+\alpha/2}p$		$\frac{\alpha/2}{1+\alpha/2}p$	

The standard equilibrium constant in terms of degree of dissociation is

$$K_p^\circ = \frac{(p_{H_2}/p^\circ)\sqrt{(p_{O_2}/p^\circ)}}{(p_{H_2O}/p^\circ)} = \frac{\left(\left(\dfrac{\alpha}{1+\alpha/2}\dfrac{p}{p^\circ}\right)\left(\dfrac{\alpha/2}{1+\alpha/2}\dfrac{p}{p^\circ}\right)\right)^{1/2}}{\left(\dfrac{(1-\alpha)}{1+\alpha/2}\dfrac{p}{p^\circ}\right)}$$

Simplifying, we get

$$K_p^\circ = \frac{\alpha^{3/2}(p/p^\circ)^{1/2}}{\sqrt{2}(1-\alpha)(1+\alpha/2)^{1/2}} \tag{7.14.1}$$

This equation is slightly difficult to solve for α. However, a simplified expression can be used since the value of K_p for the reaction is very small as can be seen from the following calculations.

From the table of $\Delta_f G^\circ$ we can calculate $\Delta_r G^\circ_{298K}$ of the reaction.

$$H_2O(g) = H_2(g) + \tfrac{1}{2}\ O_2(g)$$

$$\Delta_r G^\circ_{298K} = \Delta_f G^\circ(H_2,\ g) + \tfrac{1}{2}\ \Delta_f G^\circ(O_2,\ g) - \Delta_f G^\circ(H_2O,\ g)$$

$$= 0 + 0 - (-\ 228.57\ \text{kJ mol}^{-1})$$

$$= 228.57\ \text{kJ mol}^{-1}$$

Since $\quad \Delta_r G^\circ = -RT \ln K_p^\circ$

$$\log K_p^\circ = -\frac{\Delta_r G^\circ}{2.303\,RT} = -\frac{(228\,570\ \text{J mol}^{-1})}{(2.303)(8.314\ \text{J K}^{-1}\ \text{mol}^{-1})(298\,\text{K})}$$

$$= -40.059$$

Hence $\quad K_p^\circ = 8.73 \times 10^{-41}$

As K_p° is small, we can approximate

$$1 - \alpha \simeq 1 \quad \text{and} \quad 1 + \alpha/2 \simeq 1$$

Thus

$$K_p^\circ = \frac{\alpha^{3/2}(p/1\ \text{bar})^{1/2}}{\sqrt{2}} \tag{7.14.2}$$

or

$$\alpha = \left(\frac{\sqrt{2}\,K_p^\circ}{(p/1\ \text{bar})^{1/2}}\right)^{2/3} = \left[\frac{1.414 \times 8.73 \times 10^{-41}}{\left\{\left(\dfrac{27}{760} \times 1.013\,25\ \text{bar}\right)/1\ \text{bar}\right\}^{1/2}}\right]^{2/3}$$

$$= 7.51 \times 10^{-27}$$

We can see that $\alpha \ll 1$, therefore, our approximation is justified.

SYNTHESIS OF AMMONIA

Thermodynamic Feasibility of the Reaction

Before making attempts to carry out a given reaction, it is worthwhile to consider its feasibility using the thermodynamic principle. Nernst and Haber from the following thermodynamic data showed that the synthesis of ammonia based on the reaction,

$$\tfrac{1}{2}\ N_2(g) + \tfrac{3}{2}\ H_2(g) \rightleftharpoons NH_3(g)$$

in principle, is feasible.

	$\dfrac{\Delta_f H^\circ_{298K}}{kJ\ mol^{-1}}$	$\dfrac{S^\circ_{298K}}{J\ K^{-1}\ mol^{-1}}$	$\dfrac{C_{p,m}}{J\ K^{-1}\ mol^{-1}}$
$N_2(g)$	—	191.62	29.12
$H_2(g)$	—	130.68	28.83
$NH_3(g)$	−46.11	192.45	35.06

$$\Delta_r S^\circ_{298K} = \left\{192.45 - \tfrac{1}{2}(191.62) - \tfrac{3}{2}(130.68)\right\}\ J\ K^{-1} mol^{-1} \tag{7.14.3}$$

$$= -\ 99.38\ J\ K^{-1}\ mol^{-1}$$

$$\Delta_r G^\circ_{298K} = \Delta_r H^\circ_{298K} - T\ \Delta_r S^\circ_{298K}$$

$$= -46\ 110\ J\ mol^{-1} - (298\ K)\ (-\ 99.38\ J\ K^{-1}\ mol^{-1})$$

$$= -16\ 494.8\ J\ mol^{-1} \tag{7.14.4}$$

The negative value of Gibbs function of reaction indicates that the reaction is feasible.

Evaluation of Equilibrium Constant

The equilibrium constant of the reaction is computed as follows:

$$\log K^\circ_{298K} = -\frac{\Delta_r G^\circ_{298K}}{2.303\ RT} = \frac{(16\ 494.8\ J\ mol^{-1})}{2.303 \times (8.314\ J\ K^{-1}\ mol^{-1})\ (298\ K)}$$

$$= 2.89$$

$$K^\circ_{298K} = 7.78 \times 10^2 \tag{7.14.5}$$

Though the reaction is possible with such a large value of K°_p, its rate of combination is very slow. We can increase the rate by increasing temperature. This will decrease the value of K°_p, since the reaction is exothermic, but it is still helpful. The rate can also be increased by using a suitable catalyst. The latter will not change the value of $\Delta_r G^\circ$ or K°_p for the above reaction, but will help in increasing the rate of the reaction by following an alternative path.

Evaluation of Equilibrium Concentrations of Species

Now we shall consider how to calculate the value of K°_p or equilibrium concentrations of N_2, H_2 and NH_3 under a given condition of the system. We write the reaction as

$$\tfrac{1}{2}N_2(g) + \tfrac{3}{2}H_2(g) \rightleftharpoons NH_3(g)$$

with
$$K^\circ_p = \frac{(p_{NH_3}/p^\circ)}{(p_{N_2}/p^\circ)^{1/2}(p_{H_2}/p^\circ)^{3/2}}$$

For the sake of convenience, we start with the dissociation of NH_3. The reaction will be just reverse of the above reaction, i.e.

$$NH_3(g) \rightleftharpoons \tfrac{1}{2}\ N_2(g) + \tfrac{3}{2}\ H_2(g)$$

Its equilibrium constant is

$$K_p^{\circ\prime} = \frac{(p_{N_2})^{1/2}(p_{H_2})^{3/2}}{p_{NH_3}} \frac{1}{p^\circ}$$

Obviously $K_p^{\circ\prime} = 1/K_p^\circ$

Let α be the degree of dissociation of NH_3. We then have

	$NH_3(g)$	$\rightleftharpoons$	$\frac{1}{2}N_2(g)$	+	$\frac{3}{2}H_2(g)$	*Total amount*
At $t = 0$	n		0		0	n
At equilibrium	$n(1-\alpha)$		$(1/2)\, n\, \alpha$		$(3/2)\, n\, \alpha$	$n(1+\alpha)$
Mole fraction	$\frac{1-\alpha}{1+\alpha}$		$\frac{\alpha}{2(1+\alpha)}$		$\frac{3\alpha}{2(1+\alpha)}$	
Partial pressure	$\frac{1-\alpha}{1+\alpha}p$		$\frac{\alpha}{2(1+\alpha)}p$		$\frac{3\alpha}{2(1+\alpha)}p$	

Thus $$K_p^{\circ\prime} = \frac{(p_{N_2})^{1/2}(p_{H_2})^{3/2}}{p_{NH_3}} \frac{1}{p^\circ}$$

$$= \frac{\left(\frac{\alpha}{2(1+\alpha)}p\right)^{1/2}\left(\frac{3\alpha}{2(1+\alpha)}p\right)^{3/2}}{\left(\frac{1-\alpha}{1+\alpha}p\right)} \frac{1}{p^\circ} = \frac{3\sqrt{3}\,\alpha^2}{4(1-\alpha^2)}\left(\frac{p}{p^\circ}\right) \qquad (7.14.6)$$

and $$K_p^\circ = \frac{4(1-\alpha^2)}{3\sqrt{3}\,\alpha^2}\left(\frac{p^\circ}{p}\right) \qquad (7.14.7)$$

Example 7.14.1 In a system, the equilibrium reaction

$$(1/2)\, N_2(g) + (3/2)H_2(g) \rightleftharpoons NH_3(g)$$

was studied starting from pure NH_3. It is found that at 10.1325 bar pressure and 673 K, the gaseous mixture contains 3.85 mol % of NH_3. Calculate K_x, K_p° and α.

Solution The gaseous mixture contains 3.85 mol % of NH_3. Thus, its mole fraction is given by

$$x_{NH_3} = \frac{3.85}{100} = 0.038\,5$$

The total mole fraction of the remaining gases, namely N_2 and H_2, is

$$x_{N_2} + x_{H_2} = 1 - 0.038\,5 = 0.961\,5$$

Since nitrogen and hydrogen are present in one is to three ratio, therefore, their individual mole fractions are

$$x_{N_2} = (1/4) \times 0.961\,5 = 0.240\,4$$
$$x_{H_2} = 0.961\,5 - 0.240\,4 = 0.721\,1$$

$$K_x = \frac{x_{NH_3}}{(x_{N_2})^{1/2}(x_{H_2})^{3/2}} = \frac{0.038\,5}{(0.240\,4)^{1/2}(0.721\,1)^{3/2}} = 0.128$$

Their partial pressures are

$$p_{NH_3} = x_{NH_3}p = 0.038\,5 \times 10.132\,5 \text{ bar} = 0.390\,1 \text{ bar}$$
$$p_{H_2} = x_{H_2}p = 0.721\,1 \times 10.132\,5 \text{ bar} = 7.306\,5 \text{ bar}$$
$$p_{N_2} = x_{N_2}p = 0.240\,4 \times 10.132\,5 \text{ bar} = 2.435\,9 \text{ bar}$$

Substituting these in the expression of K_p° we get

$$K_p^\circ = \frac{(p_{NH_3}/p^\circ)}{(p_{H_2}/p^\circ)^{3/2}(p_{N_2}/p^\circ)^{1/2}}$$

$$= \frac{(0.390\,1 \text{ bar}/1 \text{ bar})}{(7.306\,5 \text{ bar}/1 \text{ bar})^{3/2}(2.435\,9 \text{ bar}/1 \text{ bar})^{1/2}}$$

$$= 0.012\,65$$

If α is the degree of dissociation of NH_3, then we have

	NH_3	$\rightleftharpoons$	$(1/2)N_2$	+	$(3/2)H_2$
At equilibrium	$1 - \alpha$		$\alpha/2$		$3\,\alpha/2$

$$\text{Total fraction} = 1 - \alpha + \frac{\alpha}{2} + \frac{3\alpha}{2} = 1 + \alpha$$

Thus, the mole fraction of NH_3 is

$$x_{NH_3} = \frac{1-\alpha}{1+\alpha}$$

Equating this to 0.038 5, we get

$$\alpha = 0.926$$

or, at equilibrium 92.6 mol % of NH_3 is dissociated into H_2 and N_2.

Example 7.14.2 For a reaction

$$NH_3(g) \rightleftharpoons (1/2)N_2(g) + (3/2)H_2(g)$$

$K_p^\circ = 79.1$ at 400 °C. Show that the fraction of NH_3 dissociated at a total pressure p is given by

$$\alpha = \left(1 + \frac{3\sqrt{3}}{4}\frac{(p/p^\circ)}{K_p^\circ}\right)^{-1/2}$$

Calculate the value of K_c° for the given reaction.

Solution The given reaction is

$$NH_3(g) \rightleftharpoons (1/2)N_2(g) + (3/2)H_2(g)$$

Its K_p° is given as

$$K_p^\circ = \frac{(p_{N_2}/p^\circ)^{1/2}(p_{H_2}/p^\circ)^{3/2}}{(p_{NH_3}/p^\circ)} = 79.1$$

The expression of K_p° in terms of degree of dissociation (α) of NH_3 as given by Eq. (7.14.6) is

$$K_p^\circ = \frac{3\sqrt{3}\,\alpha^2(p/p^\circ)}{4(1-\alpha)(1+\alpha)}$$

Rearranging this, we get

$$4(1-\alpha^2)K_p^\circ - 3\sqrt{3}\,\alpha^2(p/p^\circ) = 0$$

or $$\alpha^2\{4K_p^\circ + 3\sqrt{3}\,(p/p^\circ)\} = 4K_p^\circ$$

Thus $$\alpha = \left(\frac{4K_p^\circ}{4K_p^\circ + 3\sqrt{3}\,(p/p^\circ)}\right)^{1/2} = \left(\frac{1}{1+\dfrac{3\sqrt{3}}{4}\dfrac{(p/p^\circ)}{K_p^\circ}}\right)^{1/2}$$

$$= \left(1+\frac{3\sqrt{3}}{4}\frac{(p/p^\circ)}{K_p^\circ}\right)^{-1/2}$$

Now $$K_c^\circ = K_p^\circ\left(\frac{c^\circ RT}{p^\circ}\right)^{-\Delta\nu_g}$$

For the given reaction, $\Delta\nu_g = +1$

Thus $$K_c^\circ = (79.1)\left[\frac{(1\ \text{mol dm}^{-3})(8.314\ \text{kPa dm}^3\ \text{K}^{-1}\ \text{mol}^{-1})(673\ \text{K})}{100\ \text{kPa}}\right]^{-1} = 1.414$$

Example 7.14.3 If in the system given in Example 7.14.1 the pressure is changed to 5 066.25 kPa, the mixture was found to contain 15.11 mol % of NH_3. Calculate the degree of dissociation.

Solution In the present case, we have

$$x_{NH_3} = 0.151\,1 \qquad p_{NH_3} = 0.151\,1 \times 5\,066.25\ \text{kPa} = 765.51\ \text{kPa}$$
$$x_{N_2} = 0.212\,2 \qquad p_{N_2} = 0.212\,2 \times 5\,066.25\ \text{kPa} = 1\,075.06\ \text{kPa}$$
$$x_{H_2} = 0.636\,7 \qquad p_{H_2} = 0.636\,7 \times 5\,066.25\ \text{kPa} = 3\,225.68\ \text{kPa}$$

Thus $$K_x = \frac{0.1511}{(0.212\,2)^{1/2}(0.636\,7)^{3/2}} = 0.643$$

$$K_p^\circ = \frac{(765.51\ \text{kPa}/100\ \text{kPa})}{(1\,075.06\ \text{kPa}/100\ \text{kPa})^{1/2}\ (3\,225.68\ \text{kPa})\ (100\ \text{kPa})^{3/2}} = 0.012\,74$$

It is to be noted that K_p° is the same since it is independent of pressure but K_x depends on the value of pressure.

The degree of dissociation as calculated from the expression

$$x_{NH_3} = \frac{1-\alpha}{1+\alpha} = 0.151\,1$$

is $$\alpha = 0.736$$

Thus, it can be seen that the degree of dissociation has decreased as the pressure in increased, as would be expected from the Le Chatelier's principle.

DISSOCIATION OF DINITROGEN TETROXIDE

The following is the Table containing information needed to deal with the dissociation of dinitrogen tetroxide.

	$N_2O_4(g)$	$\rightleftharpoons$	$2NO_2(g)$	Total amount
Initial amount	n		0	0
Amount at time t	$n(1-\alpha)$		$2n\alpha$	$n(1+\alpha)$
Mole fraction	$\frac{1-\alpha}{1+\alpha}$		$\frac{2\alpha}{1+\alpha}$	
Partial pressure	$\frac{1-\alpha}{1+\alpha}p$		$\frac{2\alpha}{1+\alpha}p$	
$\Delta_f H°/\text{mol}^{-1}$	9.16 kJ		33.18 kJ	

Degree of Dissociation from Density Measurement at Constant Pressure

Let at $t = 0$, V_0 be the volume of the gas at a pressure p. Using ideal gas equation, the density of the gas is given by

$$pV_0 = nRT = \frac{m}{M}RT$$

or

$$\rho_0 = \frac{m}{V_0} = \frac{pM}{RT}$$

Let at equilibrium, the total amount of the gases be $n(1 + \alpha)$ and let V_1 be its volume.

Then $\quad pV_1 = n(1+\alpha)RT \quad$ or $\quad V_1 = \frac{n(1+\alpha)RT}{p}$

Density of the gas at equilibrium will be given by

$$\rho_1 = \frac{n\times M}{V_1} = \frac{nMp}{n(1+\alpha)RT}$$

or

$$\rho_1 = \frac{\rho_0}{1+\alpha}$$

Rearranging this, we have

$$\alpha = \frac{\rho_0 - \rho_1}{\rho_1} \tag{7.14.8}$$

Degree of Dissociation from Pressure Measurement at Constant Volume

When the reaction is carried out under constant volume condition, the pressure of the system increases since on dissociation, amount of the gas increases from n to $n(1 + \alpha)$. If the ideal gas behaviour is assumed, then

$$p_0 = \frac{nRT}{V} \quad \text{and} \quad p_1 = n(1+\alpha)\frac{RT}{V}$$

Thus $\quad p_1 = (1+\alpha)p_0 \quad$ or $\quad \alpha = \frac{p_1 - p_0}{p_0} \qquad (7.14.9)$

Thus, the value of α can be calculated at any stage of dissociation from the pressure measurements.

Expression of Equilibrium Constant

The expression for K_p° is

$$K_p^\circ = \frac{(p_{NO_2}/p^\circ)^2}{(p_{N_2O_4}/p^\circ)} = \frac{\left(\dfrac{2\alpha}{1+\alpha}\dfrac{p}{p^\circ}\right)^2}{\left(\dfrac{1-\alpha}{1+\alpha}\dfrac{p}{p^\circ}\right)} = \frac{4\alpha^2}{1-\alpha^2}\left(\frac{p}{p^\circ}\right) \quad (7.14.10)$$

Experimentally, it is found that at room temperature and 1 bar total pressure, the degree of dissociation of N_2O_4 at equilibrium is 0.19, the numerical value of the standard equilibrium constant will be

$$K_p^\circ = \frac{4\times 0.19^2}{1-0.19^2}(1\text{ bar}/1\text{ bar}) = 0.149\,8$$

Thermodynamic Data for the Reaction

With K_p°, the value of $\Delta_r G^\circ$ can be calculated as follows.

$$\begin{aligned}\Delta_r G^\circ_{298\text{K}} &= -2.303\ RT \log K_p^\circ \\ &= -(2.303)(8.314\ \text{J K}^{-1}\ \text{mol}^{-1})(298\ \text{K}) \log (0.149\,8) \\ &= 4\ 704\ \text{J mol}^{-1} \simeq 4.70\ \text{kJ mol}^{-1}\end{aligned}$$

By measuring the percentage dissociation as a function of temperature, we can determine $\Delta_r G^\circ$ as a function of temperature. Then with the help of Gibbs-Helmholtz equation, we can obtain enthalpy of the reaction $\Delta_r H_{298\text{K}}$.

Effect of Pressure on Degree of Dissociation

The degree of dissociation of N_2O_4 does not depend upon the amount of N_2O_4 taken to start with. However, the degree of dissociation of N_2O_4 does depend upon pressure. The expression relating these two can be obtained by rearranging Eq. (7.14.9). Thus, we have

$$\alpha = \left[\frac{K_p^\circ}{K_p^\circ + 4(p/p^\circ)}\right]^{1/2} \quad (7.14.11)$$

As pressure increases towards infinity, α approaches zero. On the other hand, as the pressure approaches zero, α approaches one. These changes are exactly what would be expected from the Le Chatelier's principle.

7.15 MISCELLANEOUS NUMERICALS

1. (a) For the reaction

$$C_2H_4(g) + H_2(g) \rightleftharpoons C_2H_6(g)$$

calculate K_p° at 25 °C. Given are the standard free energies of formation at 298 K:

$$\Delta_f G^\circ(C_2H_4) = 68.124\ \text{kJ mol}^{-1}; \qquad \Delta_f G^\circ(C_2H_6) = -32.886\ \text{kJ mol}^{-1}$$

(b) For the reaction given in part (a), standard enthalpies of formation and standard entropies of various species are given below. Calculate K_p°.

$$\Delta_f H^\circ(C_2H_4) = 52.283 \text{ kJ mol}^{-1}; \qquad \Delta_f H^\circ(C_2H_6) = -84.667 \text{ kJ mol}^{-1}$$

$$S^\circ(C_2H_4) = 219.45 \text{ J K}^{-1} \text{ mol}^{-1}; \qquad S^\circ(C_2H_6) = 229.49 \text{ kJ mol}^{-1};$$

$$S^\circ(H_2) = 130.58 \text{ J K mol}^{-1};$$

(c) Also calculate K_c° for the reaction given in part (a).

Solution **(a)** The standard free energy change of the reaction

$$C_2H_4(g) + H_2(g) \rightarrow C_2H_6(g)$$

is given by

$$\Delta_r G^\circ = \Delta_f G^\circ(C_2H_6) - \Delta_f G^\circ(C_2H_4)$$

Thus substituting the given data at 298 K, we get

$$\Delta_r G^\circ_{298K} = (-32.886 - 68.124) \text{ kJ mol}^{-1} = -101.01 \text{ kJ mol}^{-1}$$

Now since $\quad \log K_p^\circ = \dfrac{-\Delta_r G^\circ}{RT \times 2.303}$

we get $\quad \log K_p^\circ = \dfrac{(101.01 \times 10^3 \text{ J mol}^{-1})}{(8.314 \text{ J K}^{-1} \text{ mol}^{-1})(298 \text{ K})(2.303)} = 17.704$

or $\quad K_p^\circ = 5.06 \times 10^{17}$

(b) $\Delta_r H^\circ$ of the given the reaction is given by

$$\Delta_r H^\circ = \Delta_f H^\circ(C_2H_6) - \Delta_f H^\circ(C_2H_4)$$

$$= (-84.667 - 52.283) \text{ kJ mol}^{-1} = -136.95 \text{ kJ mol}^{-1}$$

$\Delta_r S^\circ$ of the reaction is

$$\Delta_r S^\circ = S^\circ(C_2H_6) - S^\circ(C_2H_4) - S^\circ(H_2)$$

$$= (229.49 - 219.45 - 130.58) \text{ J K}^{-1} \text{ mol}^{-1} = -120.54 \text{ J K}^{-1} \text{ mol}^{-1}$$

Now since $\quad \Delta_r G^\circ = \Delta_r H^\circ - T\Delta_r S^\circ$

we get $\quad \Delta_r G^\circ = -136.95 \times 10^3 \text{ J mol}^{-1} - (298 \text{ K})(-120.54 \text{ J K}^{-1} \text{ mol}^{-1})$

$$= -101.029 \times 10^3 \text{ J mol}^{-1} = -101.029 \text{ kJ mol}^{-1}$$

Calculating K_p° from the reaction

$$\log K_p^\circ = -\frac{\Delta_r G^\circ}{2.303\, RT}$$

we get $\quad \log K_p^\circ = \dfrac{101.029 \times 10^3 \text{ J mol}^{-1}}{2.303(8.314 \text{ J K}^{-1} \text{ mol}^{-1})(298 \text{ K})} = 17.707$

or $\quad K_p^\circ = 5.06 \times 10^{17}$

(c) The change in the stoichiometry number of gaseous species in a reaction is given by

$$\Delta\nu_g = \underset{\text{(products)}}{\sum_{p}} \nu_p - \underset{\text{(reactants)}}{\sum_{r}} |\nu_r|$$

Thus, for the given reaction, we get

$$\Delta\nu_g = -1$$

Substituting this along with K_p° at 298 K in the relation

$$K_c^\circ = K_p^\circ \left(\frac{c^\circ RT}{p^\circ}\right)^{-\Delta\nu_g}$$

we get $$K_c^\circ = (5.06 \times 10^{17})\left(\frac{(1\ \text{mol dm}^{-3})(8.314\ \text{J K}^{-1}\ \text{mol}^{-1})(298\ \text{K})}{100\ \text{kPa}}\right)^{-(-1)}$$

$$= 1.254 \times 10^{19}$$

2. What is the partial pressure of HCl(g) above a solution that is 2 mol dm^{-3} in H$^+$ and 2 mol dm^{-3} in Cl$^-$? Given: $\Delta_f G^\circ$(HCl, g) = −95.27 kJ mol^{-1} and $\Delta_f G^\circ$(Cl$^-$) = −131.17 kJ mol^{-1}.

Solution To calculate the partial pressure of HCl above the solution containing H$^+$ and Cl$^-$, we consider the reaction

	H^+	+	Cl^-	$\rightleftharpoons$	HCl(g)
$\Delta_f G^\circ$/kJ mol^{-1}	0		−131.17		−95.27

Thus $\Delta_r G^\circ = -95.27\ \text{kJ mol}^{-1} - (-131.17\ \text{kJ mol}^{-1}) = 35.9\ \text{kJ mol}^{-1}$

$$\log K^\circ = \frac{-\Delta_r G^\circ}{2.303\ RT} = \frac{-35.9 \times 10^3\ \text{J mol}^{-1}}{2.303(8.314\ \text{J K}^{-1}\ \text{mol}^{-1})(298\ \text{K})} = -6.29$$

Hence $K^\circ = 5.1 \times 10^{-7}$

Since $$K^\circ = \frac{(p_{\text{HCl}}/1\ \text{bar})}{([\text{H}^+]/\text{mol dm}^{-3})([\text{Cl}^-]/\text{mol dm}^{-3})} = 5.1 \times 10^{-7}$$

we get $p_{\text{HCl}}/1\ \text{bar} = 2 \times 2 \times 5.1 \times 10^{-7}$

or $p_{\text{HCl}} = (2.04 \times 10^{-6})(1\ \text{bar}) = 2.04 \times 10^{-6}\ \text{bar} = 0.204\ 6\ \text{Pa}$

3. In a study of the water-gas reaction

$$CO_2(g) + H_2(g) \rightleftharpoons CO(g) + H_2O(g)$$

a mixture of CO_2 and H_2 initially containing 42.4 mol % H_2 was brought to equilibrium in a closed vessel at 1 259 K. The system was then found to contain 15.2 mol % H_2. Calculate K_p° and $\Delta_r G^\circ$ for the reaction at 1 259 K.

Solution At $t = 0$, H_2 is equal to 42.4 mol %, and therefore CO_2 will be 100 − 42.4 = 57.6 mol %. At equilibrium, H_2 is found to be 15.2 mol %, i.e. 42.4 − 15.2 = 27.2 mol % has combined with CO_2 to give CO(g) and H_2O(g). It is clear from the stoichiometry that the same amount of CO or H_2O is formed. The mole percentage of CO_2 will be equal 57.6 − 27.2 = 30.4. Thus at equilibrium, the amount per cents of various species are

$$\underset{30.4\%}{CO_2(g)} + \underset{15.2\%}{H_2(g)} \rightleftharpoons \underset{27.2\%}{CO(g)} + \underset{27.2\%}{H_2O(g)}$$

Now since for the reaction

$$\Delta\nu_g = 0$$

therefore $$K_p^\circ = K_c^\circ = K_n^\circ = \frac{(n_{CO}/n^\circ)(n_{H_2O}/n^\circ)}{(n_{CO_2}/n^\circ)(n_{H_2}/n^\circ)};$$ (where n° = 1 mol)

Substituting the value, we get

$$K_p^\circ = K_n^\circ = \frac{27.2 \times 27.2}{30.4 \times 15.2} = 1.602$$

Since $\Delta_r G^\circ = -RT \ln K_p^\circ$

we get $\Delta_r G^\circ = -(8.314\ \text{J K}^{-1}\ \text{mol}^{-1})(1\ 259\ \text{K})(2.303) \times \log(1.602)$

$$= -4\ 933.65\ \text{J mol}^{-1}$$

4. H_2S dissociates according to the equation

$$2H_2S(g) \rightleftharpoons 2H_2(g) + S_2(g)$$

At 1 125 °C and at a total pressure of 101.325 kPa, the degree of dissociation of H_2S is 0.305. Calculate K_p° for this reaction at 1 125 °C.

Solution

If α is the degree of dissociation of H_2S, then the amounts of species involved in the reaction are

$$\begin{array}{ccccc} 2H_2S(g) & \rightleftharpoons & 2H_2(g) & + & S_2(g) \\ n(1-\alpha) & & n\alpha & & n(\alpha/2) \end{array}$$

Total amount of species = $n(1 - \alpha) + n\alpha + n(\alpha/2) = n(1 + \alpha/2)$

The partial pressures of the species are

$$p_{H_2S} = x_{H_2S}p = \frac{n(1-\alpha)}{n(1+\alpha/2)}p = \frac{1-\alpha}{(1+\alpha/2)}p$$

$$p_{H_2} = x_{H_2}p = \frac{n\alpha}{n(1+\alpha/2)}p = \frac{\alpha}{1+\alpha/2}p$$

$$p_{S_2} = x_{H_2}p = \frac{n\alpha/2}{n(1+\alpha/2)}p = \frac{\alpha/2}{1+\alpha/2}p$$

Substituting these in the expression of K_p°, we get

$$K_p^\circ = \frac{\left\{\frac{\alpha}{(1+\alpha/2)}(p/100\text{ kPa})\right\}^2\left\{\frac{\alpha/2}{(1+\alpha/2)}(p/100\text{ kPa})\right\}}{\left\{\frac{(1-\alpha)}{(1+\alpha/2)}(p/100\text{ kPa})\right\}^2}$$

$$= \frac{0.5\,\alpha^3\,(p/100\text{ kPa})}{(1-\alpha)^2\,(1+0.5\,\alpha)}$$

Substituting the given value α, we get

$$K_p^\circ = \frac{0.5\times(0.305)^3\times 1.013\,25}{(0.695)^2\,(1+0.152\,5)} = 0.025\,8$$

---

5. $COCl_2$ gas dissociates according to the equation

$$COCl_2(g) \rightleftharpoons CO(g) + Cl_2(g)$$

When $COCl_2$ is heated to 724 K at 101.325 kPa, density of the gas mixture at equilibrium is 1.162 g dm^{-3}. Calculate (a) the degree of dissociation, (b) K_p° and (c) $\Delta_r G^\circ$ for the reaction at 724 K.

Solution

(a) Let α be the degree of dissociation of $COCl_2$. The amount of various species involved in the given reaction are

$$\begin{array}{lccccc} & COCl_2 & \rightleftharpoons & CO & + & Cl_2 \\ \text{at } t=0 & n & & 0 & & 0 \\ \text{and at } t_{eq} & n(1-\alpha) & & n\alpha & & n\alpha \end{array}$$

Total amount of the gases = $n(1 + \alpha)$

Volume of the flask, $V = \dfrac{n(1+\alpha)RT}{p}$

Density of the mixture, $\rho = \dfrac{nM_{COCl_2}}{V} = \dfrac{nM_{COCl_2}p}{n(1+\alpha)RT}$

or $$\alpha = \frac{pM_{COCl_2}}{\rho RT} - 1$$

Substituting the given values of ρ, p, M_{COCl_2} and T, we get

$$\alpha = \frac{(101.325\text{ kPa})(99\text{ g mol}^{-1})}{(1.162\text{ g dm}^{-3})(8.314\text{ kPa dm}^3\text{ K}^{-1}\text{ mol}^{-1})(724\text{ K})} - 1$$

$$= 1.435 - 1 = 0.435$$

(b) The partial pressures of the species involved in the reaction are

$$p_{COCl_2} = \frac{n(1-\alpha)}{n(1+\alpha)}p = \frac{1-\alpha}{1+\alpha}p; \qquad p_{CO} = \frac{n\alpha}{n(1+\alpha)}p = \frac{\alpha}{1+\alpha}p$$

$$p_{Cl_2} = \frac{n\alpha}{n(1+\alpha)}p = \frac{\alpha}{1+\alpha}p$$

Substituting these in the expression

$$K_p^\circ = \frac{(p_{CO}/p^\circ)(p_{Cl_2}/p^\circ)}{(p_{COCl_2}/p^\circ)}$$

we get $$K_p^\circ = \frac{\left(\dfrac{\alpha}{1+\alpha}\dfrac{p}{p^\circ}\right)^2}{\left(\dfrac{1-\alpha}{1+\alpha}\dfrac{p}{p^\circ}\right)} = \frac{\alpha^2}{1-\alpha^2}\left(\frac{p}{p^\circ}\right)$$

Hence $$K_p^\circ = \frac{(0.435)^2}{1-(0.435)^2} \times \left(\frac{101.325\text{ kPa}}{100\text{ kPa}}\right) = \frac{0.189\,2}{1-0.189\,2}(1.0132\,5)$$

$$= 0.236\,4$$

(c) Since $\Delta_r G^\circ = -RT\ln K_p^\circ$, we get

$$\Delta_r G^\circ = -(8.314\text{ J K}^{-1}\text{ mol}^{-1})(724\text{ K})(2.303) \times \log(0.236\,4)$$

$$= 8\,682.8\text{ J mol}^{-1}$$

---

6. N_2O_4 dissociates according to the equation

$$N_2O_4(g) \rightleftharpoons 2NO_2(g)$$

When 0.578 g of N_2O_4 was introduced into a 1 dm^3 flask maintained at 308 K, the equilibrium pressure was 24.12 kPa. Calculate (a) the degree of dissociation and, (b) K_p° at this temperature.

Solution Let α be the degree of dissociation of N_2O_4. The amounts of the species involved in the reaction are

$$\mathrm{N_2O_4(g)} \rightleftharpoons \mathrm{2NO_2(g)}$$

At t_{eq} $\quad n(1-\alpha) \qquad 2n\alpha$

where $n = (0.578\ \mathrm{g})/M_{\mathrm{N_2O_4}} = (0.578\ \mathrm{g})/(92\ \mathrm{g\ mol^{-1}}) = 0.006\ 282\ \mathrm{mol}$

Total amount of species $= n(1-\alpha) + 2n\alpha = n(1+\alpha)$

Since $pV = nRT$, for this case

$$n' = n(1+\alpha) = \frac{pV}{RT} = \frac{(24.12\ \mathrm{kPa})\,(1\ \mathrm{dm^3})}{(8.314\ \mathrm{kPa\ dm^3\ K^{-1}\ mol^{-1}})\,(308\ \mathrm{K})}$$

$$= 9.42 \times 10^{-3}\ \mathrm{mol}$$

Solving for α, we get

$$\alpha = \frac{0.009\ 42}{0.006\ 282} - 1 = 0.5$$

Substituting α in the expression of K_p° we get

$$K_p^\circ = \frac{(p_{\mathrm{NO_2}}/p^\circ)^2}{(p_{\mathrm{N_2O_4}}/p^\circ)} = \frac{\left(\dfrac{2\alpha}{1+\alpha}(p/p^\circ)\right)^2}{\left(\dfrac{1-\alpha}{1+\alpha}(p/p^\circ)\right)} = \frac{4\alpha^2(p/p^\circ)}{1-\alpha^2}$$

$$= \frac{4\times 0.5^2}{1-0.5^2}\left(\frac{24.12\ \mathrm{kPa}}{100\ \mathrm{kPa}}\right)$$

$$= 0.322$$

---

7. (a) PCl_5 dissociates according to the equation

$$\mathrm{PCl_5(g) \rightarrow PCl_3(g) + Cl_2(g)}$$

When 0.03 mol of PCl_5 was brought to equilibrium at 502 K and 101.325 kPa the volume of the system was 2.09 dm^3. Calculate (i) the degree of dissociation and (ii) K_p°

(b) What will be the degree of dissociation when 0.20 mol of PCl_5 is brought to equilibrium in a 3 dm^3 vessel at 502 K?

Solution

Let α be the degree of dissociation of PCl_5. The amount of various species involved in the reaction are

$$\mathrm{PCl_5(g) \rightleftharpoons PCl_3(g) + Cl_2(g)}$$

	$PCl_5(g)$	$PCl_3(g)$	$Cl_2(g)$
At $t = 0$	n	0	0
At t_{eq}	$n(1-\alpha)$	$n\alpha$	$n\alpha$

Total amount of species at equilibrium $= n\ (1+\alpha)$

(a) In this case

$$n = 0.03\ \mathrm{mol}$$

Thus from the ideal gas equation, we have

(i) $$(0.03\ \mathrm{mol})\,(1+\alpha) = \frac{pV}{RT} = \frac{(101.325\ \mathrm{kPa})\,(2.09\ \mathrm{dm^3})}{(8.314\ \mathrm{kPa\ dm^3\ K^{-1}\ mol^{-1}})\,(502\ \mathrm{K})}$$

$$= 0.050\ 7\ \mathrm{mol}$$

or $$\alpha = \frac{(0.050\ 7\ \mathrm{mol})}{(0.03\ \mathrm{mol})} - 1 = 0.69$$

(ii) The partial pressure of the species are

$$p_{PCl_5} = \frac{n(1-\alpha)}{n(1+\alpha)}p = \frac{1-\alpha}{1+\alpha}p; \qquad p_{PCl_3} = \frac{n\alpha}{n(1+\alpha)}p = \frac{\alpha}{1+\alpha}p$$

$$p_{Cl_2} = \frac{n\alpha}{n(1+\alpha)}p = \frac{\alpha}{1+\alpha}p$$

Substituting these in the expression of K_p°, we get

$$K_p^\circ = \frac{(p_{PCl_3}/p^\circ)(p_{Cl_2}/p^\circ)}{(p_{PCl_5}/p^\circ)} = \frac{\left(\frac{\alpha}{1+\alpha}(p/p^\circ)\right)^2}{\left(\frac{(1-\alpha)}{(1+\alpha)}(p/p^\circ)\right)} = \frac{\alpha^2(p/p^\circ)}{1-\alpha^2}$$

Substituting the value of α, we get

$$K_p^\circ = \frac{0.69^2}{1-0.69^2}\left(\frac{101.325}{100\ \text{kPa}}\right) = 0.920\ 4$$

(b) We have

$$K_p^\circ = \frac{\alpha^2(p/p^\circ)}{1-\alpha^2}$$

Replacing p in terms of total amount of gases, we get

$$K_p^\circ = \frac{\alpha^2}{1-\alpha^2}\frac{n'RT}{Vp^\circ} = \frac{\alpha^2}{1-\alpha^2}\frac{n(1+\alpha)RT}{Vp^\circ}$$

$$= \frac{\alpha^2}{1-\alpha}\frac{nRT}{Vp^\circ}$$

or $$\frac{\alpha^2}{1-\alpha} = K_p^\circ\frac{Vp^\circ}{nRT}$$

For the given data, we have

$$\frac{\alpha^2}{1-\alpha} = \frac{(0.920\ 4)\ (3\ \text{dm}^3)\ (100\ \text{kPa})}{(0.2\ \text{mol})\ (8.314\ \text{kPa dm}^3\ \text{K}^{-1}\ \text{mol}^{-1})\ (502\ \text{K})} = 0.33$$

or $$\alpha^2 + 0.33\ \alpha - 0.33 = 0$$

Solving for α, we get

$$\alpha = \frac{-0.33 \pm \sqrt{(0.33)^2 + 4(0.33)}}{2} = \frac{-0.33 \pm 1.195}{2}$$

$$= -0.762 \text{ (not possible) or } 0.432$$

8. PCl_5(g) dissociates according to the reaction

$$PCl_5(g) \rightleftharpoons PCl_3(g) + Cl_2(g)$$

At 523 K, $K_p^\circ = 1.80$. Find the density of the equilibrium mixture in g dm^{-3} at a total pressure of 100 kPa.

Solution Let α be the degree of dissociation of PCl_5. The amount of various species at equilibrium are

$$\begin{array}{llll} & PCl_5 & \rightleftharpoons PCl_3 & + \; Cl_2 \\ \text{At } t_{eq} & n(1-\alpha) & n\alpha & n\alpha \end{array}$$

Total amount of the gases = $n\,(1 + \alpha)$

The value of α can be determined from K_p°

$$K_p^\circ = \frac{(p_{PCl_3}/p^\circ)\,(p_{Cl_2}/p^\circ)}{(p_{PCl_5}/p^\circ)} = \frac{\left(\frac{\alpha}{1+\alpha}(p/p^\circ)\right)^2}{\left(\frac{1-\alpha}{1+\alpha}(p/p^\circ)\right)} = \frac{\alpha^2(p/p^\circ)}{1-\alpha^2}$$

(where p° = 100 kPa)

or $\alpha^2 = K_p^\circ(1-\alpha^2)$ (since p = 100 kPa)

or $\alpha = \sqrt{K_p^\circ/(1+K_p^\circ)}$

Substituting the value of K_p° we get

$$\alpha = \sqrt{180/2.80} = 0.80$$

Now using ideal gas equation, we have

$$pV = n'RT = n(1+\alpha)\,RT = (m/M_{PCl_5})\,(1+\alpha)\,RT$$

$$\rho = \frac{m}{V} = \frac{pM_{PCl_5}}{(1+\alpha)RT} = \frac{(100\text{ kPa})\,(102\text{ g mol}^{-1})}{(1.8)\,(8.314\text{ J K}^{-1}\text{ mol}^{-1})\,(523\text{ K})}$$

$$\rho = 1.30\text{ g dm}^{-3}$$

---

9. Solid NH_4HS dissociates according to the equation

$$NH_4HS(s) \rightleftharpoons NH_3(g) + H_2S(g)$$

The dissociation pressure of solid NH_4HS is 66.87 kPa at 288 K. (a) Calculate K_p° for this reaction. (b) What fraction of the solid will dissociate when 0.1 mol of NH_4HS is introduced into a 1 dm^3 flask at 298 K? (c) What fraction will dissociate when 0.1 mol of NH_4HS is introduced into a 1 dm^3 flask that contains NH_3 at 20.26 kPa and 298 K?

Solution

(a) Dissociation pressure = 66.87 kPa

Since NH_3 and H_2S are present in equal amounts, we have

$$p_{NH_3} = p_{H_2S} = 33.435\text{ kPa}$$

Thus $$K_p^\circ = (p_{NH_3}/p^\circ)\,(p_{H_2S}/p^\circ) = \left(\frac{33.435\text{ kPa}}{100\text{ kPa}}\right)^2 = 0.111$$

(b) Let α be the degree of dissociation. Therefore, at equilibrium, we have

$$\begin{array}{llll} & NH_4HS(s) \rightleftharpoons & NH_3(g) & + \; H_2S(g) \\ \text{At } t_{eq} & (0.1\text{ mol})(1-\alpha) & (0.1\text{ mol})\alpha & (0.1\text{ mol})\alpha \end{array}$$

$$p_{NH_3} = p_{H_2S} = \frac{nRT}{V} = \frac{(0.1\,\alpha\text{ mol})\,(8.314\text{ kPa dm}^3\text{ mol}^{-1}\text{ K}^{-1})\,(298\text{ K})}{1\text{ dm}^3}$$

$$= (247.76\,\alpha)$$

Thus $K_p^\circ = (p_{NH_3}/100 \text{ kPa})\,(p_{H_2S}/100 \text{ kPa}) = (2.477\,6\ \alpha)^2$

or $\alpha = \dfrac{\sqrt{K_p^\circ}}{2.477\,6}$ which gives $\alpha = \dfrac{\sqrt{0.111}}{2.477\,6} = 0.135$

(c) Let α' be the degree of dissociation. Therefore, we have

$$\begin{array}{lccc} & NH_4HS(s) \rightleftharpoons & NH_3(g) \quad + & H_2S(g) \\ \text{At } t_{eq} & (0.1 \text{ mol})(1-\alpha') & (0.1 \text{ mol})\alpha' & (0.1 \text{ mol})\alpha' \end{array}$$

$$p_{NH_3} = 20.26 \text{ kPa} + \frac{(0.1\,\alpha' \text{ mol})\,(8.314 \text{ kPa dm}^3 \text{ mol}^{-1} \text{ K}^{-1})(298 \text{ K})}{1 \text{ dm}^3}$$

$$= 20.26 \text{ kPa} + 247.76\ \alpha' \text{ kPa}$$

$$p_{H_2S} = \frac{(0.1\,\alpha' \text{ mol})\,(8.314 \text{ kPa dm}^3 \text{ mol}^{-1} \text{ K}^{-1})(298 \text{ K})}{1 \text{ dm}^3}$$

$$= 247.76\ \alpha' \text{ kPa}$$

$$K_p^\circ = (p_{NH_3}/p^\circ)\,(p_{H_2S}/p^\circ)\,; \qquad (p^\circ = 100 \text{ kPa})$$

$$= (0.202\,6 + 2.477\,6\ \alpha')\,(2.477\,6\ \alpha')$$

or $(2.477\,62\ \alpha')^2 + 0.202\,6 \times 2.477\,6\ \alpha' - K_p^\circ = 0$

$$6.139\ \alpha'^2 + 0.502\ \alpha'^2 - 0.111 = 0$$

Solving for α', we get

$$\alpha' = \frac{-0502 + \sqrt{0.502^2 + 4(6.139)(0.111)}}{2 \times 6.139} = \frac{-0.502 + 1.726}{12.278} = 0.10$$

---

10. The density of an equilibrium mixture of N_2O_4 and NO_2 at 101.325 kPa is 3.62 g dm^{-3} at 288 K and 1.84 g dm^{-3} at 348 K. What is the enthalpy of reaction for

$$N_2O_4(g) \rightleftharpoons 2NO_2(g)$$

Solution Let n be the amount of N_2O_4 at $t = 0$ and α be the degree of dissociation of N_2O_4 at equilibrium. Then

$$\begin{array}{ll} N_2O_4(g) \rightleftharpoons & 2NO_2(g) \\ n(1-\alpha) & n(2\alpha) \end{array}$$

Total amount of species at equilibrium $= n(1 + \alpha)$

Now employing the ideal gas equation, we get

$$pV = n(1+\alpha)\,RT = \frac{m}{M}\,(1+\alpha)RT$$

$$pM = \rho(1+\alpha)RT$$

or $\alpha = \dfrac{pM}{\rho RT} - 1$

At 288 K, $\alpha = \dfrac{(101.325 \text{ kPa})\,(92 \text{ g mol}^{-1})}{(3.62 \text{ g dm}^{-3})\,(8.314 \text{ kPa dm}^3 \text{ K}^{-1} \text{ mol}^{-1})\,(288 \text{ K})} - 1$

$$= 1.076 - 1 = 0.076$$

At 348 K, $\alpha' = \dfrac{(101.325\ \text{kPa})\ (92\ \text{g mol}^{-1})}{(1.84\ \text{g dm}^{-3})\ (8.314\ \text{kPa dm}^3\ \text{K}^{-1}\ \text{mol}^{-1})\ (288\ \text{K})} - 1$

$= 1.752 - 1 = 0.752$

Now $$K_p^\circ = \frac{(p_{NO_2}/p^\circ)^2}{(p_{N_2O_4}/p^\circ)} = \frac{\left(\dfrac{2\alpha}{1+\alpha}(p/p^\circ)\right)^2}{\left(\dfrac{1-\alpha}{1+\alpha}(p/p^\circ)\right)} = \frac{4\alpha^2}{1-\alpha^2}(p/p^\circ)$$

Therefore K_p° at 288 K $= \dfrac{4\times 0.076^2}{1-0.076^2}\left(\dfrac{101.325\ \text{kPa}}{100\ \text{kPa}}\right) = 0.023\,6$

and K_p° at 348 K $= \dfrac{4\times 0.752^2}{1-0.752^2}\left(\dfrac{101.325\ \text{kPa}}{100\ \text{kPa}}\right) = 5.275$

Now $$\ln\frac{K^\circ_{p(T_2)}}{K^\circ_{p(T_1)}} = -\frac{\Delta_r H^\circ}{R}\left[\frac{1}{T_2}-\frac{1}{T_1}\right] = -\frac{\Delta_r H^\circ (T_1 - T_2)}{RT_1T_2}$$

For the given two temperatures, this gives

$$2.303\log\left(\frac{5.275}{0.023\,6}\right) = \frac{\Delta_r H^\circ\ (60\ \text{K})}{(8.314\ \text{J K}^{-1}\ \text{mol}^{-1})\ (288\ \text{K})\ (348\ \text{K})}$$

Therefore $$\Delta_r H^\circ = \frac{(2.303)\ (8.314\ \text{J K}^{-1}\ \text{mol}^{-1})\ (288\ \text{K})\ (348\ \text{K})}{(60\ \text{K})}\log\left(\frac{5.275}{0.023\,6}\right)$$

$= 75.14 \times 10^3\ \text{J mol}^{-1} = 75.14\ \text{kJ mol}^{-1}$

---

11. HgO dissociates according to the equation

$$HgO(s) \rightleftharpoons Hg(g) + \tfrac{1}{2}\,O_2(g)$$

This dissociation pressure is 51.56 kPa at 693 K and 108.02 kPa at 723 K. For this reaction, calculate (a) $\Delta_r H$ and (b) $\Delta_r S$ at 723 K.

Solution

Pressure p at 693 K = 51.56 kPa

Pressure p at 723 K = 108.02 kPa

Stoichiometry of the given reaction indicates that

$$n_{Hg(g)} = 2n_{O_2}(g)$$

Therefore, it follows that

$$p_{Hg(g)} = 2p_{O_2}(g)$$

The partial pressures of Hg(g) and O_2(g) at the given temperatures are:

At 693 K $\quad p_{Hg(g)} = \dfrac{2}{3}\times 51.56\ \text{kPa} = 34.373\ \text{kPa}$

and $\quad p_{O_2(g)} = \dfrac{1}{3}\times 51.56\ \text{kPa} = 17.187\ \text{kPa}$

At 723 K $\qquad p_{Hg(g)} = \frac{2}{3} \times 108.02 \text{ kPa} = 72.013 \text{ kPa}$

and $\qquad p_{O_2}(g) = \frac{1}{3} \times 108.02 \text{ kPa} = 36.007 \text{ kPa}$

Substituting these in the expression of K_p° we get

$$K_p^\circ \text{ at } 693 \text{ K} = (p_{Hg}/p^\circ)(p_{O_2}/p^\circ)^{1/2} = (0.343\,7)(0.171\,9)^{1/2} = 0.142\,5$$

$$K_p^\circ \text{ at } 723 \text{ K} = (p_{Hg}/p^\circ)\,(p_{O_2}/p^\circ)^{1/2} = (0.7201)(0.360\,1)^{1/2} = 0.432\,1$$

Now since $\qquad \ln \dfrac{K^\circ_{p(T_2)}}{K^\circ_{p(T_1)}} = -\dfrac{\Delta_r H^\circ}{R}\left(\dfrac{1}{T_2} - \dfrac{1}{T_1}\right)$

we get $\qquad \Delta_r H^\circ = \dfrac{2.303 \times R \times T_1 T_2}{T_2 - T_1} \log \dfrac{K^\circ_{p(T_2)}}{K^\circ_{p(T_1)}}$

For the given two temperatures, this yields

$$\Delta_r H^\circ = \frac{2.303\,(8.314 \text{ J K}^{-1} \text{ mol}^{-1})\,(693 \text{ K})\,(723 \text{ K})}{(30 \text{ K})} \log\left(\frac{0.432\,1}{0.142\,5}\right)$$

$$= 154\,061 \text{ J mol}^{-1} = 154.061 \text{ kJ mol}^{-1}$$

Now $\qquad \Delta_r G^\circ \text{ at } 723 \text{ K} = -RT \ln K_p^\circ$

$$= -\,(8.314 \text{ J K}^{-1} \text{ mol}^{-1})\,(723 \text{ K}) \times 2.303 \log 0.432\,1$$

$$= 5\,044.7 \text{ J mol}^{-1} = 5.045 \text{ kJ mol}^{-1}$$

Thus $\qquad \Delta_r S^\circ = \dfrac{\Delta_r H^\circ - \Delta_r G^\circ}{T} = \dfrac{154\,061 \text{ mol}^{-1} - 5\,044.7 \text{ J mol}^{-1}}{723 \text{ K}}$

$$= 206.11 \text{ J K}^{-1} \text{ mol}^{-1}$$

---

12. The value of K_p° for the reaction

$$\tfrac{1}{2}\, H_2(g) + \tfrac{1}{2}\, I_2(g) \rightleftharpoons HI(g)$$

is 8.32 at 873 K and 1 bar pressure. Calculate K_p°, K_c°, and K_x for

(i) $\qquad 2HI(g) \rightleftharpoons H_2(g) + I_2(g)$

(ii) $\qquad H_2(g) + I_2(g) \rightleftharpoons 2HI(g)$

(iii) $\qquad HI(g) \rightleftharpoons \tfrac{1}{2} H_2(g) + \tfrac{1}{2} I_2(g)$

Solution

For the reaction

$$\tfrac{1}{2} H_2(g) + \tfrac{1}{2} I_2(g) \rightleftharpoons HI(g)$$

we have $\qquad K_p^\circ = \dfrac{(p_{HI}/p^\circ)}{(p_{H_2}/p^\circ)^{1/2}(p_{I_2}/p^\circ)^{1/2}} = 8.32$

(i) For the reaction $2HI(g) \rightleftharpoons H_2(g) + I_2(g)$, we have

$$K_p^{\circ\prime} = \frac{(p_{H_2}/p^\circ)(p_{I_2}/p^\circ)}{(p_{HI}/p^\circ)^2} = \frac{1}{K_p^{\circ 2}} = \frac{1}{(8.32)^2} = 0.014\,45$$

(ii) For the reaction

$$H_2(g) + I_2(g) \rightleftharpoons 2HI(g)$$

we have $K_p^{\circ\prime\prime} = \dfrac{(p_{HI}/p^\circ)^2}{(p_{H_2}/p^\circ)(p_{I_2}/p^\circ)} = K_p^{\circ 2} = (8.32)^2 = 69.22$

(iii) For the reaction

$$HI(g) \rightleftharpoons \tfrac{1}{2}H_2(g) + \tfrac{1}{2}I_2(g)$$

we have $K_p^{\circ\prime\prime\prime} = \dfrac{(p_{H_2}/p^\circ)^{1/2}(p_{I_2}/p^\circ)^{1/2}}{(p_{HI}/p^\circ)} = \dfrac{1}{K_p^\circ} = \dfrac{1}{(8.32)} = 0.012$

Since for these reactions $\Delta\nu_g = 0$, it follows that for all the reactions

$$K_p^\circ = K_x = K_c^\circ$$

---

13. Assume that the decomposition of HNO_3 can be represented by the following equation

$$4HNO_3(g) \rightleftharpoons 4NO_2(g) + 2H_2O(g) + O_2(g)$$

and that at a given temperature and pressure the reaction approaches equilibrium. Show that if we start with pure HNO_3, then at equilibrium

$$K_p^\circ = \frac{1024\,(p_{O_2})^7}{(p - 7p_{O_2})^4}\frac{1}{(p^\circ)^3}$$

Solution The reaction is

$$4HNO_3(g) \rightleftharpoons 4NO_2(g) + 2H_2O(g) + O_2(g)$$

Its K_p° is

$$K_p^\circ = \frac{(p_{NO_2})^4\,(p_{H_2O})^2\,p_{O_2}}{(p_{HNO_3})^4}\frac{1}{(p^\circ)^3}$$

It is obvious from the reaction that

$$p_{NO_2} = 4p_{O_2}$$
$$p_{H_2O} = 2p_{O_2}$$
$$p_{HNO_3} = p - p_{NO_2} - p_{H_2O} - p_{O_2} = p - 7p_{O_2}$$

Thus substituting these in the above expression, we get

$$K_p^\circ = \frac{1024(p_{O_2})^7}{(p - 7p_{O_2})^4}\frac{1}{(p^\circ)^3}$$

---

14. Consider the dissociation of $N_2O_4(g)$ at 25 °C. Suppose one mole of $N_2O_4(g)$ is confined in a vessel under 101.325 kPa pressure. The standard free energies of formation of $N_2O_4(g)$ and $NO_2(g)$ are respectively 97.89 and 51.31 kJ mol^{-1}. (a) Calculate K_p°, K_x and K_c°. (b) Calculate the amount of N_2O_4 dissociated. (c) If 5 mol of argon are introduced and the mixture confined under 101.325 kPa total pressure, what is the amount of N_2O_4 dissociated? (d) If the volume of the vessel, determined by conditions in (a), is kept constant and 5 mol of argon are introduced, what will be the amount of N_2O_4 dissociated?

Solution

(a) Given that

$$N_2O_4(g) \rightleftharpoons 2NO_2(g)$$

$\Delta_f G^\circ$/kJ mol^{-1} 97.89 51.31

The free energy change of the reaction is

$$\Delta_r G^\circ = 2\Delta_f G^\circ(NO_2) - \Delta_f G^\circ(N_2O_4)$$

$$= (102.62 - 97.89)\ \text{kJ mol}^{-1} = 4.73\ \text{kJ mol}^{-1} = 4\ 730\ \text{J mol}^{-1}$$

Since $\Delta_r G^\circ = -RT \ln K_p^\circ$, therefore

$$\log K_p^\circ = \frac{-\Delta_r G^\circ}{2.303\,RT} = \frac{-4\ 730\ \text{J mol}^{-1}}{2.303\,(8.314\ \text{J K}^{-1}\ \text{mol}^{-1})\,(298\ \text{K})}$$

$$= -0.829\ 0$$

$$K_p^\circ = 0.148$$

Also $$K_p^\circ = \frac{(p_{NO_2}/p^\circ)^2}{(p_{N_2O_4}/p^\circ)} = \frac{\{x_{NO_2}(p/p^\circ)\}^2}{x_{N_2O_4}(p/p^\circ)}$$

$$= \frac{x_{NO_2}^2}{x_{N_2O_4}}(p/p^\circ) = K_x(p/p^\circ)$$

For p = 101.325 kPa, we have

$$K_x = \frac{K_p^\circ}{(p/p^\circ)} = \frac{0.148}{1.0132\ 5} = 0.146$$

$$K_c^\circ = \frac{(c_{NO_2}/c^\circ)^2}{c_{N_2O_4}/c^\circ} = \frac{(p_{NO_2}/RT)^2}{p_{N_2O_4}/RT}\frac{1}{c^\circ}$$

$$= \frac{p_{NO_2}^2}{p_{N_2O_4}}\frac{1}{c^\circ RT} = \frac{K_p}{c^\circ RT} = \frac{K_p^\circ p^\circ}{c^\circ RT}$$

$$= \frac{(0.148)\,(100\ \text{kPa})}{(1\ \text{mol dm}^{-3})\,(8.314\ \text{kPa dm}^{-3}\ \text{mol}^{-1}\ \text{K}^{-1})\,(298\ \text{K})}$$

$$= 5.97 \times 10^{-3}$$

(b) If α is the amount of N_2O_4 dissociated, then

$n(N_2O_4) = (1\ \text{mol} - \alpha)$ and $n(NO_2) = (2\alpha)$

Total amount of gases = 1 mol + α

The mole fractions are

$$x_{N_2O_4} = \frac{1\ \text{mol} - \alpha}{1\ \text{mol} + \alpha} \quad \text{and} \quad x_{NO_2} = \frac{2\alpha}{1\ \text{mol} + \alpha}$$

$$K_p^\circ = \frac{(p_{NO_2}/p^\circ)^2}{(p_{N_2O_4}/p^\circ)} = \frac{\{x_{NO_2}(p/p^\circ)\}^2}{x_{N_2O_4}(p/p^\circ)} = \frac{\left\{\dfrac{2\alpha}{1\ \text{mol} + \alpha}(p/p^\circ)\right\}^2}{\left\{\dfrac{1\ \text{mol} - \alpha}{1\ \text{mol} + \alpha}(p/p^\circ)\right\}}$$

$$= \frac{4\alpha^2(p/p^\circ)}{(1\ \text{mol})^2 - \alpha^2}$$

Therefore $\alpha/\text{mol} = \sqrt{\dfrac{K_p^\circ}{4(p/p^\circ)+K_p^\circ}} = \sqrt{\dfrac{0.148}{4\times 1.0132\,5+0.148}} = 0.187\,7$

The amount of N_2O_4 dissociated is

$$\alpha = 0.187\,7 \text{ mol}$$

(c) Let α' be the amount of N_2O_4 dissociated in the presence of 5 mol of Argon, then

$$n(N_2O_4) = 1 \text{ mol} - \alpha' \quad \text{and} \quad n(NO_2) = 2\alpha'$$

and the total amount will be

$$5 \text{ mol} + (1 \text{ mol} + \alpha') = 6 \text{ mol} + \alpha'$$

Substituting these in the expression of K_p°, we get

$$K_p^\circ = \frac{\left\{\dfrac{2\alpha'}{6\text{ mol}+\alpha'}(p/p^\circ)\right\}^2}{\left\{\dfrac{(1\text{ mol}-\alpha')}{6\text{ mol}+\alpha'}(p/p^\circ)\right\}} = \frac{4\alpha'^2}{(1\text{ mol}-\alpha')\,(6\text{ mol}+\alpha')}(p/p^\circ)$$

Rearranging this, we get

$$(\alpha'/\text{mol})^2\,\{4(p/p^\circ) + K_p^\circ\} + 5K_p^\circ\,(\alpha'/\text{mol}) - 6K_p^\circ = 0$$

Substituting p = 101.325 kPa and K_p° = 0.148, we get

or $\quad 4.201\,(\alpha'/\text{mol})^2 + 0.74\,(\alpha'\text{ mol}) - 0.888 = 0$

Solving for α', we get

$$\alpha'/\text{mol} = \frac{-0.74+\sqrt{(0.74)^2+4\times 4.201\times 0.888}}{2\times 4.201} = \frac{-0.74+3.933}{8.402}$$

$$= 0.38$$

(d) Volume of the vessel, $V = \dfrac{nRT}{p} = \dfrac{(1.187\,7\text{ mol})RT}{101.325\text{ kPa}}$

Pressure of the system after introducing 5 mol of Argon is given as

$$p = \frac{nRT}{V} = \frac{(6\text{ mol}+\alpha'')RT}{(1.187\,7\text{ mol})\,RT/101.325\text{ kPa}}$$

$$= \frac{(6\text{ mol}+\alpha'')(101.325\text{ kPa})}{(1.187\,7\text{ mol})}$$

where α'' is the new amount of N_2O_4 dissociated.

Since $\quad K_p^\circ = \dfrac{(4\alpha''^2)(p/p^\circ)}{(1\text{ mol}-\alpha'')(6\text{ mol}+\alpha'')}$

Substituting the value of p in terms of α'', we have

$$K_p^\circ = \frac{4\alpha''^2}{(1\text{ mol}-\alpha'')\,(6\text{ mol}+\alpha'')}\,\frac{(6\text{ mol}+\alpha'')}{(1.187\,7\text{ mol})}\,\frac{(101.325\text{ kPa})}{(100\text{ kPa})}$$

$$= \frac{4\alpha''^2}{(1\,\text{mol} - \alpha'')\,(1.172\,2\,\text{mol})}$$

or $$4(\alpha''/\text{mol})^2 + 1.172\,2\,K_p^\circ\,(\alpha''/\text{mol}) - 1.172\,2\,K_p^\circ = 0$$

or $$4(\alpha''/\text{mol})^2 + 0.173\,5\,(\alpha''/\text{mol}) - 1.173\,5 = 0 \qquad (\text{since } K_p^\circ = 0.148)$$

Solving for α'', we get

$$\alpha''/\text{mol} = \frac{-0.173\,5 + \sqrt{(0.173\,5)^2 + 4\times 4\times 0.1735}}{2\times 4} = \frac{-0.173\,5 + 1.675}{8}$$

$$= 0.188$$

---

15. At high temperature CO_2 dissociates according to the equation

$$2CO_2(g) \rightleftharpoons 2CO(g) + O_2(g)$$

At 101.325 kPa pressure the percentage of oxygen at equilibrium is 2.05×10^{-5} at 1 000 K and at 1 400 K is 1.27×10^{-2}. Assuming the enthalpy change $\Delta_r H$ of the reaction to be independent of temperature, calculate the standard free energy change and the standard entropy change at 1 000 K.

Solution At equilibrium, we have

$$\begin{matrix} 2CO_2(g) & \rightleftharpoons & 2CO(g) & + & O_2(g) \\ n(1-2\alpha) & & n(2\alpha) & & n\alpha \end{matrix}$$

Total amount of gases $= n(1 - 2\alpha) + 2n\alpha + n\alpha = n(1 + \alpha)$

Percentage of oxygen at equilibrium, $x = \dfrac{\alpha}{1+\alpha} \times 100$

Since α is expected to be very small, we may write $x = \alpha \times 100$ or $\alpha = x/100$

Thus, At 1 000 K, $\alpha = 2.05 \times 10^{-5}/100 = 2.05 \times 10^{-7}$

At 1 400 K, $\alpha = 1.27 \times 10^{-2}/100 = 1.27 \times 10^{-4}$

The partial pressures of CO_2, CO and O_2 are

$$p_{CO_2} = x_{CO_2} p = \frac{n(1-2\alpha)}{n(1+\alpha)} p = \frac{1-2\alpha}{1+\alpha} p$$

$$p_{CO} = x_{CO} p = \frac{n(2\alpha)}{n(1+\alpha)} p = \frac{2\alpha}{1+\alpha} p$$

$$p_{O_2} = x_{O_2} p = \frac{n\alpha}{n(1+\alpha)} p = \frac{\alpha}{1+\alpha} p$$

Substituting these in the expression

$$K_p^\circ = \frac{(p_{CO}/p^\circ)^2 (p_{O_2}/p^\circ)}{(p_{CO_2}/p^\circ)^2}$$

we get $$K_p^\circ = \frac{\left(\frac{2\alpha}{1+\alpha}(p/p^\circ)\right)^2\left(\frac{\alpha}{1+\alpha}(p/p^\circ)\right)}{\left(\frac{1-2\alpha}{1+\alpha}(p/p^\circ)\right)^2} = \frac{4\alpha^3}{(1-2\alpha)^2(1+\alpha)}\left(\frac{p}{p^\circ}\right)$$

Assuming α to be negligible in comparison to 1, we get

$$K_p^\circ = 4\alpha^3(p/p^\circ)$$

Thus
$$K^\circ_{p(1000\ \mathrm{K})} = 4 \times (2.05 \times 10^{-7})^3\ (1.013\ 25) = 32.42 \times 10^{-21}$$
$$K^\circ_{p(1400\ \mathrm{K})} = 4 \times (1.27 \times 10^{-4})^3\ (1.013\ 25) = 8.30 \times 10^{-12}$$
$$\Delta_r G^\circ = -RT \ln K_p^\circ$$
$$= -(8.314\ \mathrm{J\ K^{-1}\ mol^{-1}})\ (1\ 000\ \mathrm{K}) \times 2.303 \log (32.42 \times 10^{-21})$$
$$= 373\ 095\ \mathrm{J\ mol^{-1}} = 373.10\ \mathrm{kJ\ mol^{-1}}$$

since $$\log\left\{\frac{K^\circ_{p(T_2)}}{K^\circ_{p(T_1)}}\right\} = \frac{\Delta_r H^\circ}{2.303\,R}\left(\frac{1}{T_1} - \frac{1}{T_2}\right)$$

$$\Delta_r H^\circ = \frac{2.303(8.314\ \mathrm{J K^{-1}\ mol^{-1}})\ (1\ 000\ \mathrm{K})\ (1\ 400\ \mathrm{K})}{(400\ \mathrm{K})}\log\left(\frac{8.30\times10^{-12}}{32.42\times10^{-21}}\right)$$
$$= 563\ 482.2\ \mathrm{J\ mol^{-1}} = 563.48\ \mathrm{kJ\ mol^{-1}}$$

$$\Delta_r S^\circ = \frac{\Delta_r H^\circ - \Delta_r G^\circ}{T} = \frac{(563.48\times10^3 - 373.10\times10^3)\ \mathrm{J\ mol^{-1}}}{(1\ 000\ \mathrm{K})}$$
$$= 190.38\ \mathrm{J\ K^{-1}\ mol^{-1}}$$

---

16. PCl_5 vapour decomposes on heating according to

$$PCl_5(g) \rightleftharpoons PCl_3(g) + Cl_2(g)$$

The density of a sample of partially dissociated PCl_5 at 101.325 kPa and 503 K was found to be 4.8 g dm^{-3}. Calculate the degree of dissociation and $\Delta_r G^\circ$ for the reaction at 503 K.

Solution Let α be the degree of dissociation of PCl_5. At equilibrium, we will have

$$\begin{array}{ccccc} PCl_5 & \rightleftharpoons & PCl_3 & + & Cl_2 \\ n(1-\alpha) & & n\alpha & & n\alpha \end{array}$$

Total amount of gases = $n(1 + \alpha)$

Now employing ideal gas equation, we get

$$pV = n(1 + \alpha)\ RT$$

or $$1+\alpha = \frac{pV}{nRT} = \frac{pV}{(m/M_{PCl_5})RT} = \frac{pM_{PCl_5}}{\rho RT}$$

Thus $$\alpha = \frac{pM_{PCl_5}}{\rho RT} - 1$$

For the given value of ρ, we get

$$\alpha = \frac{(101.325\ \mathrm{kPa})\ (208.5\ \mathrm{g\ mol^{-1}})}{(4.8\ \mathrm{g\ dm^{-3}})\ (8.314\ \mathrm{kPa\ dm^3\ mol^{-1}\ K^{-1}})\ (503\ \mathrm{K})} - 1$$
$$= 1.053 - 1 = 0.053$$

Now $$K_p^\circ = \frac{(p_{PCl_3}/p^\circ)(p_{Cl_2}/p^\circ)}{(p_{PCl_5}/p^\circ)} = \frac{\left(\frac{\alpha}{1+\alpha}(p/p^\circ)\right)\left(\frac{\alpha}{1+\alpha}(p/p^\circ)\right)}{\left(\frac{1-\alpha}{1+\alpha}(p/p^\circ)\right)}$$

$$= \frac{\alpha^2}{1-\alpha^2}\left(\frac{p}{p^\circ}\right)$$

Substituting $\alpha = 0.053$, we get

$$K_p^\circ = \frac{0.053^2}{1-0.053^2}(1.013\ 25) = 2.854 \times 10^{-3}$$

$$\Delta_r G^\circ = -RT \ln K_p^\circ$$

$$= -(8.314\ \text{J K}^{-1}\ \text{mol}^{-1})(503\ \text{K}) \times 2.303 \times \log(2.854 \times 10^{-3})$$

$$= 24\ 506\ \text{J mol}^{-1} = 24.506\ \text{kJ mol}^{-1}$$

---

17. The equilibrium pressure at 25 °C for the reaction

$$CuSO_4 \cdot 3H_2O(s) \rightleftharpoons CuSO_4 \cdot H_2O(s) + 2H_2O(g)$$

is 0.746 8 kPa. The standard enthalpy change at this temperature is 112.968 kJ mol^{-1}.

(a) What is the value of K_p° ?

(b) Determine whether the reaction is spontaneous at 298 K and 101.325 kPa.

(c) Calculate $\Delta_r S^\circ$ for the reaction.

(d) Give a physical explanation for the result in (c).

(e) Explain the result in (b) using the values of $\Delta_r H^\circ$ and $\Delta_r S^\circ$.

Solution

(a) For the reaction

$$CuSO_4 \cdot 3H_2O(s) \rightleftharpoons CuSO_4 \cdot H_2O(s) + 2H_2O(g)$$

we have $$K_p^\circ = (p_{H_2O}/p^\circ)^2 = (0.746\ 8\ \text{kPa}/100\ \text{kPa})^2$$

$$= (7.468 \times 10^{-3})^2 = 5.577 \times 10^{-5}$$

(b) To predict the nature of reaction, we calculate $\Delta_r G^\circ$ of the reaction

$$\Delta_r G^\circ = -RT \ln K_p^\circ$$

$$= -(8.314\ \text{J K}^{-1}\ \text{mol}^{-1})(298\ \text{K}) \times 2.303 \times \log(5.577 \times 10^{-5})$$

$$= 24\ 270\ \text{J mol}^{-1} = 24.27\ \text{kJ mol}^{-1}$$

Since $\Delta_r G^\circ$ is positive, the reaction will not be spontaneous.

(c) $$\Delta_r S^\circ = \frac{\Delta_r H^\circ - \Delta_r G^\circ}{T} = \frac{112.968\ \text{kJ mol}^{-1} - 24.27\ \text{kJ mol}^{-1}}{(298\ \text{K})}$$

$$= 0.298\ \text{kJ K}^{-1}\ \text{mol}^{-1} = 298\ \text{J K}^{-1}\ \text{mol}^{-1}$$

(d) Since the number of gaseous substance on the product side is larger than in the reactant side, the reaction as written will be attended to by a large increase in entropy.

(e) Large positive value of change of enthalpy factor dominates over $T\Delta_r S^\circ$ term, and thus overall $\Delta_r G^\circ$ is positive.

---

18. The following data for the equilibrium composition of the reaction

$$2Na(g) \rightleftharpoons Na_2(g)$$

at 1.013 MPa pressure and 1 482.53 K have been obtained.

mass % Na (monomer gas) = 71.3

mass % Na_2 (dimer gas) = 28.7

Calculate the standard equilibrium constant K_p°.

Solution At equilibrium, we have

$$\begin{array}{ccc} 2Na(g) & \rightleftharpoons & Na_2(g) \\ 71.3\text{ g} & & 28.7\text{ g} \\ 3.1\text{ mol} & & 0.624\text{ mol} \end{array}$$

Total amount of species = 3.1 mol + 0.624 mol = 3.724 mol

The partial pressures are

$$p_{Na} = x_{Na}p = \frac{3.1}{3.724} \times 1.013\text{ MPa} = 0.843\text{ MPa}$$

$$p_{Na_2} = x_{Na_2}p = \frac{0.624}{3.724} \times 1.013\text{ MPa} = 0.170\text{ MPa}$$

$$K_p^\circ = \frac{p_{Na_2}/p^\circ}{(p_{Na}/p^\circ)^2} = \frac{(170\text{ kPa}/100\text{ kPa})}{(843\text{ kPa}/100\text{ kPa})^2} = 0.023\,9$$

---

19. Given are the following standard free energies of formation at 298 K.

	CO(g)	CO_2(g)	H_2O(g)	H_2O(l)
$\Delta_f G/\text{kJ mol}^{-1}$	−137.17	−394.36	−228.57	−237.13

(a) Find $\Delta_r G^\circ$ and the standard equilibrium constant K_p° at 298 K for the reaction

$$CO(g) + H_2O(g) \rightleftharpoons CO_2(g) + H_2(g)$$

(b) Find the vapour pressure of pure water at 298 K.

(c) If CO, CO_2 and H_2 are mixed so that the partial pressure of each is 101.325 kPa and the mixture is brought into contact with excess of liquid water. What will be the partial pressure of each gas when equilibrium is attained at 298 K. The volume available to the gases is constant.

Solution **(a)**

$$\Delta_r G^\circ = \Delta_f G^\circ(CO_2, g) + \Delta_f G^\circ(H_2, g) - \Delta_f G^\circ(CO, g) - \Delta_f G^\circ(H_2O, g)$$

$$= [-394.36 + 0 + 137.17 + 228.57]\text{ kJ mol}^{-1}$$

$$= -28.62\text{ kJ mol}^{-1}$$

Since $\Delta_r G^\circ = -RT \ln K_p^\circ$

therefore $$\log K_p^\circ = -\frac{\Delta_r G^\circ}{2.303\,RT} = \frac{(28\,620\text{ mol}^{-1})}{2.303\,(8.314\text{ J K}^{-1}\text{ mol}^{-1})(298\text{ K})}$$

$$= 5.02$$

Thus $K^\circ = 1.047 \times 10^5$

(b) For the determination of vapour pressure of water, we consider the equilibrium

$$H_2O(1) \rightleftharpoons H_2O(g)$$

The free energy change of the reaction is

$$\Delta_r G^\circ = \Delta_f G^\circ(H_2O, g) - \Delta_f G^\circ(H_2O, 1)$$

$$= -228.57\text{ kJ mol}^{-1} + 237.13\text{ kJ mol}^{-1} = 8.56\text{ kJ mol}^{-1}$$

Now employing the reaction

$$\log K_p^\circ = -\frac{\Delta_r G^\circ}{2.303\,RT}$$

we get $$\log K_p^\circ = -\frac{(8.56\times 10^3\ \text{J mol}^{-1})}{(2.303)\,(8.314\ \text{J K}^{-1}\ \text{mol}^{-1})\,(298\ \text{K})} = -1.500$$

or $$K_p^\circ = 3.16 \times 10^{-2}$$

Now, since $K_p^\circ = p_{H_2O(g)}/p^\circ$ the vapour pressure of water will be equal to 3.16 kPa.

(c) Since now $H_2O(g)$ will remain in equilibrium with $H_2O(1)$, the partial vapour pressure of the former will be given by

$$p_{H_2O} = 3.16\ \text{kPa}$$

At equilibrium, the partial vapour pressure of CO(g) will decrease from the given value of 101.325 kPa as it will combine with $H_2O(g)$ to give products. Let x be the decrease in the pressure of CO, then

$$p_{CO_2} = p_{H_2} = 101.325\ \text{kPa} + x$$

Substituting these in the expression

$$K_p^\circ = \frac{(p_{CO_2}/p^\circ)\,(p_{H_2}/p^\circ)}{(p_{CO}/p^\circ)\,(p_{H_2O}/p^\circ)} = \frac{p_{CO_2}\,p_{H_2}}{p_{CO}\ p_{H_2O}}$$

we get $$K_p^\circ = \frac{(101.325\ \text{kPa} + x)^2}{(101.325\ \text{kPa} - x)\,(3.16\ \text{kPa})} = 1.047\times 10^5$$

$$= \frac{(1 + x/101.325\ \text{kPa})^2}{(1 - x/101.325\ \text{kPa})\,(0.03119)} = 1.047\times 10^5$$

or $$1 + (x/101.325\ \text{kPa})^2 + 2(x/101.325\ \text{kPa}) = 3\,266 - 3\,266\,(x/101.325\ \text{kPa})$$

or $$(x/101.325\ \text{kPa})^2 + 3\,268\,(x/101.325\ \text{kPa}) - 3\,265 = 0$$

Solving for $x/101.325$ kPa, we get

$$x/101.325\ \text{kPa} = \frac{-3\,268 + \sqrt{3\,268^2 + 4\times 3\,265}}{2} = \frac{-3\,268 + 3\,270}{2}$$

$$x/101.325\ \text{kPa} \simeq 1$$

Therefore the partial pressures of various species at equilibrium will be

$$p_{CO_2} = p_{H_2} = 202.65\ \text{kPa}$$

$$p_{H_2O} = 3.16\ \text{kPa}$$

$$p_{CO} = \frac{p_{CO_2}\,p_{H_2}}{p_{H_2O}\,K_p^\circ} = \frac{(202.65\ \text{kPa})\,(202.65\ \text{kPa})}{(3.16\ \text{kPa})\,(1.047\times 10^5)} = 0.124\ \text{kPa}$$

---

20. A container whose volume is V contains an equilibrium mixture that consists of 2 mol each of PCl_5, PCl_3 and Cl_2 (all as gases). The pressure is 3 bar and temperature is T. A certain amount of $Cl_2(g)$ is now introduced, keeping the pressure and temperature constant, until the equilibrium volume is $2V$. Calculate the amount of Cl_2 that was added and the value of K_p°.

Solution

At equilibrium, we have

$$\begin{array}{ccccc} PCl_5 & \rightleftharpoons & PCl_3 & + & Cl_2 \\ 2\ \text{mol} & & 2\ \text{mol} & & 2\ \text{mol} \end{array}$$

Total amount of gases = 6 mol

Thus $$K_p^\circ = \frac{(p_{PCl_3}/p^\circ)(p_{Cl_2}/p^\circ)}{(p_{PCl_5}/p^\circ)} = \frac{\left(\frac{2}{6}p/p^\circ\right)^2}{\left(\frac{2}{6}p/p^\circ\right)}$$

Substituting p = 3 bar, we get

$$K_p^\circ = 1$$

Let x be the amount of PCl_3 that combines when the amount y of chlorine is added keeping p and T constant. Thus, the amounts of PCl_3, Cl_2 and PCl_5 become

$$n(PCl_3) = 2\text{ mol} - x$$

$$n(Cl_2) = y + 2\text{ mol} - x$$

$$n(PCl_5) = 2\text{ mol} + x$$

Since the final volume after the addition of Cl_2 is twice the initial volume, it follows that the total amount of gases in $2V$ is 2 × 6 mol = 12 mol. Since $n(PCl_3) + n(PCl_5)$ is 4 mol, the total amount of chlorine is 8 mol.

$$\text{Total amount of gases} = y + 6\text{ mol} - x = 12\text{ mol}$$

Their partial pressures are

$$p_{PCl_3} = \frac{2\text{ mol} - x}{12\text{ mol}}p = \frac{2\text{ mol} - x}{12\text{ mol}} \times 3\text{ bar}$$

$$p_{Cl_2} = \frac{8\text{ mol}}{12\text{ mol}}p = \frac{8\text{ mol}}{12\text{ mol}} \times 3\text{ bar} = 2\text{ bar}$$

$$p_{PCl_5} = \frac{2\text{ mol} + x}{12\text{ mol}}p = \frac{2\text{ mol} + x}{12\text{ mol}} \times 3\text{ bar}$$

Substituting these in the expression

$$K_p^\circ = \frac{(p_{PCl_3}/p^\circ)(p_{Cl_2}/p^\circ)}{(p_{PCl_5}/p^\circ)} \qquad \text{(where } p^\circ = 1\text{ bar)}$$

we get $$\frac{\left(\frac{2\text{ mol} - x}{4\text{ mol}}\right)(2)}{\left(\frac{2\text{ mol} + x}{4\text{ mol}}\right)} = \frac{(2\text{ mol} - x)(2)}{(2\text{ mol} + x)} = 1 \qquad (\text{as } K_p^\circ = 1)$$

or $$4 - 2\,(x/\text{mol}) = 2 + (x/\text{mol})$$

or $$3\,(x/\text{mol}) = 2$$

$$x/\text{mol} = \frac{2}{3} = 0.67$$

Therefore, the amount of Cl_2 added

$$y = 6\text{ mol} + x = 6.67\text{ mol}$$

21. For the reaction

$$C_2H_6(g) \rightleftharpoons C_2H_4(g) + H_2(g)$$

K_p° is 0.05 and $\Delta_r G^\circ$ is 22.384 kJ mol^{-1} at 900 K. If an initial mixture comprising 20 mol of C_2H_6 and 80 mol of N_2(inert) is passed over a dehydrogenation catalyst at 900 K, what is the equilibrium percentage composition of the effluent gas mixture? The total pressure is kept at 0.5 bar. Given: $\Delta_r S^\circ = 135.143$ J K^{-1} mol^{-1} at 300 K. Calculate $\Delta_r G^\circ$ at 300 K. (Assume $\Delta_r C_p = 0$.)

Solution

Let x be the amount out of 20 mol of C_2H_6 that dissociate to give C_2H_4 and H_2. Thus, we have

$$\begin{array}{cccc} C_2H_6(g) & \rightleftharpoons C_2H_4(g) & + & H_2(g) \\ 20\text{ mol} - x & x & & x \end{array}$$

Total amount of gases = 80 mol + (20 mol – x + x + x) = 100 mol + x

Thus $$K_p^\circ = \frac{\left(\dfrac{x}{100\text{ mol} + x}(p/p^\circ)\right)\left(\dfrac{x}{100\text{ mol} + x}(p/p^\circ)\right)}{\left(\dfrac{20\text{ mol} - x}{100\text{ mol} + x}(p/p^\circ)\right)}$$

$$= \frac{x^2(p/p^\circ)}{(100\text{ mol} + x)(20\text{ mol} - x)}$$

$$= \frac{(x/\text{mol})^2(p/p^\circ)}{\{100 + (x/\text{mol})\}\{20 - (x/\text{mol})\}}$$

Substituting the value of p and K_p° we have

$$0.05 = \frac{0.5\,(x/\text{mol})^2}{2\,000 - 80(x/\text{mol}) - (x/\text{mol})^2} \qquad \text{(where } p^\circ = 1 \text{ bar)}$$

or $$11(x/\text{mol})^2 + 80(x/\text{mol}) - 2\,000 = 0$$

$$(x/\text{mol}) = \frac{-80 + \sqrt{80^2 + 4(11)(2\,000)}}{2 \times 11} = \frac{-80 + 307.3}{22} = \frac{227.3}{22}$$

$$= 10.33$$

Therefore $n(C_2H_6) = 20$ mol – 10.33 mol = 9.67 mol

$n(H_2) = n(C_2H_4) = 10.33$ mol

$n(N_2) = 80$ mol

Total amount of gases = 100 mol + 10.33 mol = 110.33 mol

$$\text{Amount \% of } C_2H_6 = \frac{9.67\text{ mol}}{110.33\text{ mol}} \times 100 = 8.765$$

$$\text{Amount \% of } H_2 \text{ and } C_2H_6 \text{ separately} = \frac{10.33\text{ mol}}{110.33\text{ mol}} \times 100 = 9.363$$

$$\text{Amount \% of } N_2 = \frac{80\text{ mol}}{110.33\text{ mol}} \times 100 = 72.509$$

Since $\Delta_r C_p$ is zero, therefore, $\Delta_r H°$ and $\Delta_r S°$ will be independent of temperature, i.e.

$$\Delta_r H°_{300\text{K}} = \Delta_r H°_{900\text{K}}$$

$$\Delta S°_{300\text{K}} = \Delta S°_{900\text{K}}$$

Now

$$\Delta_r G°_{900\text{K}} = \Delta_r H°_{900\text{K}} - (900\text{ K})\,\Delta_r S°_{900\text{K}}$$

$$\Delta_r G°_{300\text{K}} = \Delta_r H°_{300\text{K}} - (300\text{ K})\,\Delta_r S°_{300\text{K}}$$

$$\Delta_r G°_{300\text{K}} - \Delta_r G°_{900\text{K}} = (600\text{ K})\,\Delta_r S° = 600 \times 135.143\text{ J mol}^{-1}$$

$$= 81.085.8\text{ J mol}^{-1} \approx 81.086\text{ kJ mol}^{-1}$$

$$\Delta_r G°_{300\text{K}} = \Delta_r G°_{900\text{K}} + 81.086\text{ kJ mol}^{-1}$$

$$= 22.384\text{ kJ mol}^{-1} + 81.086\text{ kJ mol}^{-1} = 103.47\text{ kJ mol}^{-1}$$

---

22. A 2 dm^3 flask maintained at 700 K contains 0.1 mol of CO and a catalyst for the reaction

$$CO(g) + 2H_2(g) \rightleftharpoons CH_3OH(g)$$

Hydrogen is introduced until the equilibrium total pressure is 7 bar at which point 0.06 mol of methanol is formed. (a) Calculate $K_p°$. (b) What would be the final pressure if the same amounts of CO and H_2 are used, but no catalyst is present such that no reaction occurs?

Solution

(a) Let y be the amount of H_2 that is the introduced and let the amount of methanol formed be equal to x. At equilibrium, we will have

$$\begin{array}{ccccc} CO & + & 2H_2 & \rightleftharpoons & CH_3OH \\ 0.1\text{ mol} - x & & y - 2x & & x \end{array}$$

Total amount, $n = (0.1\text{ mol} - x) + (y - 2x) + x = 0.1\text{ mol} + y - 2x$

The total amount as given by the ideal gas equation is

$$n = \frac{pV}{RT} = \frac{(700\text{ kPa})(2\text{ dm}^3)}{(8.314\text{ kPa dm}^3\text{ K}^{-1}\text{ mol}^{-1})(700\text{ K})} = 0.240\,6\text{ mol}$$

$$\text{Amount of } H_2 \text{ introduced} = y = n - 0.1\text{ mol} + 2x$$

Now it is given that $x = 0.06$ mol. Thus, we have

$$y = 0.240\,6\text{ mol} - 0.1\text{ mol} + 0.12\text{ mol} = 0.260\,6\text{ mol}$$

$$K_p° = \left(\frac{(p_{CH_3OH}/p°)}{(p_{CO}/p°)(p_{H_2}/p°)^2}\right) = \frac{\left(\frac{x}{n}(p/p°)\right)}{\left(\frac{0.1\text{ mol} - x}{n}(p/p°)\right)\left(\frac{y-2x}{n}(p/p°)\right)^2}$$

$$= \frac{\left(\frac{0.06}{0.240\,6}\times 7\right)}{\left(\frac{0.04}{0.240\,6}\times 7\right)\left(\frac{0.140\,6}{0.240\,6}\times 7\right)^2} = \frac{1.746}{(1.164)(16.733)} = \frac{1.722}{19.581}$$

$$= 0.089\,6$$

(b) Total amount of CO and H_2 if no reaction occurs
$= 0.1 \text{ mol} + 0.260\,6 \text{ mol} = 0.360\,6 \text{ mol}$

The pressure corresponding to this would be

$$p = \frac{0.360\,6 \times 7 \text{ bar}}{0.240\,6} = 10.49 \text{ bar}$$

---

23. For the reaction

$$CO_2(g) + H_2(g) \rightleftharpoons CO(g) + H_2O(g)$$

K_p° has the value 10^{-5} at 298 K and $\Delta_r S^\circ = -41.84 \text{ J K}^{-1} \text{ mol}^{-1}$. ($\Delta_r H^\circ$ and $\Delta_r S^\circ$ do not change with temperature). One mole of CO, 2 mol of H_2 and 3 mol of CO_2 are introduced into a 5 dm^3 flask at 298 K. Calculate: (a) $\Delta_r G^\circ$ at 298 K, (b) the equilibrium pressure, (c) the amount of $H_2O(g)$ present at equilibrium, and (d) K_p° at 373 K.

Solution Let x be the amount of CO_2 that combines at equilibrium. Thus, we have

$$\begin{array}{ccccccc} CO_2(g) & + & H_2(g) & \rightleftharpoons & CO(g) & + & H_2O(g) \\ 3 \text{ mol} - x & & 2 \text{ mol} - x & & 1 \text{ mol} + x & & x \end{array}$$

(a) $\Delta_r G^\circ = -RT \ln K_p^\circ$
$= -(8.314 \text{ J K}^{-1} \text{ mol}^{-1})(298 \text{ K}) \times 2.303 \log(10^{-5})$
$= 28\,529 \text{ J mol}^{-1} = 28.529 \text{ kJ mol}^{-1}$

(b) Total amount of gases $= (3 \text{ mol} - x) + (2 \text{ mol} - x) + (1 \text{ mol} + x) + x = 6 \text{ mol}$

$$p_{eq} = \frac{nRT}{V} = \frac{(6 \text{ mol})(8.314 \text{ kPa dm}^3 \text{ K}^{-1} \text{ mol}^{-1})(298 \text{ K})}{(5 \text{ dm}^3)}$$

$$= 2\,973.09 \text{ kPa}$$

(c) $$K_p^\circ = \frac{(p_{CO}/p^\circ)(p_{H_2O}/p^\circ)}{(p_{CO_2}/p^\circ)(p_{H_2}/p^\circ)} = \frac{\left(\dfrac{1 \text{ mol} + x}{6 \text{ mol}}(p/p^\circ)\right)\left(\dfrac{x}{6 \text{ mol}}(p/p^\circ)\right)}{\left(\dfrac{3 \text{ mol} - x}{6 \text{ mol}}(p/p^\circ)\right)\left(\dfrac{2 \text{ mol} - x}{6 \text{ mol}}(p/p^\circ)\right)}$$

$$= \frac{x(1 \text{ mol} + x)}{(3 \text{ mol} - x)(2 \text{ mol} - x)} = \frac{(x/\text{mol})(1 + x/\text{mol})}{(3 - x/\text{mol})(2 - x/\text{mol})}$$

$$= 10^{-5}$$

Since K_p° is very small, therefore, the value of x/mol is expected to be small. Thus

$$K_p^\circ = \frac{(x/\text{mol})(1)}{(3)(2)} = 10^{-5}$$

or $x/\text{mol} = 6.0 \times 10^{-5}$

(d) $\Delta_r G^\circ_{373\text{K}} = \Delta_r H^\circ_{373\text{K}} - T\,\Delta_r S^\circ_{373\text{K}}$

$\Delta_r G^\circ_{298\text{K}} = \Delta_r H^\circ_{298\text{K}} - T\,\Delta_r S^\circ_{298\text{K}}$

$\Delta_r G^\circ_{373\text{K}} - \Delta_r G^\circ_{298\text{K}} = \Delta_r S^\circ (298 \text{ K} - 373 \text{ K})$;

(since $\Delta_r H^\circ$ and $\Delta_r S^\circ$ are independent of temperature)

$\Delta_r G^\circ_{373\text{K}} = \Delta_r G^\circ_{298\text{K}} - (75 \text{ K})\,\Delta_r S^\circ$

$$= 28\ 529\ \text{J mol}^{-1} + (75\ \text{K})\ (41.84\ \text{J K}^{-1}\ \text{mol}^{-1})$$

$$= 31\ 667\ \text{J mol}^{-1} = 31.667\ \text{kJ mol}^{-1}$$

Now since $\Delta_r G° = -RT \ln K_p°$, we get

$$\log K_p° = -\frac{\Delta_r G°}{2.303\ RT} = -\frac{31\,667\ \text{J mol}^{-1}}{(2.303)\ (8.314\ \text{J K}^{-1}\ \text{mol}^{-1})\ (373\ \text{K})} = -\ 4.434$$

$$K_p° = 3.681 \times 10^{-5}$$

Alternatively, since

$$\Delta_r G° = \Delta_r H° - T\,\Delta_r S°$$

therefore,

$$\Delta_r H° = \Delta_r G° + T\,\Delta_r S°$$

$$= 28\ 529\ \text{J mol}^{-1} + (298\ \text{K})\ (-\ 41.84\ \text{J K}^{-1}\ \text{mol}^{-1})$$

$$= 16\ 060.68\ \text{J mol}^{-1}$$

$$\log \frac{K°_{p(T_2)}}{K°_{p(T_1)}} = -\frac{\Delta_r H°}{2.303\ R}\left[\frac{1}{T_2} - \frac{1}{T_1}\right]$$

Thus

$$\log \frac{K°_{p(373\text{K})}}{10^{-5}} = \frac{(16\,060.68\ \text{J mol}^{-1})}{(2.303)\ (8.314\ \text{J K}^{-1}\ \text{mol}^{-1})}\left[\frac{75\ \text{K}}{(373\ \text{K})\ (298\ \text{K})}\right]$$

or

$$\log \frac{K°_{p(373\text{K})}}{10^{-5}} = 3.681$$

$$K°_{p(373\text{K})} = 3.681 \times 10^{-5}$$

REVISIONARY PROBLEMS

7.1 (a) What do you understand by the term extent of reaction?

(b) If the reaction is carried out starting only from the reactants, show that the change in the amount of each component is related to its stoichiometry number by the expression

$$\frac{dn_1}{\nu_1} = \frac{dn_2}{\nu_2} = \frac{dn_3}{n_3} = \cdots = d\xi$$

where νs have negative values for reactants and positive values for products.

7.2 (a) The change in the free energy of a chemical reaction or due to changes in temperature and pressure is

$$dG = -S\ dT + V\ dp + \sum_{\text{B}} \mu_{\text{B}}\ dn_{\text{B}}$$

Express this in terms of extent of reaction and show that the reaction potential $\Delta_r G$ is given by

$$\Delta_r G = \sum_{\text{B}} \mu_{\text{B}} \nu_{\text{B}} = (\partial G/\partial \xi)_{T,\,p}$$

(b) Derive the following conditions regarding the nature of the reaction

$dG = \Delta_r G\ d\xi < 0$ spontaneous

$dG = \Delta_r G\ d\xi > 0$ not possible

$dG = \Delta_r G\ d\xi = 0$ equilibrium

7.3 (a) Show that the homogeneous ideal gas reactions, the reaction potential is given as

$$\Delta_r G = \Delta_r G° + RT \ln \{\prod_{B} (p_B / p°)^{\nu_B}\}$$

where the symbol Π stands for multiplication to be carried over all the species of the reaction with ν_Bs negative for reactants and positive for products.

(b) Comment upon the statement that 'the relative magnitudes of the $\Delta_r G°$ term and the logarithm term in the equation of part (a) determines the thermodynamic progress of the reaction'.

(c) What do you understand by the term reaction at equilibrium? State the conditions when the reaction is at equilibrium? The value of ξ at equilibrium is known as extent of reaction at equilibrium, ξ_{eq}. Can it have a value of zero or one?

(d) Equation of part (a) can be rewritten as

$$\Delta_r G = \Delta_r G° + RT \ln Q_p^\circ$$

where Q_p° is the standard reaction quotient. Answer the following:

(i) What will be the value of $\Delta_r G$ at equilibrium?

(ii) Q_p° at equilibrium is written as K_p°. What is the name given to K_p°?

(e) Show that at equilibrium

$$K_p^\circ = 10^{-\Delta_r G°/(2.303\, RT)}$$

From this relation, justify that 'K_p° of a reaction depends only on temperature and is independent of partial pressures of the components and that of the total pressure of the system'.

(f) Show that the equation of part (a) can be written as

$$\Delta_r G = RT \ln \frac{Q_p^\circ}{K_p^\circ}$$

Discuss the nature of the reaction when

(i) $Q_p^\circ > K_p^\circ$; (ii) $Q_p^\circ = K_p^\circ$ and (iii) $Q_p^\circ < K_p^\circ$

7.4 For a chemical reaction at equilibrium involving only gases, derive the following expressions:

(i) $K_p^\circ = K_c^\circ (c°RT / p°)^{\Delta\nu_g}$ (ii) $K_p^\circ = K_x (p/p°)^{\Delta\nu_g}$

(iii) $K_p^\circ = K_n^\circ \left(\dfrac{p_{total} / p°}{n_{total}/n°}\right)^{\Delta\nu_g}$

where K_c°, K_x and K_n° have their usual meanings.

7.5 (a) What do you understand by the term 'reaction at equilibrium'? Is it a static equilibrium?

(b) Show that the reaction at equilibrium can be characterized by an equilibrium constant, defined as

$$K = \frac{k_f}{k_b}$$

where k_f and k_b are the rate constants for the forward and the backward reactions? Using the law of mass action of Guldberg and Waage, show that the equilibrium constant of a reaction

$$a\,A + b\,B \rightleftharpoons c\,C + d\,D$$

can be written as

$$K = \frac{[C]^c [D]^d}{[A]^a [B]^b}$$

7.6 (a) State the principal of Le Chatelier and Braun as applicable to the chemical reactions. With the help of this rule predict qualitatively the effect of

(i) changing the pressure

(ii) changing the temperature, and

(iii) addition of one of the components of the reaction.

(b) The quantitative explanation of Le Chatelier principle can be provided by thermodynamic criterion for equilibrium with the aid of which following expression can be derived

$$\left(\frac{\partial \xi_{eq}}{\partial T}\right)_p = \frac{\Delta_r H}{T G''_{eq}} \quad \text{and} \quad \left(\frac{\partial \xi_{eq}}{\partial p}\right)_T = -\frac{\Delta_r V}{G''_{eq}} = -\frac{(\Delta \nu_g)RT}{p G''_{eq}}$$

where $G''_{eq} = \{(\partial^2 G/\partial \xi^2)_{T,\,p}\}_{eq}$. Derive these relations.

(c) Using the relation of part (b), discuss the effects of temperature and pressure on the reaction.

$$\frac{1}{2}N_2(g) + \frac{3}{2}H_2(g) \rightleftharpoons NH_3(g) \qquad \Delta_r H^\circ = -46.11 \text{ kJ mol}^{-1}$$

$$H_2(g) + Cl_2(g) \rightleftharpoons 2HCl(g) \qquad \Delta_r H^\circ = -184.61 \text{ kJ mol}^{-1}$$

$$N_2O_4(g) \rightleftharpoons 2NO_2(g) \qquad \text{Endothermic}$$

Show that the above effects are consistent with the Le chatelier's principle.

7.7 Discuss the effect of pressure on each of the following equilibrium constants.

K_p, K_x, K_c and K_n

7.8 (a) Derive the relation

$$\frac{d \ln K_p^\circ}{dT} = \frac{\Delta_r H^\circ}{RT^2}$$

or
$$\ln K_p^\circ = -\frac{\Delta_r H^\circ}{RT} + I$$

or
$$\ln \frac{K^\circ_{p(T_2)}}{K^\circ_{p(T_1)}} = -\frac{\Delta_r H^\circ}{R}\left[\frac{1}{T_2} - \frac{1}{T_1}\right]$$

(b) Give schematic sketches of plots between log K_p° ant $1/T$ for an endothermic and an exothermic reaction. What will be their slopes?

(c) With the relation given in part (a), derive the following conclusions.

$K^\circ_{p(T_2)} > K^\circ_{p(T_1)}$ if $\Delta_r H^\circ$ is positive

and $K^\circ_{p(T_2)} < K^\circ_{p(T_1)}$ if $\Delta_r H^\circ$ is negative

where $T_2 > T_1$.

(d) Derive the following relation

$$\frac{d \ln K_c^\circ}{dT} = \frac{\Delta_r U^\circ}{RT^2}$$

7.9 Taking the typical reaction of

$$SO_2(g) + \frac{1}{2}O_2(g) \rightleftharpoons SO_3(g)$$

discuss the effect of an inert gas on equilibrium.

7.10 (a) Explain why it is permissible to omit the activity terms of pure solids and liquids while writing the K_p° expression for a heterogeneous reaction involving gaseous and pure condensed phases.

(b) Take the following typical heterogeneous reaction

$$A(1) \rightleftharpoons A(g) \quad \text{or} \quad A(s) \rightleftharpoons A(g)$$

show that $$\ln\frac{p_2}{p_1} = -\frac{\Delta_r H^\circ}{R}\left[\frac{1}{T_2} - \frac{1}{T_1}\right]$$

7.11 Taking the following typical reactions along with the data provided, write down the expressions for the equilibrium constants K_p°, K_x and K_c°.

(i)	A(g)	+	B(g)	⇌	C(g)	+	D(g)
$t = 0$	a		b		0		0
At t_{eq}	?		?		x		?

(ii)	A(g)	⇌	B(g)	+	C(g)
$t = 0$	a		0		0
At t_{eq}	?		x		?

(iii)	B(g)	+	C(g)	⇌	A(g)
$t = 0$	a		b		0
At t_{eq}	?		?		x

(iv)	A(g)	⇌	B(g)	+	2C(g)
$t = 0$	a		0		0
At t_{eq}	?		x		?

(v)	A(g)	+	2B(g)	⇌	C(g)
$t = 0$	a		b		0
At t_{eq}	?		?		x

(vi)	A(g)	+	B(g)	⇌	C(g)
$t = 0$	a		b		c
At t_{eq}	?		?		$c + x$

Discuss qualitatively the effects of pressure, volume and inert gas on the value of x. Are these effects consistent with the Le Chatelier's principle?

7.12 On the basis of Le Chatelier's principle, discuss the effects of temperature, pressure and addition of common substance on the following typical examples:

(i) $\Delta\nu_g = 0$; exothermic (ii) $\Delta\nu_g = 0$; endothermic

(ii) $\Delta\nu_g = +ve$; exothermic (iv) $\Delta\nu_g = +ve$; endothermic

(v) $\Delta\nu_g = -ve$; exothermic (vi) $\Delta\nu_g = -ve$; endothermic

7.13 For the reaction

$$N_2O_4(g) \rightleftharpoons 2NO_2(g)$$

(i) Derive the following relations for one mole of gaseous mixture:

$$G_{pure} = x_D\mu_D^\circ + x_M\mu_M^\circ + RT\ln(p/p^\circ)$$

$$\Delta_{mix}G = RT(x_D\ln x_D + x_M\ln x_M)$$

where subscript D stands for dimer and M for monomer.

(i) Show that for this reaction ξ/mol is equal to α.

(ii) Show that G_{pure} and $\Delta_{mix}G$ can be written in terms of ξ' (= ξ,/ mol) as follows:

$$G_{pure}/\text{mol} = \mu_D^\circ + (2\mu_M^\circ - \mu_D^\circ)\,\xi' + [1 + \xi']\,RT \ln (p/p^\circ)$$

$$= \mu_D^\circ + \Delta_r G\,\xi' + [1 + \xi']\,RT \ln (p/p^\circ)$$

$$\Delta_{mix}G/\text{mol} = RT\,[(1 - \xi')\ln (1 - \xi') + 2\xi' \ln (2\xi') - (1 + \xi')\ln (1 + \xi')]$$

(iv) If $\mu_D^\circ = 97.89$ kJ mol^{-1}, $\mu_M^\circ = 51.31$ kJ mol^{-1} and $\Delta_r G^\circ = 4.73$ kJ mol^{-1}.
Calculate the values of G_{pure}, $\Delta_{mix}G$ and G_{total} for the following values of ξ' and then plot these against ξ'.

$$\xi' = 0.0, 0.1, 0.15, 0.2, 0.4, 0.6, 0.8, 1.0$$

(v) Show that at equilibrium $\Delta_{mix}G$ decreases as rapidly of G_{pure} increases, i.e.

$$\left(\frac{\partial G_{pure}}{\partial \xi}\right)_{T,p} = -\left(\frac{\partial(\Delta_{mix}G)}{\partial \xi}\right)_{T,p}$$

(vi) Show that

$$\left(\frac{\partial G_{pure}}{\partial \xi}\right)_{T,p} < -\left(\frac{\partial(\Delta_{mix}G)}{\partial \xi}\right)_{T,p}$$

before equilibrium

$$\left(\frac{\partial G_{pure}}{\partial \xi}\right)_{T,p} > -\left(\frac{\partial(\Delta_{mix}G)}{\partial \xi}\right)_{T,p}$$

beyond equilibrium.

7.14 Consider the reaction

$$H_2(g) + I_2(g) \rightleftharpoons 2HI(g)$$

(a) If there are one mole of H_2, one mole of I_2 and zero mole of HI present before the reaction advanced, express the free energy of the reaction in terms of extent of reaction ξ.

(b) What form would the expression for the G have if the iodine were present as solid?

7.15 Define the term standard free energy of formation of a compound. What reference is chosen in defining the standard free energy of formation of ions? Explain, how with the help of standard free energies of formation, $\Delta_r G^\circ$ of a given reaction can be calculated.

7.16. Explain:

(i) Why is the synthesis of NH_3 preferably carried out at low temperature and high pressure?

(ii) Why is the manufacture of O_3 carried out at high temperature and high pressure?

TRY YOURSELF PROBLEMS

7.1 What is the effect of a catalyst on the $\Delta_r G^\circ$ of a chemical reaction?

7.2 Prove that for a chemical reaction

$$\ln K_p^o = \frac{\Delta_r S^\circ}{R} - \frac{\Delta_r H^\circ}{RT} \quad \text{and} \quad \Delta_r G = -T\,\Delta S_{universe}$$

7.3 The reaction

$$\frac{1}{2}N_2(g) + \frac{3}{2}H_2(g) = NH_3(g)$$

was started with initial amounts as

$$n(N_2, g) = 1 \text{ mol}; \quad n(H_2, g) = 2 \text{ mol}; \quad \text{and} \quad n(NH_3, g) = 0.5 \text{ mol}.$$

What would be the extent of reaction at the stage when

(a) $n(N_2, g) = 0.5$ mol, (b) $n(H_2, g) = 1.2$ mol, and (c) $n(NH_3, g) = 1.0$ mol?

(*Ans.* (a) 1 mol, (b) 0.533 mol, and (c) 0.5 mol)

7.4 Suppose the reaction $N_2O_4(g) \rightleftharpoons 2NO_2(g)$ is started with initial amounts as

$$N_2O_4(g) = n \quad \text{and} \quad NO_2(g) = n'$$

Show that the extent of the reaction is related to the degree of dissociation through the relation

$$\xi = n\alpha$$

and hence show that if $n = 1$ mol then $\xi / \text{mol} = \alpha$.

7.5 The maximum percentage yield of SO_3 obtained from the reaction of SO_2 and O_2 is greater at 25 °C than at 250 °C. Predict the sign of $\Delta_r H°$.

7.6 The free energy of a reaction at progress may be taken as a function of T, p and ξ, that is

$$G = f(T, p, \xi)$$

Show that

(a)

$$dG = -S\,dT + V\,dp + \Delta_r G\,d\xi$$
$$dA = -S\,dT - p\,dV + \Delta_r G\,d\xi$$
$$dH = T\,dS + V\,dp + \Delta_r G\,d\xi$$
$$dU = T\,dS - p\,dV + \Delta_r G\,d\xi$$

(b)

$$\Delta_r G = \left(\frac{\partial G}{\partial \xi}\right)_{T,p} = \left(\frac{\partial A}{\partial \xi}\right)_{T,V} = \left(\frac{\partial H}{\partial \xi}\right)_{S,p} = \left(\frac{\partial U}{\partial \xi}\right)_{S,V}$$

(c)

$$\left(\frac{\partial G}{\partial T}\right)_{p,\xi} = \left(\frac{\partial A}{\partial T}\right)_{V,\xi} = -S; \qquad \left(\frac{\partial G}{\partial p}\right)_{T,\xi} = \left(\frac{\partial H}{\partial p}\right)_{S,\xi} = V$$

$$\left(\frac{\partial A}{\partial V}\right)_{T,\xi} = \left(\frac{\partial U}{\partial V}\right)_{S,\xi} = -p; \qquad \left(\frac{\partial H}{\partial S}\right)_{p,\xi} = \left(\frac{\partial U}{\partial S}\right)_{V,\xi} = T$$

NUMERICAL PROBLEMS

Equilibrium Constant

7.1 (a) For ozone at 298 K $\Delta_f G° = 163.43$ kJ mol^{-1}, compute the standard equilibrium constant K_p° for the reaction

$$\frac{3}{2} O_2(g) \rightleftharpoons O_3(g)$$

(b) The above reaction is studied starting from 3 mol of O_2 at 298 K. Assuming that the degree of dissociation at equilibrium is very much less than unity, show that

$$\xi = \frac{3}{2}\sqrt{K_p\, p^{1/2}}$$

(*Ans.* 2.25×10^{-29})

7.2 0.1 mol of H_2 and 0.2 mol of CO_2 are introduced in an evacuated flask at 723 K and the following reaction occurs to give an equilibrium pressure of 50.662 5 kPa.

$$H_2(g) + CO_2(g) \rightleftharpoons H_2O(g) + CO(g) \qquad (1)$$

Analysis of the mixture shows that it contains 10 mol % of H_2O.

A mixture of CoO(s) and Co(s) is then introduced such that the additional equilibria (2) and (3) are established

$$CoO(s) + H_2(g) \rightleftharpoons Co(s) + H_2O(g) \qquad (2)$$

$$CoO(s) + CO(g) \rightleftharpoons Co(s) + CO_2(g) \qquad (3)$$

Analysis of the new equilibrium mixture thus obtained is found to contain 30 mole % of H_2O.

Calculate the three standard equilibrium constants K_{c1}, K_{c2} and K_{c3} for the reaction (1), (2) and (3), respectively. (*Ans.* 0 075 63, 9, 119.048)

7.3 Consider the equilibrium

$$2NO(g) + Cl_2(g) \rightleftharpoons 2NOCl(g)$$

At 298 K for NOCl(g), $\Delta_f G° = 66.358$ kJ mol^{-1}. The $\Delta_f G°$ for NO(g) is 86.688 kJ mol^{-1}. If NO and Cl_2 are mixed in the molar ratio 2 : 1 show that at equilibrium

$$x_{NO} = \left(\frac{2}{p\,K_p}\right)^{1/3} \quad \text{and} \quad x_{NOCl} = 1 - \frac{3}{2}\left(\frac{2}{p\,K_p}\right)^{1/3}$$

Assume that $x_{NOCl} = \simeq 1$. Note how each one of these quantities depends on pressure. Evaluate x_{NO} at 101.325 kPa and 1 013.25 kPa. (*Ans.* 5.3×10^{-3} , 2.4×10^{-3})

7.4 For the reaction

$$MnCO_3(s) \rightleftharpoons MnO(s) + CO_2(g)$$

$$\Delta_r G°/\text{kJ mol}^{-1} = 115.729 - 0.059\,2\,(T/\text{K}) \log (T/\text{K})$$

determine the temperature at which the dissoiciation pressure of CO_2(g) will be 50.662 5 kPa. (*Ans.* 668.85 K)

7.5 Calculate the equilibrium pressure of oxygen over CuO and Cu at 25 °C. Given $\Delta_f G°(\text{CuO}) = -127.2$ kJ mol^{-1}. (*Ans.* 2.62×10^{-43} kPa)

7.6 In the reaction $XY_2 \rightleftharpoons X + 2Y$, all the three substance are ideal gases. A 10.0 dm^3 flask contains initially 0.40 mol of XY_2. A catalyst for dissociation is then introduced and when the equilibrium is attained, the pressure of the mixture is 121.59 kPa. The temperature is 300 K. Find the standard equilibrium constant for the given reaction. (*Ans.* 0.005 79)

7.7 For the reaction

$$H_2(g) + I_2(g) \rightleftharpoons 2HI(g)$$

$K_p^\circ = 54.8$ at 417 °C. (a) When a mixture initially containing 8 mol of HI, 2 mol of H_2 and 0.5 mol of I_2 is heated to 417 °C, will more HI be formed? (b) What is the maximum amount of HI that can be formed when 1 g of H_2 and 100 g of I_2 are heated to 417 °C in a 2 dm^3 flask? (c) A mixture of 1 mol each of H_2, I_2 and HI was brought to equilibrium at 417 °C. What was the composition of the system?

(*Ans.* (a) no more HI formation, but some HI will decompose;
(b) 0.677 mol;
(c) $n(H_2) = 0.319$ mol, $n(I_2) = 0.319$ mol, $n(\text{HI}) = 2.362$ mol)

7.8 Consider the reaction for manufacture of NO from NH_3

$$4NH_3(g) + 5O_2(g) = 4NO(g) + 6H_2O(g) + \text{heat}$$

Suppose 4 mol of NH_3 and 5 mol of O_2 are mixed at 773 K and 1.013 MPa pressure and equilibrium is attained.

(a) Explain whether this much information is sufficient for the calculation of concentrations of all species.
(b) Discuss the change in relative amounts that would follow if temperature is raised while pressure is constant.
(c) What change in relative amounts will take place if pressure is decreased to 0.101 3 MPa keeping T constant?

(d) How will the value of K_p° for the equilibrium at 773 K compare with that of K_p° at 873 K?

(e) How will the value of K_p° at 1.013 MPa compare with the value of K_p° at 0.101 3 MPa.

(*Ans.* (a) Since K_p is not known, equilibrium concentrations cannot be determined,
(b) equilibrium will be shifted to left,
(c) equilibrium will be shifted to right,
(d) $\ln (K^\circ_{p(873\text{K})} / K^\circ_{p(773\text{K})}) = (100\ \text{K})\ \Delta_r H^\circ / \{R(773\ \text{K})\ (873\ \text{K})\}$
(e) remains same)

Thermodynamic Feasibility

7.9 It is required to pass CO and H_2O vapour at 1.013 MPa and 0.506 5 MPa, respectively, into a reaction chamber at 973 K and to withdraw CO_2 and H_2 at partial pressures of 1.595 MPa. Is this theoretically possible? K_p° for the reaction

$$CO(g) + H_2O(g) \rightleftharpoons CO_2(g) + H_2(g)$$

is 0.71.

7.10 A chemist claims that the following reaction occurs at 298 K

$$SF_6(g) + 8HI(g) \rightleftharpoons H_2S(g) + 6HF(g) + 4I_2(s)$$

Is it thermodynamically possible? Given

$\Delta_f G^\circ(SF_6, g) = -991.608\ \text{kJ mol}^{-1}$; $\Delta_f G^\circ(H_2S, g) = -33.012\ \text{kJ mol}^{-1}$

$\Delta_f G^\circ(HI, g) = 1.297\ \text{kJ mol}^{-1}$; $\Delta_f G^\circ(HF, g) = -270.73\ \text{kJ mol}^{-1}$

7.11 The $\Delta_f G^\circ$ of $CCl_4(1)$ is – 68.618 kJ mol^{-1}. Is it thermodynamically possible to prepare $CCl_4(1)$ from C(s) and Cl_2(g) at 298 K? Are there any conditions assumed in your answer?

7.12 A chemist wants to find a catalyst for the formation NO at 101.325 kPa from O_2 and N_2 at room temperature. Is this possible?

7.13 It is possible to prepare essentially pure HCl by combination of elements. This is not so for HI. Explain how you would arrive at this conclusion thermodynamically.

Temperature Variations

7.14 For the reaction

$$FeCl_3(g) \rightleftharpoons FeCl_2(g) + \tfrac{1}{2} Cl_2(g)$$

K_p° is 7.4×10^{-13} at 1 473 K and 1.2×10^{-11} at 1 773 K. Calculate $\Delta_r H^\circ$ and $\Delta_r S^\circ$ for the reaction at 1 773 K and 101.325 kPa. (Let initially the amount of $FeCl_3$ be one mole and those of Cl_2 and $FeCl_2$ be zero.) (*Ans.* –201.64 kJ mol^{-1}, 95.34 J K^{-1})

7.15 Consider the reaction

$$Ag_2O(s) \rightleftharpoons 2Ag(s) + \tfrac{1}{2} O_2(g)$$

for which

$$\Delta_r G^\circ/\text{J mol}^{-1} = 32\ 384 + 17.32(T/\text{K}) \log (T/\text{K}) - 116.48(T/\text{K})$$

(i) At what temperature will the equilibrium pressure of O_2 be 101.325 kPa?

(ii) Express $\log K_p^\circ$ $\Delta_r H^\circ$ and $\Delta_r S^\circ$ as a function of temperature

[*Ans.* (i) 460 K

(ii) $\log K_p^\circ = -\dfrac{1691}{T} - 0.9045 \log T + 6.0834$;

$\Delta_r H^\circ/\ \text{J mol}^{-1} = 32\ 384 - 7.531\ (T/\text{K})$

$\Delta_r S^\circ/\ \text{J K}^{-1}\ \text{mol}^{-1} = 108.95 - 17.32 \log (T/\text{K})$]

7.16 The following table gives the standard Gibbs free energy of formation of Cl(g) at three temperatures:

T/K	$\Delta_f G°$/ kJ mol^{-1}
100	115.19
1 000	62.19
3 000	–56.43

(a) For the reaction $\frac{1}{2}\,Cl_2(g) \rightleftharpoons Cl(g)$, determine the standard equilibrium constant K_p° at each of these temperatures. Determine the mean $\Delta_r H°$ and then find $\Delta_r S°$ at each of these temperatures.

(b) At 1 700 K and 111.46 kPa, one gram of chlorine occupies 19.6 dm^3. What are K_p° K_x and K_c° for the following reaction?

$$\tfrac{1}{2}\,Cl_2(g) \rightleftharpoons Cl(g)$$

Assume ideal gas behaviour.

(c) At 1 600 K, the degree of dissociation of chlorine is 0.071 at 101.325 kPa. What are K_p°, K_x and K_c° for the following reaction?

$$Cl_2(g) \rightleftharpoons 2Cl(g)$$

Assume ideal gas behaviour.

7.17 (a) The equilibrium constant for the dissociation

$$2H_2S(g) \rightleftharpoons 2H_2(g) + S_2(g)$$

is K_p° = 0.0118 at 1338 K while the enthalpy of dissociation is $\Delta H°$ = 177.40 kJ mol^{-1}. Find the standard equilibrium constant of the reaction at 1 200 K.

(*Ans.* 0.0738)

(b) If the degree of dissociation of the reaction of part (a) at 1 023 K and 101.325 kPa is 0.055, calculate K_p° of the reaction. What will be the effect of (i) increase in pressure, and (ii) the addition of an inert gas in the reaction mixture?

(*Ans.* 0.009 186)

7.18 For the reaction

$$H_2(g) + I_2(g) \rightleftharpoons 2HI(g)$$

K_p° = 50.0 at 721 K and K_c° = 669 at 623 K. Find $\Delta_r U°$ and $\Delta_r H°$ of the reaction.

(*Ans.* 98.84 kJ mol^{-1}, 98.84 kJ mol^{-1})

7.19 The standard equilibrium constant for the reaction

$$Ba^{2+} + Y^{4-} \rightleftharpoons BaY^{2-}$$

(where H_4Y represents ethylendiamine tetraacetic acid)
is $10^{8.01}$ at 0 °C and $10^{7.68}$ at 30 °C Calculate the enthalpy change and the entropy change for the reaction. Assume $\Delta H°$ to be independent of temperature.

(*Ans.* –17.422 kJ mol^{-1}, 89.55 J K^{-1} mol^{-1})

Annexure Chemical Equilibrium in an Ideal Solution

An ideal solution is defined as the one whose constituents follow Raoult's law ($p_B = x_B\, p_B^*$) in the entire range of concentration. The expression for the chemical potential of constituents in an ideal solution can be obtained as follows.

Thermodynamic criterion of equilibrium for a constituent in solution and vapour phase is

$$\mu_{\mathrm{B(soln)}} = \mu_{\mathrm{B(vap)}} \tag{1}$$

If the vapour is assumed to be ideal, we write:

$$\mu_{\mathrm{B(vap)}} = \mu^{\circ}_{\mathrm{B(vap)}} + RT\,(p_{\mathrm{B}}/p^{\circ}) \tag{2}$$

where p° is the standard-state unit pressure. With this, Eq. (1) becomes

$$\mu_{\mathrm{B(soln)}} = \mu^{\circ}_{\mathrm{B\,(vap)}} + RT \ln\,(p_{\mathrm{B}}/p^{\circ}) \tag{3}$$

Making use of the Raoult's law, we have

$$\mu_{\mathrm{B(soln)}} = \left\{\mu^{\circ}_{\mathrm{B(vap)}} + RT \ln (p^*_{\mathrm{B}} / p^{\circ})\right\} + RT \ln x_{\mathrm{B}}$$

$$= \mu^*_{\mathrm{B\,(l)}} + RT \ln x_{\mathrm{B}} \tag{4}$$

where $\mu^*_{\mathrm{B(l)}}$ is the chemical potential of the pure constituent B in liquid phase. The general condition of chemical equilibrium is

$$\Delta_{\mathrm{r}} G = \Sigma_{\mathrm{B}}\, \nu_{\mathrm{B}} \mu_{\mathrm{B}} = 0 \tag{5}$$

where νs are the stoichiometric coefficients which carry positive and negative signs for the products and reactants, respectively. Substitution of Eq. (4) in Eq. (5) gives

$$\Sigma_{\mathrm{B}} \nu_{\mathrm{B}} \left[\mu^*_{\mathrm{B(l)}} + RT \ln (x_{\mathrm{B}})_{\mathrm{eq}}\right] = 0 \tag{6}$$

or $$\Sigma_{\mathrm{B}}\, \nu_{\mathrm{B}} \mu^*_{\mathrm{B\,(l)}} = -RT\, \Sigma_{\mathrm{B}}\, \nu_{\mathrm{B}} \ln\,(x_{\mathrm{B}})_{\mathrm{eq}} = -RT \ln \left\{\prod_{\mathrm{B}} (x_{\mathrm{B}})^{\nu_{\mathrm{B}}}_{\mathrm{eq}}\right\}$$

or $$\Delta_{\mathrm{r}} G^{\circ} = -RT \ln K_x \tag{7}$$

The above treatment may be extended to ideal dilute solution where the following expressions hold good.

Solvent (Raoult's law); $p_1 = x_1 p_1^*$ (8)

Solute (Henry's law) ; $p_{\mathrm{B}} = x_{\mathrm{B}} k_{\mathrm{H,B}}$ (9)

where $k_{\mathrm{H,B}}$ is the Henry's law constant for the *i*th species in the solution.† Equation (7) still holds good for an equilibrium reaction involving solutes in an ideal dilute solution with the following expression of $\mu^*_{\mathrm{B\,(l)}}$.

$$\mu^*_{\mathrm{B(l)}} = \mu^{\circ}_{\mathrm{B(vap)}} + RT \ln\,(k_{\mathrm{H,B}}/p^{\circ}) \tag{10}$$

The term $\mu^*_{\mathrm{B(l)}}$ will be equal to the chemical potential of the solute B at $x_{\mathrm{B}} = 1$ if Henry's law is applicable over the entire range from $x_{\mathrm{B}} = 0$ to $x_{\mathrm{B}} = 1$.

Quite often, the equilibrium constant is written in terms of molarities (or molalities) of the constituents of reaction at equilibrium. The thermodynamic relations involving these concentration terms can be derived as follows.

Equilibrium Constant in terms of Molalities

If n_1 and n_B are the amounts of solvent and the solute B present in the solution, then

$$x_B = \frac{n_B}{n_1 + \sum_B n_B} \quad \text{and} \quad m_B = \frac{n_B}{n_1 M_1} \tag{11}$$

For a dilute solution where $\sum_B n_B << n_1$, we have

$$x_B \simeq \frac{n_B}{n_1} = m_B M_1 \tag{12}$$

With this, Eq. (6) becomes

$$\sum_B \nu_B \left[\mu^*_{B(1)} + RT \ln (m_B M_1)_{eq}\right] = 0$$

or $$\sum_B \nu_B \left[\mu^*_{B(1)} + RT \ln\left(\frac{M_1}{\text{kg mol}^{-1}}\right)\right] = -RT \ln\left[\prod_B \left(\frac{m_B}{\text{mol kg}^{-1}}\right)^{\nu_B}_{eq}\right]$$

or $$\Delta_r G'^{\circ} = -RT \ln K^{\circ}_m \tag{13}$$

Equilibrium Constant in terms of Molarities

If c_B is the molarity (i.e. amount of substance concentration), then

$$c_B = \frac{n_B}{V} \equiv \frac{n_B \rho}{(V\rho)} = \frac{n_B \rho}{m_{soln}} = \frac{n_B \rho}{n_1 M_1 + \sum_B n_B M_B}$$

For a dilute solution where $\sum_B n_B M_B << n_1 M_1$, we have

$$c_B = \frac{n_B \rho}{n_1 M_1} = x_B \frac{\rho}{M_1} \quad \text{or} \quad x_B = c_B \frac{M_1}{\rho} \tag{14}$$

With this, Eq. (6) becomes

$$\sum_B \nu_B \left[\mu^*_{B(1)} + RT \ln (c_B M_1 / \rho)\right] = 0$$

or $$\sum_B \nu_B \left[\mu^*_{B(1)} + RT \ln\left(\frac{M_1/\rho}{\text{dm}^3 \text{ mol}^{-1}}\right)\right] = -RT \ln\left\{\prod_B \left(\frac{c_B}{\text{mol dm}^{-3}}\right)^{\nu_B}_{eq}\right\}$$

or $$\Delta_r G''^{\circ} = -RT \ln K^{\circ}_c \tag{15}$$

† See Section 4.3 of Volume 3.

Appendix I Values of Thermodynamic Properties

SELECTED VALUES OF THERMODYNAMIC PROPERTIES AT 298.15 K AND 1 BAR

Substance	$\frac{\Delta_f H^\circ}{\text{kJ mol}^{-1}}$	$\frac{\Delta_f G^\circ}{\text{kJ mol}^{-1}}$	$\frac{S^\circ}{\text{J K}^{-1}\text{ mol}^{-1}}$	$\frac{C_{p,m}}{\text{J K}^{-1}\text{ mol}^{-1}}$
	Inorganic Compounds			
NaCl(s)	–411.15	–384.14	72.13	50.50
NaBr(s)	–361.06	–348.98	86.82	51.38
NaI(s)	–287.78	–286.06	98.53	52.09
Na_2CO_3(s)	–1130.68	–1044.44	134.98	112.30
$NaHCO_3$(s)	–950.81	–851.0	101.7	87.61
Na_2SO_4(s)	–1387.08	–1270.16	149.58	128.20
NaOH(s)	–425.61	–379.49	64.46	59.54
KCl(s)	–436.75	–409.14	82.59	51.30
KBr(s)	–393.80	–380.66	95.90	52.30
KI(s)	–327.90	–324.89	106.32	52.93
$KMnO_4$(s)	–837.2	–737.6	171.71	117.57
$K_2Cr_2O_7$(s)	–2061.5	–1881.6	291.2	219.24
KOH(s)	–424.76	–379.08	78.9	64.9
MgO(s)	–601.70	–569.43	26.94	37.15
$MgCO_3$(s)	–1095.8	–1012.1	65.7	75.52
$MgCl_2$(s)	–641.32	–591.79	89.62	71.38
$Mg(OH)_2$(s)	–924.54	–833.51	63.18	77.03
CaO(s)	–635.09	–604.03	39.75	42.80
$Ca(OH)_2$(s)	–986.09	–898.49	83.39	87.49
CaC_2(s)	–59.8	–64.9	69.96	62.72
$CaCO_3$(calcite)	–1206.92	–1128.79	92.9	81.88
$BaCO_3$(s)	–1216.3	–1137.6	112.1	85.35
BaO(s)	–553.5	–525.1	70.42	47.78
Fe_2O_3(s, haematite)	–824.2	–742.2	87.40	103.85
Fe_3O_4, (s, magnetite)	–1118.4	–1015.4	146.4	143.43
FeS(s, α)	–100.0	–100.4	60.29	50.54
FeS_2(s)	–178.2	–166.9	52.93	62.17
ZnO(s)	–348.28	–318.30	43.64	40.25
CuO(s)	–157.3	–129.7	42.63	42.30
$CuSO_4$(s)	–771.36	–661.8	109	100.0
$CuSO_4 \cdot H_2O$(s)	–1085.8	–918.11	146.0	134
$CuSO_4 \cdot 5H_2O$(s)	–2279.7	–1879.7	300.4	280
CuS(s)	–53.1	–53.6	66.5	47.82

(Contd...)

Substance	$\Delta_f H°$ / kJ mol^{-1}	$\Delta_f G°$ / kJ mol^{-1}	$S°$ / J K^{-1} mol^{-1}	$C_{p,m}$ / J K^{-1} mol^{-1}
AgCl(s)	–127.07	–109.79	96.2	50.79
$AgNO_3$(s)	–124.39	–33.41	140.92	93.05
PbO(s, red)	–218.99	–188.93	66.5	45.81
PbO(s, yellow)	–217.32	–187.89	68.70	45.77
PbO_2(s)	–277.4	–217.33	68.6	64.64
Pb_3O_4(s)	–718.4	–601.2	211.3	146.9
Al_2O_3(s. α)	–1675.7	–1582.3	50.92	79.04
SiO_2(s, α)	–910.94	–856.64	41.84	44.43
CdO(s)	–258.2	–228.4	54.8	43.43
$CdSO_4$.8/3 H_2O(s)	–1729.4	–1465.14	229.63	213.26
HgO(s)	–90.83	–58.54	70.29	44.06
Hg_2Cl_2(s)	–265.22	–210.75	192.5	
SnO(s)	–285.8	–256.9	56.5	44.31
SnO_2(s)	–580.7	–519.6	52.3	52.59
O_3(g)	142.7	163.2	238.93	39.20
H_2O(g)	–241.82	–228.57	188.83	33.58
H_2O(l)	–285.83	–237.13	69.91	75.29
H_2O_2(l)	–187.78	–120.35	109.6	89.1
HF(g)	–271.1	–273.2	173.78	29.13
HCl(g)	–92.31	–95.30	186.91	29.12
Br_2(g)	30.91	3.11	245.46	36.02
HBr(g)	–36.40	–53.45	198.70	29.14
I_2(g)	62.44	19.32	260.69	36.90
HI(g)	26.48	1.70	206.59	29.16
S(monoclinic)	0.33	0.1	32.6	23.6
SO_2(g)	–296.83	–300.19	248.22	39.87
SO_3(g)	–395.72	–371.06	256.76	50.67
H_2S(g)	–20.63	–33.56	205.79	34.23
NO(g)	90.25	86.57	210.76	29.84
NO_2(g)	33.18	51.31	240.06	37.20
N_2O(g)	82.05	104.20	219.85	38.45
N_2O_4(l)	–19.50	97.54	209.2	142.7
N_2O_4(g)	9.16	97.89	304.29	77.28
N_2O_5(s)	–43.1	113.9	178.2	143.1
N_2O_5(g)	11.3	115.1	355.7	84.5
NH_3(g)	–46.11	–16.45	192.45	35.06
N_2H_4(l)	50.63	149.43	121.21	139.3
NH_3NO_3(s)	–365.56	–183.87	151.08	84.1
NH_4Cl(s)	–314.43	–202.87	94.6	
PCl_3(g)	–287.0	–267.8	311.78	71.84
PCl_5(g)	–374.9	–305.0	364.58	112.8
C(diamond)	1.90	2.90	2.38	6.11
CO(g)	–110.53	–137.17	197.67	29.12
CO_2(g)	–393.51	–349.36	213.74	37.11
HCN(g)	135.1	124.7	201.78	35.86
CS_2(l)	89.70	65.27	151.34	75.7

(Contd...)

Substance	$\Delta_f H°$ / kJ mol^{-1}	$\Delta_f G°$ / kJ mol^{-1}	$S°$ / J K^{-1} mol^{-1}	$C_{p,m}$ / J K^{-1} mol^{-1}
		Organic Compounds		
CH_4(g)	–74.81	–50.72	186.26	35.31
C_2H_2(g)	226.73	209.20	200.94	43.93
C_2H_4(g)	52.26	68.15	219.56	43.56
C_2H_6(g)	–84.68	–32.82	229.60	52.63
HCHO(g)	–117	–113	91.2	
CH_3OH(l)	–238.66	–166.27	126.8	81.6
CH_3CHO(l)	–192.38	–128.12	160.2	
CH_3CO_2H(l)	–484.5	–389.46	178.7	
C_2H_5OH(l)	–277.69	–174.78	160.7	111.46
C_3H_8(g)	–103.89	–23.38	270.02	73.51
C_4H_{10}(g)	–126.15	–17.03	310.23	97.45
C_6H_6(g)	–82.93	129.72	269.31	81.67
		Ions in Solutions		
OH^-	–229.99	–157.24	–10.75	–148.5
F^-	–332.63	–278.79	–13.8	–106.7
Cl^-	–167.16	–131.23	56.5	–136.4
Br^-	–121.55	–103.96	82.4	–141.8
I^-	–55.19	–51.57	111.3	–142.3
S^{2-}	33.1	85.8	–14.6	
HS^-	–17.6	–12.08	62.8	
SO_4^{2-}	–909.27	–744.53	2.01	–293
NO_3^-	–205.0	108.74	146.4	–86.6
NH_4^+	–132.51	–79.31	113.4	79.19
CO_3^{2-}	–677.14	–527.81	–56.9	
HCO_3^-	–691.99	–586.77	91.2	
CN^-	150.6	172.4	94.1	
BrO_3^-	–67.6	18.60	161.71	
IO_3^-	–221.3	–128.0	118.4	
MnO_4^-	–541.4	–447.2	191.2	
MnO_4^{2-}	–653	–500.7	59	
Zn^{2+}	–153.89	–147.06	–112.1	46
Cu^{2+}	–64.77	65.49	–99.6	
Ag^+	105.58	77.11	72.68	21.8
Fe^{2+}	–89.1	–78.90	–137.7	
Fe^{3+}	48.5	–4.7	–315.9	
Mg^{2+}	–466.85	–454.8	–138.1	
Li^+	–278.49	–293.31	13.4	68.6
Na^+	–240.12	–261.91	59.0	46.4
K^+	–252.38	–283.87	102.5	21.8

(*Contd...*)

Substance	$\frac{\Delta_f H^\circ}{\text{kJ mol}^{-1}}$	$\frac{\Delta_f G^\circ}{\text{kJ mol}^{-1}}$	$\frac{S^\circ}{\text{J K}^{-1}\text{ mol}^{-1}}$	$\frac{C_{p,m}}{\text{J K}^{-1}\text{ mol}^{-1}}$
Cr^{3+}	–1999.1			
		Gaseous Atoms		
H	217.97	203.25	114.71	20.78
O	249.17	231.73	161.06	21.91
F	78.99	61.91	158.75	22.74
Cl	121.68	105.68	165.20	21.84
Br	111.88	82.40	175.02	20.79
I	106.84	70.25	180.79	20.79
S	278.81	238.25	167.82	23.67
N	472.70	455.56	153.30	20.79
C	716.68	671.26	158.10	20.84
		Elements in their Stable States of Aggregation		
H_2(g)	0	0	130.68	28.82
O_2(g)	0	0	205.14	29.36
F_2(g)	0	0	202.78	31.30
Cl_2(g)	0	0	223.07	33.91
Br_2(l)	0	0	152.23	75.69
I_2(s)	0	0	116.14	54.44
S(rhombic)	0	0	31.80	22.64
N_2(g)	0	0	191.61	29.13
C(graphite)	0	0	5.74	8.53
Al(s)	0	0	28.33	24.35
Zn(s)	0	0	41.63	25.40
Cu(s)	0	0	33.15	24.43
Ag(s)	0	0	42.55	25.35
Fe(s)	0	0	27.28	25.10
Mg(s)	0	0	32.68	24.89
Li(s)	0	0	29.12	24.77
Na(s)	0	0	51.21	28.24
K(s)	0	0	64.18	29.58

ENTHALPY OF FORMATION OF SOLUTES AND SOLUTIONS AT 298.15 K AND 1 BAR PRESSURE

		$\Delta_f H^\circ$ / kJ mol^{-1}
NaOH(s)		–425.61
	in 100 H_2O	–469.65
	in 200 H_2O	–469.61
	∞ H_2O	–470.11
NaCl(s)		–411.15
	in 100 H_2O	–407.07
	in 200 H_2O	–406.92
	∞ H_2O	–407.27
KOH(s)		–424.76
	in 100 H_2O	–481.64
	in 200 H_2O	–481.74
	∞ H_2O	–482.37
KCl(s)		–436.75
	in 100 H_2O	–419.32
	in 200 H_2O	–419.19
	∞ H_2O	–419.53
HCl(g)		–92.31
	in 100 H_2O	–165.93
	in 200 H_2O	–166.27
	∞ H_2O	–167.16

APPROXIMATE BOND ENTHALPIES (IN kJ mol^{-1}) AT 298.15 K

	H	C	N	O	F	Cl	Br	I	S	Si
H—	436	412	388	463	565	431	366	299	338	
C—		348			484	338	276	238	259	
C=		612								
N—		305	163		270	200				
N=		613	409							
N≡		890	945							
O—		360		146	185	203				
O=		743		497						
F—	565				155					
Cl—	431					254(242)				
Br—	366					219	193			
I—	99					210	178	151		
S—						250	212		264	
Si—										176

MOLAR HEAT CAPACITIES OF GASES AT CONSTANT PRESSURE AS A FUNCTION OF TEMPERATURE (298 K TO 1 000 K)

$$C_{p,\,m}/\text{J K}^{-1}\,\text{mol}^{-1} = a + b\,(T/\text{K}) + c(T/\text{K})^2$$

Gas	*a*	$b \times 10^3$	$c \times 10^6$
H_2	28.166	2.3372	–0.29816
N_2	29.295	–2.2634	5.6649
O_2	29.648	9.55	–1.328
CO	28.993	–0.80728	5.0568
CO_2	22.299	57.303	–25.294
H_2O	39.074	–17.586	19.870
CH_4	19.408	55.349	–2.9618
C_2H_4	7.3138	138.07	–51.482
C_2H_6	7.3230	167.59	–52.194
C_6H_6	–27.867	422.69	184.95
CH_3OH	17.173	97.03	–24.754
HCl	29.698	–3.5048	5.4351
Cl_2	29.190	19.249	–11.001
NO	29.712	–1.1792	5.4541
NO_2	25.036	45.550	–18.420
NH_3	25.627	34.821	–3.9571
N_2O_4	36.208	160.98	–77.984

Appendix II Units and Conversion Factors

CGS UNITS VIS-A-VIS UNITS

Physical quantity	*CGS units*		*SI units*	
	Name	*Symbol*	*Name*	*Symbol*
Length	centimetre	cm	metre	m
	Angstrom (10^{-8} cm)	Å		
mass	gram	g	kilogram	kg
time	second	s	second	s
temperature	Celsius	°C	kelvin	K
	kelvin	°K		
energy	calorie	cal	joule	J
	kilocalorie	kcal	kilojoule	kJ
	litre-atmosphere	L-atm		
	ergs	erg		
electric current	ampere	A	ampere	A

UNITS DERIVED FROM THE BASE SI UNITS

Physical quantity	*Name of unit*	*Symbol and definition*
force	newton	$N = kg\ m\ s^{-2}$ or $J\ m^{-1}$
energy	joule	$J = kg\ m^2\ s^{-2}$ or $N\ m$
electric charge	coulomb	$C = A\ s$
potential difference	volt	$V = kg\ m^2\ s^{-3}\ A^{-1}$ or $J\ A^{-1}\ s^{-1}$
resistance	ohm	$\Omega = kg\ m^2\ s^{-3}\ A^{-2}$ or $V\ A^{-1}$
frequency	hertz = cycle per second	$Hz = s^{-1}$
area	square metre	m^2
volume	cubic metre	m^3
density	kilogram per cubic metre	$kg\ m^{-3}$
velocity	metre per second	$m\ s^{-1}$
angular velocity	radian per second	$rad\ s^{-1}$
acceleration	metre per $(\text{second})^2$	$m\ s^{-2}$
pressure	newton per square metre or pascal	$Pa = N\ m^{-2}$
conductivity	siemen	$S = \Omega^{-1}$
magnetic field	tesla	$T = Wb\ m^{-2} = V\ s\ m^{-2}$
electric capacitance	farad	$F = C\ V^{-1}$
magnetic flux	weber	$Wb = V\ s$
inductance	henry	$H = Wb\ A^{-1}$

CONVERSION OF CGS UNITS TO SI UNITS

Quantity	*Unit*	*Equivalent**
length	Å	10^{-10} m = 10^{-1} nm = 10^2 pm
volume	litre	10^{-3} m = 1 dm^3
force	dyne	10^{-5} N
energy	erg	10^{-7} J
	cal	4.184 8 J
	eV	$1.602\ 1 \times 10^{-19}$ J
	eV/mole	98.484 kJ mol^{-1}
pressure	atmosphere	101.325 kN m^{-2}
	mmHg (or Torr)	133.322 N m^{-2}
	bar (10^6 dyn/cm^2)	10^5 N m^{-2}
viscosity	poise	10^{-1} kg m^{-1} s^{-1}

* Symbols used for fractions and multiples are

Fraction	*Prefix*	*Symbol*	*Multiple*	*Prefix*	*Symbol*
10^{-1}	deci	d	10	deka	da
10^{-2}	centi	c	10^2	hecto	h
10^{-3}	milli	m	10^3	kilo	k
10^{-6}	micro	μ	10^6	mega	M
10^{-9}	nano	n	10^9	giga	G
10^{-12}	pico	p	10^{12}	tera	T
10^{-15}	femto	f	10^{15}	peta	P

VALUES OF SOME PHYSICO-CHEMICAL CONSTANTS

Constant	*SI units*
Acceleration of gravity, g	9.806 65 m s^{-2}
Avogadro constant, N_A	$6.022\ 05 \times 10^{23}$ mol^{-1}
Bohr magneton	$9.274\ 09 \times 10^{-24}$ A m^2
Bohr radius, a_0	$5.291\ 77 \times 10^{-11}$ m
Boltzmann constant, k	$1.380\ 66 \times 10^{-23}$ J K^{-1}
Elementary charge, e	$1.602\ 19 \times 10^{-19}$ C
Electronic rest mass, m_e	$9.109\ 53 \times 10^{-31}$ kg
Faraday constant, F	$9.648\ 46 \times 10^4$ C mol^{-1}
Gas constant, R	$8.205\ 75 \times 10^{-2}$ dm^3 atm K^{-1} mol^{-1}
	8.314 41 J K^{-1} mol^{-1}
	8.314 41 Pa m^3 K^{-1} mol^{-1}
	8.314 41 kPa dm^3 K^{-1} mol^{-1}
	8.314 41 MPa cm^3 K^{-1} mol^{-1}
Molar volume of ideal gas at 0 °C and 1 atm, V_m	$2.241\ 4 \times 10^{-2}$ m^3 mol^{-1}
Planck's constant, h	$6.626\ 18 \times 10^{-34}$ J s
Proton rest mass, m_p	1.672 65 10^{-27} kg
Vacuum speed of light, c_0	$2.997\ 925 \times 10^8$ m s^{-1}
Standard atmospheric pressure	101.325 kN m^{-2} = 101.325 kPa

Index